LEARNING A[...]K

INVENTOR 2008

A Process-Based Approach

by

Thomas Short
Munro & Associates, Inc.
Troy, Michigan

Anthony Dudek
A.F. Dudek & Associates, Inc.
Chicago, Illinois

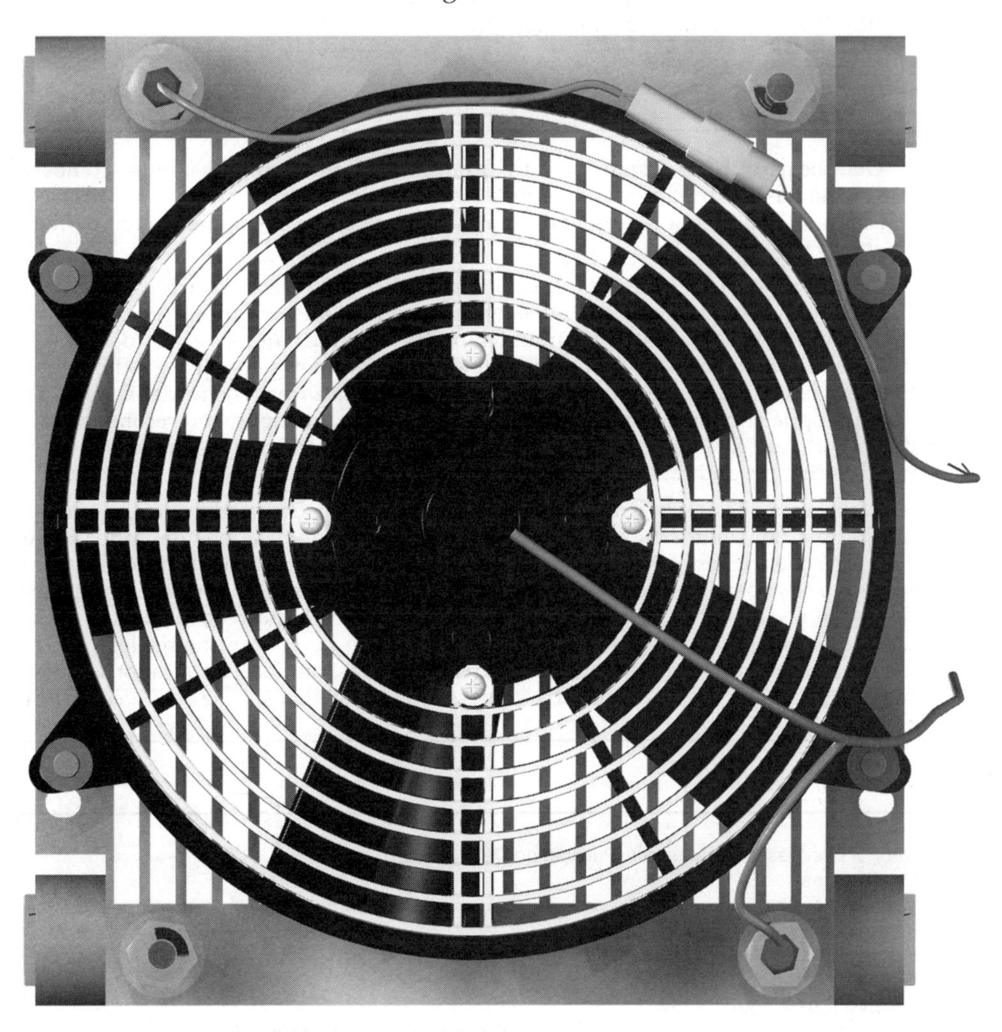

Publisher
The Goodheart-Willcox Company, Inc.
Tinley Park, Illinois
www.g-w.com

Manufactured in the United States of America.

Library of Congress Catalog Card Number 2007036141

ISBN: 978-1-59070-864-4

1 2 3 4 5 6 7 8 9 – 08 – 12 11 10 09 08 07

Library of Congress Cataloging-in-Publication Data
Short, Thomas
Learning Autodesk Inventor : a process-based approach/by Thomas Short, Anthony Dudek. — 4th ed.
p. cm.
Includes index.
ISBN-13: 978-1-59070-864-4
1. Engineering graphics. 2. Engineering models—Data processing. 3. Autodesk Inventor (Electronic resource) I. Dudek, Anthony. II. Title.
T353.S4664 2007
604.2—dc22 2007036141

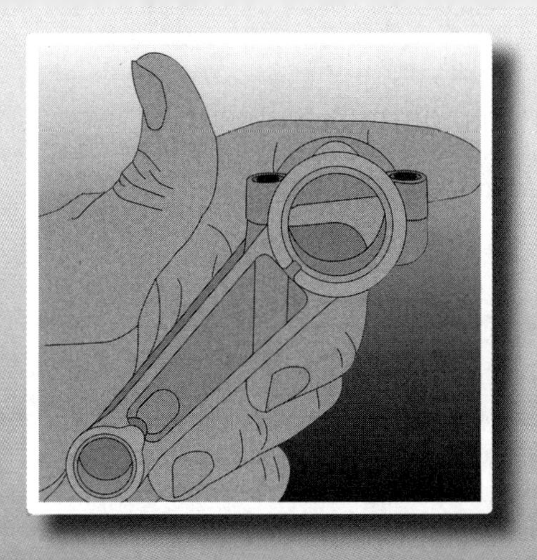

Introduction

Learning Autodesk Inventor is a text designed with the student as well as the engineering professional in mind. You will find it is presented in a manner designed to facilitate learning—practical examples and clear instructions. If you are looking for loads of theory, you will not find it in this text. The intention of this text is to recreate the actual workflow experienced by professionals as they use the software. After all, this text was written by those very same professionals. By the time you have finished this text, you will have a keen understanding of the methods used to produce a viable solid model part or assembly in Inventor.

Parametric design is very important in Inventor. You will encounter this a great deal. Throughout this text you will find example after example of parametric design principles. It cannot be stressed enough. The fact that your solid model should not be a static part, but a dynamic part able to withstand revision after revision is of great importance. You will hear this principle echoed throughout the text. The software was designed for this capability, so take advantage of it.

The goal of *Learning Autodesk Inventor* is to present a process-based approach to the Inventor tools, options, and techniques. Each topic is presented in a logical sequence where they naturally fit in the design process of real-world products. In addition, this text offers the following features.

- Inventor tools are introduced in a step-by-step manner.
- Easily understandable explanations of how and why the tools function as they do.
- Numerous examples and illustrations to reinforce concepts.
- Professional tips explaining how to use Inventor effectively and efficiently.
- Practices involving tasks to reinforce chapter topics.
- Chapter tests for reviewing tools and key Inventor concepts.
- Chapter exercises to supplement each chapter.

Fonts Used in This Text

Different typefaces are used throughout the chapters to define terms and identify Inventor tools. Important terms always appear in ***bold-italic, serif*** type. Inventor menus, tools, dialog box names, and button names are printed in **bold-face, sans serif** type. File names, folder names, paths, and selections in drop-down lists appear in Roman, sans serif type. Keyboard keys are shown inside of brackets [ ] and appear in Roman, sans serif type. For example, [Enter] means to press the enter (return) key.

Flexibility in Design

Flexibility is the key word when using *Learning Autodesk Inventor.* This text is an excellent training aid for individual as well as for classroom instruction. *Learning Autodesk Inventor* teaches you to apply Inventor to real-world problems. It is also a useful resource for professionals using Inventor in the work environment.

Notices

There are a variety of notices you will see throughout the text. These notices consist of technical information, hints, and cautions that will help you develop your Inventor skills. The notices that appear in the text are identified by icons and rules around the text.

PROFESSIONAL TIP

These are ideas and suggestions aimed at increasing your productivity and enhancing your use of Inventor tools and techniques.

NOTE

A note alerts you to important aspects of the tool or activity that is being discussed.

CAUTION

A caution alerts you to potential problems if instructions are not followed or tools are used incorrectly, or if an action could corrupt or alter files, folders, or disks. If you are in doubt after reading a caution, consult your instructor or supervisor.

Reinforcement and Evaluation

The chapter examples, practices, tests, and exercises are set up to allow you to select individual or group learning goals. Thus, the structure of *Learning Autodesk Inventor* lends itself to the development of a course devoted entirely to Inventor training. *Learning Autodesk Inventor* offers several ways for you to evaluate performance. Included are:

- **Examples.** The chapters include tutorial examples that offer step-by-step instructions for producing Inventor drawings, parts, and/or assemblies. The examples not only introduce topics, but also serve to help reinforce and illustrate principles, concepts, techniques, and tools.
- **Practices.** Chapters have short sections covering various aspects of Inventor. A practice composed of several instructions is found at the end of many sections. These practices help you become acquainted with the tools and techniques just introduced. They emphasize a specific point.
- **Chapter Tests.** Each chapter also includes a written test. Questions may require you to provide the proper tool, option, or response to perform a certain task.
- **Chapter Exercises.** A variety of drawing exercises follow each chapter. The exercises are designed to make you think and solve problems. They are used to reinforce the chapter concepts and develop your skills.

Student CD

Each chapter consists of examples, practices, and exercises. Most of these activities require a file that has been created and supplied to you on the Student CD. This CD is packaged with the text. At various points throughout this text, you will be instructed to open or access a file. The file will be used to develop, emphasize, and reinforce Inventor concepts and techniques.

In addition, the chapter practices and exercises are on the Student CD. The Student CD must be installed to your hard drive and the installation program is automatically started when the CD is inserted. After installation, use the menu that runs when the Student CD icon (on your desktop or in the Start menu) is selected to access the appropriate material. Once you have the instructions displayed on screen, use the [Alt][Tab] key combination to switch between the instructions and Inventor. The appendix is also included on the Student CD.

The CD installation is set up so each chapter has a file folder with specific chapter-related files. A comprehensive list of the files needed for that chapter is also included in the folder.

About the Authors

The authors have been using the software since its infancy. They are professionals with experience in design and training using Inventor. Their wide range of skills help make this book successful.

Thomas Short

Thomas Short is a nationally recognized expert in Inventor and 3D solid/surface modeling. He is a registered mechanical engineer in Michigan and has his B.S. and M.S. in mechanical engineering. He was a faculty member in the Mechanical Engineering Department at Kettering University (formerly General Motors Institute). He is also a member of the Society of Manufacturing Engineers and the Society of Automotive Engineers.

Tom has been using, teaching, and consulting in CAD since 1975, and AutoCAD since 1983. In 1984, he founded CommandTrain, Inc., which was a Premier Authorized AutoCAD Training Center. He is now a consulting engineer for Munro & Associates in Troy, Michigan. He is a Certified AutoCAD Instructor and Certified Technical Instructor. He has written for, and taught, many courses in Autodesk software including AutoCAD, AutoCAD 3D, AutoLISP, and Inventor. He has taught at every Autodesk University. He has also taught AutoCAD classes in the United States, Mexico, Brazil, Canada, and England.

As a CAD consultant and trainer he has worked for many companies, including Ford Motor Company, General Motors, Visteon, 3M, and McDonnell Douglas. He has helped several tooling and manufacturing companies in the Detroit area implement successful strategies for using Inventor as their solid modeling system.

Anthony Dudek

Anthony Dudek has been working in the mechanical engineering industry since 1985, using both AutoCAD and Inventor. He is now a nationally recognized expert in Inventor and 3D solid/surface modeling. Anthony is a Certified Autodesk Instructor and Autodesk Training Specialist. For the past several years, he has been a speaker at Autodesk University.

Tony has been a mechanical designer, 3D modeler, CAD manager, programmer, project manager, and management/networking consultant. For over 10 years, he has run A.F. Dudek & Associates, which is a CAD consulting firm in the Chicago area. He is also an Inventor instructor at Moraine Valley Community College's Premier Authorized AutoCAD Training Center, as well as other venues. He is the author of numerous articles and training materials. He is currently working with BP as CAD and Document Management Project Manager.

Notice to the User

This text is designed as a complete entry-level Inventor teaching tool. The authors present a typical point of view. Users are encouraged to explore alternative techniques for using and mastering Inventor. The authors and publisher accept no responsibility for any loss or damage resulting from the contents of information presented in this text. This text contains the most complete and accurate information that could be obtained from various authoritative sources at the time of production. The publisher cannot assume responsibility for any changes, errors, or omissions.

NOTE

It is assumed that **Content Center** is installed. You will need to use **Content Center** throughout this book. Refer to the Inventor documentation for instructions on installing **Content Center**, if it was not included in the original installation.

Also, the screen captures and descriptions of screen colors in this book are based on the Presentation color scheme. In some instances, the color of a particular item is based on the current color scheme. Changing color schemes is discussed in Chapter 2.

Acknowledgments

The authors and publisher would like to thank the following individuals for their assistance and contributions.

Mike Berna
David Boomer
Mary Dudek
Fern Espino
Jim Irvine
J.C. Malitzke
Lawrence Maples
Jerry McNaughton
Doug Montgomery
Rick Oprisu
Eliza Perry
Del Radloff

Cover image based on a design modeled by Jerry McNaughton.

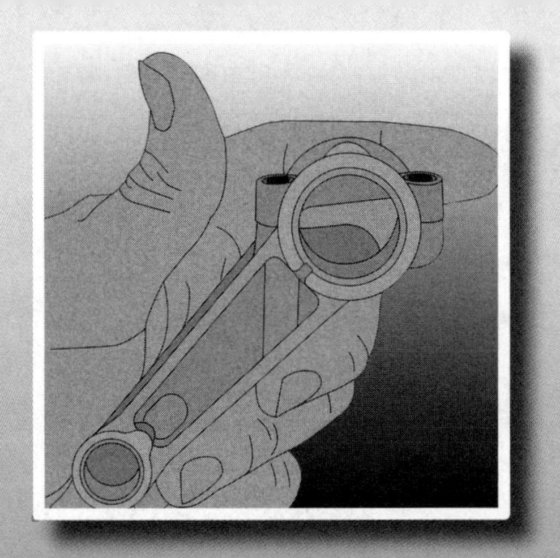

Brief Contents

Chapter 1: Introduction to Autodesk Inventor 17
Chapter 2: User Interface 25
Chapter 3: Sketching, Constraints, and the Base Feature 65
Chapter 4: Complex Sketching, Constraints, Formulas, and the Construction Geometry 87
Chapter 5: Secondary Sketches and Work Planes 107
Chapter 6: Adding Features 131
Chapter 7: Adding More Features 159
Chapter 8: Creating Part Drawings 187
Chapter 9: Dimensioning and Annotating Drawings. 211
Chapter 10: Sweeps and Lofts. 235
Chapter 11: Building Assemblies with Constraints. 251
Chapter 12: Working with Assemblies 273
Chapter 13: Motion Constraints and Assemblies. 293
Chapter 14: iParts and Factories. 303
Chapter 15: Parameters in Assemblies 321
Chapter 16: Surfaces. 331
Chapter 17: Assembly Drawings 347
Chapter 18: Presentation Files. 371
Chapter 19: Sheet Metal Parts. 391
Chapter 20: Inventor Studio 425
Index 451

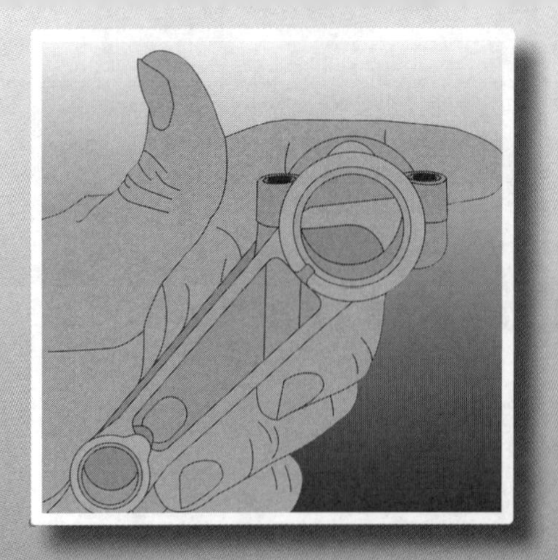

Expanded Contents

Chapter 1

Introduction to Autodesk Inventor 17

A Journey Begins 17
Feature-Based Modeling 18
Sketched Features 19
Placed Features 19
Work Features 19
Parametric Modeling 19
Assembly Modeling 21
Modeling Motion 22
2D Drawings 22
Presentations 22
Engineer's Notebook 23
Design Assistant 24

Chapter 2

User Interface 25

Opening an Existing Part File 25
User Interface Overview 26
Inventor Standard Toolbar 27
New Flyout Menu 27
Open Button 27
Save Button 27
Undo and Redo Buttons 30
Select Flyout 30
Return Button 30
Sketch Button 30
Update Button 31
Zoom All Button 32
Zoom Window Button 32

Zoom Button 32
Pan Button 32
Zoom Selected Button 32
Rotate Button 32
Look At Button 36
Display Flyout 36
Camera Flyout 37
Shadow Flyout 37
Component Opacity Flyout 38
Analyze Faces Button 38
Panel Bar 38
Panels 39
Resizing the Panel Bar 41
Browser 41
Changing Dimensions 43
Renaming Features 43
Pop-Up Menu 43
Working with Features 45
Menu Bar 46
File Pull-Down Menu 46
View Pull-Down Menu 47
Insert Pull-Down Menu 47
Format Pull-Down Menu 47
Tools Pull-Down Menu 53
Convert Pull-Down Menu 57
Applications Pull-Down Menu 57
Window Pull-Down Menu 57
Web Pull-Down Menu 58
Help Pull-Down Menu 58
Engineer's Notebook 59
Projects 60
Selecting a Project 61
Creating a Project 63

Chapter 3

Sketching, Constraints, and the Base Feature 65

Process for Creating a Part 65
Sketching 68
Constraints 70
Dimensions 71
Applying Fix Constraints 72
Extruding the Part 73
Editing the Feature and the Sketch 74
Circles, Tangent and Horizontal Constraints, and Trimming 76
Arcs and More Constraints 77
Drawing an Arc from within the Line Tool 79
Things That Can Go Wrong with Sketches 80
Review of All Constraints 83

Chapter 4

Complex Sketching, Constraints, Formulas, and the Construction Geometry

Complex Sketching, Constraints, Formulas, and the Construction Geometry . . . 87
Creating Complex (Ambiguous) Profiles . . . 87
Two Point Rectangle Tool and Precise Input . . . 88
Using the d0 Model Parameters in Equations in Dimensions . . . 89
More on Equations . . . 92
Construction Geometry . . . 93
Sketch Mirror Tool . . . 96
Revolve Tool . . . 97
Inventor Precise Input Toolbar for 2D Sketches . . . 99

Chapter 5

Secondary Sketches and Work Planes

Secondary Sketches and Work Planes . . . 107
Creating Secondary Sketch Planes and Features . . . 107
Operation and Extents Options of the Extrude Tool . . . 110
Projecting Silhouette Curves to the Sketch Plane . . . 114
Default Work Planes and Midplane Construction . . . 117
Intersect Option of the Extrude Tool . . . 121
Creating and Using New Work Planes . . . 122
  Offset from an Existing Face or Work Plane . . . 123
  Offset through a Point . . . 124
  Angled from a Face or Existing Work Plane . . . 124
  Midway between Two Parallel Faces . . . 125
  Perpendicular to a Line at Its Endpoint . . . 126
  Parallel to a Face or Plane and Tangent to a Curved Face . . . 126
  Tangent to a Curved Face through an Edge . . . 126
  Through Two Parallel Work Axes . . . 126
  Perpendicular to an Axis and through a Point . . . 126
  Between Edges or Points . . . 126

Chapter 6

Adding Features

Adding Features . . . 131
Adding Nonsketch Features to the Part . . . 131
Hole Tool . . . 132
  Threaded Holes . . . 134
  Clearance Holes . . . 135
  Taper Thread Holes . . . 136
  Counterbored and Countersunk Holes . . . 136
  Creating Holes without a Sketch Plane . . . 137
Thread Tool . . . 139
Fillets and Rounds . . . 141
  Applying Fillets and Rounds . . . 141
  Filleting and Rounding All Edges . . . 143
  More Fillet Tool Options . . . 145
  Variable-Radius Fillets and Setbacks . . . 147
  Full Round Fillets . . . 149
Adding Chamfers . . . 150

Rectangular and Circular Patterns. . . . 153
Feature Patterns 153
Pattern the Entire Part 154
Sketch Patterns 155
Mirror Feature Tool. . . . 156

Chapter 7

Adding More Features. . . . 159
Shell Tool 159
Rib Tool. . . . 163
Creating Solids from Open Sketches 167
Moving the Coordinate System on a Sketch Plane. . . . 167
Text and Emboss Tools 169
Decal Tool. . . . 171
Face Drafts 176
Fixed Edge 177
Fixed Plane 178
Other Face Drafts 178
Bend Part Tool 180
Split Tool. . . . 180
Using the Split Tool 181
Splitting a Helical Gear 181

Chapter 8

Creating Part Drawings. . . . 187
Create a Layout Drawing 187
Creating the Drawing Views 189
Creating a Base View 189
Projecting Views. . . . 191
Creating Other Views 192
Full Section View 192
Half-Section View. . . . 192
Aligned Section View 194
Section Views with Depth. . . . 195
Section Views Based on Sketch Geometry 196
Isometric Views. . . . 196
Auxiliary Views 197
Detail Views 198
Broken Views. . . . 198
Draft Views 201
Editing the Drawing Views 201
Repositioning Annotations. . . . 203
Editing a Drawing View's Lines. . . . 203
Changing the Model. . . . 205
Changing the View Orientation 205
Inventor's Options for Drawings 206
Working with AutoCAD Drawings 207
Opening an AutoCAD Drawing 207
Saving an Inventor Drawing as an AutoCAD Drawing 208

Chapter 9

Dimensioning and Annotating Drawings 211

Preparing to Annotate a Drawing Layout 211
Drafting Standards 211
Layers 213
Centerlines 214
Adding Centerlines to the Top View 214
Adding Centerlines to the Section Views 215
Adding Centerlines to the Auxiliary View 216
Dimensioning in Inventor 218
Parametric Dimensions 218
Dimension Styles 219
Dimensioning the Top View 220
Dimensioning the View SECTION A-A 220
Dimensioning the Auxiliary View 220
Adding Hole or Thread Notes 221
Surface Texture Symbols 222
Feature Control Frames 223
Adding Text 226
Sketched Symbols 227
Revision Table and Revision Tag 228
Weld Symbols 230

Chapter 10

Sweeps and Lofts 235

Sweeps and Lofts 235
Creating Sweep Features 236
Creating a Sweep Path 236
Creating a Cross-Sectional Profile 236
Creating the Sweep 236
Profile not Perpendicular to the Path 237
Closed Paths 237
Practical Example of Sweeps 238
Creating the Sweep Path 238
Creating the Cross Section 239
Completing the Part 240
3D Sweeps 241
First Example of a 3D Sketch 241
Second Example of a 3D Sketch 242
Inline Work Features 244
Editing a 3D Sweep 246
Lofts 246
Basic Loft 246
Closed-Loop Loft 248
Editing a Loft Feature 249

Chapter 11

Building Assemblies with Constraints ... 251

Building an Assembly File ... 251
Creating a Project with Folders ... 252
Building the Assembly ... 254
Removing Degrees of Freedom ... 256
Constraining Centerlines ... 257
Using the Insert Constraint ... 259
Placing the Bearing Blocks ... 259
Placing the Bushings ... 260
Determining the Cap Screw Specifications ... 261
Placing Standard Parts ... 262
Constraining Edges of Parts ... 265
Tangent Constraint ... 266
Derived Parts ... 268
Derived Gears ... 268
Derived Mold ... 270

Chapter 12

Working with Assemblies ... 273

Setting Additional Paths in Projects ... 273
Creating Parts in the Assembly View ... 275
Contact Solver ... 277
Angle Constraint ... 278
Driving Constraints ... 279
Constraining Work Planes and Axes ... 280
Assembly Constraints and Sketches ... 282
Adaptive Parts ... 284
Adaptive Location ... 284
Adaptive Size ... 286
Visibility and Design Views ... 288

Chapter 13

Motion Constraints and Assemblies ... 293

Motion Constraint ... 293
Rotation Constraint ... 294
Opposite Direction of Rotation ... 294
Same Direction of Rotation ... 295
Rotation-Translation Motion Constraint ... 296
Collision Detector ... 298
Rotation-Translation Constraint—Second Solution ... 299
Transitional Motion Constraint ... 300

Chapter 14

iParts and Factories....303

Assembling iParts....304
Placing an iPart....304
Constraining the iPart....306
Placed iPart....307
Constraining iMates....307
Placing the Remaining Parts....310
Creating an iPart from Scratch....311
Modeling the Part....311
Creating iMates....311
iPart Authoring....313
Testing the iPart in the Factory....318

Chapter 15

Parameters in Assemblies....321

Process of Working with Parameters....321
Controlling a Part with a Spreadsheet....322
Using a Spreadsheet to Control an Assembly....324
Editing the Part....326
Flanged Pipe Run—An Advanced Example....328
Summary....330

Chapter 16

Surfaces....331

Surface Display....332
Extruded Surfaces....333
Revolved Surfaces....333
Lofted Surfaces....333
Swept Surfaces....333
Thicken/Offset Tool....334
Replace Face....335
Surfaces as Construction Geometry....337
Stitched Surfaces....337
Boundary Patch....338
Surfaces in 2D Drawings....339
Analyzing Faces and Surfaces....340

Chapter 17

Assembly Drawings....347

Views....347
Section Views....349
Sliced and Offset Section Views....350
Break Out Views....351
Annotations....355
Centerlines....355
Dimensions....355

Leader Text ... 359
Balloons ... 359
Parts Lists ... 364
Changing Standards ... 366
Printing the Drawing ... 367
Creating Drawing Views Using Design View Representations ... 367
Overlay Views ... 369

Chapter 18

Presentation Files

Presentation Files ... 371
Presentation Files ... 371
Adding Tweaks ... 374
Animate Tool and Editing Tweaks ... 376
Creating Multiple Exploded Views ... 378
Complex and Rotational Tweaks ... 382
Tweaks ... 383
Views ... 384
Visibility ... 385
Styles ... 385
Part Colors in Assemblies ... 386
Lighting ... 386
Colors ... 388

Chapter 19

Sheet Metal Parts

Sheet Metal Parts ... 391
Sheet Metal Parts ... 391
Style Tool ... 393
Face Tool ... 395
Adding a Second Face ... 395
Adding a Third Face ... 397
Overriding Styles ... 398
Creating a Flat Pattern ... 399
Cut Tool ... 400
Projecting a Flat Pattern into a Sketch ... 401
Flanges ... 402
More on Flanges ... 404
Applying a Flange to a Portion of an Edge ... 405
Rounds and Chamfers ... 407
Hems ... 408
Holes and Cuts across Bends ... 409
Contour Flanges ... 412
Creating a Contour Flange ... 412
Extending a Contour Flange from an Edge ... 413
Complex Patterns as Contour Flanges ... 414
Fold Tool ... 415
Corner Seam Tool ... 418
Punch Tools ... 420

Chapter 20

Inventor Studio

Inventor Studio . . . 425
Rendering Process . . . 426
Animation Process . . . 426
Rendering Components . . . 427
Creating a Camera . . . 428
Creating a Lighting Style . . . 429
Moving Lights and Cameras . . . 431
Rendering the Scene . . . 432
Creating a Surface Style . . . 434
Rendering a Scene Containing Threads . . . 436
Animating Components . . . 436
Creating a Camera . . . 437
Entering Animation Mode . . . 438
Setting the Animation Time . . . 438
Animating Components . . . 439
Animating Constraints . . . 442
Creating a Camera . . . 442
Adding Lighting . . . 442
Animating the Constraint . . . 442
Animating a Camera . . . 444
Animating Fade . . . 445
Rendering an Animation . . . 446
Animating Parameters . . . 447
Adaptive Spring . . . 447
Robot Hand . . . 448
Index . . . 451

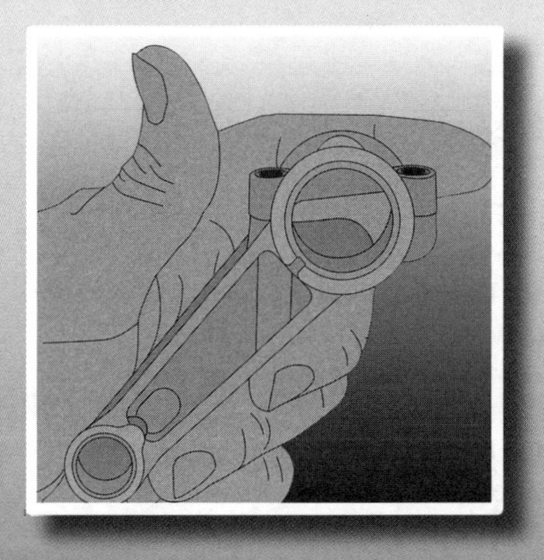

Chapter 1

Introduction to Autodesk Inventor

Objectives

After completing this chapter, you will be able to:

- Define a feature in Inventor.
- List the types of features in Inventor.
- Explain how to edit a part.
- Define an assembly in Inventor.
- Explain how to model motion in Inventor.
- Explain the engineer's notebook.

A Journey Begins

Inventor is a mature parametric solid modeling program now in its twelfth release. It is built for mechanical design and is in wide use throughout the world. With this program, you can design 3D models of complex parts, find their engineering properties, and create dimensioned detail drawings. Then, you can combine the parts into assemblies along with standard components, such as fasteners, from a built-in library. Assemblies can be animated to study their motion and to check for part interference. Assembly drawings with parts lists and balloons are easy to create and, like part drawings, are directly related to the model. In Inventor Studio, you can create realistic renderings and video files of animated assemblies.

This book uses a process-based approach. This offers a measured pace for the learner. A wealth of examples and exercises, along with the intuitive nature of Inventor, will allow you to become proficient and confident using Inventor with a moderate amount of study and practice.

Feature-Based Modeling

To explore the *feature-based* aspect of Inventor, refer to **Figure 1-1.** This is a part tree in the **Browser Bar**, or **Browser**, for a completed part. The **Browser** is an important part of the user interface. The ***part tree*** in the **Browser** lists the pieces and processes that make up the part. Inventor refers to these pieces and processes as ***features.*** If you think about any "single" part, it is really the end result of several features. The .ipt file extension is used for Inventor part files.

The part tree in **Figure 1-1** shows several features that together form the actual solid geometry of the final part. These features are named and have a colored icon next to their name that represents the type of feature. In this example, features include Hole for Outlet, Cleanup, Bolt Hole, and Bolt Pattern. These descriptive names were entered by the designer. By default, a feature is created with a generic name that represents the feature, such as Extrusion1 for an extrusion or Revolution2 for a revolution.

Work features are used for construction purposes. When created, they have names with the word Work in them, such as Work Plane2 and Work Plane3. The designer may elect to give work features descriptive names, such as Work Plane for Extrusion. Work features are not typically displayed in the graphics window, except when needed. Features that are not visible in the graphics window are grayed out in the **Browser**. The feature can be made visible as needed.

All of the features are listed in the part tree in the order in which they were created, with the first feature at the top. Features may be reordered by simply picking and dragging them to a new position in the part tree. Reordering features may alter the part. In addition, features cannot be moved above other features on which they are based. For example, if a hole has a fillet applied to it, the fillet feature cannot be moved above the hole feature. The last feature to be listed in any Inventor part tree is the End of Part marker. Its icon is a red sphere with a white X. This serves as an "end-of-file" marker for the part.

Features can be edited on an individual basis to change their size, shape, and, in some cases, location on the part. Features can also be suppressed. This means that the features are still in the part tree, but their effects on the part are not applied. To permanently remove a feature from the part, it is simply deleted. Features can even be exported for use in other parts.

The part that you are creating is never really edited; its *features* are edited. Since the part is made up of features, as the features are edited, the part is altered. Inventor is truly a feature-based solid modeling program and features form the heart of the system.

Inventor's features can be broken down into three categories—sketched, placed, and work. These categories are introduced in the next sections. As you work through this book, you will become very familiar with features and the tools used to create and edit them.

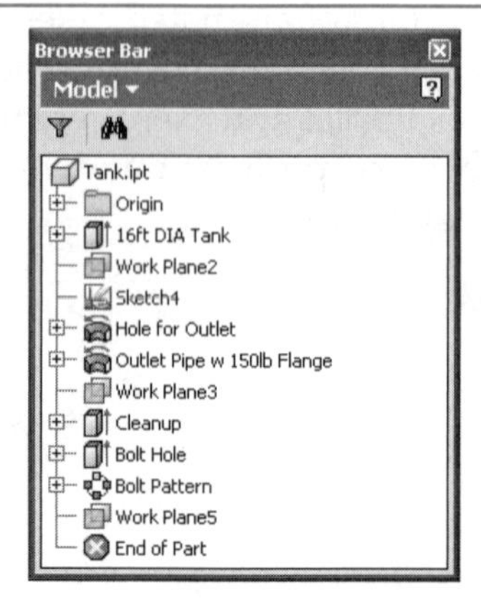

Figure 1-1.
The part tree is displayed in the **Browser Bar**, or **Browser**. The tree contains all of the features that make up the part.

Figure 1-2.
This is a fully dimensioned and constrained 2D sketch.

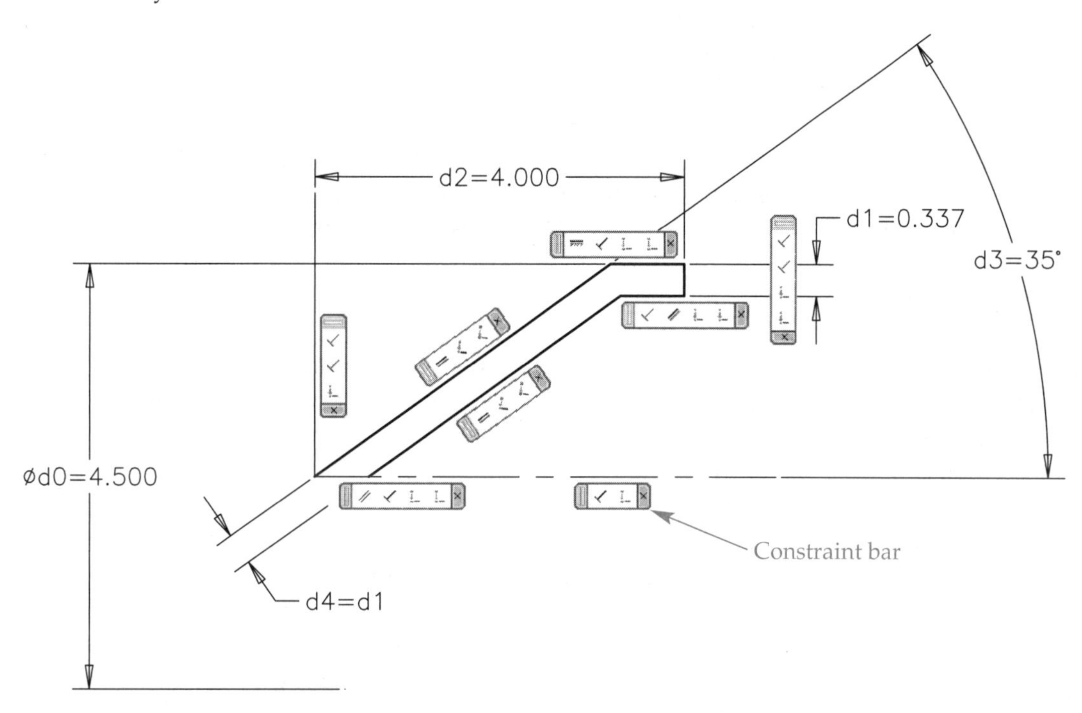

Sketched Features

A 2D ***sketch*** forms the basis for a sketched feature. This sketch can be drawn, or sketched, as any 2D geometry from lines, arcs, circles, and so on. The sketch is then dimensioned to exact size. Furthermore, the geometric relationship of the sketch geometry is constrained to a final shape. See **Figure 1-2.** A ***fully constrained sketch*** completely describes the size, shape, and location of the sketched geometry so Inventor cannot inadvertently change the design in future operations. You will learn much more about sketches and constraints in later chapters.

Placed Features

Placed features are not based on a sketch. Instead, the designer uses tools to "place" features onto an existing part, which is a collection of features. Placed features are similar to machining processes and can be thought of as operations performed on an existing part. Examples of placed features include fillets, chamfers, and holes.

Work Features

Work features are used for construction purposes. Work features include work planes, work axes, and work points. In a 3D environment, sometimes there is nothing on which to base a 2D sketch except a work plane. The work plane serves as the "drafting table" for the sketch's "paper." A work axis can serve as a centerline of revolution. A work point can serve as an anchor point for a 3D path.

Parametric Modeling

As you have seen, an Inventor part is made up of features. This is why Inventor is considered a *feature-based* solid modeling program. However, Inventor is also a *parametric* solid modeling program. A ***parametric model*** contains parameters, or dimensions, that define the model. In Inventor, the features that make up a part have dimensions that control the size, shape, and location of the features. By altering the dimensions (parameters), the features are altered.

For example, suppose you need to design a rectangular plate. The preliminary design indicates the plate is 8″ × 4″ × .25″. To create the plate, you first sketch a rectangle that is roughly 8″ × 4″ on a sketch plane (the "paper"). Then, you apply a horizontal and a vertical dimension to the sketch and edit the values to 8″ and 4″. See **Figure 1-3.** The sketch completely describes the shape and size of the plate's top view. Next, you extrude the sketch .25″ to fully describe the part in three dimensions. See **Figure 1-4.** Since Inventor is a parametric modeler, you can now alter any or all of the three dimensions (parameters) to change the part.

Another powerful aspect of Inventor's parametric modeling is the ability to establish relationships between dimensions and features. For example, suppose the plate described above will always be half as wide as it is long, no matter what the length dimension is. An equation can be entered as a dimension to accomplish this. Every dimension, or parameter, in Inventor has a unique name. By default, the first dimension name is d0, the second is d1, and so on. As you will learn, these can be renamed to descriptive names as needed. Now, since the vertical distance (d1) needs to be one-half of the horizontal distance (d0), instead of entering a number for the dimension, the simple equation d0/2 is entered for the value of d1. See **Figure 1-5.** As d0 (the horizontal distance) is altered, d1 changes so the vertical distance is one-half of the horizontal distance. This is known as *specifying design intent.* It is the intention of the designer that this plate always maintain its aspect ratio of 2:1 length to width. The sketch is thus dimensioned to reflect this design intent.

Figure 1-3.
A rectangle is sketched and dimensioned.

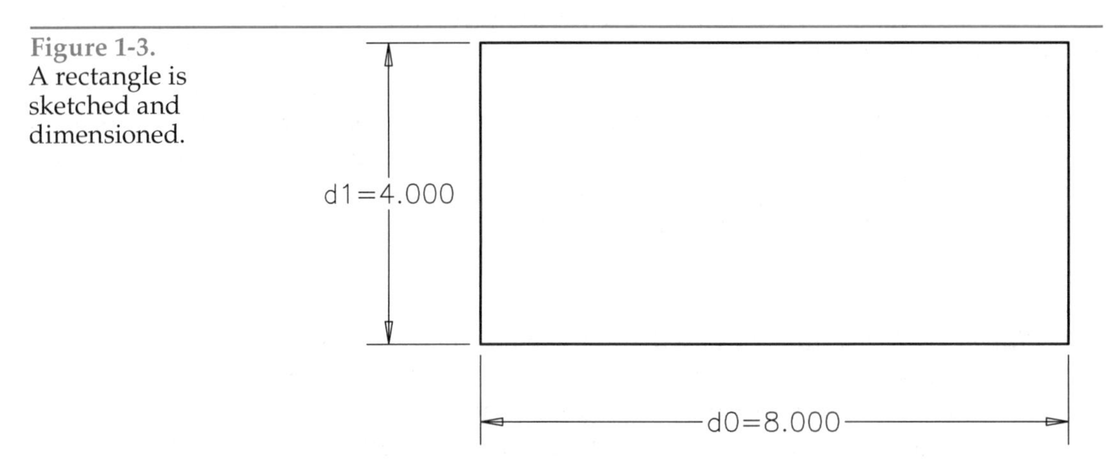

Figure 1-4.
The sketch from Figure 1-3 is extruded .25″ to create a solid part.

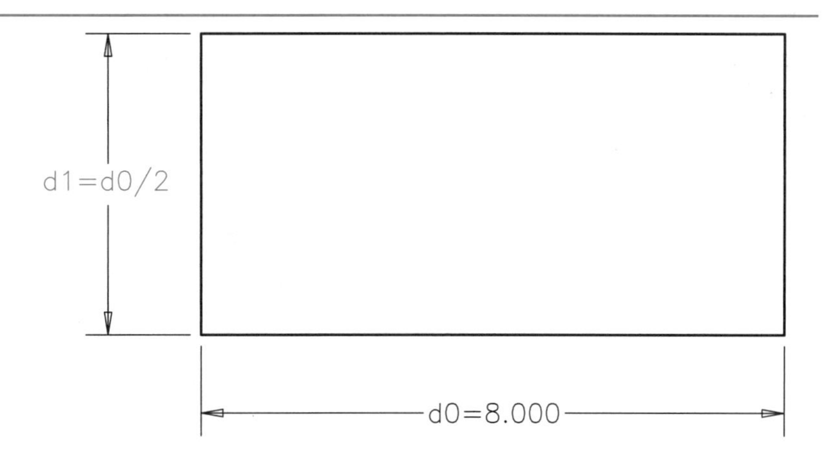

Figure 1-5.
An equation that is based on the length dimension is entered for the width dimension.

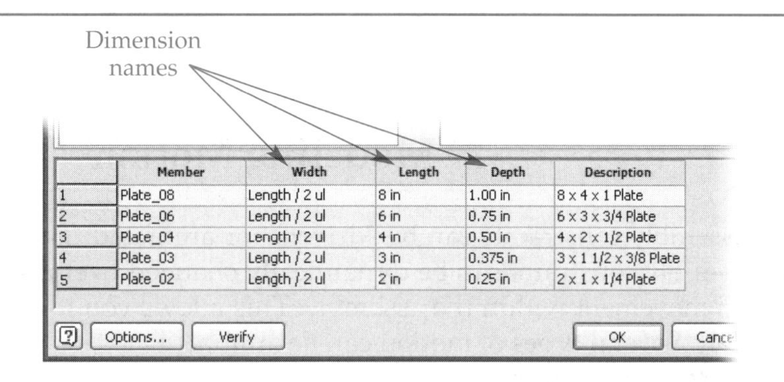

	Member	Width	Length	Depth	Description
1	Plate_08	Length / 2 ul	8 in	1.00 in	8 x 4 x 1 Plate
2	Plate_06	Length / 2 ul	6 in	0.75 in	6 x 3 x 3/4 Plate
3	Plate_04	Length / 2 ul	4 in	0.50 in	4 x 2 x 1/2 Plate
4	Plate_03	Length / 2 ul	3 in	0.375 in	3 x 1 1/2 x 3/8 Plate
5	Plate_02	Length / 2 ul	2 in	0.25 in	2 x 1 x 1/4 Plate

Figure 1-6.
By using descriptive names, you can build a spreadsheet that is used to "manufacture" different versions of the part.

As mentioned, the default dimension names (d0, etc.) can be renamed. Descriptive names, such as Length and Width, are meaningful to the drafter and designer. A year after the part is created, anybody can open the part file and instantly know what the dimension controls. Also, designers and drafters are almost always part of a team. Using descriptive names allows other team members to interpret your design intent. This is further enhanced by adopting a standard naming convention in your department or company.

Another use of named dimensions is to "manufacture" different versions of a part based on data in a spreadsheet. See **Figure 1-6.** As the part is "manufactured," the designer is prompted to enter values for various parameters. The prompts are based on the dimension names. Therefore, Pipe Length is a meaningful name, where d0 is not.

Assembly Modeling

Inventor does a great job modeling parametric parts. However, one of the most powerful aspects of Inventor is the ability to create assemblies. As a part is a collection of features, an assembly is a collection of parts. See **Figure 1-7.** Each part is created and saved. Then, the parts are placed into an assembly file. Parts are only referenced in the assembly. The part files remain separate. Changes made to the part file are reflected in the assembly. Finally, the spatial relationships between parts are defined, or constrained, within the assembly file. An assembly file has an .iam file extension.

A partially-constrained part can be dynamically dragged to analyze its movement within the assembly. The extents of its movement is the part's ***work envelope*** in the assembly. You can also move parts in an assembly and then measure distances to determine design data for additional parts.

Figure 1-7.
This is a complex assembly. A number of individual parts were placed into the assembly and constrained to finish the assembly.

Modeling Motion

Assembly constraints can be "driven," or animated. The numeric values used in the assembly constraint can be dynamically changed over a specified range to model a part's movement within the assembly. This allows you to animate the motion of an assembly. Several types of motion can be animated:

- Rotational (gears).
- Rotational-translational (rack and pinion).
- Translational (cam and cam follower).

2D Drawings

Prints of 2D drawings are always required for the machinists, assemblers, and other workers in the shop. These drawings must follow accepted drafting conventions for lineweight, linetype, and symbol use. Inventor provides tools for creating 2D drawings of parts and assemblies. Inventor drawings have an .idw file extension. Parts and assemblies are referenced into the drawing and displayed using orthographic projection rules. Changes made to the part are automatically reflected in the drawing.

Presentations

If a picture is worth a thousand words, then how many words is a movie worth? A 2D drawing will always be required. However, you can capture to digital video the process for building an assembly. This is called a ***presentation.*** See **Figure 1-8.** Inventor uses an .ipn file extension for presentation files.

The assembly file is referenced into the presentation and always reflects the current state of the assembly. The animated presentation can be distributed to others or saved as an AVI file to be viewed on workstations not equipped with Inventor. Presentations provide a very effective form of communication.

Figure 1-8.
An animated assembly presentation, such as this one, can be saved as an AVI file and played in Windows Media Player.

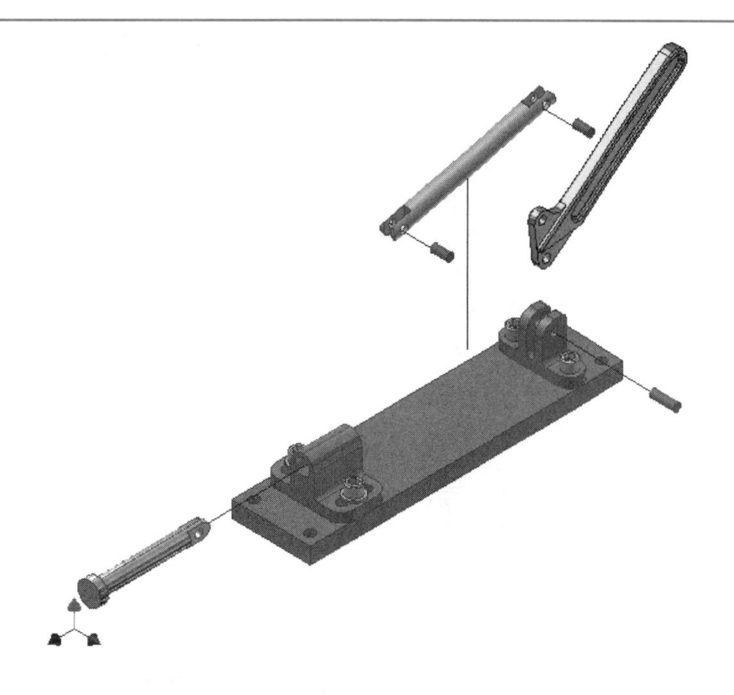

Engineer's Notebook

There is a handy little utility in Inventor that allows you to annotate a part in a notebook-style environment. This allows you to add notes to a part or feature that are not intended to be part of the final 2D drawing. See **Figure 1-9.** The engineer's notebook is meant to act as a communication tool between those collaborating on the design. Multiple notations may be made on a single part or assembly. The notes appear in the **Browser** in a separate branch in the part or assembly tree. An icon also appears in the graphics window near the annotated part. The engineer's notebook is discussed in detail in Chapter 2.

Figure 1-9.
The engineer's notebook can be used to share comments that are not actual drawing annotations.

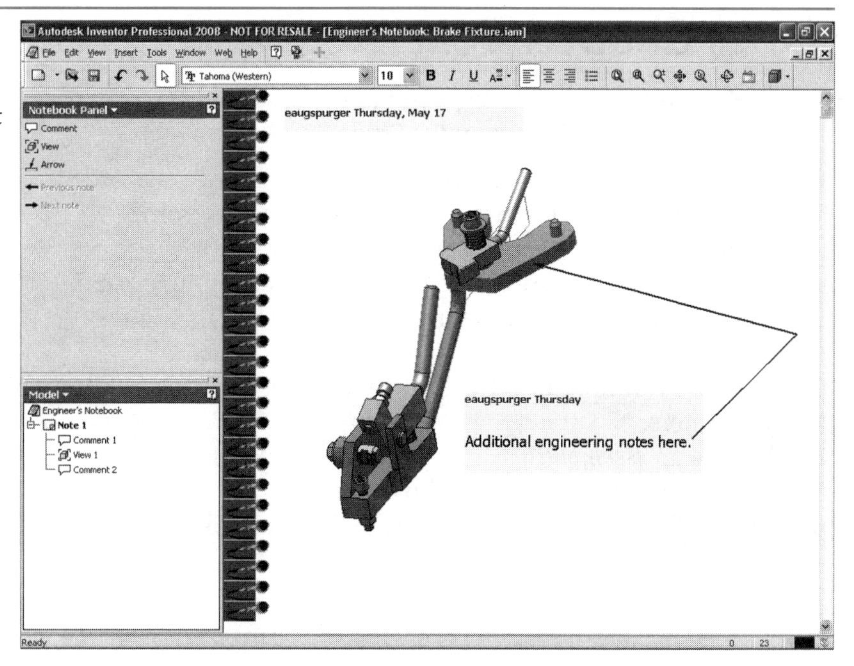

Figure 1-10.
Design Assistant serves as a document management system.

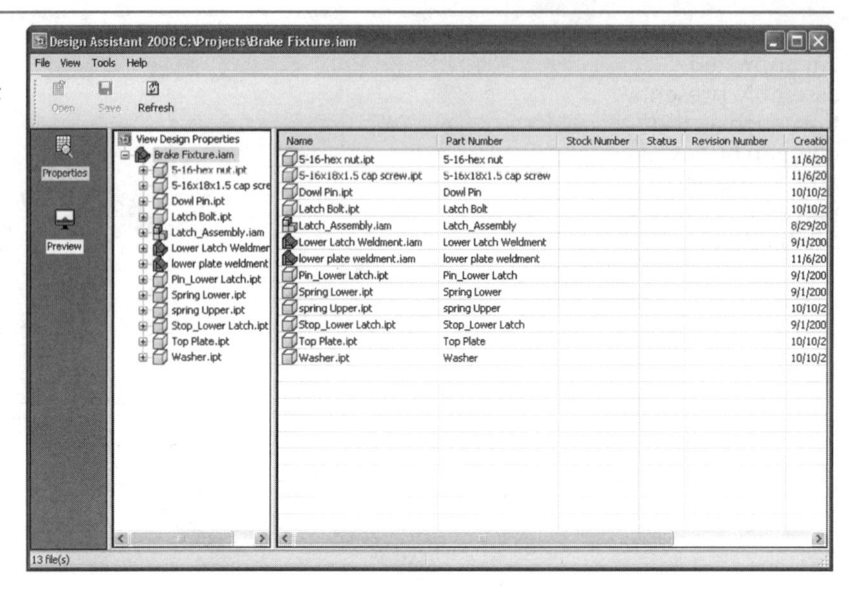

Design Assistant

In the process of creating parts, subassemblies, and an assembly, you may end up with hundreds of files. **Design Assistant** is a utility that acts as a document management system, **Figure 1-10.** It can track items such as revision number, status of design, and iProperties. When accessed through Windows Explorer, **Design Assistant** is a great utility for managing Inventor files.

Another very useful utility, called **Pack and Go**, allows you to select an assembly and it will find all of the parts, subassemblies, design views, etc., in the assembly and copy them to a destination folder. Then, you can zip all of the files and e-mail the zipped file.

Chapter Test

Answer the following questions on a separate sheet of paper or complete the electronic chapter test on the Student CD.

1. What are *features* in Inventor?
2. How is Inventor a feature-based modeling program?
3. What are the three types of features in Inventor?
4. Which type of feature is used for construction purposes?
5. What is the basic process for editing a part?
6. How is Inventor a parametric modeling program?
7. Give one example of where you would establish a relationship between a part's parameters.
8. What is an *assembly?*
9. How can motion be modeled in Inventor?
10. What is the purpose of the engineer's notebook?

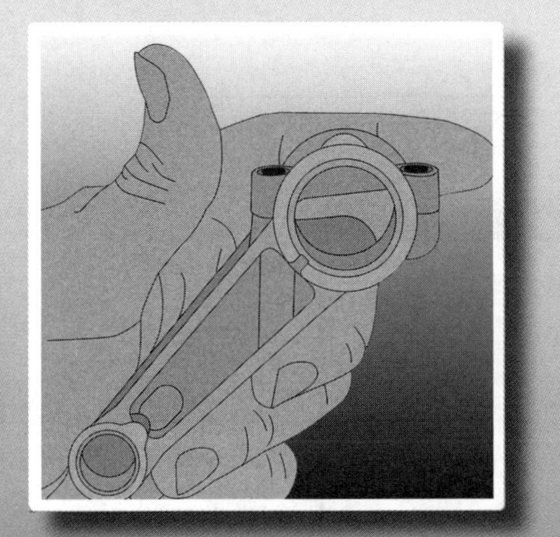

Chapter 2

User Interface

Objectives

After completing this chapter, you will be able to:

- Explain the various components of Inventor's user interface.
- Locate the various components of the user interface.
- Open an existing file.
- Create a new file.
- Create and edit solid model geometry.

User's Files

The Student CD included with this text contains several files required for this chapter. Refer to the file File List.txt *in the* \Ch02 *folder for the comprehensive list.*

Opening an Existing Part File

By default, the **Open** dialog box is displayed when Inventor is started. See **Figure 2-1.** At any time during the Inventor session, this dialog box is accessed by selecting **Open** in the **File** pull-down menu, picking the **Open** button on the **Inventor Standard** toolbar, or pressing the [Ctrl][O] key combination. The **Open** dialog box is used to open various types of files in Inventor.

Navigate to the folder containing the examples for Chapter 2. Pick once on the Example_02_01.ipt file in the list box. A preview of the part is shown in the lower-left corner of the dialog box. To open the file, pick the **Open** button at the bottom of the dialog box with the file name highlighted or double-click on Example_02_01.ipt in the list box.

Figure 2-1.
The **Open** dialog box.

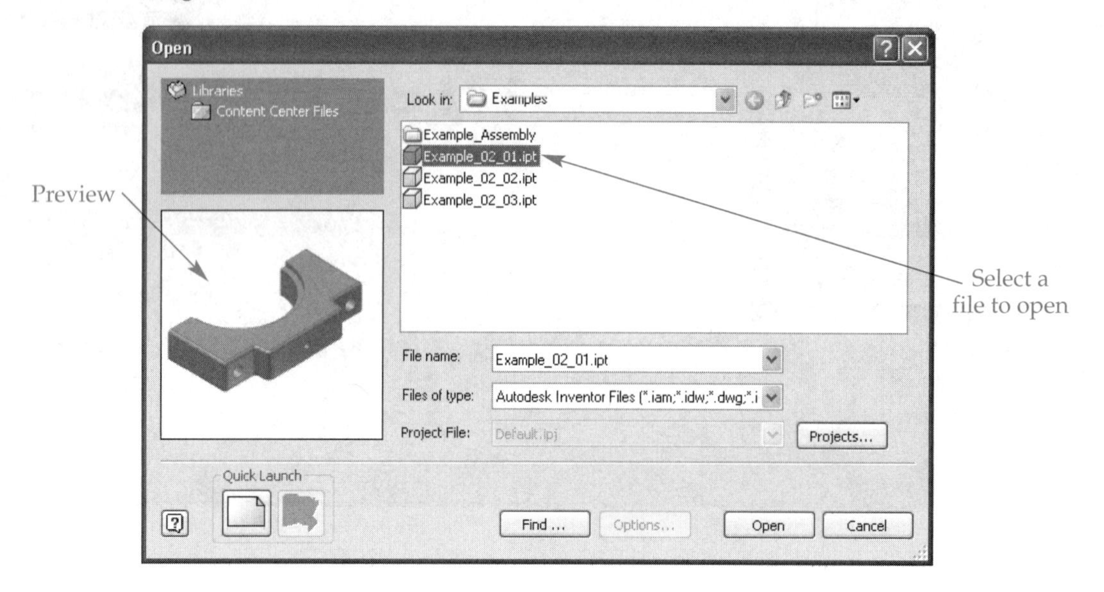

User Interface Overview

Once the file is opened, look at the user interface. See **Figure 2-2.** At the top of the screen are the **Menu Bar** and the **Inventor Standard** toolbar. Along the left side of the screen are the **Panel Bar** and the **Browser**. These components are all context sensitive; that is, they change as you work in different modes. The **Panel Bar** will drastically change, displaying completely different tools, and the others will slightly change.

The modes for part modeling are: sketch, 3D sketch, part (or features), sheet metal, and solids. In the first part of this book, the sketch and part modes will be used. Solids mode is used for parts imported from other CAD systems, and 3D sketch mode is for

Figure 2-2.
The default appearance of Inventor with a part file open.

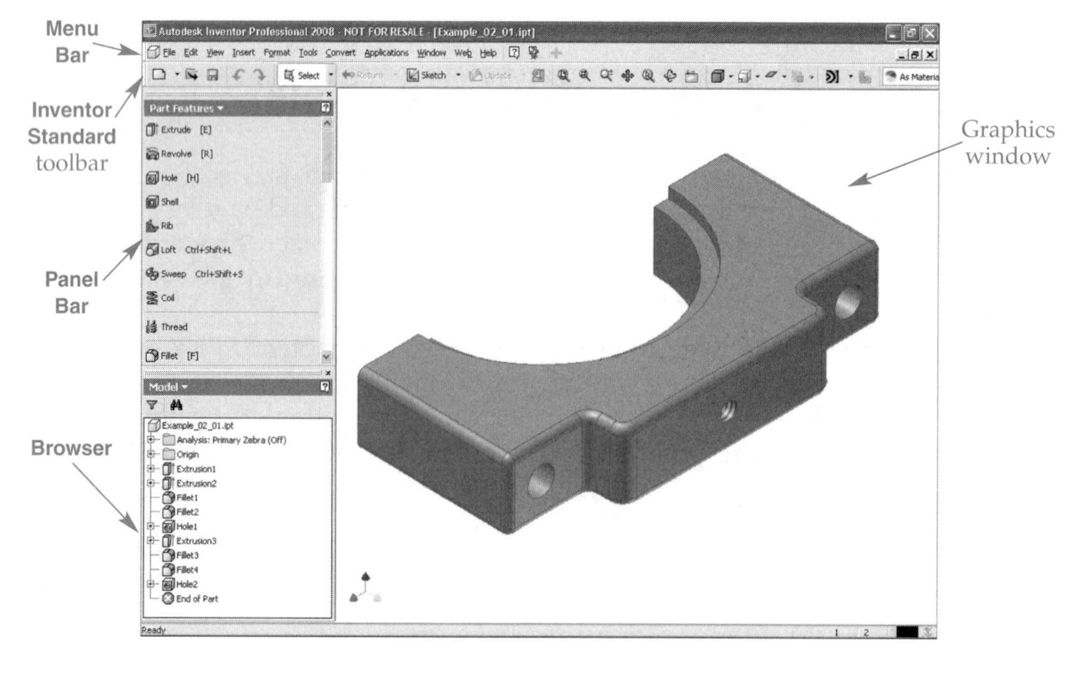

creating paths based on existing geometry for extruding sweep features, such as pipes or tubes. When Example_02_01.ipt is opened, Inventor is currently in part mode.

Inventor Standard Toolbar

The **Inventor Standard** toolbar is the main toolbar used when working in Inventor. See **Figure 2-3.** All of the common utility functions (**Save**, **Undo**, **Zoom Extents**, etc.) are located on this toolbar.

New Flyout Menu

There are four options in the **New** flyout menu. Pick the arrow next to the **New** button to display the flyout menu. See **Figure 2-4.** These options allow you to start a new file in one of four Inventor file types—assembly, drawing, part, and presentation. Other Inventor file types are not available in the flyout.

An assembly file is a collection of part files that are related to one another to create a completed product. Selecting **Assembly** in the **New** flyout menu creates a new assembly file. The default name is Assembly*x*, where *x* is a number. When the file is saved, you have the opportunity to give it a more meaningful name. Assembly files are saved with a .iam file extension. A new assembly file is based on the template corresponding to the default unit of measure, which was specified when Inventor was installed.

Drawing files are used to create 2D orthographic views of parts and assemblies. Selecting **Drawing** in the **New** flyout menu creates a new drawing file. The default name is Drawing*x*, where *x* is a number. Drawing files are saved with a .idw file extension.

Part files are the basis of everything in Inventor. They are used to create assemblies as well as drawings. Selecting **Part** in the **New** flyout menu creates a new part file. The new part file is based on the default template for the default unit of measure. The default name is Part*x*, where *x* is a number. Part files are saved with a .ipt file extension.

Figure 2-3.
The **Inventor Standard** toolbar, shown here floating and resized.

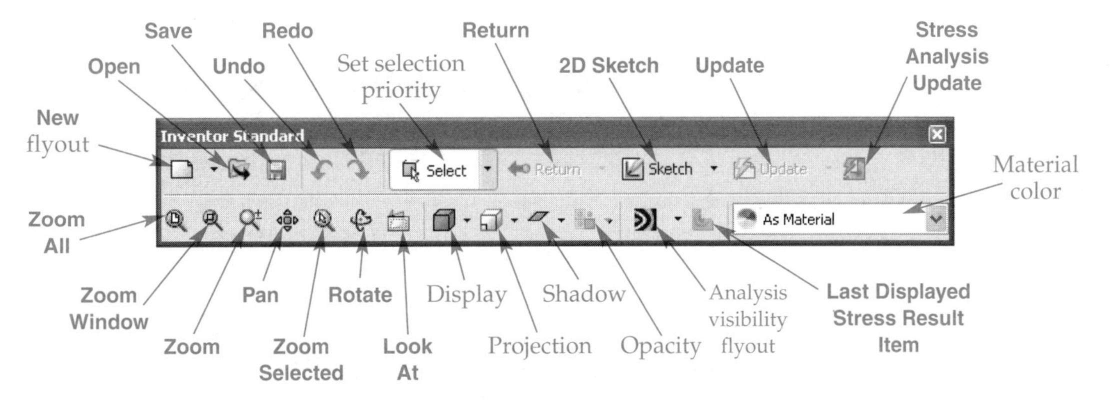

Figure 2-4.
The **New** flyout menu.

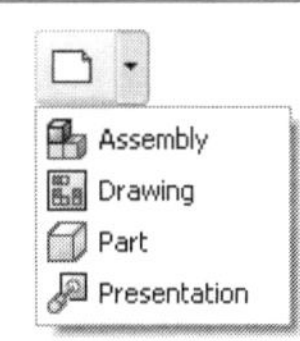

Presentation files are used to document the steps in the assembly or disassembly of an assembly file. Selecting **Presentation** in the **New** flyout menu creates a new presentation file. The default name is Presentation*x*, where *x* is a number. Presentation files are saved with a .ipn file extension.

NOTE

If you pick the **New** button, not the arrow next to the **New** button, the **New File** dialog box is displayed. This dialog box allows you to choose any of Inventor's file types for the new file.

Open Button

The **Open** button displays the **Open** dialog box. This dialog box is used to find and open an existing Inventor file. Navigate to files and folders as you would in a standard Windows "open" dialog box.

Near the bottom of the dialog box is the **Files of type:** drop-down list. This list includes all of the file types that can be opened in Inventor. The Inventor file types are IAM, IDW, IPT, IPN, and IDE. The DWG file type is the native format from AutoCAD and DXF is used to exchange 3D information with AutoCAD. The IGES, SAT, and STEP file types are for translation of 3D geometry from other CAD systems.

At the bottom of the **Open** dialog box is the **Find...** button. Pick this button to display the **Find** dialog box. See **Figure 2-5.** This dialog box can be used to search the target location for files that meet specified criteria.

Save Button

When working on a new file that has yet to be saved and the **Save** button is picked, the **Save As** dialog box is displayed. See **Figure 2-6A.** This dialog box is not displayed when saving a file that has previously been saved. The **Save As** dialog box is a standard Windows "save" dialog box. Use the **Save in:** drop-down list to navigate to the folder where the file is to be saved. When specifying a filename, you do not need to type the extension. Inventor knows the file type, which will be the only type available in the **Save as type:** drop-down list.

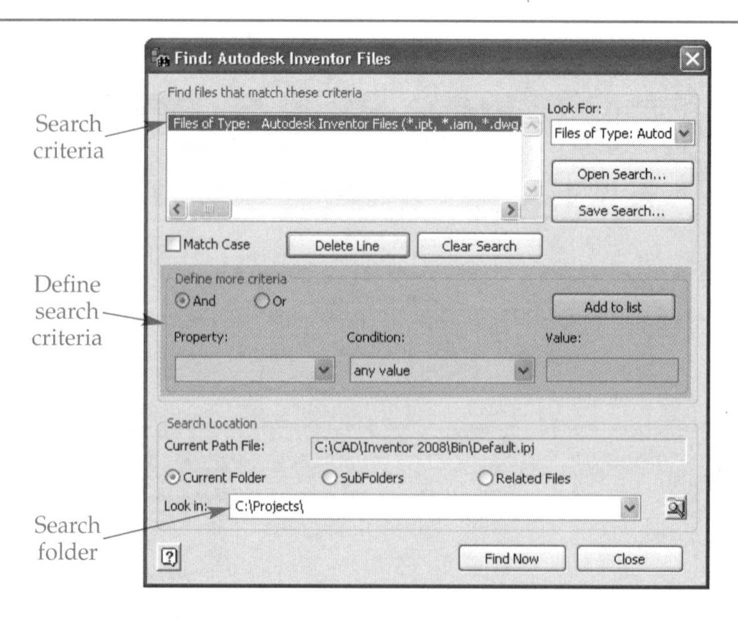

Figure 2-5.
The **Find** dialog box is used to search a target location for files that meet specified criteria.

Picking the **Options...** button in the **Save As** dialog box displays the **File Save Options** dialog box. See **Figure 2-6B.** Check the **Save Preview Picture** check box to save a thumbnail image with the file. The thumbnail is displayed in Windows Explorer when it is set to view thumbnails. When saving thumbnails, you can select one of the following for the thumbnail image source.

- Select **Active Component Iso View on Save** to capture an isometric view regardless of what is displayed in the active window.
- Select **Active Window on Save** if you wish to capture whatever is in the active window.
- Select **Active Window** if you wish to capture the active window. This image will be used from then on whenever **Save** is pressed. When this option is selected, the **Capture** button is used to capture whatever is in the active window.
- Select **Import From File** if you wish to use an external, 120 pixel × 120 pixel bitmap image as the thumbnail image. When this option is selected, the **Import** button is used to select the bitmap.

Figure 2-6.
A—The **Save As** dialog box. B—The **File Save Options** dialog box contains options for saving a thumbnail image with the saved file.

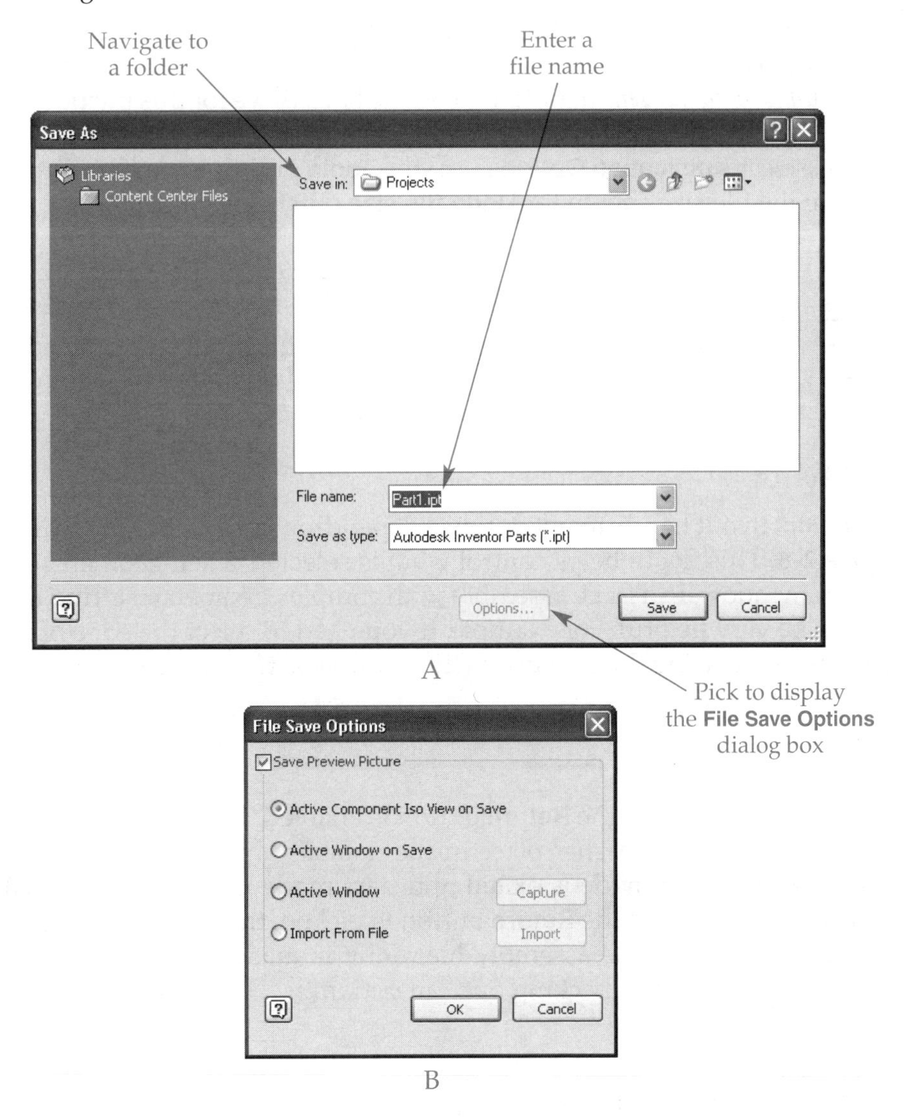

Figure 2-7.
The size of the undo temporary file can be increased in the **General** tab of the **Options** dialog box.

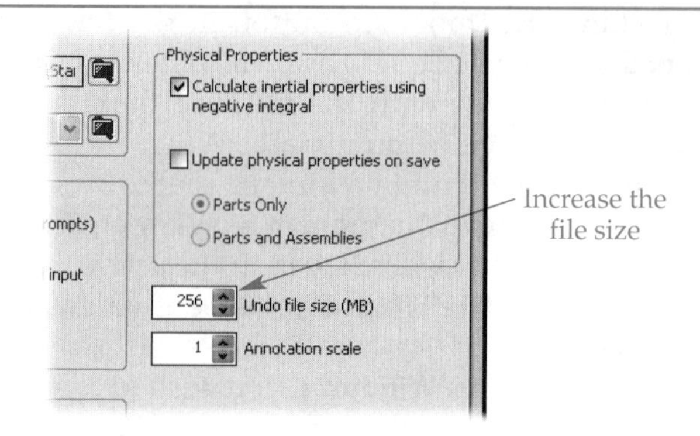

Undo and Redo Buttons

Selecting the **Undo** button will undo the last action, except view-related operations such as **Zoom** or **Pan**. Undo also acts on assemblies that have multiple part files open. Selecting the **Redo** button reverses the effect of the **Undo** button. A redo basically undoes the undo. As with undo, view-related operations are not affected by a redo. Both buttons can be picked multiple times to step backward or forward through undo and redo operations.

Undo is extremely powerful. It will correctly undo changes made to the possibly hundreds of part files that may have been changed due to a change in the assembly. There is an *undo temporary file* maintained on the hard drive for this purpose. Increasing its size may help performance issues when working with a large assembly. To increase the file size, select **Application Options...** in the **Tools** pull-down menu. On the **General** tab, increase the file size shown for **Undo file size (MB)**. See **Figure 2-7.**

NOTE

If the **Undo** button is selected enough times, the opened file will close. Selecting the **Redo** button will open that file again.

Select Flyout

The **Select** flyout has different options, depending on which type of file is current. See **Figure 2-8.** This flyout helps control what is selected when geometry is picked in the graphics window. In a large assembly with complex geometry, setting the selection priority can be very helpful. For example, if you need to select the edge of a part in an assembly, but are getting the part itself when you pick, try changing the select priority to **Select Faces and Edges**.

Return Button

When editing a sketch, the **Return** button is enabled. In general, the **Return** button allows you to return to a higher place in the database hierarchy. For example, if the button is picked, sketch mode is exited and part mode is made current. Also, when editing a subassembly and the **Return** button is picked, the subassembly is exited and the next assembly up in the assembly hierarchy is made current. Using the **Return** button can be thought as a "backing out" operation.

Sketch Button

The **Sketch** button is used to create a sketch on a work plane or the face of a part. While in part mode, pick the **Sketch** button and then select a planar face on a part. This starts a new sketch, which appears in the **Browser**, and activates sketch mode. The

Figure 2-8.
The **Select** flyout options help control what is selected when geometry is picked in the graphics window. Which options are displayed depend on the current mode. A—Part. B—Assembly. C—Drawing. D—Presentation.

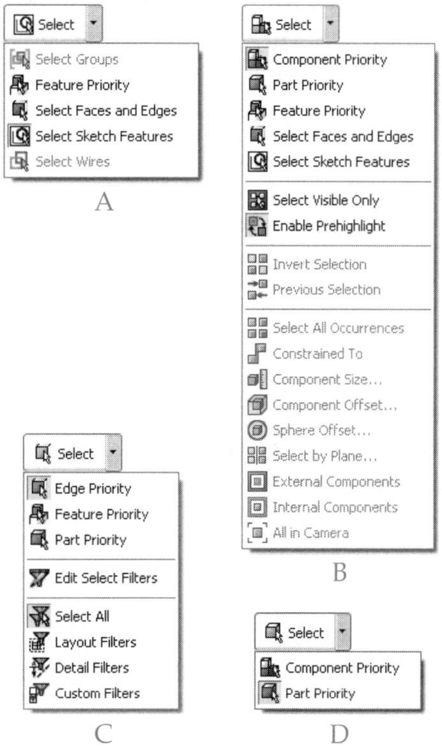

Select
Select Groups
Feature Priority
Select Faces and Edges
Select Sketch Features
Select Wires
A

Select
Component Priority
Part Priority
Feature Priority
Select Faces and Edges
Select Sketch Features
Select Visible Only
Enable Prehighlight
Invert Selection
Previous Selection
Select All Occurrences
Constrained To
Component Size...
Component Offset...
Sphere Offset...
Select by Plane...
External Components
Internal Components
All in Camera
B

Select
Edge Priority
Feature Priority
Part Priority
Edit Select Filters
Select All
Layout Filters
Detail Filters
Custom Filters
C

Select
Component Priority
Part Priority
D

Panel Bar also changes to show the 2D sketch tools. Try it now by picking the **Sketch** button and then picking the top face on Example_02_01.ipt. You should be in sketch mode and a new sketch should be listed at the bottom of the **Browser** immediately above the End of Part feature. In sketch mode, you can draw a new 2D shape. This shape can be constrained and extruded into a new 3D feature added onto the part. This is covered in greater detail in Chapter 3. To exit sketch mode and return to part mode, pick the **Return** button or right-click and select **Finish Sketch** from the pop-up menu.

Update Button

Inventor is a parametric modeler. Parts and assemblies designed in Inventor can be changed time and time again. The **Update** flyout menu is used after a part has been modified so the part reflects the changes. See **Figure 2-9.**

With Example_02_01.ipt open, right-click on Extrusion1 in the **Browser** and select **Edit Sketch** from the pop-up menu. The dimensions for the extrusion and the sketch on which it is based appear in the graphics screen. Double-click on the radius dimension (33), change its value to 35 in the **Edit Dimension** dialog box, and pick the check mark button in the dialog box or press [Enter]. The dimension of the feature has been changed; however, the part is not yet updated to reflect the change. Pick the **Update** button to update the part. After examining the change, pick the **Undo** button until the change is reversed.

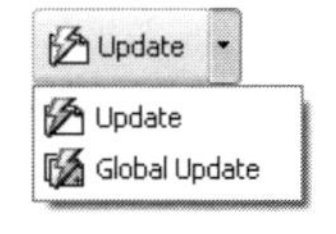

Figure 2-9.
The **Update** flyout is used to update the part to reflect the changes just made.

One of the main uses for the **Update** flyout menu is for updating parts in assembly mode when you are working on a large assembly with subassemblies. The **Update** button is for the subassembly you are currently editing. The **Global Update** button is to force an update of the main assembly and all subordinate assemblies.

Zoom All Button

The **Zoom All** button is used to fill the graphics window with all of the part or assembly. The display will zoom out or in as needed.

Zoom Window Button

The **Zoom Window** button is used to fill the graphics window with a selected rectangular area of the display. After picking the button, select two points to define the window. The two points are defined with two mouse picks, or a pick-and-drag method can be used. Inventor zooms to fill the graphics screen with that window. Pressing [F5] function key redisplays the previous view. Remember, **Undo** does not undo view-related operations.

Zoom Button

The **Zoom** button is used to magnify or reduce the display in the graphics windows. After picking the button, the zooming cursor appears in the graphics window. You can also press the [F3] function key and hold it down to activate the **Zoom** tool. To zoom in, press and hold the left mouse button and drag the mouse toward you. To reduce the size of the image in the graphics window, drag the mouse away from you. When done, press [Esc] or right-click in the graphics window and select **Done** from the pop-up menu. Zooming can be used at any time, even within a command. This is known as zooming transparently.

PROFESSIONAL TIP

A great trick is to simply use the mouse's roller wheel to zoom—roll the wheel toward you to magnify or away from you to reduce the image.

Pan Button

The **Pan** button is used to shift the display within the graphics window. After pressing the **Pan** button, the panning cursor appears in the graphics window. The [F2] function key can also be held down to activate the **Pan** tool. To pan around the graphics window, press and hold the left mouse button and drag to obtain the desired display. When done, press [Esc] or right-click in the graphics window and select **Done** from the pop-up menu. Panning can be done at any time, even within a command.

Zoom Selected Button

The **Zoom Selected** button is used to zoom in on a selected feature or face of a part. To use this tool, first pick a feature or a face of a part. Then, pick the **Zoom Selected** button. The graphics window is zoomed so the selected item fills the display.

Rotate Button

The **Rotate** button may be the most useful viewing tool in Inventor. It is sometimes called *3D rotate.* With this tool, the viewpoint can be moved around the model in 3D

Figure 2-10.
The **Rotate** tool is used to rotate the view. The rotate circle appears in the graphics window when the tool is active.

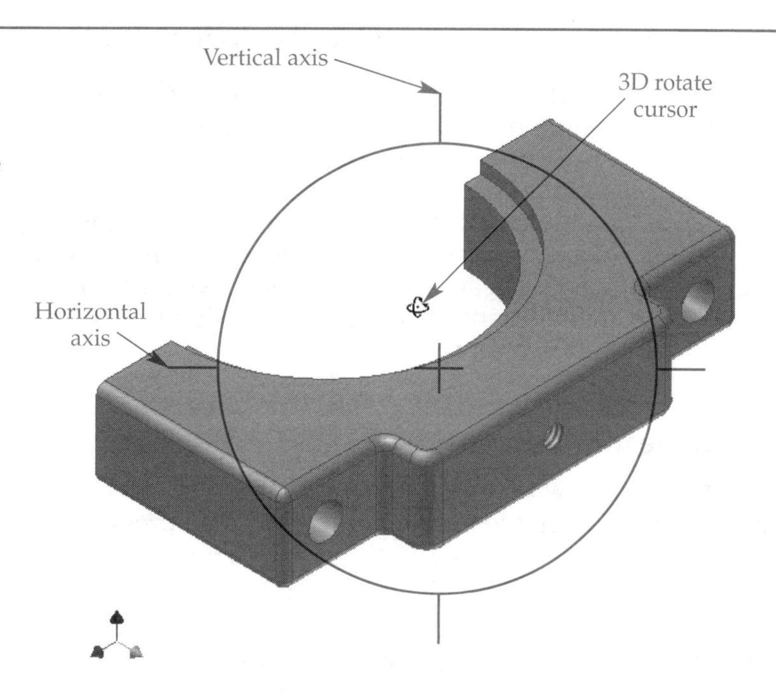

space. Once the **Rotate** button is picked, the rotate circle and 3D rotate cursor appear on screen. See **Figure 2-10.** The [F4] function key can also be held down to activate the tool.

Keep in mind as you use the **Rotate** tool that what you are actually doing is manipulating the point in space from which you are viewing the part—your viewpoint. This tool is an extremely effective visualization tool. It will be used every day you use Inventor, so take the time to master its use. To revert to the default 3D view after using the tool, right-click with no tool active and select **Isometric view** from the pop-up menu.

Position the 3D rotate cursor inside of the circle. Press and hold the left mouse button and move the mouse around. Your viewpoint should be rotating around the part. This is known as ***free-rotate mode.***

Release the mouse button and position the cursor outside of the circle. Press and hold the left mouse button. Move the mouse in a large circle around the screen. This rotates the view about the view axis, which is the imaginary line coming straight out of the monitor. This is similar to a wheel rotating on an axis.

Release the mouse button and position the cursor over the vertical axis near the top of the screen. Press and hold the left mouse button. Move the mouse straight down. The part rotates about an imaginary horizontal axis. This is similar to a barbecue spit turning over the grill.

Release the mouse button and position the cursor over the horizontal axis near the left-hand edge of the screen. Press and hold the left mouse button. Move the mouse straight across the screen toward the right. The part rotates about an imaginary vertical axis.

Finally, release the mouse button and position the cursor inside of the circle. Then, pick the point by pressing the left mouse button. The view is panned to center the point in the view screen. You can also pick outside of the circle to pan as long as the cursor is not the return cursor. The return cursor appears as an arrow pointing to the left. Picking when this cursor is displayed ends the **Rotate** tool.

PROFESSIONAL TIP

A little-known trick when using the **Rotate** tool is to create a continuous orbit. Pick the **Rotate** button and position the cursor inside of the circle. Hold down the [Shift] key and the left mouse button. Then, "stroke" across the graphics screen in any direction and release the mouse button. The display spins in the direction of your mouse stroke. The length of the stroke determines the spin speed. To cancel the spin and leave the **Rotate** tool active, left-click in the graphics window. To exit the operation, pick the **Rotate** button or press the [Esc] key.

Common view option

Another option with the 3D rotate tool is the ***common view.*** To display the **Common View** tool, pick the **Rotate** button and then press the spacebar. A translucent cube with green arrows appears in the graphics window. See **Figure 2-11.** This is the **Common View** tool. To change the operation back to the 3D rotate view, simply press the spacebar again.

The green arrows represent view directions. If you pick on a green arrow, then that view of the part is shown in the graphics window. Picking an arrow at a corner of the cube will result in the isometric view from that direction. For example, if you select the arrow at the top, back corner of the cube, the view rotates so that isometric view is displayed. The green arrows on the flat faces of the cube correspond to the six standard orthographic views.

Notice the coordinate system indicator in the lower-left corner of the graphics window as you pick different arrows. The common view cube and arrows are oriented along the XYZ axes of the coordinate system. Pick the arrow for the top of the part. This displays the top, or plan, view of the part. If you pick the same arrow again, the bottom view is displayed. You can do this for any of the arrows. Picking twice on any view direction displays the first view and then its opposite view.

Pick the top arrow to give a plan view of the part. Now, hover over one of the edges of the cube. Notice that half of the line turns red. Pick on the lower "half edge" and the view is rotated 90° to the right (counterclockwise). Pick on the upper half edge and the view is rotated 90° to the left (clockwise).

To exit the common view, press the [Esc] key or right-click in the graphics window and select **Done** from the pop-up menu. The **Rotate** button works as a toggle between

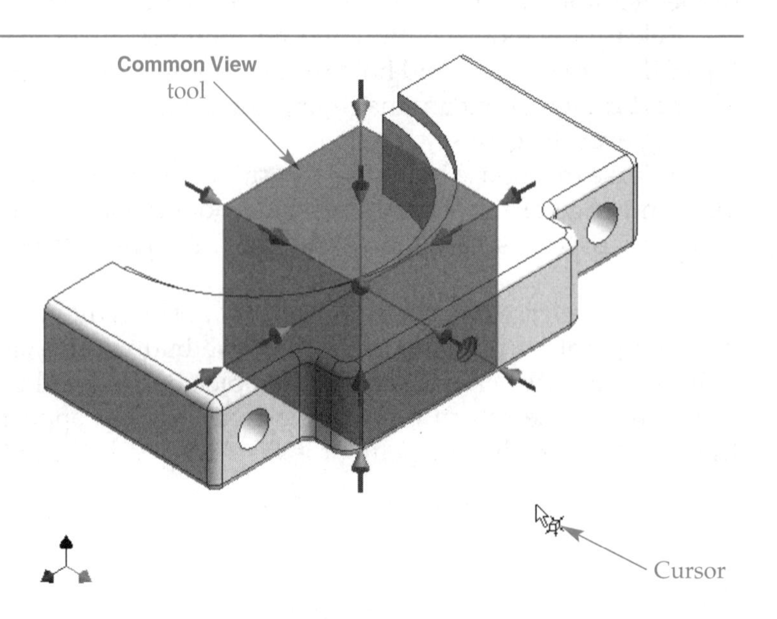

Figure 2-11.
The **Common View** tool is used to display plan and isometric views of the part.

the common view and the 3D rotate view. Whichever view was last used—common view or 3D rotate view—is activated the next time the **Rotate** button is picked.

Redefining the isometric view

Inventor has a default isometric view. To display this view, press the [F6] key or right-click in the graphics window and select **Isometric View** from the pop-up menu. This view is always displayed when the isometric view is restored. However, the isometric view can be redefined.

To redefine the isometric view, first display the common view. Then, using the green arrows at the corner of the cube, display the new isometric view. The new view must be an isometric view, not an orthographic view or one created in the 3D rotate view. Finally, right-click in the common view and pick **Redefine Isometric** from the pop-up menu.

Open the Example_02_02.ipt file. Right-click and choose **Isometric View**. Examine the orientation of the part in this view. Next, pick the **Rotate** button and, if needed, press the spacebar to activate the common view. Pick the green arrow pointing to topmost corner of the cube. Refer to **Figure 2-12.** After the view rotates to the new viewpoint, right-click in the graphics window and select **Redefine Isometric** from the pop-up menu. Then, press the [Esc] key to exit the rotate operation. Press and hold the [F4] key to activate the 3D rotate view and then change the viewpoint. Now, right-click in the graphics window and select **Isometric View** from the pop-up menu. The viewpoint rotates to the new isometric view, not the one that was initially defined. Close the part file without saving changes.

Rotating around a large part

Sometimes, when rotating a zoomed-in display of a large part or assembly, the currently displayed area rotates way out of the view. To avoid this problem, initiate the 3D rotate view. Then, use the left-click pan method to place the area you want to view in the center of the circle. This redefines the center of the operation. Then, rotate the view as needed.

Open Dogging_Assembly.ipn found in the \Ch02\Example_Assembly folder. Make sure you open the IPN file, not the IAM file. Zoom in on the assembly until it is displayed at approximately twice the original size. Some of the assembly will not be visible in the display. Then, pick the **Rotate** button and rotate the viewpoint. To swing

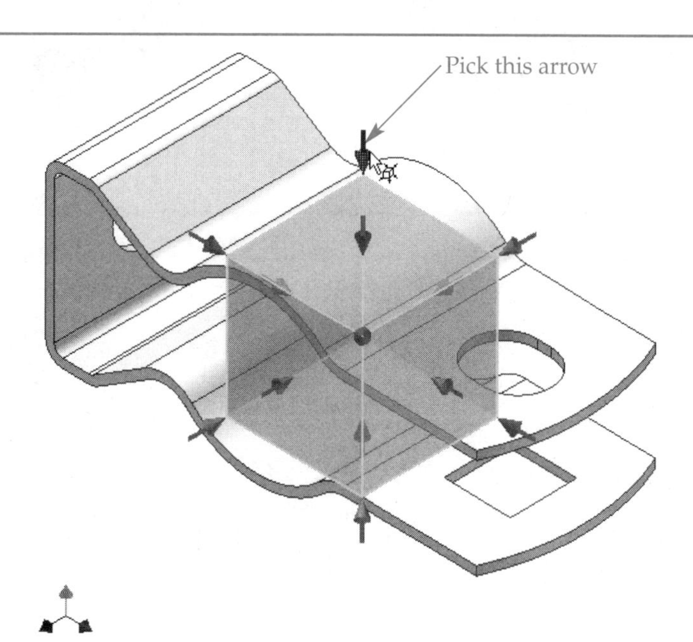

Figure 2-12.
To redefine the isometric view, first display the new isometric view to be the default. Then, right-click and select **Redefine Isometric** from the pop-up menu.

the viewpoint around the bronze-colored component on the left of the assembly, left-click to pan it into the center of the rotate circle. You may need to pick more than once if it is not currently visible in the display. Now, rotate the view again. The bronze-colored component remains in the center of the view as you rotate the view. Close the file without saving it.

Look At Button

The **Look At** tool is used to display a plan view of a selected feature. This tool is useful when working on sketches to switch from an isometric to a plan view of the sketch.

Open the file Example_02_01.ipt, if it is not already open. Pick the **Look At** button and then a face on the part. A plan view, or "straight on" view, of that particular face is displayed. Display the isometric view and try using the tool on the other part faces.

Sometimes when the **Look At** tool is used, the part turns around leaving it upside down. Each face has its own coordinate system and Inventor is orienting that one with the screen.

Display Flyout

The buttons in the display flyout are used to change the shading of the part. The **Shaded Display** button displays the part as a shaded object. The **Hidden Edge Display** button displays the part as a shaded object with hidden lines shown. The **Wireframe Display** button displays the part as a wireframe with all lines shown. See **Figure 2-13.**

Figure 2-13.
The different shading modes in Inventor. A—**Shaded Display**. B—**Hidden Edge Display**. C—**Wireframe Display**.

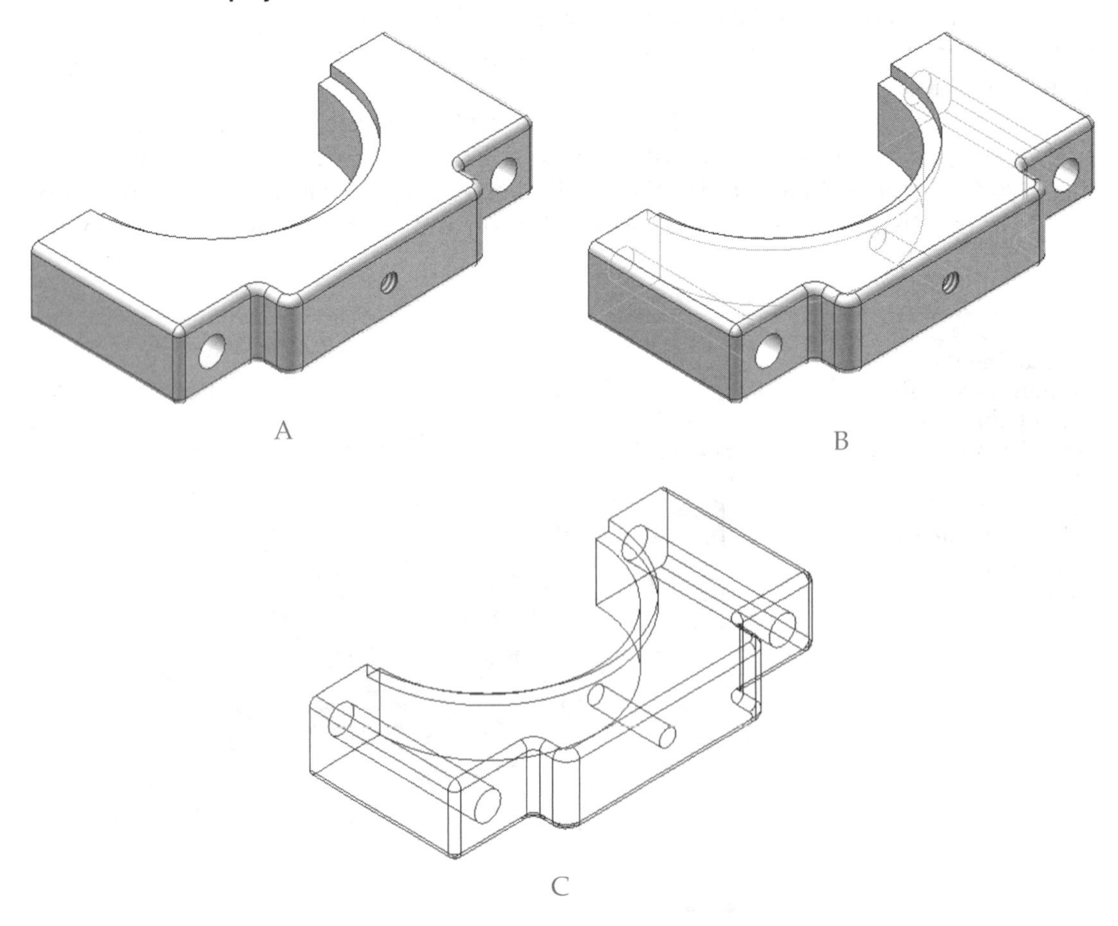

Figure 2-14.
The view in Inventor can be based on orthographic or perspective projection.

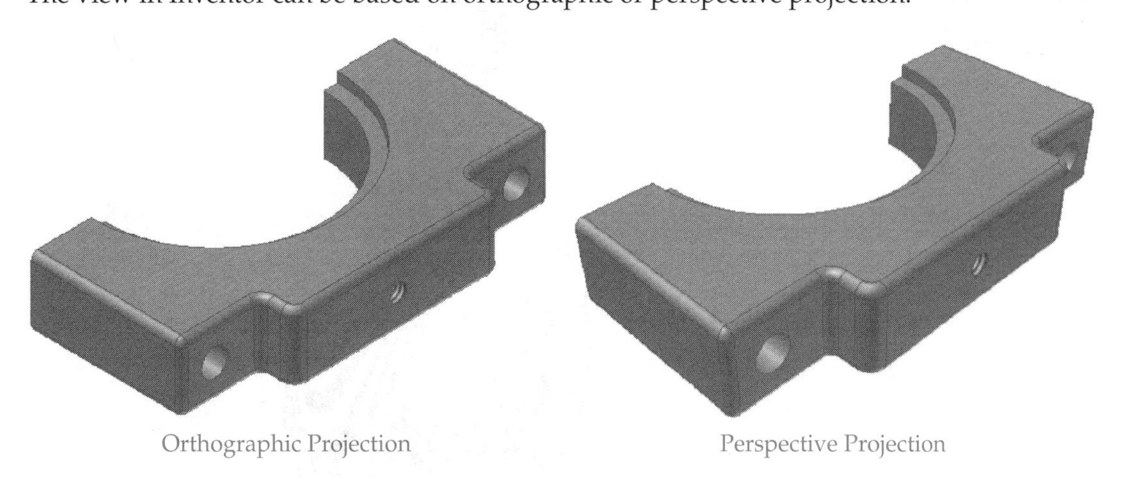

Camera Flyout

An imaginary camera generates the view of the model. The viewpoint determines the location of the camera. The view generated by the camera is based on one of two methods of projection. See **Figure 2-14.** ***Orthographic projection*** shows the part so all of its points project along parallel lines to their positions on the screen. This is the mode used most often. It is set current by picking the **Parallel** button in the projection flyout. In ***perspective projection,*** parallel lines recede to a common point in the distance. This mode is very useful for presentations or whenever a more realistic picture of the part is desired. It provides a sense of how parts and assemblies would appear to the human eye when actually manufactured. Projection mode is set current by picking the **Perspective** button in the projection flyout.

Shadow Flyout

The buttons in the shadow flyout determine if a shadow of the part is displayed in the graphics screen and, if so, the type of shadow. See **Figure 2-15.** Shadows are useful in combination with perspective view for realistic images.

Figure 2-15.
Ground shadow effects. A—**Ground Shadow**. B—**X-Ray Ground Shadow**.

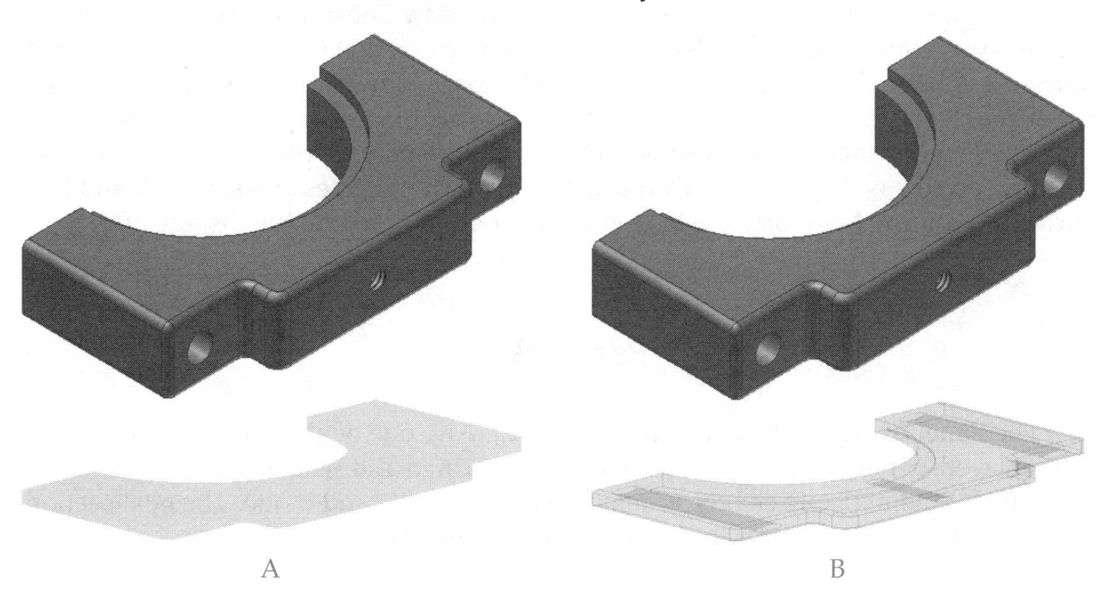

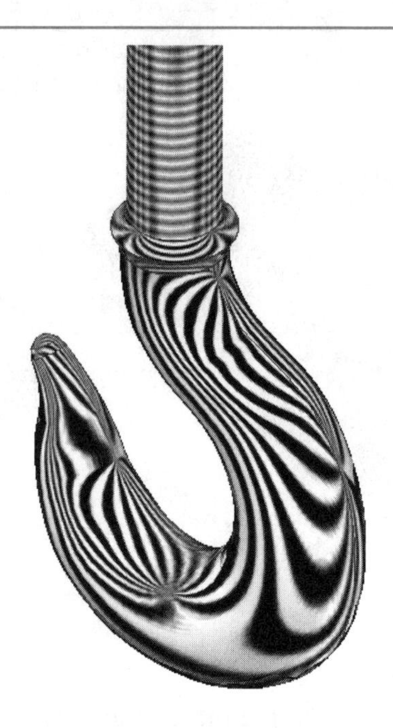

Figure 2-16.
A part showing the results of analyzed surfaces. This is a zebra analysis.

When the **No Ground Shadow** button is selected, no shadow is shown. A simple shadow under the part is shown when the **Ground Shadow** button is selected. When the **X-Ray Ground Shadow** is selected, a shadow is displayed with hidden lines within the shadow. It is a nice effect, but not very realistic.

Component Opacity Flyout

The buttons in the opacity flyout turn component opacity on and off. Component opacity refers to the transparency of assembly components above the currently edited subassembly in the assembly hierarchy. It is used when editing assemblies and is covered in greater detail in later chapters of this textbook.

Analyze Faces Button

The buttons in the analysis visibility flyout display the part with an analysis of the surfaces for continuity, curvature, cross sections, or valid draft angles. An example of continuity known as zebra analysis is shown in **Figure 2-16.** The button in the flyout acts as a toggle, turning the analysis on or off.

Open the Example_02_03.ipt file and pick the **New Zebra Analysis** button in the analysis visibility flyout. In the **Zebra Analysis** dialog box that is displayed, pick the **OK** button. A zebra stripe pattern is displayed on the part. Now, rotate the part to examine the surface continuity. Pick the **New Surface Analysis** button in the analysis visibility flyout. In the **Surface Analysis** dialog box that is displayed, pick the **OK** button. A pattern of colors is displayed on the part. This pattern indicates areas of high and low surface curvature. Now, pick the button on the toolbar to toggle the display off.

Panel Bar

The **Panel Bar** is the main Inventor interface. Any one of a number of panels may be displayed, depending on the type of file being edited and what mode Inventor is in. The panels automatically change to match the current file and mode. The next section discusses the panels that may be displayed in the **Panel Bar**.

Figure 2-17.
The **2D Sketch** panel in expert mode.

Some of the buttons in the panel bars have submenus, or flyouts, that are denoted by a small black arrow next to the button name. If you left-click on one of these arrows, a submenu is accessed.

Inventor has an "expert mode" for the panels in the **Panel Bar**. In expert mode, the explanatory text for each button is hidden; only the icon is displayed. **Figure 2-17** shows the **2D Sketch Panel** in expert mode. Notice the text is gone and only the icon is shown on the button. To enter expert mode, right-click on the background of a panel bar and uncheck **Display Text with Icons** in the pop-up menu. To exit expert mode, select **Display Text with Icons** from the pop-up menu so it is checked.

Panels

With Example_02_03.ipt open, notice the **Part Features** panel is displayed in the **Panel Bar**. See **Figure 2-18A.** This indicates Inventor is in part mode. The **Part Features** panel contains feature tools needed to create a part.

If you select the top face of the hook and pick the **Sketch** button on the **Inventor Standard** toolbar, sketch mode is entered and the **2D Sketch Panel** is displayed in the **Panel Bar**. See **Figure 2-18B.** This panel contains the tools needed to draw and constrain a sketch. If you pick the **Return** button on the **Inventor Standard** toolbar, you are returned to part mode and the **Part Features** panel is again displayed in the **Panel Bar**.

Inventor has a built-in sheet metal modeling module. To start a new sheet metal file, pick the **New** button on the **Inventor Standard** toolbar. Then, select Sheet Metal.ipt in the **New File** dialog box. The new file is opened in sketch mode, as indicated by the **2D Sketch Panel** in the **Panel Bar**. Pick the **Return** button on the **Inventor Standard** toolbar to enter sheet metal mode. The **Sheet Metal Features** panel is then displayed in the **Panel Bar**. See **Figure 2-18C.** This panel contains the tools necessary to complete a sheet metal part.

Inventor also has the ability to create 2D drawing views. To start a new drawing file, pick the **New** button on the **Inventor Standard** toolbar. Then, select Standard.idw in the **New File** dialog box. The new drawing file is opened in drawing mode. The **Drawing Views Panel** is displayed in the **Panel Bar**. See **Figure 2-18D.** This panel contains the tools necessary to create different types of 2D views.

To start a new assembly file, pick the **New** button on the **Inventor Standard** toolbar. Then, select Standard.iam in the **New File** dialog box. Alternately, you can pick **Assembly** in the **New** flyout, as described earlier. The **Assembly Panel** is displayed in the **Panel Bar**. See **Figure 2-18E.** This panel is used to import, manipulate, and constrain part components into an assembly.

To start a new presentation file, pick the **New** button on the **Inventor Standard** toolbar. Then, select Standard.ipn in the **New File** dialog box. Alternately, you can pick **Presentation** in the **New** flyout, as described earlier. The **Presentation Panel** is displayed in the **Panel Bar**. See **Figure 2-18F.** This panel is used to import and manipulate components into a presentation.

Figure 2-18.
The panel displayed in the **Panel Bar** is dependent on the current mode. A—**Part Features** panel. B—**2D Sketch Panel**. C—**Sheet Metal Features** panel. D—**Drawing Views Panel**. E—**Assembly Panel**. F—**Presentation Panel**.

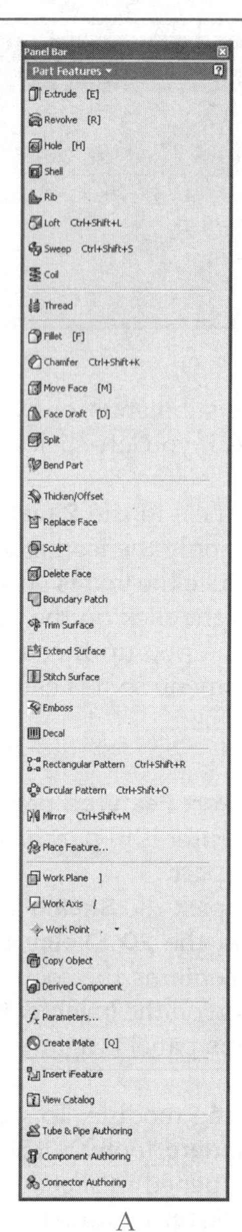

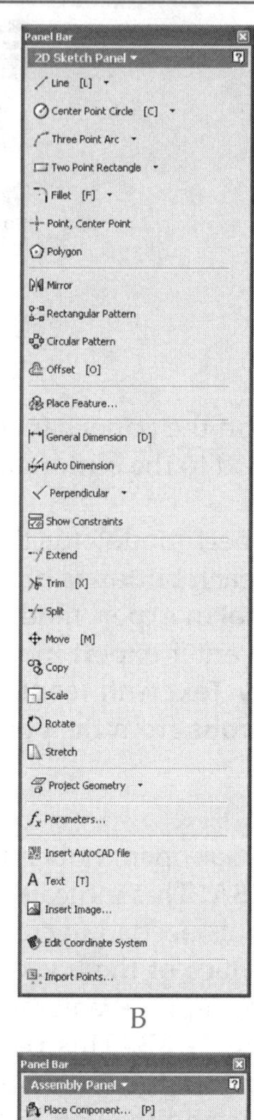

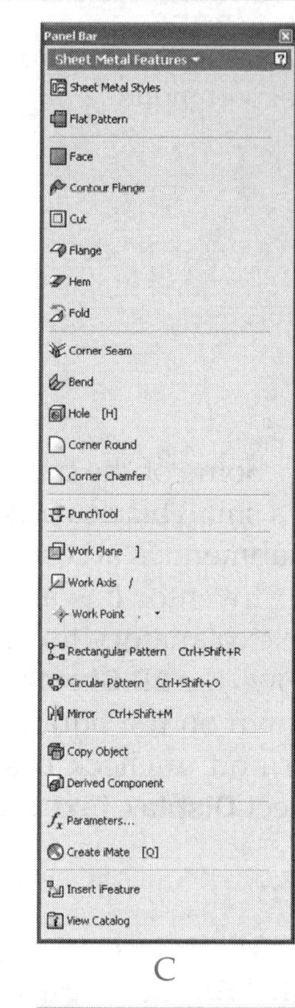

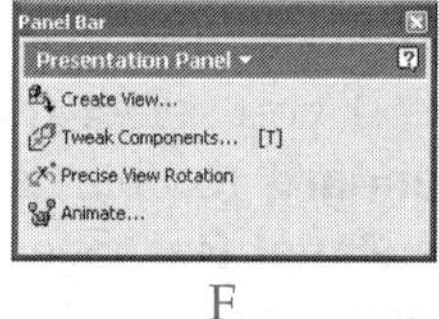

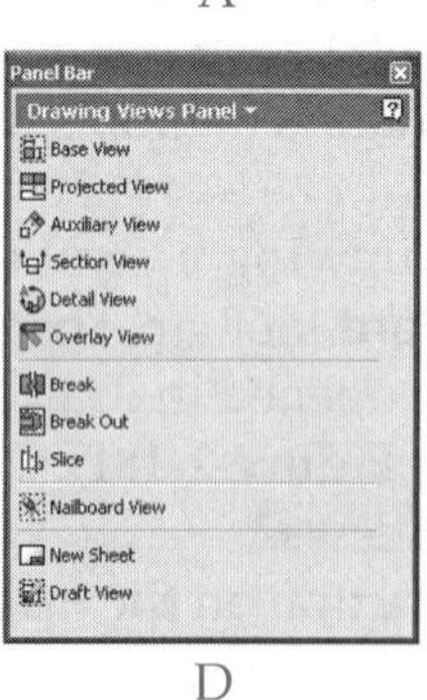

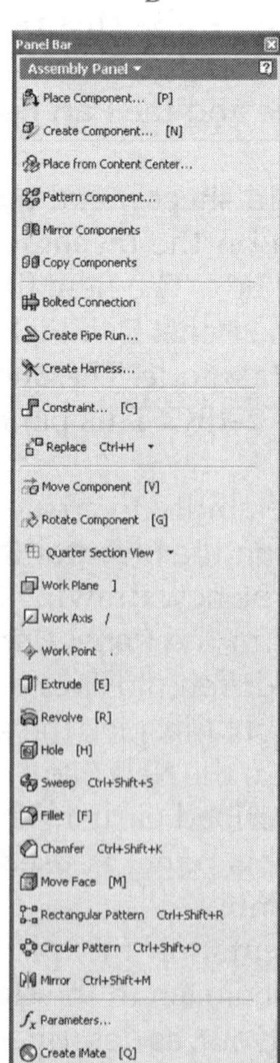

Resizing the Panel Bar

The **Panel Bar** can be resized by hovering over its border until the standard Windows resizing cursor appears. Then, pick, hold, and drag the **Panel Bar** to the new size.

The **Panel Bar** can also be floated by picking and holding on the two bars at the top of the panel and dragging away from the docked location. To dock the floating **Panel Bar**, pick and hold on the title bar and drag it to a docked position.

The **Panel Bar** can be closed by picking the **X** in the upper-right corner of the panel. To redisplay the **Panel Bar**, pick **Panel Bar** in the **Toolbar** cascading menu of the **View** pull-down menu. You can also right-click on a toolbar and select **Panel Bar** in the pop-up menu so it is checked.

Browser

The **Browser Bar**, or **Browser**, is the heart of Inventor. You will spend a great deal of time using it to manage features. Inventor creates parametric parts that are based on parametric features. The features are, in turn, based on sketches. All of the items used to build a part are listed in the **Browser** as they are created. The **Browser** can be floated or docked and resized in the same manner as the **Panel Bar**.

Figure 2-19 shows the **Browser** listing for Example_02_01.ipt. If you move your cursor over the name of a feature, a red box appears around the feature name. Also, the feature is highlighted in red in the graphics window as long as the cursor is over the feature name. See **Figure 2-20.** If you pick a feature in the **Browser**, such as Extrusion1, it stays highlighted in the graphics window.

More than one feature can be selected in the **Browser**. To select consecutively listed features, pick the first one in the **Browser**, hold down the [Shift] key, and pick the last one. To select individual features, hold down the [Ctrl] key and pick the features in the **Browser**.

In most CAD systems, geometry is selected in the graphics window only. In Inventor, geometry can be selected in the graphics window or the **Browser**. This is extremely useful because in some situations you can *only* select items from the **Browser**.

With the file Example_02_01.ipt open, pick the plus sign next to Extrusion1 in the **Browser**. The tree is expanded to show the sketch that is the basis for that feature. Features defined by a sketch, such as the extrusions, are called ***sketched features.*** Some of the features in the **Browser**, such as the fillets, do not have a plus sign next to them. These are called ***placed features*** and are not defined by a sketch. Rather, they are "placed" on other features.

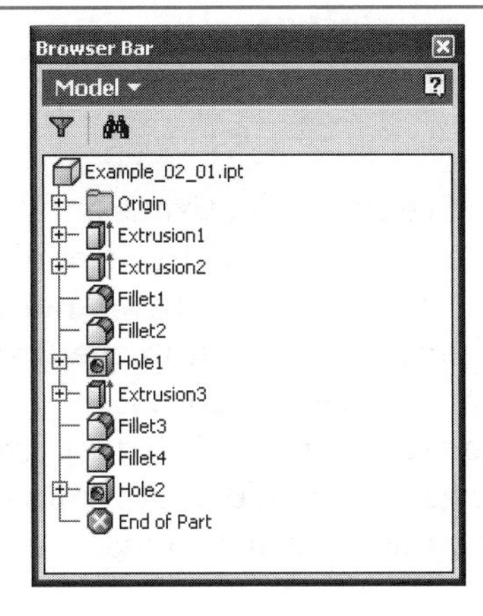

Figure 2-19.
The **Browser** shows the tree, or hierarchy, of a part.

Figure 2-20.
Highlighting a feature in the **Browser** also highlights it in the graphics window.

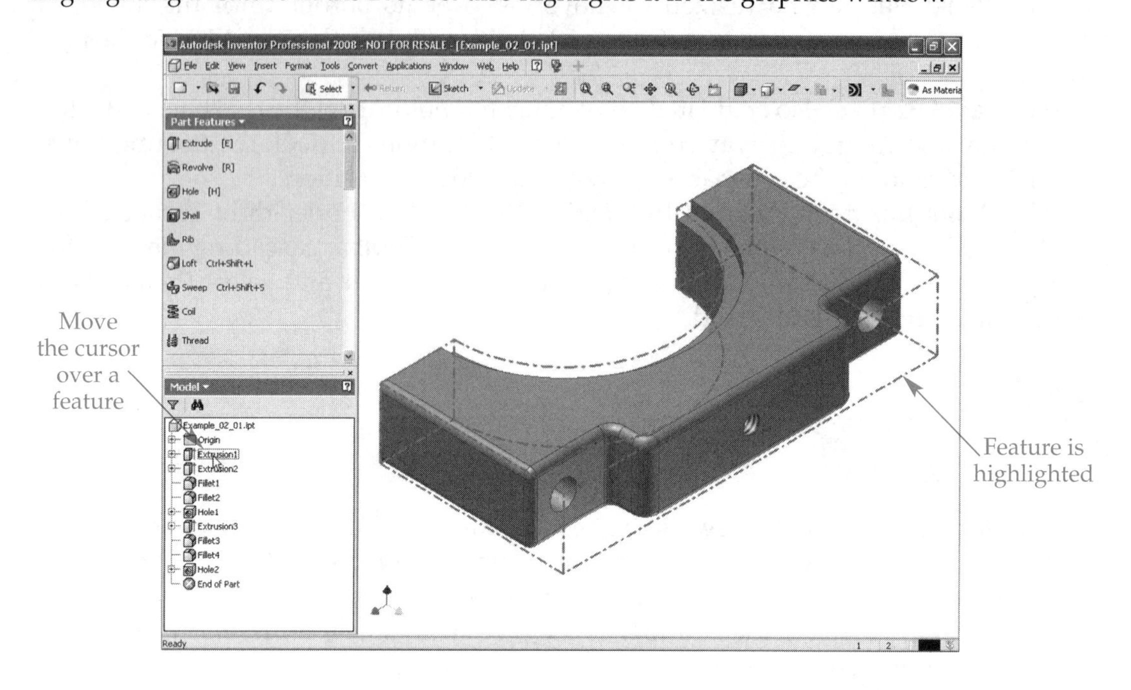

Figure 2-21.
Double-click on the sketch name in the **Browser** to change to sketch mode.

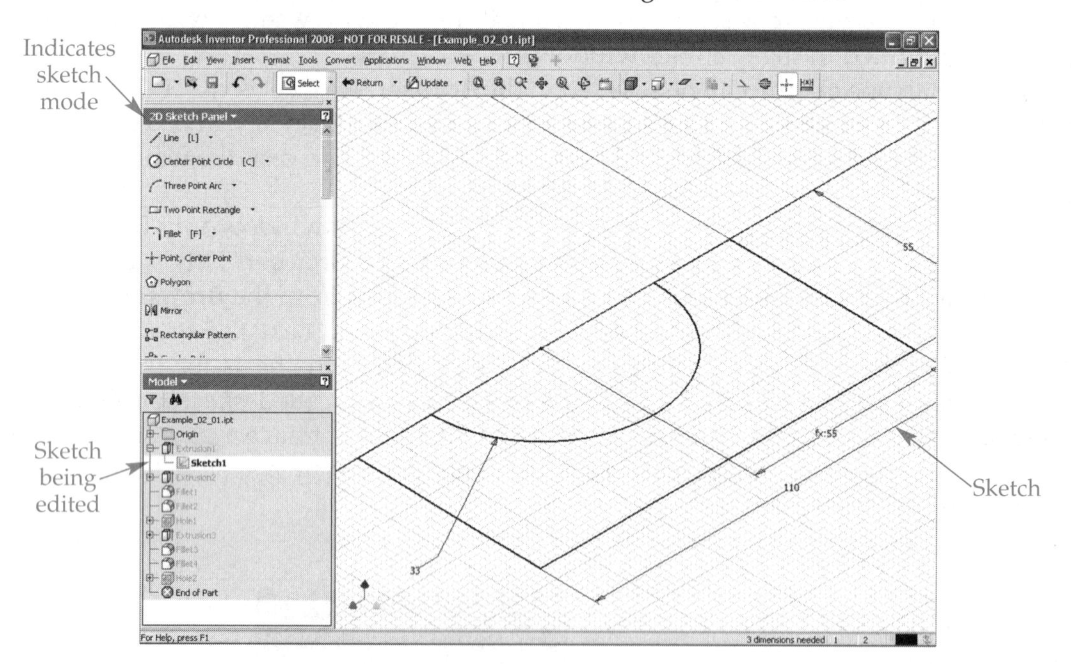

Double-click on Sketch1 under Extrusion1. Your screen should look like **Figure 2-21.** Sketch mode has been entered and the graphics window displays the 2D sketch that is the basis of Extrusion1. Notice that the **Panel Bar** displays the **2D Sketch Panel**. Also, notice that everything is grayed out in the **Browser** except for Sketch1. This signifies that the other features are unavailable.

At this point the sketch can be edited—perhaps the dimensioning scheme needs to be changed or some geometry added or removed. However, do not use this method to change the numeric values of dimensions, even though you could. The proper method is discussed in the next section.

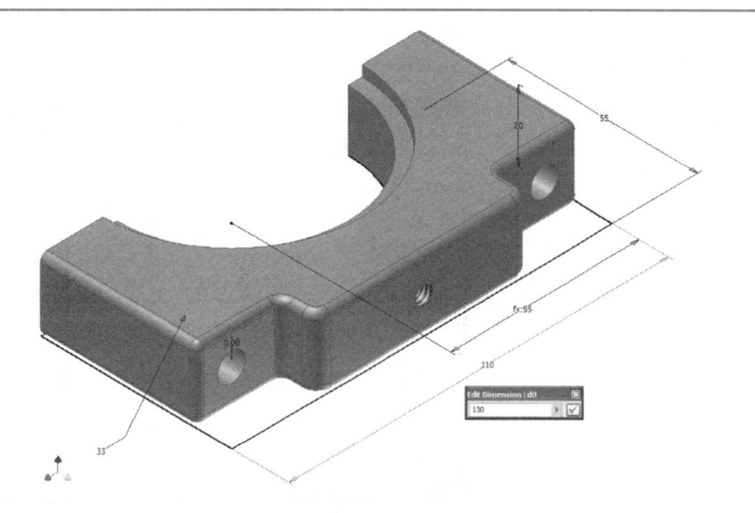

Figure 2-22.
Editing dimensions of the feature in part mode.

Exit sketch mode by picking the **Return** button on the **Inventor Standard** toolbar or by right-clicking in the graphics window and picking **Finish Sketch** in the pop-up menu. You are now back in part mode. Notice the change in the **Panel Bar** and the **Browser**.

Changing Dimensions

To change the dimensions controlling a feature, while in part mode, right-click on the feature in the **Browser** and pick **Show Dimensions** in the pop-up menu. For example, with Example_02_01.ipt open, right-click on Extrusion1 and pick **Show Dimensions**. In the graphics window, find the 110 mm dimension and double-click on it. The **Edit Dimension** dialog box is opened with the current dimension value displayed. Change the value to 130 mm and press [Enter] or pick the green check mark. See **Figure 2-22.**

The dimension changes, but the feature does not change. To update the part, pick the **Update** button on the **Inventor Standard** toolbar. The feature is changed to reflect the new dimension. This example illustrates the parametric capabilities of Inventor.

Renaming Features

As features are created, Inventor automatically names them. The default feature names, such as Extrusion1, are not very descriptive. The **Browser** can be used to rename features. With Example_02_01.ipt open, left-click two single times on Extrusion1 in the **Browser**; do not double-click. See **Figure 2-23.** The name is replaced by a text box and the current name is highlighted. Now, type a new name, such as Base Feature, and press [Enter].

Renaming features is very important. For example, the hole features in Example_02_01.ipt should be renamed to be more descriptive—perhaps revealing the size of the hole or if it is tapped. You do not want to have a finished part containing 50 features with names Extrusion1 through Extrusion50. The little extra time required to rename features can save hours of headaches later trying to figure out which feature is which.

Pop-Up Menu

Right-clicking on a sketched feature displays a pop-up menu. See **Figure 2-24.** The options in this menu are:

- **3D Grips.** Displays the model in grip mode, which is a wireframe with grips. By dragging the grips, the size of the feature can be changed.
- **Move Feature.** Allows a feature to be moved using a triad; similar to grip mode.

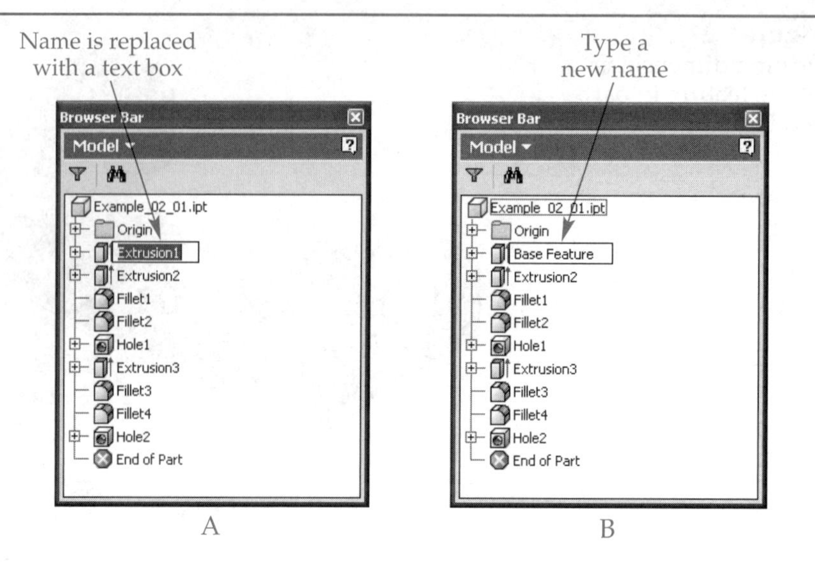

Figure 2-23.
Renaming a feature in the **Browser**. A—Left-click two single times on the name. B—The renamed feature.

- **Copy.** Places a copy of the feature onto the Windows clipboard for use in another part.
- **Delete.** Removes the feature and any dependent features from the part. Dependency is covered in more detail later in the book.
- **Show Dimensions.** Displays the dimensions of the sketch on which the feature is based. To turn off the display of the dimensions, pick in the graphics window to deselect the feature and then pick a feature in the **Browser**.
- **Edit Sketch.** Displays in sketch mode the sketch on which the feature is based. This is an alternative to double-clicking on the feature's sketch in the **Browser**.
- **Edit Feature.** Allows you to edit more than just the feature's dimensions. Its use is covered in later chapters.
- **Infer iMates.** Used to create iMate definitions for the feature.
- **Create Note.** Activates the engineer's notebook, which is covered later in this chapter.
- **Suppress Features.** Simplifies the part or temporarily removes a feature from the part. This is covered in detail in later chapters.

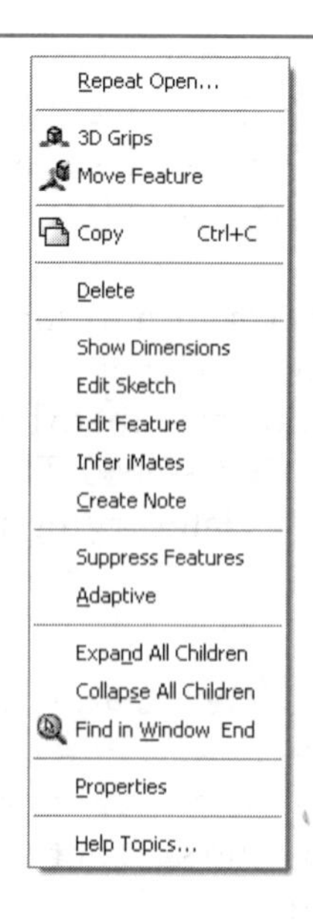

Figure 2-24.
The pop-up menu displayed by right-clicking on a sketched feature.

- **Adaptive.** Turns on adaptivity for that feature. This is an extremely powerful capability of Inventor. When a feature is adaptive, it does not have a fixed size, but is free to adapt its size to other parts in the assembly. This is covered in more detail in the assembly chapters of this book.
- **Expand All Children.** Expands all branches in the **Browser** revealing sketches and other subdata forms.
- **Collapse All Children.** Collapses all branches in the **Browser** hiding sketches and other subdata forms.
- **Find in Window.** Zoom the graphics screen to a close-up of the feature.
- **Properties.** Accesses the **Features Properties** dialog box where the feature can be renamed and suppressed; the adaptivity changed to **Sketch**, **Parameters**, or **From/To Planes**; or the color of the feature changed.

Working with Features

The top branch of the **Browser** tree is the name of the currently open document. If the file has been saved, the file extension is shown, which helps identify which type of file is open. The icon next to the name also identifies the type of file. There is a different icon for each type of Inventor file.

Immediately below the name is a branch containing a folder called Origin. This is an extremely useful feature found in every part and assembly. It contains three work planes, three work axes, and one work point. See **Figure 2-25.** The work point, named Center Point, is at the coordinate system's origin (0,0,0). The work axes lie along the coordinate system's three axes (X, Y, and Z). The work planes correspond to the planes that are formed by the coordinate system's X, Y, and Z axes. These work features are important in part modeling, as well as constraining assemblies. Throughout this book there are references to these work planes, work axes, and work point.

The last branch of the **Browser** tree is a special feature labeled End of Part. This denotes the end of the feature listing, but it also has specific uses. For example, with Example_02_01.ipt open, pick on End of Part in the **Browser**, drag it up the tree, and drop it below the Origin branch. Notice that all features are grayed out in the **Browser** and are no longer displayed in the graphics window. The End of Part feature represents the end of the features list—the end of the database. After the move, it is higher in the tree than any of the features, so the features are basically no longer in the part.

Now, drag-and-drop the End of Part branch below the first part feature, Extrusion1 (or Base Feature, if you renamed it earlier). The basic shape that was used to start this part is displayed in the graphics window and is no longer grayed out in the **Browser**. Continue to drag-and-drop the End of Part feature down one feature at a time. This reveals the method(s) used to construct this part. This is a great way to see how the part was built.

Figure 2-25.
Every part and assembly contains the features indicated here.

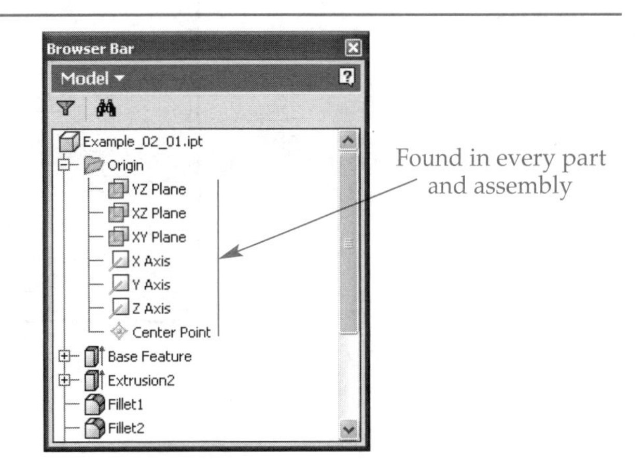

Menu Bar

By default, the **Menu Bar** is docked above the **Inventor Standard** toolbar. It contains eleven pull-down menus—**File**, **Edit**, **View**, **Insert**, **Format**, **Tools**, **Convert**, **Applications**, **Window**, **Web**, and **Help**. They contain important tools with which you need to become familiar. The next sections discuss the pull-down menus.

File Pull-Down Menu

There are four options for saving found in the **File** pull-down menu. If **Save** is selected while working on a new file that has not been saved, the **Save As** dialog box is displayed. Otherwise, the file is simply saved. The **Save As...** option is used to save the current file under a new name. In essence, the software simply renames the current file. The **Save As** dialog box is displayed when this option is selected.

Save Copy As is different than **Save As...** in that the current file does not change. A copy is saved with a new name. However, the copy does not become the current document.

Save All saves the active file and all of its dependent files. For example, suppose you are working on an assembly that is composed of dozens, perhaps hundreds, of parts. Picking **Save All** saves the assembly and all of its component parts.

Picking **iProperties...** opens the **Properties** dialog box. Most of the tabs are used to enter text information that will be saved along with the part. Some of this information will be used in the title block of a 2D drawing that is created for the part. The **Physical** tab is used for the physical calculations of the part. See **Figure 2-26.** If you select Very High for accuracy and pick the **Update** button in the tab, the physical properties of the part are displayed. The material from which the part will be manufactured can be set in the **Material** drop-down list. The property values automatically change to reflect the material. This topic is covered in more detail in Chapter 14.

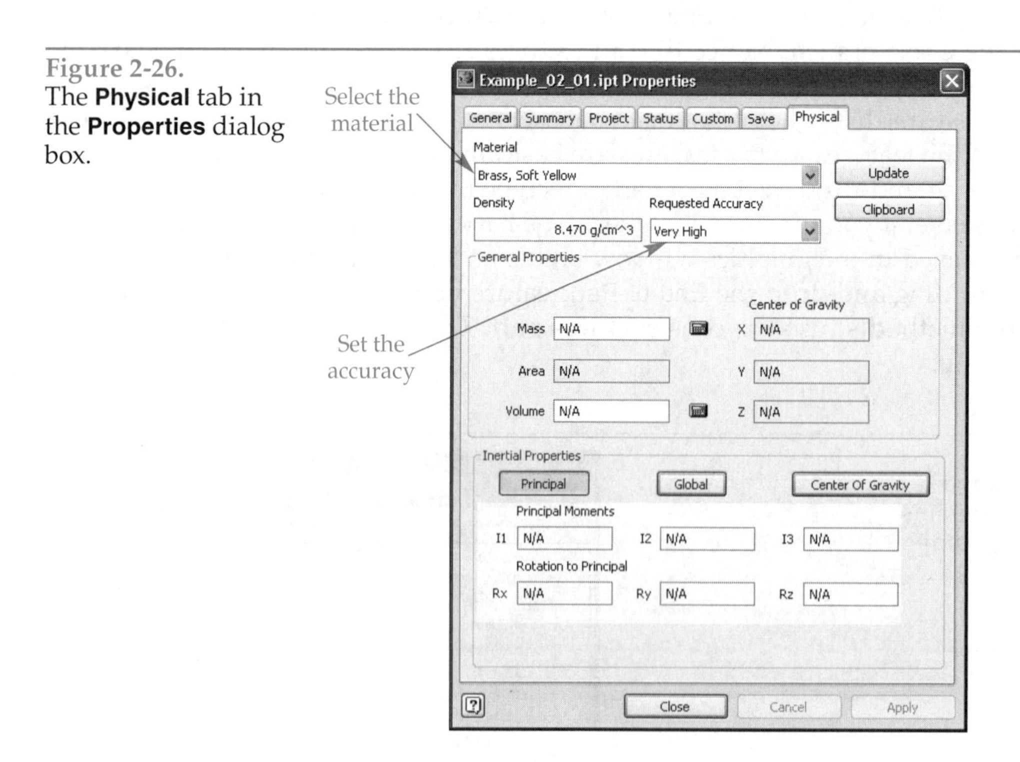

Figure 2-26.
The **Physical** tab in the **Properties** dialog box.

PROFESSIONAL TIP

The **Save Copy As** option in the **File** pull-down menu can be used to save the file in one of several formats, such as JPEG, DWF, or STL. In the **Save Copy As** dialog box, select the appropriate file type in the **Save as type:** drop-down list. Then, pick the **Options...** button and specify the settings for the file type. For example, to create a high-resolution image—one that will print well on a large-format plotter—select an image file type, such as JPEG, and specify a size of at least 4000 pixels × 4000 pixels. Adjust the pixel count to find the balance of resolution and file size that works for you.

PRACTICE 2-1 Complete the practice problem on the Student CD.

View Pull-Down Menu

Selecting **Toolbar** in the **View** pull-down displays a cascading menu containing options for showing or hiding the toolbars that are available in the current mode. For example, while in the part mode, the **Panel Bar, Browser, Inventor Standard**, and **Sketch Properties** toolbars are available. On the other hand, when in sketch mode, the cascading menu also contains **2D Sketch Panel** and **Inventor Precise Input** toolbars.

The **Status Bar** option in the **View** pull-down menu toggles the display of the one-line bar at the bottom of the screen. As you work with Inventor, prompts appear in this area indicating what information is needed or what to do next.

There are many other options in the **View** pull-down menu. These are covered throughout the text as needed.

Insert Pull-Down Menu

There are three options in the **Insert** pull-down menu. **Object...** is used to link or embed external files, like Microsoft Word or Excel documents, in an Inventor model or drawing. **Insert Image...** is used to bring in a raster image file for use in a sketch. It is only available when in sketch mode. **Import...** is used to bring in a 3D geometry translation file, such as IGES or SAT, and is only available when in part mode. These types of files are used to exchange 3D geometry with other CAD systems.

Format Pull-Down Menu

The **Format** pull-down menu contains options for creating, editing, and activating styles. A ***style*** consists of settings for lighting, material, or color that are applied to the part. Inventor has three types of styles—color, lighting, and material. The following sections explain each of the options in the **Format** pull-down menu.

Active standard

Picking the **Active Standard...** option in the **Format** pull-down menu displays the **Document Settings** dialog box with the **Standard** tab current. See **Figure 2-27.** This dialog box can also be opened by selecting **Document Settings...** in the **Tools** pull-down menu. The **Standard** tab contains two drop-down lists: **Material** and **Active Lighting Style**. Selecting a style from the list makes that style the default. The two styles are independent of each other.

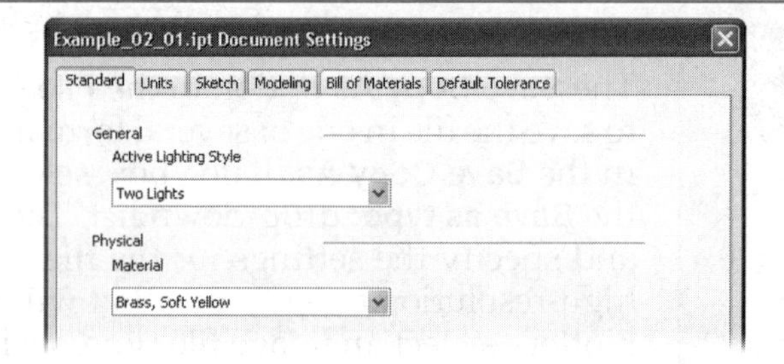

Figure 2-27.
The **Standard** tab of the **Document Settings** dialog box is used to select the active styles.

Styles editor

Selecting the **Style and Standard Editor...** option in the **Format** pull-down menu displays the **Style and Standard Editor** dialog box. See **Figure 2-28.** This is where the designer can access and modify color, lighting, and material styles. The style definitions can be stored either in the part file or in style libraries.

Along the top of the dialog box are four buttons and a drop-down list. Picking the **Back** button displays the previous content in the list below the buttons. Picking the **New...** button creates a new style based on the currently displayed style. Picking the **Save** button saves changes to the currently displayed style. Picking the **Reset** button prior to saving reverts to the previously saved version of the currently displayed style. The setting in the **Filter Styles** drop-down list determines which styles are available. If Local Styles is selected, only the styles that have been applied to the part or copied into the current document from the style library are available. If All Styles is selected, all of the styles defined in the global styles library are available.

On the left side of the **Styles and Standards Editor** dialog box is a tree. The top branches in the tree are Color, Lighting, and Material. These represent the three types of

Figure 2-28.
The **Styles and Standards Editor** dialog box is where styles are created and managed.

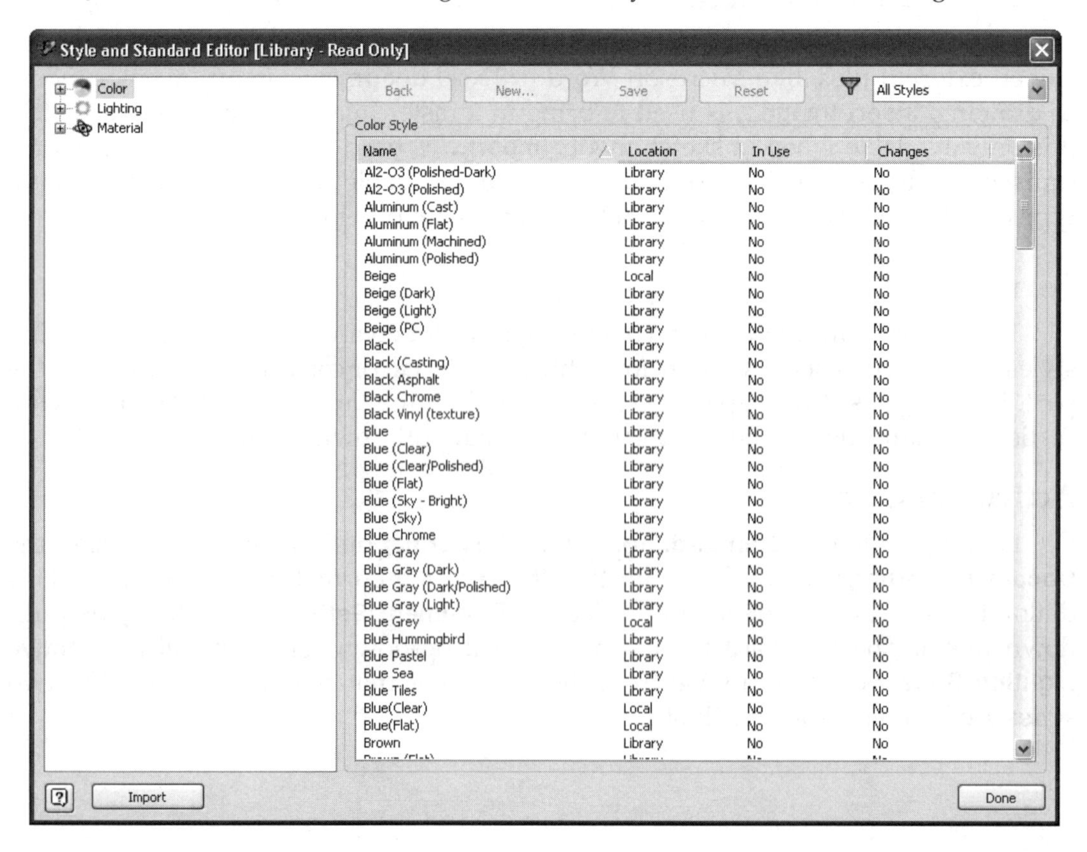

styles. Expanding one of the main branches displays the available styles of that type. Highlighting one of the styles in the tree displays the settings for the style on the right-hand side of the dialog box, where they can be edited. Editing a styles is described in the following sections.

PROFESSIONAL TIP

To quickly jump to a style or texture within a long list, pick in the list to activate it and then type the first letter of style or texture you are seeking.

Editing a Color Style. With a color style selected, the **Color Style** area is displayed on the right-hand side of the **Style and Standard Editor** dialog box. See **Figure 2-29.** The **Colors** area displays four boxes that determine the overall color of the style. To modify a color component, pick the color swatch and select a new color. If black is selected, then color component has no effect on the model. The color components are:

- **Diffuse.** The color of the surface when seen in directed light. All parts are seen in directed light by default.
- **Emissive.** The color of the surface if the part is lighted from within. In the absence of directed lights or ambient lighting this color appears.
- **Specular.** The color of highlighted points, such as edges or corners.
- **Ambient.** The color of the part when lighted only by ambient light. This color appears on the areas of a part's surface in shadow.

The **Appearance** area contains two sliders that adjust the shininess and transparency of the color style. The **Shiny** setting determines the size and quality of the lighted points. Highlights on a shiny surface are small and sharp. Translucent (partially transparent) color styles have an **Opaque** setting of less than 100%. Take the time to compare the values for the Chrome, Lexan-Clear, and Metal-Steel styles. Notice the **Shiny** value for Chrome and notice the **Opaque** value for Lexan-Clear as compared to Metal-Steel.

Figure 2-29.
Editing a color style.

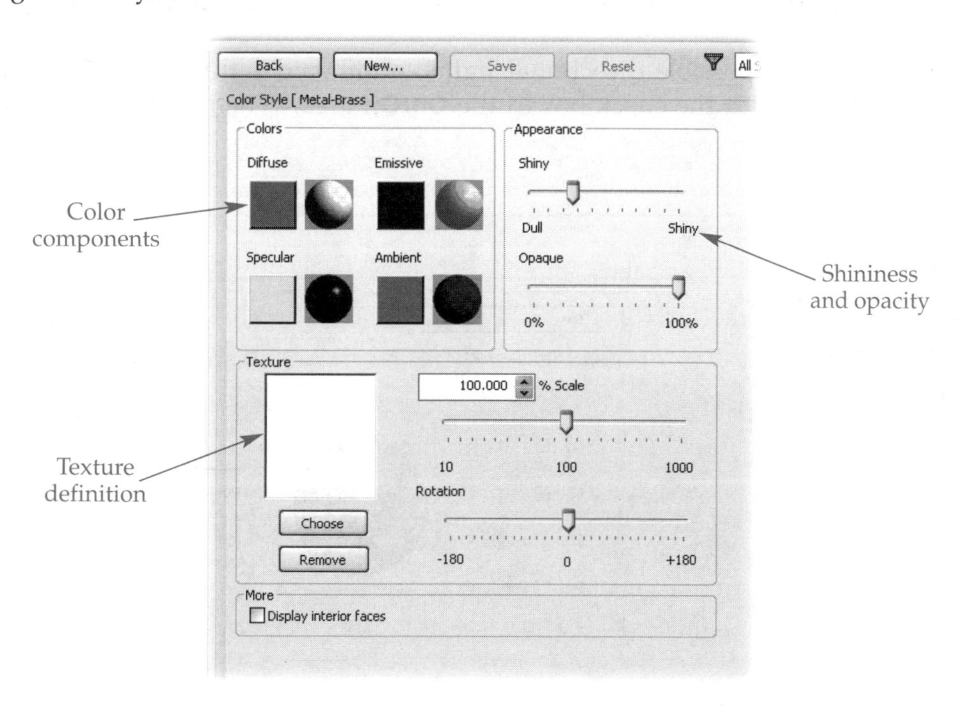

The **Texture** area allows you to specify a texture map to be included in the color style. For example, a color style meant to simulate wooden planking would use one of the wood grain textures defined in the texture database.

Now, create a new color style. First, select the Yellow (Flat) color style. Then, pick the **New...** button, change the name to Wood, and press [Enter]. The new color appears in the tree. Change the **Diffuse** color to red 180, green 160, and blue 5 by picking the color swatch and then the **Define Custom Colors>>** button in the **Color** dialog box. Then, enter the values in the appropriate text boxes and pick the **Add to Custom Colors** button. Finally, select the new color and pick the **OK** button. Change **Specular** to red 250, green 250, and blue 135 in a similar manner. Also, change **Ambient** to red 150, green 100, and blue 0. Now, change the **Shiny** setting to the dull end of the scale—wood is not very shiny unless it has been polished. Next, pick the **Choose** button in the **Texture** area and select Wood_5 in the **Texture Chooser** dialog box. Pick the **OK** button to close the **Texture Chooser** dialog box. Finally, pick **Save** to save the changes and **Done** to close the **Styles and Standards Editor** dialog box.

To display the part in your new color style, pick the materials drop-down list on the far right of the **Inventor Standard** toolbar. Then, select Wood (your new style) in the drop-down list. The part is displayed in the selected color style. Experiment by selecting other color styles from the drop-down list. Before continuing, revert to the previous setting by selecting As Material from the drop-down list.

Currently, the new style is only available for this part. Making styles available for other parts is discussed in the Importing and Exporting Styles section of this chapter.

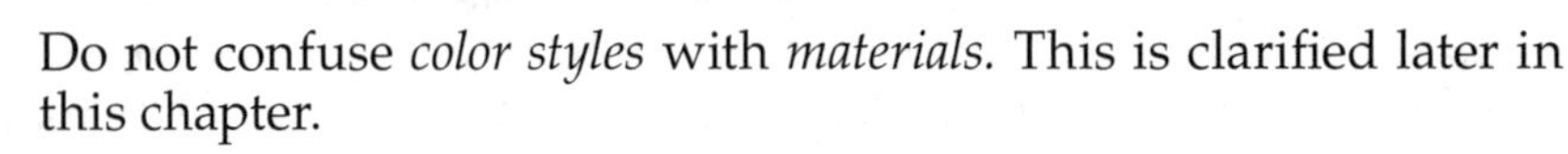

PROFESSIONAL TIP

Do not confuse *color styles* with *materials.* This is clarified later in this chapter.

Editing a Lighting Style. With a lighting style selected, the **Lighting Style** area is displayed on the right-hand side of the **Styles and Standards Editor** dialog box. See **Figure 2-30.** This is where you can define the brightness, color, and position of one to four directed lights. This panel also controls the level of the ambient light, which is the light that comes from all directions. You can create styles and turn each of the directed lights on or off. By simply dragging a slider, dramatic changes in the lighting can be made.

There are four possible directional lights, numbered 1 through 4. To turn any of these on or off, pick the check box in the **On/Off** area. When a light's check box is checked, it is on.

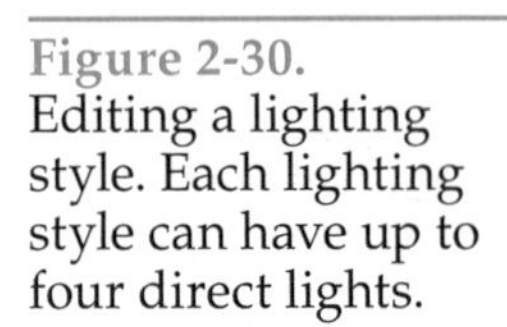

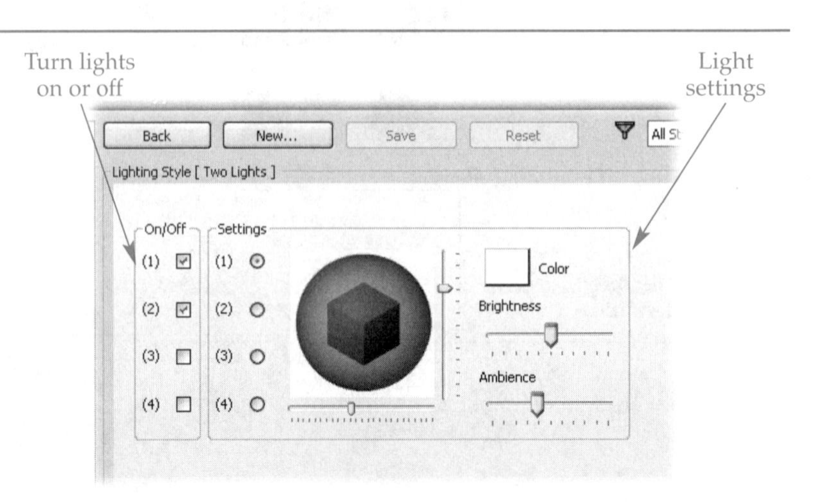

Figure 2-30.
Editing a lighting style. Each lighting style can have up to four direct lights.

The **Brightness** slider in the **Settings** area controls the level of intensity for *all* of the directed lights. Although you cannot control the brightness of each directed light by using the slider, you can affect the brightness by adjusting the color of each directed light. A darker color provides less illumination than a lighter color. The **Ambience** slider controls the ambient lighting, which in turn determines the difference between the lighted and unlighted areas of the model. The slider to the right of the 3D cube controls the height of the current light source and the slider below moves the light source left or right.

To create a new lighting style, select a style in the tree on which to base the new style. Then, pick the **New...** button, type a new name in the **New Style Name** dialog box, and press [Enter]. The new lighting style appears in the tree. Make changes to the directed lights and the **Brightness** and **Ambience** slider bars. Then, pick the **Save** button.

Editing a Material Style. When a material style is selected, the **Material Style** area is displayed on the right-hand side of the **Style and Standard Editor** dialog box. See **Figure 2-31.** All of the information in this area is needed to properly define any material. These numeric values determine the results in the **Physical** tab of the **Properties** dialog box. For example, the density value is used to calculate the weight of the part. A thorough explanation of the values is beyond the scope of this text. However, notice the **Color Style** drop-down list. The color style selected in this list determines how the part is shaded in the graphics window.

To create a new material style, select a style in the tree on which to base the new style. Then, pick the **New...** button. Type a name for the new style in the **New Style Name** dialog box, such as Test Substance, and pick the **OK** button. The new material style appears in the tree. Make any changes to the material properties and pick the **Save** button. To apply the material style to the part, open the **Properties** dialog box and display the **Physical** tab. Then, select the style in the **Material** drop-down list.

Reloading a Style. A style can be reloaded from the global styles library. Right-click on the style name in the tree in the **Style and Standard Editor** dialog box and select **Update Style** from the pop-up menu. The settings for the style are changed to match the settings in the global styles library. For example, with Example_02_01.ipt open, display the **Style and Standard Editor** dialog box. Then, expand the Material branch and select ABS Plastic to display its properties on the right-hand side of the dialog box. Notice that the color style is set to Bubble Gum. Right-click on the style name and select **Update Style** from the pop-up menu. The color style for the material style now matches the definition in the global styles library, which is Plastic (White).

Figure 2-31.
Editing a material style. Notice how a material style has a color style associated with it.

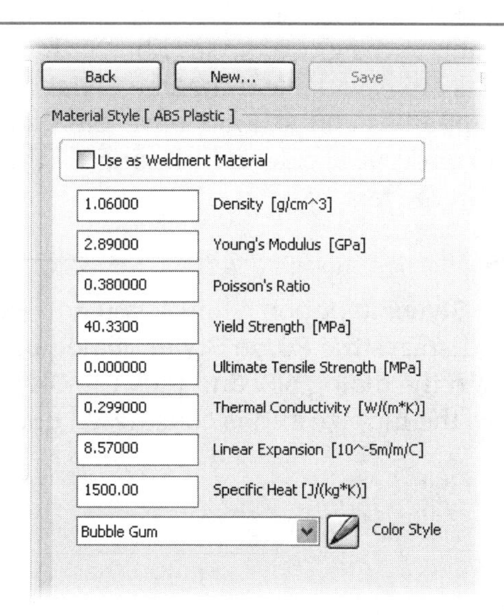

Caching a Style. All of the material styles in the current part and the global styles library are available in the **Material** drop-down list in the **Properties** dialog box. Assigning a material style to the part automatically loads it into the current file. However, a material style can also be loaded from the global styles library without assigning it to a part. This is called ***caching*** the style. To cache a style, select All Styles in the **Filter Styles** drop-down list in the **Style and Standard Editor** dialog box. Then, right-click on the style to cache in the tree and select **Cache in Document** from the pop-up menu.

For example, select Local Styles in the **Filter Styles** drop-down list. Notice that there is not a material style with the name Silver. Now, select All Styles in the **Filter Styles** drop-down list. Since there is a Silver material style in the global styles library, it appears in the tree. Right-click on Silver and choose **Cache in Document** from the pop-up menu. The style is now available from within the part file. Verify this by selecting Local Styles in the **Filter Styles** drop-down list.

Importing and Exporting Styles. A style can be exported to a file by right-clicking on the style name in the **Style and Standard Editor** dialog box. Then, select **Export...** from the pop-up menu. In the **Export style definition** dialog box that is displayed, navigate to the folder where the file is to be stored and enter a file name. The exported file has a .styxml extension. The STYXML file can then be copied to another computer or e-mailed to another designer, for example.

To import a saved style, pick the **Import** button in the lower, left-hand corner of the **Style and Standard Editor** dialog box. In the **Import style definition** dialog box that is displayed, navigate to the folder where the STYXML file is saved and select it.

Update styles

Picking the **Update Styles...** option in the **Format** pull-down menu opens the **Update Styles** dialog box. This dialog box lists all of the local styles with settings that are different from those in the global styles library. The Update? column initially displays No for all of the styles. To update specific local styles, pick on No for the style to change it to Yes. You can turn all of the No settings to Yes by picking the **Yes to All** button. Picking the **No to All** button changes all of the Yes settings to No. When done making Yes and No settings, pick the **OK** button to update the styles and close the dialog box.

Save styles to style library

The **Save Styles to Style Library...** option in the **Format** pull-down menu is used to save the locally defined styles to the global styles library. This allows the local styles to be available for use in other parts or assemblies. Use this option with caution because you may be saving to the global styles library that others use.

The option is only available if the **Use Styles Library** setting for the current project is Yes. Most companies and schools have the **Use Styles Library** setting for projects as Read Only to prevent accidental altering of the global styles library. Projects are discussed later in the text.

Purge styles

The **Purge Styles...** option allows you to remove any unused styles from the current part. It displays the **Purge Styles** dialog box. Change the setting for the styles to purge to Yes in the dialog box and pick the **OK** button. Purging unused styles from the part reduces the file size and is considered good housekeeping.

Tools Pull-Down Menu

The options in the **Tools** pull-down menu have many uses. Most of the options in the pull-down menu are covered in later chapters. The focus in this section is on the **Document Settings...** and **Application Options...** options.

Document settings

Selecting **Document Settings...** in the **Tools** pull-down menu opens the **Document Settings** dialog box. The settings in this dialog box affect the current document, whether it is a part, drawing, or assembly. The dialog box contains six tabs: **Standard**, **Units**, **Sketch**, **Modeling**, **Bill of Materials**, and **Default Tolerance**.

The **Standard** tab contains two drop-down lists. These are used to set the active material and lighting styles. To set an active style, simply select it from the appropriate drop-down list.

The unit of measure and the appearance of the dimension labels are set in the **Units** tab. See **Figure 2-32A.** In the **Units** area, set the unit of measure for length, angle, time, and mass. In the **Modeling Dimension Display** area, set the precision for linear and angular dimensions and how the dimension is displayed. The **Display as expression** setting will be the most useful.

Settings pertinent to the sketch environment are set in the **Sketch** tab. See **Figure 2-32B.** The snap spacing, increment for grid lines, and auto-bend radius can be set.

The **Modeling** tab contains settings related to the part modeling environment. See **Figure 2-32C.** When the **Compact Model History** check box is checked, the feature history of the part is condensed, which reduces the file size. However, the features have to be "rebuilt" when the part is opened and this takes time. Therefore, only use this option when needed. The setting in the **Tapped Hole Diameter** drop-down list determines how the diameter of a tapped hole is applied. This should be set to Minor for most applications. Checking the **Apply to Legacy Tapped Holes** check box applies the setting to existing holes. For most work, the **Participate in Assembly and Drawing Sections** check box should be checked, which allows the part to be sectioned when creating a section view in a 2D drawing (IDW).

Inventor supports the use of tolerancing in part design. The **Default Tolerance** tab is used to set up default tolerances. See **Figure 2-32D.** Add default tolerances only if your dimensioning scheme includes tolerances.

Application options

Selecting **Application Options...** in the **Tools** pull-down menu opens the **Options** dialog box. The settings in this dialog box are applied to Inventor itself and affect all files opened. The 13 tabs in this dialog box are covered in the next sections.

General Tab. The **General** tab contains general utility settings. The value in the **Locate tolerance** text box in the **Selection** area is the maximum number of pixels an object can be from the pick point to have Inventor select it. The value in the **Undo file size (MB)** text box is the size of the undo file, as discussed earlier. The entry in the **User name:** text box appears in the title block of a 2D drawing. Unchecking the **Start-up action** check box prevents the specified action (**Open** dialog box, **New** dialog box, or new based on a template) from being initiated when Inventor is launched.

Save Tab. The **Save** tab contains various settings for saving files when minor changes have been made that do not affect the design intent of the part or assembly. In a multiuser environment, you may not want to save the affected files because it will require that you check out the files in order to save them. This may interfere with another team member's design work. In this tab, you can determine for which changes the user will be prompted. The user is always prompted to save changes that affect the design intent.

Figure 2-32.
A—The **Units** tab is used to change the units used for the part. B—The **Sketch** tab is used to change settings for the sketch environment. C—The **Modeling** tab is used to change settings related to the part modeling environment. D—The **Default Tolerance** tab is used to set up default tolerances.

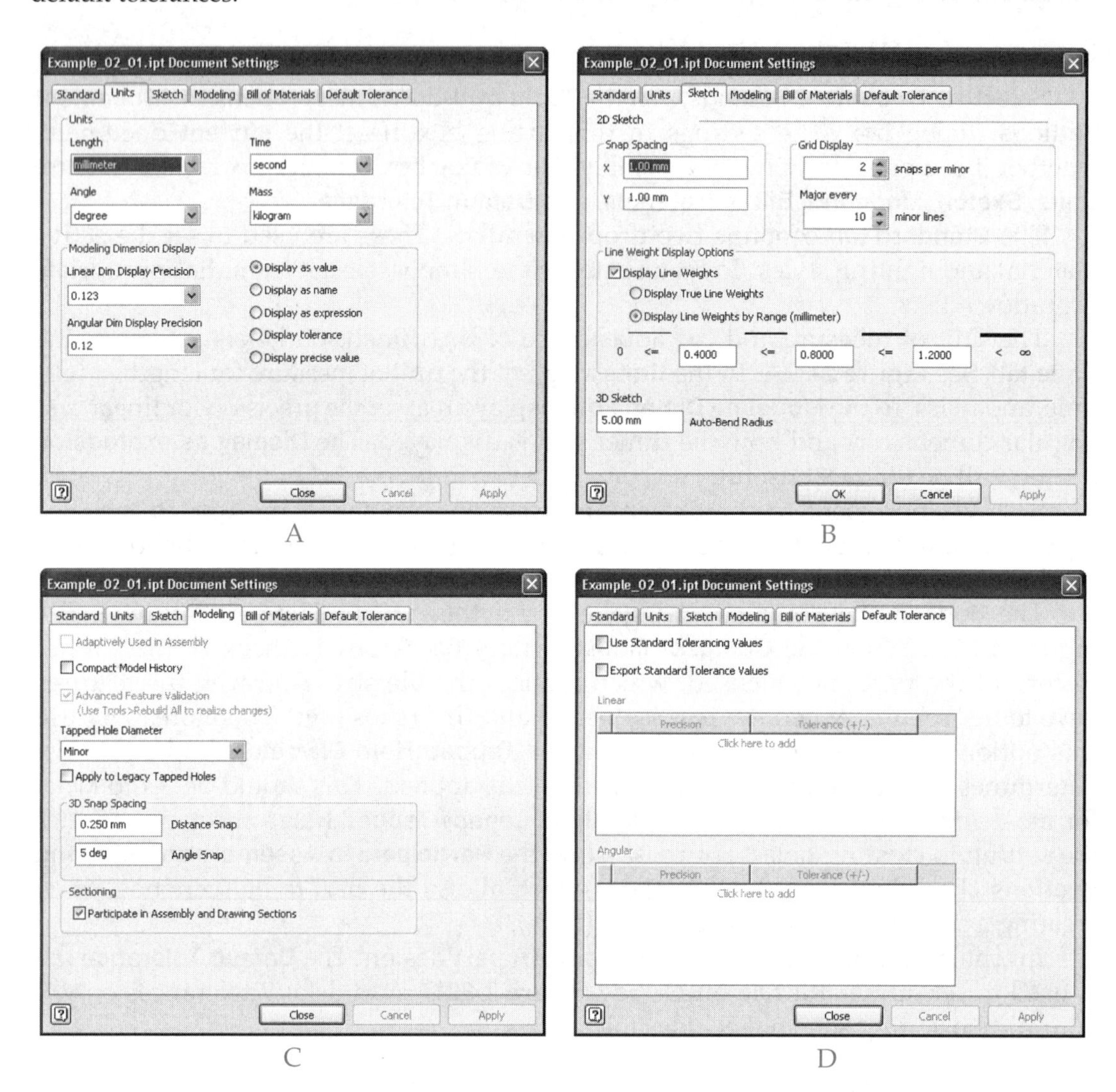

File Tab. Inventor depends on many files as it runs. This tab specifies where the needed files are located on the hard disk. To specify a new location, enter the path in the appropriate text box or pick the browse button to the right of the text box and manually locate the folder.

Colors Tab. The settings in the **Colors** tab control the appearance of the Inventor screen. See **Figure 2-33.** A different color scheme can be set for the design environment and the drafting environment by picking the corresponding button in the preview area. To change the color scheme, select a different scheme from the list in the **Color Scheme** area.

Display Tab. The **Display** tab contains many of the settings that affect how Inventor uses the screen. See **Figure 2-34.** The settings in the **Wireframe display mode** area are used when the part is displayed in wireframe. The settings in the **Shade display modes** area are used when the part is displayed shaded.

In the **Display Quality** area, you can select one of three settings for the display quality. Smooth quality takes longer to regenerate. The default **Medium** setting can be used in most situations.

Figure 2-33.
The **Colors** tab is used to control the appearance of the Inventor screen.

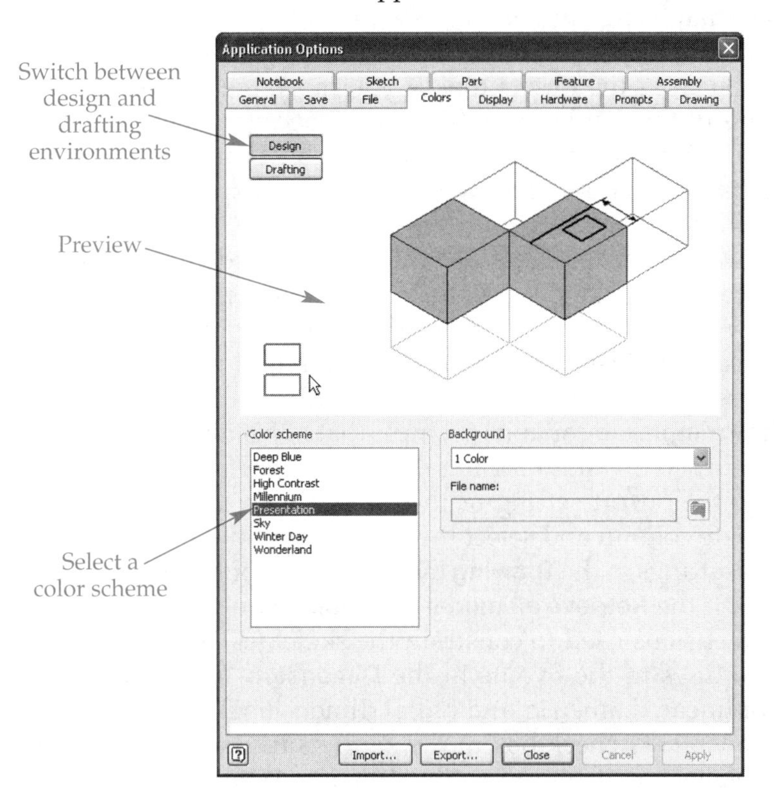

Figure 2-34.
The **Display** tab is used to control wireframe and shaded displays.

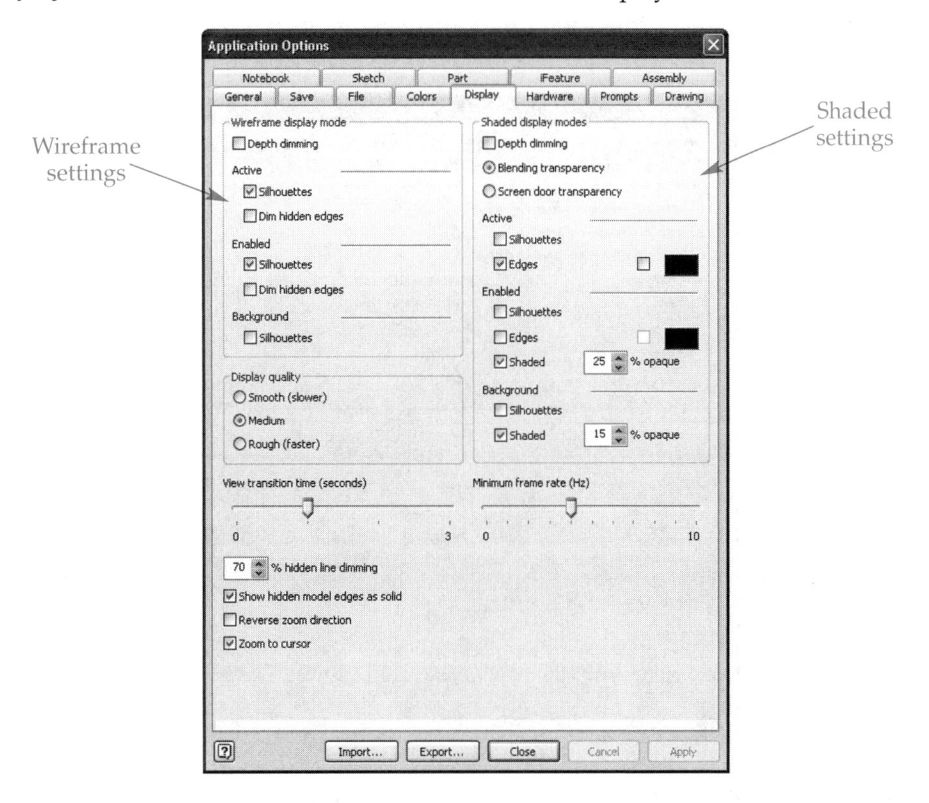

The **View transition time (seconds)** slider determines the amount of time Inventor takes to switch to a new view. If this slider is set to zero, the view is changed instantaneously rather than being animated.

The **Minimum frame rate (Hz)** slider controls the display of the model as certain display tools are used, such as the **Rotate** tool. If you move the cursor too fast when using this tool, portions of the display may "drop out" because the screen cannot be redrawn fast enough. This will happen in a larger assembly or if your computer has a slower video card. Setting the slider higher will prevent the display drop out, but will also slow down the tool.

Hardware Tab. Inventor makes great demands on the video capabilities of the computer. The **Hardware** tab is used to tweak the computer's performance or diagnose what is causing any problems. A major cause of computer problems is the use of incorrect device drivers. The video card, mouse, monitor, and motherboard all have device drivers. The correct one must be used for best performance.

Prompts Tab. The **Prompts** tab contains prompts that appear in many dialog boxes. These prompts can be turned on and off using this tab. The default response can also be changed. To change the on/off setting or the default response, right-click on the prompt in the appropriate column and select the new setting in the pop-up menu.

Drawing Tab. The settings in the **Drawing** tab control various aspects of a drawing (IDW). See **Figure 2-35.** If the **Retrieve all model dimensions on view placement** check box is checked, the dimensions used to constrain the sketch(es) are placed on the 2D drawing views. The settings in the flyouts in the **Dimension Type Preferences** area determine the style for linear, diametric, and radial dimensions.

If the **Display Line Weights** check box in the settings in the **Line Weight Display** area is checked, lineweights are displayed on the drawing. When unchecked, all lines in the drawing are displayed at the same lineweight. To change lineweight settings,

Figure 2-35.
The **Drawing** tab is used to set dimension types, lineweight options, and the location of the title block.

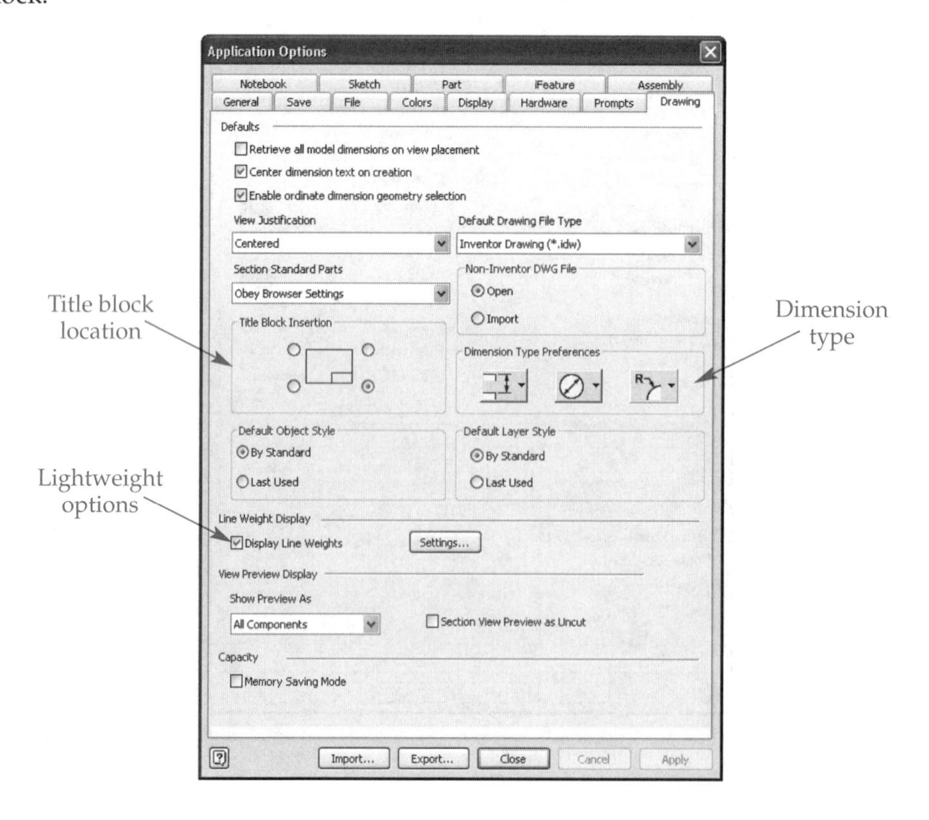

pick the **Settings...** button and make the changes in the **Line Weight Settings** dialog box that is displayed.

Notebook Tab. The settings in the **Notebook** tab control the appearance of the engineer's notebook and whether or not notes appear in the model. The engineer's notebook is a place for making notes to communicate information to others. It is covered in detail later in this chapter.

Sketch Tab. The **Sketch** tab contains many options for working in sketch mode, such as the display of grid lines and snap to grid. This tab should not be confused with the **Sketch** tab in the **Document Settings** dialog box. Remember, the settings in the **Options** dialog box control Inventor on a global basis. The choices in the **Document Settings** dialog box control only the current document.

When the **Edit dimension when created** check box is checked, the **Edit Dimension** dialog box is displayed immediately after a dimension is placed. Since Inventor is based on sketching, the measured distance almost always needs to be changed and this can speed up the process. When the **Autoproject edges for sketch creation and edit** check box is checked, a new sketch can be automatically located in reference to an existing part's edges or work features. This is checked by default.

Part Tab. The **Part** tab contains a few of the options for part mode. The default sketch plane for a new part can be set in the **Sketch on new part creation** area. The settings in the **3D grips** area pertain to using 3D grips.

iFeature Tab. The settings in the **iFeature** tab specify the file locations for iFeatures. iFeatures are 3D features, such as slots, tabs, etc., contained within a symbol library, or catalog.

Assembly Tab. The **Assembly** tab contains many options controlling how Inventor acts when creating assemblies. Many of these settings are covered in later chapters.

When the **Defer update** check box is unchecked, updates to the assembly occur automatically whenever individual parts are changed. Updating an assembly may take a long time, especially if the assembly is large. Deferring the update allows changes to be made to the parts without the assembly updating until the **Update** button is picked.

Convert Pull-Down Menu

The **Convert** pull-down menu is used to switch between part and sheet metal modes. The current mode is indicated in the pull-down menu with a check. Also, the **Panel Bar** displays the corresponding panel.

Applications Pull-Down Menu

The **Applications** pull-down menu is used to switch between part (or sheet metal) mode and Inventor Studio. Inventor Studio is a rendering application for Inventor. When in Inventor Studio, the **Inventor Studio** panel is displayed in the **Panel Bar**.

Window Pull-Down Menu

Several files can be open at the same time in Inventor. The display of these files is managed via the **Window** pull-down menu. Switching between the open files is as simple as selecting the file name in the pull-down menu. One of the advantages of having multiple files open is that geometry can be copied from one into another.

The **New Window** option opens a copy of the current file in a new window. Think of it as a clone. You can use the two windows to display different views of the same model. Changes made in one window are applied in the other window as well.

The **Cascade** option arranges all open files like a deck of cards. See **Figure 2-36A.** The **Arrange All** option arranges all open files in a tiled configuration. See **Figure 2-36B.**

Figure 2-36.
A—The **Cascade** option in the **Window** pull-down menu displays all open files in overlapping windows. B—The **Arrange All** option displays all open files in tiled windows.

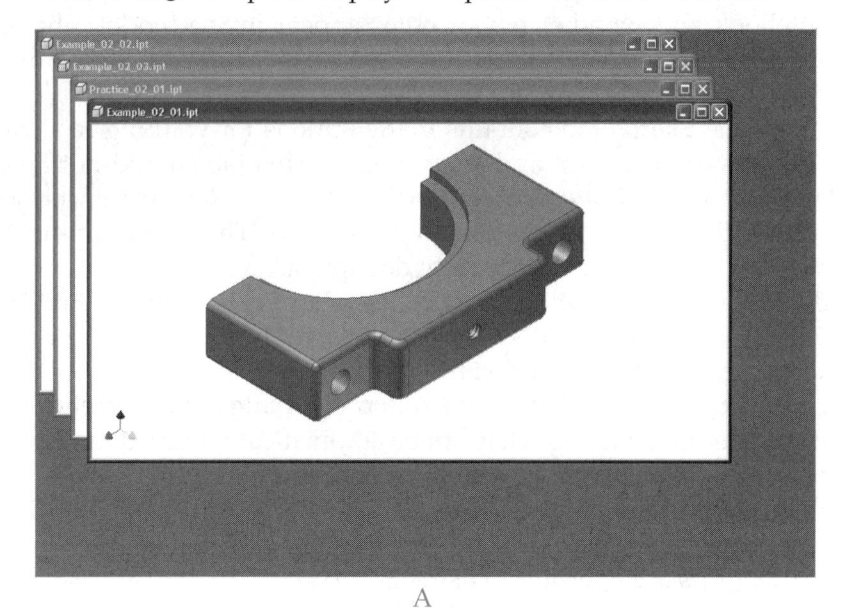

A

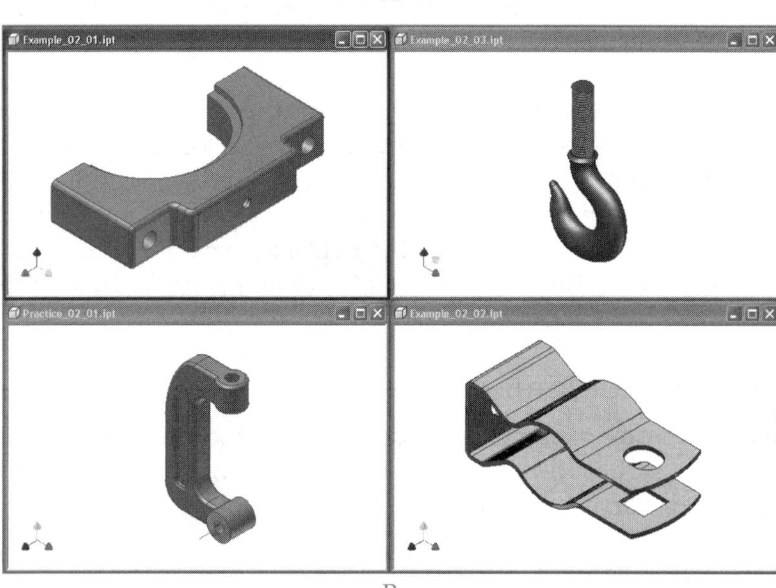

B

Web Pull-Down Menu

The **Web** pull-down menu provides access to Inventor-specific websites. The **Autodesk Inventor** option accesses the Inventor page on the Autodesk website. The **Manufacturing Community** option accesses the manufacturing community page on the Autodesk website. The **Streamline** option accesses the Autodesk Streamline page on the Autodesk website. Autodesk Streamline is an on-line collaboration tool for workgroups and vendors. The **Team Web** option opens a sample "team web" web page that allows you to sample the iDrop technology of drag-and-drop geometry.

Help Pull-Down Menu

This pull-down menu provides access to a comprehensive help system. It gives a user a complete guide to the software. The options are:

- **Help Topics.** Opens the Autodesk Inventor documentation. The documentation contains an index, word search, links to tutorials, and other helpful features.

- **What's New.** Offers information on the new features in this release of the software.
- **Tutorials.** Provides samples procedures for working with Inventor tools and other options.
- **Skill Builders.** This provides access to exercises designed to help you build your Inventor skills.
- **Shortcut Quick Reference.** Selecting this displays the **Shorcut/Alias Quick Reference** dialog box. All of the shortcut keys are displayed in this dialog box.
- **Programming Help.** Offers Visual Basic information for programming macros in Inventor.
- **Graphics Drivers.** Opens the Autodesk inventor hardware website, which provides information on video card drivers. Many of these drivers are downloadable here.
- **Customer Involvement Program**. The customer involvement program, when enabled, allows Autodesk to collect anonymous information about how you use Inventor. The purpose of this is to determine how the software is being used so future releases can be improved.
- **About Autodesk Inventor.** Provides information about the program's release number, build number, serial number, etc.

There are four other options in the **Help** pull-down menu—**Subscription e-Learning Catalog**, **Create Support Request**, **View Support Request**, and **Edit Subscription Center Profile**. These options are used for subscription licenses of Inventor.

Engineer's Notebook

Have you ever wanted to make a note concerning a step in the design process? The engineer's notebook lets you do just that. Open Example_02_01.ipt if it is not open. In the **Browser**, right-click on Extrusion1 (this was renamed Base Feature earlier in the chapter) and choose **Show Dimensions** from the pop-up menu. In this way, the dimensions will be displayed in the engineer's notebook along with the part. Next, pick anywhere in the graphics window to deselect the extrusion. Then, right-click on the 130 mm dimension and choose **Create Note** from the pop-up menu. The engineer's notebook is displayed. Notice how the dimensions show up in the notebook graphic. In the yellow comment box, enter the text shown in **Figure 2-37.** This documents the change to the 110 mm dimension.

There are several view controls. Right-click on the view to display the pop-up menu containing these controls. Several of these, such as **Rotate**, work the same in the engineer's notebook as they do in the regular Inventor graphics window. **Delete** removes the view from the notebook. **Freeze** locks the view at the current design phase. The **Display** flyout offers the three standard shading modes. **Restore Camera** acts as a "previous view." **Help Topics...** opens the help file.

The pull-down menus along the top of the screen contain many of the same items found in the standard Inventor pull-down menus. However, some pull-down menus and options are specific to the engineer's notebook. The **Inventor Standard** toolbar in the engineer's notebook contains common tools used for working with text entry, such as font choice, point size, format (bold and italic), justification, and bullets. Also on this toolbar are buttons for pan, zoom, undo, and redo operations.

There are five buttons in the **Notebook Panel**. The **Comment** button allows you to add additional comment boxes. The **View** button is used to add an additional view. The **Arrow** button allows you to add an arrow in the view. The **Previous Note** and **Next Note** buttons are used to "tab" between notes (not comments). When finished, exit the engineer's notebook by using the "document close" **X** in the upper-right corner.

Figure 2-37.
Adding a note in the engineer's notebook.

Type the note

Once the engineer's notebook is closed, a note icon appears in the graphics window. If the note is applied to a feature, it is listed under the feature. The note icon is a yellow icon displayed near the feature. Hover the cursor over the icon and the note is displayed next to the cursor. The visibility of this icon and the text can be controlled in the **Notebook** tab in the **Options** dialog box.

PROFESSIONAL TIP

It is possible to create a note by right-clicking on a face or edge of a part and selecting **Create Note** from the pop-up menu. The engineer's notebook is displayed as normal. A note added this way is listed in the **Browser** after the Origin branch. Inventor does this to signify that an edge or face was selected and not a feature. The note added to the dimension, as described above, is not listed in the **Browser**.

Projects

Many of the Inventor file types depend on other files. For example, an assembly file (IAM) is composed of part files (IPT). When the assembly is opened for editing, Inventor must be able to locate those parts on the hard disk. As another example, when a drawing file (IDW) is opened, the part or assembly that it depicts must be located. If the part or assembly cannot be located, then you have a serious problem. You may end up with nothing shown on the screen. Inventor uses a concept called ***projects*** to solve the problem of locating dependent files. A project file (IPJ) is a text file that stores the paths of folders for the project.

If you do not set up a project beforehand and try to open a file that depends on other files, you will encounter the **Resolve Link** dialog box. See **Figure 2-38.** This dialog box is asking you to specify the location of the needed file(s). You could use this dialog box to find the file for Inventor, but this is tedious and unnecessary. Using projects is much more efficient.

Figure 2-38.
The **Resolve Link** dialog box is displayed if a file that depends on other files is opened before a project is set up.

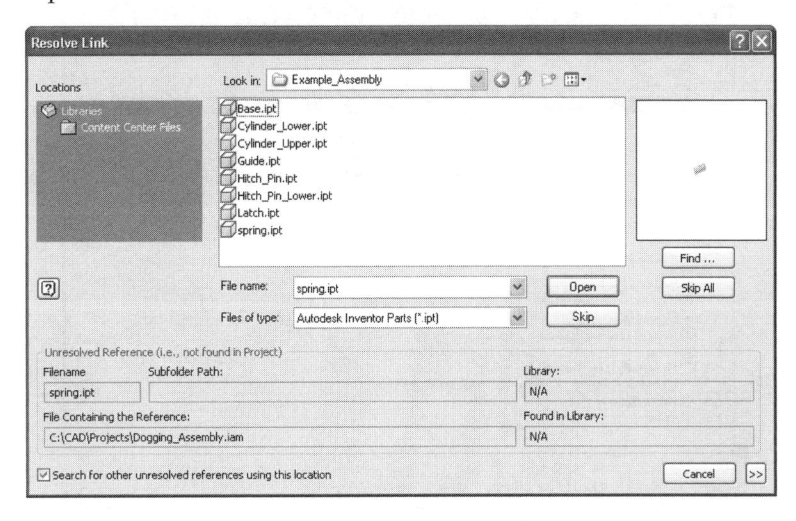

Selecting a Project

You will see the benefits of using a project by examining one that has been created for this chapter. If you have not installed the files from the Student CD, do so now. During the installation, folders are created to organize all of the files required for a project. Refer to **Figure 2-39** for an example of this layout.

A project cannot be activated with Inventor files open, so close all files. Next, pick **Projects...** from the **File** pull-down menu to display the **Projects** dialog box. Pick the **Browse...** button at the bottom of the dialog box. In the **Choose project file** dialog box that is displayed, navigate to \Ch02 folder. Select the project file Chapter02.ipj and pick the **Open** button.

The **Projects** dialog box should look similar to **Figure 2-40.** Select the Chapter02 project in the top half of the dialog box, if it is not already highlighted. Details about the selected project are shown in the lower half of the dialog box. Notice the path displayed next to Location = . This is the project's main folder where the IPJ file is located. Within this folder are subfolders that contain files for the project. One of the subfolders, \Examples\Special_Hardware, is listed under Frequently Used Subfolders.

Figure 2-39.
The folder structure set up for all of the files needed for Chapter 2.

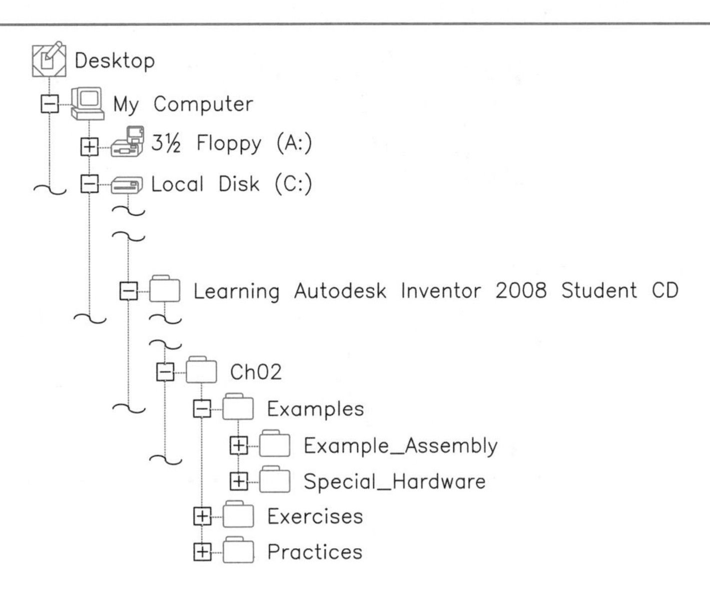

Figure 2-40.
The Chapter02 project is loaded.

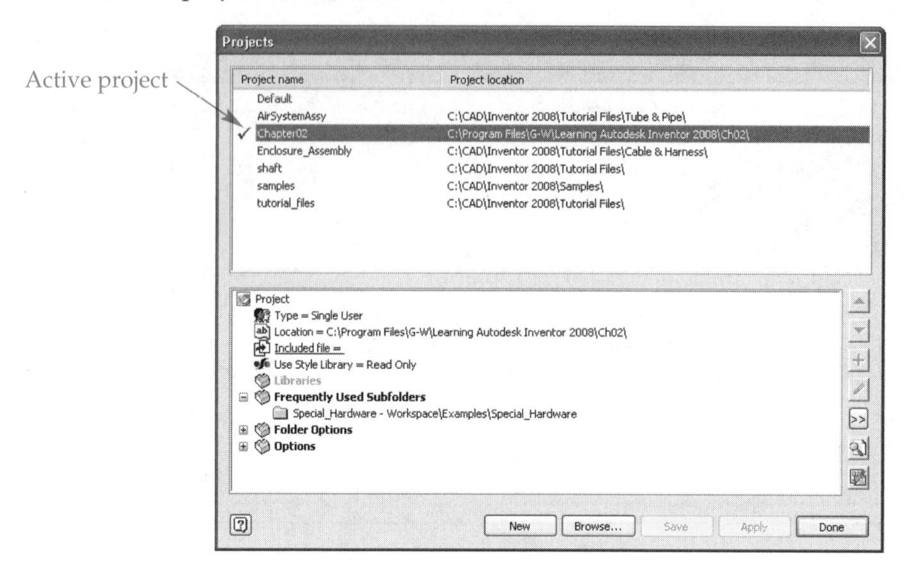

This shortcut was created by right-clicking on the Frequently Used Subfolders heading, selecting **Add Path** from the pop-up menu, and adding the path.

Now that you have explored the project, look at how this can help you locate files. If the check mark in the upper half of the **Projects** dialog box is not next to the Chapter02 project, double-click on the project name. The check mark indicates the active project. Next, pick the **Done** button to close the dialog box. Then, pick the **Open** button on the **Inventor Standard** toolbar to display the **Open** dialog box. The dialog box should look similar to **Figure 2-41.** Notice that the folder listed in the drop-down list at the top of the dialog box is the project's main folder, \Ch02, and its subfolders are listed in the file area. As long as this project is active, this will be the default view in the **Open** dialog box when opening a file. Pick on Special Hardware listed under Frequently Used Subfolders in the left of the dialog box. The contents of the \Special_Hardware folder are displayed in the file area (there are currently no files in the folder). Picking Workspace again displays the contents of the project's main folder.

Figure 2-41.
Opening a file with the Chapter02 project active.

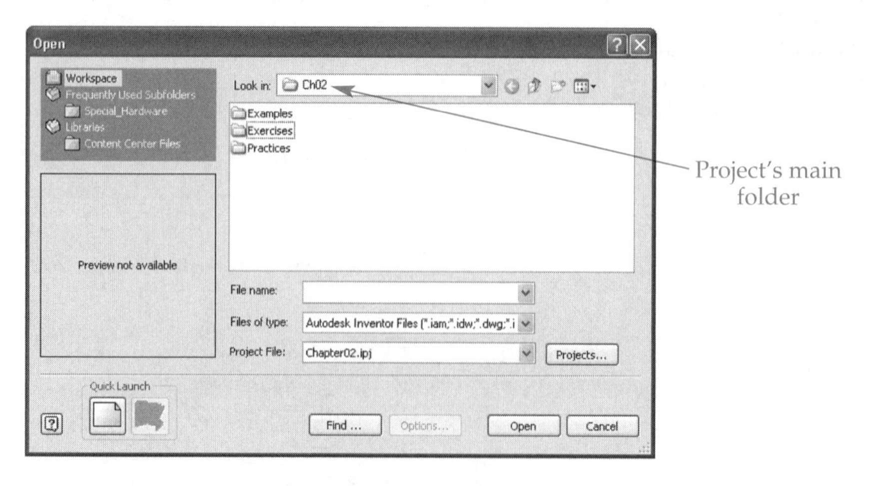

Creating a Project

You can create your own projects. This allows you the freedom to organize your work along project lines. Close all files, because you cannot create a new project with files open in Inventor. Then, select **Projects...** in the **File** pull-down menu to display the **Projects** dialog box. Pick the **New** button located near the bottom of the dialog box. The first page of the **Inventor project wizard** is displayed. On this page, you select the type of project, such as a Vault project, shared project, or single-user project. If Vault is not installed, you can only create single-user projects. After selecting the project type, pick the **Next** button to display the second page of the wizard. See **Figure 2-42A.** Specify the name of the project file. Also, specify a project folder. This is the location on the hard drive where the Inventor files that will be a part of the project are (or will be) stored.

At this point, you can pick the **Finish** button to create the project. However, you can also pick the **Next** button to continue to an optional final step. See **Figure 2-42B.** This page of the wizard allows you to specify the location of any library that may be needed for the project.

Figure 2-42.
A—Enter the name and folder location for the project. B—Include any library that may be needed in the course of working on the project.

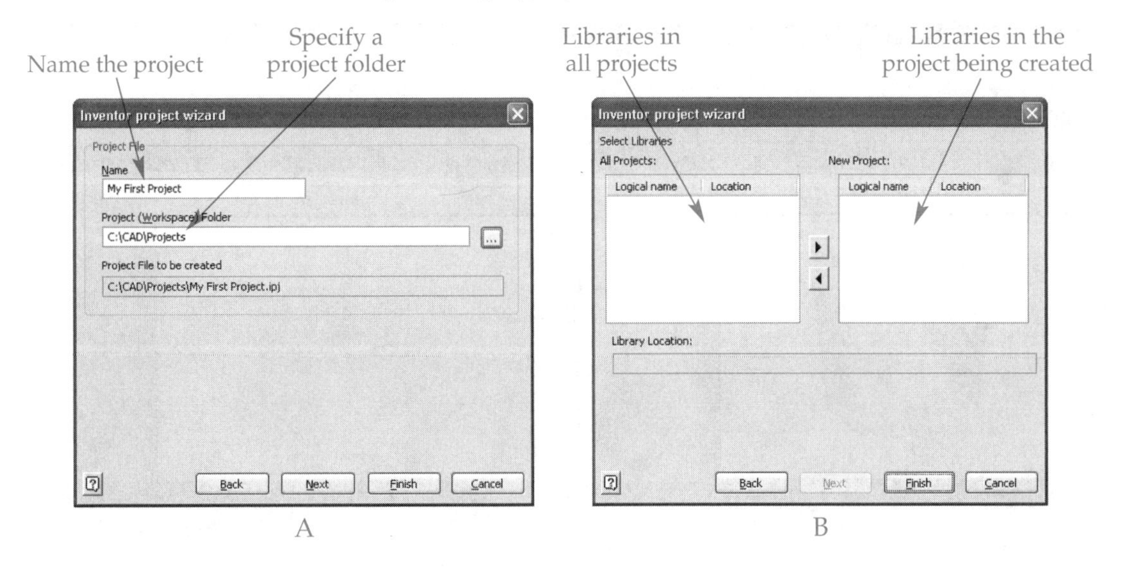

Chapter Test

Answer the following questions on a separate sheet of paper or complete the electronic chapter test on the Student CD.

1. In which dialog box is the option for saving a thumbnail preview located and how is it displayed?
2. Which operations cannot be undone with the **Undo** button?
3. In general, what is the function of the **Return** button?
4. Briefly explain how to redefine the isometric view.
5. Define *sketched feature.*
6. Define *placed feature.*
7. How can features be renamed in the **Browser**?
8. What are the three types of styles in Inventor?

9. Settings made in the ______ dialog box are applied to the current file, while settings made in the ______ dialog box are applied to Inventor itself and affect all files.
10. What is the purpose of the engineer's notebook?
11. Briefly describe the purpose of a project.

Chapter Exercises

Exercise 2-1. Ring Gear.

Complete the exercise on the Student CD

Exercise 2-2. Swivel Yoke.

Complete the exercise on the Student CD

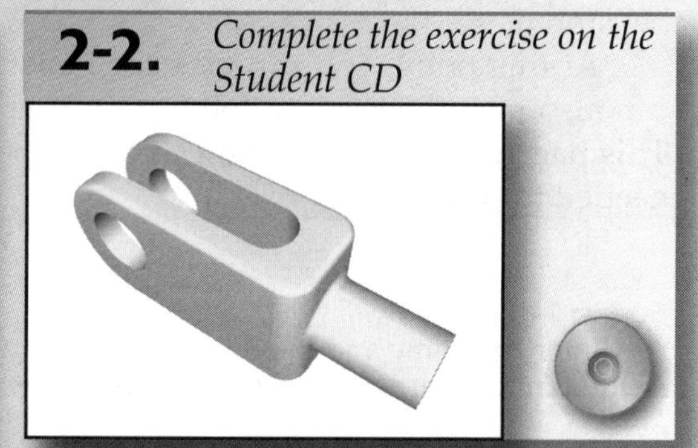

Exercise 2-3. Piston.

Complete the exercise on the Student CD

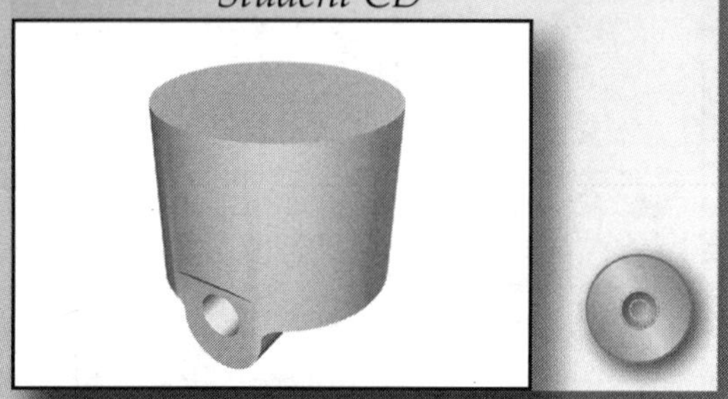

Exercise 2-4. Gold Contact Bracket.

Complete the exercise on the Student CD

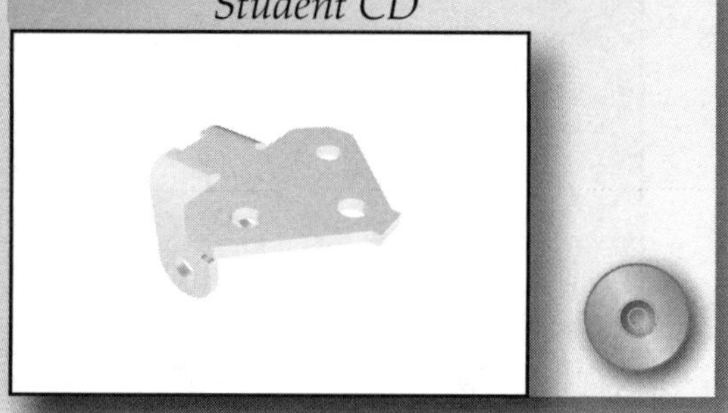

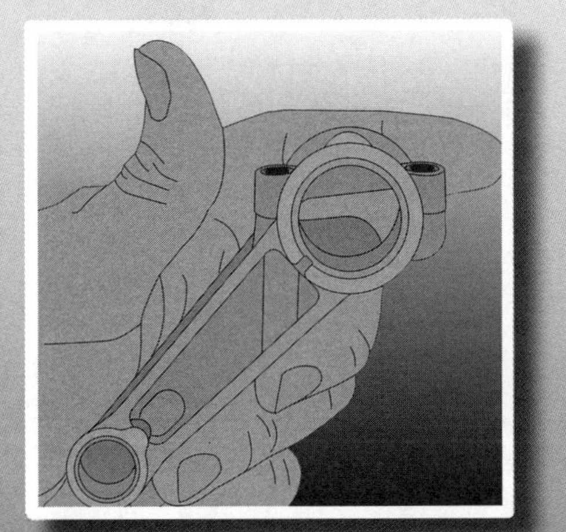

Chapter 3

Sketching, Constraints, and the Base Feature

Objectives

After completing this chapter, you will be able to:

- Describe the procedure for creating a base sketch.
- Sketch curves, including lines and arcs.
- Explain the geometric constraints that Inventor can apply.
- Apply and display geometric constraints.
- Add dimensions to constrain a sketch.
- Extrude solid parts from a sketch.

User's Files

The Student CD included with this text contains several files required for this chapter. Refer to the file File List.txt *in the* \Ch03 *folder for the comprehensive list.*

Process for Creating a Part

All parts in Inventor start with a base 2D profile that is extruded, revolved, or swept into a solid. The profile is constructed in sketch mode, so it is called a ***sketch.*** A sketch is constructed from geometric shapes: lines, arcs, circles, and splines. It must be closed; an open sketch will extrude into a surface, not a solid. The closed sketch cannot cross over itself, but it can have interior islands or separated closed shapes. Valid and invalid shapes for solid profiles are shown in **Figure 3-1.** In this chapter, you will learn how to create single, closed shapes. These are called *unambiguous profiles.* You will also learn how to extrude these profiles into solid parts.

An Inventor sketch made up of six lines is shown in **Figure 3-2.** All lines, arcs, and circles in Inventor are considered curves. The curves (lines) in **Figure 3-2** are called ***intelligent objects*** because they each contain, or "know," unique information about themselves and each other. For example, line AB was horizontally constructed and Inventor remembers this by automatically attaching a horizontal constraint to the line. Line BC knows that it is perpendicular to line AB and the perpendicular symbol

Figure 3-1.
The shapes on the left are valid 2D profiles for extruding. The shapes on the right are invalid.

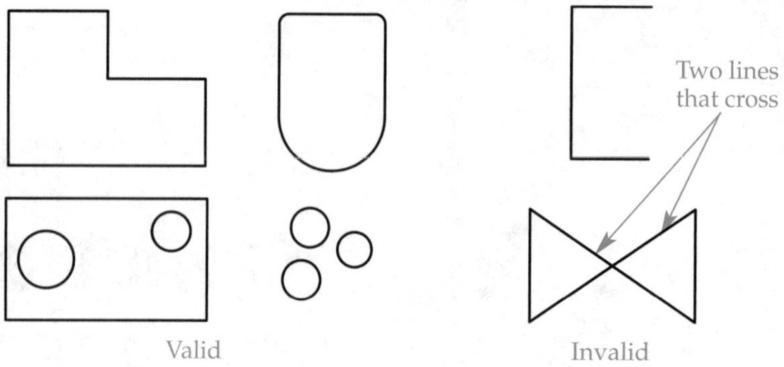

is displayed on both lines. Line BC also knows that it is connected to lines AB and CD and those symbols (coincident) are displayed on both pairs of lines. Line CD knows that it is parallel to line AB.

Constraints define and maintain the relationships between the geometry in the sketch. By default, Inventor applies parallel and perpendicular constraints whenever possible in place of horizontal and vertical constraints. Horizontal, perpendicular, coincident, and parallel are all geometric constraints; four of the 12 total. You can also apply dimension constraints.

The automatic constraint symbols for all six lines are displayed in **Figure 3-2.** The display of constraints can be turned on and off. Constraints can be manually controlled as well. Displaying, adding, and deleting constraints are discussed later in this chapter.

The sketch in **Figure 3-2** was drawn in metric units. When line AB was constructed, it was not drawn to the exact length of 170 mm. Although it is possible to construct precisely, it is not necessary. After the sketch is drawn, dimensions can be added. Adjusting a dimension value changes the associated geometry. The fully dimensioned sketch is shown in **Figure 3-3.**

Even though you do not need to construct precisely, you do want to be reasonably close. Large changes in part size made by editing dimensions may distort the sketch beyond repair. You may be forced to undo the changes and start again. Based on these ideas, here are some sketching guidelines:

- Sketch the size reasonably close.
- Sketch so the constraints are exact.
- Precisely dimension the objects.

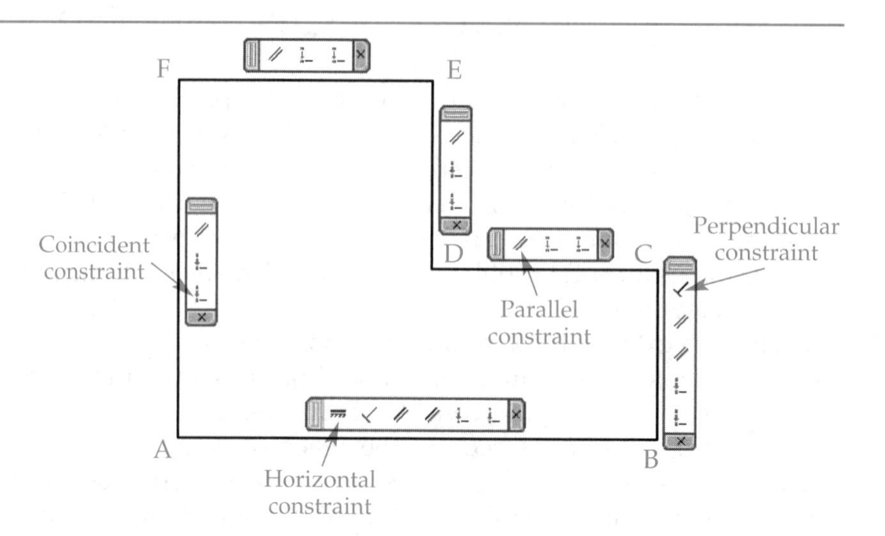

Figure 3-2.
This is a sketch created in Inventor. All lines, arcs, and circles in Inventor are called curves.

Figure 3-3.
This is the sketch from Figure 3-2 fully dimensioned and constrained.

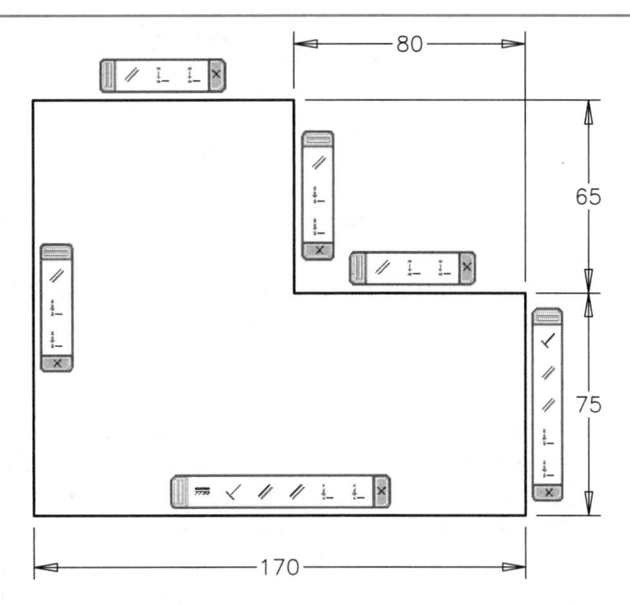

In addition, make the base sketch representative of the fundamental feature of the part, yet as simple as possible. This will give you control of all of the added features for making quick changes. For example, a bracket and its base sketch are shown in **Figure 3-4.** The fillets, cutouts, and holes were all added as features.

In some cases, you may not add the dimensions. For example, this may be a preliminary design study and the dimensions may not be known. Of course, the dimensions can be added later. In another example, you may want to control the size of the part (its dimensions) by the position of other parts in the assembly. This is called an *adaptive part* and is discussed in Chapter 12.

Figure 3-4.
A—This is the completed part.
B—The base sketch used to create the part (shown in an isometric view).

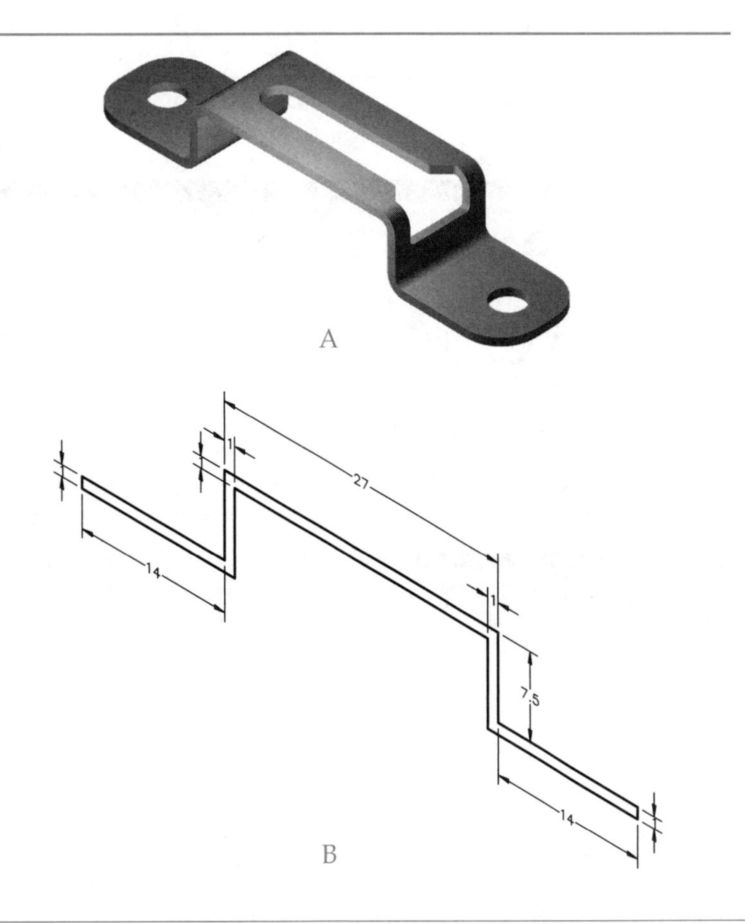

Sketching

Now you will learn how to create a part. First, you will sketch the 2D profile shown in **Figures 3-2** and **3-3.** Then, you will extrude this sketch to create a part. Begin by starting a new metric part file:

1. Select **New...** from the **File** pull-down menu, pick the **New** button on the **Inventor Standard** toolbar, or press [Ctrl][N].
2. Select the **Metric** tab in the **New File** dialog box that is displayed.
3. Select the Standard (mm).ipt template file and pick the **OK** button.

This brings up the graphics window with a grid and a coordinate system indicator (CSI) displayed in the lower-left corner of the grid. See **Figure 3-5.** The CSI is not, however, at the origin (0,0). A CSI can be placed at the origin by checking the **Coordinate system indicator** check box in the **Sketch** tab of the **Application Options** dialog box. With the Presentation color scheme set current, the X and Y axes are shown as black lines. The major horizontal and vertical grid lines are shown as medium gray lines. The major grid lines are 20 mm apart. The minor gird lines are shown as light gray lines and 2 mm apart; this is considered a 2 mm grid.

The default name of the file is Part1 (or Part2, Part3, etc.). Later, when the file is saved, it can be given a more logical name. You cannot save in sketch mode, which is the current mode. Sketch mode is indicated by the **2D Sketch Panel** in the **Panel Bar**. Also, Sketch1 is highlighted in the **Browser**. If you try to save the file now (after sketching geometry), Inventor will ask if you want to exit this mode.

Since the part is 170 mm wide, zoom out to display more of the grid. Pick the **Zoom** button on the **Inventor Standard** toolbar. Then, pick anywhere in the graphics window and drag the cursor up until ten major grid lines are displayed. Now, pick the **Pan** button on the **Inventor Standard** toolbar and pan the intersection of the X and Y axes to the lower-left corner of the screen so there are nine major grid lines to the right of the

Figure 3-5.
The Inventor screen after starting a new part.

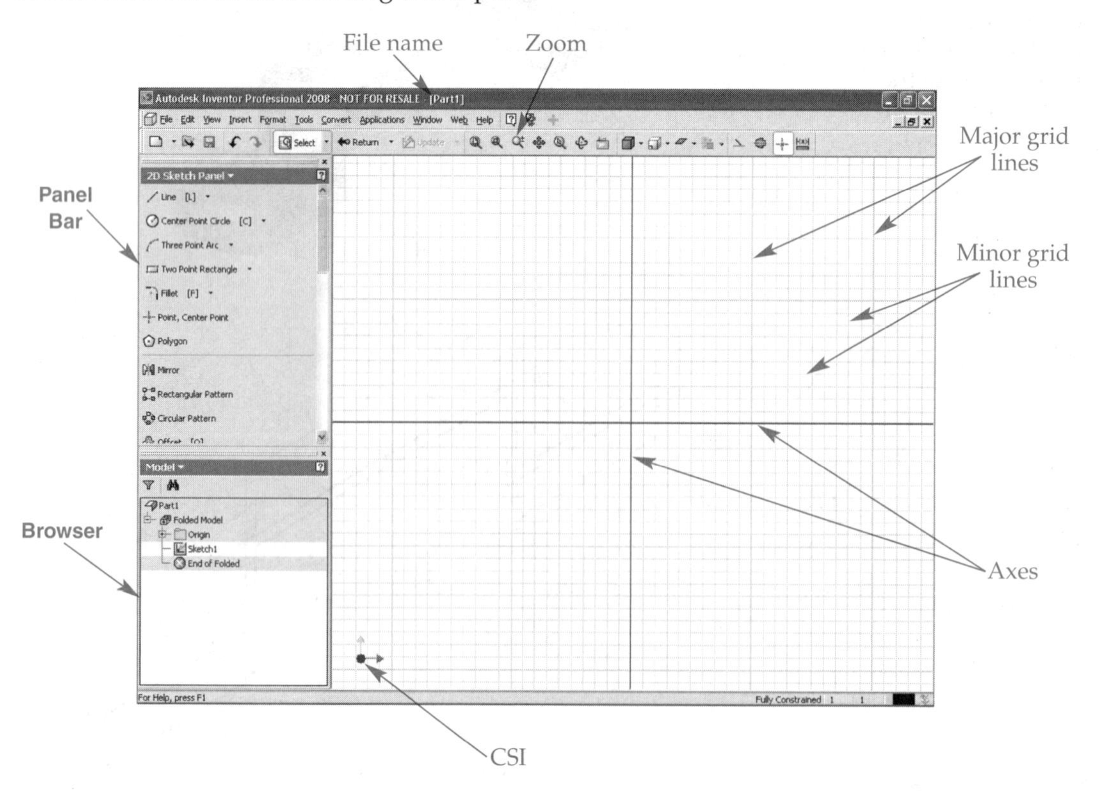

Y axis. Press [Esc] to exit the **Pan** tool. If you have a roller-wheel mouse, you can roll the wheel to zoom or, if the wheel also acts as a button, hold the wheel down to pan.

The intersection of the axes is called the ***origin*** and has the coordinates 0,0. In this example, you will start drawing the sketch at X=0, Y=0. Starting at the origin is not necessary, but you will find it very useful when using work planes, which is introduced in Chapter 5. The default work point, named Center Point, located in the Origin branch in the **Browser** is at the origin. This point can be projected as geometry into the sketch to help start drawing at the origin. The **Project Geometry** tool discussed in Chapter 5 is used to project geometry.

Since Inventor is in sketch mode, the 31 sketching tools with their names and buttons are displayed in the **2D Sketch Panel** in the **Panel Bar**. If you hover the cursor over a button, a tooltip is displayed next to the cursor and a help string is displayed at the bottom of the screen. Note that several of the buttons are flyouts. For example, the **Center Point Circle** button is a flyout and contains the **Tangent Circle** and **Ellipse** buttons as well. The button you pick from a flyout remains "on top" as the default button after you are done using the tool. Several of the tools have default hot keys; the **Line** tool is [L], the **General Dimension** tool is [D], and the **Circle** tool is [C]. Notice that the key is displayed in square brackets next to the tool name. Pressing the key on the keyboard activates the tool.

Now, it is time to draw the first line, line AB. Refer to **Figures 3-2** and **3-3.** First, activate the **Line** tool by picking the **Line** button in the **2D Sketch Panel** or pressing the [L] key on the keyboard. The **Line** button is illuminated (depressed) to indicate the tool is active. Also, the message Select start of line, drag off endpoint for tangent arc appears at the bottom-left of the screen. If tooltips have been enabled by checking the **Show command prompting (Dynamic Prompts)** check box in the **General** tab of the **Application Options** dialog box, then a tooltip is also displayed next to the cursor.

Move the cursor to the origin and pick to set the first endpoint of the line. As you move the cursor, the coordinates are shown at the bottom-right of the screen. Now, move the cursor to the right. A line appears attached to the cursor and the first endpoint. A yellow dot appears under the cursor, which represents the second endpoint. Also, if you move the cursor horizontally, a small horizontal line appears under the cursor. This symbol indicates that the automatic horizontal constraint will be applied. If you move the cursor vertically, a small vertical line appears.

Notice that there are three windows in the lower-right corner of the screen. See **Figure 3-6.** The left-hand window displays the coordinates of the cursor position. The

Figure 3-6.
As you draw, various information is displayed at the bottom of the screen. Also, notice the constraint symbol that is displayed next to the cursor.

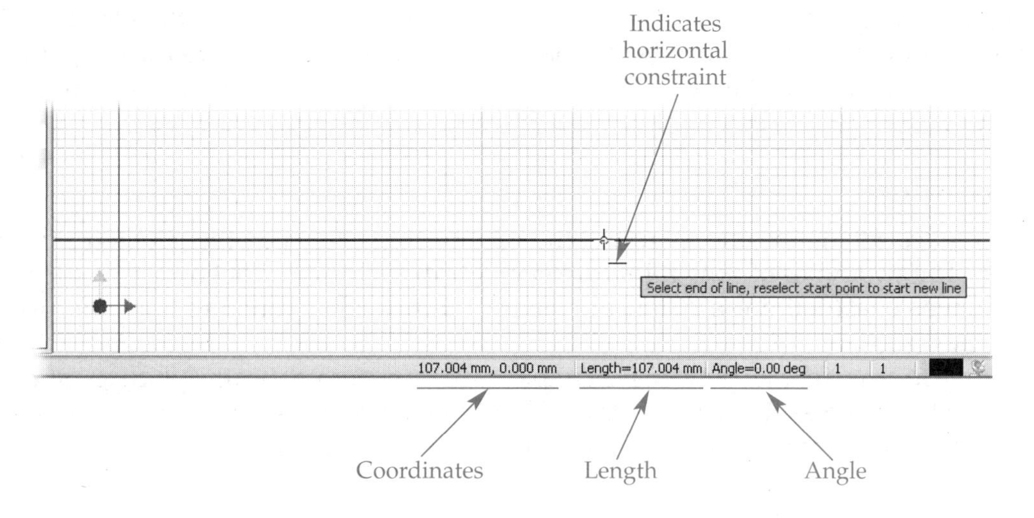

middle displays the length of the line from the first endpoint to the cursor location. The right-hand window displays the angle from the positive X axis to the line.

Move the cursor to the right so the line is horizontal and about 170 mm in length; *about* means between 160 and 180. Pick to set the second endpoint of the first line, line AB. When the cursor is directly over the endpoint, the dot is gray. When you move the cursor off of the endpoint, the dot is yellow.

Now, draw the other five lines in the sketch using the following steps. Remember, the lengths do not need to be exact at this point, rather *about* the correct length.

1. Move the cursor straight up. Notice that a perpendicular symbol appears next to the cursor and under line AB. This indicates that the automatic perpendicular constraint will be applied. When the line is perpendicular and about 60 mm in length, pick to set the second endpoint of line BC.
2. Draw line CD by moving the cursor to the left. Notice that a parallel symbol appears next to the cursor and under line AB. This indicates that the automatic parallel constraint will be applied. When the line is parallel and about 80 mm in length, pick to set the second endpoint of line CD.
3. Move the cursor straight up. Notice that the automatic parallel constraint will be applied. Inventor prefers the parallel constraint over the perpendicular constraint. Therefore, line DE will be constrained parallel to line BC, not perpendicular to line AB. When the line is parallel and about 65 mm in length, pick to set the second endpoint of line DE.
4. Move the cursor to the left. Notice that the automatic parallel constraint (parallel to line AB) will be applied. When you move the cursor above the first endpoint of line AB, a dotted vertical line appears indicating what Inventor thinks is a logical endpoint for the next line. This helps you locate the second endpoint of line EF. When the dotted line appears and line EF is parallel, pick to set the second endpoint of line EF. Note: The dotted line may not be visible because it is on top of the Y axis line.
5. Now, move the cursor to the first endpoint of line AB, which is the origin in this case. The dotted line appears as you move the cursor straight down (which is hidden by the Y axis in this example). Also, the automatic parallel constraint (parallel to line BC) will be applied. When the cursor is directly on top of the first endpoint of line AB, the yellow dot changes to green. Also, the coincident constraint symbol appears next to the cursor that indicates the two endpoints will be connected. Pick on the first endpoint of line AB to draw line FA and close the sketch.
6. Exit the **Line** tool by pressing the [Esc] key or right-clicking in the graphics window and selecting **Done** from the pop-up menu.

If you make a mistake while drawing the profile, pick the **Undo** button on the **Inventor Standard** toolbar, press the [Ctrl][Z] key combination, or select **Undo** from the **Edit** pull-down menu. This deletes the last line segment and cancels the **Line** tool. To continue drawing the profile, pick the **Line** button, move the cursor to the endpoint of the last line, and pick when the yellow dot becomes green. Then, finish sketching the profile.

Constraints

Constraints define and maintain the relationships between the geometry in the sketch. The two types of constraints are geometric (such as parallel or perpendicular) and dimensional (defining a distance or length). Dimensional constraints are discussed in the next section.

As you saw in the previous example, you can automatically place geometric constraints on the sketch as you draw. If for some reason you want to prevent the automatic constraint from being applied, hold down the [Ctrl] key as you draw.

To show how the geometry is constrained, pick the **Show Constraints** button in the **2D Sketch Panel**. Then, pick on the geometry. The constraints for that geometry appear in a small bar, called a ***constraint bar***, next to the geometry. Review the constraints by placing the cursor over any of the constraint symbols. The constraint is highlighted in yellow and the geometry to which the constraint applies is highlighted in red. To exit the **Show Constraints** tool, press the [Esc] key.

The order in which symbols appear in the constraint bar, from left to right, is the order of precedence for the constraints. To delete a constraint, right-click on the symbol in the constraint bar and select **Delete** from the pop-up menu. To move the constraint bar, move the cursor to the double line at the left-hand end or top of the bar. The "move" cursor appears. Pick and drag the bar to a new location. To turn off the display of (close) a constraint bar, pick the X at the right-hand end or bottom of the bar.

You can show the constraints for all geometry by right-clicking in the graphics window when no tool is active and selecting **Show All Constraints** from the pop-up menu. The [F8] key can also be used to show all constraints. To turn off the display of all constraint bars, right-click in the graphics window and select **Hide All Constraints** from the pop-up menu. The [F9] key can also be used to close all constraint bars.

Dimensions

Since there are currently no dimensions on the sketch, you can dynamically change the geometry in the sketch. Select a line or an intersection of two lines, hold down the mouse button, and then drag the geometry to the new location. Try this by selecting any corner in your sketch and moving it to a new location. However, notice that as you move the corner, the two lines attached to it remain constrained by the geometric constraints that were automatically placed as you drew the profile. This is a useful technique for preliminary design work.

You have two choices for applying the dimensions needed to accurately size the part. You can apply automatic dimensions, in which case Inventor decides where the dimensions should be placed. Alternately, you can manually apply and place each dimension.

To manually apply a dimension, pick the **General Dimension** button in the **2D Sketch Panel** or press the [D] key. Then, pick the geometry to which you want the dimension applied. Finally, drag the dimension to the desired location. To cancel the **General Dimension** tool, press the [Esc] key.

For this example, first apply automatic dimensioning. Pick the **Auto Dimension** button in the **2D Sketch Panel**. The **Auto Dimension** dialog box shown in **Figure 3-7** is opened. Notice the dialog box reports that six dimensions are required. This means six dimension constraints are required to fully constrain the sketch. The number of dimensions needed to fully constrain the sketch is based on the geometric constraints currently applied to the sketch. If you remove some of the geometric constraints, additional dimensions may be needed. Now, pick the **Apply** button in the **Auto Dimension** dialog box and Inventor constructs *four* dimensions.

Why four? Notice the dialog box reports that two dimensions are still required. These two dimensions are required to constrain the sketch to a location in 2D space.

Figure 3-7.
You can automatically dimension a sketch using this dialog box.

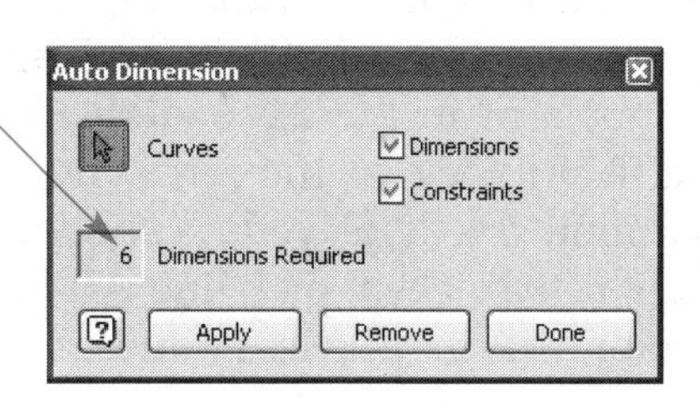

Figure 3-8.
By editing a dimension, you can alter the associated geometry. A—Double-clicking on the dimension displays the **Edit Dimension** dialog box. B—A new dimension value is entered.

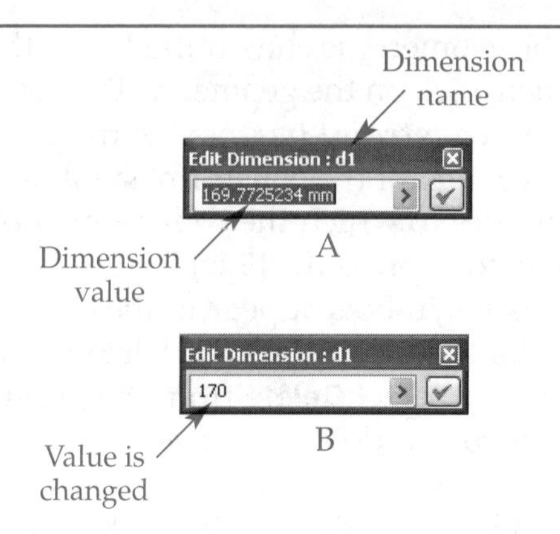

They cannot be automatically applied. For this example, you will apply geometric constraints rather than locational dimensions. The next section details the procedure.

Dimensions are what allow you to draw the sketch at its approximate size. Once placed, a dimension value can be edited to an exact value. Since the associated geometry is constrained by the dimension, it is updated to match the exact value. First, close the **Auto Dimension** dialog box by picking the **Done** button. Then, double-click on the dimension across the bottom that is supposed to be 170 mm. The **Edit Dimension** dialog box shown in **Figure 3-8** is displayed. Each dimension has a name that is, initially, assigned by Inventor. In this case, the name is d1, as reflected in the title bar. Depending on how you drew your sketch, this name may be different, but it will start with the letter d. In the next chapter, you will learn how to rename dimensions and relate dimensions to each other. For now, type the exact value of 170 in the field and pick the check mark button to set the dimension and close the dialog box. Edit the remaining dimensions to the exact values shown in **Figure 3-3.**

You do not need to enter the "mm" for millimeters, as this is a metric part. If other units are entered, you need to include the unit abbreviation, such as 6.6929 in, 0.558 ft, or 17 cm. Any of these entries will change the dimension to 170 mm. Regardless of the units entered, the dimension on the sketch will always display the millimeter value.

Once a dimension is placed, you can reposition it. Simply pick on the number when the "move" cursor is displayed. Then, drag it to a new location. The **Auto Dimension** and **General Dimension** tools must be inactive to do this.

PROFESSIONAL TIP

You can use the **Auto Dimension** tool at any time to check for the number of dimension constraints that are still required to fully constrain the sketch.

Applying Fix Constraints

The sketch is now dimensionally and geometrically constrained. However, it is not locked to the coordinate system. This means that the geometry can be moved around in the XY plane of the coordinate system. You can fix the sketch to the coordinate system in order to take advantage of the existing work planes. To fix this sketch, you will apply a fix geometric constraint to line AB and line AF.

The default constraint in the **2D Sketch Panel** is the perpendicular constraint. However, this is a flyout. Pick on the arrow next to the button name. Buttons for all 12 geometric constraints are displayed, as shown in **Figure 3-9.** Pick the **Fix** button in the flyout to make it active. Then, pick line AB to apply the fix constraint. Notice that the line changes color (this may be less noticeable in certain color schemes). Also, line

Figure 3-9.
The constraint flyout contains the twelve geometric constraints.

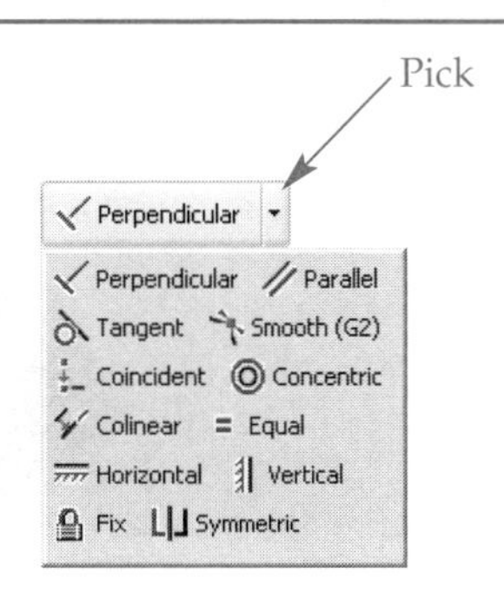

CD and line EF change color. This is because they are constrained parallel to line AB, which is now constrained to the Y axis.

Exit the **Fix** tool by pressing the [Esc] key. You can now pick on the sketch and move it left and right. However, it cannot be moved up and down. This is because the Y values of line AB are constrained, or locked, to the coordinate system.

Pick the **Fix** button again and pick line AF. Now, all of the lines in the sketch have changed color, indicating that the sketch is fully constrained. Press the [Esc] key to exit the **Fix** tool. Now, none of the lines can be moved.

Extruding the Part

Now that the sketch is fully constrained, it can be extruded into a solid. First, display the isometric view by right-clicking in the graphics window and selecting **Isometric View** from the pop-up menu or pressing the [F6] key. It is easier to see the extrusion direction choices in this view. Right-click again and select **Finish Sketch** from the pop-up menu, or press the [S] key. You can now save the file as Example_03_01.ipt; remember, you cannot save in sketch mode.

Once you have "finished" the sketch, the **2D Sketch Panel** in the **Panel Bar** is replaced by the **Part Features** panel. There are 41 tools in the **Part Features** panel. The one you will use is **Extrude**, which has a hot key of [E]. This tool does several things:

- Since there is only one unconsumed and unambiguous sketch on the screen, it selects the sketch as the profile and shades it.
- It displays the **Extrude** dialog box shown in **Figure 3-10.**
- It displays a wireframe preview of the profile extruded with the default settings.

The **Shape** tab of the **Extrude** dialog box contains three buttons in the middle of the tab. These are **Join**, **Cut**, and **Intersect**. Since this is the first profile, only **Join** is allowed. The other two buttons are grayed out.

The middle text box on the right side of the **Shape** tab has a value of 10 mm by default. This is the thickness of the extrusion. For this example, change the value to 60 mm; do not press [Enter]. The wireframe preview changes to reflect the value. Now, in the graphics window, pick one of the top edges of the wireframe and drag it up and

Figure 3-10.
Extruding a sketch into a solid.

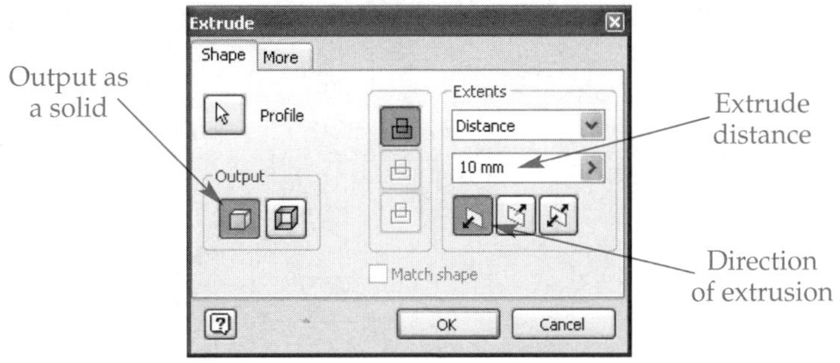

Figure 3-11.
The completed part. Notice the taper angle.

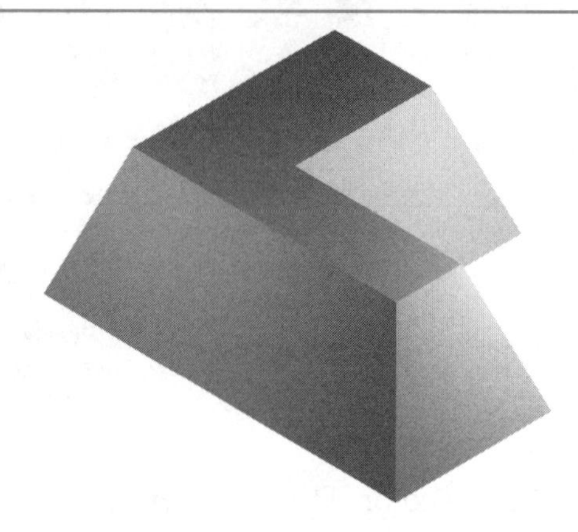

down (or back and forth). The distance value in the **Extrude** dialog box changes with the wireframe movement. Set the value back to 60 mm.

Change the direction of the extrusion to the negative Z axis by picking the center button of the direction indicators in the **Extrude** dialog box. Change it to mid-plane with the third choice. In this case, one-half of the extrusion distance is applied to each side of the sketch.

The extrusion is by, default, a solid object. You can extrude to a surface object by picking the **Surface** button in the **Output** area of the **Shape** tab. However, most often the **Solid** button is used.

Pick the **More** tab and type –20 in the **Taper** text box; do not press [Enter]. This is a degree value for a taper angle. A negative angle makes the extrusion smaller as it is extruded, a positive angle makes it larger. The wireframe changes to reflect the setting.

When you are finished making settings, pick the **OK** button. The part is extruded with a taper angle on all faces, as shown in **Figure 3-11.**

Editing the Feature and the Sketch

The tree for Part1, or Example_03_01.ipt if you saved the file, in the **Browser** now includes Extrusion1. If you expand this branch, Sketch1 appears below Extrusion1. Right-click on Extrusion1 to display a pop-up menu with many options, **Figure 3-12.** There are four options that allow you to change the size and shape of the part: **Show Dimensions**, **Edit Sketch**, **Edit Feature**, and **3D Grips**.

Figure 3-12.
This pop-up menu is displayed when you right-click on the extrusion name in the part tree in the **Browser**.

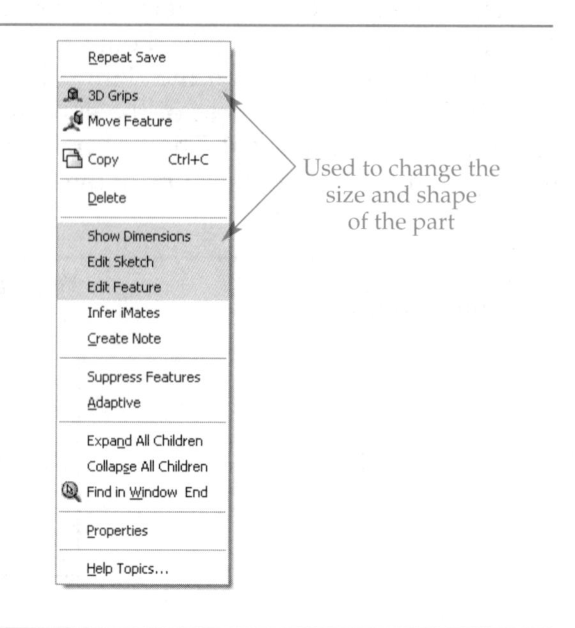

Figure 3-13.
Updating the part.

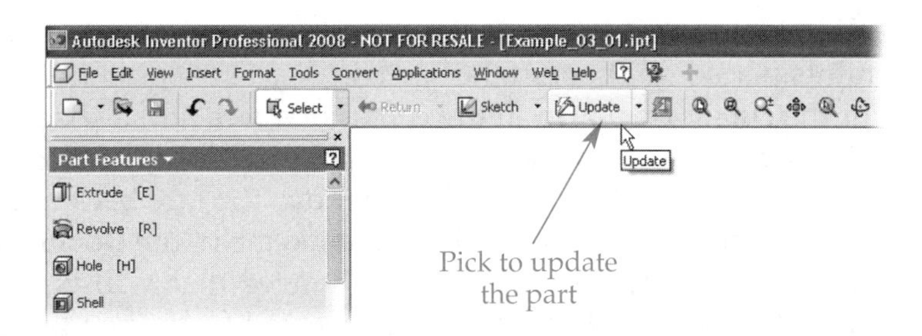

Select **Show Dimensions** from the pop-up menu and all of the dimensions are displayed on the part, including the extrusion height and angle. Double-click on the 170 mm dimension and change it to 150 mm in the dialog box that appears. When you pick the check mark or press [Enter], the dimension changes but the part does not. You can change as many dimensions as you wish. To update the part, pick the **Update** button on the **Inventor Standard** toolbar, **Figure 3-13.**

The second option that allows you to modify the part is **Edit Sketch**. Select this option to return to sketch mode. Now, change the 150 mm dimension back to 170 mm. To complete the edit, right-click and select **Finish Sketch** from the pop-up menu. This applies the change, exits sketch mode, and displays the updated, extruded part.

The third option that allows you to modify the part is **Edit Feature**. This option displays the **Extrude** dialog box in which changes can be made to the extruded distance, direction, taper angle, and sketch dimensions. Change the taper to 0° in the **More** tab. Then, pick the **OK** button and the part is updated.

The fourth choice is **3D Grips**. Selecting this option displays the sketch dimensions, which can be edited, and a number of small circles on the part called ***grips,* Figure 3-14.** The grips turn red if you hover the cursor over them. If you hover over the grip in the middle of a face, it is replaced by a 3D arrow. Pick and hold on one of the midpoint grips of the vertical lines and drag that line around the screen to change the part.

Figure 3-14.
A part can be edited using 3D grips.

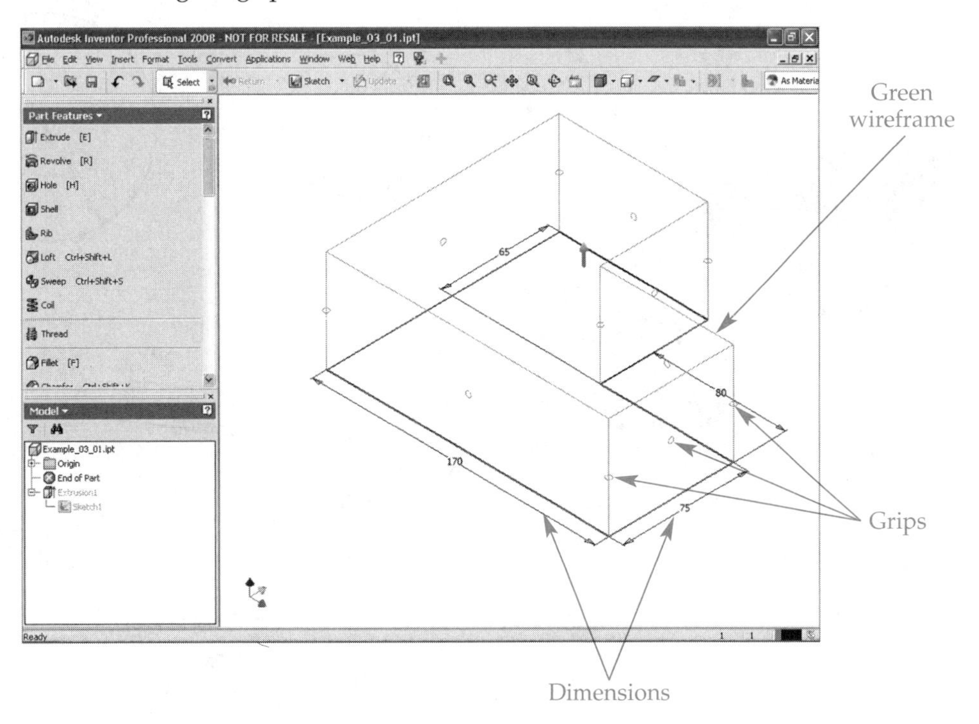

The vertical line at the origin will not move because the two lines touching it have fix constraints placed on them in the sketch.

Put the cursor on a grip in the center of a face and a shaded arrow appears. There is also a shaded arrow for the extrusion height. When the cursor is over the arrow, the arrow turns red. Pick, hold, and drag to move the face back and forth in increments of 0.25 mm. The increment is set in the **Modeling** tab of the **Document Settings** dialog box.

The green wireframe reflects the changes that will be made to the part when the **Update** button is picked or **Done** is selected from the pop-up menu. To cancel the changes, press the [Esc] key or select **Cancel** from the pop-up menu.

Circles, Tangent and Horizontal Constraints, and Trimming

In this section, you will use the **Center point circle** and **Line** tools to construct the part shown in **Figure 3-15A.** You will also apply tangent and horizontal constraints to the sketch before extruding the profile. Start a new Standard (mm).ipt file. With the **Center point circle** tool, construct two circles as shown in **Figure 3-15B.** Pick once to place the center of the circle. Place the center of the small circle on the origin of the coordinate system. Remember, the default work point can be projected into the sketch. Projecting geometry is discussed in Chapter 5. Then, move the cursor and pick to set the diameter. The radius of the circle is displayed at the bottom-right corner of the screen where the length and angle of a line are displayed. Remember, you can draw the circles to an approximate size and edit the dimensions to exact values.

Next, apply dimensions and edit them to the exact values. To put a diameter dimension on a circle, pick the **General Dimension** button in the **2D Sketch Panel.** Then, pick anywhere on the circumference of the circle and drag the dimension to the desired location. You can pick on either the circumferences or the center points of the circles to place the 55 mm linear dimension.

Figure 3-15.
A—The completed part. B—Start the sketch used to create the part by drawing two circles. C—The sketch is completed and fully constrained.

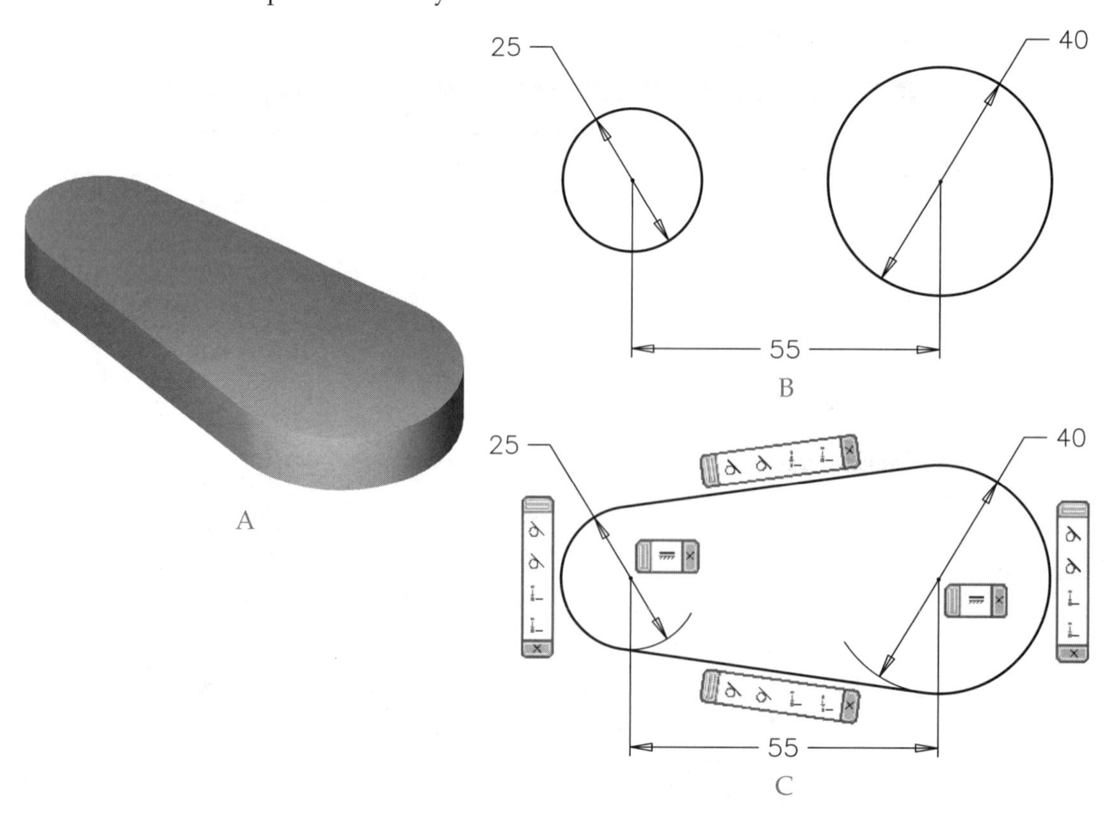

Now, place a horizontal constraint between the center points of the two circles to vertically align them. Pick the **Horizontal** button in the constraint flyout in the **2D Sketch Panel**. Then, pick the center point of the small circle. Finally, pick the center point of the large circle. If the circles were drawn so their centers are not horizontal (different Y values), the large circle moves to reflect the constraint. Think of this constraint acting as if a horizontal line is drawn between the two circle centers.

To complete the sketch, you need to draw two lines and trim the circles. Pick the **Line** button in the **2D Sketch Panel**. Then, pick the first endpoint of the line anywhere on the top of the smaller circle. The cursor will show the coincident constraint symbol. Now, put the cursor on the larger circle and move it around until you see both the coincident and tangent constraint symbols. Pick this point to finish the line.

The line is tangent to the large circle, but not to the small circle. To constrain the line tangent to the small circle, pick the **Tangent** button in the constraint flyout. Then, pick the line and the small circle. The first endpoint of the line moves to a location so that the line is tangent to the small circle. The second endpoint of the line also moves slightly so the line remains tangent to the large circle.

Use a similar process to draw a line across the bottoms of the circles. Then, display the constraints for the sketch. With no tool active, right-click in the graphics window and select **Show All Constraints** from the pop-up menu or press [F8]. Both lines have two coincident and two tangent symbols displayed in their constraint bar. This is because each endpoint is on (coincidental to) a circle and each line is tangent to each circle.

Currently, the sketch is an ambiguous profile. Ambiguous profiles are discussed in the next chapter. To make the sketch unambiguous so it will be automatically selected for extrusion, the inner portions of the circles need to be removed. The **Trim** tool is used to do this. Pick the **Trim** button in the **2D Sketch Panel** or press the [X] key. Then, place the cursor over the inside edge of the small circle. The portion that will be removed by the tool is displayed as a dashed line. When you pick that portion of the circle, it is trimmed (removed). Also, trim the inside edge of the large circle, **Figure 3-15C.**

After using the **Trim** tool, you may still see a portion of an arc. This arc is actually the dimension line for the diameter dimension. Since the dimension is not part of the geometry, this does *not* create an ambiguous sketch.

The sketch is complete, fully constrained, and unambiguous. You can now extrude it into a solid part. Display the isometric view by right-clicking anywhere in the graphics window and selecting **Isometric View** from the pop-up menu. Then, finish the sketch by right-clicking again and selecting **Finish Sketch** from the pop-up menu. Finally, extrude the profile 10 mm with a 0° taper using the **Extrude** tool in the **Part Features** panel. Save the file as Example_03_02.ipt.

Arcs and More Constraints

Five lines and an arc are required to construct the part shown in **Figure 3-16A.** Three constraints will be used to locate the arc: equal, colinear, and coincident. Start a new English Standard (in).ipt file. Using the **Line** tool, draw the five lines shown in **Figure 3-16B.** Make sure to leave the gap between lines AB and CD. Apply a fix constraint to lines AB and AF, which will keep them from moving as other lines are dynamically moved. Now, dimension the sketch as shown in **Figure 3-16B.**

Show all constraints. Notice that the only relationship between lines AB and CD is that they are parallel. In fact, if you pick on line CD, you can drag it independent of line AB and line DE. By adding a colinear constraint to lines AB and CD, the lines are constrained to the same Y values. If the lines were vertical, they would be constrained to the same X values. If the lines are angled, they would be constrained to the same "slope" line. Pick the **Colinear** button in the constraint flyout. Then, pick line AB and line CD. In this case, the order is not important. However, the second line picked is constrained colinear to the first line picked, so in the case of angled lines, the order is important.

Figure 3-16.
A—The completed part. B—Start the sketch used to create the part by drawing the line segments shown here. Be sure to leave the gap between points B and C.

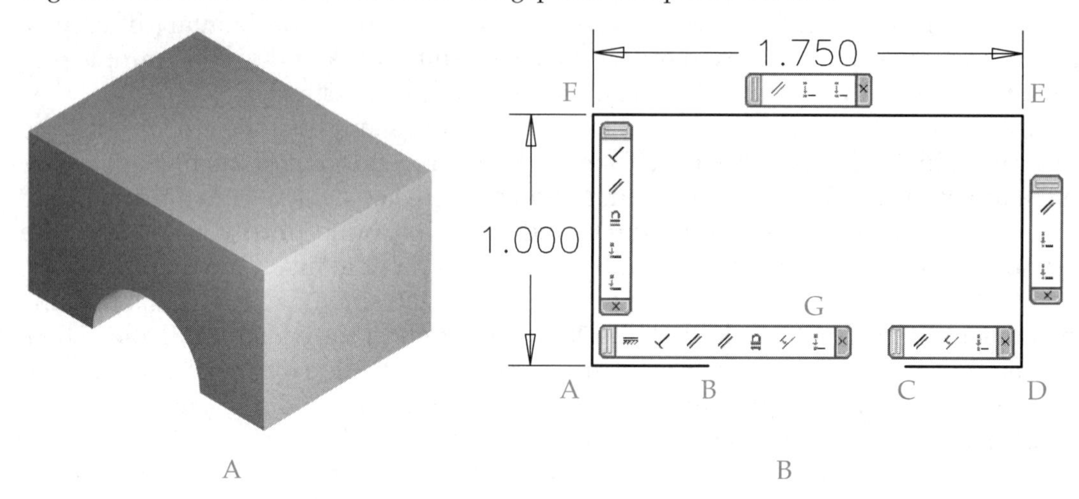

Now, the arc needs to be added to the sketch using the **Three Point Arc** tool. Pick the **Three Point Arc** button in the **2D Sketch Panel**. Three point arcs are drawn by selecting the start point, endpoint, and then middle point (not center point). Pick the right endpoint of line AB; the yellow dot will change to green. Then, pick the left endpoint of line CD. Finally, pick at approximately point G (shown in **Figure 3-16**). The Y coordinate of the center of the arc is approximately the same as the Y coordinate for lines AB and CD.

An equal constraint applied to lines AB and CD will horizontally center the arc between the lines. This is because the endpoints of the arc are constrained to the endpoints of the lines. Pick the **Equal** button in the constraint flyout, pick line AB, and then pick line CD.

Add another equal constraint to the two vertical lines, AF and DE. You should receive a message indicating that applying the constraint will over-constrain the sketch. See **Figure 3-17.** Inventor does not allow you to over-constrain a sketch. Line ED is already constrained by the 1.000 dimension, the coincident constraints, and the colinear constraint of AB and CD. Pick the **Cancel** button in the message box to close it. Press the [Esc] key to cancel the **Equal** constraint tool. Because the first equal constraint is valid, it remains to control lines AB and CD.

The arc now needs to be vertically constrained. This can be done by applying a coincident constraint to the center of the arc and line CD (or line AB). Pick the **Coincident** button in the constraint flyout. Then, pick the center of the arc (the dot) and line CD. Make sure you pick when line CD is highlighted.

Figure 3-17.
If you attempt to apply a dimension or geometric constraint that will over-constrain the sketch, a warning appears. Inventor will not allow you to over-constrain a sketch.

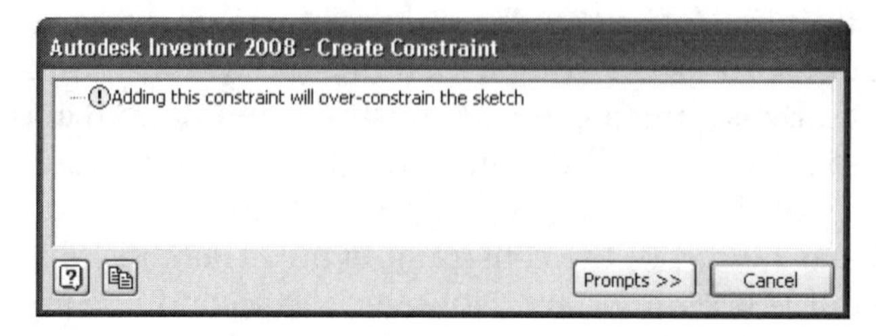

Figure 3-18.
Lines and constraints can be added by right-clicking in the graphics window while in sketch mode and selecting the tool in the pop-up menu.

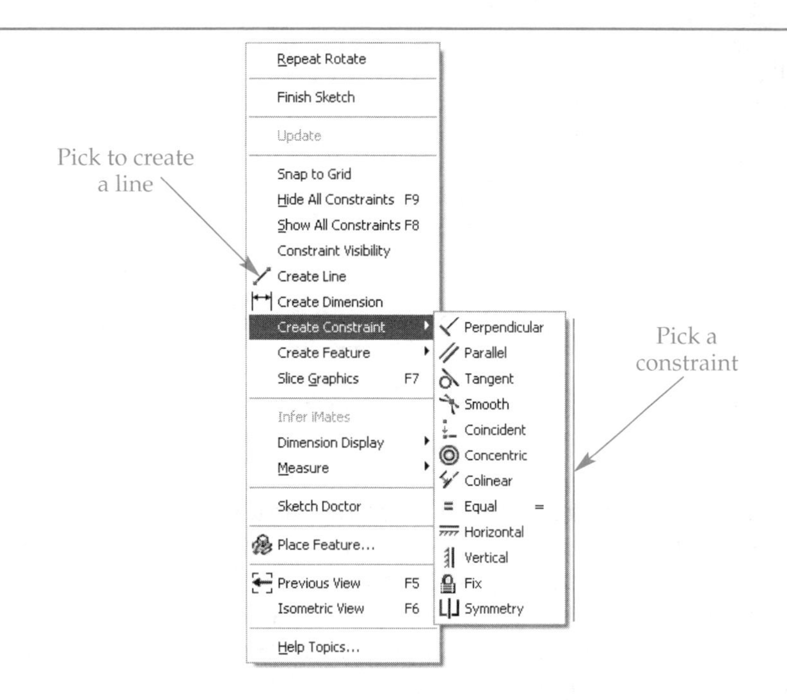

The last constraint that needs to be added to the sketch is a dimension constraint for the size of the arc. This will constrain not only the size of the arc, but the length of lines AB and CD. Pick the **General Dimension** button in the **2D Sketch Panel**. Then, pick the arc and drag the dimension to the desired location. Finally, edit the dimension and enter .40 for the radius value.

The sketch is complete, fully constrained, and unambiguous. Display the isometric view and finish the sketch. Also, save the file as Example_03_03.ipt. Complete the part by extruding the sketch 1.25″ with a 0° taper.

PROFESSIONAL TIP

The **Line** tool and the 12 constraints can also be accessed through the pop-up menu when in sketch mode. Press the [Esc] key to cancel any tools you may have selected. Right-click on an empty area of the graphics window to display the pop-up menu, **Figure 3-18.** Notice the **Line** tool is available. Select **Create Constraint** to display a cascading menu with all of the constraints that are available.

PRACTICE 3-1 Complete the practice problem on the Student CD.

Drawing an Arc from within the Line Tool

Tangent and perpendicular arcs can also be constructed from *within* the **Line** tool. For example, the sketch shown in **Figure 3-19** can be constructed, not including dimensions, using only one session of the **Line** tool. This is a little tricky, but once mastered, you will find it very useful. For AutoCAD users, the procedure is similar to constructing polyarcs within the **POLYLINE** command, but without the typing.

Start a new metric Standard (mm).ipt file. Draw a line from point A to point B about 170 mm long. Now, move the cursor over point B and let it "hover." The yellow dot will change to gray. Pick and drag the cursor to the right and up, as if you were sketching an

Figure 3-19.
All curves in this sketch can be drawn in a single session of the **Line** tool.

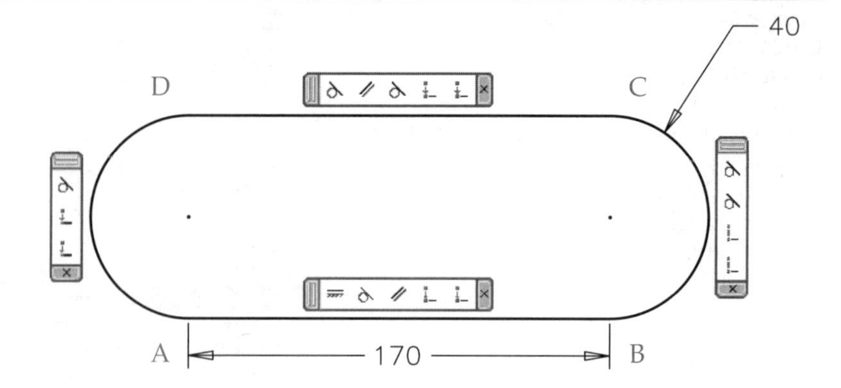

arc. Release the mouse button when the radius is about 40, as shown in the lower-right corner of the screen. The arc segment is constructed. Then, draw line CD normally. Finally, use the same "drag off" method to draw the second arc from point D to point A.

Show all constraints. Make sure both arcs are tangent to both lines. If not, add the necessary tangent constraint. Now, if you were going to finish the part, you can add dimensions and adjust their values. Then, finish the sketch and extrude the part.

PRACTICE 3-2 Complete the practice problem on the Student CD.

Things That Can Go Wrong with Sketches

There are several mistakes you can make in the construction of sketches that will generate errors when you attempt to use the **Extrude** tool. The two most common errors are having gaps at corners and having overlapping lines. For example, open the Example_03_04.ipt file. There is a very small gap between the two lines in the upper-right corner, as shown in **Figure 3-20.**

Pick the **Extrude** button in the **Part Features** panel. In the **Extrude** dialog box, notice that the **Surface** button is on in the **Output** area. Because of the gap, Inventor assumes this sketch is to be extruded into a surface. However, you want this to be a solid, not a surface, so pick the **Solid** button. A new button—with a red cross—appears below the **Output** area. See **Figure 3-21.** This button indicates a solid cannot be created because a problem exists. If you hover the cursor over the button, the tooltip is **Examine Profile Problems**.

Figure 3-20.
At a "normal" zoom level, the sketch on the left appears to be closed. However, if you zoom in on the upper-right corner of the sketch, there is a small gap, as shown in the detail.

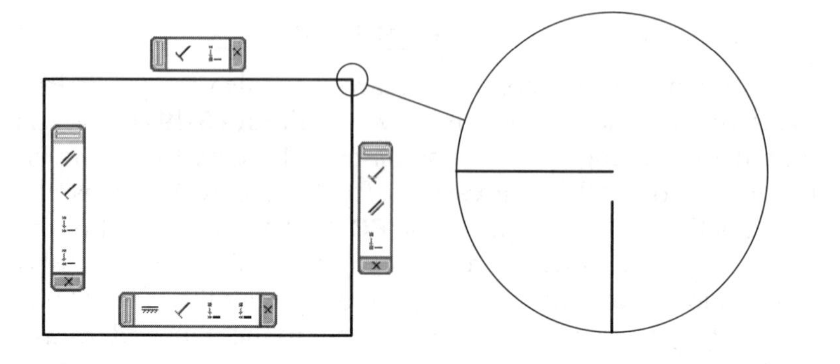

Figure 3-21.
If you attempt to extrude an open sketch into a solid, the **Examine Profile Problems** button appears in the **Extrude** dialog box. Pick this button to open the **Sketch Doctor** wizard.

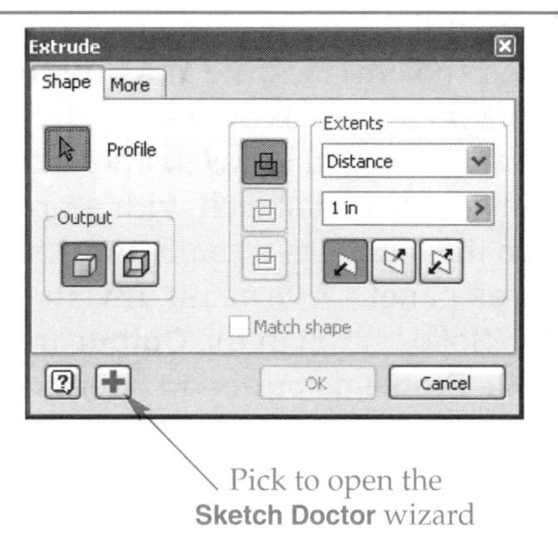

Now, pick the **Examine Profile Problems** button and the **Sketch Doctor** wizard appears, **Figure 3-22.** A description of the problem and possible solutions is provided on the first page of the wizard. In this case, the problem is indicated as an open loop. Also, the problem is indicated on the sketch by a green dot.

Pick the **Next>** button to display the next page of the wizard. On this page, you are given options for fixing the problem, **Figure 3-23.** Inventor refers to these as "treatments." Highlight the **Close Loop** treatment and pick the **Finish** button. In this case, a message box appears indicating there is a gap and asking if you would like Inventor to close the

Figure 3-22.
The first page of the **Sketch Doctor** wizard

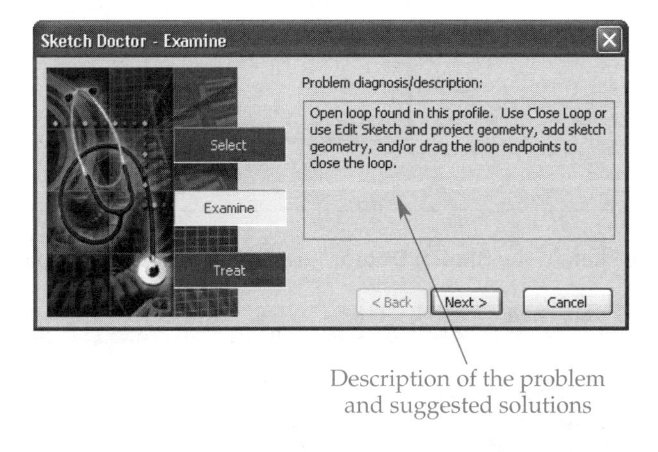

Figure 3-23.
Selecting a treatment option in the **Sketch Doctor** wizard.

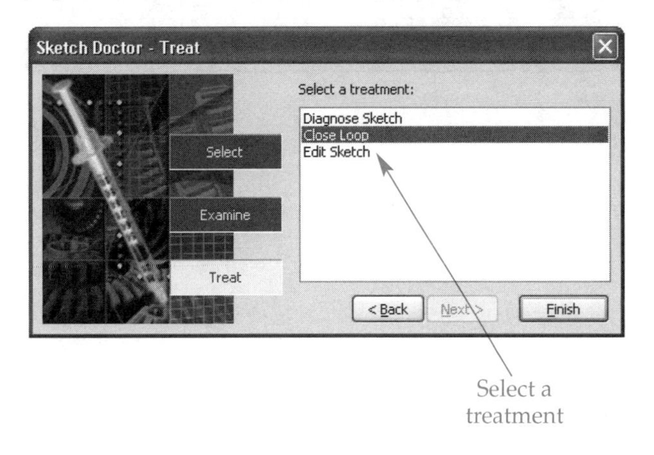

gap between the highlighted lines. In other instances, this message may be different, depending on the problem. Pick the **Yes** button and Inventor will automatically close the sketch. Another message box appears indicating that the problem was successfully fixed. Now, you can "finish" the sketch and extrude it into a solid.

For another example of a sketch with a problem, open the Example_03_05.ipt file. The bottom line in this sketch is actually two lines that overlap. Pick the **Extrude** button in the **Part Features** panel. Once again, Inventor defaults to creating a surface instead of a solid. Pick the **Solid** button in the **Output** area of the **Extrude** dialog box. Then, pick the **Examine Profile Problems** button to open the **Sketch Doctor** wizard. The first page of the wizard indicates the problem. Pick the **Next>** button to continue.

On the second page of the wizard, highlight the **Diagnose the Sketch** treatment. Then, pick the **Finish** button. The **Diagnose Sketch** dialog box appears, **Figure 3-24.** This dialog box allows you to choose which tests to perform. Check all of the tests and pick the **OK** button to continue.

The sketch is analyzed and the **Sketch Doctor** wizard is redisplayed. There are three problems listed in the wizard—a missing constraint, overlapping lines, and a gap (open loop). See **Figure 3-25.** In addition, as you highlight a problem in the wizard, that problem is highlighted on the sketch. In this case, the multiple problems make it hard for Inventor to automatically fix the sketch. The easiest fix is to cancel the wizard, delete the bottom lines, and properly draw a new line.

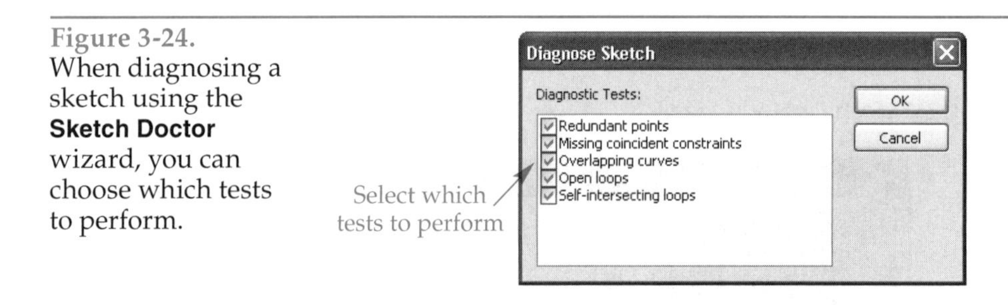

Figure 3-24.
When diagnosing a sketch using the **Sketch Doctor** wizard, you can choose which tests to perform.

Figure 3-25.
After diagnosing the sketch, the **Sketch Doctor** wizard displays the problems it found.

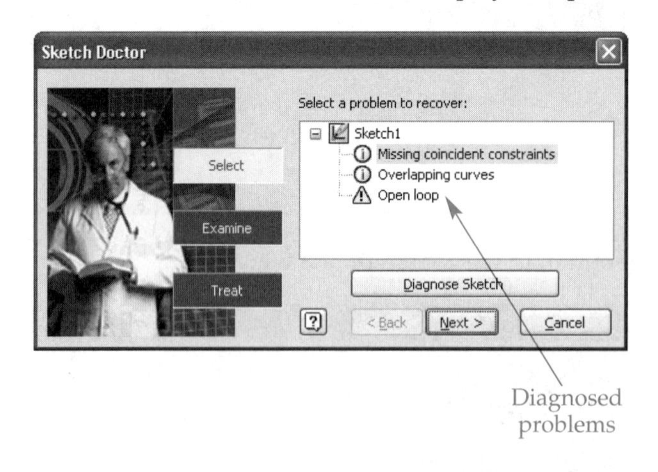

Review of All Constraints

There are 12 geometric constraints available in the constraints flyout in the **2D Sketch Panel**. The types of objects or "picks" required when applying each constraint varies. In addition, the symbol that appears in the constraint flyout is the same symbol that appears in the constraint bar for geometry. The chart in **Figure 3-26** reviews the choices for all constraints. In order to become more familiar with the available constraints, copy this chart by hand to a blank sheet of paper. Be neat so you can use the sheet as a reference as you work in Inventor.

Figure 3-26.
Review of constraints.

Name	1st Choice	2nd Choice	Symbol
Perpendicular	Line	Line	
Parallel	Line	Line	
Tangent	Line, arc, or circle	Line, arc, or circle	
Smooth (G2)	Spline	Line, arc, or circle	
Coincident	Center of arcs or circles Endpoints of lines or arcs	Centers of arcs or circles; endpoints of lines Endpoints of lines or arcs	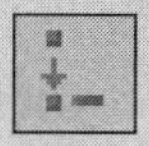
Concentric	Circle or arc	Circle or arc	
Colinear	Line	Line	
Equal	Line, arc, or circle	Line, arc, or circle	 =
Horizontal	Line Center, endpoint, or midpoint	*(none)* Center, endpoint, or midpoint	
Vertical	Line Center, endpoint, or midpoint	*(none)* Center, endpoint, or midpoint	
Fix	Line, arc, circle, or spline	*(none)*	
Symmetric	Line, arc, circle, or spline	Line (Normal, Construction, or Center)	

Chapter Test

Answer the following questions on a separate sheet of paper or complete the electronic chapter test on the Student CD.

1. Define *sketch* in Inventor.
2. All lines, arcs, and circles in Inventor are considered ______ objects.
3. What is a *constraint* in Inventor?
4. By default, in which mode is Inventor when a new part file is started?
5. What is the hot key for the **Line** tool?
6. When using the **Line** tool, what information is provided in the three windows at the lower-right corner of the Inventor screen?
7. What are the two basic types of constraints?
8. How do you display the constraints for all geometry in the sketch?
9. When constraints are displayed for geometry, how are they represented on-screen?
10. If you have drawn a line that is 150 mm in length, but it should be 148 mm in length, how can you correct this problem in Inventor?
11. Which tool allows you to manually apply a dimension?
12. What function does the fix constraint serve?
13. Briefly describe how to extrude a fully constrained, unambiguous sketch into a solid part.
14. Once you have extruded a sketch into a solid part, how can you edit the original sketch, in sketch mode, to change the part?
15. Briefly describe how to draw an arc from within the **Line** tool. Assume you have drawn one straight line segment.

Chapter Exercises

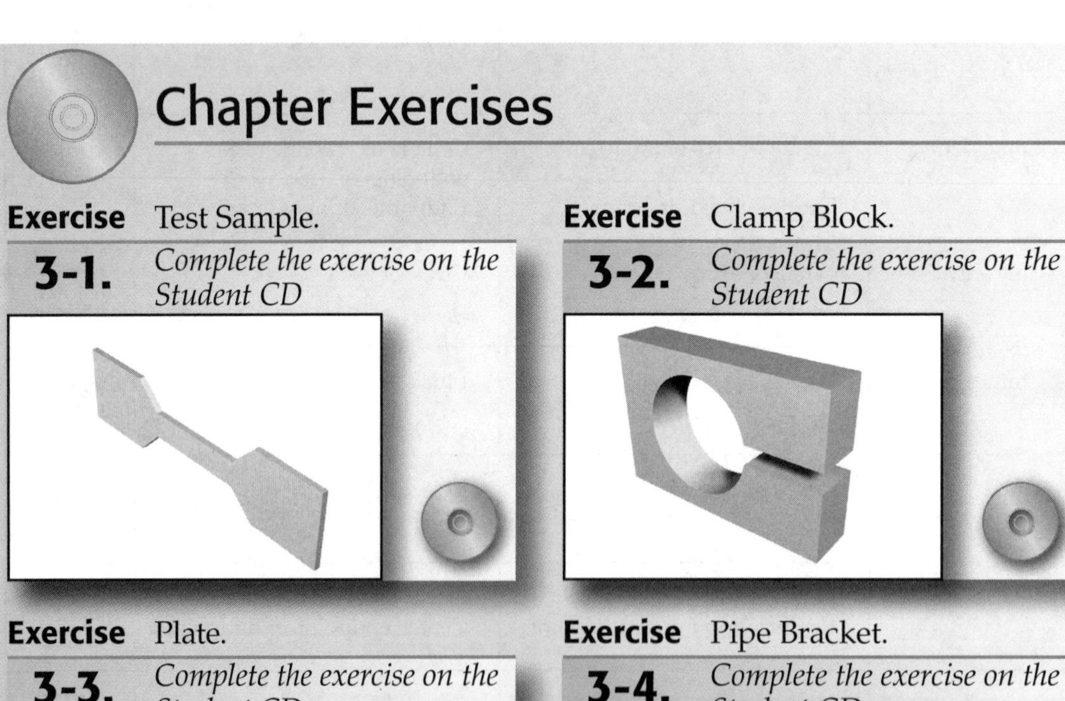

Exercise 3-1. Test Sample.
Complete the exercise on the Student CD

Exercise 3-2. Clamp Block.
Complete the exercise on the Student CD

Exercise 3-3. Plate.
Complete the exercise on the Student CD

Exercise 3-4. Pipe Bracket.
Complete the exercise on the Student CD

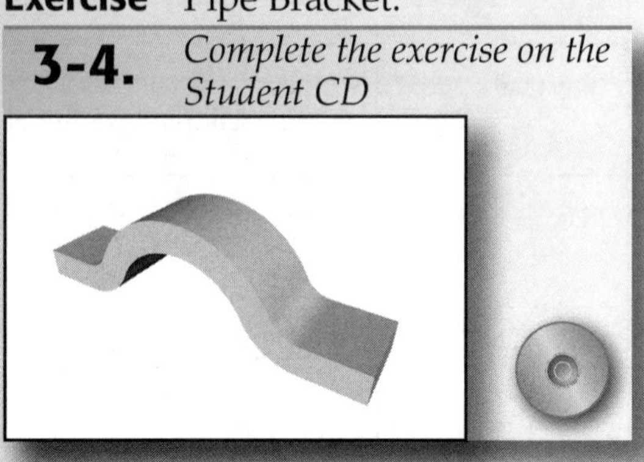

Exercise 3-5. Shear Blade.

Complete the exercise on the Student CD

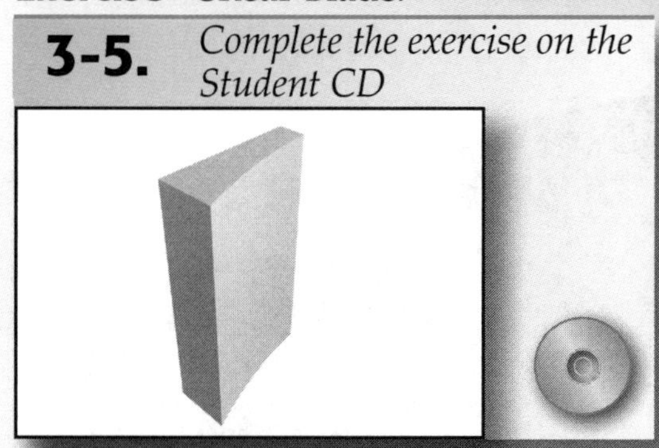

Exercise 3-6. Taper Wedge.

Complete the exercise on the Student CD

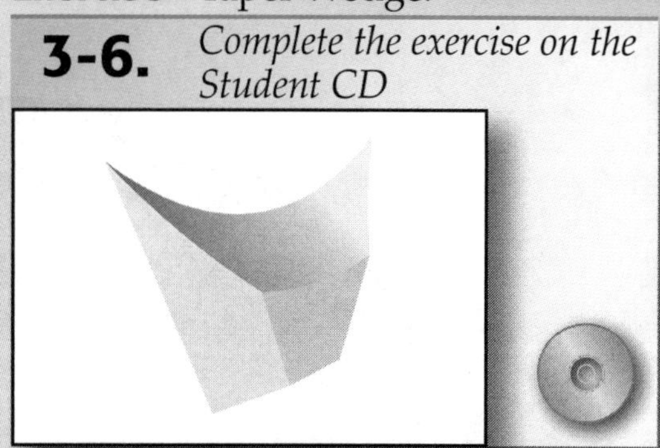

Exercise 3-7. U-Bracket.

Complete the exercise on the Student CD

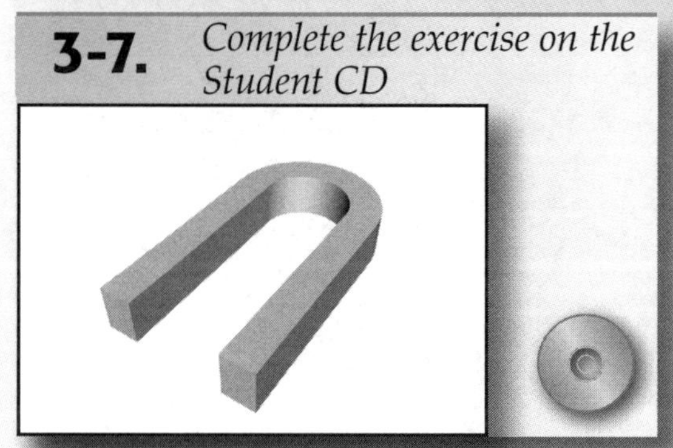

Exercise 3-8. Cover Plate.

Complete the exercise on the Student CD

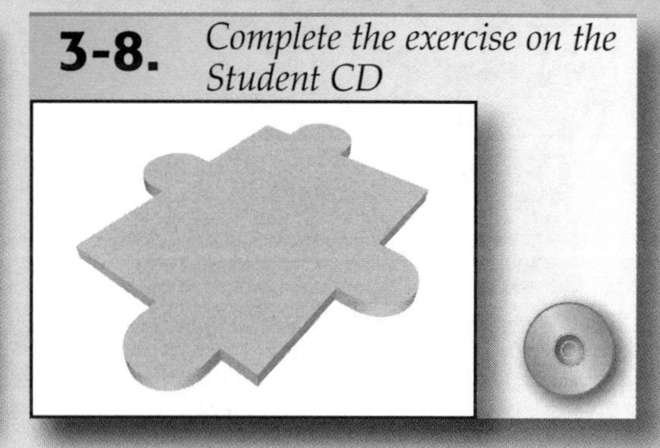

Exercise 3-9. Block.

Complete the exercise on the Student CD

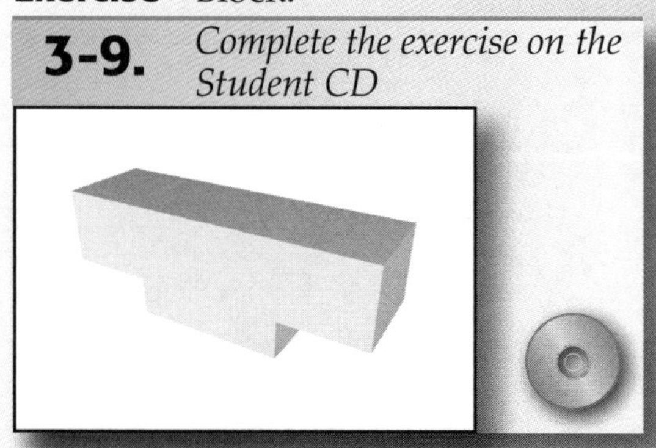

Many sketches were required to create this model of a cover. (Model courtesy of Autodesk, Inc.)

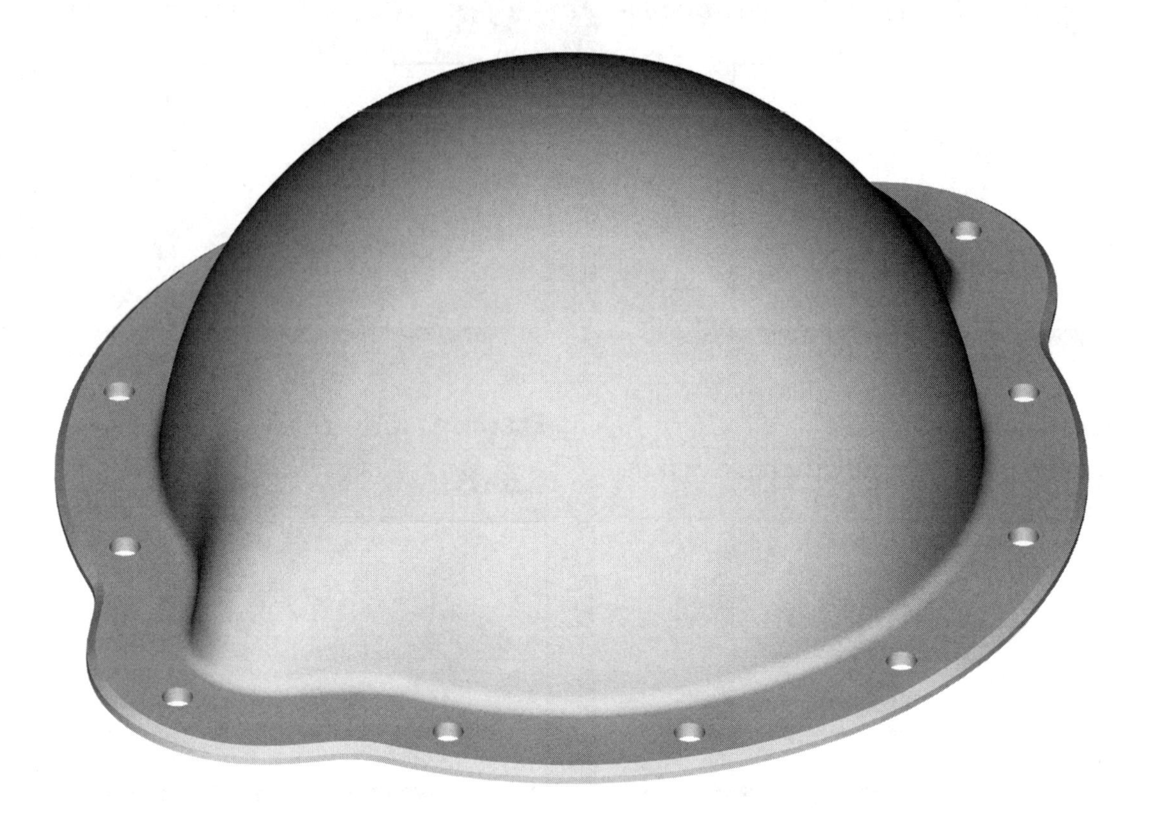

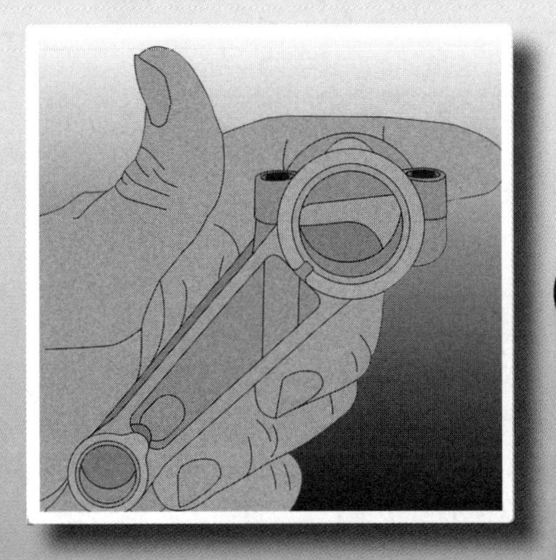

Chapter 4

Complex Sketching, Constraints, Formulas, and the Construction Geometry

Objectives

After completing this chapter, you will be able to:

- Explain how to create and select ambiguous profiles.
- Use dimension names to create dimensions with equations.
- Create and use construction geometry and centerlines to locate sketch objects.
- Use the **Mirror** tool in a sketch.
- Use the **Revolve** tool and explain its features.
- Explain and use the **Inventor Precise Input** toolbar.

User's Files

The Student CD included with this text contains several files required for this chapter. Refer to the file File List.txt *in the* \Ch04 *folder for the comprehensive list.*

Creating Complex (Ambiguous) Profiles

The sketches you created in Chapter 3 had no interior geometry, or "islands." There was only one choice for the profile; therefore, they were *unambiguous.* When you initiated the **Extrude** tool, you did not have to select the profile. Inventor selected the only possible profile for you. In this chapter, you will work with sketches that have several possible profiles. These are called ***ambiguous profiles.*** You must select the specific profile to extrude. The islands in the sketch will extrude as ***cutouts.*** They are not called holes because there is a **Hole** tool, which is discussed in the next chapter.

An Inventor sketch made up of four lines and two circles is shown in **Figure 4-1.** This geometry presents seven possible profiles that can be extruded, which are shown in the figure. You can select the profile to extrude by picking the **Profile** button in the **Extrude** dialog box, then pick within the area(s) you want to extrude. As you move the cursor over an area, the area that will be added to the selection is highlighted in red. The total area in the selection is highlighted in blue. This represents the profile that will be extruded. You can remove an area from the selection by holding down the

Figure 4-1.
There are seven possible profiles from the sketch shown on the left. The possible profiles are shown on the right. The area that will be extruded is represented by the hatch lines.

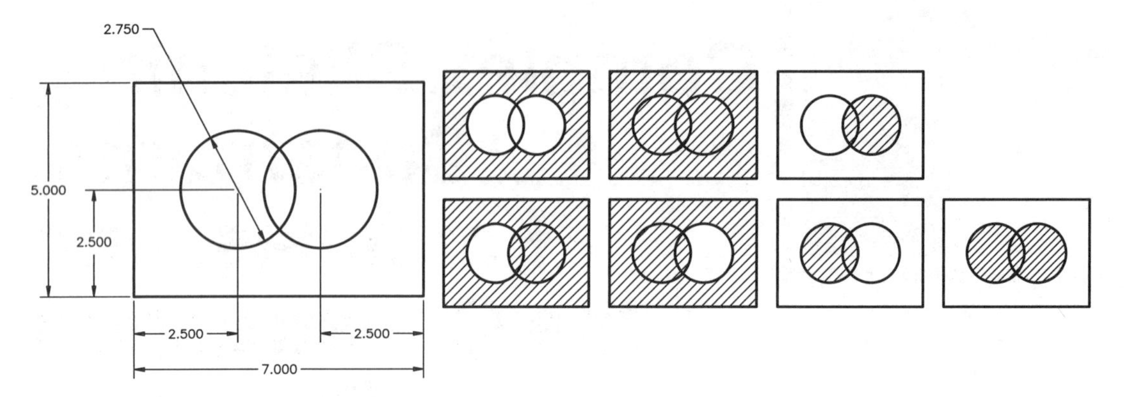

[Shift] key and picking in that area again. The area that will be removed is highlighted in red before you pick. To see that the area has been removed, move the cursor off of the sketch. Only the profile that will be extruded is highlighted.

Open the file Example_04_01.ipt. Then, pick the **Extrude** button in the **Part Features** panel. In the **Extrude** dialog box, notice that the **Profile** button is selected (depressed). However, no part of the sketch is highlighted; the sketch is ambiguous. With the button selected, move the cursor over the sketch and pick the appropriate areas to create the 1″ thick part shown in **Figure 4-2A.** Create the part. Then, right-click on Extrusion1 in the **Browser** and select **Edit Feature** from the pop-up menu. Pick the **Profile** button in the **Extrude** dialog box and select or deselect the areas to create the extruded part shown in **Figure 4-2B.** Pick the **OK** button to update the part.

Two Point Rectangle Tool and Precise Input

Start a new English Standard (in).ipt file. Now, you will create a part that is similar to the one in the last section. You will use the **Rectangle** tool, precise size input, and

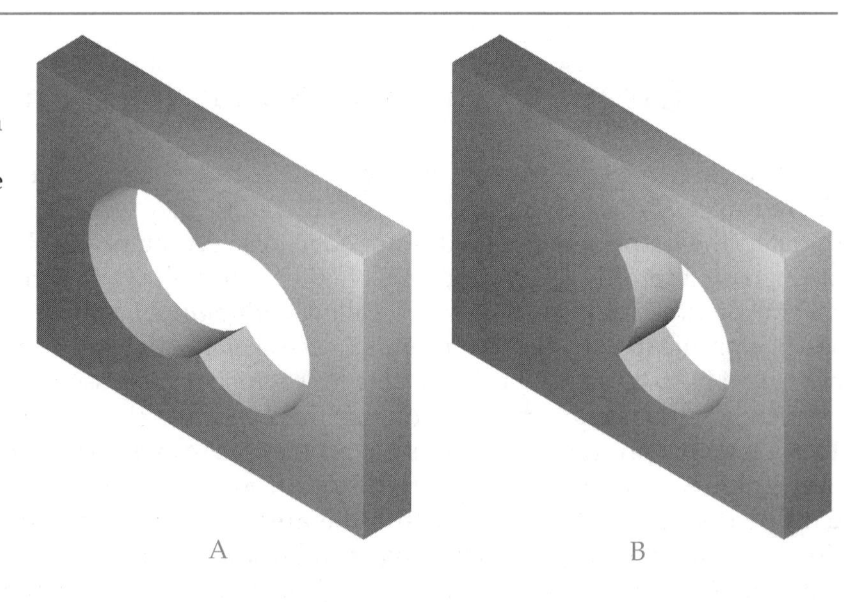

Figure 4-2.
A—One possible part produced by extruding the sketch shown in Figure 4-1.
B—Another possible part.

Figure 4-3.
You can use the **Inventor Precise Input** toolbar to enter exact coordinates.

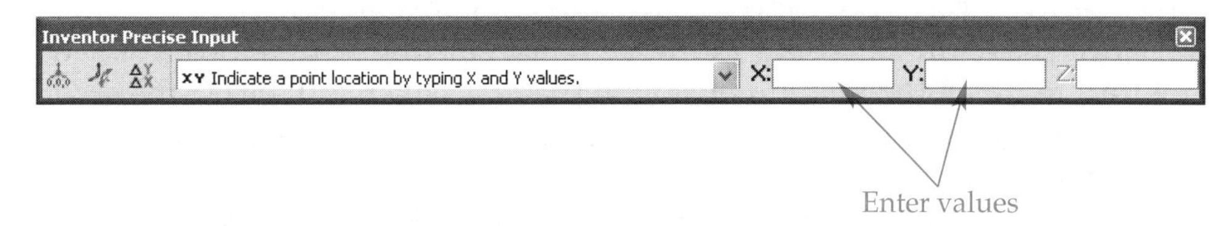

equations in the dimensions. You can input precise sizes using the **Inventor Precise Input** toolbar. Display this toolbar by selecting **Toolbar** from the **View** pull-down menu. Then, select **Inventor Precise Input** from the cascading menu. You must be in sketch mode for this option to appear. You can also right-click on a blank area of any toolbar and select **Inventor Precise Input** from the pop-up menu.

Now, pick the **Two point rectangle** button in the **2D Sketch Panel**. Pick the first point of the rectangle at the origin. The **Inventor Precise Input** toolbar will look like **Figure 4-3** with the vertical blinking bar in the **X:** text box. You can now type a value of 5 for the X coordinate of the second corner of the rectangle. Then, either press the [Tab] key to move to the **Y:** text box or pick in that text box with the cursor. Now, enter a value of 4 for the Y coordinate of the second corner of the rectangle. As you enter values, the second corner of the rectangle moves in the graphics window. The toolbar also has a **Z:** text box, but it is grayed out for 2D sketches. When the coordinates are correct, press the [Enter] key to finish the rectangle.

Pick the **Zoom All** button to see the entire rectangle. Then, pick the **Center point circle** button in the **2D Sketch Panel**. Using the **Inventor Precise Input** toolbar, enter X=2.5 and Y=2.0 for the center point. Enter X=3.5 and Y=2.0 for a point on the circumference of the circle to set the size. Close the **Inventor Precise Input** toolbar and end the **Circle** tool.

Now, you can finish the sketch. Using the **Extrude** tool, select the proper profile and create the 1″ solid part.

Precise input is a nice technique that you might find useful for establishing the start of a sketch. For this example, using precise input to draw the rectangle may have been useful. However, using precise input to construct the circle was probably counterproductive.

PROFESSIONAL TIP

The options on the **Inventor Precise Input** toolbar are reviewed at the end of this chapter.

Using the d0 Model Parameters in Equations in Dimensions

In Chapter 3, you learned how to add dimensions to constrain a sketch. Now, you will learn how to relate dimensions to each other. For example, in the previous section you drew a circle in the center of a rectangle. Using dimensions, you can constrain the circle so it is always in the horizontal center of the rectangle.

With the extruded part from the previous section still open, expand Extrusion1 in the **Browser** and right-click on Sketch1. Select **Edit Sketch** in the pop-up menu. Now, using the **General Dimension** tool, place a dimension across the bottom of the part. If the X and Y values were correctly input in the previous section, the value will be 5.000 inches. Then, place a dimension from the center of the circle to the left-hand edge of the part. Press the [Esc] key to cancel the **General Dimension** tool.

Every dimension in a part has a unique name starting with a lowercase "d" and ending with a number; the first dimension in the first sketch will be d0, the second d1, and so on. These names will be used to locate the center of the circle at the horizontal midpoint of the rectangle, regardless of the rectangle's width.

With the dimension tool still active, pick on the 2.500 dimension to open the **Edit Dimension** dialog box. While the dimension value is highlighted, pick on the 5.000 dimension in the sketch. The name of the 5.000 dimension appears in the dialog box in place of the dimension value. This is d2 if you drew the rectangle described above starting in a new session of Inventor. Type /2 after the d2 so the equation reads d2/2, **Figure 4-4.** This equation means that the value of d3 (the dimension from the center to the left edge) is the value of d2 divided by 2. In other words, the value of dimension d3 will always be one-half of the value of dimension d2.

Once you press [Enter] or pick the check button to update the dimension, the value is shown as fx: 2.500. The fx: indicates that the dimension is a function of (related to) another dimension or is driven by an equation.

Now, change dimension d2 (the 5.000 dimension) to a value of 10. The circle remains horizontally centered because dimension d3, which horizontally constrains the center of the circle, is defined as one-half of d2. Edit dimension d2 to a value of 5 and the circle returns to its original position.

Using a similar method, vertically constrain the center of the circle. Place a dimension on the left side of the part. Then, place a dimension from the center of the circle to the bottom edge of the part. Now, pick the vertical dimension to the center of the circle to edit it. The value should be 2.000 inches. With the dimension value highlighted, pick the overall vertical dimension on the sketch. The highlighted dimension value in the **Edit Dimension** dialog box is replaced with the name of the overall vertical dimension, which should be d4. Type /2 after d4 and press [Enter]. The circle's vertical dimension is now constrained to one-half of the overall vertical dimension.

Now, you will constrain the size of the circle based on the overall size of the part. First, place a diameter dimension on the circle. Then, edit the dimension value and enter the equation d4*0.50, where d4 is the name of the overall vertical dimension. See **Figure 4-5.** This equation is the same as d4/2. Now, any time the overall vertical dimension is changed, the circle remains vertically centered *and* the diameter of the circle is one-half of the vertical distance.

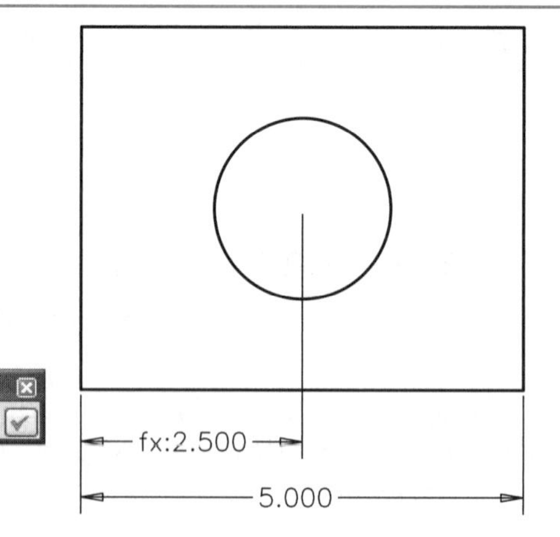

Figure 4-4.
By entering the equation in place of the 2.500 value, the center of the circle will always be at the horizontal center of the part, even if the 5.000 dimension is changed.

Figure 4-5.
Entering an equation to constrain the diameter of the circle to one-half of the overall vertical dimension.

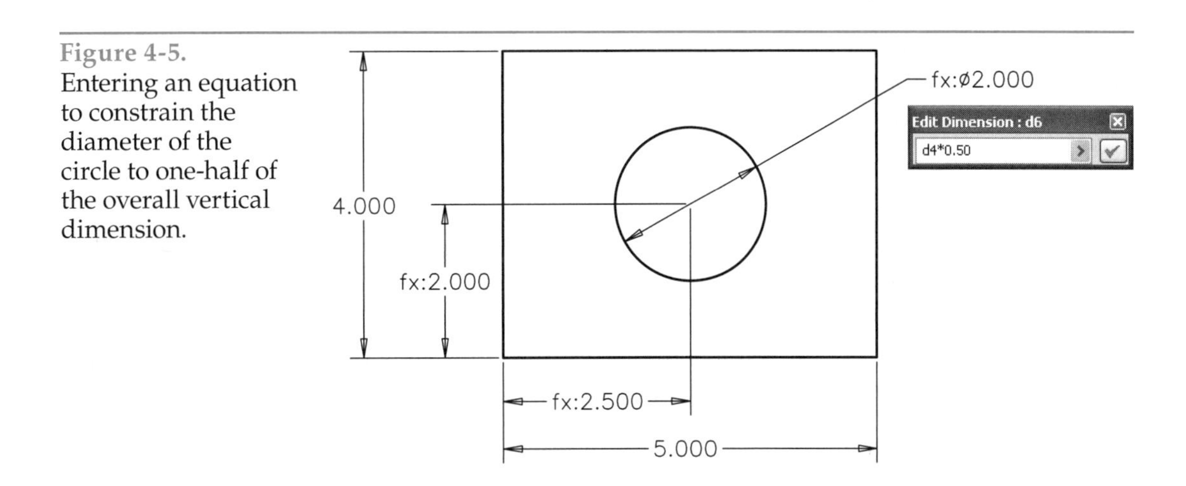

If you right-click on a dimension and select **Dimension Properties...** in the pop-up menu, the **Dimension Properties** dialog box is displayed. This dialog box contains two tabs, one for the dimension you selected and the other for the document, in this case the sketch. See **Figure 4-6.** Here, you can set the number of decimal places and the way the dimension is displayed. For example, if you select **Show Expression** in the **Modeling Dimension Display:** drop-down list in the **Document Settings** tab and pick **OK**, the three dimensions that are functions are displayed as equations. These display options can also be set by right-clicking in the sketch and selected **Dimension Display** in the pop-up menu. Then, select the display option in the cascading menu that is displayed.

Finish the sketch. With the dimensions related to each other through the use of equations, the part can easily be updated in the future. You could even take it one step further and constrain the overall vertical dimension to 80% of the overall horizontal

Figure 4-6.
A—The **Dimension Properties** dialog box can be used to show functions as equations instead of fx:. B—The various display options.

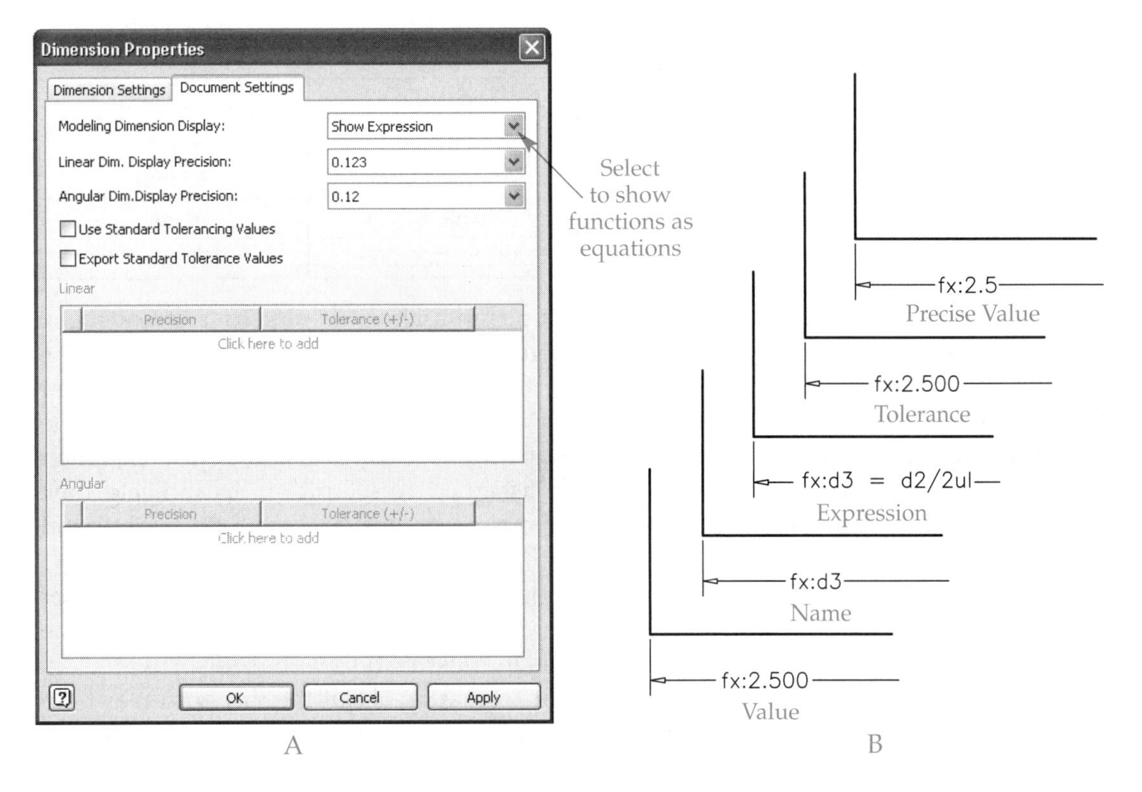

dimension, if needed. The value entered in the **Edit Dimension** dialog box can vary from a plain number to an arithmetic, trigonometric, or algebraic equation with the dimension names used as variables.

More on Equations

Inventor expects the result of an equation to have the correct units. For example, the equation d0/d2 would not be accepted for the diameter of the circle. This equation results in a ***unitless*** solution and Inventor expects a solution in the units you are working in (inches or millimeters, for example). The solution is unitless because d0 is inches and d2 is inches. When inches are divided by inches, the units cancel each other out and the result has no unit. However, the equation (d0/d2)*1 in is acceptable because the solution has units. An unacceptable equation is highlighted in red in the **Edit Dimension** dialog box.

In the previous section, you entered the equation d4*0.5 for the dimension value of the circle diameter. Verify this equation by editing the dimension. If you are not in sketch mode, right-click on the extrusion name in the **Browser** and select **Show Dimensions**. Then, double-click on the diameter dimension to open the **Edit Dimension** dialog box. Notice that Inventor has rewritten the equation you entered to d4/0.50 ul. Inventor has guessed that the constant 0.50 is *unitless* and automatically indicates this by adding the "unit" ul. Since the dimension d4 is in inches, the result of the equation is in inches and the equation is acceptable.

For the following simple equations, assume that dimension d3 has a value of 175 mm and dimension d4 has a value of 50 mm. The five arithmetic operators available in Inventor, and some sample equations, are:

Operation	Operator	Equation	Result
Addition	+	d3+d4	225 mm
		d3+100ul	275 mm
Subtraction	–	d3–d4	125 mm
Multiplication	*	0.5ul*d4	25 mm
		d3*d4	Error
Division	/	d4/10ul	5.0 mm
		(d3/d4)*1mm)	3.5 mm
Exponent	^	d3^2	Error
		(d4^2)/100mm	25 mm

To see all of the dimensions and their values in a chart similar to a spreadsheet, select **Parameters...** from the **Tools** pull-down menu. The **Parameters** dialog box is shown, **Figure 4-7.** The dimension names are shown on the left side in the **Parameter Name** column. The unit that Inventor expects is listed in the **Unit** column. The value of the dimension, whether a numeric value or an equation, is listed in the **Equation** column. Notice that all constants have been assigned the "unit" ul to indicate they are unitless. You can also add comments about a dimension using the right-hand column.

The width of a column can be widened or narrowed. To do so, move the cursor to the line separating the column headings; a double-arrow cursor appears. Drag the column border left or right. This is the same as in most Windows-based software.

Hovering the cursor over a row provides information about the dimension as help text. You can edit the dimension name, value, or comment for a dimension by picking

Figure 4-7.
The **Parameters** dialog box can be used to rename dimensions and enter equations.

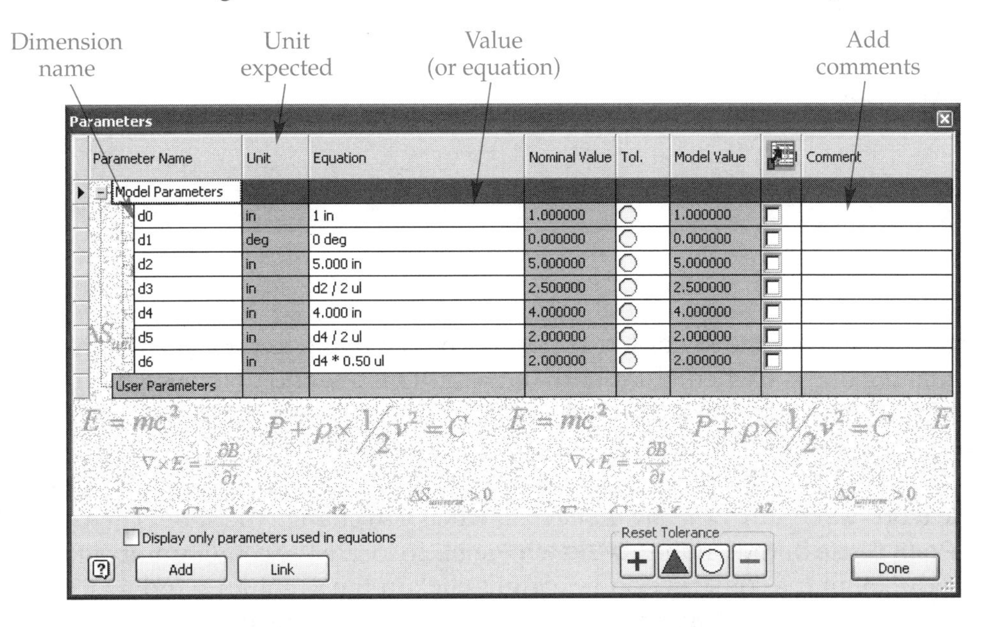

in that cell and typing. When done typing, press [Enter]. For example, change the name of d2 to Long. Notice that after you press [Enter] all equations that referenced d2 are updated to reference Long. The dimension name is case sensitive; LONG is different than Long and long. You can use some symbols, such as the underscore, letters, and numbers. You cannot use spaces, arithmetic operators, such as the plus sign, or symbols that may be used in an equation, such as a period. If you make a mistake, Inventor will give an error message: Variable name contains invalid characters. Now, add Length of part to the Comment cell of the Long dimension.

Notice in **Figure 4-7** that there are seven dimensions and the sketch only had five. The dimensions d0 and d1 are related to the extrusion and were automatically added after using the **Extrude** tool. Change the name of dimension d0 to Extrusion_Height and dimension d1 to Taper_Angle. If you like, you can add comments to these dimensions as well.

Construction Geometry

Lines can be sketched in different styles. So far, you have only sketched in the Normal style. This style creates solid (continuous) curves that are evaluated as part of the profile when the sketch is "finished." Two other available styles are Construction and Centerline. Geometry created in the Construction style can be used with constraints to locate geometry created in the Normal style without the use of dimensions. Construction geometry is not part of the profile and disappears from the model view when the profile is extruded. In this section, you will locate a circle at the center of a rectangle using a line drawn in the Construction style. The Centerline style is discussed in the next section.

Start a new metric Standard (mm).ipt file. Construct a 100 mm × 70 mm rectangle with one corner at the origin. Dimension the bottom and the left side. Next, look at the buttons near the right end of the **Inventor Standard** toolbar. See **Figure 4-8.** The **Construction** and **Centerline** buttons turn on and off the Construction and Centerline styles. These buttons are toggles. When one is pressed, it stays on until pressed again. Press the **Construction** button to toggle it on.

Figure 4-8.
The Construction and Centerline styles can be set current using the buttons on the **Inventor Standard** toolbar.

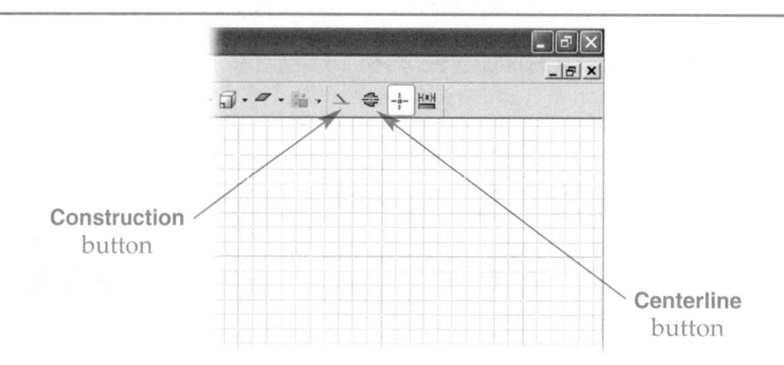

Draw a line from the lower-left corner of the rectangle to the upper-right corner. Notice that the line is dashed, thinner than the other lines, and displayed in a different color from the lines drawn in the Normal style. These differences indicate that the line is in the Construction style. Press [Esc] to exit the **Line** tool.

Change the current line style back to Normal by pressing the **Construction** button to toggle it off. Make sure nothing is selected when you change the style. Then, pick the **Center Point Circle** button in the **2D Sketch Panel**. To locate the center point of the circle at the midpoint of the construction line, right-click in the graphics window to display the pop-up menu. Select **Midpoint** to activate the midpoint object snap. Then, pick the construction line. You could also find the midpoint without using the snap by sliding the cursor along the construction line until the green dot is displayed at the midpoint and picking. The center of the circle is placed at the midpoint of the construction line. Since the construction line bisects the rectangle, this point is also the geometric center of the rectangle. Now, pick a point for the diameter. Then, dimension the circle's diameter and change the value to 40 mm.

Change the length of the part from 100 mm to 150 mm. Notice that the circle stays in the center of the rectangle. Change the dimension back to 100 mm; the circle moves to stay in the center of the rectangle. Pick the **Show Constraints** button in the **2D Sketch Panel**. Then, pick the construction line, circle, and center of the circle. See **Figure 4-9.** Press [Esc] to exit the **Show Constraints** tool. The reason the circle stays at the center of the rectangle is that there is a coincident constraint automatically applied between the center of the circle and the midpoint of the construction line. The endpoints of the construction line are similarly constrained to the corners of the rectangle. Therefore, as the corner of the rectangle is moved, the endpoint of the construction line moves. In turn, the center of the circle follows the midpoint of the construction line.

To complete this example, you will use the circle to construct a five-sided (pentagon) shape. Since the circle will not be part of the profile, it can be changed to the Construction style. First, select the circle. Selected curves are displayed in blue

Figure 4-9.
The constraints are shown for the construction line, circle, and circle center.

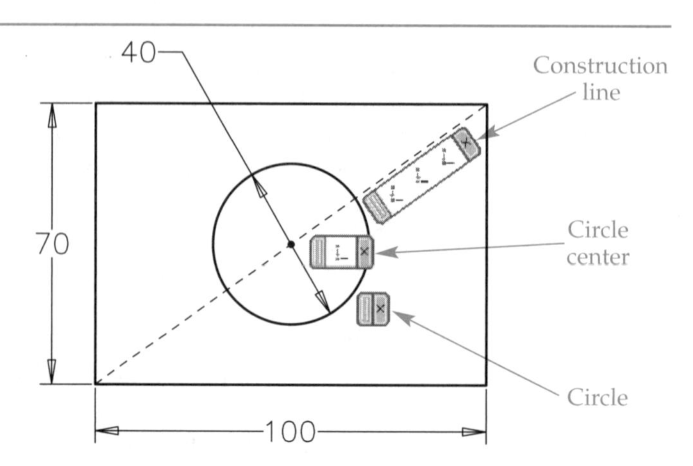

(depending on the current display color scheme). With the circle selected, press the **Construction** style button. The style of the circle is now Construction. Pick anywhere in the graphics window to deselect the circle. The circle is displayed as a thin, dashed line like the line you drew from corner to corner. Also note that the **Construction** button does not stay on.

Now, you can construct the pentagon. Each segment of the pentagon will be constrained tangent to the circle and equal to the other segments. See **Figure 4-10A.** However, before you draw anything, make sure the line style is Normal.

With the **Line** tool, draw the left side of the pentagon vertical. Remember, the starting point and length can be "about" correct. Inventor will apply a parallel constraint between the line and the left side of the rectangle. Then, holding down the [Ctrl] key to prevent automatic constraints, sketch the other four sides at approximate sizes and locations. Make sure the last endpoint of the last segment is connected to the first endpoint of the first segment. Now, apply a tangent constraint to each of the five line segments in the pentagon so they are tangent to the circle. Then, apply an equal constraint to the second through fifth line segments in the pentagon so they are equal to the first line segment. An easier method of creating the pentagon may be to use the **Polygon** tool. Finally, "finish" the sketch and extrude the profile 10 mm. Since this is an ambiguous profile, you will need to select the proper profile to create a part with a cutout (hole). See **Figure 4-10B.**

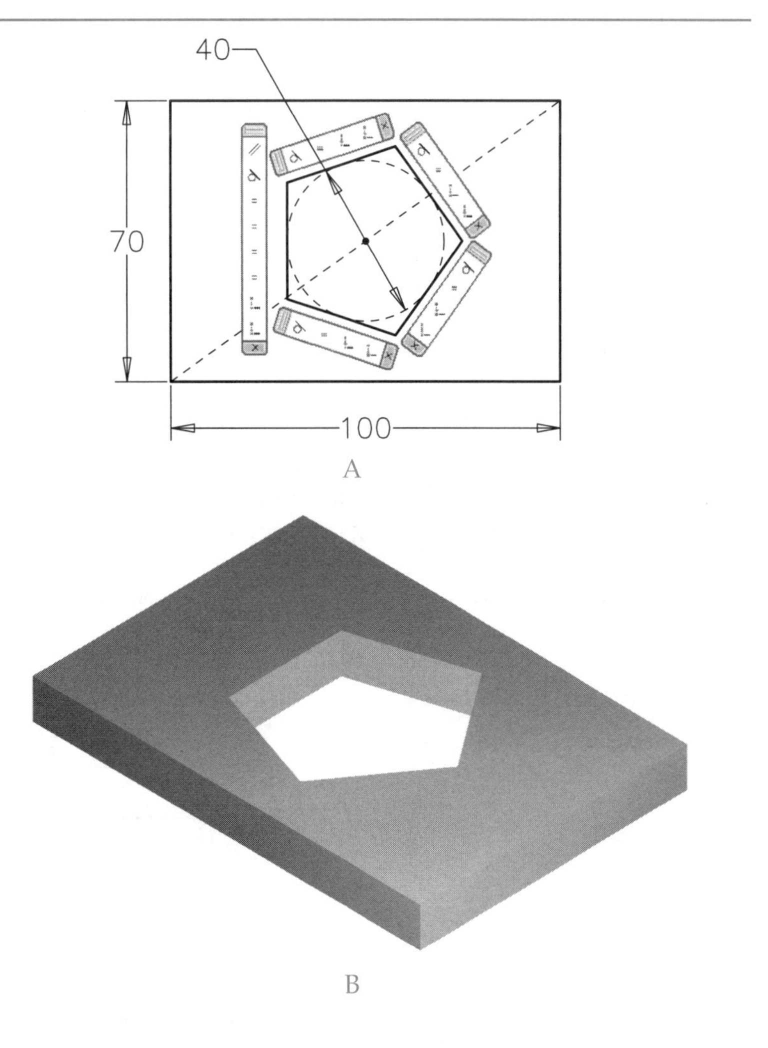

Figure 4-10.
A—Using construction lines and constraints to construct a circumscribed pentagon.
B—The sketch is extruded into a part.

PROFESSIONAL TIP

You can quickly change the style of an existing curve by selecting it and then pressing the style button. Holding the [Shift] key allows multiple curves to be selected and then changed using this method.

PRACTICE 4-1 Complete the practice problem on the Student CD.

Sketch Mirror Tool

The **Mirror** tool is used to create a mirror of geometry or objects by reflecting about a centerline. There are actually two mirror tools in Inventor. One is found in the **Part Feature** panel and the other is found in the **2D Sketch Panel**. In this section, you will look at the sketch **Mirror** tool. It is used to mirror sketched curves about any line in the sketch.

Open the file Example_04_02.ipt. Right-click on Sketch1 in the **Browser** and select **Edit Sketch** from the pop-up menu. Then, select the **Look At** button on the **Inventor Standard** toolbar and pick any line. Zoom extents if needed. See **Figure 4-11.**

In this example, the arc and two short lines will be mirrored about a line from the midpoint of line AD to the midpoint of line BC. First, using the **Line** tool, draw a line from the midpoint of line AD to the midpoint of line BC. The style can be Normal, Centerline, or Construction, but in this case, pick the **Centerline** button on the **Inventor Standard** toolbar to create a centerline. Use the midpoint object snap to locate the endpoints of the centerline.

Now, use the **Mirror** tool. Pick the **Mirror** button in the **2D Sketch Panel**. The **Mirror** dialog box is displayed. See **Figure 4-12.** If the **Select** button is not active (depressed), pick it to activate it. Then, select the two short lines and the arc on the sketch. When selected, the lines are displayed in a different color. Next, pick the **Mirror line** button in the **Mirror** dialog box. Select the centerline you just constructed on the sketch. Finally, pick the **Apply** button in the **Mirror** dialog box. The curves are mirrored. Pick the **Done** button to close the **Mirror** dialog box. Then, trim the center portion of line AB.

Figure 4-11.
The "tab" feature on this sketch will be mirrored to create a second tab.

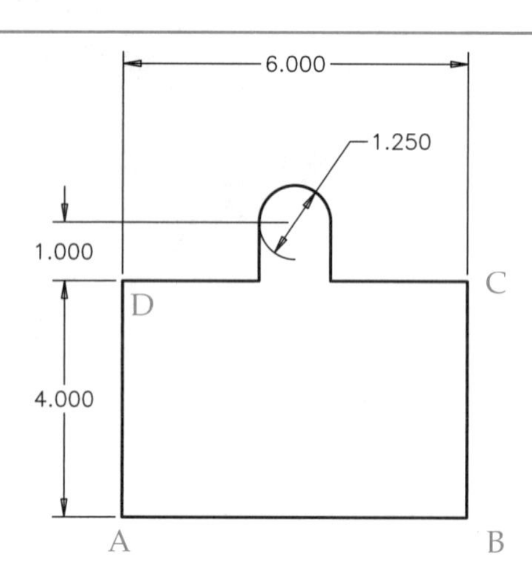

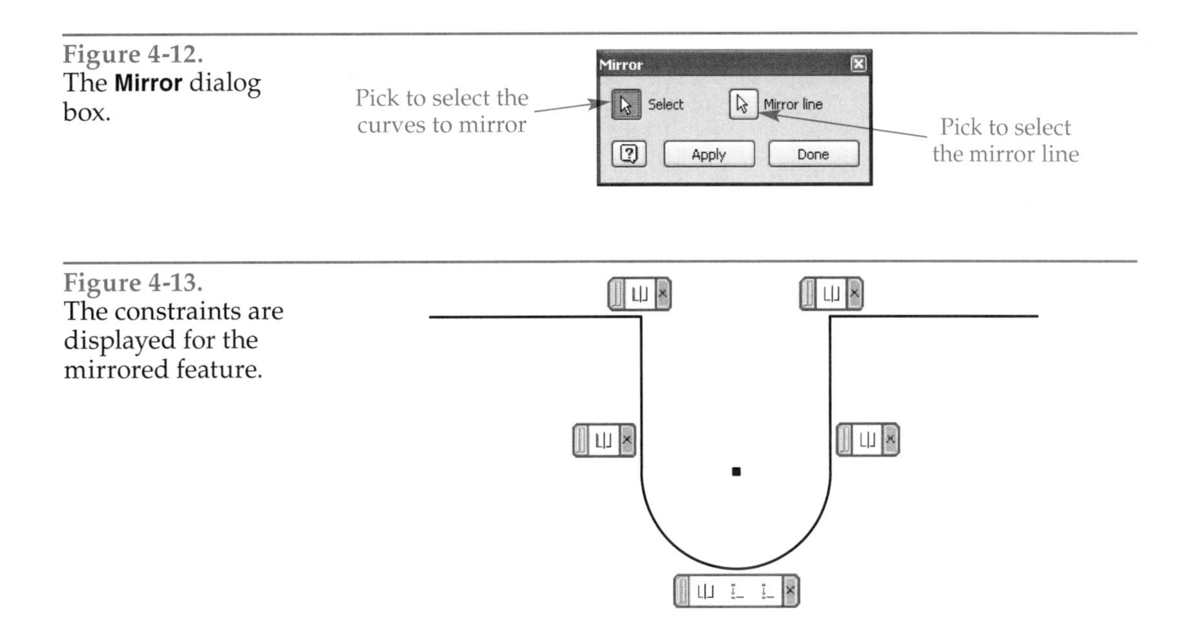

Figure 4-12.
The **Mirror** dialog box.

Figure 4-13.
The constraints are displayed for the mirrored feature.

Zoom in on the mirrored objects. Using the **Show Constraints** tool, show the constraints on the mirrored arc and the endpoints of the vertical lines, as shown in **Figure 4-13.** The **Mirror** tool applies the symmetric constraint to the arc and the endpoints of the lines that are mirrored. The constraint can be applied manually to other curves by picking the **Symmetric** button in the constraint flyout in the **2D Sketch Panel**.

Revolve Tool

The **Revolve** tool creates cylindrical parts by revolving a profile about an axis. The axis must be a straight line that, if extended indefinitely, would never cross the profile. However, the axis can touch the profile, such as an axis line that is tangent to a circle. This method produces a cylindrical model without a hole.

The style of the axis can be Normal, Construction, or Centerline. The advantage of using a centerline is that when you place dimensions in a sketch, they are placed as diametric dimensions, even though the complete diameter is not shown. ***Diametric dimensions*** measure the diameter of circular geometry. Place these dimensions by picking the geometry and the centerline. Be aware that if you pick the endpoint of the centerline, the dimension will be radial, not diametric.

Open the file Example_04_03.ipt. Edit the sketch by right-clicking on Sketch1 in the **Browser** and selecting **Edit Sketch** from the pop-up menu. Then, pick the **Look At** button on the **Inventor Standard** toolbar and select any line. Pick the **Zoom All** button. See **Figure 4-14.** Line BD is a construction line. Point B is the origin of the coordinate system. The top of the sketch is made of two separate lines, lines JD and DK.

First, apply an equal constraint to lines JD and DK, and to lines EF and GH. This horizontally centers the sketch. Now, place a dimension between line EF and the centerline (line AC). Notice how the dimension is placed as a diameter dimension. Also, the dimension does not measure from the centerline to line EF. Rather, it measures from a point on the opposite side of the centerline to line EF. This is a diametric dimension. Edit the dimension and enter a value of 4.5″. Also, place a dimension between line LM and the centerline. Edit this dimension and enter a value of 2.5″. The fully dimensioned sketch is shown in **Figure 4-15.** Display the isometric view and finish the sketch.

Figure 4-14.
This sketch will be used to create a symmetrical, circular part. Notice the centerline and the diametric dimension.

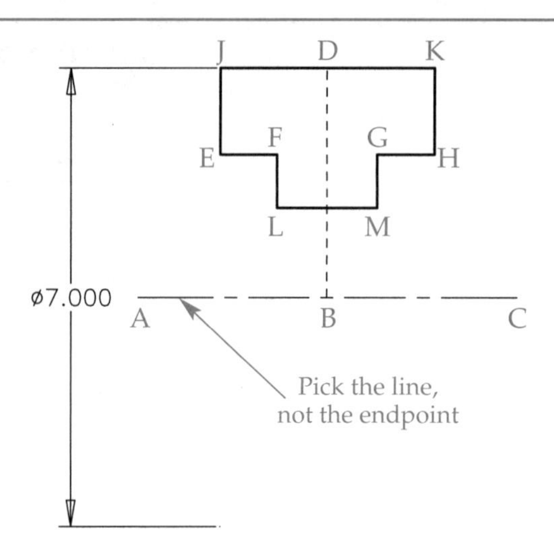

Figure 4-15.
The fully dimensioned sketch that will be revolved.

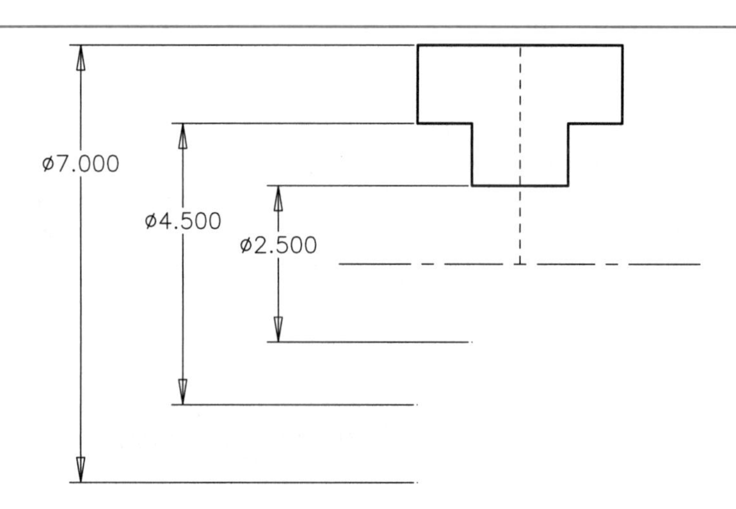

Now, you can revolve the profile about the centerline to complete the part. Pick the **Revolve** button in the **Part Features** panel or press the [R] key. The **Revolve** dialog box is displayed, **Figure 4-16.** As there is only one unambiguous, unconsumed sketch, it is selected as the profile. The area of the profile is highlighted in color. Also, the centerline in the sketch was created in the Centerline style. Since there is only one line created in this style, Inventor assumes this line will be the axis of rotation and selects it. The centerline is also displayed in color. The preview of the revolution is displayed as a green wireframe.

If you have multiple profiles in the sketch, pick the **Profile** button in the **Shape** area of the **Revolve** dialog box. Then, select the profile in the graphics area. Like the **Extrude** tool, you can select multiple profiles. If there is more than one possible axis

Figure 4-16.
The **Revolve** dialog box.

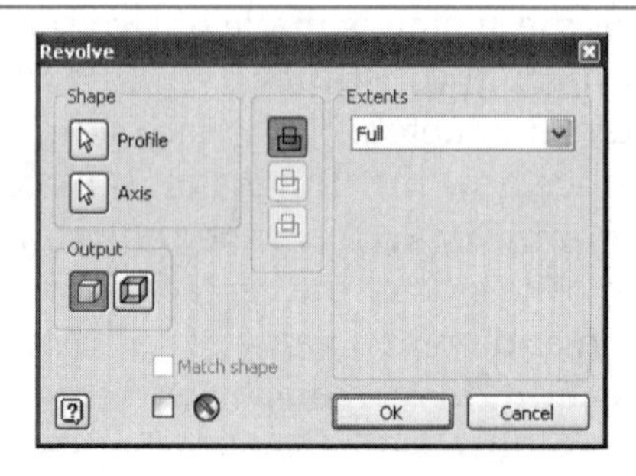

Figure 4-17.
The sketch shown in Figure 4-15 is revolved to create this part.

of revolution, or if the axis is not drawn in the Centerline style, pick the **Axis** button in the **Shape** area of the **Revolve** dialog box. Then, pick the axis in the graphics area.

Once the profile and axis are selected, pick the **OK** button in the **Revolve** dialog box. A solid model is created by revolving the profile through 360°. See **Figure 4-17.** There are other options with the **Revolve** tool that are discussed later in the book.

Inventor Precise Input Toolbar for 2D Sketches

For the most part, you will not use the **Inventor Precise Input** toolbar to help construct geometry in 2D sketches. Often, it requires more work than it saves. Remember the rules:

- Sketch the size reasonably close.
- Sketch so the constraints are exact.
- Precisely dimension the objects.

However, the toolbar can be very useful for drawing a rectangle, especially if the rectangle is the first curve sketched.

This section provides an in-depth look at the **Inventor Precise Input** toolbar. You may want to quickly read through this section and then come back to it later when you know more about Inventor and applications where precise input may be useful. The use of this toolbar for 3D sketches is discussed later in this text.

The **Inventor Precise Input** toolbar has three tools, a drop-down list, and three text boxes, **Figure 4-18.** The left-hand button, **Reset to Origin**, and the Z: text box do not work in 2D sketch mode. Therefore, they are not discussed in this section.

The toolbar has four modes, or types of input. Use the drop-down list on the toolbar to select the mode. The drop-down list indicates the mode in bold and provides a brief description. The four modes are:

- **XY.** Allows you to locate a point using the toolbar by providing an XY coordinate.
- **X∠.** Allows you to locate a point by providing its X coordinate and the angle from the X axis.

Figure 4-18.
The default settings for the **Inventor Precise Input** toolbar. Note: A drawing tool, such as the **Line** tool, must be active to enable the toolbar options.

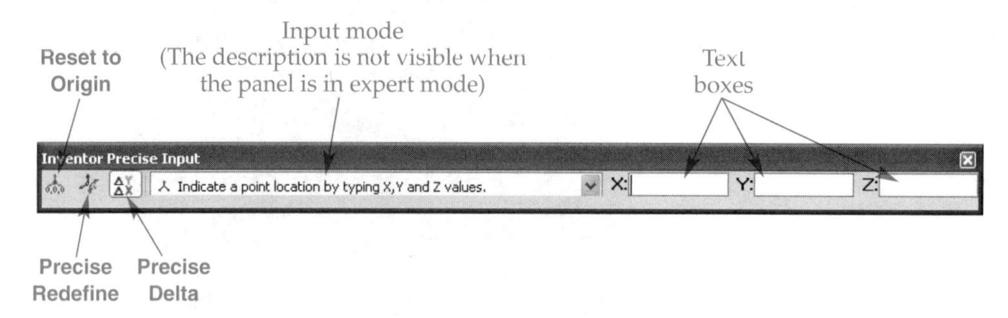

- **Y∠.** Allows you to locate a point by providing its Y coordinate and the angle from the X axis.
- **d∠.** Allows you to locate a point by providing the distance from the origin of the coordinate system and the angle from X axis.

The text boxes on the right end of the toolbar change depending on which input mode you select, **Figure 4-19.**

You can input numbers or equations in the text boxes. However, the equation used to locate a point is not saved. Rather, the equation determines a numeric value for the location. The same rules that apply to equations in dimensions apply here. You can use dimension names in the equation. Remember, the units must be correct.

To provide an example of using the **Inventor Precise Input** toolbar, start a new English Standard (in).ipt file. Then, display the **Inventor Precise Input** toolbar by right-clicking on a blank space on the **Inventor Standard** toolbar. Select **Inventor Precise Input** from the pop-up menu. The toolbar is floating by default. You can dock it or

Figure 4-19.
There are four input modes available for the **Inventor Precise Input** toolbar. A—XY. B—X∠. C—Y∠. D—d∠.

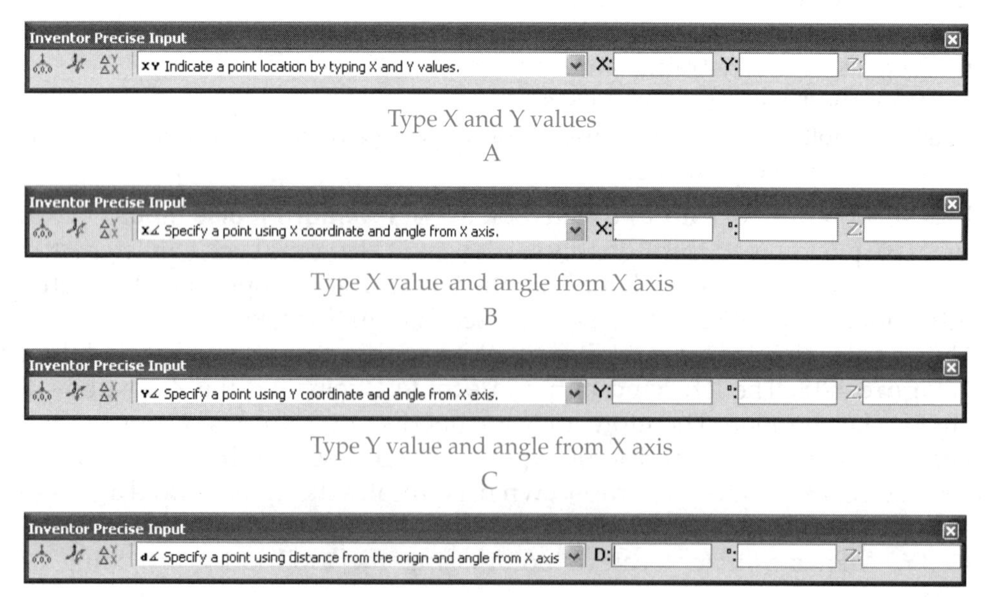

move it around the screen as you work. To prevent docking, press the [Ctrl] key as you move the toolbar. Now, use the toolbar to draw a line:

1. Pick the **Line** button in the **2D Sketch Panel**. The options in the **Inventor Precise Input** toolbar become available. Select the XY mode in the drop-down list, if it is not already selected.
2. Notice that there is a blinking, vertical bar in the **X:** text box in the toolbar. This indicates that you can type a value or equation. Type 0 in the **X:** text box.
3. Press the [Tab] key or move the cursor and pick in the **Y:** text box. Then, type 0 in the text box.
4. To place a point at the coordinates entered (0,0), press the [Enter] key.

A line now appears from the first point (0,0) to the cursor location, just as if you had picked the first point with the cursor. The CSI also appears at the origin. Continue by entering the second point:

5. The blinking, vertical bar should be in the **X:** text box. If not, pick in the text box. Then, type 0.
6. Press the [Tab] key or pick in the **Y:** text box. Type 4.25 and press [Enter] to place the second point of the line.
7. The **Line** tool remains active to draw more line segments. Press the [Esc] key to end the tool.

You have constructed a vertical line from 0,0 to 0,4.25. You can verify this by placing a dimension on the line. Also, if you show constraints on the line, it is constrained vertical.

Pick the **Line** button to start another session of the **Line** tool. In order to enable the options on the toolbar, you must first activate the **Line** tool or other drawing tool. Then, draw another line from the origin:

1. Pick the first endpoint of the line at the origin (0,0). Then, select the X∠ mode from the drop-down list on the **Inventor Precise Input** toolbar.
2. Using the **Inventor Precise Input** toolbar, set the second endpoint by typing 3 in the **X:** text box. Press the [Tab] key or pick in the **°:** text box and enter 30. The angle is measured in degrees from the X axis.
3. Press [Enter] to create the second endpoint and draw the line.
4. Press [Esc] to end the **Line** tool. Zoom extents if necessary to see both lines.

The second line is drawn from the origin to a point 3″ from the origin on the X axis and at an angle of 30°. However, it is important to note that the *length* of the line is not three inches. This can be verified with dimensions:

1. Pick the **Fix** button in the constraint flyout in the **2D Sketch Panel**. Apply a fix constraint to both lines.
2. Draw a horizontal line of any length from the origin. Use the Construction style.
3. Pick the **General Dimension** button in the **2D Sketch Panel**. To place an angular dimension, pick the construction line and then the angled line. Drag the dimension to the desired location. Notice that the dimension value is 30°. Depending on how Inventor placed constraints on the lines, you may get a message indicating the sketch will be over constrained. If this happens, pick the **Accept** button to place a reference dimension. Inventor calls reference dimensions ***driven dimensions.***
4. Place a dimension on the vertical line.
5. Now, dimension the horizontal distance between the endpoints of the angled line. Pick the line and move the cursor straight up or down. Then, drag the dimension to the desired location. The dimension value should be 3.000.
6. Finally, dimension the length of the angled line. Pick the line, move the cursor off of the line, and then pick the line again. A dimension should appear that is aligned with the line. Drag the dimension to the desired location. You will receive a warning; pick the **Accept** button to place a driven (reference) dimension. Notice that the dimension value is 3.464, which is the length of the line.

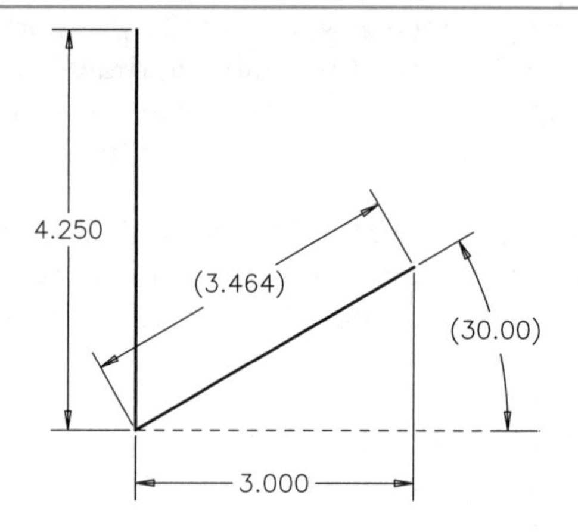

Figure 4-20.
The sketch after using the **Inventor Precise Input** toolbar to draw two lines. A construction line and dimensions have also been added.

Your sketch should look like **Figure 4-20.** When using the X∠ input mode, it is important to remember that the X value is not the length of the line. Also, certain entries are not valid. For example, if you enter 3 for the X value and attempt to enter 90 for the angle, the 90 appears in red because Inventor cannot solve this "equation." An angle of 90° places the point on the Y axis (X = 0), yet you also told Inventor that the X value is 3, not 0. Hence, the error.

You can use relative input with the **Inventor Precise Input** toolbar. With relative input, the coordinates (or angle) are based on the last point, not the origin of the coordinate system.

1. Activate the **Line** tool. Make sure the style is set to Normal. Pick the right-hand endpoint of the inclined line as the first endpoint of the new line.
2. On the **Inventor Precise Input** toolbar, pick the **Precise Delta** button. The button is depressed to indicate that the information entered will be relative to the last point (the first endpoint of the new line).
3. Set the input mode to XY.
4. Type 0 in the **X:** text box and 1.75 in the **Y:** text box.
5. Press the [Enter] key to place the second point and draw the line, **Figure 4-21.**

A vertical line segment is added to the sketch. The second endpoint of the new line is 0 units on the X axis and 1.75 on the Y axis from the first point, not from the origin of the coordinate system. Notice that the **Precise Delta** button remains active (depressed). Also, the CSI has moved to the second endpoint of the new line. This is because the second endpoint, once drawn, becomes the "last" point. With the **Line** tool and **Precise Delta** button active, continue sketching:

6. Type the equation d2/–2 in the **X:** text box and 0 in the **Y:** text box. Dimension d2 is the horizontal dimension between the endpoints of the angled line. Press [Enter] to draw the line.

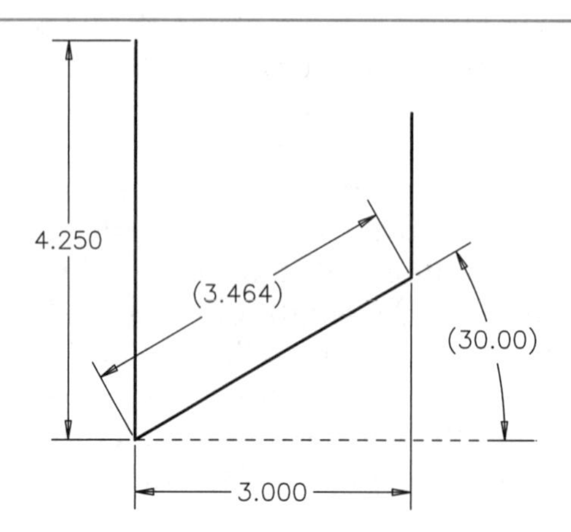

Figure 4-21.
A vertical line is added to the sketch using relative input.

7. Type 0 in the **X:** text box and .768 in the **Y:** text box. Press [Enter] to draw the line.
8. Pick the **Precise Delta** button on the **Inventor Precise Toolbar** to exit relative mode. The button is no longer depressed and the CSI is displayed at 0,0.
9. Type 0 in the **X:** text box and 4.25 in the **Y:** text box. Press [Enter] to draw the line.
10. Press [Esc] to end the **Line** tool.

Chapter Test

Answer the following questions on a separate sheet of paper or complete the electronic chapter test on the Student CD.

1. A sketch containing islands is considered a(n) ______ profile.
2. An island is also called a(n) ______ once the sketch is extruded.
3. If d3 = 2.5 mm and d2 = 1.5 mm, what is the solution calculated by Inventor for the equation d3/d2?
4. How does Inventor indicate an invalid or unacceptable equation?
5. For an item that has no units, such as a constant, how does Inventor indicate the "unit" in an equation?
6. List one advantage of using the Construction style to draw construction lines.
7. Once a curve is drawn, how can you change its line style?
8. What is the function of the sketch **Mirror** tool?
9. What is the function of the **Revolve** tool?
10. What relationship must the axis of revolution have to the profile when using the **Revolve** tool?
11. When using the **Revolve** tool, what is the advantage of drawing the axis of revolution in the Centerline style?
12. How can you select multiple profiles for extrusion?
13. How can you select multiple profiles for revolution?
14. When using the **Inventor Precise Input** toolbar, how long is an equation entered in the **X:** text box stored once the point is entered (placed)?
15. When the **Precise Delta** button on the **Inventor Precise Input** toolbar is on, from where are the coordinates you enter measured?

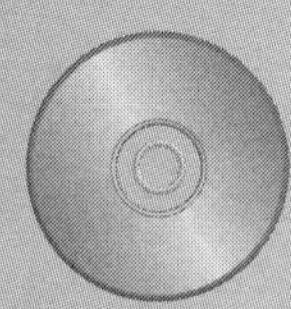

Chapter Exercises

Exercise 4-1. Link.

Complete the exercise on the Student CD

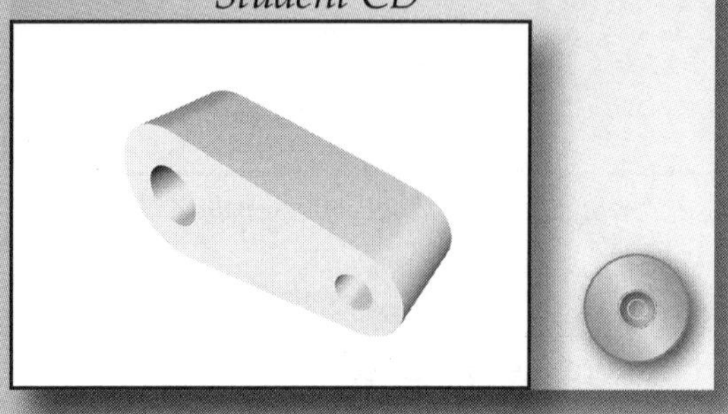

Exercise 4-2. Valve.

Complete the exercise on the Student CD

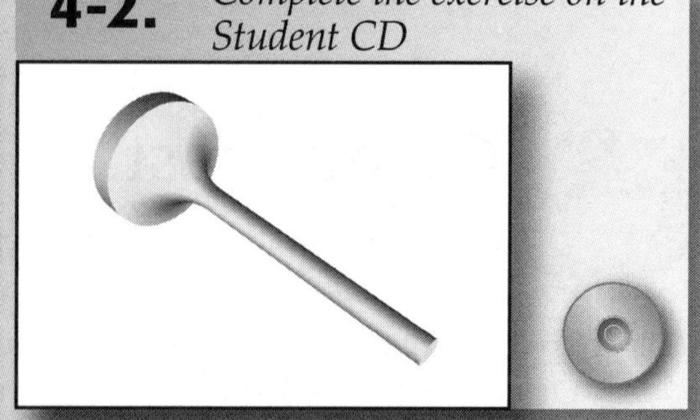

Exercise 4-3. Stamping.

Complete the exercise on the Student CD

Exercise 4-4. Mirror Part.

Complete the exercise on the Student CD

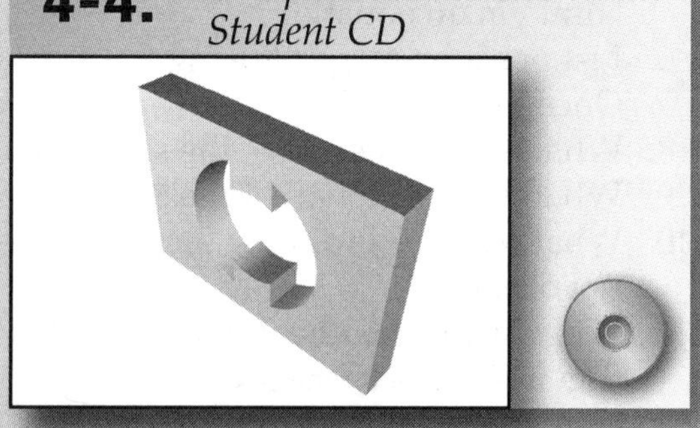

Exercise 4-5. Cap.

Complete the exercise on the Student CD

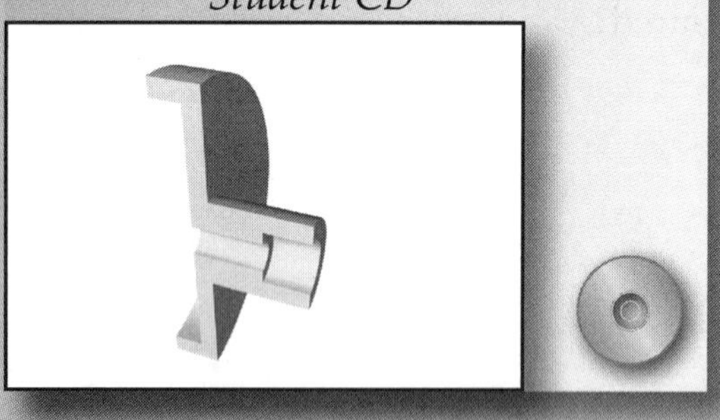

Exercise 4-6. Wheel.

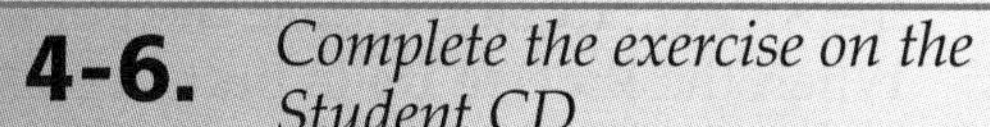

Complete the exercise on the Student CD

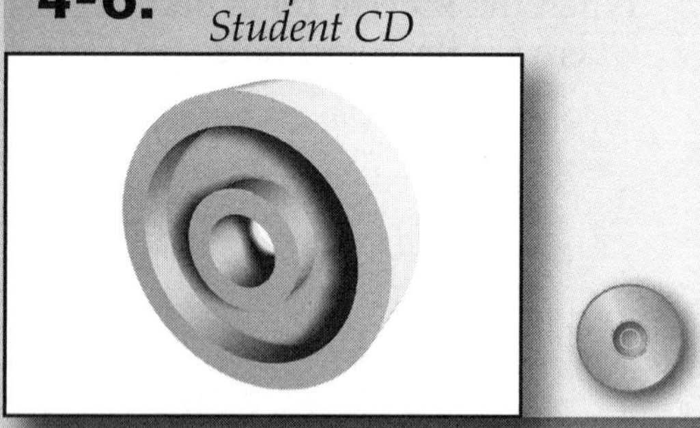

Exercise 4-7. Blanked Cam Plate.

Complete the exercise on the Student CD

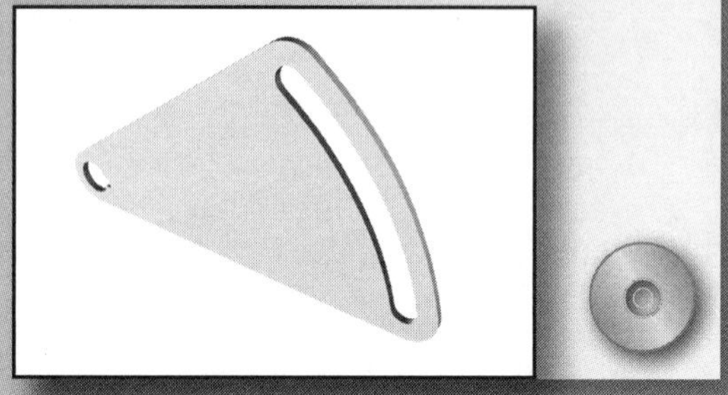

Exercise 4-8. Locking Cam.

Complete the exercise on the Student CD

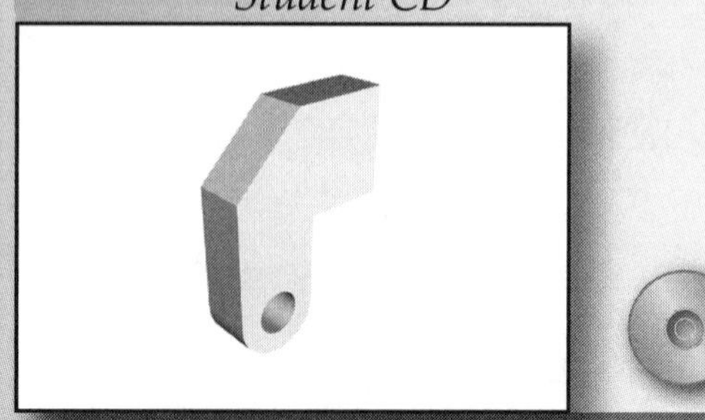

Exercise 4-9. Metering Pin.

Complete the exercise on the Student CD

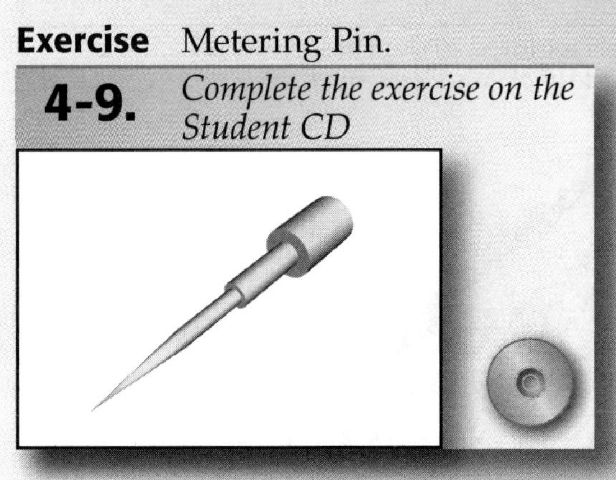

This complex model of a custom automotive wheel required 10 sketches, one of which was used to establish three work features. (Model courtesy of Autodesk, Inc.)

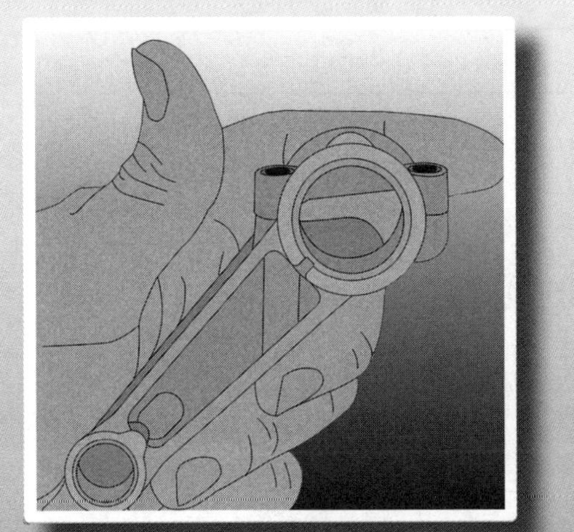

Chapter 5

Secondary Sketches and Work Planes

Objectives

After completing this chapter, you will be able to:

- Explain how to create secondary sketch features.
- Explain and use the five termination features in the **Extrude** tool.
- Project silhouette curves onto a sketch plane.
- Use the **Cut** and **Intersect** options of the **Extrude** tool.
- Display, adjust, and use the three fundamental work planes of a sketch.
- Create, display, and use additional work planes.

User's Files

The Student CD included with this text contains several files required for this chapter. Refer to the file File List.txt *in the* \Ch05 *folder for the comprehensive list.*

Creating Secondary Sketch Planes and Features

The parts you created in Chapters 3 and 4 were all based on one primary sketch and profile. In this chapter, you will look at how to create additional, secondary sketches on the faces of the part and on work planes. Open the file Example_05_01.ipt. This part has five planar (flat) faces that can be used as sketch planes. It also has one curved face that *cannot* be used as a sketch plane. See **Figure 5-1.**

In this section, you will create a boss on the top face. First, to provide a visual reference, change the color of the top face. To do this, right-click on the top face. Then, select **Properties** from the pop-up menu. The **Face Properties** dialog box is displayed, **Figure 5-2.** Select Green (Flat) from the drop-down list and pick the **OK** button. The face is now displayed in green. You can change the color of any face using this procedure.

The procedure for creating the boss is to select the planar face on which to sketch, create the sketch, and extrude the profile. This basic procedure can be used to create a

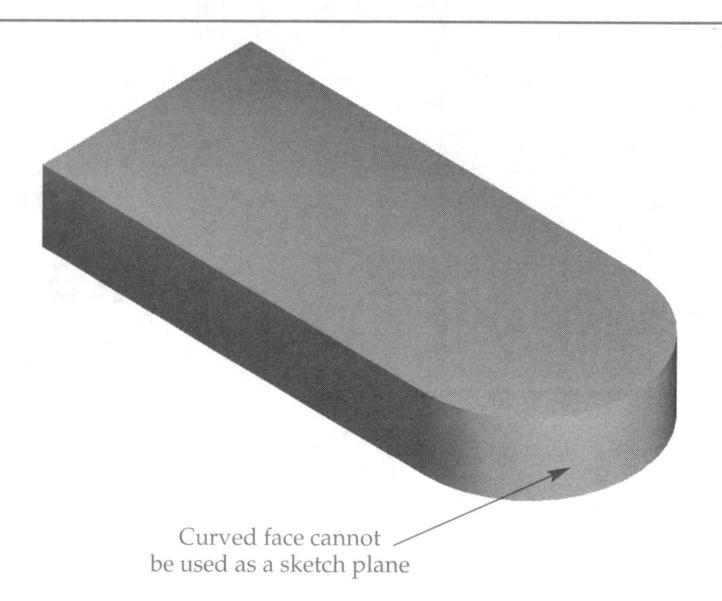

Figure 5-1.
This part has five planar faces that can be used as sketch planes. The curved face cannot be used as a sketch plane.

new feature on any planar face. The existing part edges of the face can be used as part of the sketch or as reference geometry for dimensions and constraints.

1. Pick the **Sketch** button on the **Inventor Standard** toolbar or press the [S] key to initiate sketch mode. A pencil and paper icon appears next to the cursor in the graphics window. This indicates that you need to select a sketch plane. If tooltips are turned on, this is also indicated as help text. You must select a plane on which to sketch before sketch mode is entered.
2. Move the cursor over any face on the part. The edges of the face are highlighted. If you momentarily pause, the cursor changes to the select tool. This tool allows you to choose between adjacent faces by picking the left or right arrow. To make a selection, pick the center button. To close the select tool, simply move the cursor off of the tool.
3. Move the cursor over the green face and pick to select it.

The edges of the face and the center point of the arc have been automatically projected into the sketch. They are displayed in color to indicate this. The exact color depends on which color scheme is being used. The grid is also displayed. Notice that the grid is on the selected face. This indicates that the current coordinate system coincides with the selected face. However, you do not know where the origin of the coordinate system is located. You can place a CSI at the origin as follows.

1. Select **Application Options...** from the **Tools** pull-down menu. The **Options** dialog box is displayed.
2. Pick the **Sketch** tab.
3. In the **Display** area of the **Sketch** tab, place a check in the **Coordinate system indicator** check box.
4. Pick the **OK** button to close the dialog box and apply the setting. Do not pick the Windows close button (the small X) as this cancels the settings.

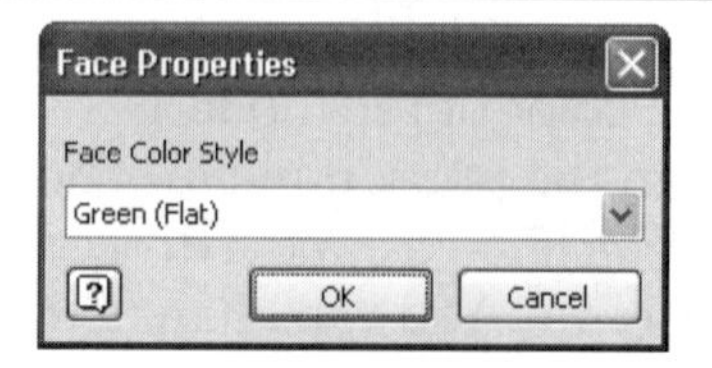

Figure 5-2.
Changing the color of a face using the **Face Properties** dialog box.

A CSI now appears at the origin of the current coordinate system. If the CSI is not immediately displayed, pick any of the edges on the part to display the CSI. The coordinate system determines horizontal and vertical directions in relation to the part. Later, you will learn how to reposition the coordinate system and why you would want to do so. Now, draw the sketch and extrude it into the boss:

1. Using the **Center Point Circle** tool, sketch a circle anywhere on the grid. For this example, do not pick the center of the circle coincident to the center of the arc.
2. Apply a concentric constraint between the circle and the arc.
3. Dimension the circle's diameter. Then, edit the dimension to a value of 25 mm.
4. Finish the sketch to return to part mode. You have now added an unconsumed profile (sketch) to the part.
5. Pick the **Extrude** button in the **Part Features** panel. Since the edges of the green face and the circle form an area that can be selected as a profile, this is an ambiguous sketch. A profile is not automatically selected; you must select the profile. With the **Profile** button active (depressed) in the **Extrude** dialog box, select the interior of the circle as the profile. The area is highlighted after it is selected and a wireframe preview of the extrusion is displayed.
6. In the **Extrude** dialog box, make sure the **Join** button is on (depressed). Then, set the distance to 22 mm. Set the extrusion direction upward by picking the left-hand button below the distance drop-down list. Also, the output should be set to a solid. See **Figure 5-3A.**
7. Pick the **OK** button to extrude the feature, **Figure 5-3B.** Notice that the extrusion is given a different name in the **Browser**, such as Extrusion2, rather than being added to the previous extrusion.

You can rename the components of the part in the **Browser**. For example, Extrusion2 is really not very descriptive. However, Boss is more meaningful as a description. To rename the boss extrusion:

1. Right-click on the extrusion name in the **Browser** and select **Properties** from the pop-up menu. The **Feature Properties** dialog box is displayed. See **Figure 5-4.**
2. In the **Name** text box, type Boss.
3. Just like individual faces, individual features can be assigned unique colors. This is usually done for emphasis or clarity. In the **Feature Color Style** drop-down list, select Chrome.
4. Pick the **OK** button. The feature is now displayed in chrome. Also, the name of the extrusion in the **Browser** is Boss.

Figure 5-3.
A—The settings for extruding the profile. B—The extrusion is added to the part.

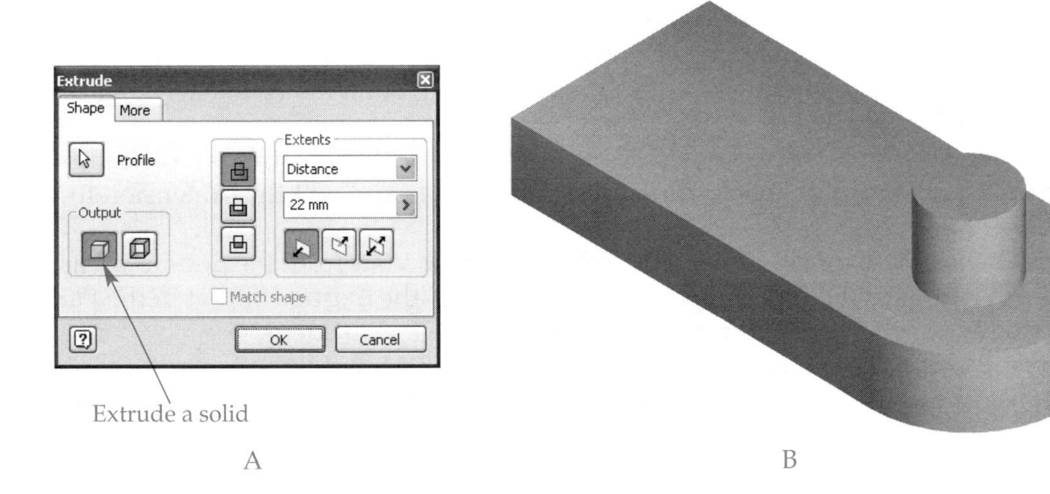

Figure 5-4.
Changing the name and color of a feature using the **Feature Properties** dialog box.

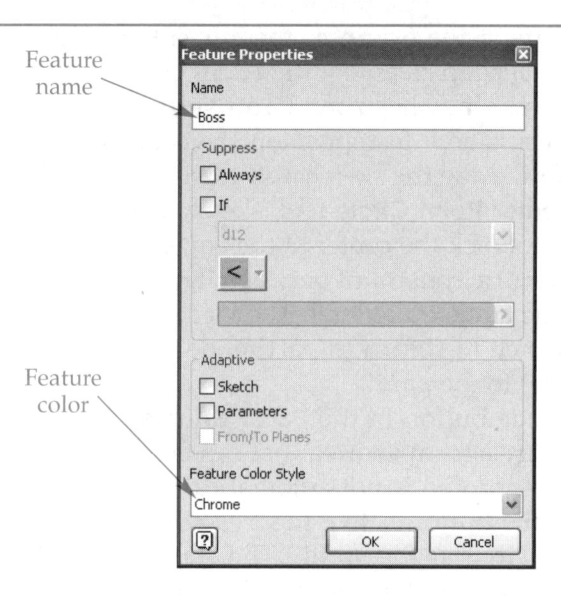

Operation and Extents Options of the Extrude Tool

There are three operation options in the **Extrude** dialog box—**Join**, **Cut**, and **Intersect**. **Join** adds the extrusion to the existing part. **Cut** removes the volume of the extrusion from the existing part. **Intersect** creates a new part based on the overlapping volume of the extrusion and the existing part.

There are also five extents options in the **Extrude** dialog box—Distance, To Next, To, From To, and All. These are used to specify how far the profile extrudes. Distance specifies an exact length for the extrusion, which is what you have used so far. To Next specifies that the extrusion will extend to the next surface of the part in the extrusion direction. To is used to pick a face to which the extrusion will extend. From To allows you to pick a face on the part from which the extrusion starts and another face to which the extrusion extends. All specifies that the extrusion goes all of the way through the part, including through all existing features.

In this next operation, you will use the All and **Cut** options. If you have not yet created the boss as described in the previous section, do so now. Then, continue as follows.

1. Pick the **Sketch** button on the **Inventor Standard** toolbar. Then, select the face on the top of the boss as the sketch plane. The edge and center point of the boss feature are automatically projected into the sketch.
2. Pick the **Look At** button on the **Inventor Standard** toolbar. Select the face on top of the boss to get a top view. This step is not necessary, but it may make locating the new sketch easier.
3. Sketch a circle with its center at the center of the boss. This constrains it concentric to the boss. Place a diameter dimension on the circle and edit the dimension to a value of 10 mm.
4. Right-click in the graphics window. Select **Isometric View** from the pop-up menu.
5. Finish the sketch to return to part mode. Then, pick the **Extrude** button in the **Part Features** panel. Select the small circle as the profile to extrude.
6. In the **Extrude** dialog box, pick the **Cut** button, which is the middle button in the center of the dialog box. Also, select All from the extents drop-down list. See **Figure 5-5A.**

Figure 5-5.
A—The settings for extruding the profile.
B—Entering a taper angle for the resulting extrusion.

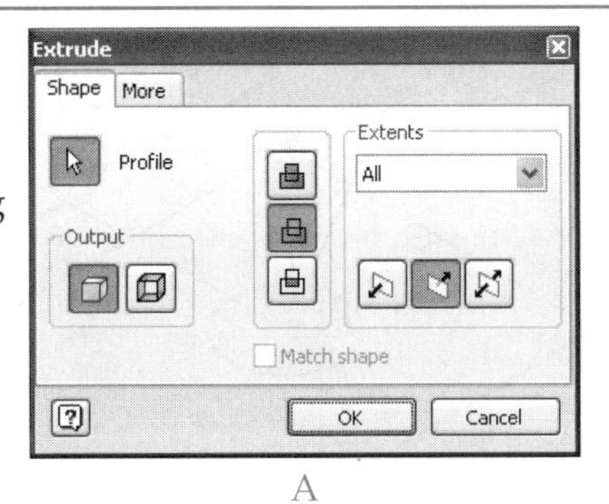

A

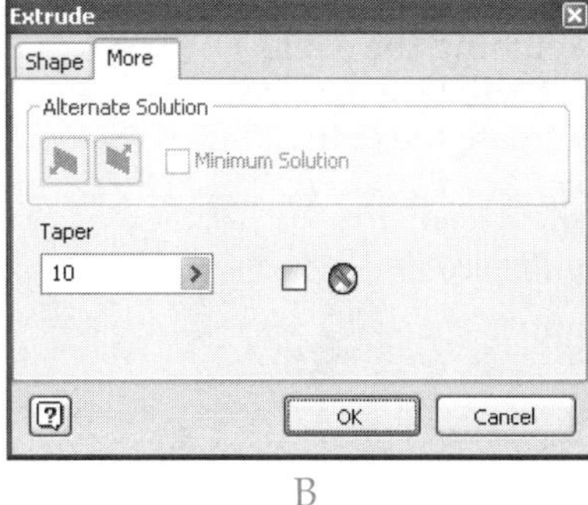

B

7. In the **More** tab of the **Extrude** dialog box, enter a taper angle of 10°. See **Figure 5-5B.** A positive taper angle makes the feature get larger as it is extruded and a negative angle makes it smaller.
8. Pick the **OK** button to extrude the profile. The finished part is shown in **Figure 5-6** with half of the part removed to show the interior extrusion.

Now, open the file Example_05_02.ipt. This is a U-shaped part, **Figure 5-7.** You will add a cutout (hole) to one leg of the part, but not the other. The To Next option of the **Extrude** tool allows you to do this by limiting the extent of the extrusion to the next face that the extrusion encounters. The "to next" face can be either flat or curved. To create the cutout:

Figure 5-6.
The extrusion is subtracted from the part. The sketch was on the top face of the boss, which is why a positive taper angle results in the bottom of the cutout being larger than the top. Note: The part is shown sliced through the middle for illustration.

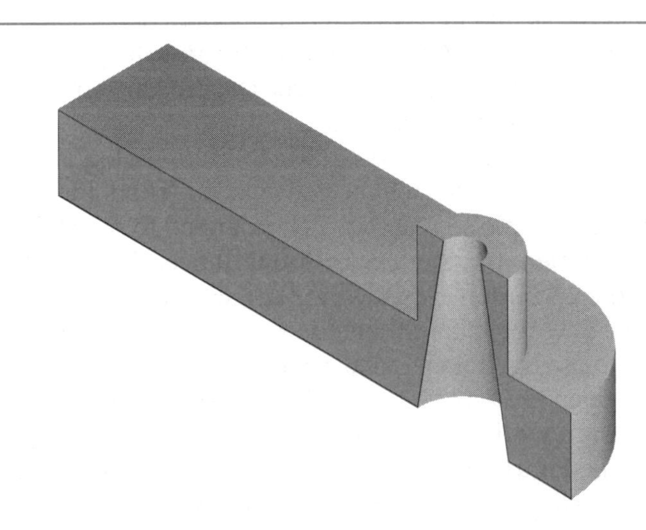

Figure 5-7.
A cutout (hole) will be added to one leg of this U-shaped part.

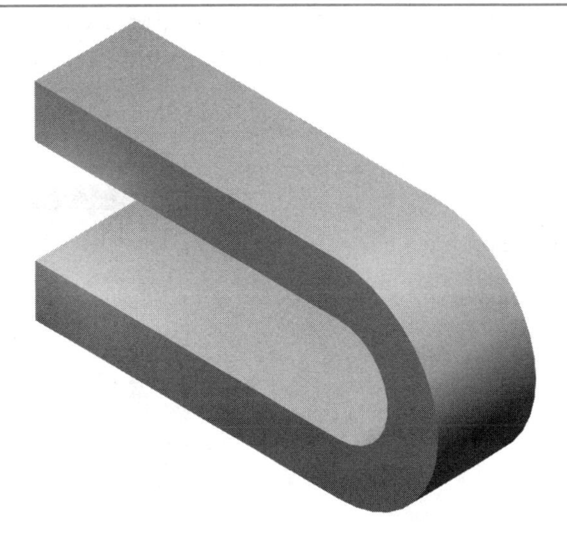

Figure 5-8.
Sketching the circle that will be extruded to create the cutout.

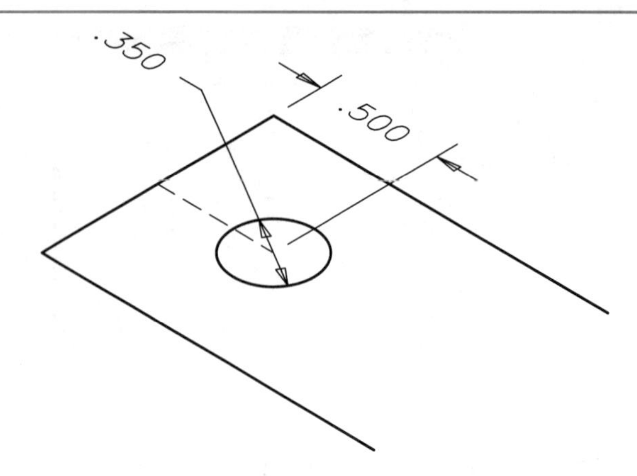

1. Pick the **Sketch** button and select the top face of the part as the sketch plane. The four edges of the rectangular face are automatically projected into the sketch.
2. Draw a construction line from the midpoint of the short, left-hand edge. If the constraint does not already exist, constrain the construction line perpendicular to the edge. Dimension the line as shown in **Figure 5-8.**
3. Turn off the **Construction** button. Then, draw a circle with its center at the endpoint of the construction line. This centers the circle in the local Y direction and constrains it coincident with the end of the line. Dimension the circle as shown in **Figure 5-8.**
4. Finish the sketch to return to part mode. Then, pick the **Extrude** button in the **Part Features** panel. Select the circle as the profile.
5. In the **Extrude** dialog box, pick the **Cut** button. Also, select To Next from the **Extents** drop-down list. See **Figure 5-9.**
6. Pick the **Terminator** button in the **Extrude** dialog box. The tooltip for this button is Select body to end the feature construction. Then, pick anywhere on the part. In this specific example, you actually do not need to select the body since it was automatically selected. In other cases, you will need to select the body.
7. Pick the **OK** button to create the extrusion. The updated part is shown in **Figure 5-10.** Notice how the cutout (hole) does not pass through both legs of the part. You may want to use the **Rotate** tool to rotate the view and better see the feature.

Now, open Example_05_03.ipt. This part is similar to the U-shaped part with which you have been working. However, one leg has a small lip projecting to the interior of the part. See **Figure 5-11.** In this example, you will add a pin to the leg that extends only as far as the lip. The To option of the **Extrude** tool allows you to do this. The circle profile that you will extrude is already drawn. Continue as follows.

1. Pick the **Extrude** button and select the circle as the profile.
2. In the **Extrude** dialog box, pick the **Join** button. Also, select To in the **Extents** drop-down list.

Figure 5-9.
The settings for extruding the profile to create a cutout (hole) in one leg only.

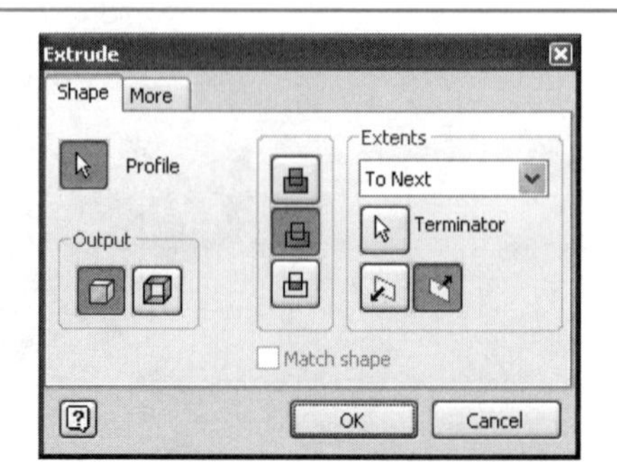

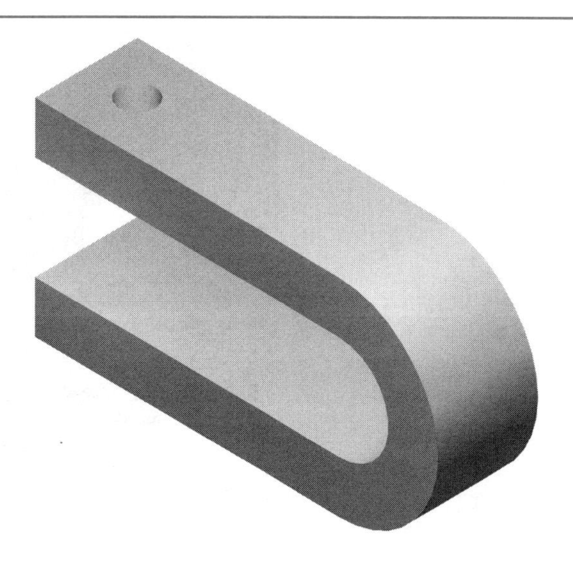

Figure 5-10.
The extrusion is subtracted from one leg only.

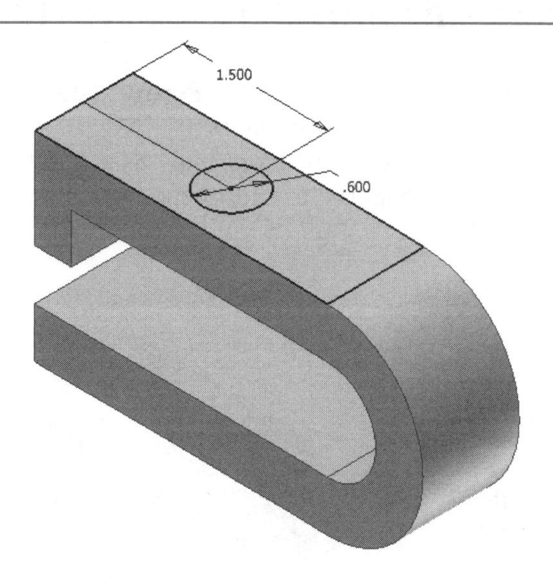

Figure 5-11.
An extrusion that extends to the same plane as the lip will be added to this part.

3. When you select To in the drop-down list, the button below the list is activated. The tooltip for this button is Select surface to end the feature creation. With the button active, pick the bottom of the lip feature. The color of this face has been changed to green in the file to help you select the correct face. Rotate the view as needed to better see the face.
4. Once you pick the face, it is highlighted. Also, a check box appears next to the button in the **Extrude** dialog box. The tooltip for this check box is Check to terminate feature on the extended face. When checked, the face is treated as an infinite plane. Because the circle is not directly above the green face, check this check box. If you do not, you will get an error.
5. Pick the **OK** button in the **Extrude** dialog box to create the new feature. See **Figure 5-12.**

PRACTICE 5-1 Complete the practice problem on the Student CD.

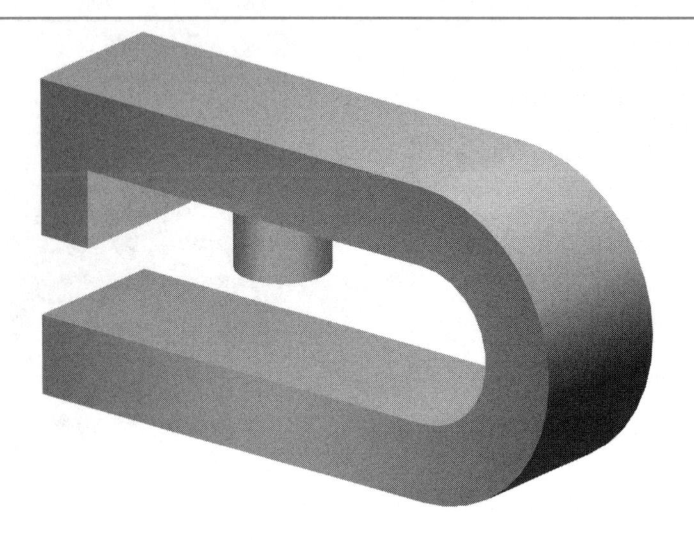

Figure 5-12.
The extrusion is added to the part. Notice how it extends to the same plane as the lip.

Projecting Silhouette Curves to the Sketch Plane

Open the file Example_05_04.ipt. **Figure 5-13A** shows the part as a solid in the isometric view. The curved surface of the part has been changed to green. **Figure 5-13B** shows the same part in the front view as a wireframe. The line on the right side of the wireframe represents the extent of the curved surface. This type of line (curve) is called a ***silhouette curve,*** but, in reality, there is no edge there. The line is displayed

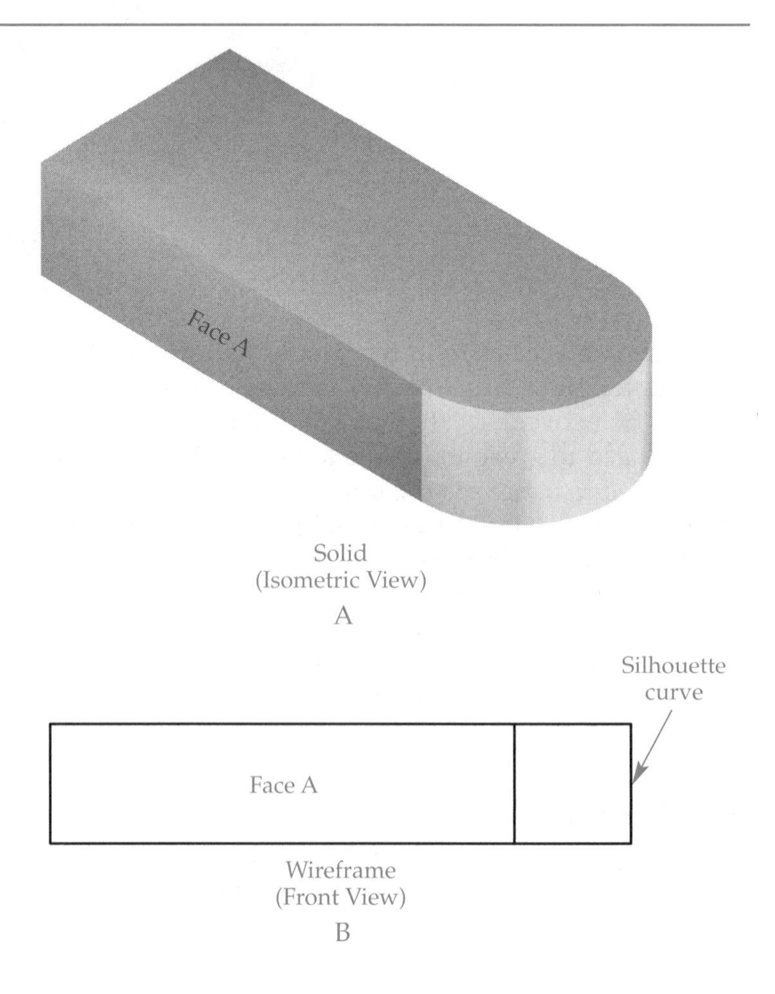

Figure 5-13.
Face A indicated here is used as the sketch plane. The silhouette curve representing the curved face will be projected onto the sketch plane.

to make the part "look right." The edges of the part can be used to define a sketch, but a special technique is necessary to use a silhouette curve. A silhouette curve must be projected onto the sketch plane. The **Project Geometry** tool is used to do this.

Now, you will construct a slot through the curved end of this part. To make visualization easier, stay in the isometric view, but change to a wireframe display. Then, continue as follows.

1. Pick the **Sketch** button. Then, select face A as the sketch plane. Face A is indicated in **Figure 5-13.** Notice that a rectangle appears in color; this is the geometry that is on the current sketch plane and automatically projected into the sketch. However, there is no line on the sketch plane that corresponds to the silhouette curve.
2. Select the **Project Geometry** button in the **2D Sketch Panel**.
3. Move the cursor around the curved (green) face until you see a vertical line at the midpoint of the curve. This line represents the silhouette curve. Pick the line and it is projected onto the sketch plane. See **Figure 5-14.**
4. Press the [Esc] key to exit the **Project Geometry** tool.
5. Sketch two construction circles, as shown in **Figure 5-15.** Constrain the large circle tangent to the top and bottom edges of the part. Constrain the small circle concentric (or coincidental) to the large circle. Dimension the small circle's diameter to a value of .2".

Figure 5-14.
Projecting the silhouette curve onto the sketch plane.

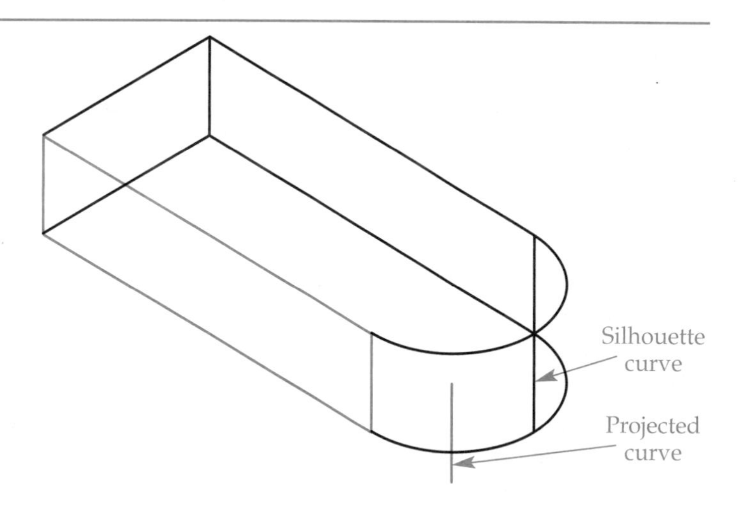

Figure 5-15.
Drawing the profile that will be extruded into the cutout. Notice how two construction circles are used to center the cutout profile.

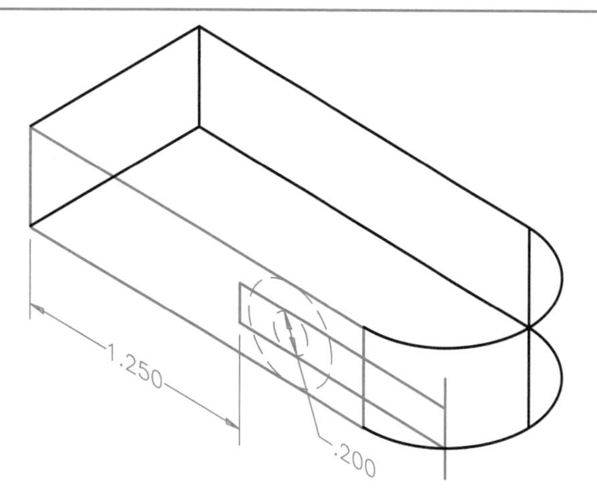

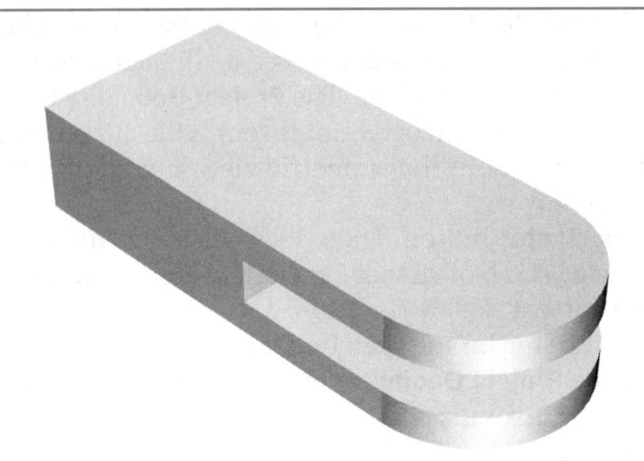

Figure 5-16.
The extrusion is subtracted from the part to create the cutout (slot).

6. Sketch the three lines forming the profile of the slot. Make sure they are not in the Construction style. The two long lines should be constrained perpendicular to the projected line and tangent to the small circle. Use a dimension to constrain the location of the short line relative to the left edge of the part, as shown in **Figure 5-15.**
7. Finish the sketch to return to part mode.
8. Pick the **Extrude** button in the **Part Features** panel. Select the rectangular area formed by the projected line and the three lines you just constructed as the profile.
9. Select the **Cut** and All options in the **Extrude** dialog box. Then, pick the **OK** button to extrude the profile and remove its volume from the part. See **Figure 5-16.**

There is a simpler way to create the above construction. Save the file as Example_05_04a.ipt and close it. Then, open Example_05_04.ipt again. Continue as follows.

1. Pick the **Sketch** button and select face A as the sketch plane.
2. Select the **Project Geometry** button in the **2D Sketch Panel** and project the silhouette curve onto the sketch plane.
3. Construct a rectangle on the face. See **Figure 5-17.**
4. Apply the two dimensions shown in the figure.
5. Constrain the midpoint of the right edge of the rectangle to the midpoint of the silhouette curve.
6. Finish the sketch and extrude the rectangle to cut the part.
7. Save the part as Example_05_04b.ipt.

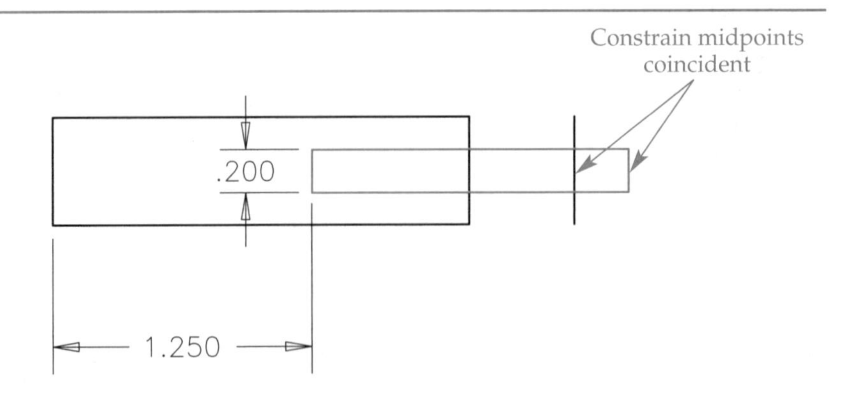

Figure 5-17.
Creating the same cutout shown in Figure 5-16 using a single rectangle.

Default Work Planes and Midplane Construction

As you saw in the previous section, the planar faces of the part can be used to define sketch planes. In addition, you can use work planes to define the sketch plane. ***Work planes*** are construction geometry that can be used to help in creating sketches and features. There are three default work planes in every Inventor part file. These work planes are aligned with the three principle planes of the coordinate system—XY, XZ, and YZ. By default, the visibility of these work planes is turned off, but they are listed in the **Browser** in the Origin branch.

In this section, you will use one of the default work planes to construct the part shown in **Figure 5-18.** Only one work plane will be used. Start a new English Standard (in).ipt file and display the isometric view. In the **Browser**, expand the Origin branch of the part tree. This branch contains the three work planes, three work axes, and one center point. As you move the cursor over the names in the **Browser**, the objects are highlighted in the graphics area.

Although its visibility is turned off, the default sketch plane is the XY work plane. Right-click on XY Plane in the **Browser** to display the pop-up menu. Currently, the **Visibility** option does not have a check mark next to it. This means that the work plane is not visible. To turn the visibility of the work plane on, pick **Visibility** in the pop-up menu. Now, if you right-click on XY Plane to display the pop-up menu, there is a check mark next to the **Visibility** option. Also, the icon next to the name in the **Browser** is no longer grayed out.

Now, hold the cursor over the work plane's name in the **Browser**. The work plane is highlighted in the graphics window. Move the cursor into the graphics window and onto the highlighted work plane. The highlight disappears until the cursor is over an edge of the work plane again. Pick the edge to select the work plane. What you have selected is an object representing an infinite plane; the edges are not true "edges."

You can move the work plane around in the graphics window. The work plane can even be resized. However, the **Auto-Resize** option must be turned off. Right-click on the work plane in the graphics window or its name in the **Browser**. Then, select the **Auto-Resize** option in the pop-up menu to uncheck it. Now, move the cursor to an edge of the work plane; the move cursor is displayed. Pick and drag the work plane to a new position. You can resize the work plane by moving the cursor to one of the four yellow

Figure 5-18.
This part is created using a single work plane.

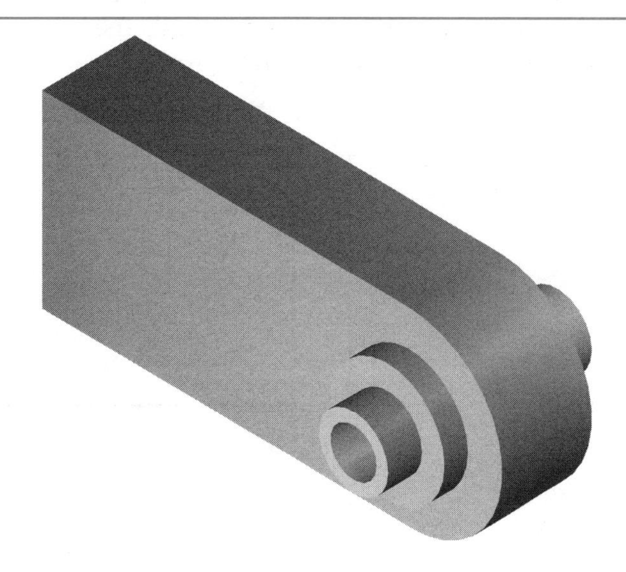

circles at the corner of the plane. When the resize cursor is displayed, pick and drag to resize the work plane. This may help if the plane is displayed inside of a shaded part and, therefore, not visible. However, since the work plane is infinite, changing the size or location has no effect on the use of the plane, just its display.

CAUTION

When moving a work plane in an isometric view, it may appear as if you are changing the elevation (local Z value) of the plane. However, you are not.

Turning on the visibility of the center point can also be useful. The center point is at the origin of the coordinate system, which is the intersection of the XY, XZ, and YZ work planes. This point is fixed; it does not move when different sketch planes are active. When its visibility is turned on, the center point is a small diamond. To turn the visibility of the center point on, right-click on Center Point in the Origin branch in the **Browser**. Then, select **Visibility** in the pop-up menu. A check mark appears next to this option when the visibility of the center point is on.

To draw the part shown in **Figure 5-17,** turn on the visibility of the XY work plane. Make sure the **Auto-Resize** option is on for the work plane. Also, turn on the visibility of the center point. Then, continue as follows.

1. Using the **Project Geometry** tool, project the center point into the sketch. This gives you a point on the sketch to which geometry can be attached. This is an excellent way to start a new part.
2. Sketch a circle centered on the projected point. This ensures the center of the circle is at the coordinate system origin for future sketches.
3. Dimension the circle diameter to 2″.
4. Construct the three straight lines and place the two dimensions shown in **Figure 5-19** (only one of the two dimensions is required; determine which one and apply it). Apply appropriate constraints. You can trim the inside portion of the circle, as shown in the figure, but this is not necessary.
5. Finish the sketch to return to part mode. Display the isometric view, if it is not already displayed.
6. Pick the **Extrude** button in the **Part Features** panel. If you did not trim the circle in the sketch, select the correct profile. In the **Extrude** dialog box, select Distance from the **Extents** drop-down list and set the distance to 1″. Also, pick the **Midplane** button, which is the right-hand button below the drop-down lists. Notice that the preview in the graphics window shows the extrusion equally divided about the work plane. Pick the **OK** button to create the extrusion.

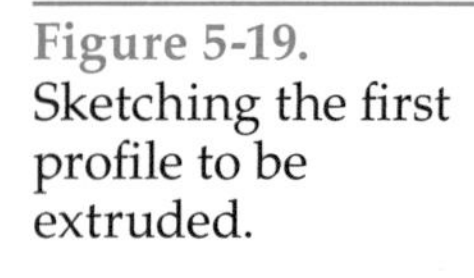

Figure 5-19.
Sketching the first profile to be extruded.

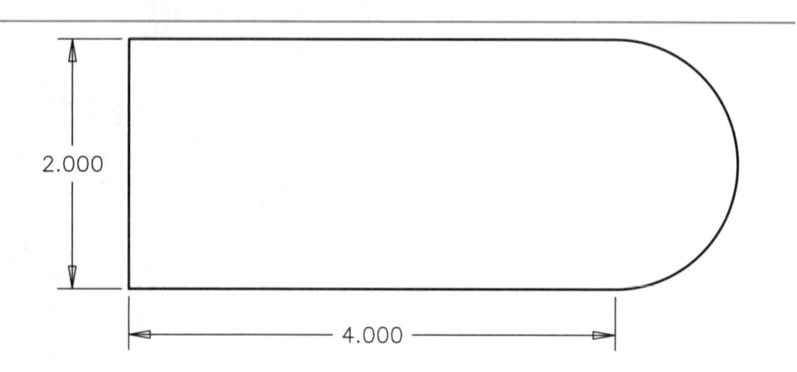

Figure 5-20.
Using the **Slice Graphics** option, the portion of the part that is above the current sketch plane (positive Z) is not displayed.

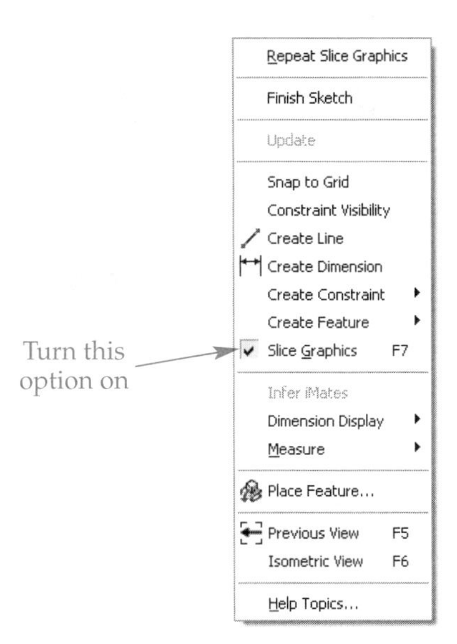

7. Pick the **Sketch** button on the **Inventor Standard** toolbar. Select the XY work plane as the sketch plane; pick an edge. This is why you turned on the visibility of the XY work plane. Otherwise, you would have to right-click on the work plane in the **Browser** and select **New Sketch** from the pop-up menu. You can also pick the **Sketch** button and then select the work plane in the **Browser**.

The sketch plane and grid pass through the part. By using the XY work plane and the **Midplane** extrusion option, you can easily create the entire part. Since the part is shaded, it would be hard to sketch within the boundary of the part. You can change to a wireframe display. However, there is another method that allows you to keep the part shaded. Right-click in the graphics window to display the pop-up menu. Then, select **Slice Graphics** from the menu. See **Figure 5-20.** The portion of the part "above" the sketch plane (positive Z axis) is hidden. Now, you can sketch within the boundary of the part and see what you are doing. Continue creating the part as follows.

8. With the **Project Geometry** tool, project the center point to the plane. Since this is a different sketch, the point needs to be projected into it.
9. Using the **Center Point Circle** tool, draw a circle on the center point.
10. Dimension the diameter and edit the value to 1.25″.
11. Finish the sketch to return to part mode. The part is no longer displayed "sliced." Pick the **Extrude** button in the **Part Features** panel.
12. In the **Extrude** dialog box, pick the **Profile** button, if it is not already active. Then, select the circle as the profile. The circle is within the shaded part, but as you move the cursor over the circle, it is highlighted.
13. In the **Extrude** dialog box, pick the **Join** button. Also, select Distance from the **Extents** drop-down list and set the distance to 1.50″. Pick the **Midplane** button and then pick **OK** to create the extrusion.
14. Using the same procedure, sketch a .75″ diameter circle and extrude it 2.25″. Make sure the **Midplane** button is selected.
15. Sketch a .50″ circle and extrude it using the **Cut** and All options.

The part is complete and should look like **Figure 5-18.** To turn off the visibility of a single work plane, right-click on it in the graphics window or on its name in the **Browser**. Then, select **Visibility** in the pop-up menu to turn off its visibility.

Open the file Example_05_05.ipt. This is a revolved part to which you will add a cutout (hole) through one side of the outer shell, as shown in **Figure 5-21.** The From To option of the **Extrude** tool will be used in conjunction with the default work planes.

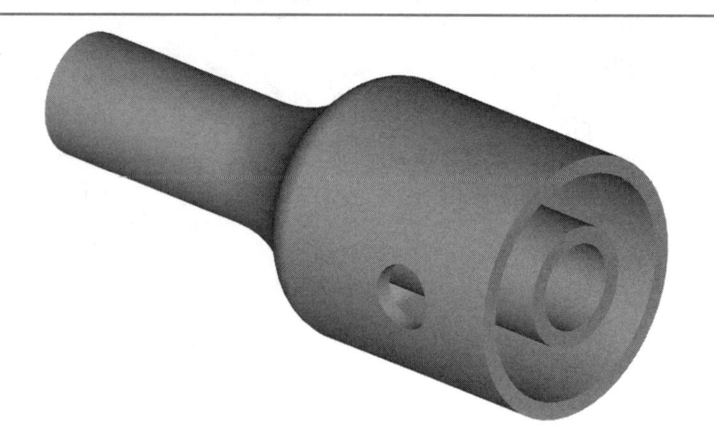

Figure 5-21.
The cutout (hole) shown here will be added to the outer shell of this part.

Expand the Origin branch of the part tree in the **Browser**. Right-click on XZ Plane and select **Visibility** in the pop-up menu. Pick the **Sketch** button on the **Inventor Standard** toolbar. Then, pick the XZ work plane as the sketch plane. Display a "sliced" part and continue as follows.

1. Pick the **Construction** button in the **Inventor Standard** toolbar. Then, pick the **Project Geometry** button in the **2D Sketch Panel** and then pick the edge of the large cylinder projecting it as a line onto the sketch plane.
2. With that tool still active, pick the **Centerline** button in the **Inventor Standard** toolbar. Then, pick on X Axis in the **Browser**, projecting it to the sketch plane.
3. Turn off the **Centerline** button and draw a circle. Dimension it as shown in **Figure 5-22.**
4. Apply a coincident constraint between the center of the circle and the projected X axis line, if it does not already exist. Finish the sketch to return to part mode.
5. Pick the **Extrude** button in the **Part Features** panel.
6. Select the circle as the profile.
7. In the **Extrude** dialog box, pick the **Cut** button and select From To in the **Extents** drop-down list.

From To means that you have to pick two work planes or faces to define the extent of the extrusion. The extruded feature does not have to touch the plane on which you sketched the profile. Once you select this option, two buttons appear below the **Extents** drop-down list in the **Extrude** dialog box. See **Figure 5-23.** Continue as follows.

8. Pick the top button below the **Extents** drop-down list. The tooltip for this button is Select surface to start the feature creation. Then, pick the outermost surface of the part. That surface is highlighted.
9. Once you select the first surface, the bottom button below the **Extents** drop-down list is activated. The tooltip for this button is Select surface to end the feature creation. With this button active, pick the inside surface of the outer shell of the part. The surface is highlighted.
10. Then, pick the **OK** button to extrude the profile.

15.000
25.000

Figure 5-22.
Drawing the profile that will be extruded to create the cutout (hole).

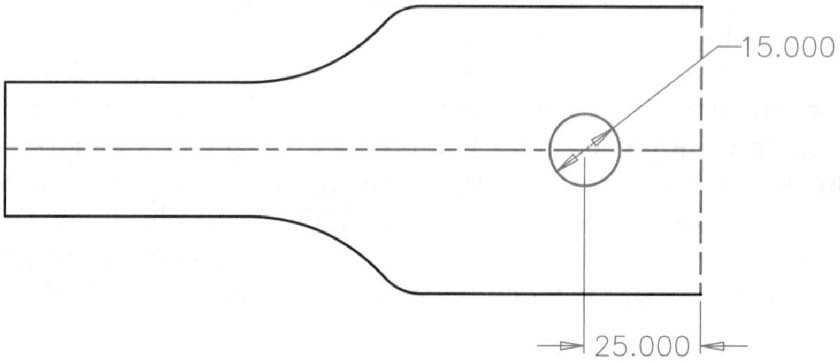

Figure 5-23.
The settings for extruding the profile.

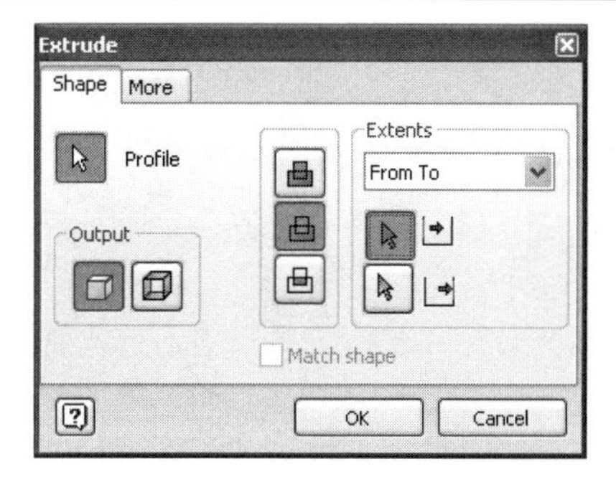

The part is completed and should look like **Figure 5-21.** If the cutout (hole) is on the opposite side of the part, right-click on the extrusion name in the **Browser**. Select **Edit Feature** from the pop-up menu. This displays the **Extrude** dialog box. Then, pick the **More** tab. In the **Alternate Solution** area of this tab are two buttons, one of which is on (depressed). These buttons determine the direction of the extrusion. Pick the button that is not currently on. Then, pick the **OK** button to complete the edit. The cutout should now be on the opposite side of the part as before.

PRACTICE 5-2 Complete the practice problem on the Student CD.

Intersect Option of the Extrude Tool

The **Intersect** option of the **Extrude** tool creates a solid of the volume that is common to the part and the new extrusion. The effects of the three options—**Join**, **Cut**, and **Intersect**—are shown in **Figure 5-24.**

In this section, you will use the **Intersect** option to create a part that would be difficult to create any other way. Open Example_05_06.ipt. See **Figure 5-25.** The YZ and the XY work planes are visible. On each plane is a sketch. If you look at the **Browser**, the part tree shows two unconsumed sketches, as indicated by the pencil and paper

Figure 5-24.
A—Extruding this profile (shown in color) can result in one of three parts, depending on the option selected. B—**Join**. C—**Cut**. D—**Intersect**.

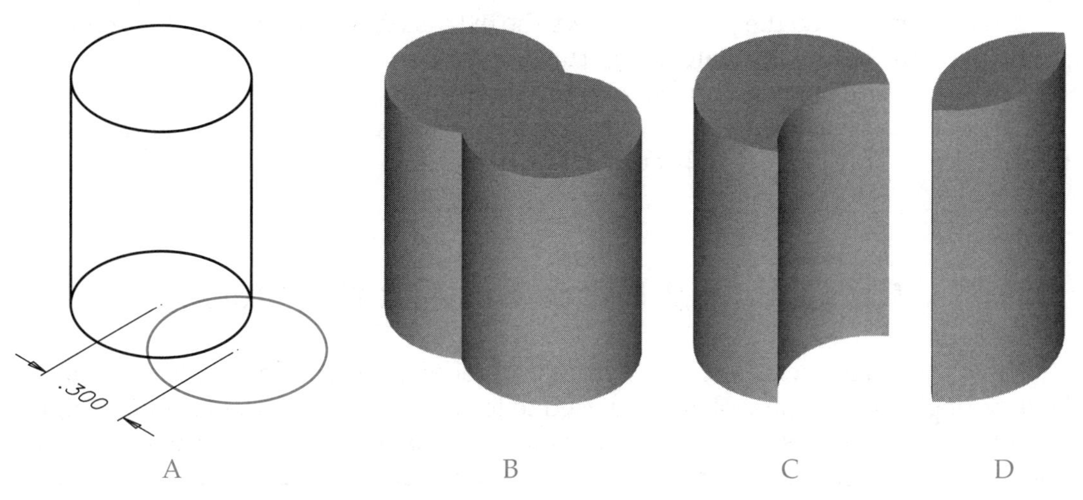

Figure 5-25.
These are two separate profiles (unconsumed sketches) that will be extruded using the **Intersect** option.

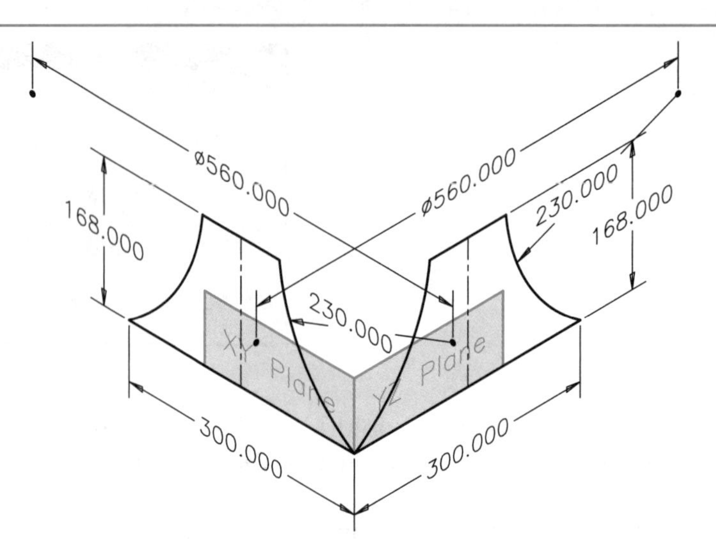

Figure 5-26.
This part is the result of extruding the profiles shown in Figure 5-25 using the **Intersect** option.

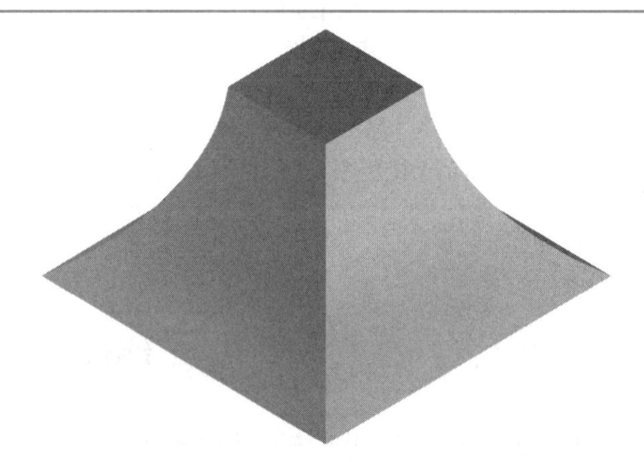

symbol next to their names. An ***unconsumed sketch*** has not yet been used for an operation, such as extrude or revolve. Consumed sketches are indicated by the same symbol, but grayed out. A ***consumed sketch*** has been used for an operation.

1. Using the **Extrude** tool, extrude the left-hand sketch 300″ to the right. The distance can also be entered as 25′. Notice that the symbol next to the name of the sketch in the **Browser** has changed to indicate a consumed sketch.
2. Pick the **Extrude** button again. Select the right-hand sketch as the profile. In the **Extrude** dialog box, select the **Distance** option and set the distance to 25′.
3. Pick the **Intersect** button in the **Extrude** dialog box. Also, pick the appropriate direction button so the preview shows the extrusion going into the existing part.
4. Pick the **OK** button to create the extrusion.
5. The result is the pagoda roof shown in **Figure 5-26.**

The **Intersect** option can be very useful in converting AutoCAD 2D drawings into Inventor parts. This process is discussed in Chapter 8.

Creating and Using New Work Planes

New work planes can be created from part faces, edges, endpoints (vertices), and the default planes. The new planes are used to terminate extrusions, to provide angled and offset sketch planes, and as references for dimensions and constraints. The latter is very useful for the 3D constraints that you will use for assemblies in Chapter 11.

The **Work Plane** tool is used to create a new plane. The basic procedure is to activate the tool, which is located in the **Part Features** panel, and then pick some combination of objects or features on the part. The position and orientation of the new plane depend on which features you pick and how you pick them. Features of a part that can be selected include faces, lines, endpoints (vertices), existing work planes, work points, and work axes. The next sections present several different applications.

Offset from an Existing Face or Work Plane

A new work plane can be created that is offset a specific distance from a face or existing work plane. This technique is often used when creating a feature attached to a cylinder. The existing XY plane needs to be moved in the Z direction so a shape can be sketched. Since it is impossible to move the existing plane, a new one is created a set distance away from the old plane. Now, sketching the required shape is possible.

For example, the part in **Figure 5-27A** is the existing part. The finished part is shown in **Figure 5-27B.** To create the new feature, you will offset the YZ work plane. Open Example_05_07.ipt and expand the Origin branch in the **Browser**. Right-click on YZ Plane and select **Visibility** from the pop-up menu to display the work plane. Then, continue as follows.

1. Pick the **Work Plane** button in the **Part Features** panel. Then, pick an edge of the YZ work plane and drag it to the right.
2. In the **Offset** dialog box that appears, enter a value of 100 mm. Then, press [Enter] or pick the check mark button in the dialog box.
3. Pick the **Sketch** button on the **Inventor Standard** toolbar. Pick the new work plane as the sketch plane.
4. With the **Project Geometry** tool, project the top and bottom of the part onto the sketch plane. Change the style of the two projected lines to Construction.
5. Draw a construction line between the midpoints of the two projected lines.
6. Draw a circle in Normal style. Dimension the diameter and edit the value to 35 mm.
7. Apply a coincident constraint between the center of the circle and the construction line you drew in Step 5.
8. Add a dimension between the projected top edge and the center of the circle. Edit the dimension to a value of 40 mm. See **Figure 5-28.**
9. Finish the sketch and pick the **Extrude** button in the **Part Features** panel.
10. Select the circle as the profile.

Figure 5-27.
A—The original part.
B—A feature is added to the part.

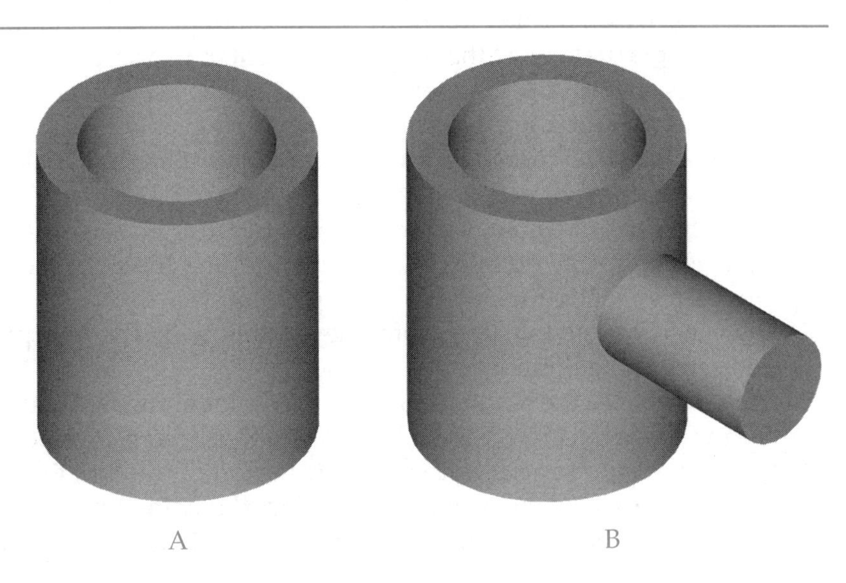

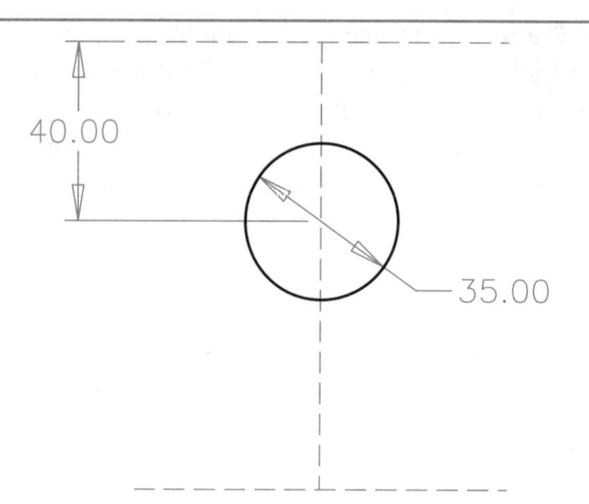

Figure 5-28.
Sketching the profile that will be extruded into the feature shown in Figure 5-27B.

11. In the **Extrude** dialog box, select the **Join** and To Next options. Pick the **Terminator** button and select the outside surface of the cylinder (in this case, the surface should be automatically selected).
12. Pick the **OK** button to create the extrusion. The new feature is added to the part.

Offset through a Point

Instead of entering an offset distance, you can select a point through which the offset work plane will pass. Using the **Work Plane** tool, pick the interior of an existing planar face or the edge of an existing work plane. Then, select an endpoint or midpoint of any edge on the part. The new work plane is created parallel to the selected face or plane and through the selected point.

Angled from a Face or Existing Work Plane

A new work plane can be created at an angle to a face or existing work plane. The new plane will pass through an edge on the part. For example, look at the existing and finished parts in **Figure 5-29.** Notice how the added feature, the pin, is not extruded perpendicular to the face. In order to do this, a work plane must be created at an angle to the face. Open Example_05_08.ipt. Then, continue as follows.

1. Pick the **Work Plane** button in the **Part Features** panel.
2. Pick face A as indicated in **Figure 5-29.**
3. Pick edge B as indicated in **Figure 5-29.** Do not pick an endpoint or the midpoint.
4. In the **Angle** dialog box that is displayed, enter an angle of 30°. The feature you are adding is angled 30° from perpendicular to the face. Then, press [Enter] or pick the check mark button to create the new work plane.
5. Pick the **Sketch** button on the **Inventor Standard** toolbar. Pick the new work plane as the sketch plane.
6. With the **Project Geometry** tool, project the four edges of face A onto the sketch plane. This can be done in one step by moving the cursor over the face until all four edges are highlighted and then picking.
7. Draw a construction line from the midpoint of the top projected line to the midpoint of the bottom projected line.
8. Draw a circle in the Normal style with its center at the midpoint of the construction line, as shown in **Figure 5-30.** Dimension the circle's diameter to 30 mm.
9. Finish the sketch and pick the **Extrude** button in the **Part Features** panel.
10. Select the circle as the profile. Also, select the **Join** and To options in the **Extrude** dialog box. Then, pick the button below the **Extents** drop-down list and select face A.
11. Pick the **OK** button to create the extrusion. The new feature is added to the part.

Figure 5-29.
A—The existing part. B—A feature is added to the part.

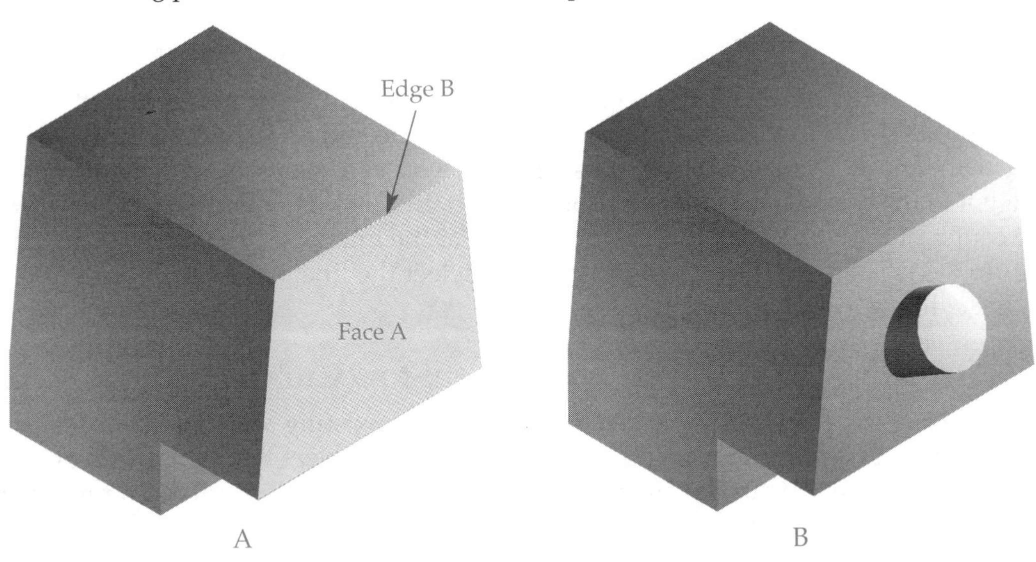

Figure 5-30.
Sketching the profile that will be extruded into the feature shown in Figure 5-29B.

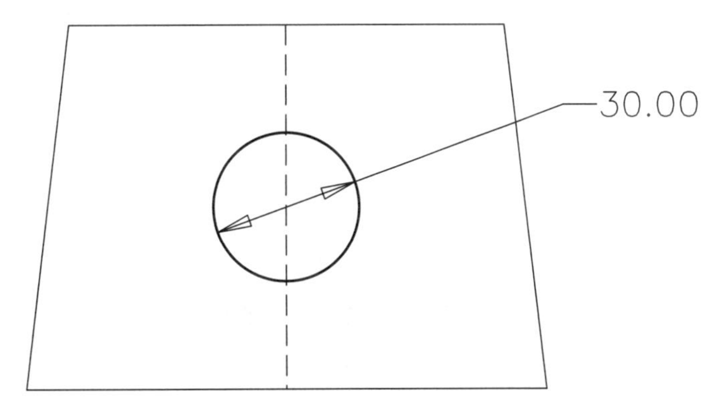

PROFESSIONAL TIP

If there is no edge on the part at the position where you want the work plane to pass through, you can sketch a construction line on an existing work plane or face. Then, select the construction line as the "edge" through which the new work plane will pass.

PRACTICE 5-3 Complete the practice problem on the Student CD.

Midway between Two Parallel Faces

You can easily construct a work plane that is midway between two parallel faces. This technique is very useful when combining parts into assemblies.

1. Open Example_05_09.ipt.
2. Pick the **Work Plane** button in the **Part Features** panel.
3. Pick the green face on the left end of the cylinder.
4. Rotate the view and pick the red face on the opposite end of the part.

A work plane is created at the midpoint of the cylinder. This plane will stay at the midpoint even if the length of the cylinder is modified.

Perpendicular to a Line at Its Endpoint

To create a work plane at the end of, and perpendicular to, a line, pick the **Work Plane** button in the **Part Features** panel. Select the line and pick the endpoint at which you want the work plane. Using this method, you can create a work plane that is tangent to the curved surface of a cylinder. To do this, you must first sketch a line from the center of the cylinder to the circumference (at the intended point of tangency). See **Figure 5-31.** Then, using the **Work Plane** tool, select the line and its endpoint that lies on the cylinder's circumference.

Parallel to a Face or Plane and Tangent to a Curved Face

A new work plane can be created parallel to an existing work plane or face and tangent to a curved surface. Using the **Work Plane** tool, select the planar face or work plane and the curved surface in either order. See **Figure 5-32.** Both faces must have their edge view in the same view.

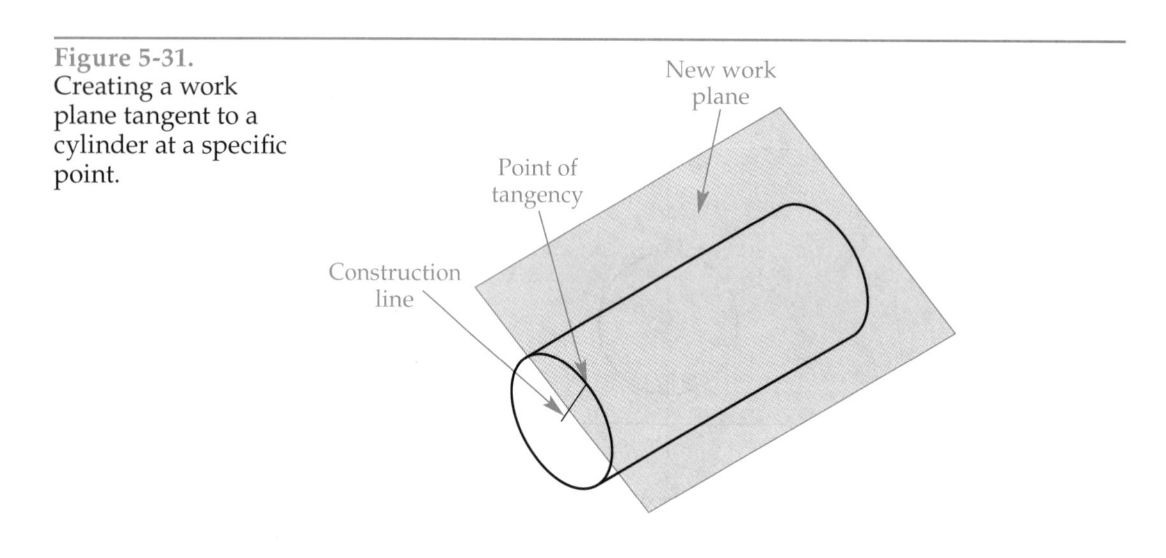

Figure 5-31.
Creating a work plane tangent to a cylinder at a specific point.

Figure 5-32.
Creating a work plane parallel to a planar face and tangent to a curved face.

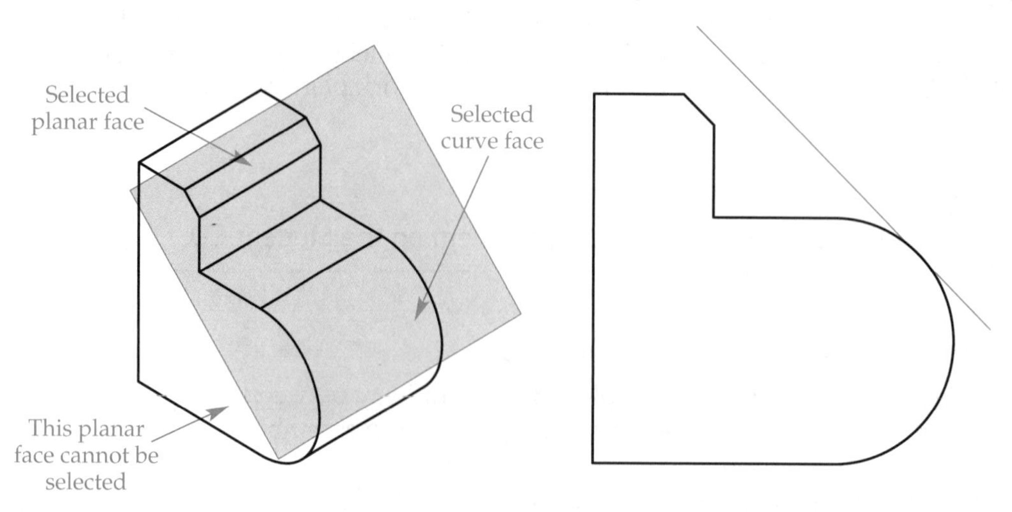

Figure 5-33.
Creating a work plane tangent to a curved face and through an edge.

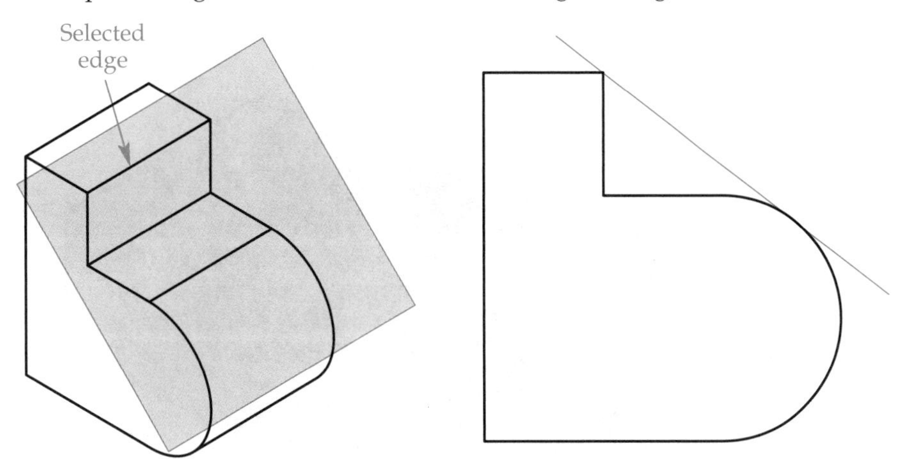

Tangent to a Curved Face through an Edge

A new work plane can be created that passes through an edge and is tangent to a curved surface. This is similar to the operation discussed in the last section. Using the **Work Plane** tool, select the curved face and the edge of a planar face in either order. See **Figure 5-33.** Note that the edge and the curved surface must appear as an edge view in the same view to create the tangent plane. The edge will appear as a point in an edge view and the curve as an arc or circle. Otherwise, the work plane will either not pass through the edge or not be tangent to the curved surface.

Through Two Parallel Work Axes

A ***work axis*** is another type of work feature. You cannot sketch on a work axis as you can a work plane. However, a work axis can be used to constrain features and create work planes. In essence, a work axis is a fixed construction line.

A work axis can be used to create a work plane. For example, you may have two cylinders and need a new work plane that passes through the center of each cylinder. To create a work axis, pick the **Work Axis** button in the **Part Features** panel. Then, pick each of the two cylinders. Do not pick the end face. You will need to restart the tool after selecting the first cylinder. Work axes are displayed in a thin line similar to a construction line, but not dashed. Now, using the **Work Plane** tool, select the two axes in any order. A work plane is created that passes through the center of each cylinder. See **Figure 5-34.**

Perpendicular to an Axis and through a Point

A work plane can be created perpendicular to a work axis. An endpoint or midpoint on a face edge is used to define the location of the work plane along the work axis. To create the work axis, use the **Work Axis** tool and select a curved surface or cylinder. Then, using the **Work Plane** tool, select the axis and the endpoint on the part in either order. See **Figure 5-35.**

Between Edges or Points

A work plane can be defined by three points that do not lie on the same line. By definition, any two points define a line. Therefore, a work plane can also be defined by an edge (line) and a point that is not on that edge. Finally, a work plane can be defined by two edges whose endpoints lie in the same plane. Using the **Work Plane** tool, select three points, an edge and a point, or two edges. The order of selection is not important. Note that selecting an edge and a point is the same as selecting a work axis and a point, as discussed in the previous section.

Figure 5-34.
Creating a work plane through the center axes of two cylinders.

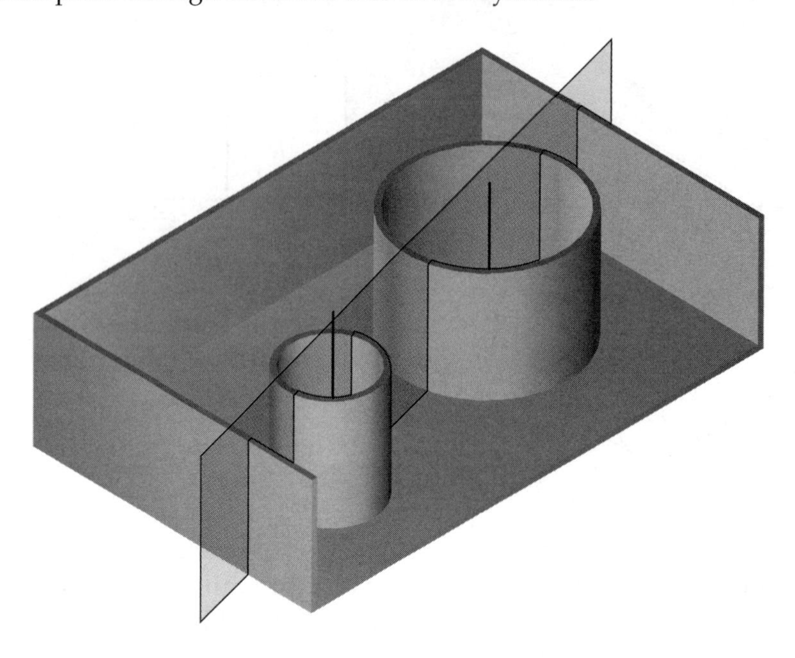

Figure 5-35.
Creating a work plane perpendicular to a work axis and through a point.

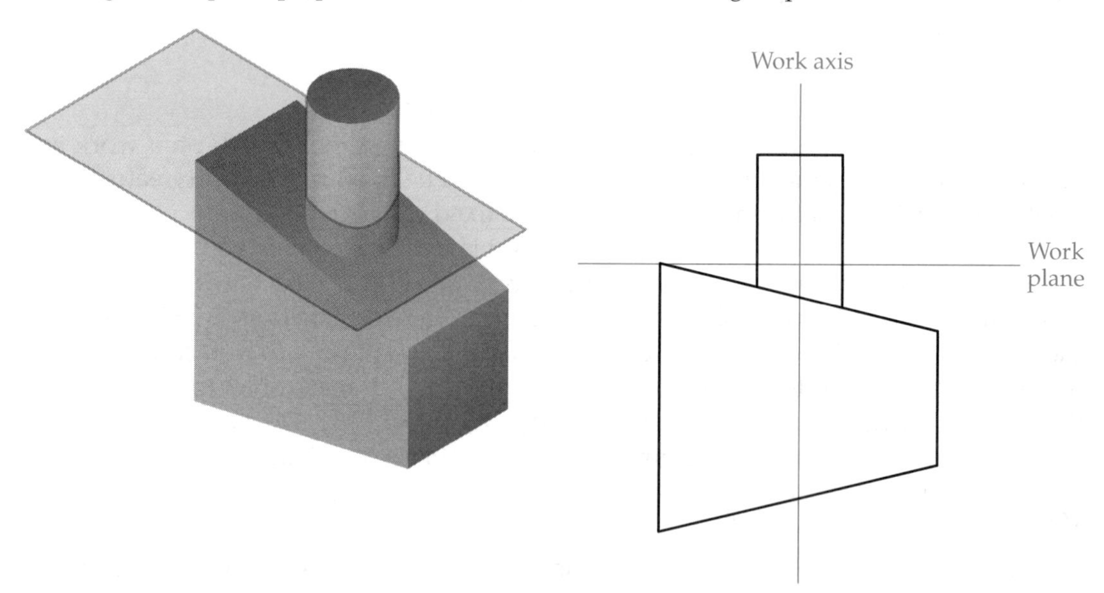

Chapter Test

Answer the following questions on a separate sheet of paper or complete the electronic chapter test on the Student CD.

1. How can you change the color of a face?
2. Which faces of a part can be used as sketch planes?
3. Which dialog box is used to turn on the display of a CSI at the origin?
4. How do you open the dialog box in Question 3?
5. Suppose you have created an extrusion that has the default name Extrusion1. How can you change the name to Pin, Locking?

6. Explain the difference between the **Join**, **Cut**, and **Intersect** options of the **Extrude** tool.
7. Explain the Distance extents option of the **Extrude** tool.
8. Explain the To Next extents option of the **Extrude** tool.
9. Explain the To extents option of the **Extrude** tool.
10. Explain the From To extents option of the **Extrude** tool.
11. Explain the All extents option of the **Extrude** tool.
12. Define *silhouette curve.*
13. How do you project a silhouette curve onto the current sketch plane?
14. Define *work plane.*
15. How many default work planes are there? Name them.
16. Describe the relationship between the default work planes and the coordinate system.
17. Explain how to turn on the visibility of a default work plane.
18. How can you remove the display of the portion of a part that is above the current sketch plane?
19. Which tool is used to create a work plane that is *not* one of the default work planes?
20. Define *consumed sketch.*
21. Define *unconsumed sketch.*
22. Name four types of items that can be used, in part, to define a new work plane.
23. Give three applications where work planes may be used.
24. Define *work axis.*
25. Suppose you have drawn two cylinders. How can you create a sketch on a plane that passes through the center of each cylinder?

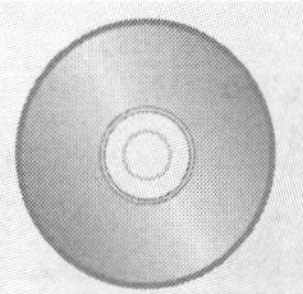

Chapter Exercises

Exercise 5-1. Boss.

Complete the exercise on the Student CD

Exercise 5-2. Bracket.

Complete the exercise on the Student CD

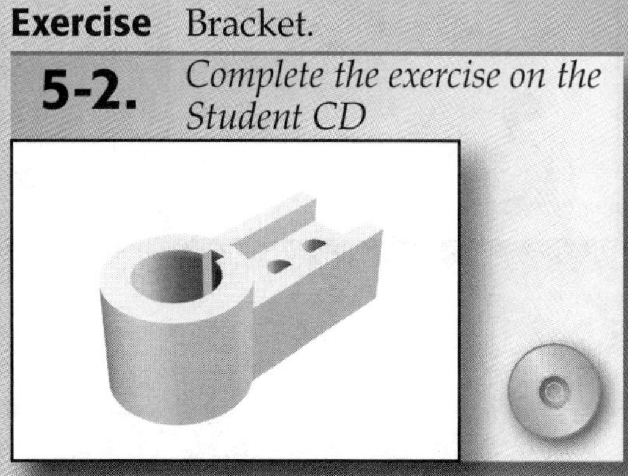

Exercise 5-3. Cutouts.

Complete the exercise on the Student CD

Exercise 5-4. Tapered Part.

Complete the exercise on the Student CD

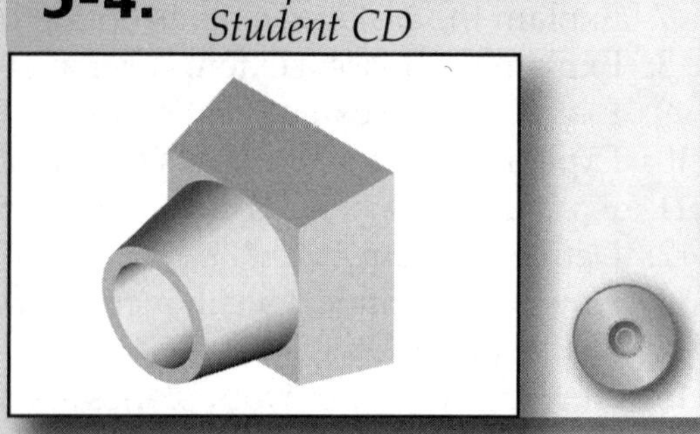

Exercise 5-5. Mirror Part.

Complete the exercise on the Student CD

Exercise 5-6. Hose Nozzle Stem.

Complete the exercise on the Student CD

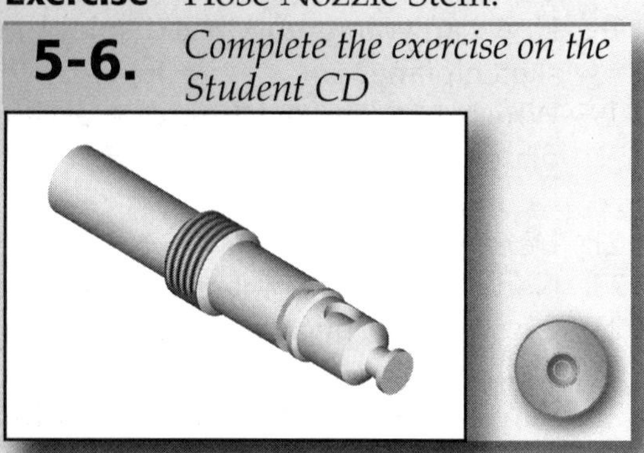

Exercise 5-7. Yoke.

Complete the exercise on the Student CD

Exercise 5-8. Brace.

Complete the exercise on the Student CD

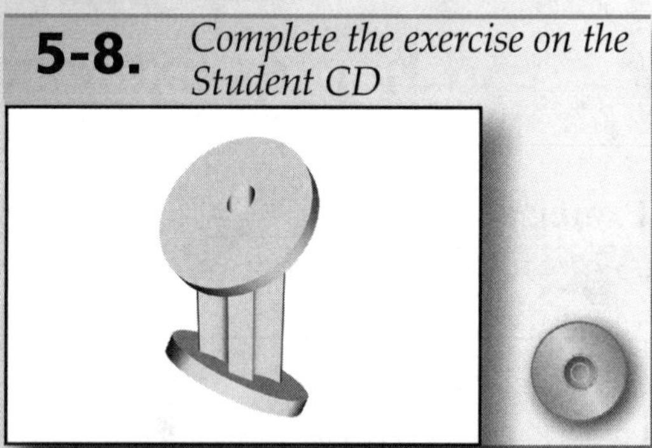

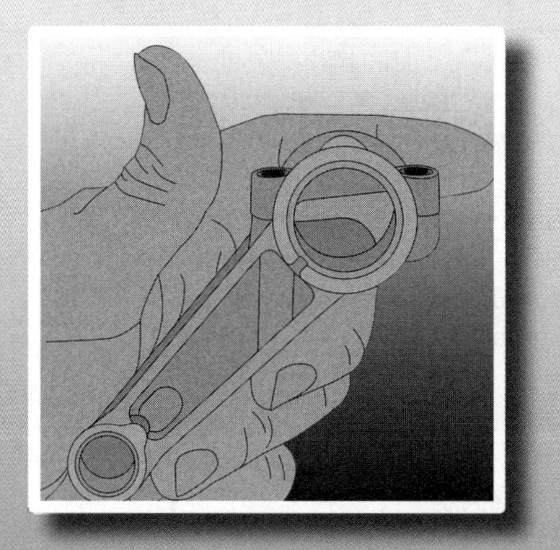

Chapter 6

Adding Features

Objectives

After completing this chapter, you will be able to:

- Add holes to a part.
- Add threads to a part.
- Fillet edges on a part.
- Chamfer edges on a part.
- Create rectangular and circular patterns.
- Mirror features.

User's Files

The Student CD included with this text contains several files required for this chapter. Refer to the file File List.txt *in the* \Ch06 *folder for the comprehensive list.*

Adding Nonsketch Features to the Part

There are several features that can be added to a part without drawing a sketch profile. For example, when using the **Fillet** tool, you need only select the edge(s) to be modified. No sketch is required. In this chapter, seven of these feature-creation tools are discussed—**Hole**, **Thread**, **Fillet**, **Chamfer**, **Mirror Feature**, **Rectangular Pattern**, and **Circular Pattern**. The rest of the feature-creation tools are discussed in the next chapter.

Hole Tool

So far, you have created holes as *cutouts* by extruding circles. However, the **Hole** tool has several advantages over extruded circles. Refer to **Figure 6-1** and the list below.

- The holes can be drilled, counterbored, or countersunk.
- Counterbores and countersinks can be automatically sized for standard fasteners.
- Thread sizes and specifications can be easily applied.
- Threads will be displayed in the shaded image.
- Threaded holes will correctly display in the part drawings.
- Hole notes can be automatically applied to the part drawings.
- Hole tables can be created in the part drawings.

Figure 6-1.
These holes were created using the **Hole** tool. When done in this manner, threads appear correctly in the part drawing and tables can be created.

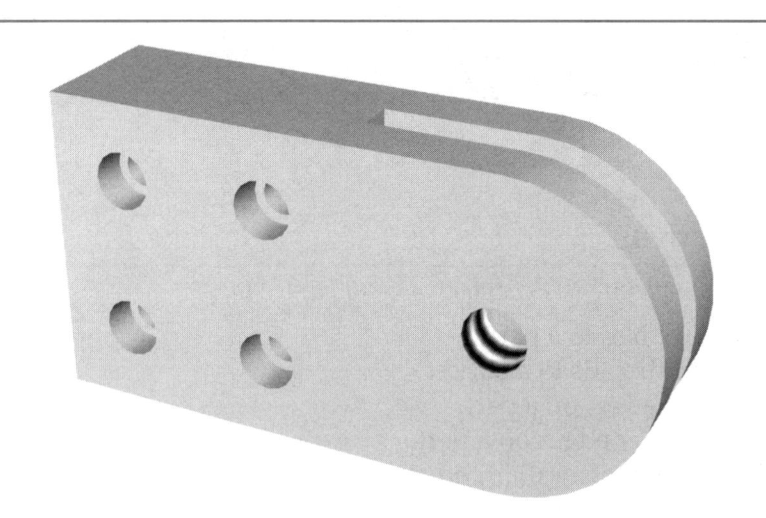

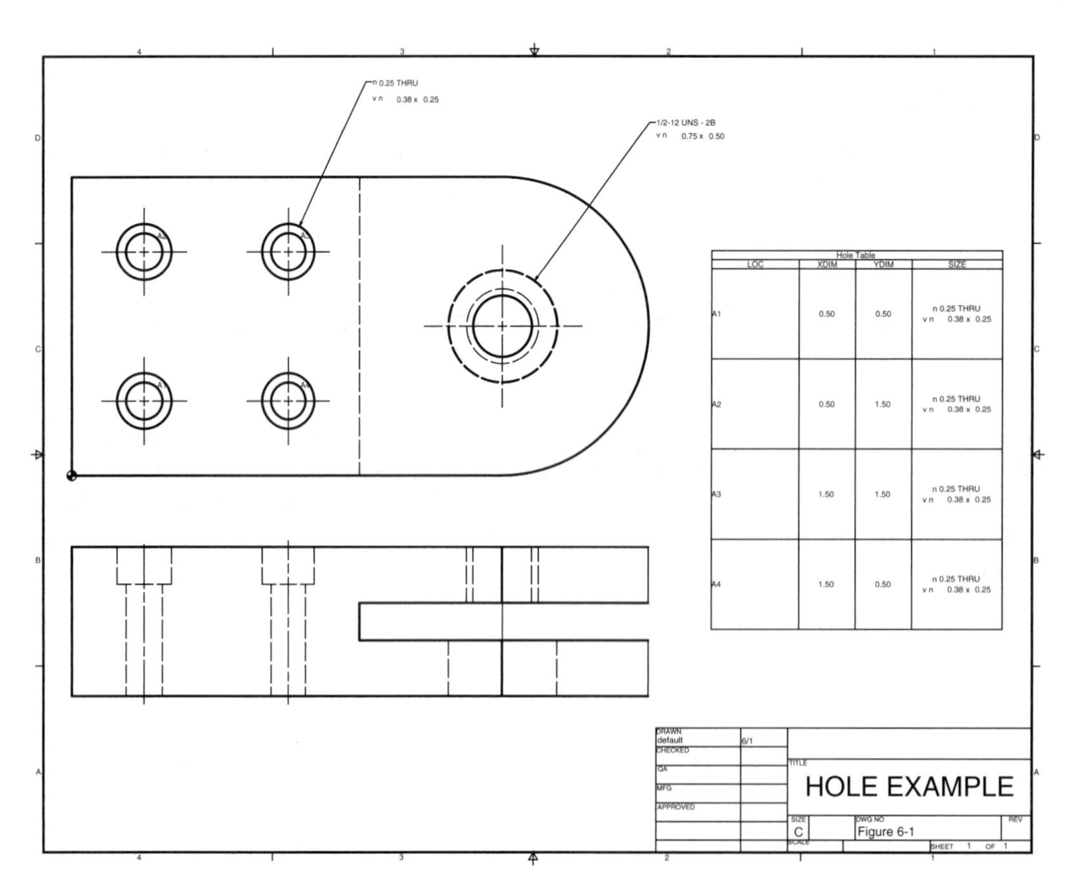

One way to use the **Hole** tool is to draw special points called ***hole centers*** on a sketch plane. The **Point, Center Point** tool is used to create these special points. The **Hole** tool is then used to specify the size and type of hole created on the hole center. By drawing multiple hole centers, multiple holes can be created in one session of the **Hole** tool. All of the holes created in one tool session have the same size and characteristics.

The **Point, Center Point** tool can create center points for holes or sketch points; this choice is controlled by the **Center Point** button on the **Inventor Standard** toolbar. When the **Center Point** button is depressed, the **Point, Center Point** tool creates crosshair-shaped hole centers. When this button is not depressed, sketch points are created.

Open Example_06_01.ipt. Pick the **Sketch** button on the **Inventor Standard** toolbar and select the large, front face as the sketch plane. Make sure the **Center Point** button is active (depressed). Next, pick the **Point, Center Point** button in the **2D Sketch Panel**. Place two hole centers anywhere on the face by picking at two locations. Apply a vertical constraint between the hole centers. This will keep them in a line parallel to the Y axis. Now, dimension the hole centers as shown in **Figure 6-2.** Then, finish the sketch to return to part mode.

In the **Part Features** panel, pick the **Hole** button or press the [H] key to activate the **Hole** tool. The hole centers in the sketch are automatically selected and the **Holes** dialog box is displayed. See **Figure 6-3.** Also, a preview of the default settings is displayed at each hole. To remove a hole center from the operation, pick the **Centers** button in the **Holes** dialog box. Then, hold down the [Shift] key and pick a hole center to deselect it. Using the same procedure, you can add a hole center to the selection set, but you do not need to hold down the [Shift] key.

There are three options in the **Termination** drop-down list in the **Holes** dialog box—Distance, Through All, and To. These are similar to the extents options for extruded profiles. Distance specifies a specific hole depth. The value is set by changing the dimension in the text box next to the image tile. Refer to **Figure 6-3.** The diameter is set by changing the dimension in the second text box. For this example, select the Distance option and change both the depth and diameter to .5″.

Holes have a direction. The direction can be reversed or "flipped" by picking the button next to the **Termination** drop-down list in the **Holes** dialog box. If the hole is

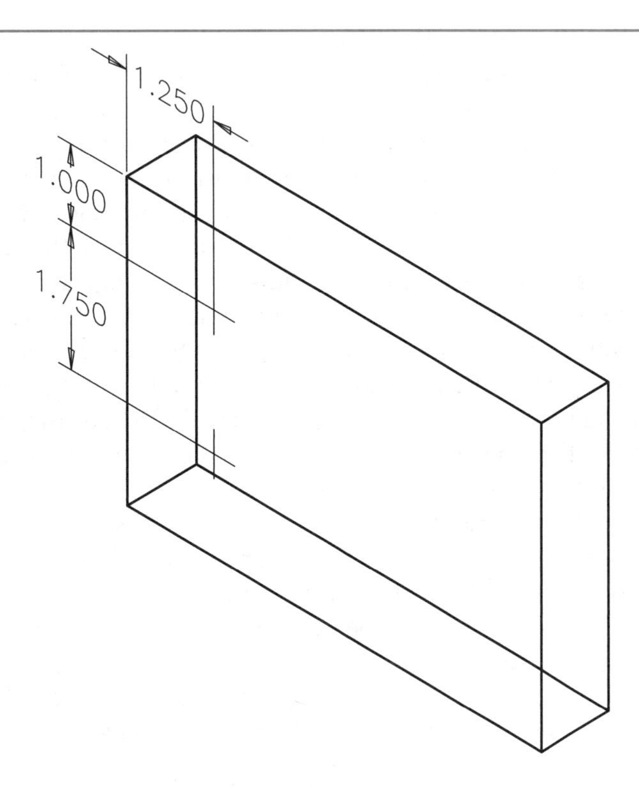

Figure 6-2.
Placing hole centers to create holes.

Figure 6-3.
The **Holes** dialog box is used to create holes.

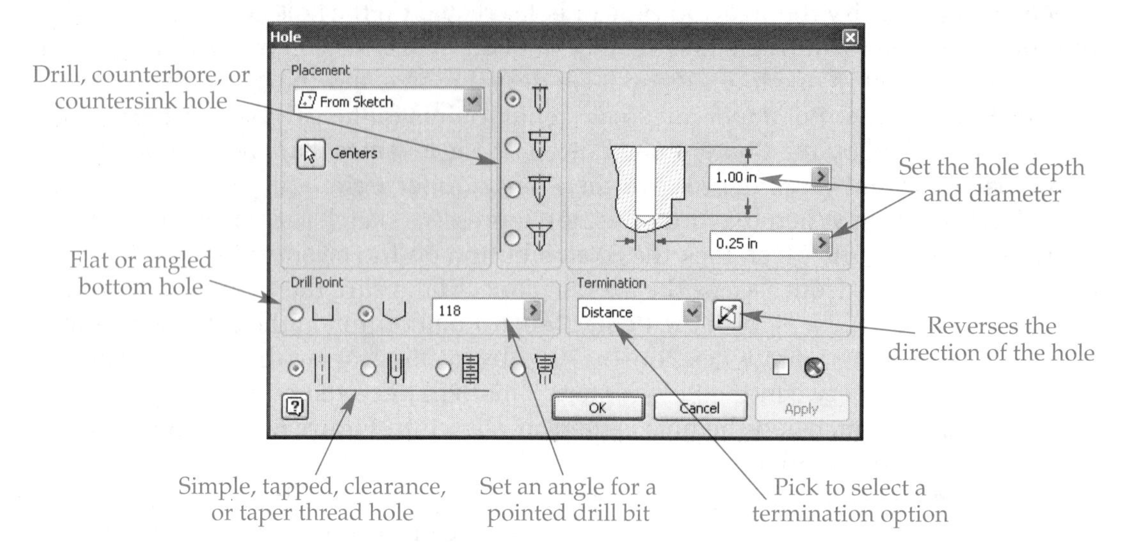

"outside" of the part, an error is generated by Inventor. Holes must be to the interior of a part. Also, there is no midplane option.

The settings in the **Drill Point** area of the **Holes** dialog box control how the bottom of the hole is created. You can set the angle of a pointed drill bit. You can also select a flat-bottom hole, as is created by an end mill. For this example, pick the **Angle** radio button and use the default angle value.

Now, pick the **OK** button in the **Holes** dialog box to create the two holes. Using the **Rotate** tool, rotate the view to see the holes. Notice how they do not pass through the part. Change to a wireframe display. Then, rotate the view so you can see the side of the holes. Notice how the bottom is tapered according to the angle set in the **Holes** dialog box.

Threaded Holes

Adding threads to the holes can be done when the holes are created or later by editing the feature. Display the isometric view of Example_06_01.ipt and shade the view. Now, right-click on the hole feature in the **Browser** and select **Edit Feature** from the pop-up menu. The **Holes** dialog box is displayed.

Select Through All from the **Termination** drop-down list. This will change the holes so they completely pass through the part. Notice how the setting is reflected in the image tile. The four radio buttons at the bottom of the dialog box are: **Simple Hole**, **Tapped Hole**, **Clearance Hole**, and **Taper Thread Hole**. These determine the type of hole. A simple hole is the type you initially created. Pick the **Tapped Hole** radio button and the **Threads** area appears in the dialog box. See **Figure 6-4.**

Next, check the **Full Depth** check box, if not already checked. This check box specifies that the hole is threaded to its full depth. When unchecked, a dimension text box appears next to the image tile that specifies the depth of thread. The value can be changed in this box.

In the **Thread Type** drop-down list, select ANSI Unified Screw Threads. Since the hole was created at .5″ diameter, the **Size** drop-down list displays 0.5 and the **Designation** drop-down list displays the standard pitches for 1/2″ ANSI standard threads. Select 1/2-20 UNF for a fine thread. Select the class of the thread as 1B. Pick the **OK** button to update the feature.

All of the data required to properly size threads are located in a spreadsheet file called Thread.xls. The only reason to modify this spreadsheet is if a custom thread is

Figure 6-4.
Adding threads to a hole.

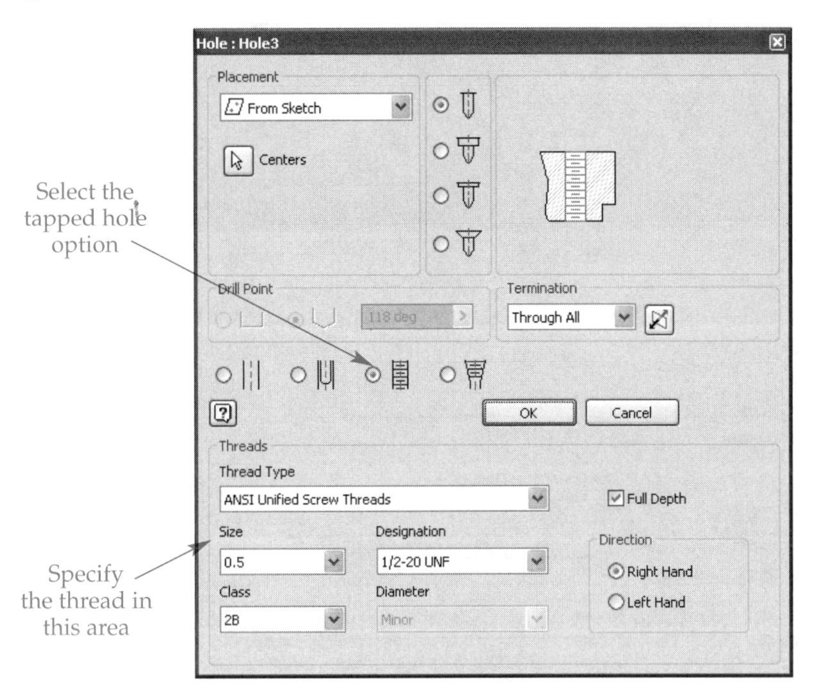

Figure 6-5.
Threads are represented with a picture. Creating the correct model geometry tremendously increases the file size.

required. Accessing and modifying this spreadsheet is beyond the scope of this book. However, there is information in Inventor's help application that may be useful.

After adding threads to the holes, a representation of the threads appears in each hole. This representation is a bitmap, or picture, of the correct thread. See **Figure 6-5.** The bitmap display is based on data in the Thread.xls spreadsheet. The bitmap is a clever technique for displaying threads and has little effect on the file size. Showing actual cut threads (geometry) in the part can be done, as discussed in Chapter 13, but it results in very large files.

Clearance Holes

The third type of hole is a clearance hole. To select this type, pick the **Clearance Hole** radio button below the **Drill Point** area of the **Holes** dialog box. See **Figure 6-6.** The size of this hole is determined by the fastener that will be located in the hole.

Figure 6-6.
Clearance holes are large enough for a fastener to pass through.

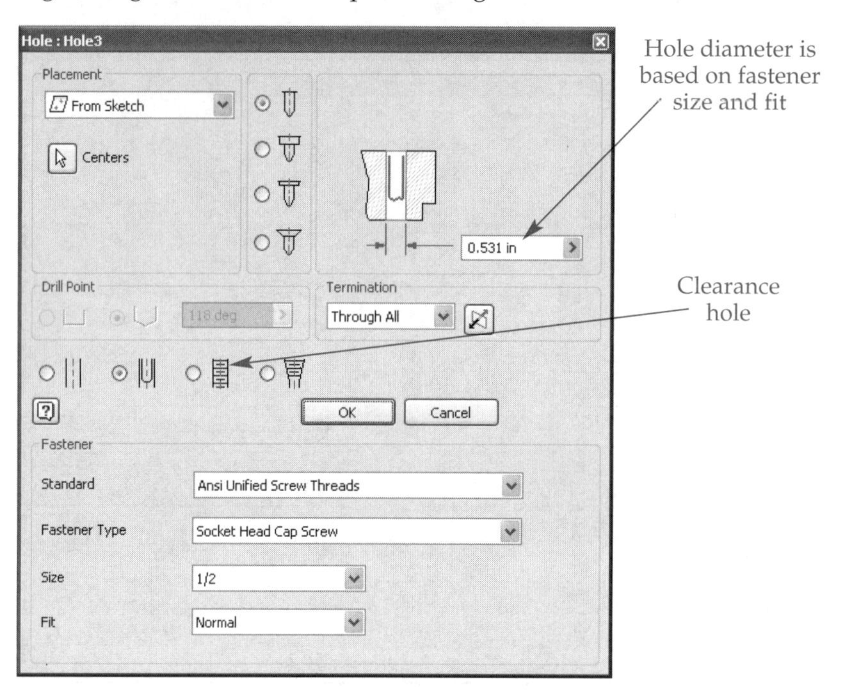

Edit the hole feature you created in Example_06_01.ipt. In the **Holes** dialog box, pick the **Clearance Hole** radio button. The area below this button changes from **Threads** to **Fastener**. Most clearance holes are made to allow a fastener to pass through the part and thread into an adjacent part. Therefore, the **Termination** should be set to **Through All** for clearance holes. Now, set the **Standard** drop-down list to ANSI Unified Screw Threads, the **Fastener Type** to Socket Head Cap Screw, the **Size** to 1/2 inch, and the **Fit** to Normal. The hole diameter next to the image tile is automatically changed to .531″. This is the proper hole size to accommodate the selected fastener. Pick the **OK** button to update the feature.

Taper Thread Holes

The fourth type of hole is a taper thread hole. This type of hole has pipe threads. Pick the **Taper Thread Hole** in the **Hole** dialog box. The area at the bottom of the dialog box changes to **Threads**. Select the type of thread in the **Thread Type** drop-down list. Then, select the diameter in the **Size** drop-down list. In the **Direction** area, you can set whether the threads are left- or right-hand threads. Finally, pick the **OK** button to create the pipe threads.

Counterbored and Countersunk Holes

You have seen drilled holes that are simple, threaded, clearance, and taper. Simple, threaded, and clearance holes can also be applied to counterbored, spotfaced, and countersunk holes. Taper thread holes can be applied spotfaced and countersunk holes. Counterbored and countersunk holes provide a clearance for the head of a fastener. Spotfaced holes provide a smooth seating surface for the fastener head.

Edit the hole feature you created in Example_06_01.ipt. At the top of the **Holes** dialog box is a column of the four hole types: **Drilled**, **Counterbore**, **Spotface**, and **Countersink**. Pick the button next to the **Counterbore** icon and review the dimensions on the right side of the dialog box. See **Figure 6-7.** These dimensions are automatically calculated based on the fastener information at the bottom of the dialog box. Change

Figure 6-7.
A counterbore clearance hole usually allows the head of the fastener to sit below the part surface.

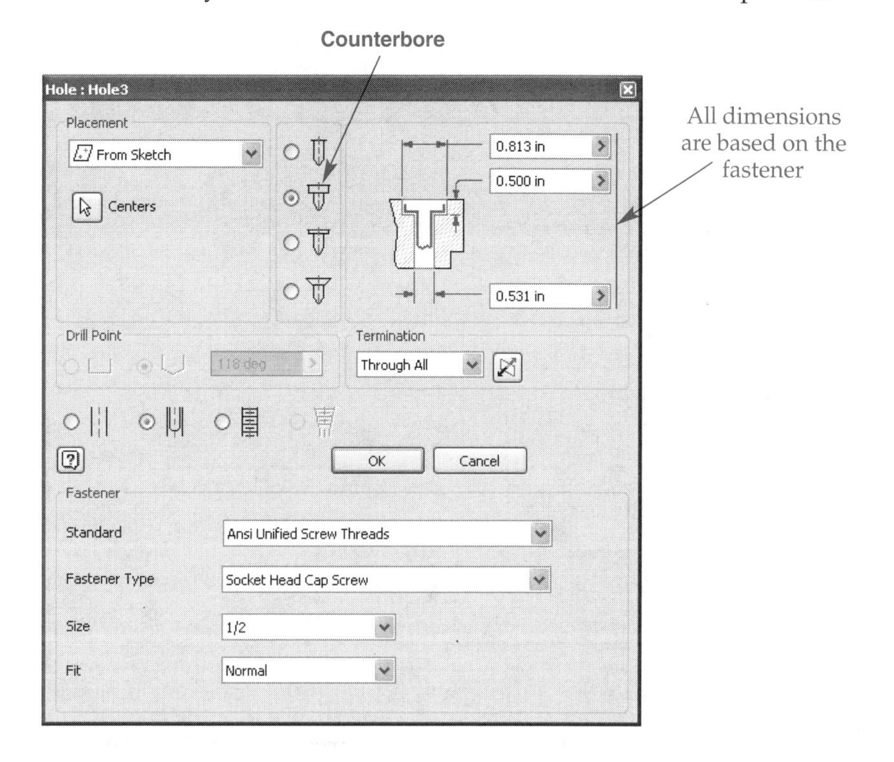

the **Fastener Type** to Hex Head Bolt and notice the counterbore dimensions change to make room for the wider bolt head.

Pick the button next to the **Countersink** icon and review the dimensions. Now, look at the fastener information at the bottom of the dialog box. See **Figure 6-8.** The **Fastener Type** drop-down list contains flat and oval head styles. This is because these are the fasteners used with a countersunk hole. Pick the **Cancel** button to close the dialog box without applying the changes.

NOTE

The head type (fastener type) can only be selected for clearance holes. If for some reason you want head clearance for tapped holes, select **Clearance Hole** first, specify the head type, and then pick the **Tapped** button.

Creating Holes without a Sketch Plane

It is not necessary to create a sketch plane and then add point hole centers in order to create a hole. Open the file Example_06_02.ipt and pick the **Hole** button in the **Part Features** panel. The **Holes** dialog box is displayed. Notice the **Placement** area of the dialog box, **Figure 6-9.** The drop-down list contains four choices: From Sketch, Linear, Concentric, On Point. The options below the drop-down list depend on which option is selected in the drop-down list.

Select Concentric from the drop-down list. The two choices below the drop-down list are **Plane** and **Concentric Reference**. Pick the **Plane** button (it should already be active) and select the face on the part labeled Front Face. Then, pick the **Concentric Reference** button, which should be activated when you select the plane, and select the circular edge of the part. Next, select Through All in the **Termination** drop-down list. Finally, set the diameter to .5 and pick the **Apply** button to apply the hole and leave the dialog box open.

Figure 6-8.
Inventor assumes that a flat head or oval head fastener will be used if you are drawing a countersink hole.

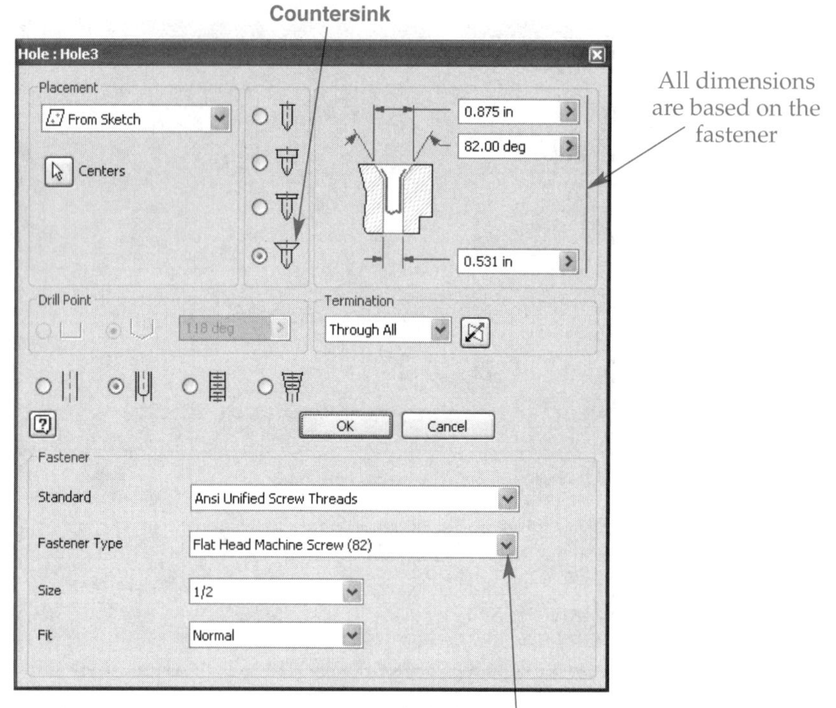

Only flat head and oval head fasteners are available for countersunk holes

Figure 6-9.
When drawing holes without using point hole centers, first select a placement option. The other options are based on the placement option.

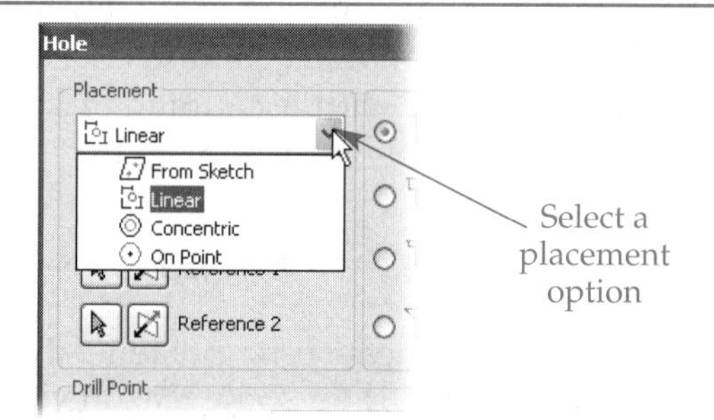

Now, select Linear from the drop-down list in the **Placement** area. With the **Face** button active, pick the face labeled Front Face. With the **Reference 1** button active, pick edge AB. With the **Reference 2** button active, pick edge AC. Change the diameter to .25, set the termination to Through All, and pick the **OK** button. The holes can be edited from the **Browser** in the normal manner. Show the dimensions for the feature and edit the **Reference 1** dimension to .5 and the **Reference 2** dimension to 1.0. Then, update the part.

PROFESSIONAL TIP

Before creating a hole without a sketch plane, check the **Edit dimension when created** check box in the **Sketch** tab of the **Options** dialog box. Then, when the **Reference 1** and **Reference 2** edges are selected, the **Edit Dimension** dialog box immediately appears and an exact value can be entered. This saves editing the dimensions after the feature is created.

Thread Tool

Threads can be applied to cylindrical features, not just to holes. The features can be external, such as a cylinder. The features can also be internal, such as cutouts created by extruding circles or revolving rectangular shapes. The **Thread** tool is used to do this.

Open Example_06_03.ipt, which is a small metric part. In the **Part Features** panel, pick the **Thread** button to activate the **Thread** tool and open the **Thread** dialog box. With the **Face** button active in the dialog box, move the cursor over the outside diameter at point A shown in **Figure 6-10.** A preview of the thread appears on the surface. Notice that the thread is applied over the entire length of that face. In the **Thread** dialog box, uncheck the **Full Length** check box. Also, type 5 in the **Length** text box. Now, pick the **Face** button and move the cursor over the outside diameter again. The preview appears on only a portion of the face, starting at the end nearest to the cursor. The length is 5 mm, as you entered in the dialog box.

Now, in the **Thread** dialog box, type 10 in the **Offset** text box. Pick the **Face** button and again move the cursor over the face. Notice that the thread preview begins 10 mm from the end nearest to the cursor. Change the offset to 0 and the length to 15. Then, pick the **Face** button and select the surface at point A. The entire face is highlighted. Finally, pick the **OK** button in the **Thread** dialog box to create the threads.

The thread specification can be changed as you create the thread. However, since you have already created the threads, right-click on the thread name in the **Browser** and select **Edit Feature** from the pop-up menu. The **Thread** dialog box is displayed. Pick the **Specification** tab, **Figure 6-11.** In this tab, there are options for setting the

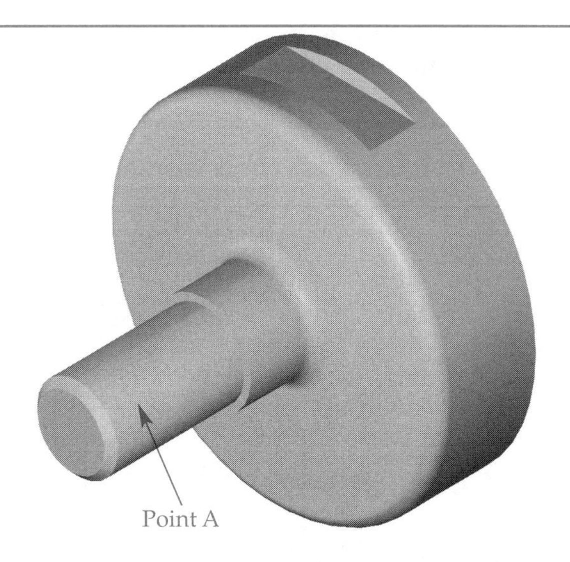

Figure 6-10.
Threads will be added to the shaft on this object.

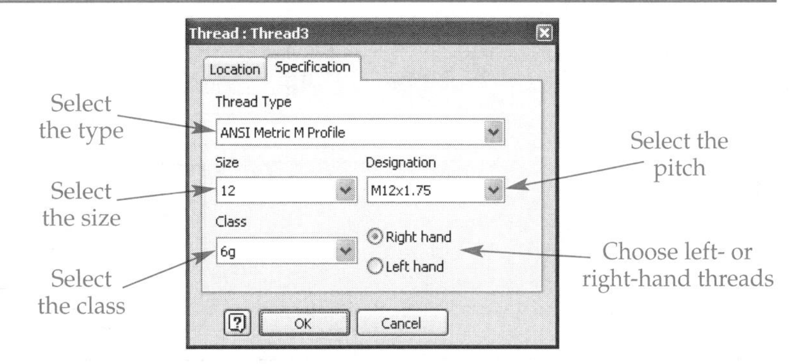

Figure 6-11.
Specifying threads using the **Thread** tool.

Figure 6-12.
An error is generated if you select a thread size that is not valid for the hole or shaft diameter.

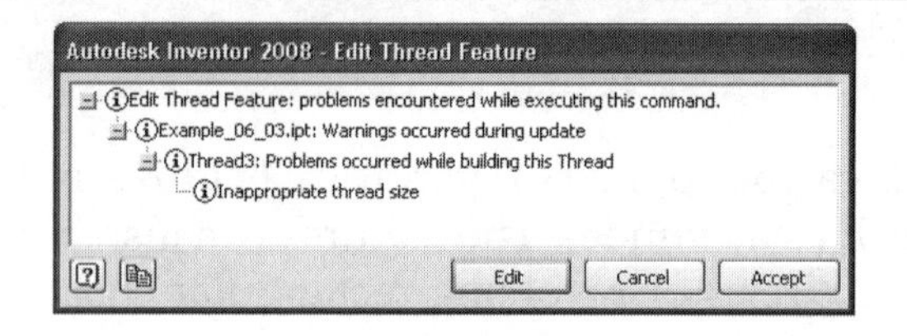

thread type, nominal size, pitch, class, and left- or right-hand threads. Use the drop-down lists to make the appropriate settings. Use the radio buttons to choose either left- or right-hand threads. However, be careful when changing the nominal size. This is a 12 mm diameter shaft and there are only two sizes of thread that can be cut on this shaft—12 mm and 14 mm. If you select any other size and pick the **OK** button, an error message is displayed, **Figure 6-12.** Pick the **Edit** button in the warning dialog box and change the thread specification as needed. When done changing the specifications, pick the **OK** button in the **Thread** dialog box to update the threads.

Now, rotate the view of the part to see the large diameter bore. On the bore, place 8 mm deep (length) threads that start at the outside face. In the **Specification** tab of the **Thread Feature** dialog box, select specifications of your choice. Then, pick the **OK** button to place the threads.

The bitmap representation of the threads approximates the coarseness of the threads. Left- and right-hand threads are also represented by the bitmaps. However, the threads will be properly displayed in the part drawing in the correct ANSI standard. See **Figure 6-13.** Creating part drawings is discussed in Chapter 8.

Figure 6-13.
Threads are represented in Inventor as a bitmap, however, the part drawing conforms to standards for representing threads.

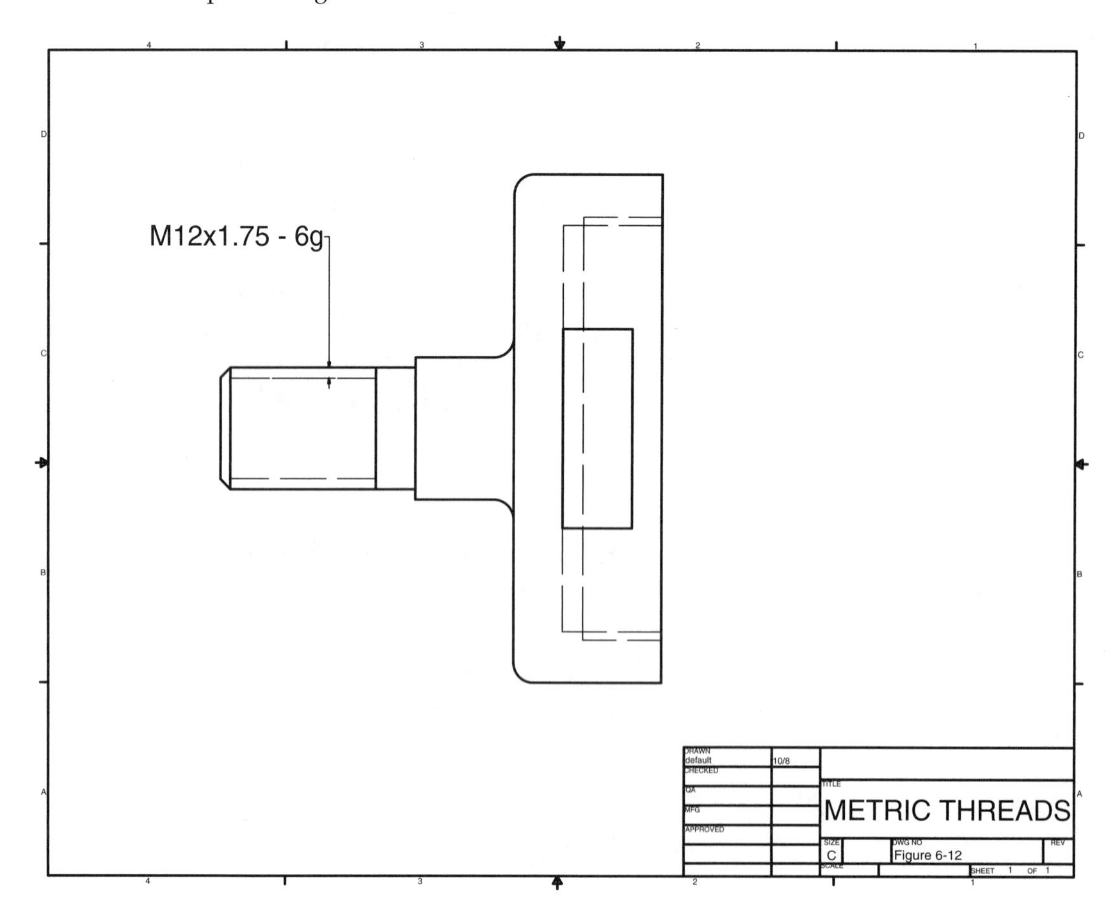

Fillets and Rounds

The **Fillet** tool puts a radius on any or all edges of a part, including intersections of part features. A radius on an inside edge adds material to the part and is called a ***fillet.*** A radius on an outside corner removes material and is called a ***round.*** See **Figure 6-14.** The terminology may be a little confusing when working in Inventor as the **Fillet** tool creates both fillets and rounds.

Applying Fillets and Rounds

Open Example_06_04.ipt. In the **Part Features** panel, pick the **Fillet** button to activate the **Fillet** tool and display the **Fillet** dialog box, **Figure 6-15.** The **Fillet** tool allows you to set several radii in the **Fillet** dialog box. You can change the fillet radius of an existing setting by selecting the value in the **Radius** column, typing the new value, and pressing [Enter]. To set a new fillet radius, pick the Click here to add entry. A new row is added and you can set the fillet radius. In this way, you can apply a variety of fillets and rounds with the same session of the **Fillet** tool. In this example, there is currently a .125″ fillet radius setting. Change this setting to .25″. Then, add another fillet radius setting and change the fillet radius to .5″.

Figure 6-14.
Fillets and rounds.

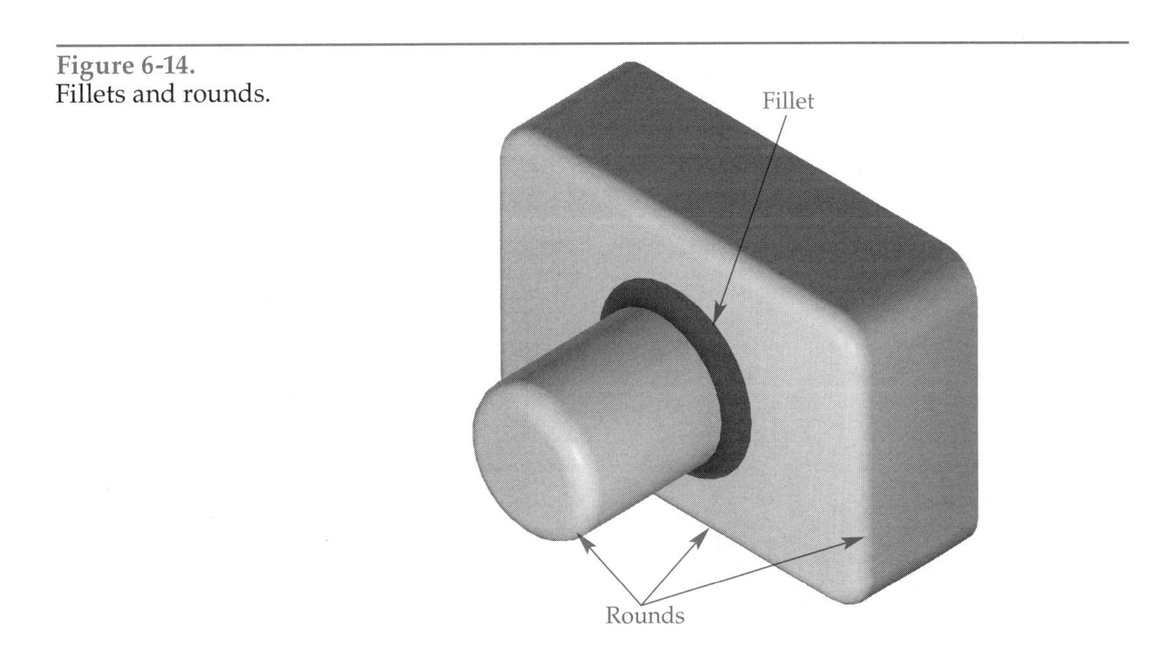

Figure 6-15.
Making settings for fillets and rounds.

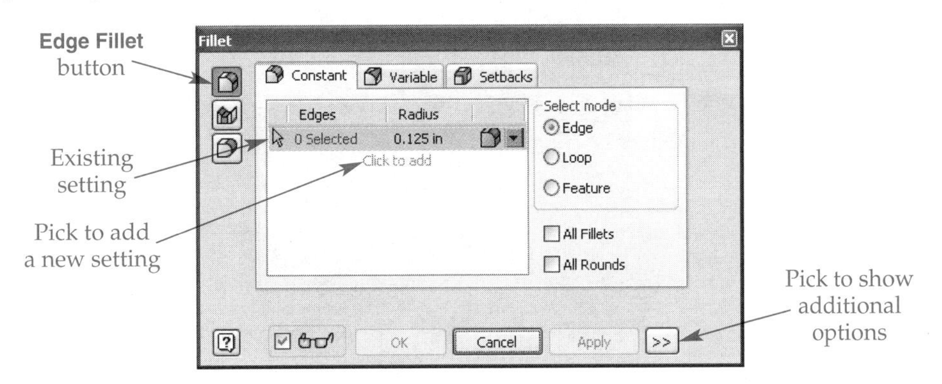

Notice that the **Edge** radio button in the **Select mode** area is on. This mode allows you to select a single edge. Also, notice in the **Edges** column that the two fillet settings all list 0 Selected. Currently, nothing is selected for any of the fillet settings. Pick the 0 Selected in the **Edges** column for the .5″ fillet setting. Then, pick the four short edges of the rectangular base. See **Figure 6-16.** As you select edges, arcs representing the round that will be applied are displayed at each edge.

Now, select the .25″ fillet radius setting in the **Fillet** dialog box and pick the **Loop** radio button in the **Select mode** area. Hover the cursor over edge A as indicated in **Figure 6-16A.** Notice that as you move the cursor, two different loops of edges are highlighted. Pick the edge when the larger front face is highlighted. After the edges are picked, look at the dialog box and verify the number of edges selected, **Figure 6-16B.** Pick **OK** and compare your part with **Figure 6-16C.**

To see how a "loop" selection works around sharp corners, edit the fillet feature you just created. In the **Fillet** dialog box, pick the .5″ radius. Now, hold down the [Shift] key and pick the four short edges of the part to remove them from the procedure. Pick **OK** to see the result.

The **Feature** radio button in the **Fillet** dialog box allows you to select a feature. Fillets and rounds will be applied to all edges on the feature, but not at intersections with other features. For example, open the **Fillet** dialog box. Add a 1″ radius fillet setting. Pick the **Feature** radio button in the **Select mode** area. Move the cursor over

Figure 6-16.
A—Pick this edge for the .25″ rounds. B—The **Fillet** dialog box. C—The completed part.

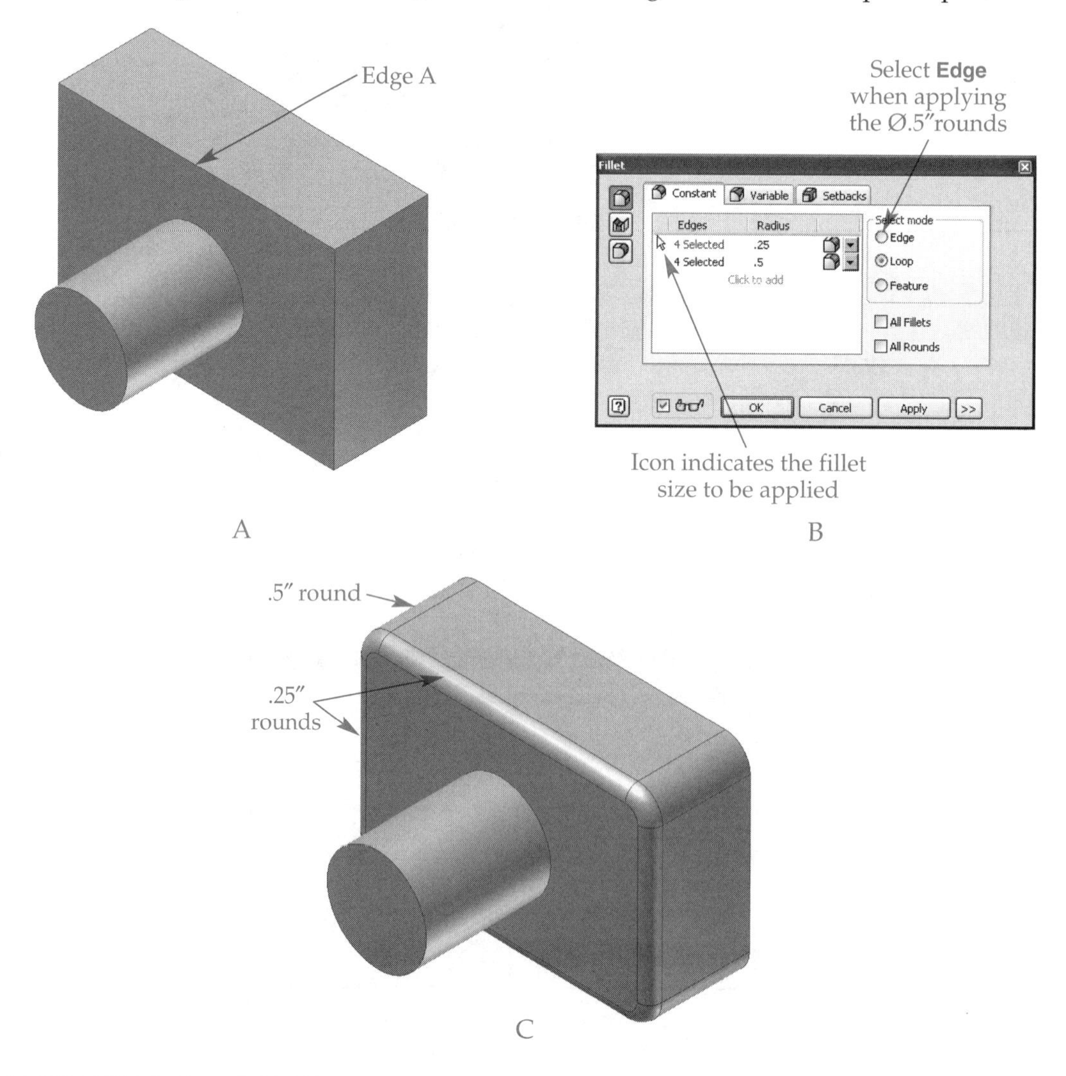

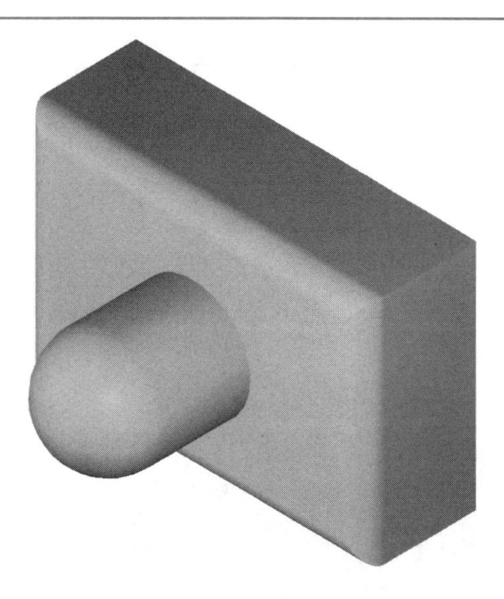

Figure 6-17.
A bullet nose round is applied to the shaft.

the part. When the cursor is over a feature, the edges of the feature that will be filleted or rounded are highlighted. Move the cursor over the post; only the circular edge on the left end is highlighted. This is because the edge on the other end is an intersection between features. Pick the post to select the feature. Then, pick the **OK** button in the **Fillet** dialog box to create the round. See **Figure 6-17.** Notice how the result is a spherical end on the cylinder. By setting the fillet radius equal to the radius of the cylinder, you can create a rounded pin. However, the fillet radius cannot be greater than the radius of the cylinder or an error is generated.

PROFESSIONAL TIP

You can change the display color of fillets and rounds for emphasis. This is done by right-clicking on the fillet name in the **Browser** and selecting **Properties** from the pop-up menu. Then, change the color in the **Feature Properties** dialog box. You can also rename the fillet in this dialog box.

Filleting and Rounding All Edges

You can have Inventor automatically apply fillets and rounds to all edges of the part. Open Example_06_05.ipt. Activate the **Fillet** tool. In the **Fillet** dialog box, change the fillet radius setting to .1″. Then, check the **All Fillets** check box. There are four inside edges on the part, so these are automatically selected. The fillets are also previewed in color on the part. Pick the **OK** button to place the fillets. See **Figure 6-18A.** Repeat the operation, this time checking the **All Rounds** check box. There are 33 external edges on the part, so these are automatically selected. The rounds are previewed on the part. Pick the **OK** button to place the rounds. See **Figure 6-18B.**

Now, there are two separate fillet operations listed in the **Browser**. Also, notice how all of the edges have been filleted and rounded, including the edges of the holes. However, holes are not normally filleted or rounded. This can be easily corrected by moving the fillet operations above the hole operations in the **Browser**. Pick one of the fillet names in the **Browser**, hold down the [Ctrl] key, and pick the other name. Both names are now selected. Drag the names up so they are above the Hole1 and Hole2 features in the part tree and release. See **Figure 6-19A.** Now, the two fillet operations are not applied to the holes, **Figure 6-19B.**

Figure 6-18.
A—Fillets are added using the **All Fillets** check box. B—Rounds are added using the **All Rounds** check box.

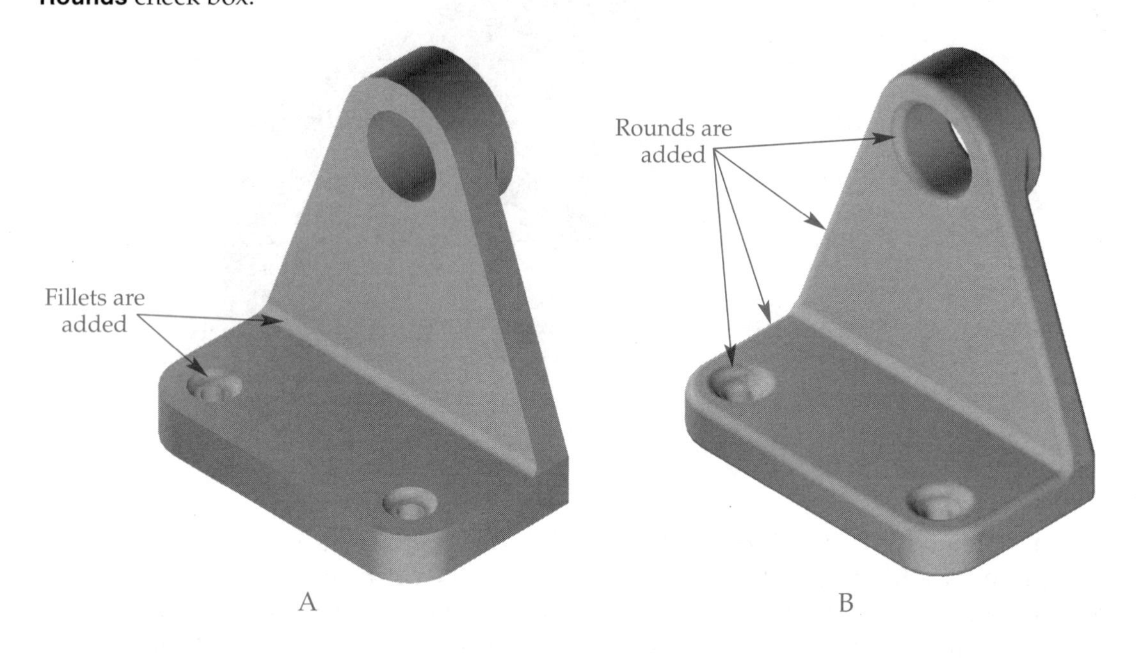

Figure 6-19.
A—Moving the fillets and rounds above the holes in the part tree. B—The holes are no longer filleted or rounded.

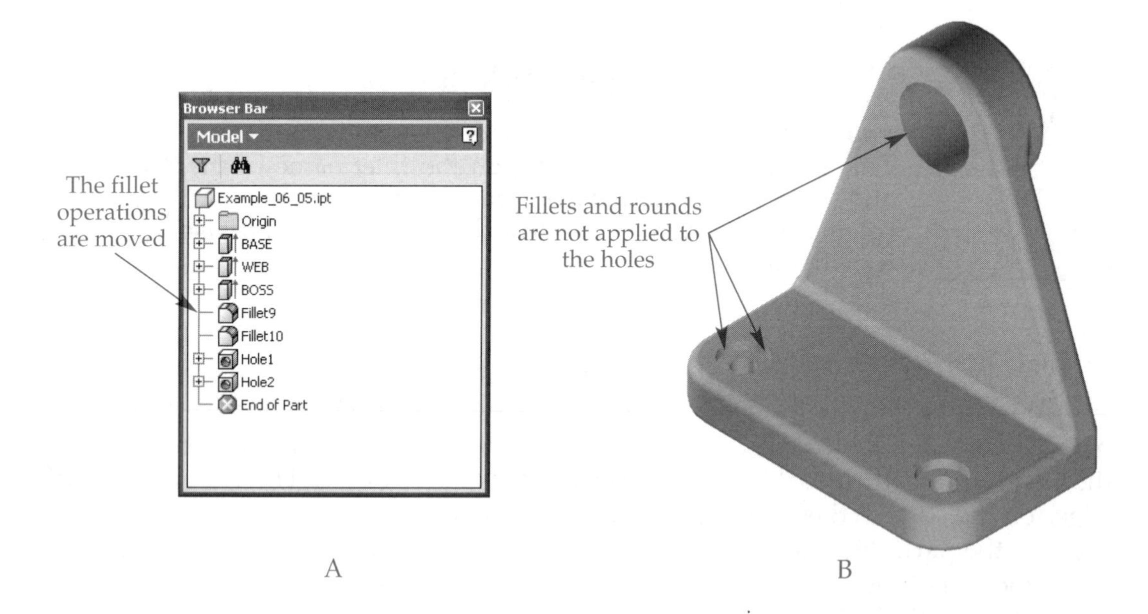

It is possible to check both **All Fillets** and **All Rounds** in the same operation. However, in this example, an error will result. The error message that is displayed refers to the inability of Inventor to perform operations with mixed convexity at a vertex. The two corners where the base meets the web will have both a concave fillet and convex rounds applied. Inventor cannot calculate the result in one operation, which is why fillets and rounds were applied in two steps for this example.

More Fillet Tool Options

You may have noticed the **>>** button at the bottom of the **Fillet** dialog box. Refer to **Figure 6-15.** Picking this button displays an expanded dialog box. See **Figure 6-20.** There are four options in the expanded portion of the dialog box—**Roll along sharp edges**, **Rolling ball where possible**, **Automatic Edge Chain**, and **Preserve All Features**.

The **Automatic Edge Chain** check box is normally checked. This results in all edges in a chain being selected by picking any edge in the chain. When unchecked, you can pick individual edges in the chain.

When the **Roll along sharp edges** check box is unchecked, a constant fillet radius is applied along the edge, even if this means extending adjacent faces. For example, open Example_06_06.ipt. Put a .75″ fillet on the intersection where the cylinder meets the base. See **Figure 6-21.** Notice how face C is extended to keep the constant radius. This is obvious in the end view. Now, edit the fillet feature and check the **Roll along sharp edges** check box. Notice how the fillet radius is now varied to keep the edge along face C a straight line. See **Figure 6-22.** This is done by varying the radius of the fillet.

Figure 6-20.
The expanded **Fillet** dialog box.

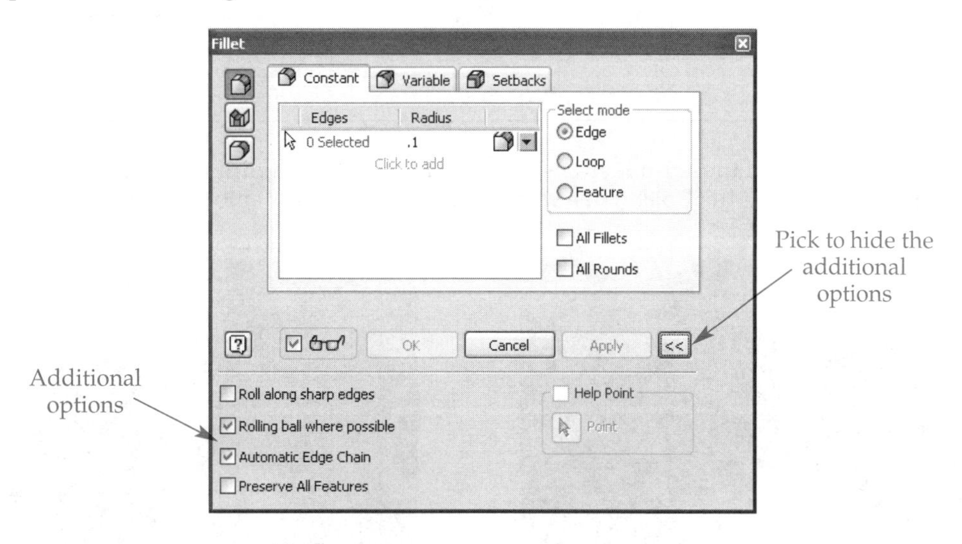

Figure 6-21.
A—Face C is extended when the **Roll along sharp edges** check box is unchecked. B—The end view of the part clearly shows how the face is extended.

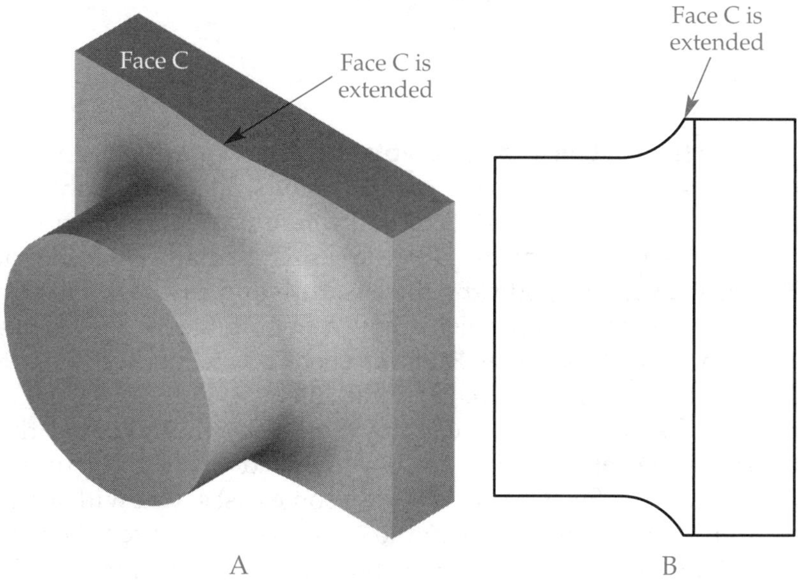

Face C

Face C is extended

Face C is extended

A

B

Figure 6-22.
A—Face C is not extended when the **Roll along sharp edges** check box is checked. B—The top view of the part clearly shows how the face is not extended.

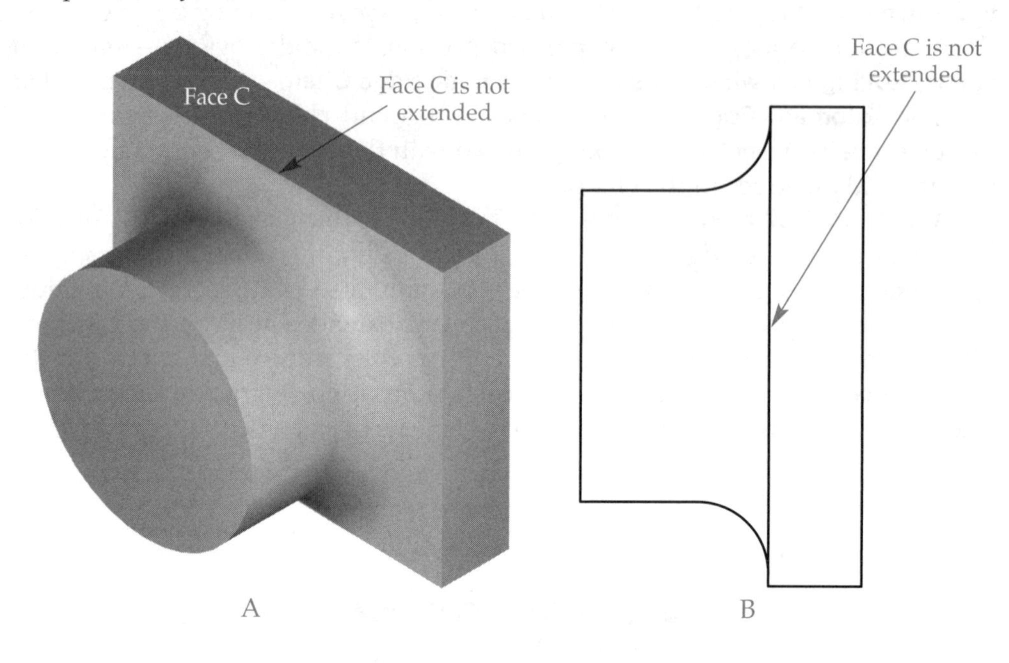

Figure 6-23.
A—The intersection of these three fillets appears as if a ball end mill was used to create the cavity. B—Unchecking the **Rolling ball where possible** check box smoothes the intersection.

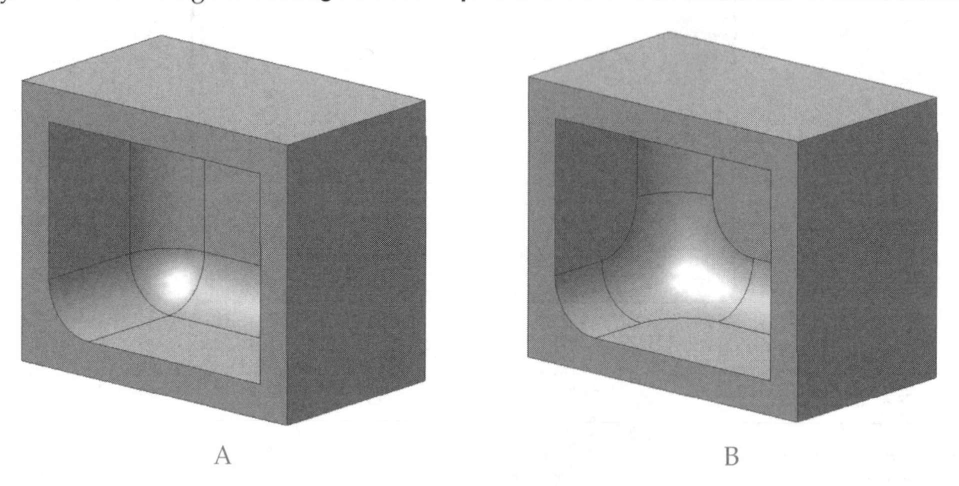

The **Rolling ball where possible** check box controls the intersection of inside edges. Open Example_06_07.ipt. Place a .75″ radius fillet on the three edges visible inside of the box. Make sure the **Rolling ball where possible** option is checked. The result of the operation at the inside corner looks as if a ball end mill machined the cavity. See **Figure 6-23A.** Now, edit the fillet feature and uncheck the option. The result of the operation at the inside corner is now a smooth blend. See **Figure 6-23B.**

When the **Preserve All Features** check box is checked, some features, such as holes, are retained when they intersect the fillet. Open Example_06_08.ipt. Place a 1″ radius round (fillet) on the short edge of the top by the hole and the post. Make sure the **Preserve All Features** check box is unchecked. When the round is applied, the hole and post disappear. See **Figure 6-24A.** In some cases, you will get an error message. Now, edit the fillet feature and check the **Preserve All Features** check box. When you pick **OK** to update the feature, the hole reappears, but not the post. See **Figure 6-24B.**

Figure 6-24.
A—When the **Preserve All Features** check box is unchecked, a fillet or round may remove some features. B—When the check box is checked, some features are retained. However, some features may still be removed by the fillet or round.

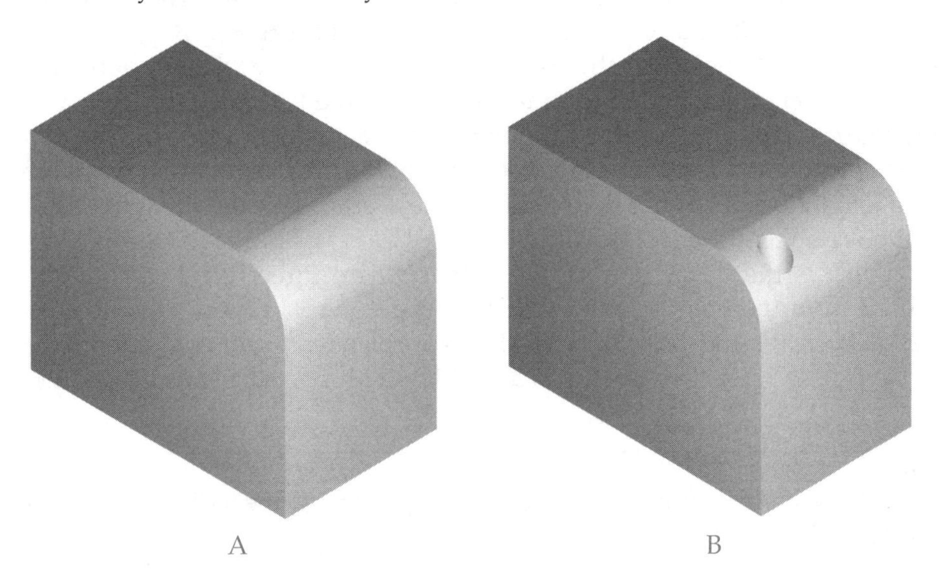

Variable-Radius Fillets and Setbacks

In some applications, the radius of a fillet must vary over the length of the fillet. This is called a ***variable-radius fillet.*** Open Example_06_09.ipt. Activate the **Fillet** tool and pick the **Variable** tab in the **Fillet** dialog box. Then, pick the long, front edge of the top face. The two endpoints are automatically selected as the Start and End points in the **Variable** tab. The point highlighted in the **Point** list in the tab is also highlighted on the part. The radius of the fillet at the highlighted point is given in the **Radius** column in the tab. Change the value by picking in the column cell and typing a new value.

Set the radius of the Start point to .1″ and the End point to 1.0″. This setting creates a smooth, variable-radius round (fillet) between the two endpoints. See **Figure 6-25A.**

Figure 6-25.
The radii are given at the transitions. A—This is a variable-radius round (fillet). B—This variable-radius round has multiple transitions.

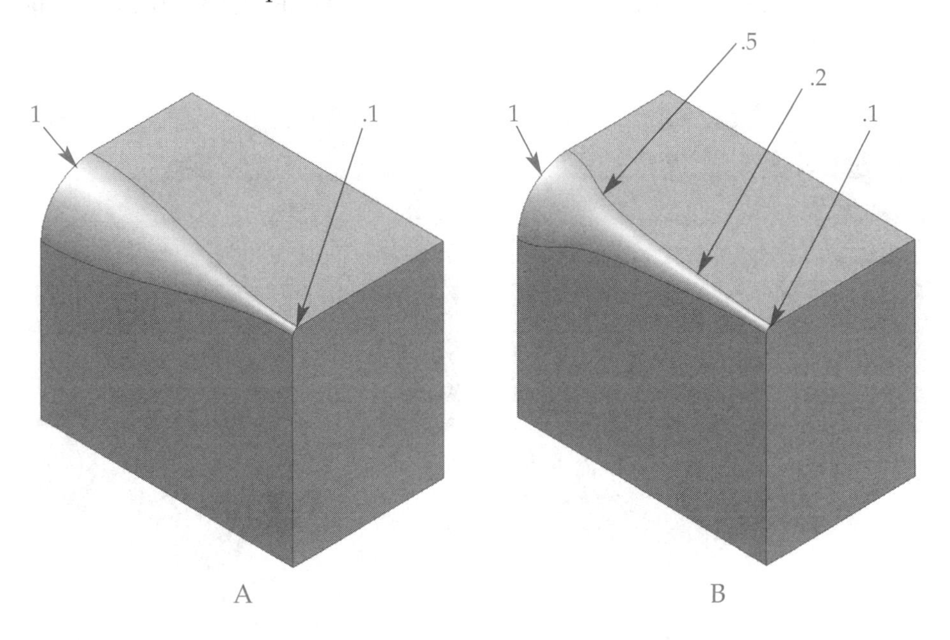

However, you can have multiple points along the length of the fillet, each with a different radius. To add points between the endpoints, simply move the cursor along the edge and pick where you want the points. For this example, add two points along the edge. Change the radius values of the two points so the fillet goes from .1, to .2, to .5, and to 1. Pick the **OK** button to apply the fillet. See **Figure 6-25B.**

Once any points between the Start and End points are picked, you can change their location along the fillet. Highlight the point in the **Variable** tab. Then, pick in the **Position** column and type a value. The value in this text box represents a percentage along the length of the fillet, where 1.00 equals 100%. You can move a point anywhere along the length, but it cannot be outside of the Start and End points, nor can it share the same position as either (1.00 and 0.00).

A ***setback*** is the distance from the corner where three fillets or rounds meet. It is measured from the point where the edges would meet to form a square corner. See **Figure 6-26.** By increasing the setback, the fillets are blended together to smooth the corner. Open Example_06_10.ipt. The block has a .5″ fillet with no setback on three intersecting

Figure 6-26.
The setback is the distance from the theoretical sharp corner to the radius. A—No setback. B—Setback of 1.00.

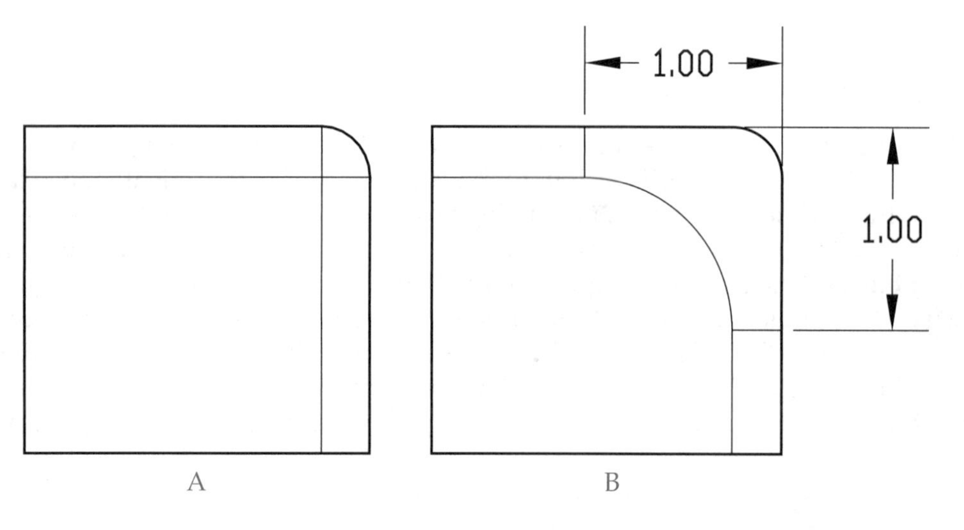

Figure 6-27.
A—The intersection of these three rounds is created without setbacks. B—Setbacks allow you to smooth the corner.

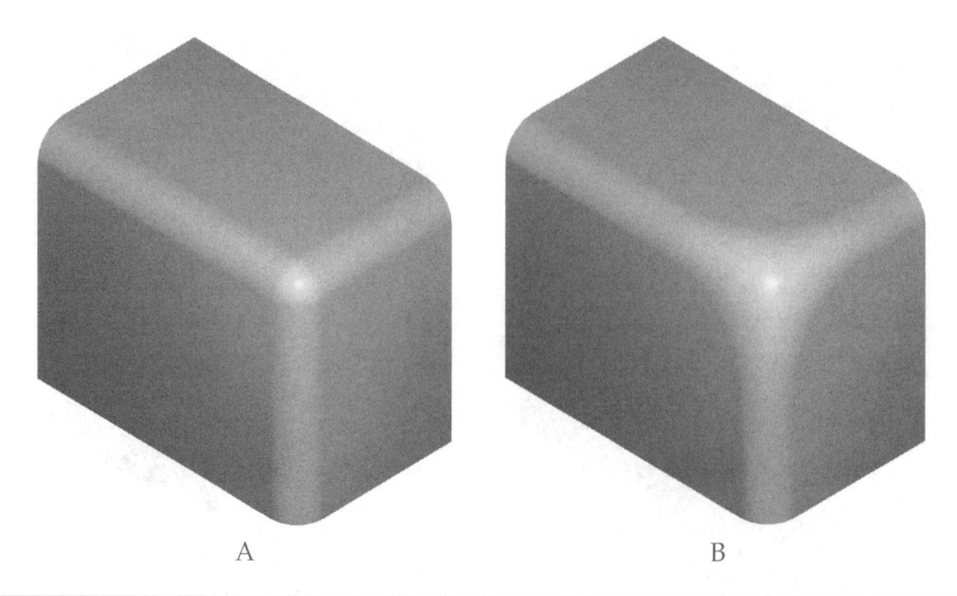

edges. See **Figure 6-27A.** Edit the fillet feature and select the **Setbacks** tab in the **Fillet** dialog box. As you move the cursor over the part, a yellow circle appears on valid vertices. Since there is only one vertex where fillets intersect, there is only one valid vertex. Pick the corner where the three fillets intersect. In the **Setback** tab, three edges are listed in the **Edge** column, each with a default setback setting of .25″. Pick in the **Setback** column for each edge and change the value to 2″. See **Figure 6-28.** Then, pick the **OK** button to update the rounds (fillets). The corner is smoothed, as shown in **Figure 6-27B.**

Full Round Fillets

In the upper-left corner of the **Fillet** dialog box are three buttons that determine if the fillet or round is applied to an edge, face, or is a full round. To this point, you have applied edge fillets and rounds with the **Edge Fillet** button on, which is the default. To apply full rounds, pick the **Full Round Fillet** button. This option requires three faces. It can be used for features such as putting a round on top of a rib.

Open the file Example_06_11.ipt. This part has a vertically tapered rib on the left side of the boss and a horizontally tapered rib on the right side of the boss. Start with the rib on the left. Pick the **Fillet** button to open the **Fillet** dialog box. Then, select the **Full Round Fillet** button and continue as follows.

1. With the **Side Face Set 1** on in the **Fillet** dialog box, pick the face on the part that is green.
2. With the **Center Set Face** button on, pick the red face on top of the rib.
3. With the **Side Face Set 2** button on, rotate the view of the part and pick the yellow face.
4. Pick the **Apply** button to create the full round fillet and leave the **Fillet** dialog box open.

The center face is replaced with a round with its radius equal to one-half of the width of the face. The effect creates a fully rounded edge on the feature. See **Figure 6-29.** Repeat the process for the rib on the right side and the center face is replaced by a variable-radius round. The edge on the feature is fully rounded, but the radius varies from one end to the other.

Figure 6-28.
Adding setbacks to the intersection of three fillets or rounds.

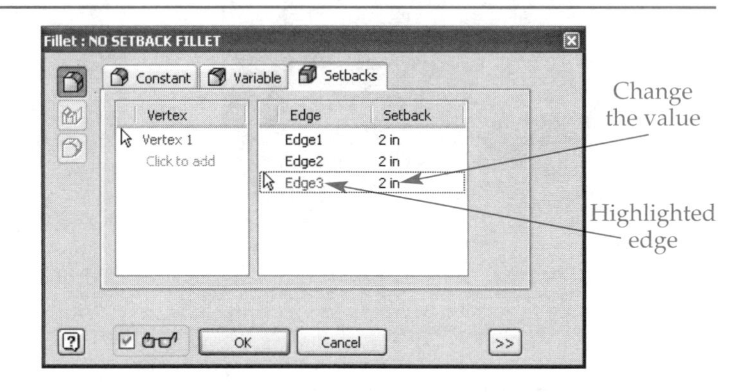

Figure 6-29.
A—The rib before the **Fillet** tool is used. B—A full round is applied to the rib.

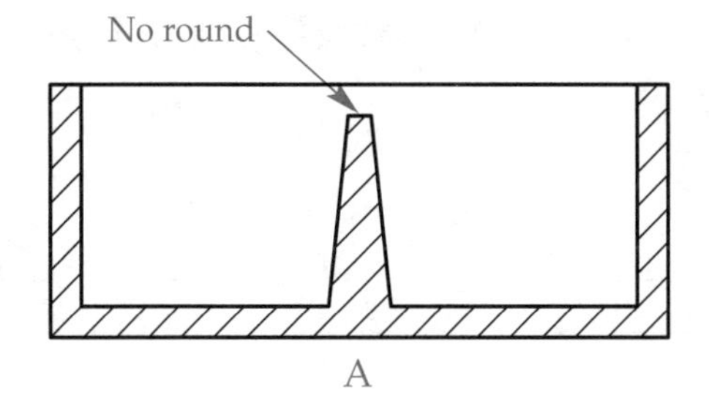

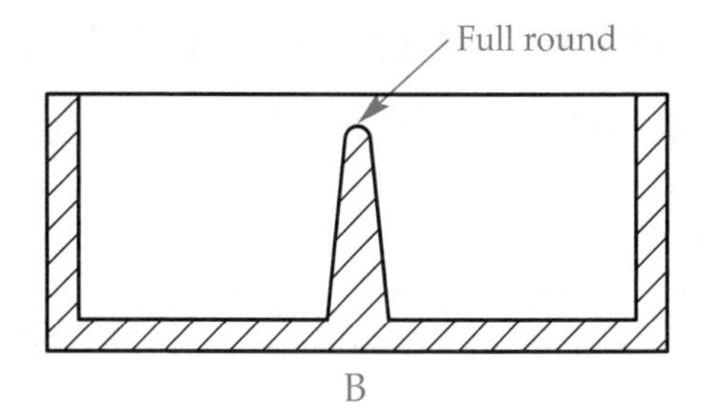

Adding Chamfers

A *chamfer* is a bevel—a flat sloping face—on the edge between two intersecting faces of the part. The faces can be flat or curved. Like fillets and rounds, chamfers add material to inside edges and remove material from outside edges. **Figure 6-30** shows chamfers applied to three types of edges.

To create a chamfer, pick the **Chamfer** button in the **Part Features** panel. The **Chamfer** dialog box is displayed. There are three methods for defining the chamfer—two equal distances, a distance and an angle, and two unequal distances. The three buttons on the left side of the **Chamfer** dialog box allow you to choose between these methods. See **Figure 6-31.**

- **Distance button.** The same distance from the edge is used on each face. This creates a 45° chamfer. With this option, specify the distance and pick the edge to chamfer.
- **Distance and Angle button.** A chamfer is created a specified distance from the edge on one face and at a specified angle to the second face. With this option, specify the distance and angle, then pick the face from which to measure the angle and the edge to chamfer. The distance is applied to the selected face.
- **Two Distances button.** Allows a different distance from the edge on each face. With this option, specify the two distances and pick the edge. Inventor automatically chooses the faces to which the distances are applied, but you can flip the order.

Figure 6-30.
A—The unchamfered part. B—Three different chamfers are applied to the part.

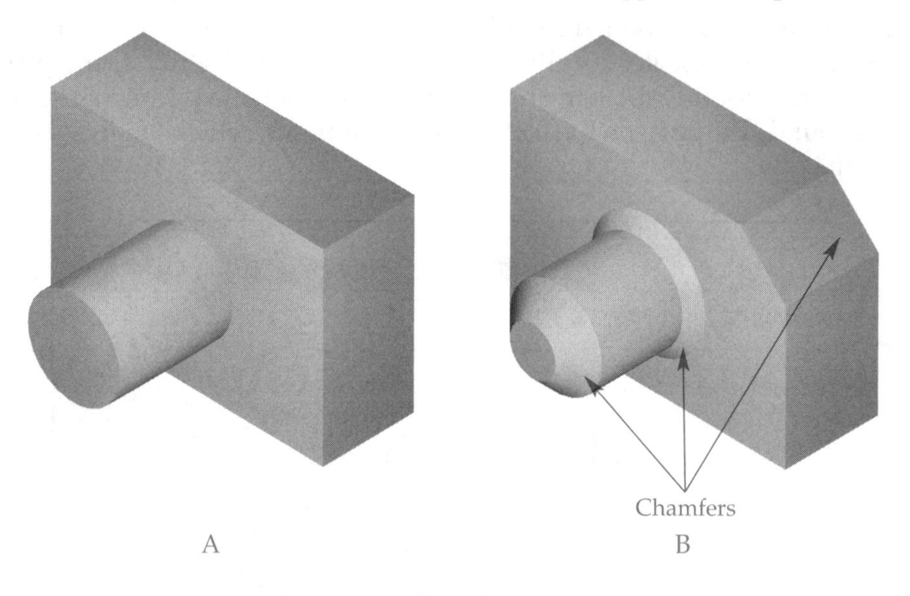

Figure 6-31.
There are three methods for defining a chamfer. A—**Distance**. B—**Distance and Angle**. C—**Two Distances**.

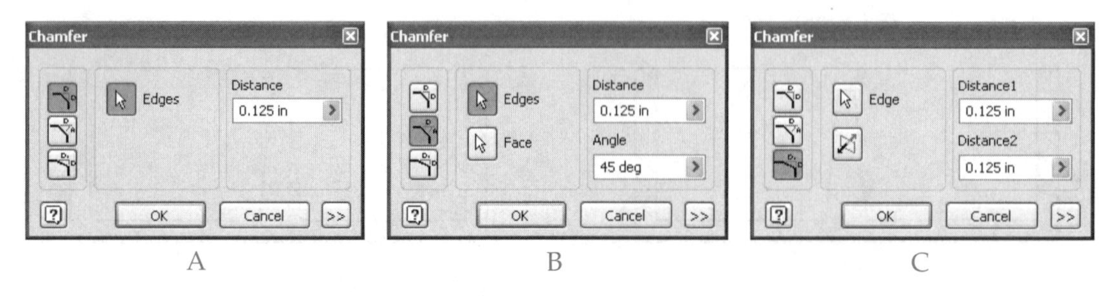

Open Example_06_12.ipt. Pick the **Chamfer** button in the **Part Features** panel. Then, continue as follows.

1. In the **Chamfer** dialog box, pick the **Distance** button.
2. Pick in the **Distance** text box and change the distance to .5″.
3. Pick the **Edges** button. Then, pick the edge at the end of the large cylinder (not the intersection with the base). A preview of the chamfer appears on the part as red lines.
4. In the **Chamfer** dialog box, pick the **OK** button to place the chamfer.
5. Pick the **Chamfer** button in the **Part Features** panel. Note: You can apply multiple chamfers of the same type in one session of the **Chamfer** tool, but you cannot apply multiple chamfers of different types.
6. In the **Chamfer** dialog box, pick the **Distance and Angle** button.
7. Set the distance to .3″ and the angle to 30°.
8. Pick the **Face** button in the **Chamfer** dialog box. Then, pick the face on the base that intersects the large cylinder.
9. The **Edge** button in the **Chamfer** dialog box should be automatically selected. If not, pick it. Then, pick the intersecting edge between the cylinder and the base. A preview of the chamfer appears on the part in red.
10. In the **Chamfer** dialog box, pick the **OK** button to place the chamfer.
11. Pick the **Chamfer** button in the **Part Features** panel.
12. In the **Chamfer** dialog box, pick the **Two Distances** button.
13. Change the value in the **Distance1** text box to .5″ and the value in the **Distance2** text box to .75″.
14. Pick the **Edge** button in the **Chamfer** dialog box and then pick the short edge just above the small cylinder.
15. In the **Chamfer** dialog box, pick the flip button so that the .75″ distance is on the vertical face.
16. Pick the **OK** button to place the chamfer. Note how the knob extends back into the chamfer. See **Figure 6-32A.**

Now, edit the last chamfer by right-clicking on its name in the **Browser** and selecting **Edit Feature** from the pop-up menu. In the **Chamfer** dialog box, change the .75″ distance to 1.5″. This will extend the chamfer down below the knob. Pick the **OK** button to update the feature. The knob disappears. See **Figure 6-32B.**

Figure 6-32.
A—The chamfer is placed and the knob is retained. B—Increasing the chamfer distance on the vertical face removes the knob.

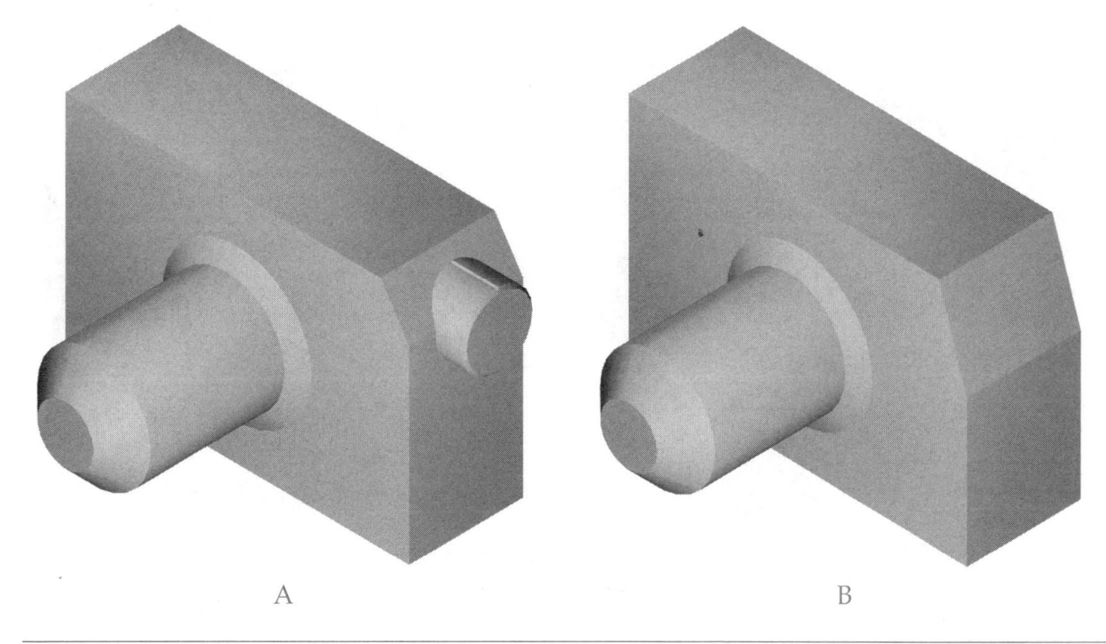

When using the **Chamfer** tool, you can select a chain or loop of edges. However, unlike fillets, they cannot have sharp corners. Also, the corner radii must be tangent to the straight edges. Pick the **>>** button in the **Chamfer** dialog box to show the expanded options. See **Figure 6-33.** There are two buttons in the **Edge Chain** area that determine if a chain of edges or a single edge is selected. Chain is the default, which has a tooltip of All tangentially connected edges.

Open Example_06_13.ipt. Face A has fillets on each corner; in other words, the face has no sharp corners. See **Figure 6-34A.** Face B, on the other hand, has three sharp corners and one smooth corner. Activate the **Chamfer** tool and expand the dialog box. Make sure the chain button is on. Then, pick the **Distance** button and set the distance to 5 mm. Pick the **Edges** button in the dialog box and move the cursor over an edge of face A and face B. Notice how all edges of face A are highlighted. However, as you move the cursor over face B, the two edges that are connected by the fillet and the other two edges are independently highlighted. If the **Single edge** button in the **Edge Chain** area of the **Chamfer** dialog box is on, only a single edge is selected. Pick the highlighted edges on face A. Also, pick the two edges on face B that are connected by the fillet. Then, pick the **OK** button to place the chamfers. See **Figure 6-34B.**

Figure 6-33.
You can determine if selecting an edge selects the entire chain or just the edge.

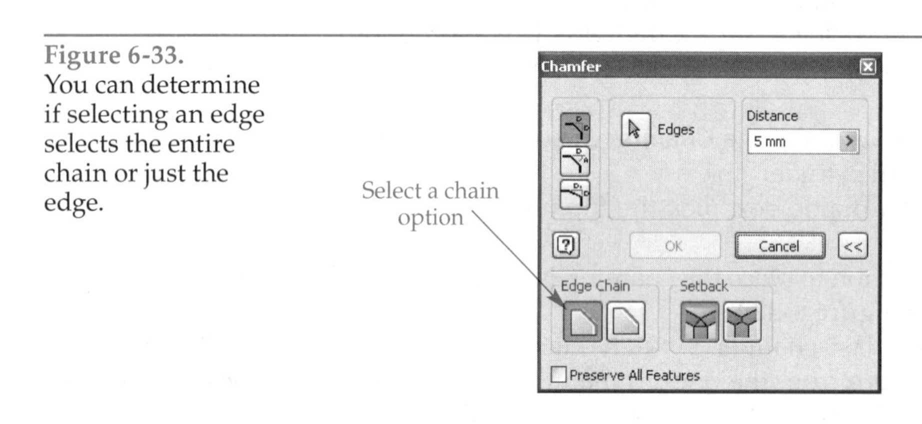

Figure 6-34.
A—The edge of face A does not have any sharp corners; the edge of face B does.
B—The chain for a chamfer stops at sharp corners, which results in the chamfer on face B.

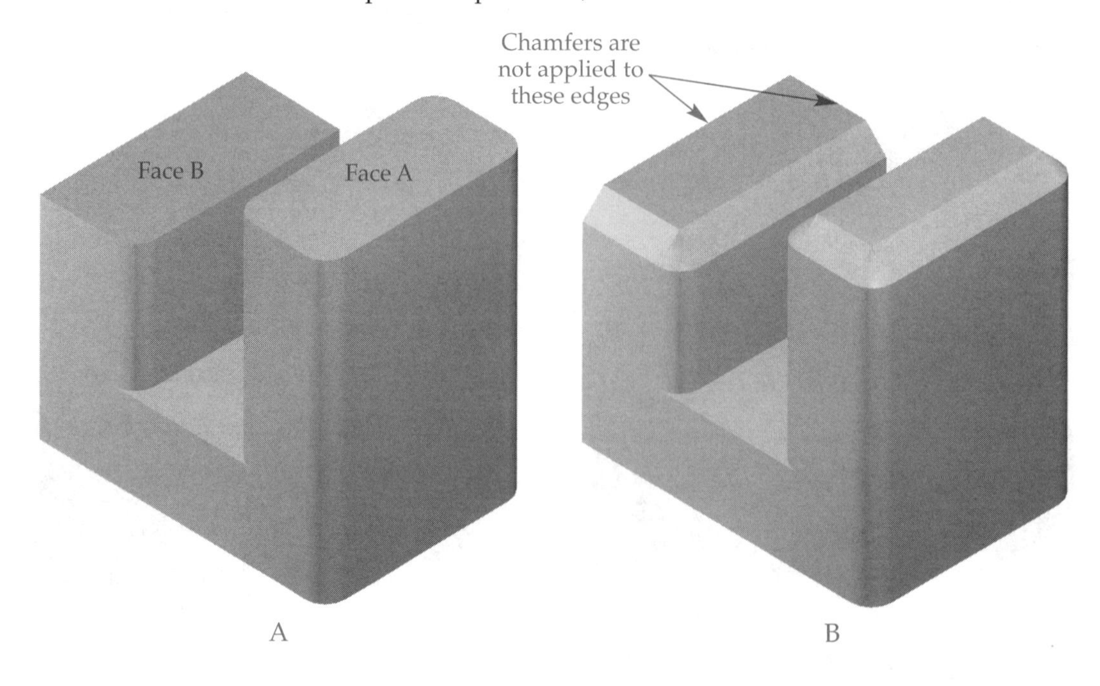

Rectangular and Circular Patterns

Patterns, or arrays, of features or sketches can be created in Inventor. The process is very similar for both features and sketches. There are two basic types of patterns—rectangular and circular. A ***rectangular pattern*** is an arrangement in rows and columns. A ***circular pattern*** is an arrangement about a center point or axis. Feature patterns are discussed first. Once you understand rectangular and circular patterns for features, you will be able to create sketch patterns with ease.

Feature Patterns

Open Example_06_14.ipt. This is a metric part with a circular sketch that has been extruded through the part to create a cutout (hole). The color of the cutout feature has been changed to blue. In the **Part Features** panel, pick the **Rectangular Pattern** button. The **Rectangular Pattern** dialog box is opened. See **Figure 6-35.**

With the **Feature** button in the dialog box active, select the blue extrusion as the feature. You can select it in the **Browser** or on the part. With the feature selected, you need to define the pattern by setting directions and values for the columns and rows.

Pick the arrow button in the **Direction 1** area of the dialog box. Then, pick near point B on edge AB. A green arrow appears on the part indicating the direction of the pattern. Next, enter 4 in the top text box in the **Direction 1** area. This is the number of columns. The tooltip for this text box is Column Count = and the current value. Also, enter 25 mm for the spacing in the middle text box. The tooltip for this text box is Column Spacing = and the current value.

In the **Direction 2** area, pick the arrow button. Then, pick near point C on edge AC. The green arrow on the part that indicates the direction of the pattern should point toward point C. If not, pick the path **Flip** button in the **Direction 2** area. Next, enter 3 for the row count and 20 mm for the spacing.

As you make changes in the **Rectangular Pattern** dialog box, a preview of the pattern is shown in green on the part. Once the pattern is defined, pick the **OK** button in the dialog box. The selected feature is arrayed in the defined pattern.

If you expand the Rectangular Pattern branch in the **Browser**, each of the holes you created is listed as an "occurrence." Right-click on the fourth one down. The pop-up menu gives you a choice of suppressing individual occurrences in the pattern.

Edges, axes, or paths can be selected for the directions. The edges can be straight or curved; the paths can be lines, arcs, or splines. Even though the **Rectangular Pattern** tool is used, these selections can generate patterns that are not rectangular. Open Example_06_15.ipt. This part is similar to the one in the previous example. However, it contains

Figure 6-35.
Defining a rectangular pattern.

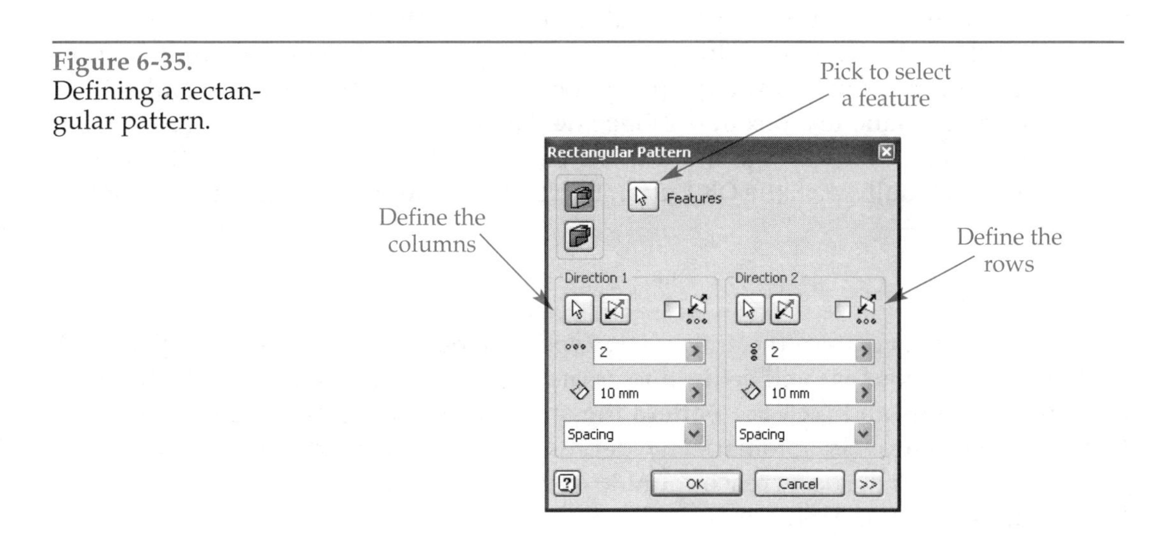

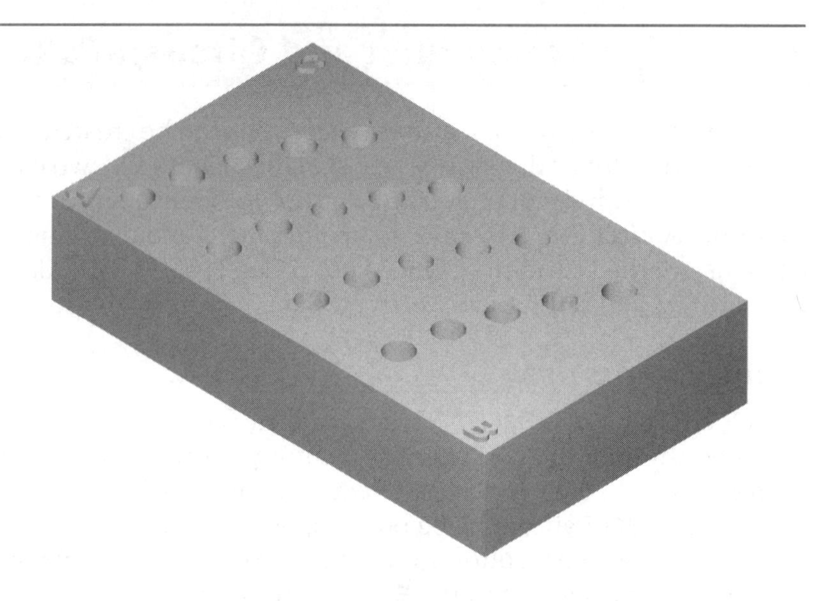

Figure 6-36.
Using a curved line as one path creates a different pattern.

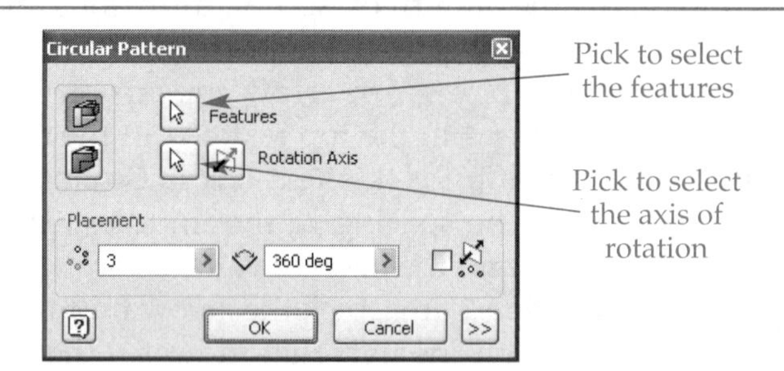

Figure 6-37.
Defining a circular pattern.

an unconsumed sketch. The curve is a spline that is part of the unconsumed sketch. Repeat the procedure in the previous example, except pick the curve for **Direction 1**, enter 5 for the number of columns, and enter 10 mm for the column spacing. Also, select edge AB for **Direction 2**, enter 4 for the number of columns, and enter 20 mm for the column spacing. See **Figure 6-36** for the result.

To create a circular pattern, you must select an axis of rotation about which the feature is arrayed. This axis can be a work axis or a circular feature. Open Example_06_16.ipt. The ear contains an extrusion, hole, and two fillets. All features that compose the ear will be arrayed about the center of the main shaft. Pick the **Circular Pattern** button in the **Part Features** panel. The **Circular Pattern** dialog box is displayed, **Figure 6-37.** With the **Features** button active, select all four features that compose the ear. Next, pick the **Rotation Axis** button in the dialog box. Since the three components of the main shaft share the same centerline, you can select any one of these features to define the axis. A preview of the array appears on the part. In the **Circular Pattern** dialog box, enter 3 in the left-hand text box in the **Placement** area to create three items. Also, enter 360 in the right-hand text box to indicate the pattern will be rotated around the entire circumference. Finally, pick the **OK** button to create the pattern. See **Figure 6-38** for the result.

Pattern the Entire Part

It is possible to pattern the entire part. Open the part Example_06_17.ipt. This part contains a work axis that will be used to create the circular pattern. Open the **Circular Pattern** dialog box and pick the **Pattern the entire solid** button. See **Figure 6-39.** The **Rotation Axis** button is automatically depressed; pick the work axis. Now, in the **Placement** area, set the occurrence count to 4 and the occurrence angle to 360 degrees. Pick **OK** to complete the pattern.

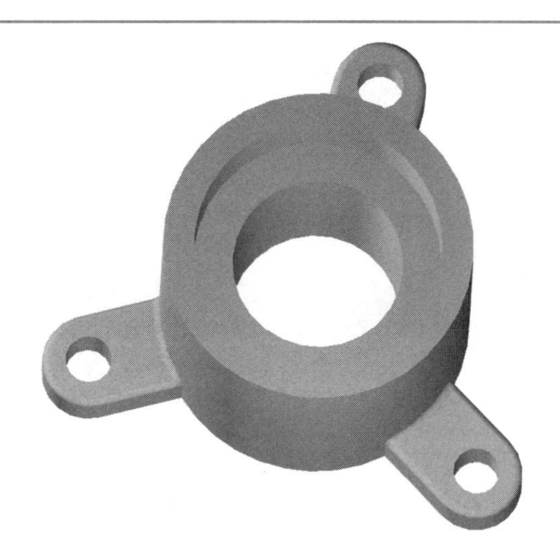

Figure 6-38.
First, a single ear was created on this part. A circular pattern was then created to place the other two ears.

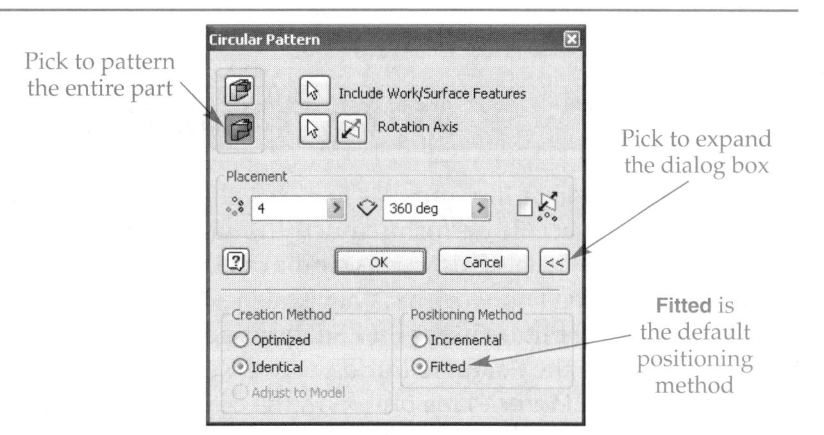

Figure 6-39.
Expand the **Circular Pattern** dialog box to modify the positioning method.

Edit the circular pattern you just created. Expand the **Circular Pattern** dialog box by picking the **>>** button. Within the **Positioning Method** area are two options—**Incremental** and **Fitted**. The default option, **Fitted**, equally spaces the occurrences of the solid throughout the occurrence angle, which is 360 degrees in this case. Pick the **Incremental** option and change the occurrence angle to 45 degrees. There are still four occurrences, but they are not spaced *within* the 45 degree angle. Instead, the solids are 45 degrees apart.

Sketch Patterns

You can also create patterns of sketches. Open Example_06_18.ipt. This part is a hinge plate with an unconsumed sketch. Edit the unconsumed sketch. Then, continue as follows.

1. Activate the **Rectangular Pattern** tool.
2. Pick all four edges of the small rectangle as the geometry.
3. Pick the **Direction 1** button and then select the edge that passes through the rectangle. Flip the direction if needed so the green arrow points to the opposite end of the large rectangle.
4. Set the number of occurrences to 4 and leave the distance as 1″.
5. Do not make any settings for **Direction 2**. In this way, only one row is created.
6. Pick the **OK** button to create the pattern.
7. Finish the sketch.
8. Pick the **Extrude** button in the **Part Features** panel. Extrude all four rectangles in the pattern using the **All** and Cut options.
9. The result is shown in **Figure 6-40.**

Figure 6-40.
A rectangle was sketched and then a rectangular pattern was created. The four profiles were then extruded to cut the part.

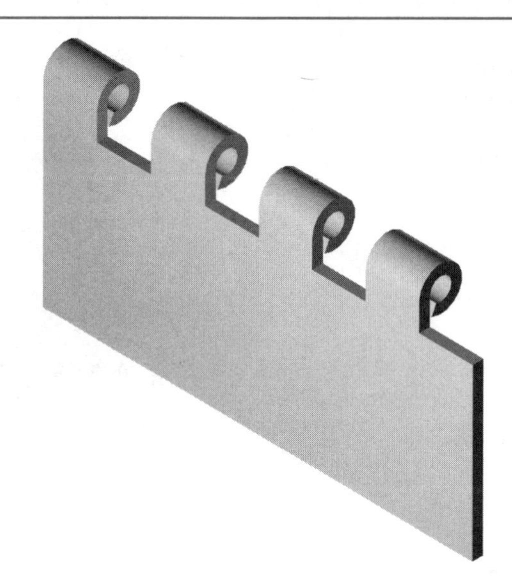

Mirror Feature Tool

The feature **Mirror** tool mirrors features about a work plane, straight edge on the part, or a planar face. It performs much the same function as the sketch **Mirror** tool. Open Example_06_19.ipt. This part is similar to the one used to create a circular pattern, except a work plane and work axis have been added.

Pick the **Mirror** button in the **Part Features** panel. The **Mirror** dialog box is displayed, **Figure 6-41.** With the **Features** button active, select all four features that compose the ear. Next, pick the **Mirror Plane** button in the dialog box. Select the work plane. Finally, pick the **OK** button in the dialog box to mirror the features. See **Figure 6-42.**

Figure 6-41.
The **Mirror Pattern** dialog box.

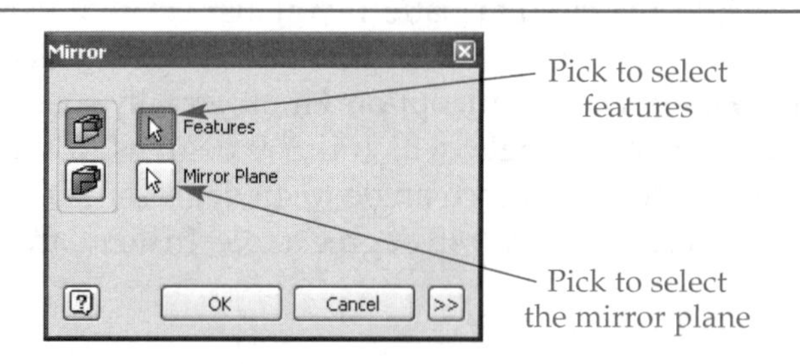

Figure 6-42.
All features in one ear were mirrored to create the second ear.

Chapter Test

Answer the following questions on a separate sheet of paper or complete the electronic chapter test on the Student CD.

1. List four advantages of creating holes using the **Hole** tool over extruding circles.
2. What are *hole centers?*
3. Which tool is used to create the object in Question 2 and where is it located?
4. What are the three termination options for a hole?
5. What are the two options for the bottom of a hole?
6. What is the purpose of a counterbore?
7. Which tool is used to create a counterbore?
8. How can you place threads in a cutout (hole) created by extruding a circle?
9. Define *fillet.*
10. Define *round.*
11. Which tool is used to create a fillet?
12. Which tool is used to create a round?
13. How can you smooth the corner where three rounds intersect?
14. What is a variable-radius fillet?
15. Define *chamfer.*
16. What are the three methods for defining a chamfer?
17. In Inventor, what is the difference between a loop for a fillet and a chamfer?
18. What are the two basic types of patterns (arrays) that can be applied to features and sketches?
19. Briefly, how can you create a pattern that has one row and four columns?
20. Which tool is used to mirror selected features of a part?

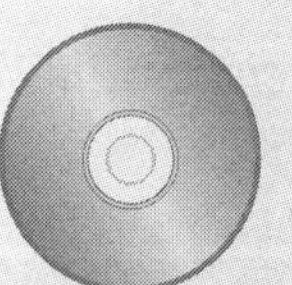

Chapter Exercises

Exercise 6-1. Sheet Metal Part.

Complete the exercise on the Student CD

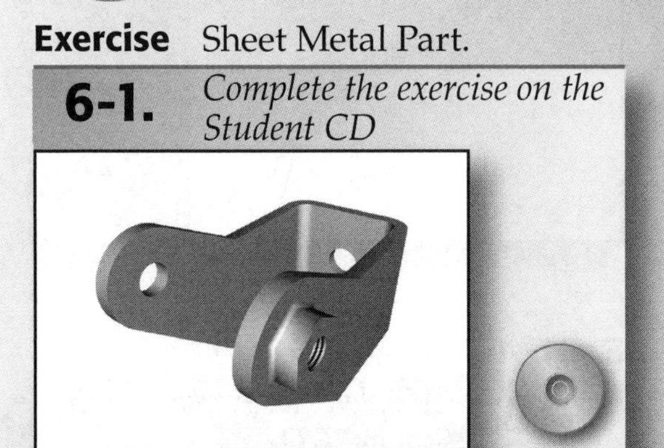

Exercise 6-2. Mounting Bracket.

Complete the exercise on the Student CD

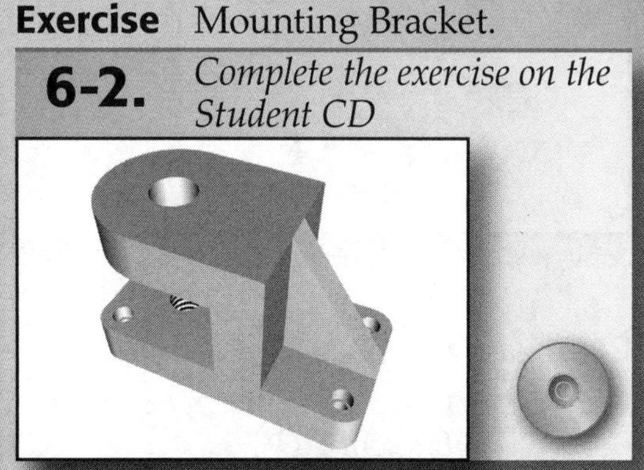

Exercise 6-3. Casting.

Complete the exercise on the Student CD

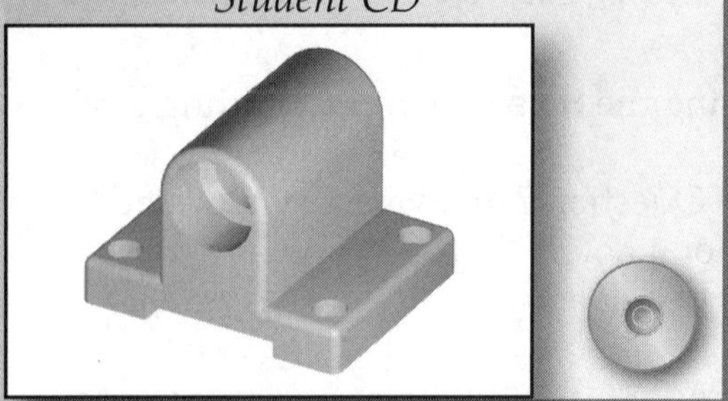

Exercise 6-4. Mounting Bracket Casting.

Complete the exercise on the Student CD

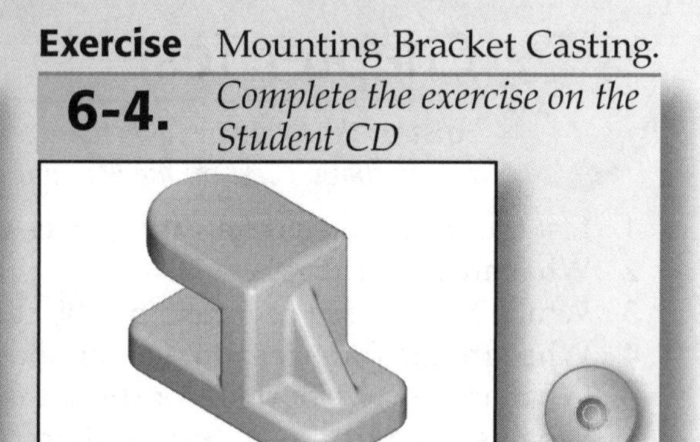

Exercise 6-5. Pin.

Complete the exercise on the Student CD

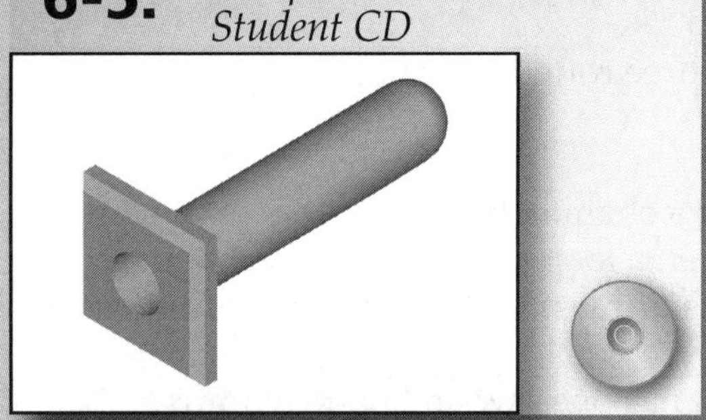

Exercise 6-6. Windshield Wiper Arm.

Complete the exercise on the Student CD

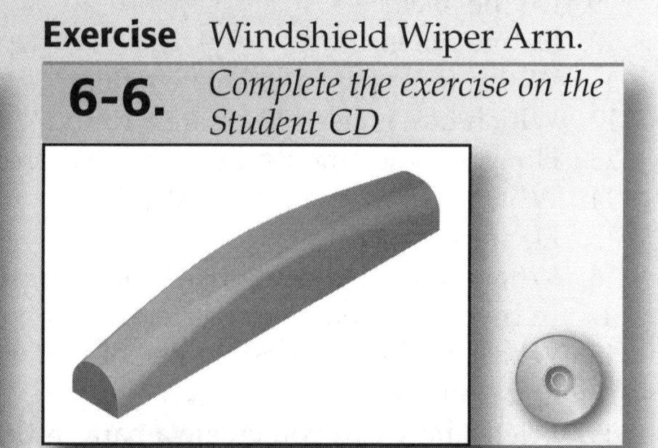

Exercise 6-7. Circular Flange.

Complete the exercise on the Student CD

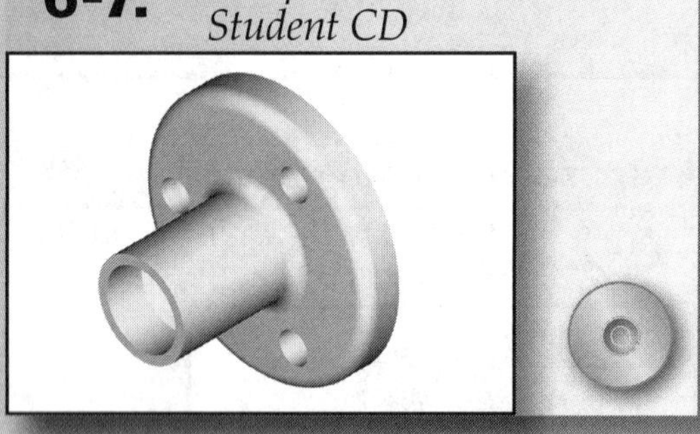

Exercise 6-8. Caulking Gun Rod.

Complete the exercise on the Student CD

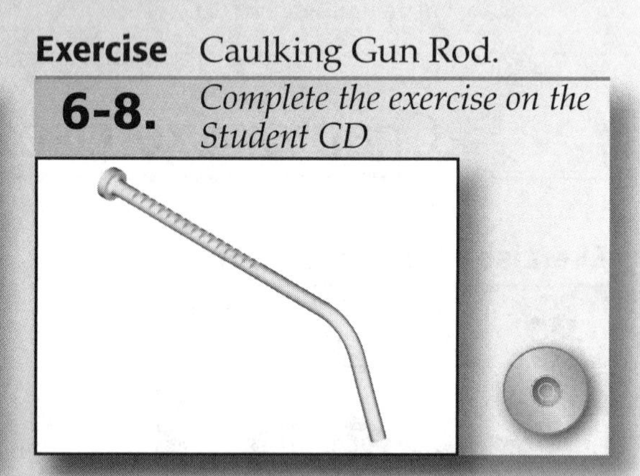

Exercise 6-9. Coupling.

Complete the exercise on the Student CD

Exercise 6-10. Flanged Yoke.

Complete the exercise on the Student CD

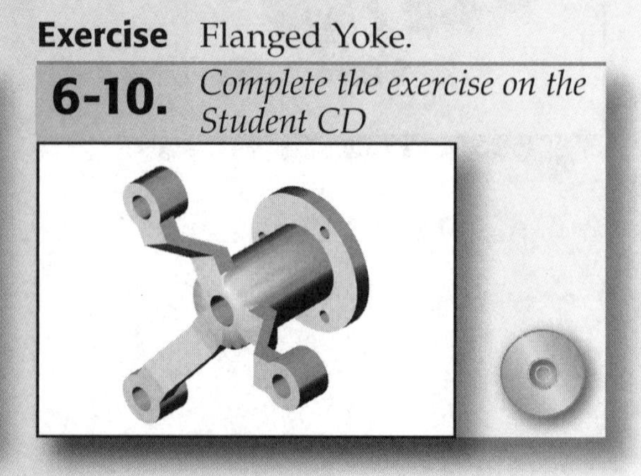

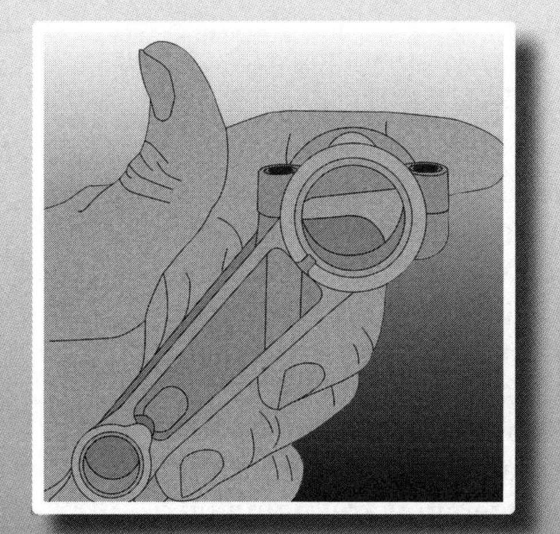

Chapter 7

Adding More Features

Objectives

After completing this chapter, you will be able to:

- Create shelled parts.
- Add ribs and webs to parts.
- Create embossed and engraved parts.
- Add decals to parts.
- Create face drafts on parts.
- Bend parts.
- Create splits.

User's Files

The Student CD included with this text contains several files required for this chapter. Refer to the file File List.txt *in the* \Ch07 *folder for the comprehensive list.*

Shell Tool

The **Shell** tool hollows out parts, turning them into thin-walled parts typical of die cast metal or injection molded plastic parts. The overall thickness of the wall is specified. In addition, individual faces can have unique thicknesses. Part faces, both flat and curved, can be removed by the shell operation to create open boxes.

Open Example_07_01.ipt. This part has eight faces, **Figure 7-1.** Four of these faces will be removed during the shell operation. First, pick the **Shell** button on the **Part Features** panel. This opens the **Shell** dialog box, **Figure 7-2.** Make sure the **Remove Faces** button in the dialog box is on (depressed). Then, move the cursor over the part. As you move the cursor, the face that will be selected is outlined. Move the cursor over face A, as indicated in **Figure 7-1,** and pick to select it. The face is highlighted in color. Then, in the **Shell** dialog box, pick the **OK** button to accept the default shell thickness of .1″. A cavity is created inside of the part. See **Figure 7-3A.** Notice how face A is removed to create an open part. If it was not removed during the operation, the cavity would be completely enclosed.

Multiple faces can be removed while using the **Shell** tool. This can be done in the initial operation or by editing the feature. To edit the feature, right-click on the shell name in the **Browser** and select **Edit Feature** from the pop-up menu. In the **Shell** dialog box, pick the **Remove Faces** button to turn it on. Then, select face B on the part. Notice how selecting face B also selects the two tangent faces. Pick the **OK** button in the dialog box to update the feature. See **Figure 7-3B.**

When editing the feature, you can also remove faces from the exclusion. In other words, you can add the face back onto the part. Edit the feature again and pick the **Remove Faces** button. The faces currently removed by the operation are highlighted in color. Hold down the [Shift] key and pick face B to deselect it. Then, release the [Shift] key and pick face C to select it. Pick the **OK** button in the dialog box to update the feature. See **Figure 7-3C.** Edit the feature again and make the changes so faces C and B are retained and faces A and D are removed. See **Figure 7-3D.** Notice the sharp chamfers created by removing face D.

The **Shell** tool has three options that determine how the cavity wall is created in relation to the base part—**Inside**, **Outside**, and **Both**. The **Inside** option keeps the outer shape of the part the same size. In other words, the cavity wall is to the inside of the part. The **Outside** option adds material to the outside of the part to create the cavity wall. The cavity has the dimensions of the original part. The **Both** option splits the difference; half of the wall thickness is to the inside and half to the outside.

Open Example_07_02.ipt. See **Figure 7-4A.** Activate the **Shell** tool and remove the large, front face. Also, set the wall thickness by picking in the **Thickness** text box and entering .5″. Make sure the **Inside** button is on and pick the **OK** button to create the shell,

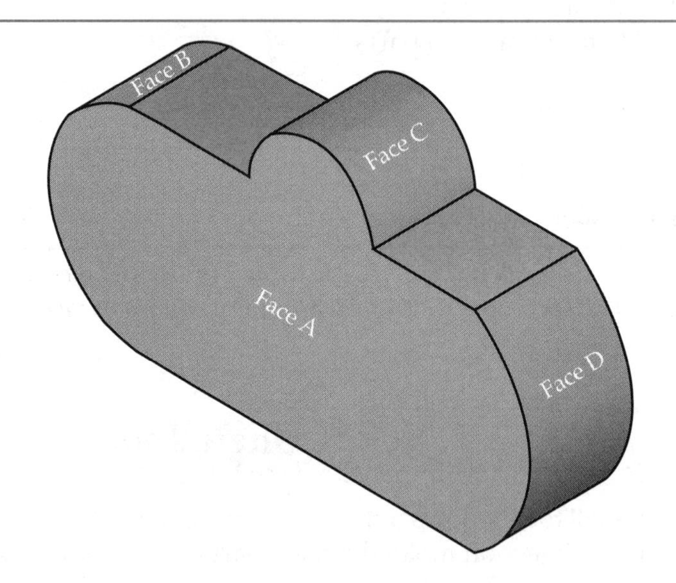

Figure 7-1.
This part will be shelled. The faces indicated here will be removed in various combinations to produce different results.

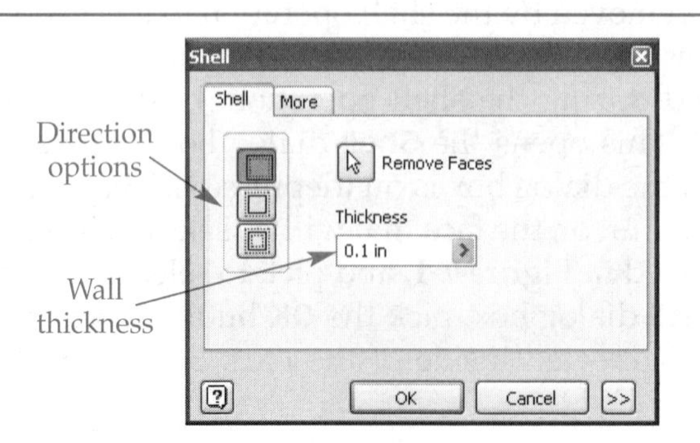

Figure 7-2.
Making settings for the shell operation.

Figure 7-3.
The shell operation with various faces removed. A—Face A. B—Faces A and B. C—Faces A and C. D—Faces A and D.

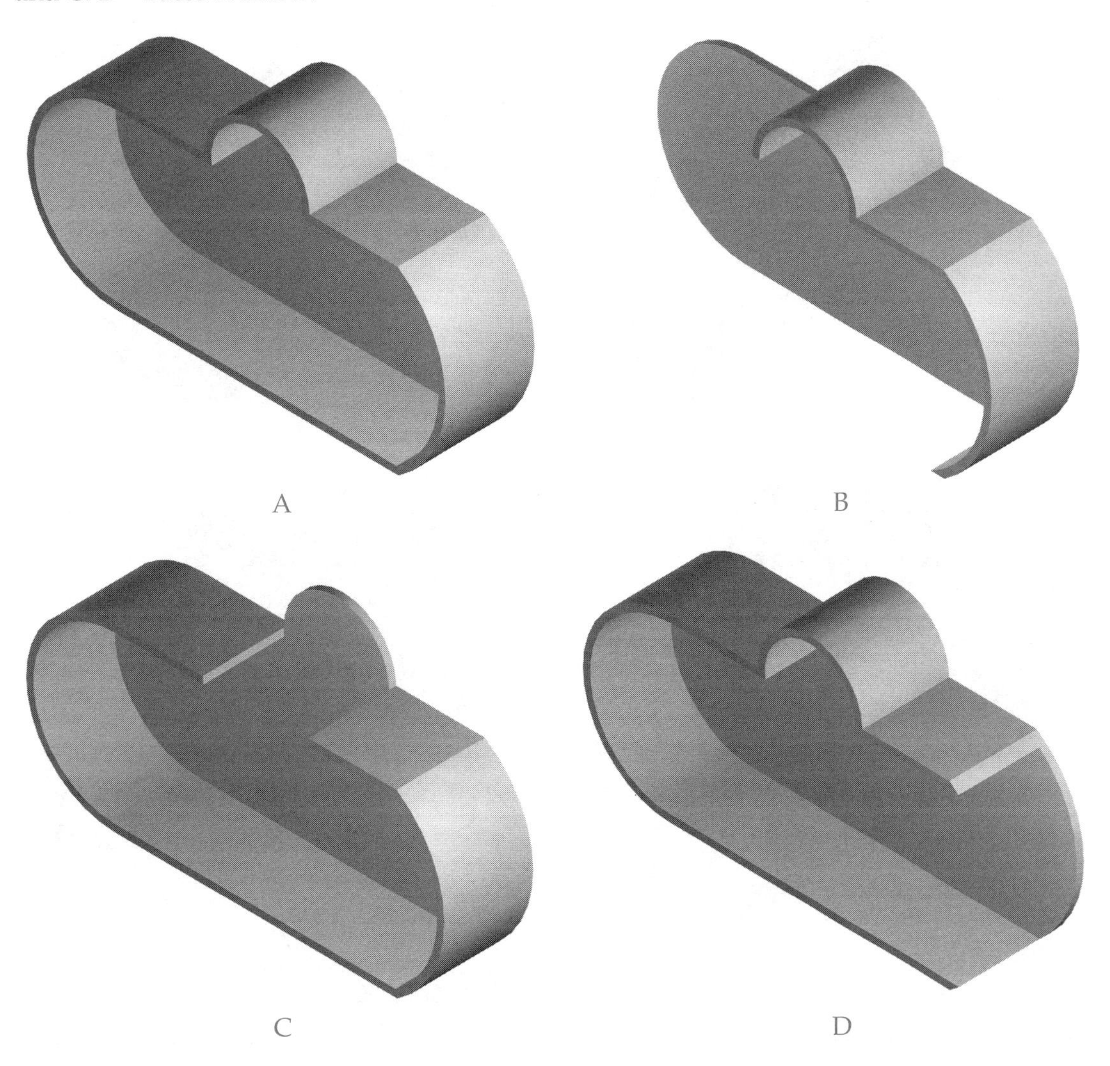

Figure 7-4B. Edit the feature and select the **Outside** button. Notice how this option makes the hole smaller. See **Figure 7-4C.** Edit the feature again and select the **Both** button. See **Figure 7-4D.**

By default, the same wall thickness is applied to all faces. However, you can apply different wall thicknesses to selected faces. In the **Shell** dialog box, pick the **>>** button to expand the dialog box. See **Figure 7-5A.** The **Unique face thickness** area is used in much the same way as setting up multiple, different-radius fillets. Open Example_07_03.ipt. This simple box was shelled with a thickness of .1″ and the front face was removed. Edit the shell feature and expand the dialog box. Pick the entry Click to Add to add a setting. Then, pick in the **Thickness** column and set the value to .3″. Pick in the **Selected** column and then select the top surface of the part. Finally, pick the **OK** button in the dialog box to update the part. The result is shown in **Figure 7-5B.** You can add multiple settings; each setting can have multiple faces if needed.

Open Example_07_04.ipt. This part was built with two extruded circles and then shelled with a 2 mm wall thickness. The side and bottom face were removed in the operation. See **Figure 7-6A.** Sketch a 70 mm diameter circle on the top of the part and extrude it 50 mm high. Shell the new feature to the inside with a thickness of 10 mm and the top face removed. The result is shown in **Figure 7-6B.**

Figure 7-4.
A—The part before shelling. B—The shelled part with the **Inside** option selected. C—The shelled part with the **Outside** option selected. D—The shelled part with the **Both** option selected.

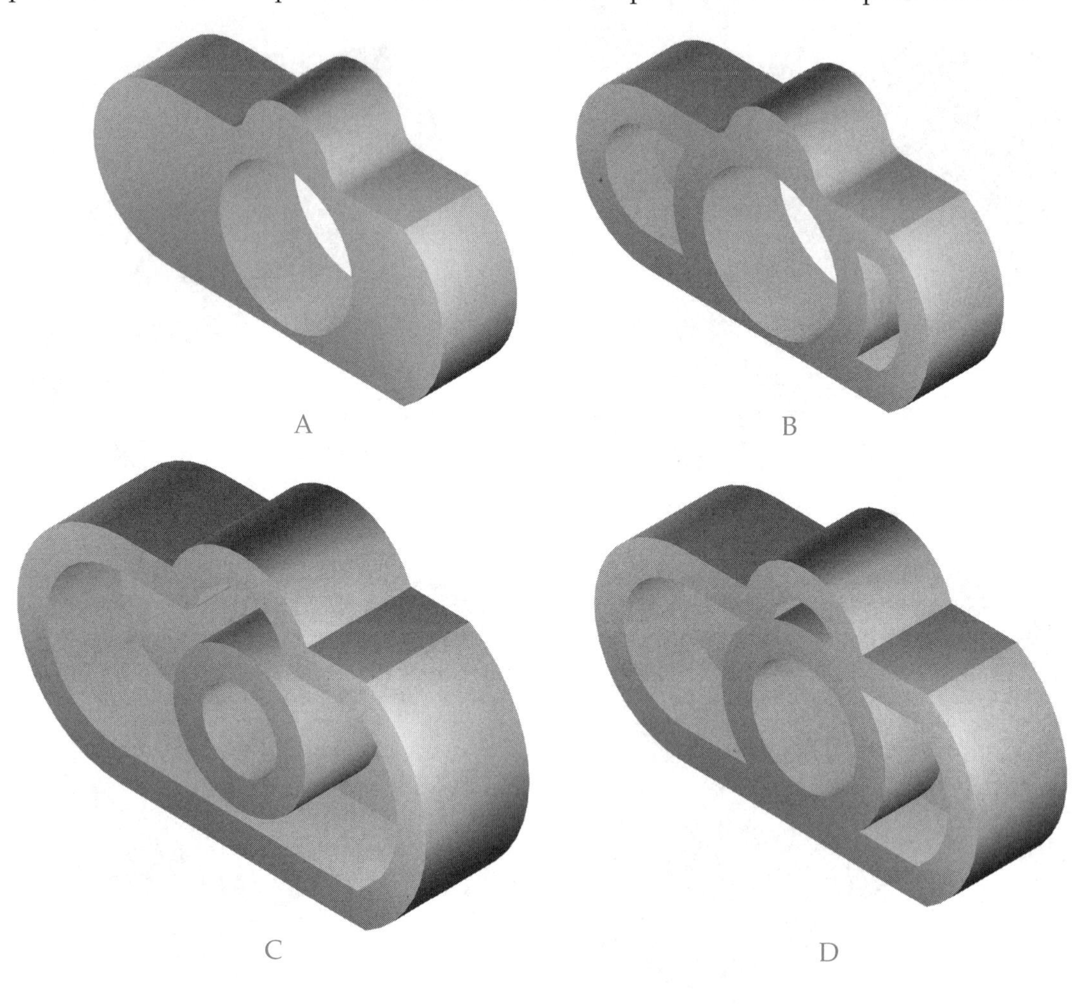

Figure 7-5.
A—The expanded **Shell** dialog box. B—The wall thickness at the top of the part is different than for the other walls.

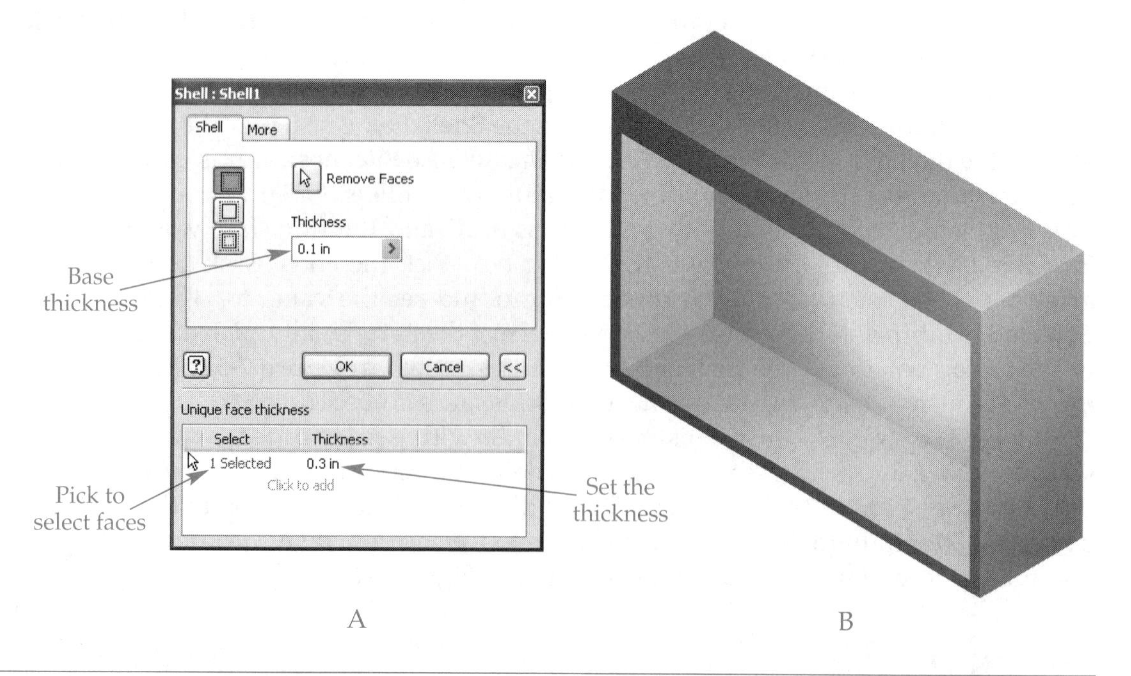

Figure 7-6.
A—The original shelled part (shown in section).
B—An extruded feature is added to the part (shown in section) and a different shell operation performed.

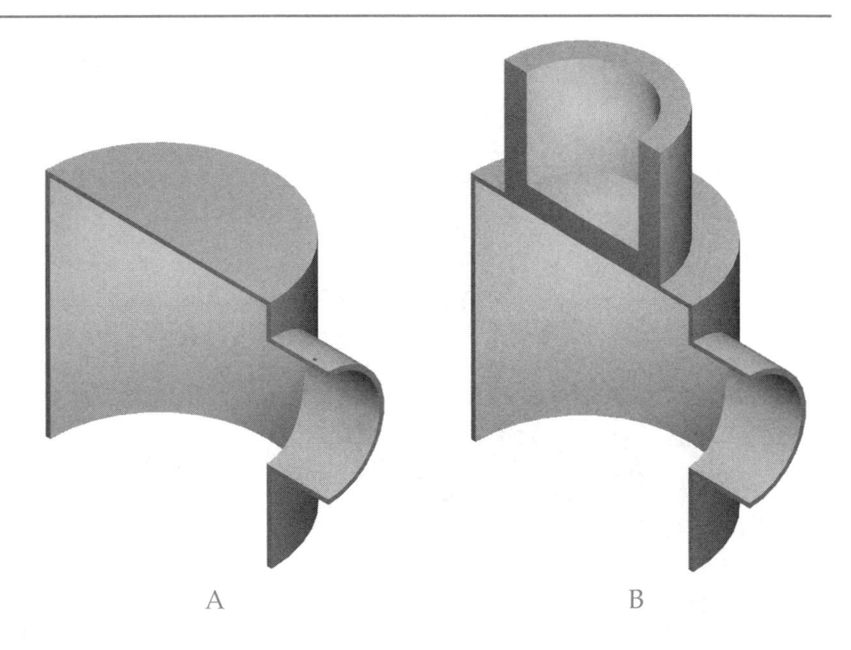

CAUTION

If you try to move the extrusion you just created above the first shell operation in the **Browser**, you will get an error message.

Rib Tool

Often, features called ribs or webs are added to parts to increase their strength or support a specific feature. The difference between a ***rib*** and ***web*** is that ribs are closed and webs are open. See **Figure 7-7.** The **Rib** tool creates both ribs and webs from a simple sketch, usually a single line.

Open Example_07_05.ipt. There is a work plane in the middle of the part. Pick the **Sketch** button and select the work plane as the sketch plane. Project the edges AB and CD onto the sketch plane, creating two points. Construct a line between the two points. This line is the profile for the rib (or web). Finish the sketch.

Pick the **Rib** button in the **Part Features** panel. This opens the **Rib** dialog box. See **Figure 7-8A.** To select a profile, pick the **Profile** button in the **Shape** area and select the profile in the graphics window. However, since there is only one unambiguous sketch, it is automatically selected as the profile. Also, the **Direction** button in the **Shape** area is automatically turned on (depressed). There are four choices for direction of the rib or web. If you look straight down the line, so it is a point, the four directions are similar to the four cardinal directions of a compass. The direction for this rib is downward toward the part. This can best be seen in a side view of the part. Use the **Look At** tool and look at the side of the part. Then, move the cursor over the line until the preview arrow points toward the part and pick. In this view, there will only be two options. Redisplay the isometric view. A preview of the rib appears on the part.

In the **Thickness** area of the **Rib** dialog box, pick in the text box and enter .375″ as the thickness. Also, select the midplane button so that the thickness is equally applied on each side of the line. In the **Extents** area of the dialog box, make sure the **To Next** button is on. This button creates a rib. If the **Finite** button is on, a web is created. Then, pick the **OK** button to create the feature. See **Figure 7-8B.**

Figure 7-7.
A—A rib is added to the angled bracket. B—A web is added to the angled bracket.

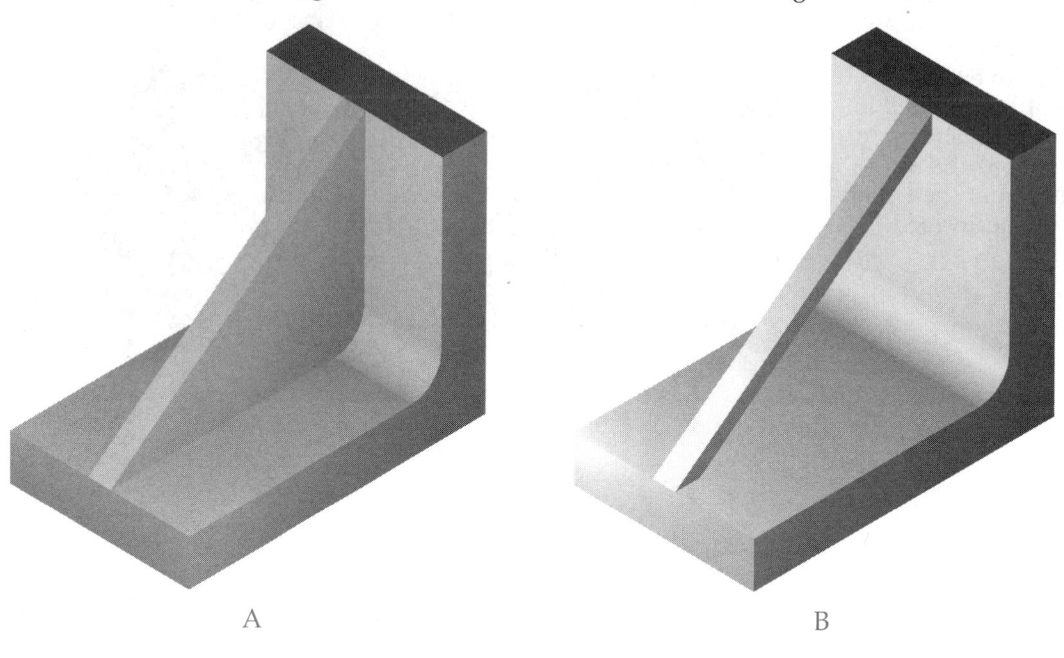

Figure 7-8.
A—Making settings for the rib operation. B—The rib is created.

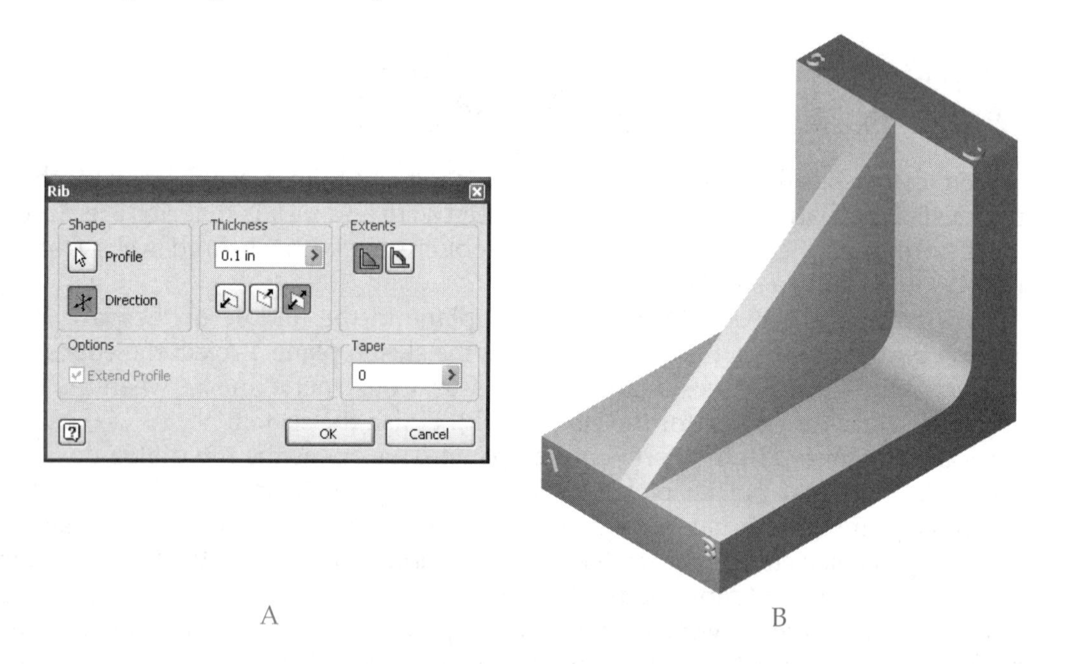

Now, edit the rib feature by right-clicking on its name in the **Browser** and selecting **Edit Feature** from the pop-up menu. In the **Extents** area of the **Rib** dialog box, pick the **Finite** button. A text box appears below the button. Pick in the text box and enter a value of 1″. Pick the **OK** button to update the part. The rib is now a web. See **Figure 7-9.** Notice that the 1″ dimension is measured perpendicular to the line.

The sketch curve (line) for a rib or web does not have to be drawn full length. By checking the **Extend Profile** check box in the **Rib** dialog box, the curve will be extended to the next faces. Open Example_07_06.ipt. The line was drawn with the top of the box as the sketch plane. The length of the line is arbitrary, but the line passes through the center of the part. Activate the **Rib** tool, pick the line as the profile, and set the thickness to 2 mm. Pick the **Direction** button and set the direction down toward the part. Note that the **Extend Profile** box is enabled and checked. Pick the **OK** button to create the rib. The rib extends from wall-to-wall. Also, notice that the rib is not created inside of the hole. See **Figure 7-10.**

Right-click on the rib in the **Browser** and select **Edit Feature** from the pop-up menu. In the **Rib** dialog box, enter 20 in the **Taper** text box. This value is used for emphasis. Pick the **OK** button to update the part. A positive taper angle creates a rib that gradually gets thicker.

Edit the rib again. Uncheck the **Extend Profile** check box and pick **OK**. The rib does not extend to the outside walls of the part. Edit the feature one more time and pick the **Finite** button. When the **Extend Profile** check box is not checked, the web does not extend to the outside walls of the part. When checked, the web extends to the outside walls. However, in both cases, notice how the web passes through the hole.

The **Rib** tool also accepts multiple lines as profiles. Open Example_07_07.ipt. Five lines are drawn through the centers of the six hole bosses. Note that the lines do not

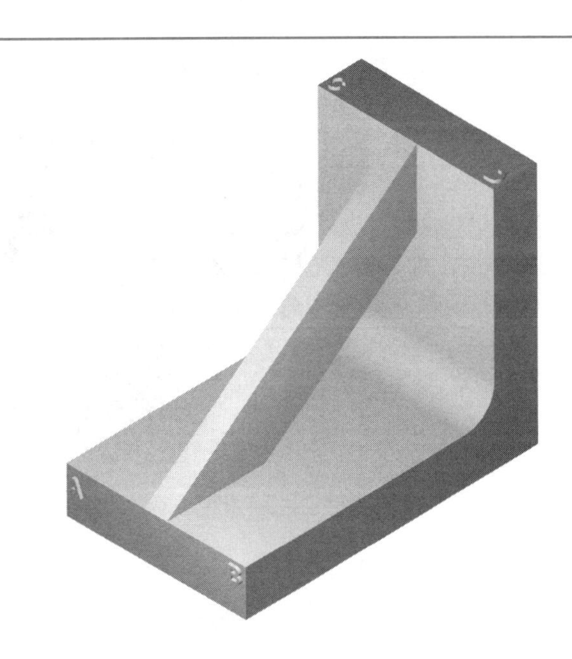

Figure 7-9.
The rib is changed into a web by editing the feature.

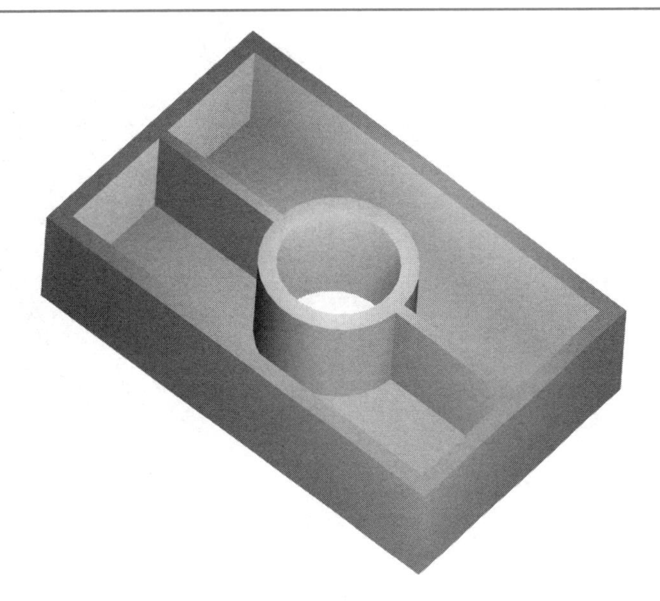

Figure 7-10.
A rib is added to the inside of the part. The rib does not pass through the hole.

touch the outside walls of the part. Activate the **Rib** tool, select all five lines as the profile, set the thickness to .1″, and pick the **To Next** button. Pick the **Direction** button and set the direction in toward the part. Then, pick the **OK** button to create the ribs. See **Figure 7-11.**

You are not limited to straight lines as the rib or web profile. Circles, arcs, and splines can also be selected as profiles. **Figure 7-12** shows a part where the ribs were created with a circle and three lines. Open Example_07_08.ipt. Study this part to see how the profiles were created. Notice that there is only one rib operation listed in the **Browser**. Select the hole and circular pattern operations and move them above the rib operation. The ribs pass through the holes. The reason you can move the hole and circular pattern above the rib operation, unlike in Example_07_06, is because the rib operation is not dependent on those two operations. In fact, you can move the rib operation anywhere in the part tree up to Extrusion2. The rib operation is dependent on Extrusion2 and all operations above it.

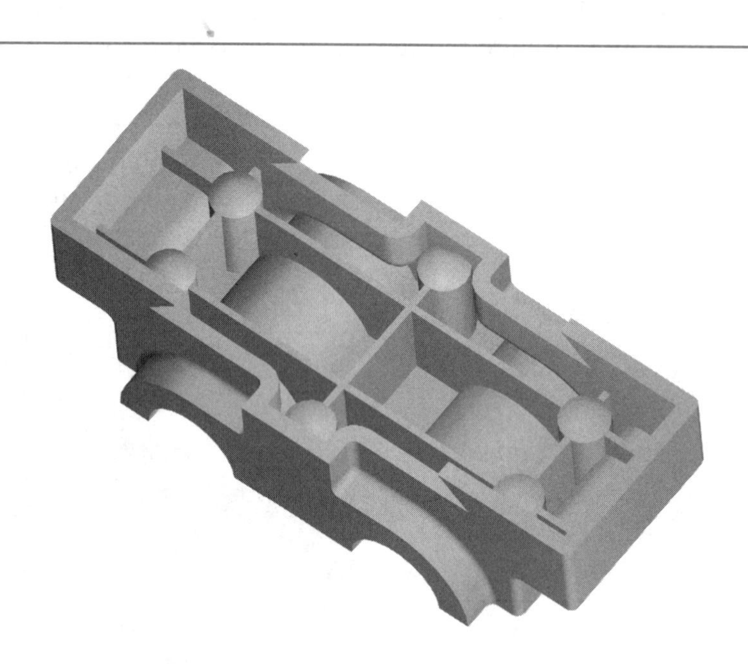

Figure 7-11.
Five lines were used to create the ribs in this part in a single operation.

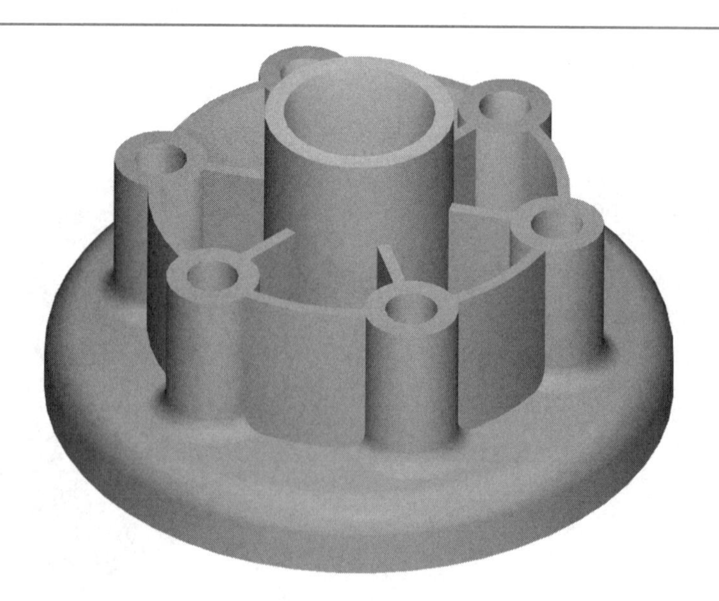

Figure 7-12.
A combination of lines and a circle were used to create the ribs on this part.

Figure 7-13.
A—When the **Match Shape** check box is unchecked, the open sketch is terminated on the existing part. B—The sketch is revolved to add a feature to the part.

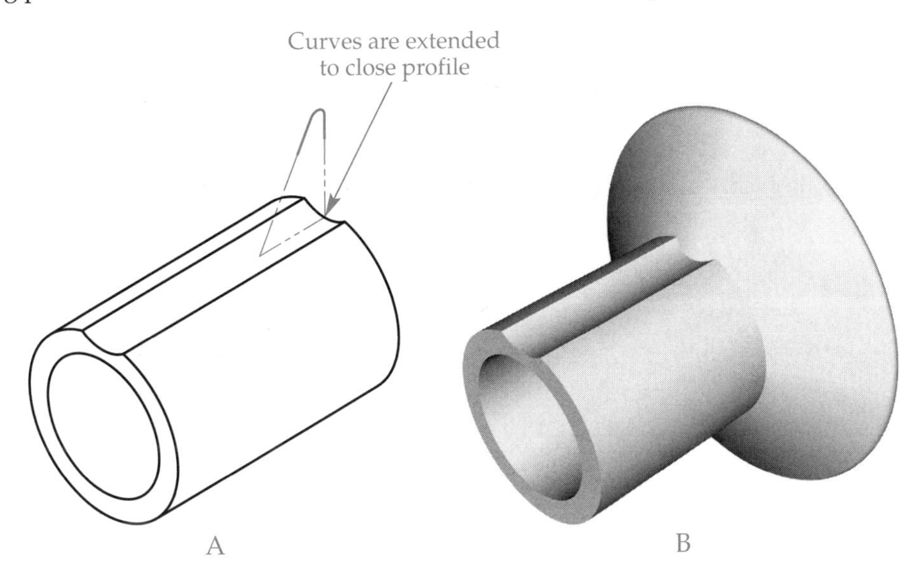

Creating Solids from Open Sketches

It is not always necessary to have a closed sketch to create a solid feature on an existing part. Sometimes, the geometry of the existing part can be used to "virtually" close the sketch. Open the Example_07_09.ipt. This part is a tube and contains a simple, open sketch of two lines and an arc.

Pick the **Revolve** button in the **Part Features** panel. Pick the two lines and the arc as the profile. In the **Revolve** dialog box, uncheck the **Match shape** check box. The profile extends, or "floods," to the edge of the part, **Figure 7-13A.** With the **Axis** button active, select the centerline as the axis of rotation. Then, select Full from the **Extents** drop-down list and pick the **Join** button. Finally, pick the **OK** button to create the feature, **Figure 7-13B.**

Moving the Coordinate System on a Sketch Plane

So far, you have changed location and orientation of the coordinate system by selecting work planes or part faces as sketch planes. The coordinate system for a sketch is called the ***sketch coordinate system (SCS),*** which is represented by the icon called the ***coordinate system indicator (CSI).*** However, you can change the coordinate system, including its orientation, using the **Edit Coordinate System** tool. The orientation of the coordinate system is important for some tools, such as the **Emboss** tool that is discussed in the next section. The text used to emboss is always aligned with the X axis of the coordinate system. Since you can change the orientation of the coordinate system, you can change the orientation of the text.

Open Example_07_10.ipt. See **Figure 7-14.** Right-click on Sketch3 in the **Browser** and select **Edit Sketch** from the pop-up menu. The top of the part was picked as the sketch plane to create Sketch3. If the CSI is not displayed at the origin, turn it on in the **Sketch** tab of the **Options** dialog box, which is accessed by selecting **Application Options...** from the **Tools** pull-down menu. The origin of the SCS was automatically placed at the intersection of the straight and arc edges when the face was selected as

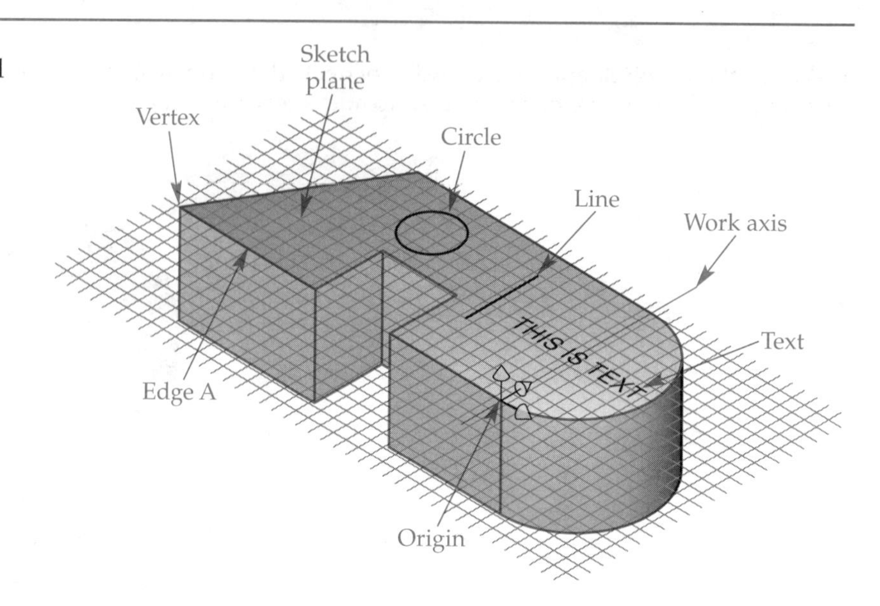

Figure 7-14.
The circle, line, and text on the top of this part were created using the default coordinate system.

the sketch plane. A line, circle, and text were then created with the appropriate sketch tools. You will learn how to create text in the next section. The Y work axis was also made visible.

To see the location and orientation of the default axes, expand the Origin branch in the **Browser**. Then, hover the cursor over an axis name and the axis is highlighted in the graphics window.

The top of the existing part has eight vertices on the sketch plane. The origin of the SCS can be moved to any of these vertices or to the center of the curved face, which is also a point on the sketch plane. In addition, the X or Y axis can be aligned with any straight edge of the part or with the work axis. The X axis is indicated by the red arrow on the CSI, the Y axis by the green arrow, and the Z axis by the blue arrow. The axes *cannot* be aligned with the sketched line. To align the Y axis of the SCS with edge A, as indicated in **Figure 7-14,** do the following.

1. Pick the **Edit Coordinate System** button in the **2D Sketch Panel**. Notice how the CSI changes to long red and green arrows and a blue dot.
2. Move the cursor over the green Y axis arrow so it is highlighted in red. Pick the arrow; it is highlighted in blue.
3. Move the cursor around the graphics window. Notice how the orientation of the long arrows change. Pick edge A. The SCS is rotated so the Y axis aligns with the edge.
4. Right-click in the graphics window and select **Done** from the pop-up menu. Do *not* press [Esc]; this cancels the operation.

The SCS is oriented so the Y axis aligns with edge A. Notice how the CSI has rotated to reflect the new SCS orientation.

Notice how the orientation of all of the sketch geometry changed with the SCS. See **Figure 7-15.** Notice that the line moved, but did not rotate. This is because a perpendicular constraint is applied to the line and one of the edges. You can relocate the sketch geometry by dragging it around and rotate geometry using the **Rotate** tool. However, you cannot change the orientation of the text. Therefore, it is good practice to orient the SCS in such a way that text will be properly aligned before you construct any sketch geometry.

Now, the orientation of the SCS is changed. However, the origin remains in the same location. As mentioned earlier, you can move the origin to any of the eight

Figure 7-15.
When the orientation of the coordinate system is changed, the geometry also rotates. The line did not rotate because of a constraint applied to it.

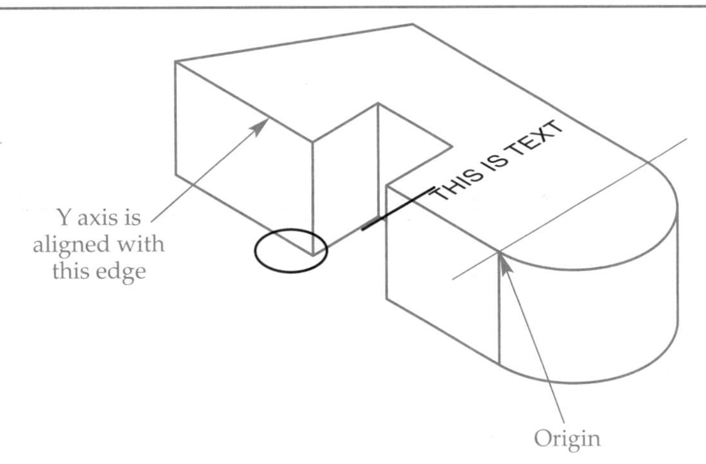

vertices or the center of the arc. Move the origin of the new SCS to the center of the arc as follows.

1. Pick the **Edit Coordinate System** button in the **2D Sketch Panel**.
2. Move the cursor to the center of the CSI. When the blue dot is highlighted in red, pick it.
3. Move the cursor around the graphics window. The X and Y axis arrows jump to the vertex closest to the cursor. Move the cursor over the arc and the arrows jump to the center of the arc.
4. When the X and Y axis arrows are at the center of the arc, pick to move the origin of the SCS.
5. Right-click in the graphics window and select **Done** from the pop-up menu. Remember, do not press [Esc].

The CSI is now displayed at the center of the arc, indicating that this is the origin of the SCS. However, notice that the orientation of the SCS has not changed. The Y axis is still aligned with edge A.

Text and Emboss Tools

To ***emboss*** text means to extrude it above a face. To ***engrave*** text means to cut it into a face. The **Emboss** tool can both emboss and engrave text created by the sketch **Create Text** tool. The face on which text is embossed or engraved can be flat or curved.

Open Example_07_11.ipt. You will now place text on the front face. Later, you will emboss this text. Pick the **Sketch** button on the **Inventor Standard** toolbar and pick the front face as the sketch plane. Then, pick the **Text** button in the **2D Sketch Panel**. Move the cursor near the top-left corner of the face and pick. The **Format Text** dialog box is displayed, **Figure 7-16.**

In the **Format Text** dialog box, pick in the large edit box and type your name. Then, using the cursor, highlight the text as you would in a text editor by picking and dragging over the text. With the text highlighted, pick the **Center Justification** and **Middle Justification** buttons at the top-left of the dialog box. You can change the font using the drop-down list below the justification buttons. If Arial is not displayed, pick the drop-down list and select it. Finally, pick in the text box to the right of the font drop-down list and type 2.8 to set the text height at 2.8″.

When you press [Enter] or pick the **OK** button, the dialog box is closed and the text is created. Now, press [Esc] to cancel the **Text** tool and then drag the text so your name is approximately centered on the face. Put a vertical constraint between the midpoint

Figure 7-16.
Adding text to a sketch.

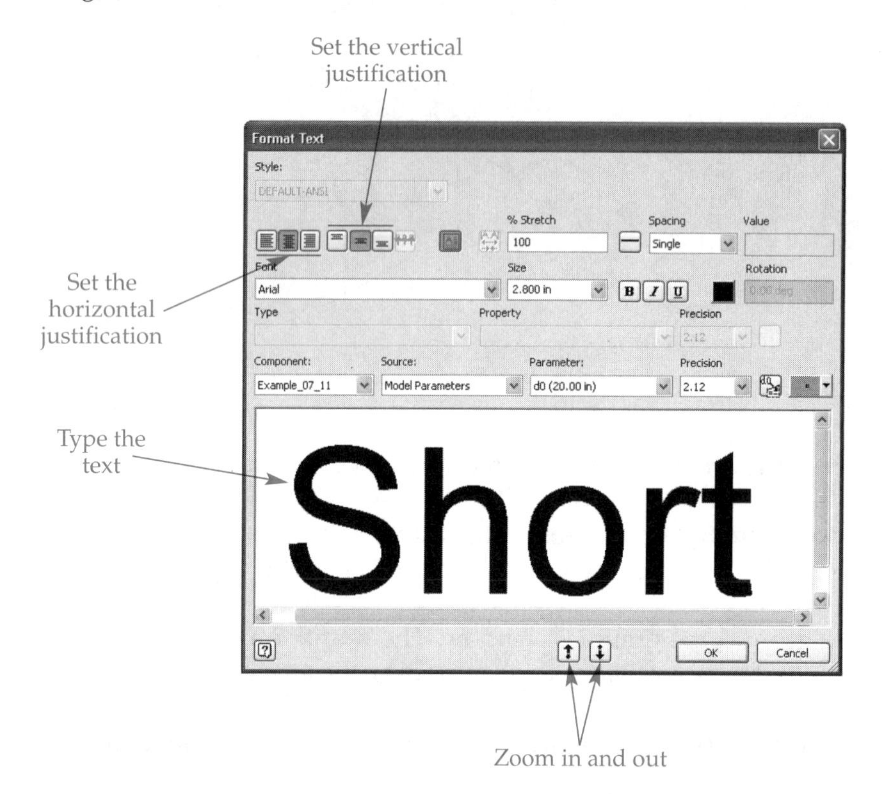

of the top of the text box and the midpoint of the top edge of the part. Finally, finish the sketch.

Pick the **Emboss** button in the **Part Features** panel. The **Emboss** dialog box is displayed. With the **Profile** button on (depressed), pick the text in the graphics window. If you have trouble picking the text, pause the cursor to display the select tool, pick an arrow to highlight the text, and pick the center button to select the text. In the dialog box, pick in the **Depth** text box and change the value to .5″. You should see a small green arrow pointing out from the text. Also, make sure the **Emboss from Face** button is on (depressed).

Notice the button below the **Depth** text box. This is the **Top Face Color** button. See **Figure 7-17.** Pick it to open the **Color** dialog box. Select Gold Metallic from the drop-down list in the **Color** dialog box and pick the **OK** button. Finally, pick the **OK** button in the **Emboss** dialog box to emboss the text. Notice how the top face is in gold instead of the part color.

Figure 7-17.
The dialog box used for the emboss operation.

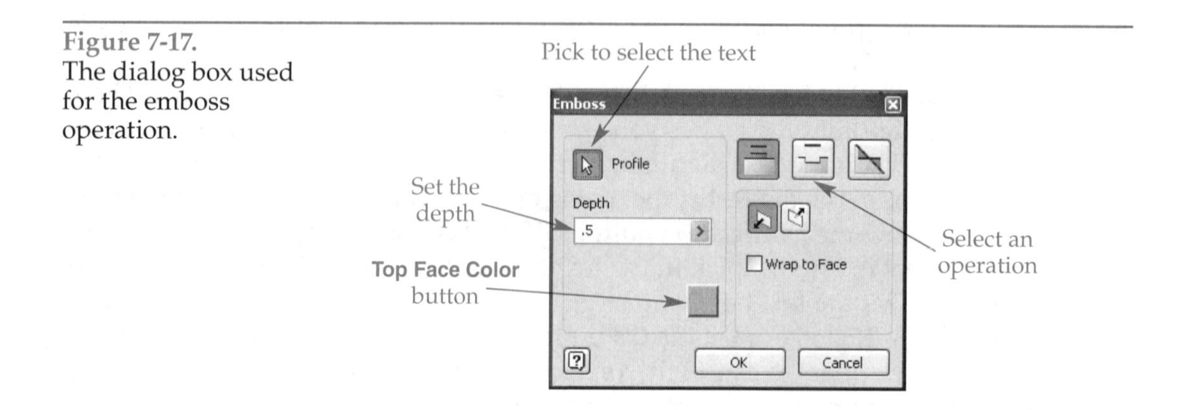

Figure 7-18.
A—Text is engraved in a curved feature, but distorted. B—In a top view, you can see that the lines are parallel, resulting in the distortion of the text.

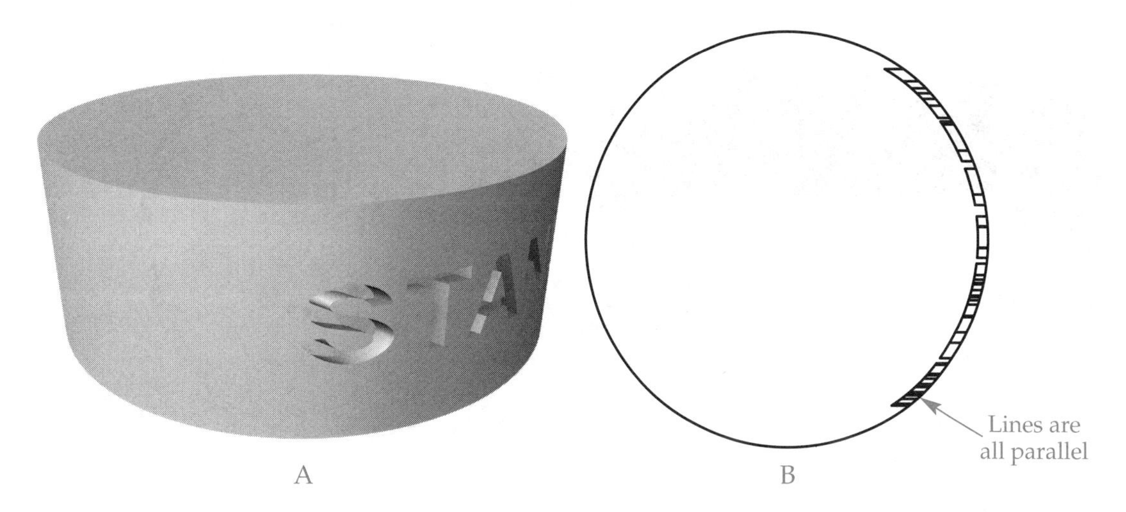

Now, right-click on the name of the emboss operation in the **Browser** and select **Properties** from the pop-up menu. In the **Feature Properties** dialog box, select Black Chrome in the **Feature Color Style** drop-down list. You can also rename the feature if you like. Then, pick the **OK** button to update the feature. Notice how the color on the top face of the embossed text *and* the color on the text sides are replaced.

Open Example_07_12.ipt. This part has text created on a work plane offset from the curved face. The text is Arial font, bold, and .5″ high. Pick the **Emboss** button in the **Part Features** panel. Then, select the text as the profile, set the height to .1″, and pick the **Engrave from Face** button. In the graphics windows, the red arrow should be pointing from the text toward the part. If not, pick the other direction button in the dialog box. Leave the **Wrap to Face** check box unchecked. Finally, pick the **OK** button to engrave the text. See **Figure 7-18.**

Notice how the text is engraved straight into the cylinder. The three letters are distorted at the beginning and end of the word. Edit the feature by right-clicking on its name in the **Browser** and selecting **Edit Feature** from the pop-up menu. In the **Emboss** dialog box, check the **Wrap to Face** check box. Then, pick the **Face** button so it is on and select the cylindrical face. Pick the **OK** button in the dialog box to update the feature. The text is now wrapped on the face before being engraved. In this manner, the letters are not distorted. See **Figure 7-19.** You can only wrap text to planar or cylindrical faces.

Decal Tool

An image can be placed on a part, much like a label on a bottle. The "image" can be any picture, drawing, or text saved as a bitmap (.bmp extension), Word file (.doc extension), or Excel file (.xls extension). Some applications include labels, logos, art, part information such as part numbers, or warranty data. You can insert an image onto a sketch plane and position it with constraints or dimensions. The **Decal** tool can then be used to attach the image to a part face.

An important property of the image to be used is its aspect ratio. The ***aspect ratio*** is the image height divided by its width. The aspect ratio cannot be changed in Inventor. However, the aspect ratio of bitmaps can be modified in graphics programs,

Figure 7-19.
A—By wrapping the text onto the curved surface, the distortion is eliminated. B—In a top view, you can see that the lines are perpendicular to the curved face.

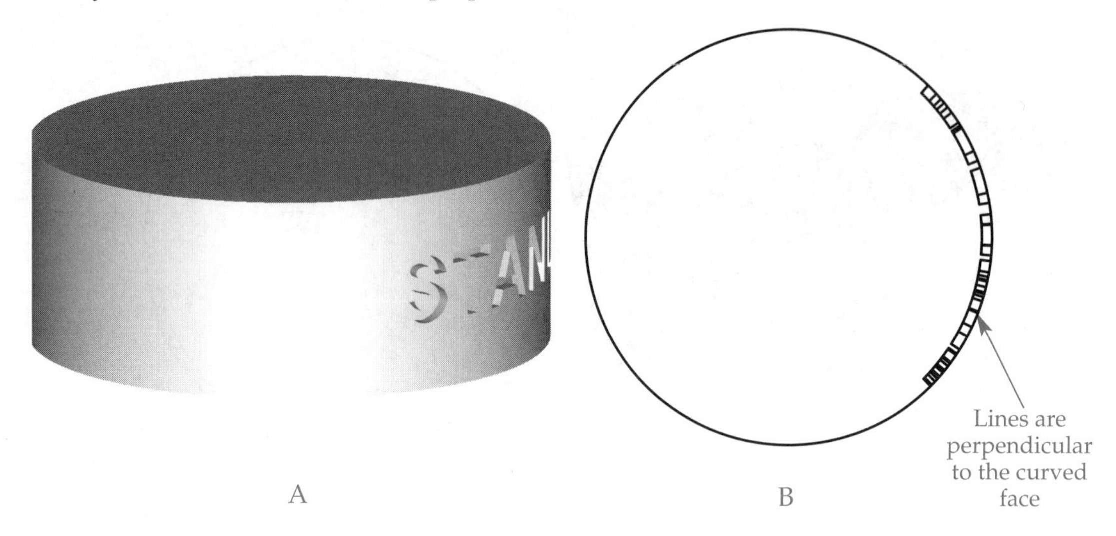

such as Adobe Photoshop. In order to keep the colors consistent when the bitmaps are inserted, the bitmaps need to be relatively small; about 400 or less pixels along the long axis.

In many industries, a label has a part number and, therefore, a drawing. A drawing of a label can be easily created by placing the image on a very thin part and creating a decal. This serves several functions:

- The label can have a part number.
- The label can have a drawing.
- Most importantly, the drawing can be used in an assembly, which is discussed in Chapter 11.

The part on which the label is placed should have the same dimensions as the label. Also, the part should be the same shape; i.e., a box for a rectangular label or a cylinder for a round label.

NOTE

If the bitmap image is displayed in Inventor as a generic icon instead of the image, a program that cannot render an image as an OLE server is set in the Windows registry to handle BMP files. Instructions for correcting this situation can be found in the support area of the Autodesk website.

Open Example_07_13.ipt. The two bitmap images you will use on this part are a photo of a brick wall, with an aspect ratio of .6, and a company logo, with a ratio of .71, **Figure 7-20.** To make this example easy, Extrusion1 has the same aspect ratio as the bricks and Extrusion2 has the same aspect ratio as the logo. First, add the brick image as follows.

1. Start a new sketch on the large front face.
2. Pick the **Insert Image...** button in the **2D Sketch Panel** or select **Insert Image...** from the **Insert** pull-down menu. The **Open** dialog box is displayed. This is a standard Windows open dialog box.
3. Navigate to the \Ch07\Examples folder, select the file bricks.bmp, and pick **Open.**

Figure 7-20.
A—This image of bricks will be used as a decal. B—This company logo will be used as a decal.

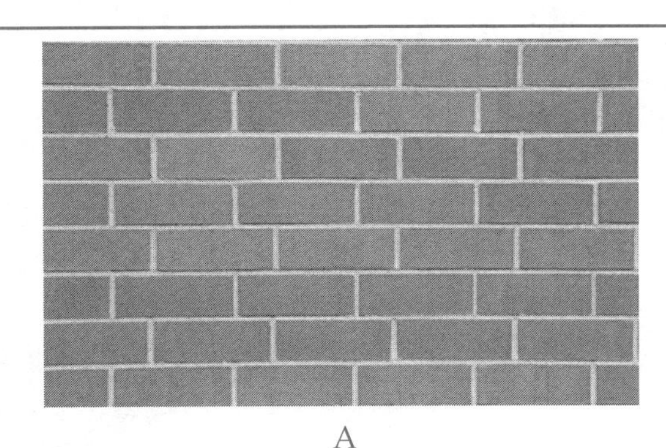

A

B

4. A small, green rectangle now appears attached to the cursor. This represents the bitmap image. Position the box at the upper-left corner of the part. When the corner of the part is highlighted as a green dot, pick to place the image.
5. Notice how the green box is still attached to the cursor. You can insert another instance of the bitmap. Since you only need one, press [Esc] to end the **Insert Image** tool.
6. Move the cursor to the lower-right corner of the image. When the corner is highlighted red, pick to select the corner. Then, drag the image corner to the lower-right corner of the part and release.
7. Finish the sketch.
8. Pick the **Decal** button in the **Part Features** panel. The **Decal** dialog box is opened, **Figure 7-21.**
9. With the **Image** button in the dialog box on (depressed), pick the brick image on the part.
10. With the **Face** button in the dialog box on, pick the front face of the part. The face is highlighted in color.
11. In the **Decal** dialog box, pick the **OK** button to apply the decal.

Notice that the decal is not applied to Extrusion2 because it was not selected as a face, which is the intended result. See **Figure 7-22.**

Using a similar process, apply the file G-W_Logo.bmp to the front face of Extrusion2. See **Figure 7-23.** You may want to use the **Slice Graphics** option to improve visibility when resizing the image. A good way to locate the bitmap is to insert it off of the part, then use the coincident constraint on the corners of the bitmap and the face.

Like the **Emboss** tool, the **Decal** tool has a **Wrap to Face** option for wrapping an image onto a curved face. Open Example_07_14.ipt. This is a model of a firecracker that needs a caution label applied. The model has the Z axis visible and a work plane offset

Figure 7-21.
Placing a decal on a part.

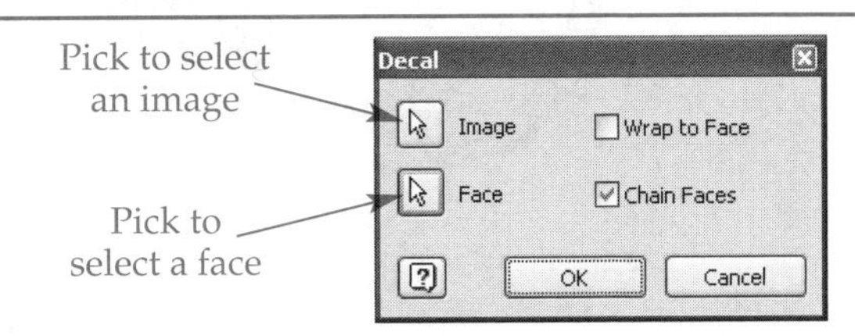

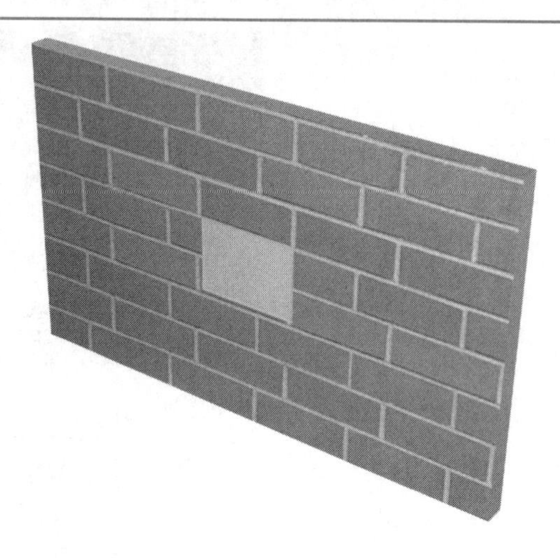
Figure 7-22.
The image of the bricks is applied to the part.

Figure 7-23.
The company logo is added to the part.

from the part. Create a new sketch on the work plane and project the Z axis onto it, as shown in **Figure 7-24.** Also, project the circular end of the part onto the sketch plane. These projection lines will be used to help align the label.

Notice the direction of the Z axis (blue) of the sketch coordinate system. This is the normal direction for the sketch plane. The destination face must point the same way or the decal will not be visible. Also, notice the orientation of the X and Y axes. The X

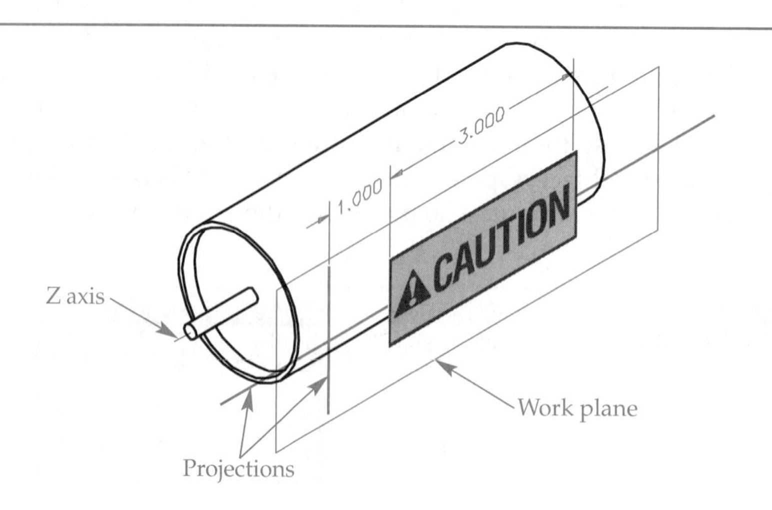

Figure 7-24.
Adding a caution label to a part.

axis of the image is aligned with the X axis of the sketch plane. The X axis of the label is along the bottom of the text. Therefore, the image will have to be rotated after it is inserted. You will use the **Rotate** tool to do this.

1. Pick the **Insert Image** button. Navigate to the \Ch07\Examples folder and select the file Caution.bmp.
2. Place the image anywhere in the graphics window. Notice how the image is rotated 90° counterclockwise to the correct orientation. Press [Esc] to end the tool.
3. Pick the **Rotate** button in the **2D Sketch Panel**. The **Rotate** dialog box is opened.
4. With the **Select** button on in the **Rotate** dialog box, pick the inserted image; select all four edges by picking in the middle of the image.
5. With the **Select** button on in the **Center Point** area of the **Rotate** dialog box, pick any corner of the image.
6. In the **Rotate** dialog box, type –90 in the **Angle** text box and press [Enter] to rotate the image. Pick the **Done** button close the dialog box (picking **Done** before pressing [Enter] cancels the operation). The image should be right reading.
7. Apply a coincident constraint between the midpoint of the left and right borders (short edges) of the image and the projected axis.
8. Dimension the sketch as shown in **Figure 7-24.** Notice how the image stays proportional.
9. Finish the sketch.
10. Pick the **Decal** button in the **Part Features** panel. Select the label as the image and the large-diameter cylinder as the face. Also, check the **Wrap to Face** check box. Pick the **OK** button to place the label on the part, **Figure 7-25.**

You can project a decal across adjacent faces using the **Chain Faces** option in the **Decal** dialog box. In **Figure 7-26A,** the **Chain Faces** check box was not checked and the decal is cut off. In **Figure 7-26B,** the check box was checked. Notice how the image is distorted as it is projected onto the adjacent face. This distortion is more pronounced the closer the face angle is to 90°, **Figure 7-27A.** This distortion can be eliminated by checking the **Wrap to Face** check box, **Figure 7-27B.** Open Example_07_15.ipt. Then, make the appropriate settings to create **Figure 7-26A.** Undo the operation and create **Figure 7-26B.** Start a new part file and create **Figure 7-27A** and **Figure 7-27B.** Use your own dimensions.

Figure 7-25.
The part with the caution label applied.

Figure 7-26.
A—When the **Chain Faces** check box is unchecked, the decal is cut off at the edge of the face.
B—When **Chain Faces** is checked, the decal wraps around the edge of the face.

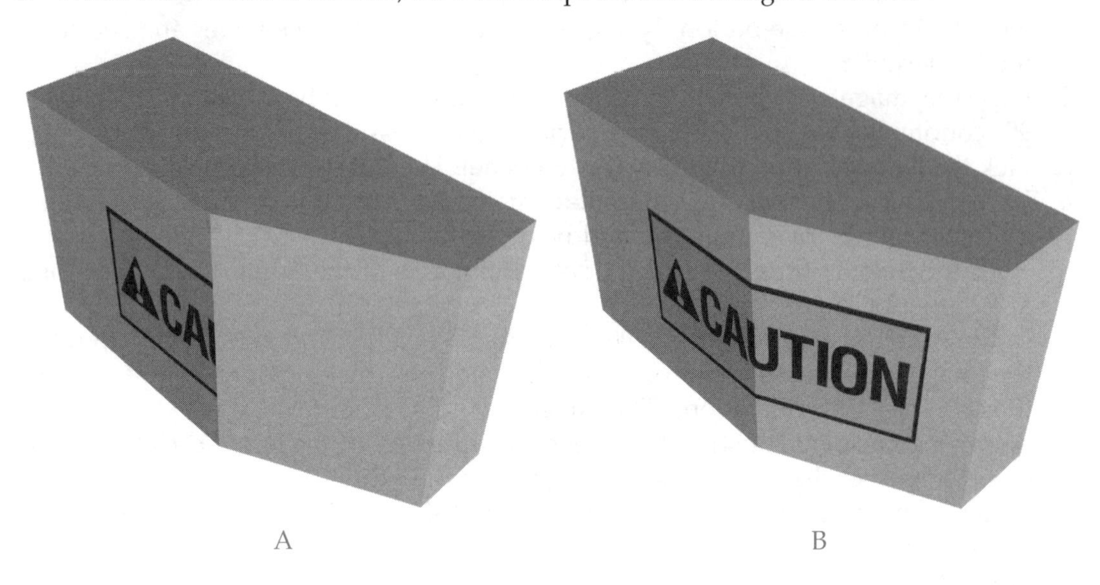

Figure 7-27.
A—When the **Chain Faces** check box is used to wrap a decal around an edge, distortion can occur.
B—Checking the **Wrap to Face** check box in addition to the **Chain Faces** corrects the distortion.

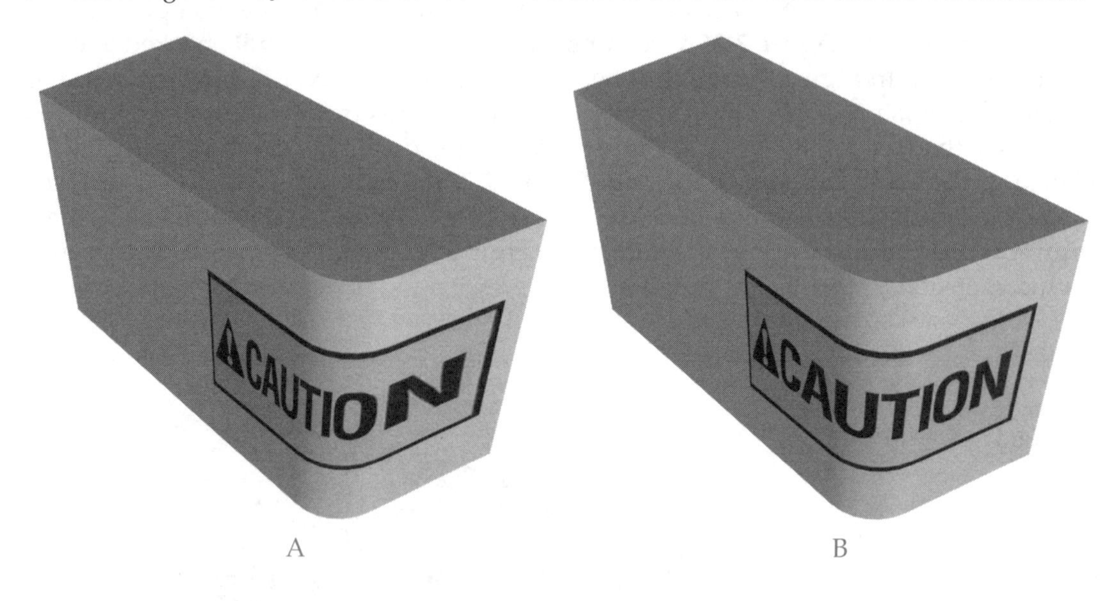

Face Drafts

Some manufacturing processes, such as casting, forging, and injection molding, require a draft angle on the part. A ***draft angle*** is an angle on the sides of a part so that the part can be removed from the forming tool. Therefore, the forming tool itself must have corresponding angles on its cavity walls. The actual feature, the angled sides, is called the ***draft.*** A draft angle can be created on a part when a sketch is extruded or later using the **Face Draft** tool. This tool can also be used to put a unique draft angle on one or more faces, simplifying the construction process.

Fixed Edge

The **Face Draft** tool requires several selections and is best explained with a simple part. Open Example_07_16.ipt. This part has two extruded blocks. The first block, **Figure 7-28A,** has two colored faces; one green face and one yellow face. The draft will be applied to the yellow face. Pick the **Face Draft** button in the **Part Features** panel. The **Face Draft** dialog box is opened, **Figure 7-29.**

The draft angle is set in the **Draft Angle** text box. Typically, a draft angle is only 2° to 3°. For this example, a large angle is used to emphasize the effect of the operation. Pick in the text box and enter 20 as the draft angle.

Pick the **Fixed Edge** button at the far-left of the dialog box. Just to the right of this button are two buttons. The **Pull Direction** flip button is currently grayed out. Pick the **Pull Direction** button (arrow) so it is on, if it is not already. Select the green face on the part. An arrow should appear on the face pointing away from the part. The draft will be at the specified angle to this arrow. If you imagine a mold over the top of this part, the pull direction is the direction the mold is "pulled" to remove the molded part. You can flip the pull direction by picking the **Pull Direction** flip button, which is now enabled.

Figure 7-28.
A—The pick point determines which edge is fixed and used to create the draft. The top edge is fixed. B—The resulting draft. C—Using a work plane to create the draft. D—The resulting draft.

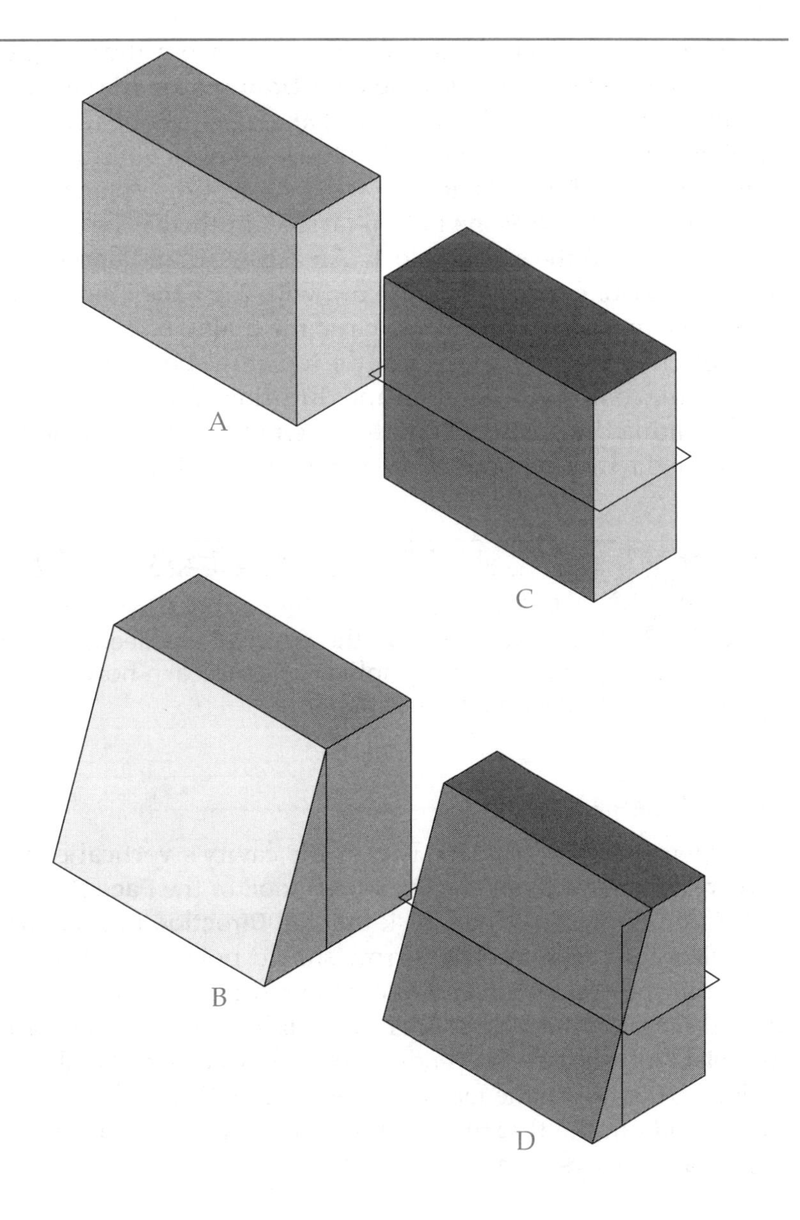

Figure 7-29.
The **Face Draft** dialog box is used to create a face draft.

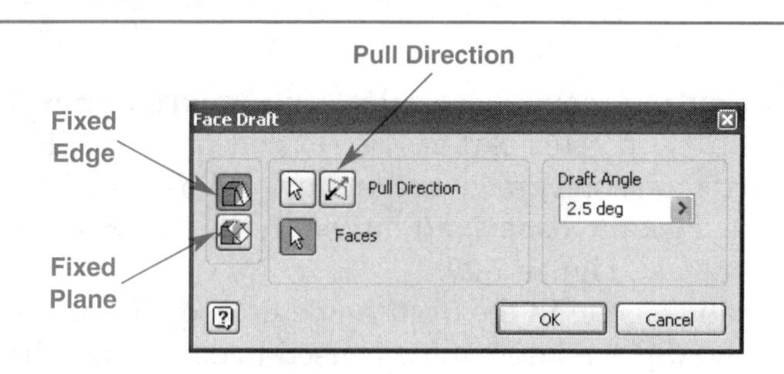

The **Faces** button is automatically selected once the pull direction is set. You will select the yellow face, however, you have two choices. If you pick near the top edge, that edge remains fixed and the part gets larger. If you pick near the bottom edge, it remains fixed and the part gets smaller. You *cannot* change this by editing the feature later. As you move the cursor over the face, a preview indicates how the face draft will be applied. Pick the yellow face near the top edge. Then, pick the **OK** button in the dialog box to create the draft angle. See **Figure 7-28B.**

Fixed Plane

The second block, **Figure 7-28C,** will have a draft angle applied using a work plane to locate the draft. Pick the **Face Draft** button in the **Part Features** panel. Pick in the text box and enter 20° as the draft angle, if it is not already entered. This time, pick the **Fixed Plane** button. Notice that the label of the buttons to the right changed from **Pull Direction** to **Fixed Plane.** You will be selecting a plane as the starting point for the draft. With the **Draft Plane** button (arrow) in the dialog box on, select the work plane in the middle of the second block. An arrow should appear on the plane and the flip button is no longer grayed out. Now, with the **Faces** button on, select the purple face to add the draft, then pick **OK** to close the dialog box.

Selecting the purple face near the top or the bottom of the block did not matter. The work plane controls the location and direction of the draft, unlike the previous method. The resulting draft is significantly different than the previous one. See **Figure 7-28D.** Compare the two methods to see where material was added and subtracted.

PROFESSIONAL TIP

Both methods of adding draft affected the entire face. If the face above or below the work plane needs to remain undrafted, the face has to be split. You will learn how to do this in the Split Tool section in this chapter.

Other Face Drafts

Open Example_07_17.ipt. Two of the cavity's vertical edges are filleted; the other two are square. Activate the **Face Draft** tool. In the **Face Draft** dialog box, set the angle to 12° (for emphasis). Then, pick the **Pull Direction** button (arrow), if it is not already on. Select the green face; the arrow should point up out of the cavity. Pick the **Faces** button in the dialog box and select the red face on the cavity wall near the top edge. When you select the face, notice how all faces connected by radii are selected, similar to the **Fillet** tool. The selection ends at the square corners. Finally, pick the **OK** button in the dialog box to create the face draft. See **Figure 7-30.** If you look at the top edge of the cavity, you can see that three of the walls have a face draft. The fourth wall, between the square corners, remains straight.

Figure 7-30.
The face draft is applied to the three walls connected by radii.

There is a second type of draft that can be created with the **Face Draft** tool—a ***shadow draft.*** This fills the shadow that a curved face projects on a flat face. Open Example_07_18.ipt. Imagine a light placed at a 50° angle to the vertical from the flat blue face, as shown in **Figure 7-31A.** The area that is not illuminated by the light, the shadow area, is filled by the **Face Draft** tool. The volume of the part is increased. See **Figures 7-31B** and **7-31C.**

Figure 7-31.
A—Imagine a light source casting a shadow on the part. B—The part before the face draft operation to create a shadow draft. C—The part with the shadow draft.

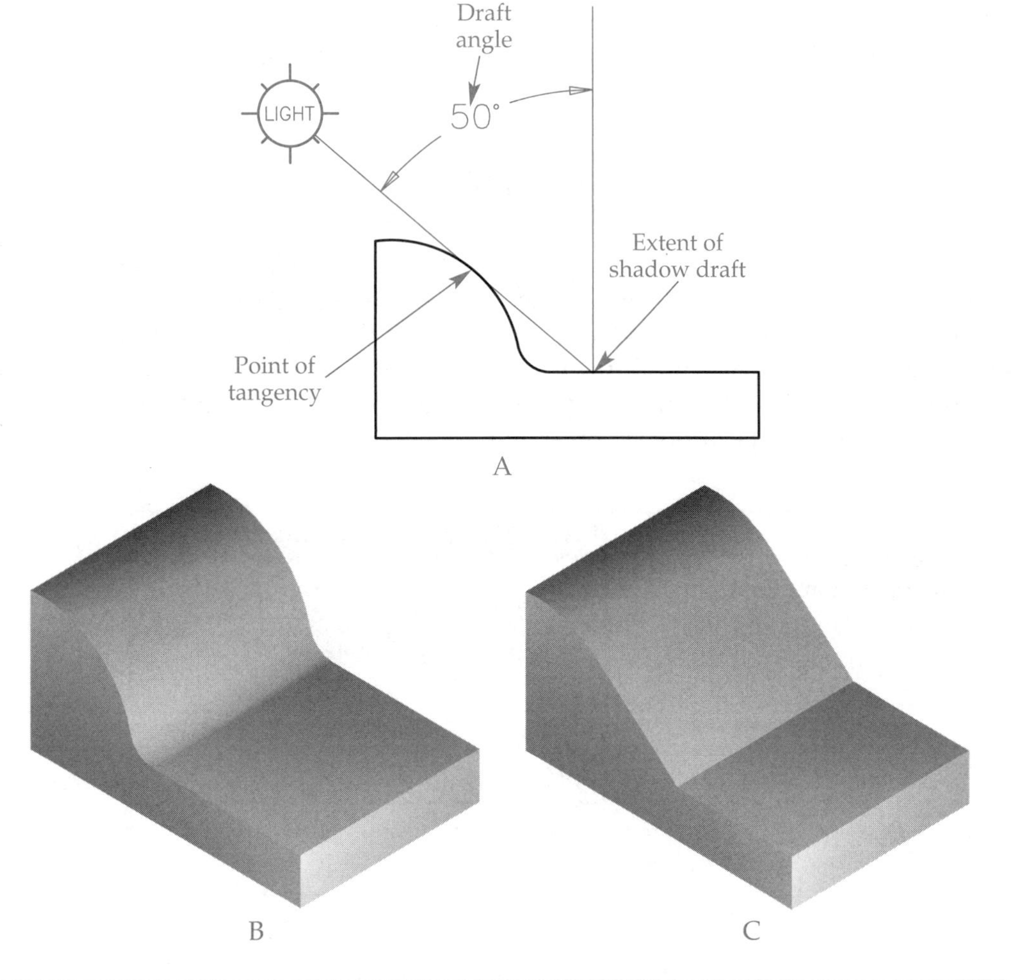

To place a shadow draft on this part, activate the **Face Draft** tool. In the **Face Draft** dialog box, set the angle to 50°. Then, pick the **Pull direction** button and select the blue face on the part. The arrow should point up away from the part. In the dialog box, pick the **Faces** button and then select the yellow face on the part. Finally, pick the **OK** button to create the shadow draft. The name of the feature in the **Browser** is FaceDraft1.

7-1 Complete the practice problem on the Student CD.

Bend Part Tool

Open Example_07_19.ipt. The **Bend Part** tool is used to bend the part about a line. There must be an unconsumed sketch containing a straight line to use the **Bend Part** tool. To use the tool, pick the **Bend Part** button in the **Part Features** dialog box. The **Bend Part** dialog box is opened, **Figure 7-32.** With the **Bend Line** button on in the dialog box, pick the sketched line. In the **Radius** text box, set the radius for the bend to .40. However, generally, the minimum inside radius for bending cold steel is the material thickness.

Picking the **Direction** button flips the bend to the opposite side of the part. Notice how the radius becomes an outside radius, as opposed to an inside radius. If the radius is set equal to the material thickness (.2 in this case) or less, the **Direction** button will *not* show a bend to the right because this generates an error. The bend angle is set in the **Angle** text box. Change the angle to 45° for this example. The **Bend Left**, **Bend Right**, and **Bend Both** buttons determine how the bend angle is applied about the line. Pick the **Bend Both** button and then pick the **OK** button to create the bend.

Split Tool

The **Split** tool is used to divide a part or face based on selected geometry. The split can be along a face, work plane, or sketched curve. When used on a part, one side of the split is discarded. When used on faces, both sides of the split are retained. Also, when used on faces you can select multiple faces or apply the split to all the faces of the part. The examples in this section demonstrate the principles of the **Split** tool. At the end of this section, a practical example is provided.

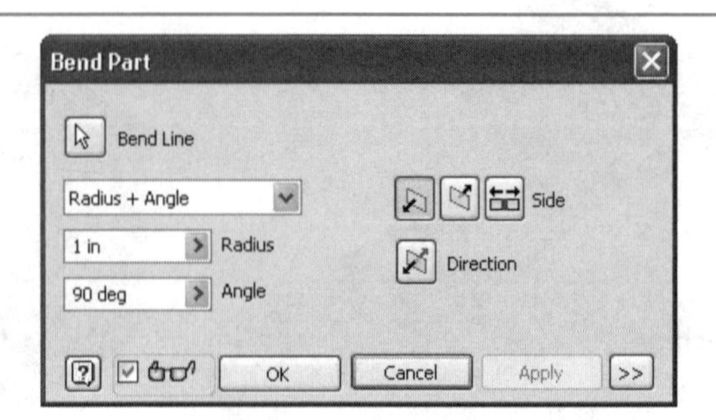

Figure 7-32.
The dialog box used for a bend operation.

Using the Split Tool

Open Example_07_20.ipt. There is an unconsumed sketch on the front face consisting of two lines and two arcs. This sketch will be used to split the part. Pick the **Split** button in the **Part Features** panel. The **Split** dialog box is opened, **Figure 7-33.** Pick the **Split Part** button in the **Method** area. This applies the operation to the part, not faces. The outline of the part is highlighted in a centerline linetype. Now, pick the **Split Tool** button, if it is not already on. Then, select the sketch in the graphics window. A preview of the split line appears in red. An arrow points in the direction of the side of the split to be removed. In the **Remove** area of the dialog box, pick one of the two buttons until the arrow points toward the top half. Then, pick the **OK** button in the dialog box to split the part. The top of the part is removed. See **Figure 7-34.**

Undo the split operation. Then, activate the **Split** tool. This time, pick the **Split Face** button in the **Method** area. With the **Split Tool** button on, pick the sketch on the part. Notice that a double arrow points perpendicular to the face. Also, the **Faces to Split** button in the **Faces** area is turned on. Select the front face on the part; it is highlighted. Finally, pick the **OK** button in the dialog box to split the face. The face is now divided into two parts. Change the color of each face by right-clicking on it and selecting **Properties** from the pop-up menu. In the **Face Properties** dialog box, select the new color. The end result is shown in **Figure 7-35.**

Splitting a Helical Gear

Open Example_07_21.ipt. This part is a gear blank with a sketch of a single gear tooth. In this example, you will sweep the sketch along a curve, create a circular

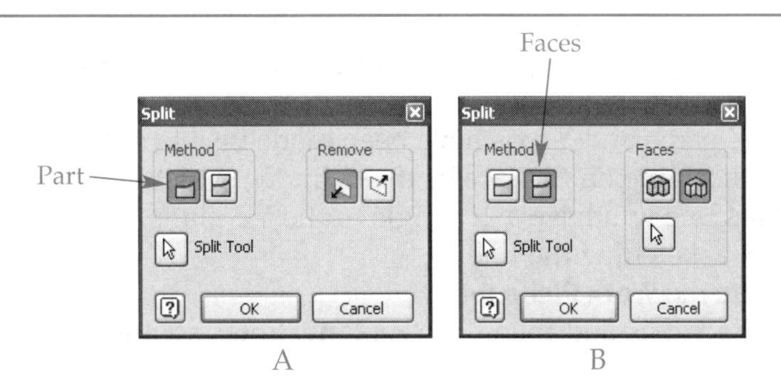

Figure 7-33.
The dialog box used for a split operation. A—Part. B—Faces.

Figure 7-34.
This part was created by splitting a block.

Figure 7-35.
The front face is split. Each face can be changed to a different color.

pattern, and trim off the excess. The end result is a helical gear. See **Figure 7-36.** Swept features are covered in detail in Chapter 10.

Gear teeth generally have a profile based on an involute curve. This complex curve can be closely approximated by calculating and creating a number of points on the profile and then drawing a spline through the points. This process, though accurate, results in large files. To show gear teeth while keeping the file small, the spline can be approximated with an arc. This process has been used by drafters for many years to show gear teeth and was used to create the tooth profile in this example. For those interested, the gear being created has a diametral pitch of five and 18 teeth, which define the diameters and profile.

The YZ plane and Y axis have been made visible in the file. These will be used to create a helical line on the face of the cylinder, which will then be used to sweep the sketch. This line will have a 21° angle to the YZ plane. You need to create a new work plane before sketching the line.

1. Activate the **Work Plane** tool.
2. Pick the Y axis and then the YZ plane. Enter an angle of –21° in the **Angle** dialog box.

Figure 7-36.
A helical gear that will be created from a "blank."

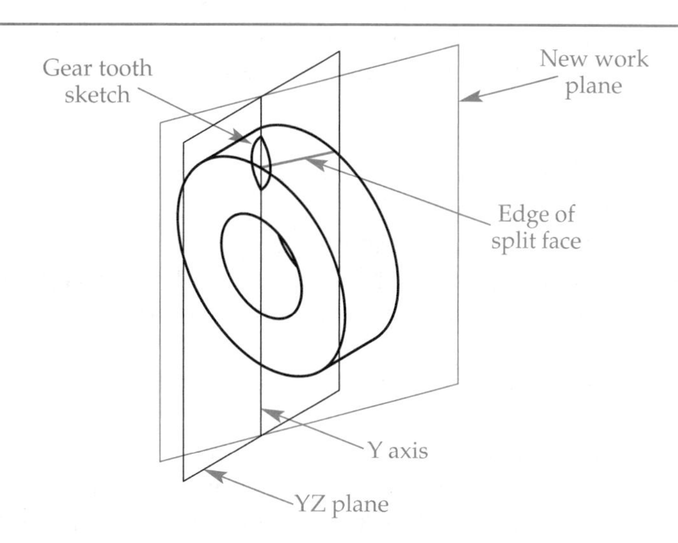

Figure 7-37.
Splitting the curved face of the blank.

3. Activate the **Split** tool. In the **Split** dialog box, pick the **Split Face** button in the **Method** area. Then, pick the **Split Tool** button and select the new work plane. Then, with the **Faces to Split** button on, select the outside curved face. Finally, pick the **OK** button to split the face. See **Figure 7-37.** Note: This actually splits the face in two places because the work plane intersects the face at the top and bottom of the part.
4. Start a new sketch on the work plane you created.
5. Use the **Project Geometry** tool to project the edge of the face split onto the sketch plane. This edge closely approximates the helical path needed to sweep the sketch.
6. Finish the sketch.
7. Pick the **Sweep** button in the **Part Features** panel. This tool is covered in detail in Chapter 10, but in this case, it is very easy to use.
8. Pick the **Profile** button in the **Sweep** dialog box and select the gear tooth sketch. Select an area as if you are extruding the profile. Then, pick the **Path** button in the dialog box and select the projection of the split face. Finally, pick the **OK** button to create the swept feature. A single tooth is created. See **Figure 7-38.**

A few more steps are required to finish this first tooth, create the pattern, and clean up the gear.

1. Using the **Fillet** tool, place a .04″ fillet on each side of the base of the tooth.
2. With the **Circular Pattern** tool, make a pattern of the tooth and the fillets. Use the cylinder for the rotation axis, and a placement of 18 at 360°.

Figure 7-38.
The profile is swept to create one tooth. The top of the tooth needs to be removed.

3. Create a sketch plane on the front of the gear. Draw a 4″ diameter circle at the gear center. In gear terminology, this dimension value is the pitch diameter plus twice the addendum.
4. Finish the sketch.
5. Using the **Split** tool, remove the portions of the gear teeth that are outside of the circle. Refer to **Figure 7-36.**

The gear is almost complete. However, use the **Rotate** tool to rotate the view and look closely at the back of the gear. A small error is present at the intersection of the tooth fillets and the back face. You may need to zoom in to see the error. To correct the error, split the gear again using a new work plane:

1. Create a work plane offset from the back face a distance of –.1″. This will put the plane "in" the part.
2. Use the **Split** tool and this work plane to remove a thin section from the back face. Make sure the direction arrow is pointing out from the back of the gear.
3. Use the **Rotate** tool to rotate the view and verify that the error has been corrected.
4. In the **Object Visibility** cascading menu in the **View** pull-down menu, select **All Work Features** to uncheck it and make the work planes and work axis invisible.

Chapter Test

Answer the following questions on a separate sheet of paper or complete the electronic chapter test on the Student CD.

1. What is the purpose of the **Shell** tool?
2. Why would faces be removed during the shell operation?
3. Which types of faces can be removed during a shell operation?
4. Give the three shell options for determining the position of the cavity wall in relation to the base part.
5. Define *web.*
6. Define *rib.*
7. Which tool is used to create a web? Which tool is used to create a rib?
8. Which type(s) of objects are used as the path by the tool(s) in Question 7?
9. What is a *sketch coordinate system (SCS)?*
10. Which tool is used to change an SCS?
11. Define *emboss.*
12. Define *engrave.*
13. Which tool is used to emboss text? Which tool is used to engrave text?
14. Which tool is used to create text in a sketch?
15. Define *aspect ratio.*
16. Which three types of "images" can be placed on a part using the **Decal** tool?
17. Define *draft angle.*
18. Which tool is used to create a draft angle on a part?
19. What is the purpose of the **Split** tool?
20. If the **Split** tool is used to split a part, what is true of one side of the split?

Chapter Exercises

Exercise 7-1. Casting.

Complete the exercise on the Student CD

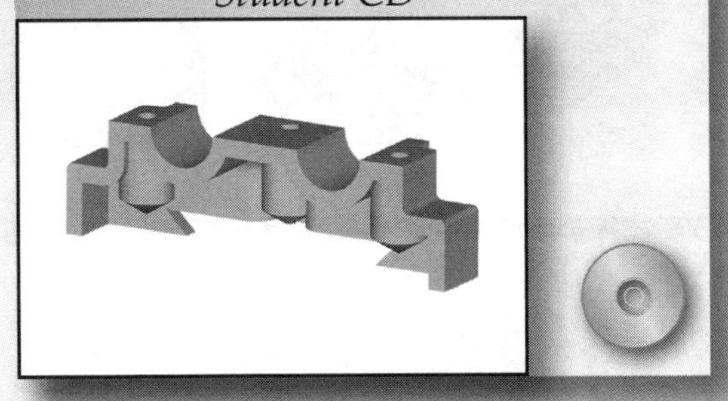

Exercise 7-2. Windshield Wiper Arm.

Complete the exercise on the Student CD

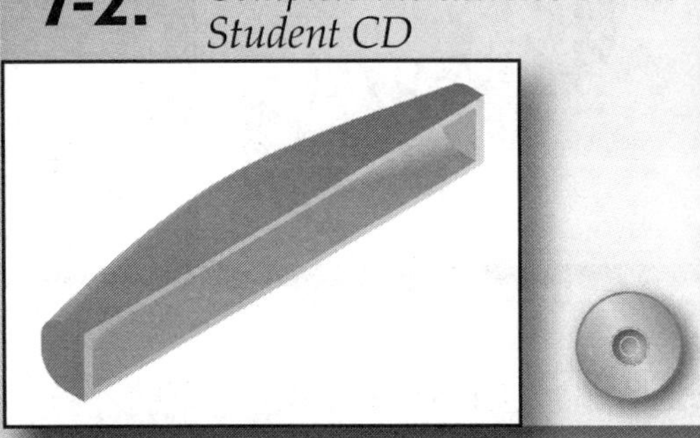

Exercise 7-3. Ribbed Collar.

Complete the exercise on the Student CD

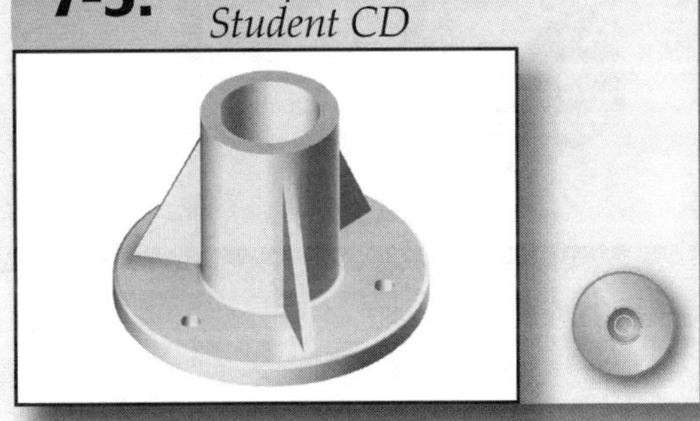

Exercise 7-4. Computer Keyboard Base.

Complete the exercise on the Student CD

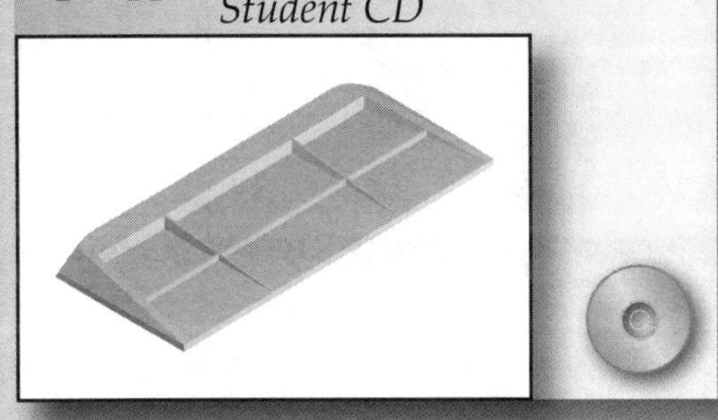

Exercise 7-5. Extruded Text.

Complete the exercise on the Student CD

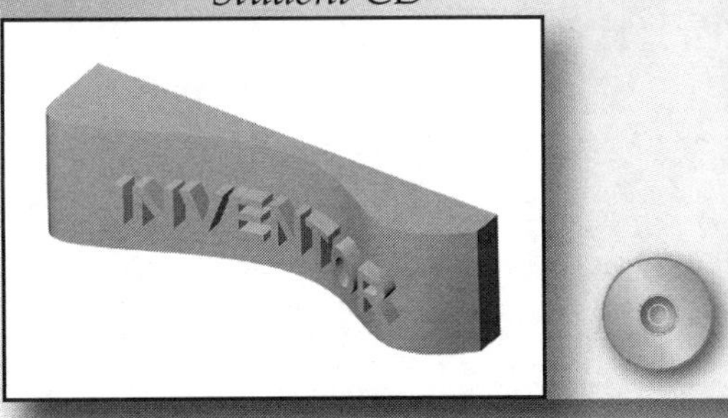

Exercise 7-6. Magnet.

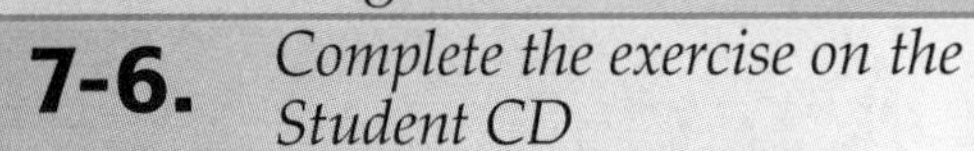

Complete the exercise on the Student CD

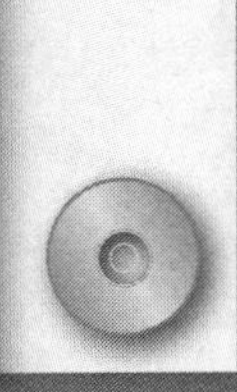

Exercise 7-7. Knurled Surface.

Complete the exercise on the Student CD

Exercise 7-8. Transformer Cover.

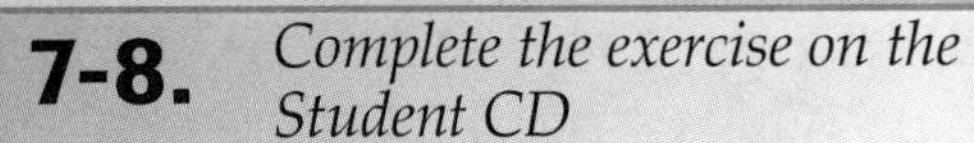

Complete the exercise on the Student CD

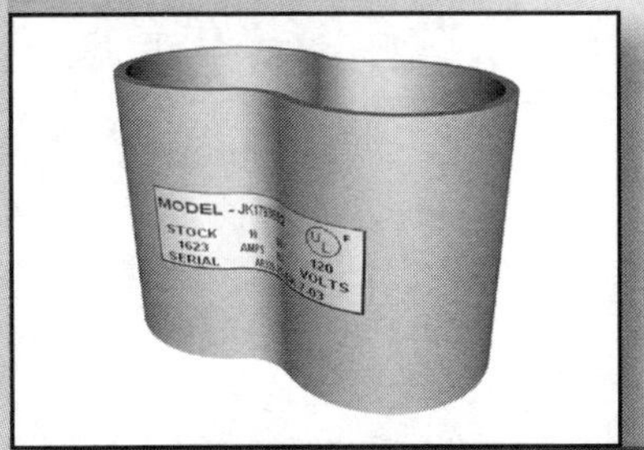

Exercise 7-9. Tapered Faces.

Complete the exercise on the Student CD

Exercise 7-10. Shaft Journal.

Complete the exercise on the Student CD

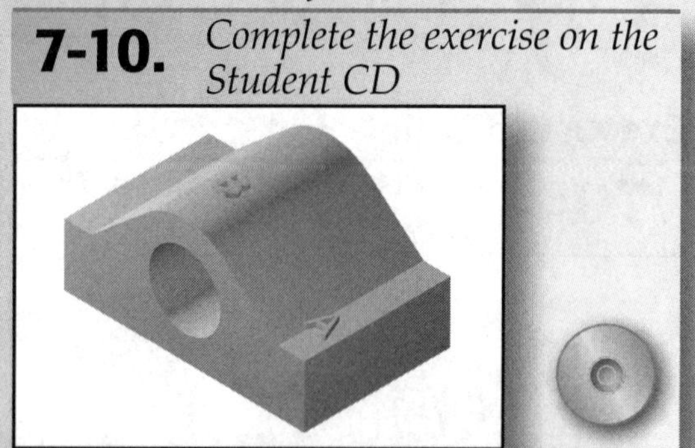

Exercise 7-11. Chisel Point.

Complete the exercise on the Student CD

Exercise 7-12. Yoke.

Complete the exercise on the Student CD

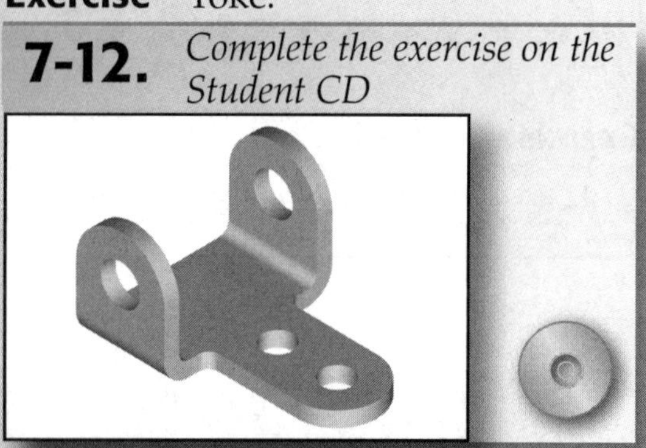

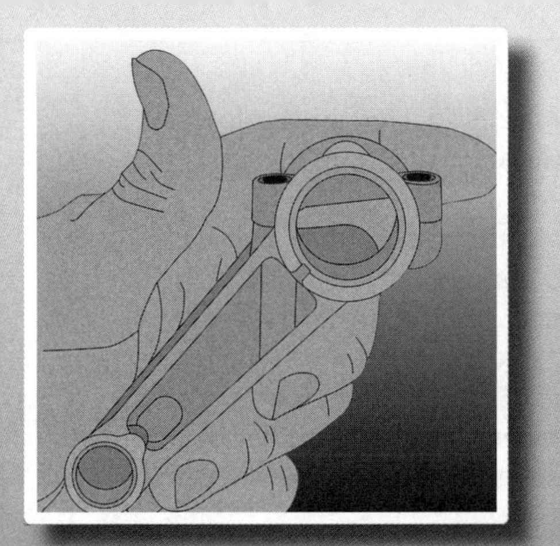

Chapter 8

Creating Part Drawings

Objectives

After completing this chapter, you will be able to:

- Create 2D drawings of Inventor solid and surface models.
- Explain the different views and how to create them.
- Specify paper size and border.
- Edit existing drawing views.
- Update drawing views when the part changes.

User's Files

The Student CD included with this text contains several files required for this chapter. Refer to the file File List.txt *in the* \Ch08 *folder for the comprehensive list.*

Creating a Layout Drawing

Engineering drawings can be directly developed from Inventor parts, surfaces, and assemblies. This chapter discusses creating the various types of view for parts. These views are directly related to the part model. Any changes in the model are automatically reflected in the drawing.

Open Example_08_01.ipt. This is a cast and machined part. Notice the part's orientation to the coordinate system. Remember, the X axis is red on the CSI, the Y axis is green, and the Z axis is blue. This will help you understand the creation of the drawing views. Pick **New...** from the **File** pull-down menu. In the **Default** tab of the **New** dialog box, notice that there are two types of drawing files: DWG and IDW. The DWG file type is native to AutoCAD. The IDW file type is the native drawing in Inventor. Double-click on the Standard.idw icon to start a new Inventor drawing.

The drawing is C-size with US customary (inch) units. The new drawing consists of a sheet, border, and title block. See **Figure 8-1.** The ***sheet*** is the beige drawing area. It determines the size of the border and the title block. For example, if you change the size of the sheet, the border and title block change with it. The ***border*** is the thick rectangle with the alphanumeric zone labels. The ***title block*** is located in the lower-right corner of the border and consists of areas for the drawing title, date, designer's initials, sheet size, etc.

Notice that this drawing's current sheet size is C. This is the default size. Suppose a D-size sheet is needed. In Inventor, the **Browser** is used to change the sheet size. Find Sheet:1 in the **Browser** and right-click on it. Select **Edit Sheet...** in the pop-up menu to open the **Edit Sheet** dialog box. In the **Size** drop-down list, select D to change the sheet size. In the **Name** text box, change the name of the sheet to D-Size. See **Figure 8-2.** Pick **OK** to exit the dialog box. In the graphics window, the larger sheet is displayed, the border expands to fill the new sheet, and the title block text indicates a D-size sheet.

Figure 8-1.
A new Inventor drawing file with the default border and title block.

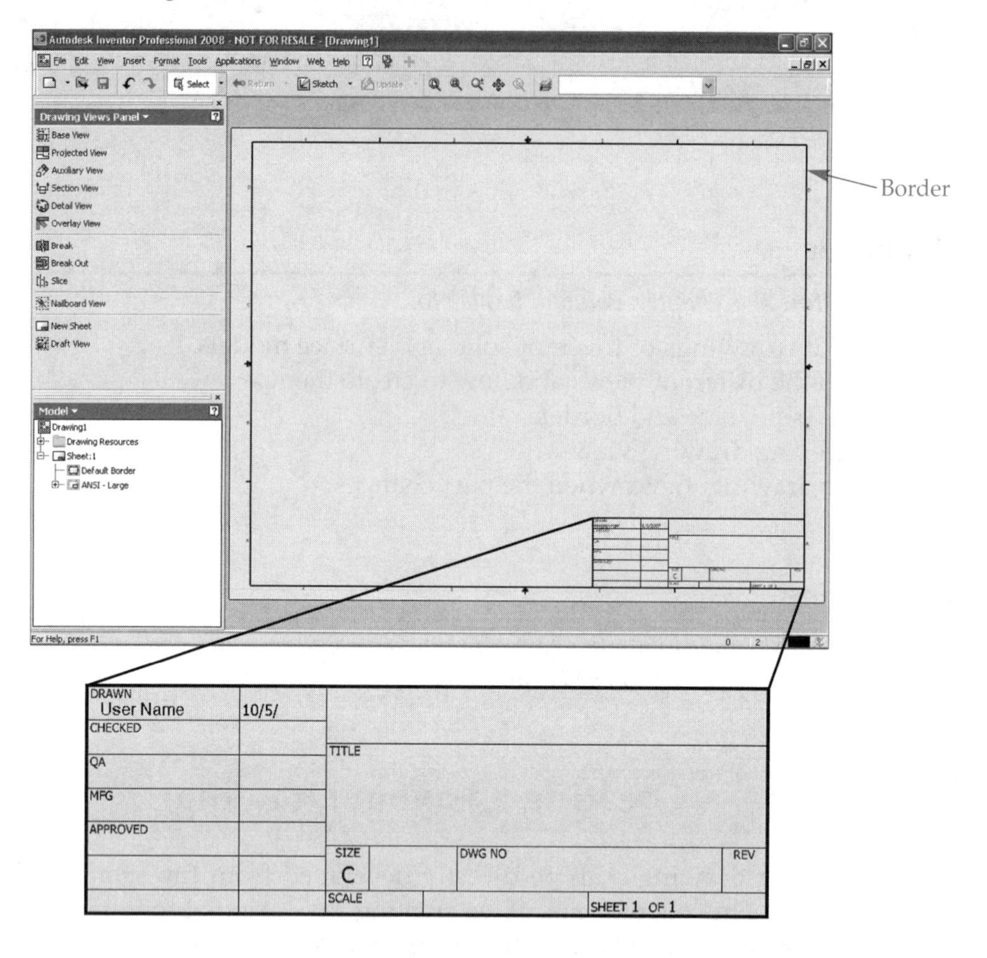

Figure 8-2.
The **Edit Sheet** dialog box is used to change the sheet size, rename the sheet, and change the sheet orientation.

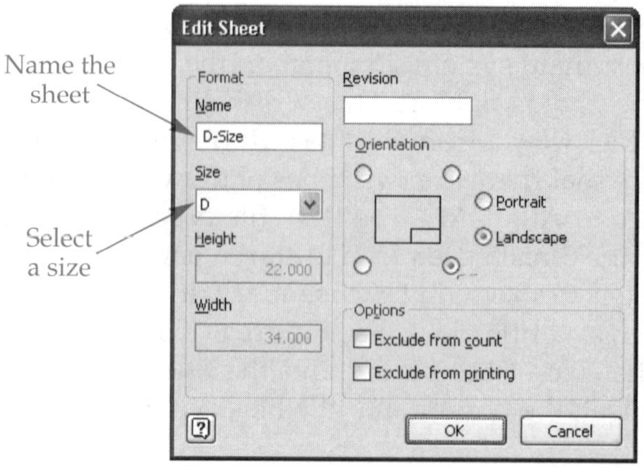

Name the sheet

Select a size

Creating the Drawing Views

As a 2D drawing layout is created, you are actually working on a separate drawing file. However, when drawing views of a part are created, that part file is linked to the drawing file. This means that any future changes made to the part are automatically updated in the drawing view. This also works the other way. Change a dimension on the drawing view and the part model updates to reflect the new values.

Save the new drawing you created as Example_08_01.idw. Now, pick the **Window** pull-down menu and notice that there are two Inventor files open. The IPT file—Example_08_01.ipt—is the part model for this example and the IDW file is the 2D drawing that you are creating.

PROFESSIONAL TIP

More than one part file can be used to create drawing views in a single drawing layout. For example, you may need to create a sheet of details that contains views of many parts.

Creating a Base View

The next step is to create drawing views. Start by creating a ***base view,*** or ***top view,*** from which all others will be generated. To create the base view, pick the **Base View** button in the **Drawing Views Panel**, select **Insert>Model Views>Base View** from the pull-down menu, or right-click on D-Size:1 in the **Browser** and choose **Base View...** from the pop-up menu. The **Drawing View** dialog box is displayed. See **Figure 8-3.**

CAUTION

Carefully follow the next step because you can do things in the graphics window while the dialog box is still open.

Before you change anything in the dialog box, move the cursor around the graphics window. Notice the cursor and image of the part in **Figure 8-3.** The image will be the base view if a point is selected on the sheet. However, do not select a point yet. If you inadvertently picked a point, just undo and start again.

The **File** drop-down list in the **Drawing View** dialog box shows the opened part model as the source for the 2D drawing layout. All open part files will be available in this drop-down list. If no part file is open, you can specify another part file as the source by picking the **Explore Directories** button next to the drop-down list.

The scale of the drawing view is entered in the **Scale** area of the dialog box. Certain standard scales can be selected by picking the arrow next to the text box. For this example, enter a value of 1.5 in the text box. All other views inherit this scale.

A name for the view is entered in the **Label** area of the dialog box. Labels are usually used for section views or detail views. By default, the view name (label) is not displayed in the drawing. To display the view name in the drawing, pick the **Toggle Label Visibility** button next to the text box in the **Label** area. The button displays a yellow lightbulb when the label will be placed on the drawing.

In the **Orientation** area of the dialog box, you can select any one of the standard orthographic or isometric drafting views for the base view. Remember, all additional views that are created will be generated from the base view, so choosing the proper start is important. For this example, select Front in the list.

There are three buttons in the **Style** area of the dialog box. These buttons determine how hidden lines are displayed and whether or not the part is shaded in the

Figure 8-3.
The **Drawing View** dialog box is used to create, scale, and label different views.

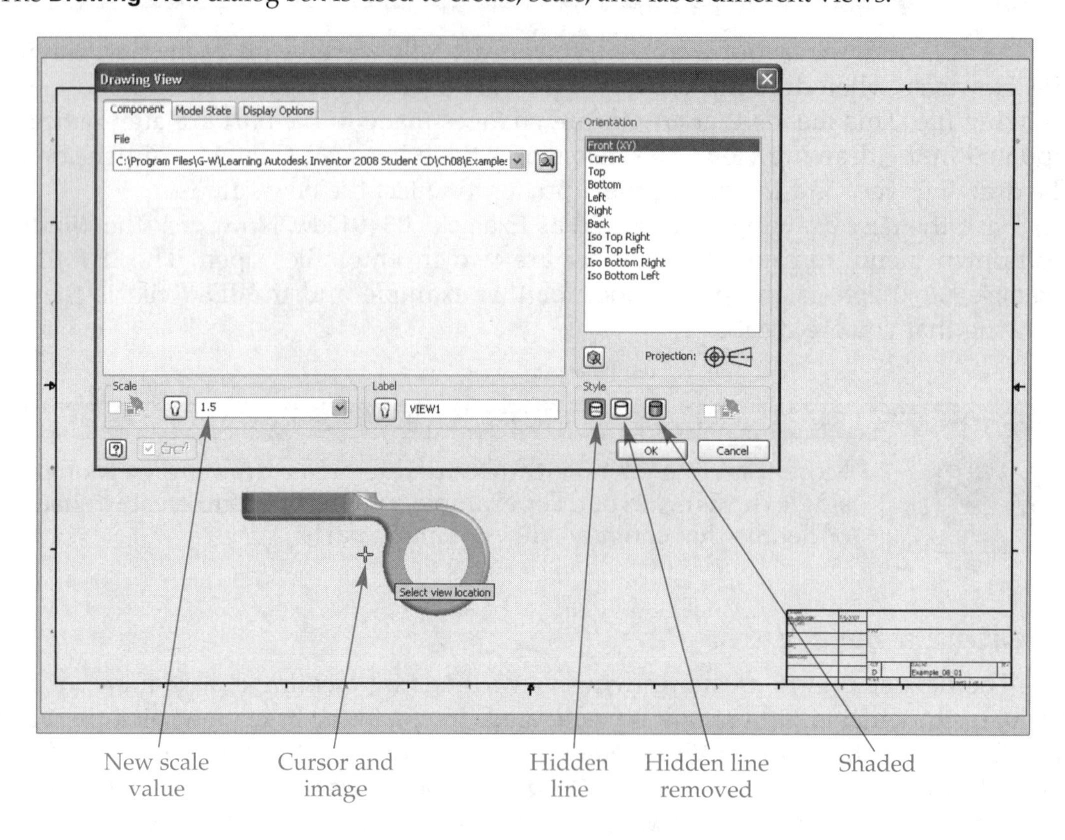

view. When the **Hidden Line** button is on, visible and hidden lines are shown. When the **Hidden Line Removed** button is on, only visible lines are shown. Either of these views can be shaded. When the **Shaded** button is on, the part is shaded. The **Shaded** button is often used for an isometric view, as you will create later in this example. A shaded view also helps in visualizing the shape of the final part. For this example, pick the **Hidden Line** button and turn the **Shaded** button off.

Finally, pick a point in the lower, left-hand corner of the drawing sheet to place the view. Move the dialog box around to make room, if needed. When you pick, the front view of the part is created and the dialog box automatically closed. What Inventor calls the front view is based on the orientation of the part to the coordinate system in the part file. If you switch back to the part file, notice that the front view in the drawing corresponds to a plan view of the XY plane. All subsequent drawing views are generated from this base view using standard orthographic projection. In the **Browser** in the drawing file, the base view is listed as VIEW*n*:*part file name*. Left-click two single times (not double-clicking) on the name in the **Browser** and rename it Front View. Notice that the part file name remains in the view name. Also, look at the title block. The drawing number is automatically entered as the part file name. Some other information, such as the drawn by name and the date, is also automatically entered.

PROFESSIONAL TIP

If you want to change the scale of the linetypes used in the drawing views, pick **Style and Standard Editor...** in the **Format** pull-down menu. Select Default Standard (ANSI) in the Standard branch. Change the value in the **Global Line Scale** text box and pick **Save** to immediately see the changes. Pick **Done** to exit the dialog box.

Projecting Views

Now, three views will be projected from the base view (front view): top, right side, and isometric. In the **Browser**, right-click on Front View to display the pop-up menu. Then, pick **Create View>Projected View**. A line appears in the graphics window that originates from the base view and follows the cursor. Pick a point directly above the front view. A rectangle representing the view appears in the drawing. Now, move the cursor to the right of the front view and pick to create a right-side view. Another rectangle appears in the drawing. Next, move the cursor to the upper-right corner of the drawing sheet and pick to create an isometric view. Finally, right-click in the drawing and select **Create** from the pop-up menu. The three rectangles are replaced by the respective views.

Isometric views generally look better if the scale is less than the corresponding orthographic views. Sometimes, they are also shaded. Move the cursor over the isometric view until the border of the view is highlighted. Right-click and select **Edit View...** from the pop-up menu. In the **Drawing View** dialog box, change the scale to 1.25 and pick the **Shaded** button to shade the view. Pick the **OK** button to close the dialog box.

Views can be moved after they have been placed. Move the cursor over the view until the border is highlighted. Then, pick and drag the view to the new location. The top and right-side views can only be moved in a straight line from the base view. However, the isometric view can be moved to anywhere on the drawing.

This drawing also has 24 ***layers***, each one for a specific type of drawing entity. These layers can be controlled in the **Style and Standard Editor**. Select **Style and Standard Editor**... from the **Format** pull-down menu. Expand the Layers branch and select any layer name. In the **Layer Styles** area on the right side of the dialog box, you can turn the layer on and off, change its color, change its lineweight, and set whether or not the layer is plotted. See **Figure 8-4.** To change one of the layer settings, pick on the

Figure 8-4.
Editing layer properties.

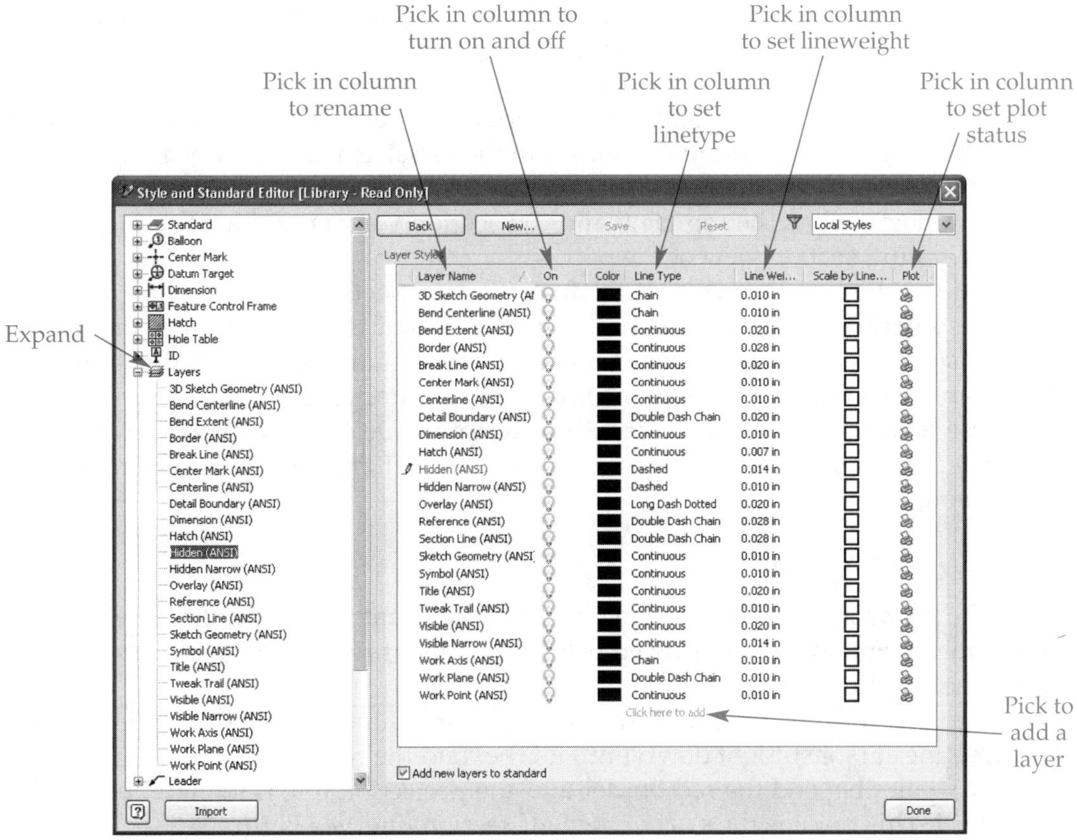

icon in the column for the layer. New layers can also be added to the standard using this dialog box.

You have now seen how to set up a basic orthographic drawing of a part. In the next section, you will learn how to create other types of drawing view. Save and close both Example_08_01.ipt and Example_08_01.idw.

Creating Other Views

Inventor can be used to create a number of other views using the base view. These views include full sections, half sections, aligned sections, isometric views, auxiliary views, detail views, broken views, and break out views. With the exception of break out views, the following sections describe these views. ***Break out views*** are used to remove covering geometry to reveal the inner geometry of a part. These are covered in detail in Chapter 17.

Full Section View

Open Example_08_02.ipt and Example_08_02.idw. Notice in the drawing file that a view is outside of the drawing sheet. Views can be "parked" outside of the sheet area. They will not print, but can be used to create other views. In the **Browser**, right-click on Top View. Pick **Create View>Section View...** in the pop-up menu. Notice that there is now a red border around the top view in the graphics window. This will be important in the next steps. Follow the steps listed below to define two points of the section line to create a full section view. Select your mouse picks carefully.

1. Hover over the center of the view until the center point is acquired, but do *not* pick.
2. Move the cursor straight to the left until it is beyond the red border. Keep a straight projection line as you do this. A dotted line gives you visual feedback that you are keeping the line straight. Once the cursor is at a point beyond the red border and the projection line is straight, pick that point. This is the start of the section line.
3. Move the cursor in a straight line to the right until it is beyond the red border on the right side of the view. Pick that point.
4. Right-click and choose **Continue** in the pop-up menu. This displays the **Section View** dialog box. Make sure A is shown in the **Label** text box and not anything else, such as A-A or B. Also, select 2:1 in the **Scale** drop-down list.
5. Pick a point below the top view and the section view is created. See **Figure 8-5.** This is a full section view of the part.

Create another section view. This time, do not select the center of the part—just sketch a straight line through the part. Then, pick and drag the cutting line. Notice how the section view is redefined. If needed, the section view can be redefined by picking and holding the cutting-plane line and then dragging it to a new location.

These two section views are not fully descriptive. Delete both of the section views just created by picking their borders and pressing the [Del] key. Then, continue to the next section.

Half-Section View

In the **Browser**, right-click on Top View and pick **Create View>Section View...** in the pop-up menu. Follow the steps listed below to define three points on a section line to create a half-section view.

1. Hover over the center of the view until the center point is acquired, but do *not* pick.
2. Move the cursor straight down until it is beyond the red border. Once your cursor is at a point beyond the red border and the projection line is straight, pick to set point 1. This point is the start of the half-section line. See **Figure 8-6.**

Figure 8-5.
The cut line and the resulting section view.

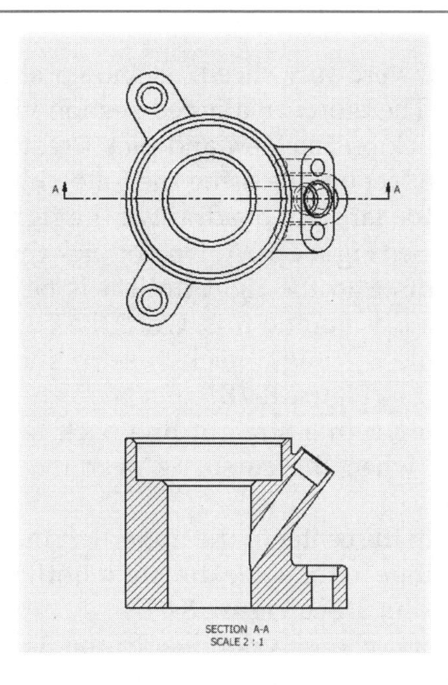

Figure 8-6.
The cut line and the resulting half-section view.

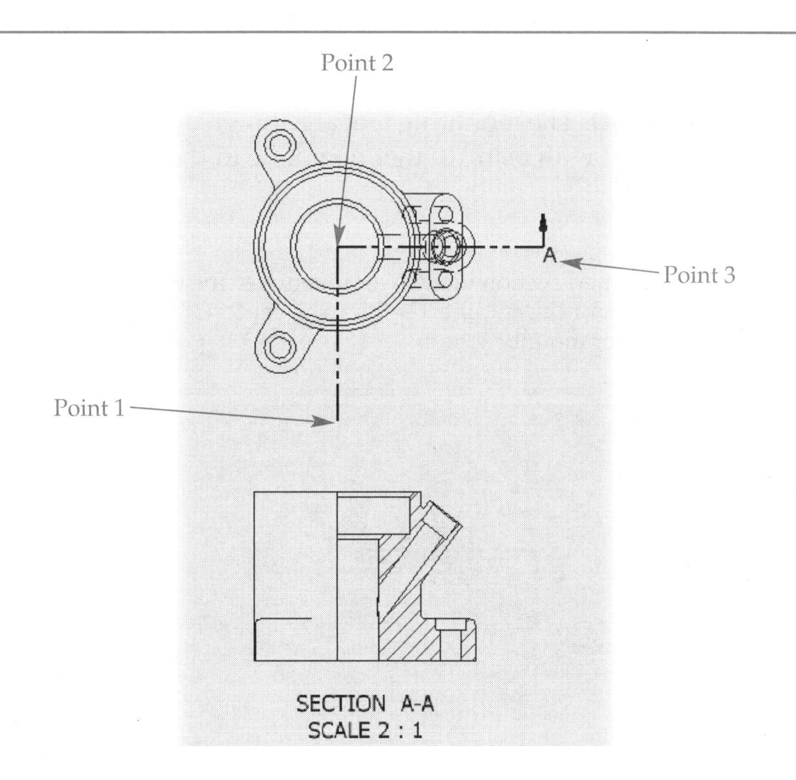

3. Move the cursor in a straight line back to the center of the view. A green dot will appear when the cursor is over the center. Pick this point, point 2.
4. Now, move the cursor to the right until it is beyond the red border on the right side of the view. Once the cursor is at a point beyond the red border and the projection line is straight, pick to set point 3.
5. Right-click and select **Continue** in the pop-up menu. In the **Section View** dialog box, make sure that the section label is A. Also, select 2:1 in the **Scale** drop-down list.
6. Pick a point on the drawing below the top view.

If you look closely, the mounting foot is not depicted very well in this new section view. The next section covers how to correct this.

Aligned Section View

If the part is vertically sliced, as shown above, the mounting feet do not lie in the same plane. Therefore, an aligned section view will best describe the part. In the **Browser**, right-click on Top View and pick **Create View>Section View...** in the pop-up menu. Create the section line using the following steps.

1. Hover over the far right quadrant of the right-hand mounting foot until the point is acquired. See **Figure 8-7A.** Do not pick this point.
2. Move the cursor to the right until it is beyond the red border. Keep a straight projection line as you do this. Once the cursor is beyond the red border and the projection line is straight, pick to set point 1. This point is the start of the half-section line. See **Figure 8-7B.**
3. Move the cursor in a straight line back to the center of the view. A green dot will appear when the cursor is over the center. Pick this point, point 2. See **Figure 8-7C.**
4. Then move to the center of the upper-left mounting foot and hover to acquire that point. Continue in the same direction until the cursor is outside of the red border. Pick to set point 3. See **Figure 8-7D.**
5. Right-click and choose **Continue**. In the **Section View** dialog box, make sure the section label is B. Also, select 2:1 in the **Scale** drop-down list.
6. Pick a point on the drawing below the first section view.

The finished section view is shown in **Figure 8-8** along with the base view and the first section view. It is clear that an aligned section view best depicts the front view of this part. The mounting feet are shown according to drafting convention for a cylindrical part with features that do not lie in the same plane.

Figure 8-7.
Creating an aligned section view. A—Hover over the edge of the mounting foot. B—Pick a point to the right of the cut line. C—Pick the center of the part. D—Pick the final point beyond the upper mounting foot.

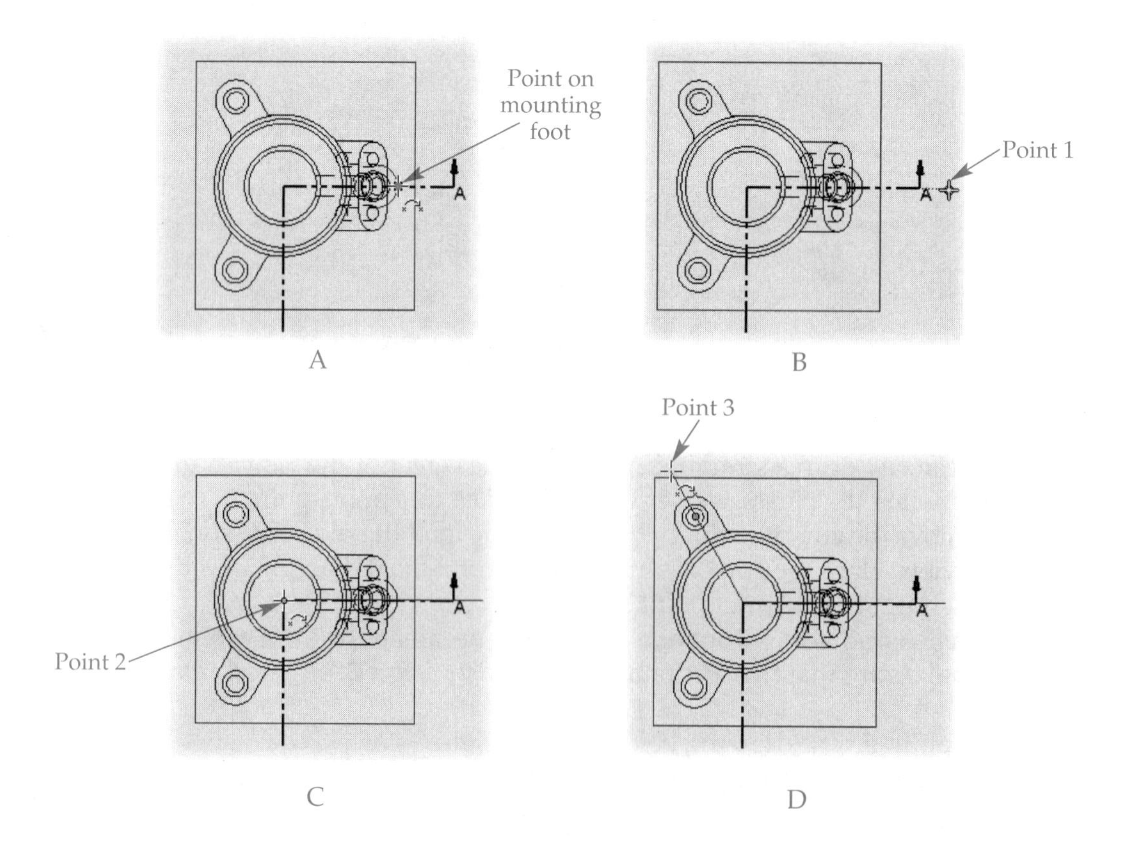

Figure 8-8.
From top to bottom: top view, half-section view, and aligned section view.

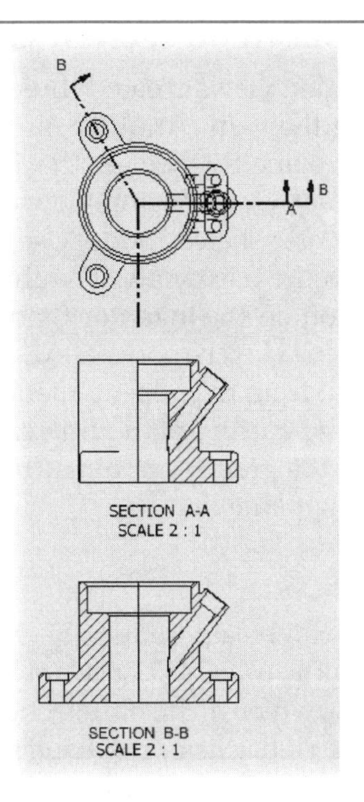

If you need to move the views to arrange them on the sheet, simply pick their edges and drag them to a new location. Notice that the projected views are constrained to the base view from which they are derived. Move the base view left or right and all other views follow.

Try moving the cutting-plane lines for these views in the top view. You will find that you will not be able to because of the way they were defined. As the points were picked, Inventor placed constraints on the lines. They are, in essence, "nailed down" to the part's features.

CAUTION

Be careful renaming the views in the **Browser** because the name affects the label shown under the section views. For example, if you rename the view to Section A-A in the **Browser**, then the label under the view reads SECTION Section A-A-Section A-A.

Section Views with Depth

The depth, or thickness, of a sectioned slice can be specified. Before placing the view, in the **Section View** dialog box, pick Distance in the drop-down list in the **Section Depth** area. Then, enter a distance in the text box below the drop-down list. As the section view is dragged in the drawing, a thick black line appears in the base view along with the cutting-plane line. This thick line represents the offset. When the section view is placed, the view depicts a slice of the part between the cutting-plane line and the thick line.

You can change the section depth at any time by right-clicking on the view on the sheet or its name in the **Browser** and choosing **Edit Section Properties...** from the pop-up menu. You also can convert it back to a full-depth section view.

Section Views Based on Sketch Geometry

For all of the section views created thus far, the cutting-plane line was specified *on the fly* before placing the view. Another method uses existing sketch geometry as the basis for the cutting-plane line. Select the Top View by picking in it once or selecting its name in the **Browser**. Once the top view is highlighted, pick **Sketch** on the **Inventor Standard** toolbar to enter sketch mode. Use the **Line** tool in the **Drawing Sketch Panel** to draw a line across the base view. The line does not need be centered on the part. Pick the **Return** button on the **Inventor Standard** toolbar or select **Finish Sketch** from the pop-up menu to exit sketch mode. Next, right-click on the sketched line and choose **Create Section View...** from the pop-up menu. Then, place the section view in the same manner as before. The cutting-plane line can be moved in the base view by picking and dragging one of the green endpoints. The section view changes to reflect the new position of the cutting-plane line.

Isometric Views

Isometric views can be automatically created in Inventor. If you have ever drawn an isometric view with AutoCAD or a drafting board, you can really appreciate this feature. SECTION A-A, which is the half-section view, will be used to create an isometric view. In the **Browser**, right-click on A:Example_08_02.ipt; this is the half-section view. In the pop-up menu, pick **Create View>Projected View**. A projection line appears from SECTION A-A to the cursor and a preview appears attached to the cursor. Pick a point to the upper right of SECTION A-A. Now, pick a second point to the lower right, thereby creating two isometric views. Right-click and choose **Create** in the pop-up menu. See **Figure 8-9.**

The two isometric views can be moved by dragging them. It is nice to keep them vertically lined up, so pick both of them at the same time by holding down the [Shift] key. Then, drag both views to the new position.

In the **Browser**, right-click on the name of one of the isometric views and choose **Edit View...** from the pop-up menu. In the **Style** area of the **Drawing View** dialog box, uncheck **Style from Base** and then pick the **Shaded** button. Also, pick the **Display Options** tab and check the **Tangent Edges** and **Foreshortened** check boxes. Finally,

Figure 8-9.
Isometric views of the part.

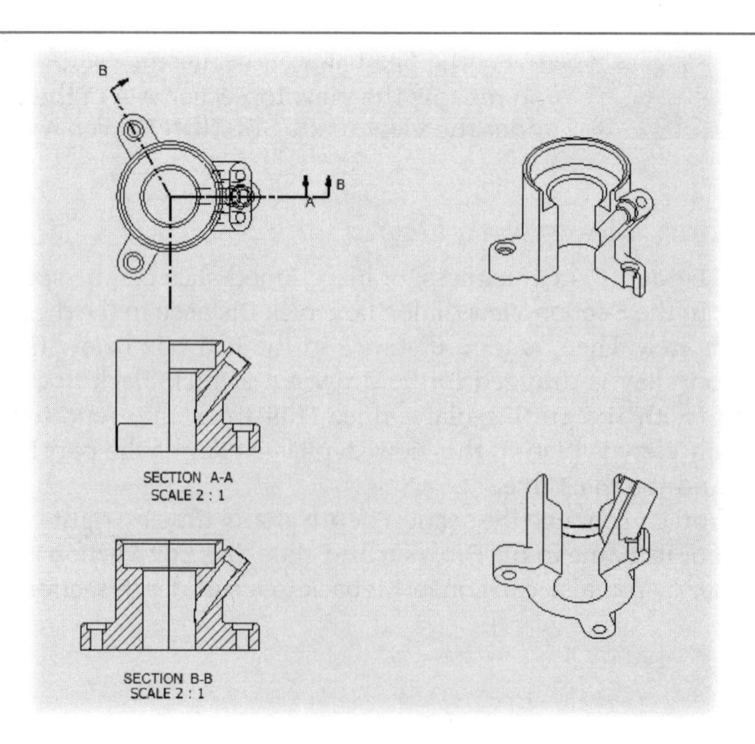

Figure 8-10.
The two isometric views are shaded.

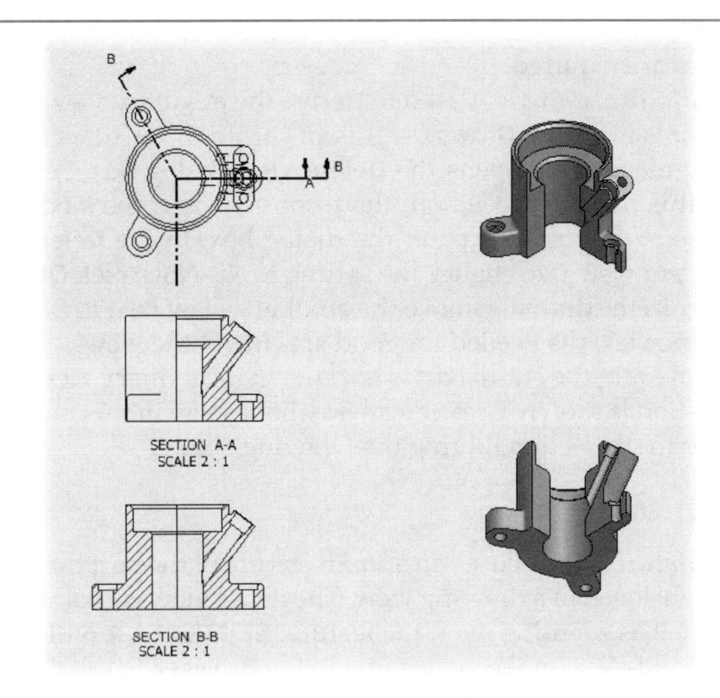

pick **OK** to exit the dialog box. The isometric view is now a shaded view. Do the same for the other isometric. See **Figure 8-10.**

Auxiliary Views

An ***auxiliary view*** can be used to get a true representation of any inclined or oblique face on a part. The angled feature on this example requires an auxiliary view to show the inclined face so it can be dimensioned. SECTION A-A will be used to create the auxiliary view. SECTION B-B is not a true representation of the part. In fact, Inventor will not allow SECTION B-B to be used to create an auxiliary view.

Right-click on Front View in the **Browser**. This is the view that is parked off of the paper. Pick **Create View**>**Auxiliary View...** in the pop-up menu. This opens the **Auxiliary View** dialog box and you are prompted to select a line. Leave the defaults in the dialog box and pick the line/edge shown in **Figure 8-11.** Move the cursor to the upper right

Figure 8-11.
Pick a point of reference for the auxiliary view.

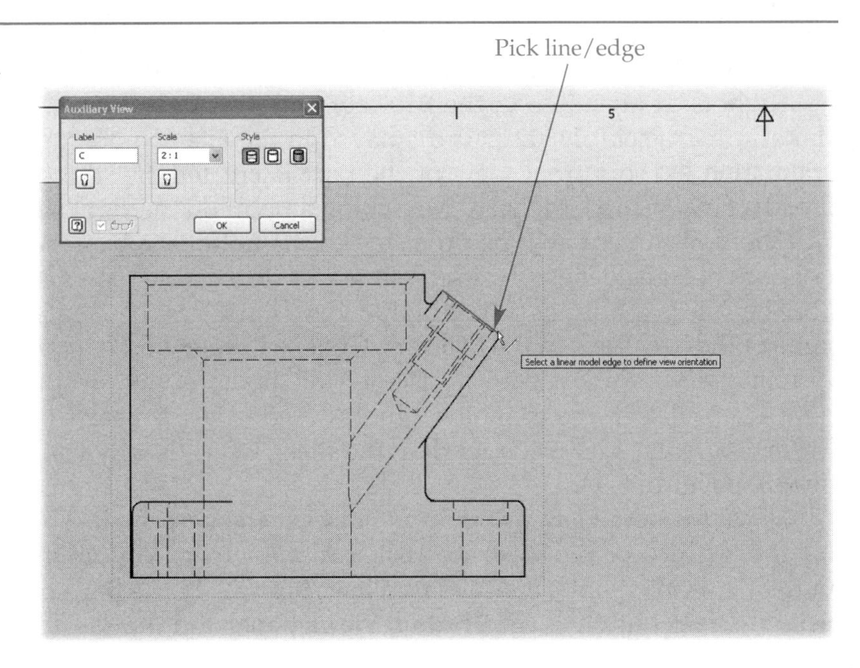

and place the view anywhere. This is a true view of the angled feature; however, some changes are required.

Now, the front view used to derive the auxiliary view can be deleted. Right-click on Front View in the **Browser**—it is in the Top View branch—and select **Delete** in the pop-up menu. This opens the **Delete View** dialog box. See **Figure 8-12.** Do not pick **OK** at this point because both the front view and the auxiliary view will be deleted. Pick the **>>** button to expand the dialog box. In the **Delete** column, pick Yes for the dependent view C to change the setting to No. Now, pick **OK** and only the front view is deleted. In the drawing, move the auxiliary view next to SECTION A-A. Also, move the isometric views as needed for good spacing of the views.

Generally, the entire part is not shown in auxiliary views. Usually, only the feature in question is shown. Creating an auxiliary view in this manner is covered later in the chapter in the section Editing the Drawing Views.

Detail Views

A ***detail view*** focuses on a small feature of the part that cannot be properly shown or dimensioned in a drawing view. The detail view is typically drawn at a large scale—think of it as zooming in on the feature. SECTION A-A will be the source for the detail view. Right-click on the view name in the **Browser**. Select **Create View>Detail View...** in the pop-up menu. The **Detail View** dialog box is opened. Changes to the scale and the view label are made using this dialog box.

Notice the prompt Select the center point of the fence. The start point, point 1, is the center of the detail circle. See **Figure 8-13.** Next, pick point 2 to define the detail circle (or fence). You are then prompted to Select a location for the view. Move the cursor to the upper right and pick a point to place the view. Make note of the label and scale of the detail view.

In the parent view (SECTION A-A), pick the circle. Select one of the green points (other than the center point) and drag the circle to a larger size to include more of the part. Notice that the detail view changes to reflect the change in the circle.

Save the drawing file. Then, close the drawing file and Example_08_02.ipt.

Broken Views

A ***broken view*** is used on long parts where a large section of the part contains no features and does not need to be shown. In effect, a broken view shortens the part so the view will fit on the drawing sheet. Open the file Example_08_03.ipt. Also, create a new drawing using the ANSI(in).idw template. Change the sheet to a D-size sheet and rename it D-Size. Save the drawing file as Example_08_03.idw.

Right-click on D-Size in the **Browser** and pick **Base View...** in the pop-up menu. Make sure Example_08_02.ipt is displayed in the **File** text box. Also, select Front in the **Orientation** list so a front view of the part is created. Set the scale to 1:1, if it is not already. Finally, place the view by picking a point on the right side of the sheet.

The broken view will be projected to the left of the base view. The **Broken View** tool does not create views by itself—it works on existing views only. In order to create a broken view, a normal orthographic view must first be created. Right-click on the name of the base view in the **Browser**. Choose **Create View>Projected View** in the pop-up menu. Then, pick a point to the left of the base view to place the orthographic view, right-click in the graphics window, and select **Create** from the pop-up menu. The orthographic view is larger than the sheet, which is why you are creating a broken view. See **Figure 8-14.**

Move the new view so that it is not over the base view. Then, pan and zoom so that the entire view and sheet are visible on the screen. Now, in the **Browser**, right-click on the new view name. Pick **Create View>Break...** in the pop-up menu. You can also pick the **Break** button in the **Drawing Views** panel and then select the view. The **Break**

Figure 8-12.
The **Delete View** dialog box is used to select which views and their dependent views will be deleted.

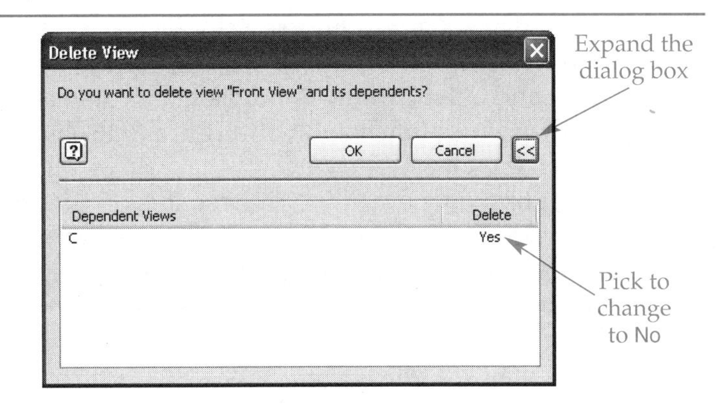

Figure 8-13.
Pick two points to define the detail view.

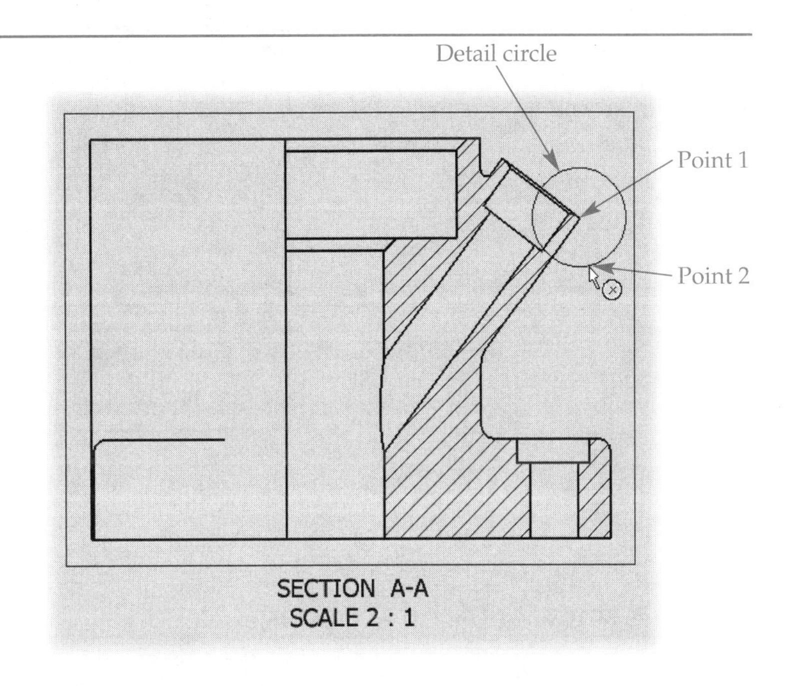

Figure 8-14.
The view runs off of the sheet before the broken view is defined.

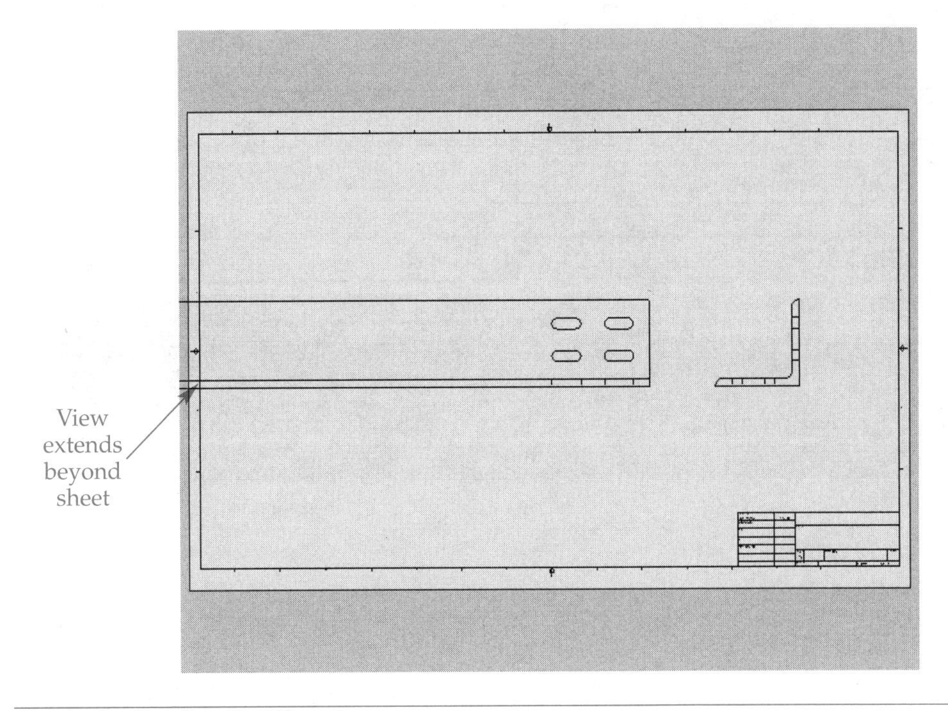

dialog box is displayed. See **Figure 8-15.** Pick the **Structural Style** button in the **Style** area, pick the **Horizontal Orientation** button in the **Orientation** area, enter 1.000 in the **Gap** text box, and enter 2 in the **Symbols** text box. Leave the slider at its default setting. Now, pick the two points shown in **Figure 8-16.** Everything *between* these points is removed from the view.

Move the broken view to the right so it is on the sheet. Now, zoom in to show the broken view. See **Figure 8-17.** Notice that there are two break symbols in each break line. Also, notice there is no specific broken view entry in the **Browser**.

To change the style of the break after the view is created, do the following. The style and display settings can be changed, but not orientation.

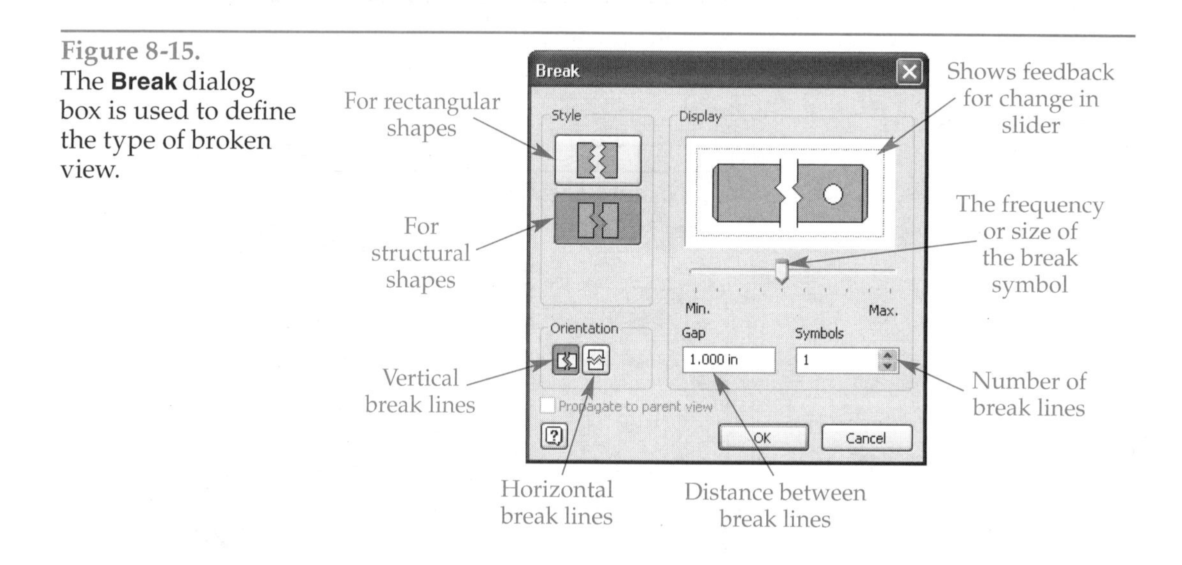

Figure 8-15.
The **Break** dialog box is used to define the type of broken view.

Figure 8-16.
Two points are used to define the break lines.

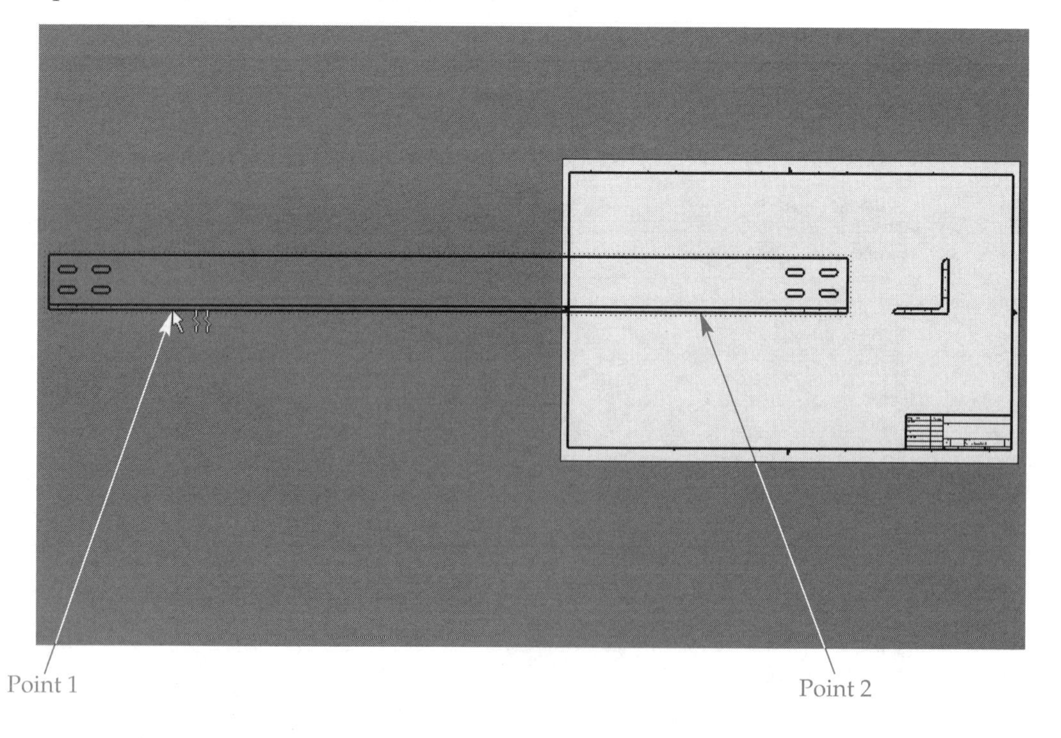

Figure 8-17.
The resulting broken view.

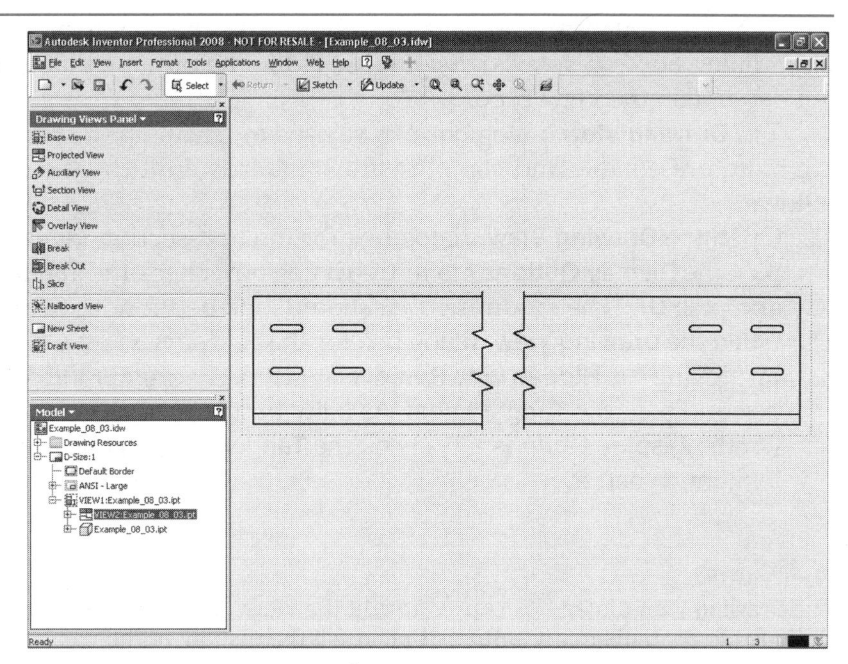

1. Hover the cursor over the break until the break is highlighted and a green dot appears.
2. Right-click and select **Edit Break...** from the pop-up menu. The **Break** dialog box is displayed.
3. Pick the **Rectangular Style** button and change the gap to 2″. The rectangular style is used for nonstructural shapes.
4. Pick **OK** to update the feature.
5. Save the drawing.

Draft Views

A draft view is created when 2D geometry needs to be created on the drawing sheet. There is no 3D model that can form the basis for the 2D geometry, so it must be manually created. To create a draft view, pick the **Draft View** button in the **Drawing Views Panel**. In the **Draft View** dialog box that appears, enter a label and scale. Then, pick **OK** to close the dialog box. Sketch mode is entered and the **Drawing Sketch Panel** is displayed. Using the tools in the **Drawing Sketch Panel**, create the needed 2D geometry. Then, right-click and select **Finish Sketch** from the pop-up menu. The 2D geometry is now in the drawing. It can be relocated on the sheet as needed. Also, notice the entry in the **Browser** for the draft view.

NOTE

Overlay views are used for assembly drawings and are discussed in Chapter 17.

Editing the Drawing Views

You now know how to create drawing views, but it is important to know how to edit them as well. Open the drawing Example_08_04.idw. This is the drawing completed earlier and consists of a base view and a number of drawing views. Right-click on a drawing view or its name in the **Browser**. Select **Edit View...** from the pop-up menu. You can also double-click on the view in the graphics window. The **Drawing View** dialog box is displayed. See **Figure 8-18.** The options that are enabled or grayed out depend on the view type.

The **Drawing View** dialog box can be used to "clean up" drawing views so that they are more presentable and comply with standards. Edit the views in this drawing as follows.

1. Open the **Drawing View** dialog box for the half-section view, SECTION A-A. Then, pick the **Display Options** tab in the dialog box, check the **Tangent Edges** check box, and pick **OK**. The mounting foot should look better now. See **Figure 8-19.**
2. Open the **Drawing View** dialog box for the auxiliary view. Change the display style by picking the **Hidden Line Removed** button. Also, enter VIEW C-C in the **Label** text box and pick the **Toggle Label Visibility** button to turn on the display of the label. On the **Display Options** tab, check the **Tangent Edges** check box. The view should look much better.

Figure 8-18.
The **Drawing View** dialog box is used to edit the view. A—The **Component** tab. B—The **Display Options** tab contains many settings that affect the display of the view.

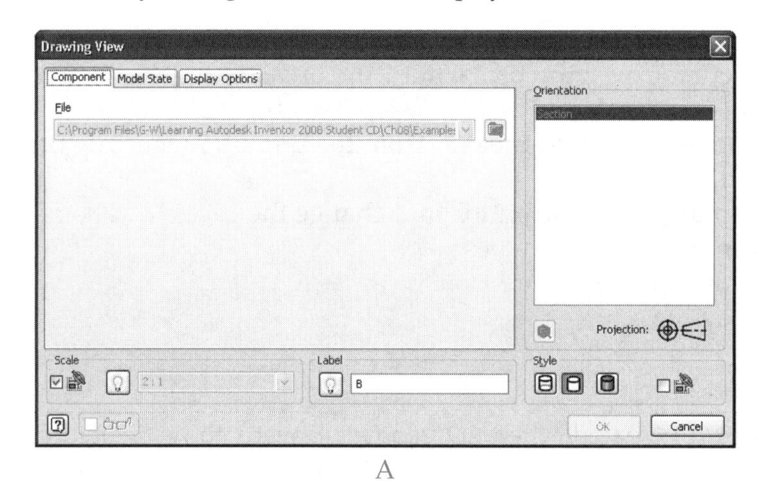

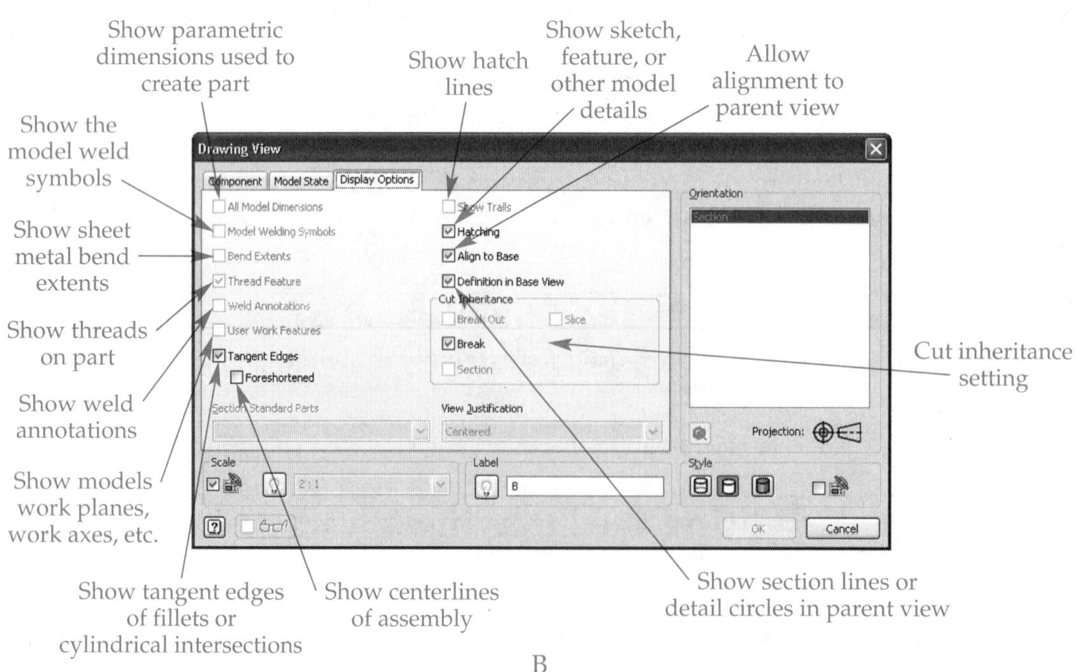

Figure 8-19.
A—The mounting foot with **Tangent Edges** check box unchecked (tangential edges not displayed).
B—The mounting foot with **Tangent Edges** check box checked.

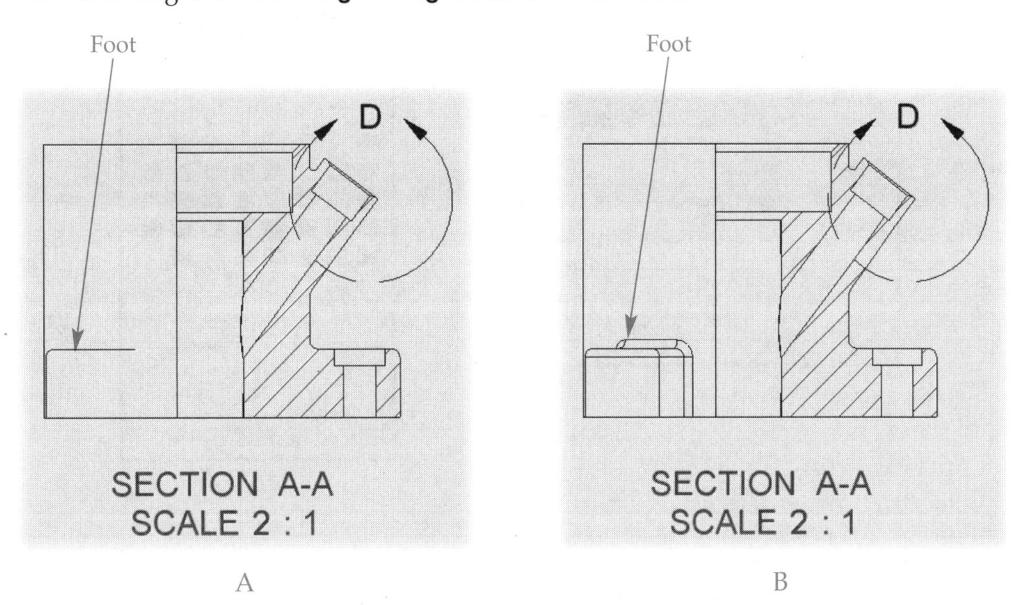

3. The hatching, or section lining, in the detail view can be confusing. Open the **Drawing View** dialog box for the detail view. On the **Display Options** tab, uncheck the **Hatching** check box and pick **OK** to exit. The detail view should be easier to read.

Repositioning Annotations

In the top view, the section view labels A and B need to be dragged into better positions. Each label should be next to the arrows on the corresponding cutting-plane line. Just pick and drag the letters into position. You can do this to any annotation in Inventor. It will retain its new position relative to the view if the view is moved.

Editing a Drawing View's Lines

In addition to changing the properties of a given view, there are other ways to edit drawing views. The lines or arcs that make up the graphics of a drawing view can be changed. The visibility, color, or linetype can be changed. To change color or linetype, select the line(s), right-click, and choose **Properties** from the pop-up menu. To select multiple lines or arcs, hold down the [Shift] key and then select them. The **Edge Properties** dialog box is displayed. See **Figure 8-20A.** In the **Edge Properties** dialog box, notice that By Layer is the default setting in the **Line Type** and **Line Weight** drop-down lists. To change the color, pick the **Color** button, uncheck the **By Layer** check box in the **Color** dialog box, and then select a new color. See **Figure 8-20B.**

To change the visibility of a particular view's lines or arcs, select the line or arc and right-click. Then choose **Visibility** in the pop-up menu, **Figure 8-21A.** It can be a bit tedious to get all of them. Do this for the auxiliary view until you get the finished view shown in **Figure 8-21B.** Save the file.

A drawing view's label can also be edited. Right-click on the label and choose **Edit View Label...** in the pop-up menu. The **Format Text** dialog box is displayed, which controls the label's properties. See **Figure 8-22.** The double chevrons (<<>>) represent the default text. While the default text cannot be deleted, its font, size, etc., can be changed. Also, more text can be added by simply typing before or after the double chevrons.

Figure 8-20.
A—Changing the style of a line (edge). B— Changing the color of a line.

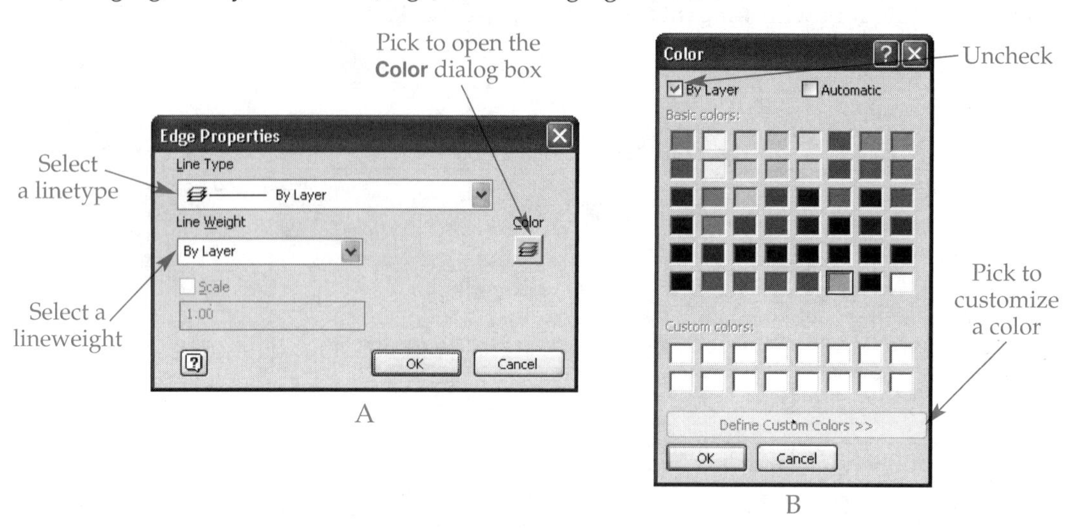

Figure 8-21.
A—Changing the visibility of a line or arc. B—The final detail view with all other lines invisible.

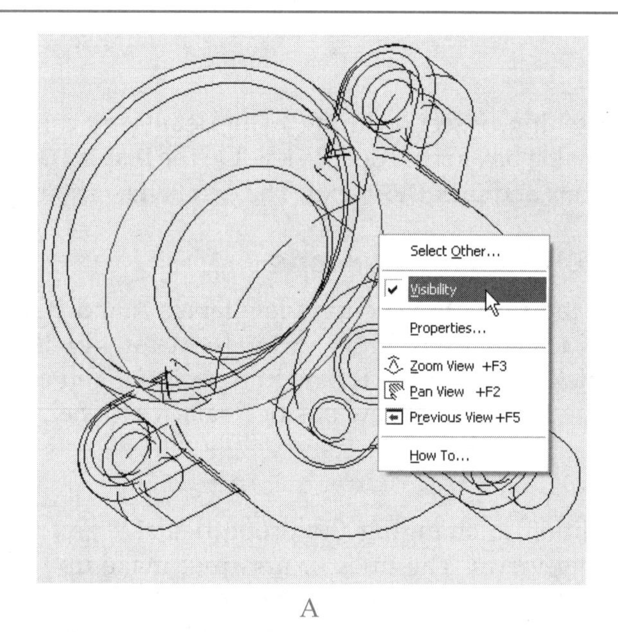

A

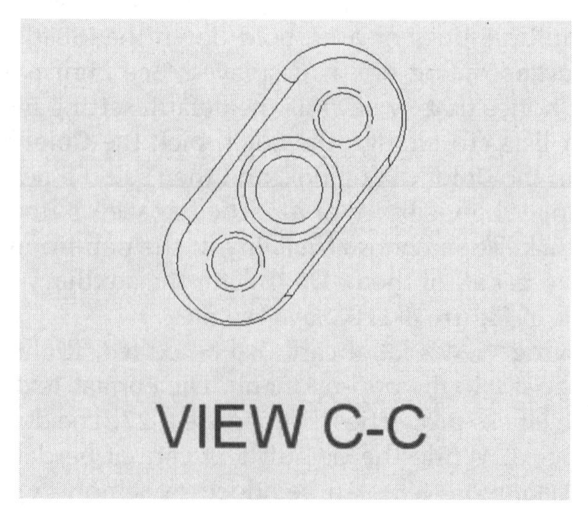

B

Figure 8-22.
Changing the properties of a label. Text can also be added before or after the default text.

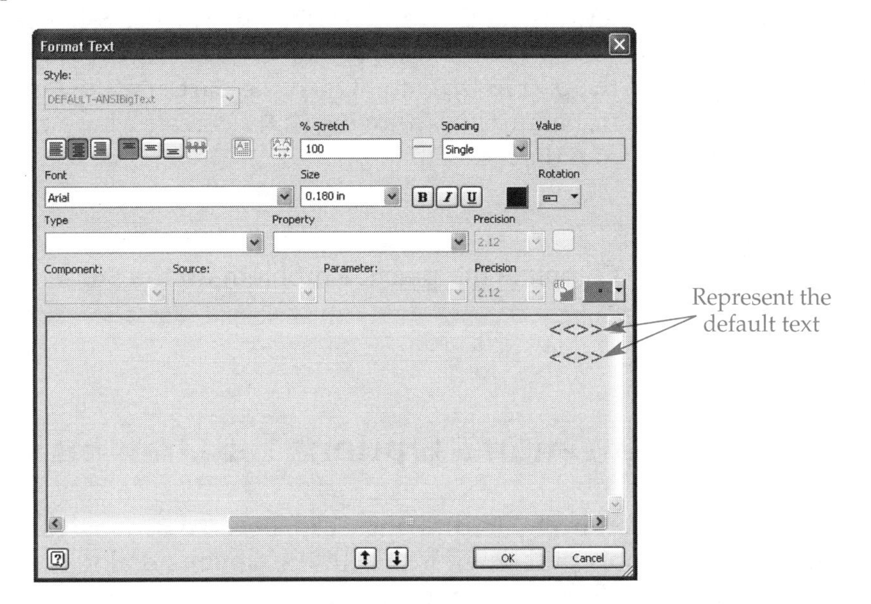

PROFESSIONAL TIP

As an alternative to changing visibility, you can create a section view with a cutting-plane line that does not pass through the part, but instead simply looks at it.

Changing the Model

Inventor is capable of bidirectional associativity. This means the drawing will follow any changes made to the part, or the part will follow any changes made to the drawing. Open Example_08_04.ipt and, if not already open, Example_08_04.idw. In the part file, change the extrusion height of **Extrusion1** to 2.5″. Without saving the part, switch to the drawing file. There will be a pause as the drawing changes to reflect the alteration made to the part. Close both files without saving the changes.

This illustrates the concept of bidirectional associativity. Conversely, you could change a model dimension in the drawing and thereby affect changes to the part. In Chapter 9, you will learn how to place dimensions and other types of annotations on a drawing.

Changing the View Orientation

Sometimes, the part is not properly oriented to Inventor's default coordinate system. At times, it will be necessary to change the view orientation using the **Drawing View** dialog box.

Inventor's default IPT template has the Z axis pointing out of the screen. This orientation is typically used for engineering analysis or in the 3D special effects industry, but is not used in the design or drafting industry.

Inventor's definition of a front view is the drafting industry's definition of a top view, or plan view of the XY plane. Instead, the drafting convention is for the Z axis to point up indicating depth. If viewed along the Z axis looking down at the part, this is referred to as the top view. All other viewpoints are derived from this top view.

Changing the view orientation, as described in Practice 8-1, is helpful in selecting the most descriptive front view. However, sometimes all that is needed is to change the current view of the part in Inventor. Open the part, use the **Rotate** tool to achieve the proper orientation, begin a new drawing (IDW), create a base view, and choose the Current orientation from the list.

Complete the practice problem on the Student CD.

Inventor's Options for Drawings

The settings for Inventor drawings can be accessed by selecting **Application Options...** in the **Tools** pull-down menu. In the **Options** dialog box, pick the **Drawing** tab. See **Figure 8-23.** Remember, these are global settings that affect all drawings.

The **Section Standard Parts** drop-down list controls the sectioning of standard parts in an assembly drawing. By default, Obey Browser Settings is selected in the drop-down list. Sectioning is off by default in the **Browser**. The other two options are Always and Never.

If the **Retrieve all model dimensions on view placement** check box is checked, any applicable model dimensions appear in the view. If unchecked, the **All Model Dimensions** check box must be checked in the **Display Options** tab of the **Drawing View** dialog box.

If the **Display Line Weights** check box is checked, lineweights will be displayed. There are two options, which are available by picking the **Settings...** button to display

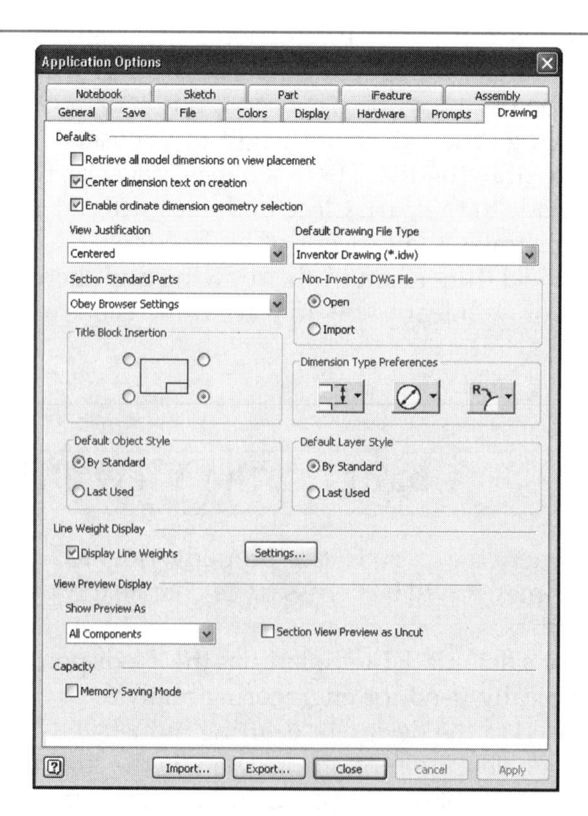

Figure 8-23.
The **Drawing** tab in the **Options** dialog box contains global settings that affect all drawings.

the **Line Weight Settings** dialog box. If the **Display true line weights** radio button is on, a line is displayed at the correct thickness regardless of zoom factor. If the **Display line weights by range (millimeter)** radio button is on, line thicknesses are displayed based on the ranges you enter in the text boxes.

Working with AutoCAD Drawings

Many designers working in Inventor have experience using AutoCAD. In addition, many existing parts have been designed in AutoCAD. Because of this, Inventor can seamlessly work with AutoCAD. In fact, Inventor can directly open AutoCAD drawing files (DWG) and can save Inventor drawings in DWG format. This section describes how to use AutoCAD 2D geometry to create an Inventor solid part. Saving an Inventor drawing as an AutoCAD drawing is also covered.

Opening an AutoCAD Drawing

In Inventor, open the file Example_08_05.dwg. This is an AutoCAD drawing file that contains 2D geometry. Notice that the graphics screen does not display the typical Inventor drawing sheet and border. However, the **Browser** is displayed and the **Drawing Review Panel** is displayed in the **Panel Bar**. See **Figure 8-24.**

Notice that all of the lines in the AutoCAD drawing are displayed. However, only the object lines displayed in blue are needed to create the solid part. The lines can be individually selected, but you can also turn off all layers other than the object layer. This can be done because the AutoCAD drawing was set up so each linetype has a separate layer. To turn off layers, select **Style and Standard Editor...** from the **Format** pull-down menu. Then, expand the Layers branch in the **Style and Standard Editor** dialog box and select any layer in the expanded tree on the left-hand side of the dialog box. All of the available layers are displayed on the right-hand side of the dialog box, including those imported with the AutoCAD drawing. Hold down the [Ctrl] key and

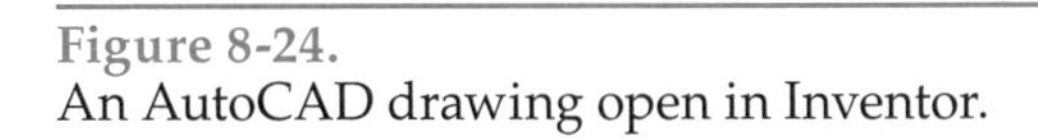
Figure 8-24.
An AutoCAD drawing open in Inventor.

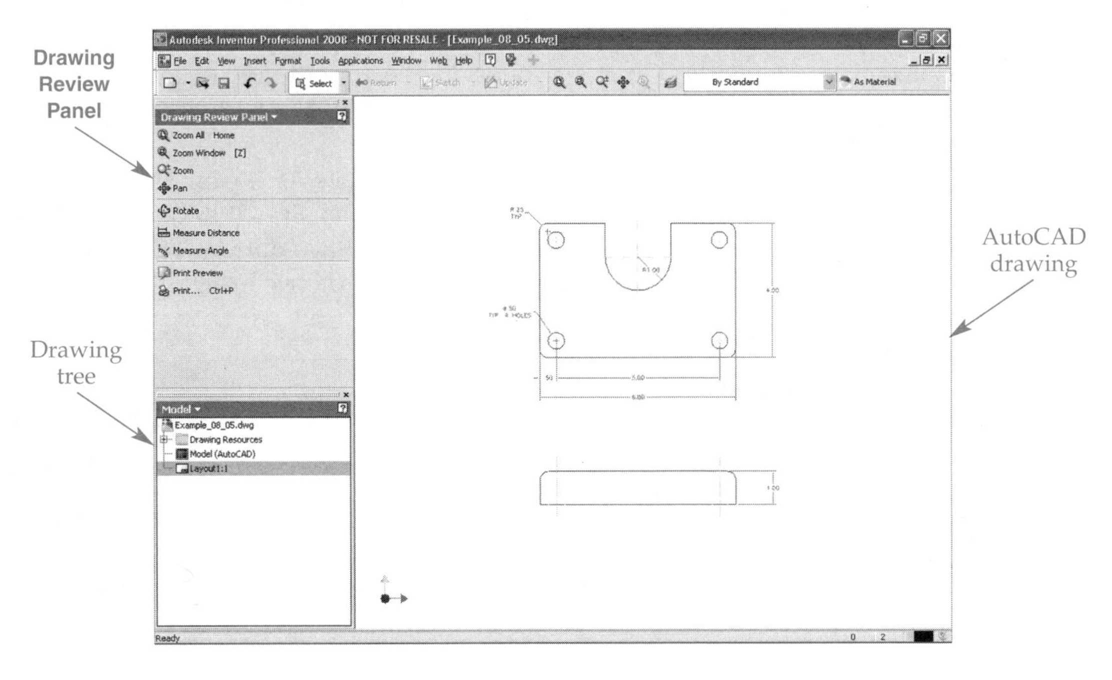

Figure 8-25.
This Inventor solid was created from 2D geometry in an AutoCAD drawing.

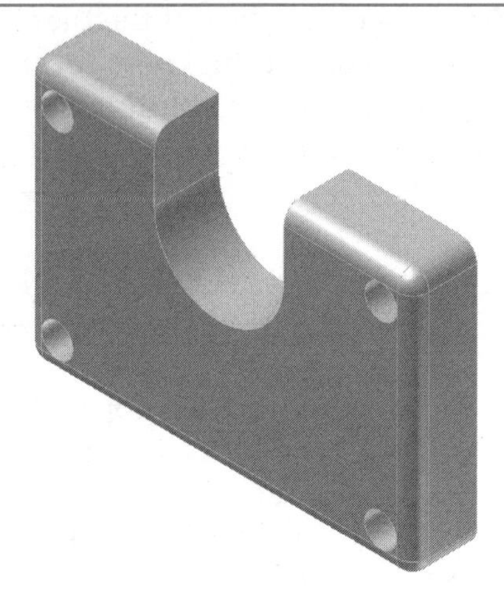

select each layer name except layer ob. Then, pick the lightbulb icon to turn it off for all selected layers. It will be gray when off; the lightbulb next to layer ob should be yellow. Finally, pick the **Done** button and then the **Save** button when prompted to save changes. Now, only the object lines are displayed in the drawing.

Drag a window around the top view to select those object lines. The lines in the front view should not be selected. Right-click and select **Copy** in the pop-up menu. Then, start a new part file based on the Standard (in).ipt template. In the sketch, right-click and select **Paste** in the pop-up menu. A rectangular outline appears attached to the cursor. Pick anywhere to place the geometry in the sketch.

Next, finish the sketch and display the isometric view. Extrude the sketch 1″. Finally, place .25″ radius rounds on the edges. See **Figure 8-25.** Save the part file as Example_08_05.ipt.

It is important to note that there is no parametric link between the final Inventor solid part and the AutoCAD 2D geometry. Also, the pasted geometry only contains coincident constraints on the line endpoints. Before extruding the profile, you should spend the time to fully constrain the sketch. Since there is no parametric link to the AutoCAD file, this includes adding dimensions.

Saving an Inventor Drawing as an AutoCAD Drawing

Start a new Inventor drawing file based on the ANSI (in).idw template. Save the drawing as Example_08_05.idw. Create a two-view orthographic drawing of the Example_08_05.ipt part. Add dimensions and any notes that are necessary to fully describe the part. Then, save the IDW file.

To save the drawing as an AutoCAD drawing, select **Save As...** from the **File** pull-down menu. In the **Save As** dialog box, select Inventor Drawing Files (*.dwg) in the **Save as type:** drop-down list. Then, name the file and pick the **Save** button. The DWG file can now be opened in AutoCAD. However, there is no parametric link to the Inventor solid part or drawing file.

NOTE

If in Windows Explorer you double-click on a DWG file that was saved from Inventor, the file is opened in Inventor, not AutoCAD. In order to open the DWG in AutoCAD, you must use the **OPEN** command from within AutoCAD. Any DWG file created in AutoCAD will open in AutoCAD when double-clicked in Windows Explorer.

Chapter Test

Answer the following questions on a separate sheet of paper or complete the electronic chapter test on the Student CD.

1. From which type of file is an Inventor drawing created?
2. In the file type(s) in question 1, which type of geometry can appear in a drawing view?
3. How can a drawing view be made so it will not print, but is still available in the drawing?
4. What is a *base view?*
5. Briefly describe how to create a base view.
6. How can a parent view be deleted without deleting the derived view?
7. What are the two display styles for a view?
8. What is the difference between a *broken view* and a *break out view?*
9. How are drawing views updated when the part is edited?
10. Briefly describe how to use 2D geometry from AutoCAD to create a 3D solid in Inventor.

Chapter Exercises

Exercise 8-1. Beam.
Complete the exercise on the Student CD

Exercise 8-2. Bevel Gear.
Complete the exercise on the Student CD

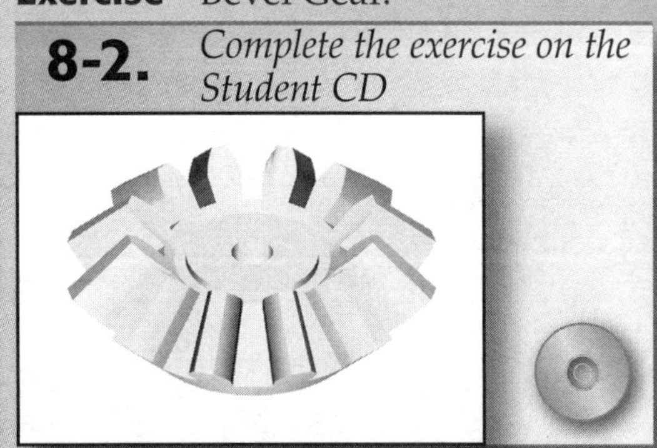

Exercise 8-3. Support Guide.
Complete the exercise on the Student CD

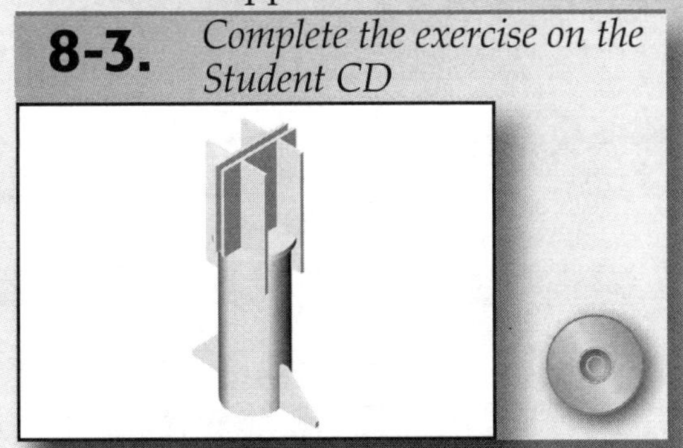

Exercise 8-4. Tubular Brace.
Complete the exercise on the Student CD

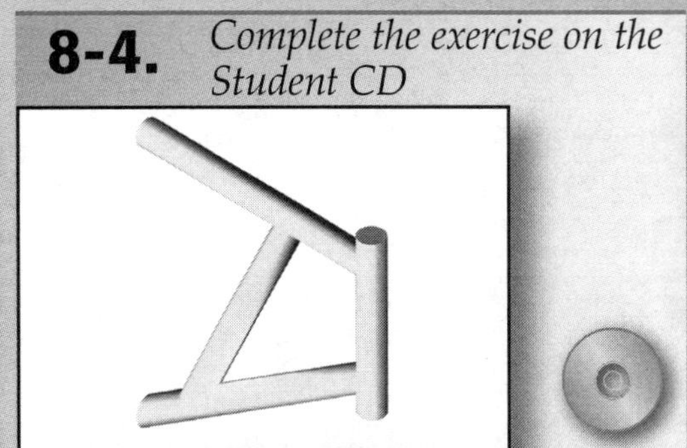

Exercise 8-5. HVAC Transition.

Complete the exercise on the Student CD

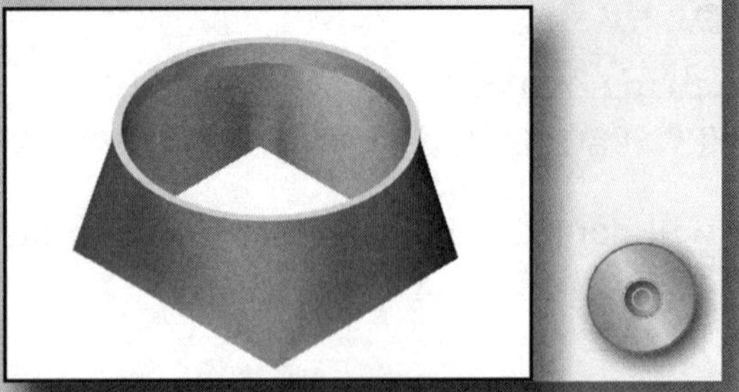

Exercise 8-6. Tubular Support.

Complete the exercise on the Student CD

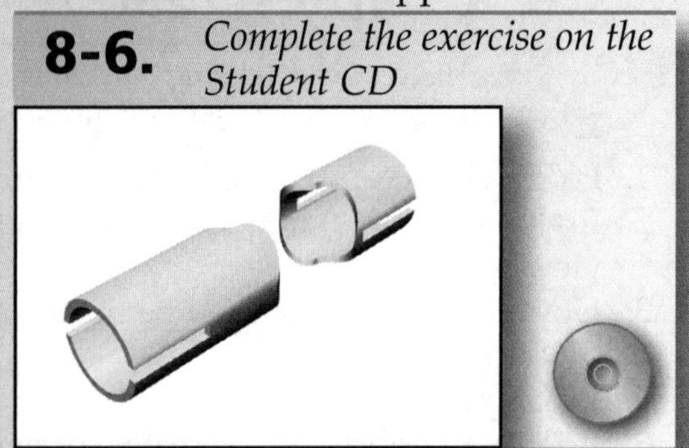

Exercise 8-7. Blow Molded Bottle.

Complete the exercise on the Student CD

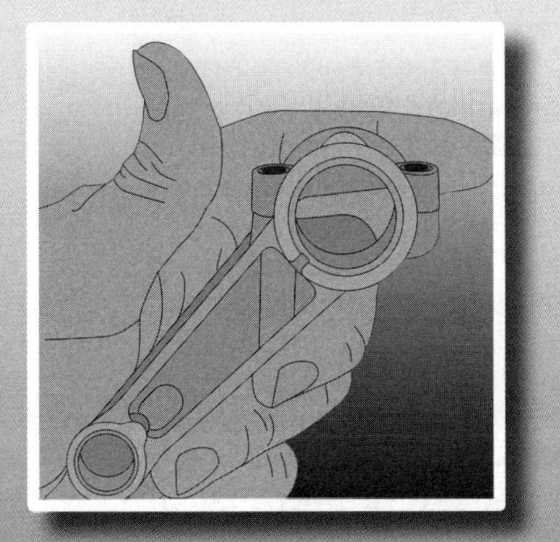

Chapter 9

Dimensioning and Annotating Drawings

Objectives

After completing this chapter, you will be able to:

- Specify a drafting standard.
- Edit a drafting standard.
- Dimension drawing views.
- Apply different annotations to an Inventor drawing.

User's Files

The Student CD included with this text contains several files required for this chapter. Refer to the file File List.txt *in the* \Ch09 *folder for the comprehensive list.*

Preparing to Annotate a Drawing Layout

Open Example_09_01.ipt and examine the part. This part file is identical to one from Chapter 8. Notice the XYZ coordinate system and the part's orientation to that coordinate system. Also, open Example_09_01.idw. This is the completed drawing file from Chapter 8. In this chapter, you will continue working on this drawing.

The **Drawing Annotation Panel** contains the tools for annotating a drawing. It is not visible by default, rather the **Drawing Views Panel** is visible in the drawing file. To switch between these two panels, with a drawing file open, pick the panel title bar or right-click on the panel. Then, select the panel to display from the menu that appears.

Drafting Standards

This section explains drafting standards and how they are implemented in Inventor. Understand that *all* of the annotations used in this chapter are governed by a drafting standard. To access the drafting standards, pick **Style and Standard**

Editor... from the **Format** pull-down menu. The **Style and Standard Editor** dialog box is displayed. See **Figure 9-1A.** This dialog box is used to make changes to a drafting standard or create a new one based on an existing standard. The current standard is displayed in bold in the Standard branch.

Be careful when making changes in this dialog box. If you make changes and save the edits, the changes will overwrite the installed standard (drafting style). You may not want to do this, but a new standard can be created based on the current one. Select the existing standard on which to base the new standard, such as ANSI, in the Standard branch. Then, pick the **New...** button to open the **New Style Name** dialog box. See **Figure 9-1B.** Enter a name for the new standard, which assumes all of the properties of the standard highlighted in the list. You can also select a different standard on which to base the new standard using the **Based On:** drop-down list in the **New Style Name** dialog box.

Figure 9-1.
A—The **Style and Standard Editor** dialog box is used to choose a drafting standard, make changes to a standard, or create a new standard based on an existing standard. B—The **New Style Name** dialog box is used to name the new style.

Figure 9-2.
The **Document Settings** dialog box can be used to select the active standard.

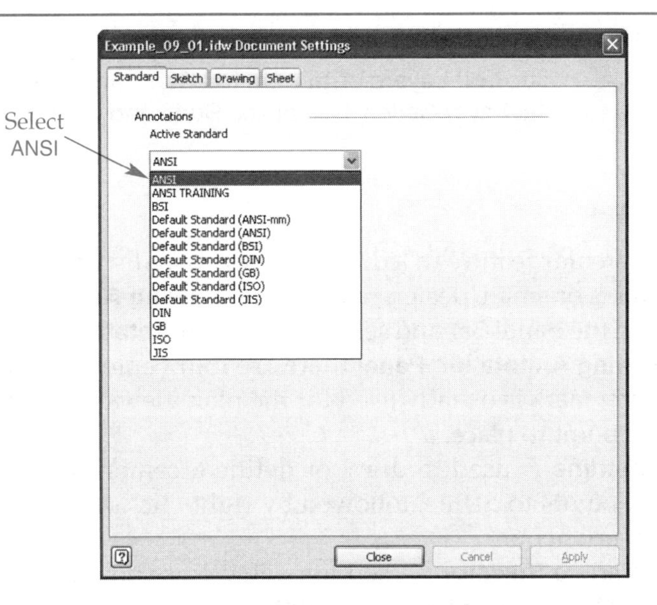

In the United States, the American National Standards Institute (ANSI) drafting standards are most commonly used. However, many engineering firms also work on international projects. These projects may require ISO standards to be in effect for all work. For this chapter, ANSI standards will be used for the most part. However, a custom standard based on ANSI will also be used.

A standard can be made active by right-clicking on its name in the Standard branch in the **Style and Standard Editor** dialog box. Then, select **Active** in the pop-up menu. A style can also be made active by selecting **Active Standard...** in the **Format** pull-down menu. Then, in the **Standard** tab of the **Document Settings** dialog box, select the standard in the drop-down list. See **Figure 9-2.** For this example, make sure ANSI is the active standard.

Layers

Inventor provides layers for organizing annotations. As you annotate the drawing views, the annotations are automatically placed on the appropriate layer. The settings for a layer can be changed.

Open the **Style and Standard Editor** dialog box. Select Active Standard in the **Filter Styles** drop-down list in the upper-right corner of the dialog box. This prevents the display of items not included in the current style. Otherwise, the items for all drafting styles will be shown in the dialog box. Notice that there are many different types of drawing objects controlled by the settings in this dialog box. You will be using this dialog box throughout this chapter. Take a moment to examine the listing before proceeding.

Now, change the color property of the Center Mark (ANSI) layer in preparation for the next section. Expand the Layers branch and select the Center Mark (ANSI) layer. On the right-hand side of the dialog box, in the **Layer Styles** area, notice that each property of the layer is available to be edited: Layer Name, On, Color, Line Type, Line Weight, Scale by Lineweight, and Plot. Pick the color swatch for the layer to display the **Color** dialog box. Select a light blue color and pick the **OK** button to change the layer color. Now, highlight the Centerlines (ANSI) layer. Change its color to the same color used for the Center Mark (ANSI) layer. Pick the **Save** button to save your changes and then pick the **Done** button to exit the dialog box.

NOTE

The **Edit Layers** button on the **Standard** toolbar provides a shortcut to the **Layer Styles** area of the **Style and Standard Editor** dialog box.

Centerlines

Every circular feature in a drawing view usually has a centerline. The only exceptions are fillets or small, design radii. If the **Drawing Annotation Panel** is not displayed, right-click in the **Panel Bar** and select **Drawing Annotation Panel** from the pop-up menu. On the **Drawing Annotation Panel** there are four centerline tools. See **Figure 9-3.**

- **Center Mark** is usually used for the plan view of a circle or arc and requires one pick point to place.
- **Centerline** is used to draw or define a centerline. This requires two or more pick points to define, followed by right-clicking and selecting **Create** from the pop-up menu.
- **Centerline Bisector** is used for side views showing the circle/arc as a cylinder. This requires just two pick points.
- **Centered Pattern** is for circular arrays/patterns, such as bolt circles. This requires many pick points to define, followed by right-clicking and selecting **Create** from the pop-up menu.

Adding Centerlines to the Top View

Zoom in on the top view in Example_09_01.idw. Pick the **Center Mark** tool and select the outermost circle of the part. A center mark appears, displayed in the light blue color you selected earlier. Next, select the *outside* arc of one of the mounting feet. Notice that only a partial center mark is created. Now, on one of the other mounting feet, select the circle that represents the hole. Notice that a full center mark is created. See **Figure 9-4.**

The center marks you just created for the mounting feet are not correct. Remove them by picking the **Undo** button twice. The **Centered Pattern** tool will place correct center marks on the feet. Pick the tool and select the outermost, large circle of the part. Next, select the outside arc of the mounting foot at the three o'clock position. Going counterclockwise, select the outside arc of the mounting foot at the ten o'clock position and then the outside arc of the mounting foot at the seven o'clock position. Right-click and select **Create** in the pop-up menu. A partial centerline circle, approximately two-thirds of a full circle, is created. Press [Esc] to exit the **Centered Pattern** tool.

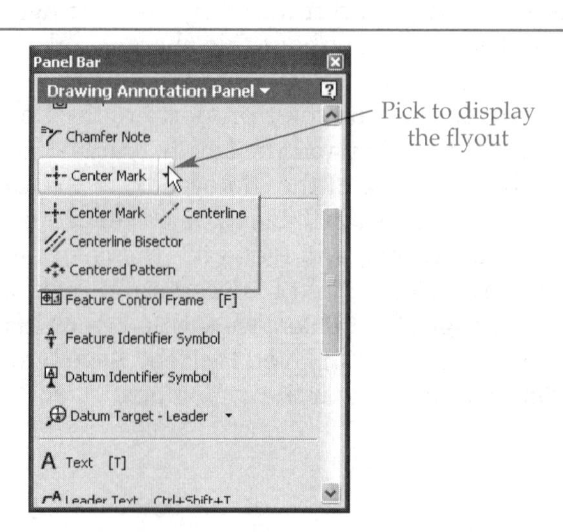

Figure 9-3.
The tools used to create centerlines and center marks.

Figure 9-4.
Center marks placed in the top view. Notice the full and partial center marks.

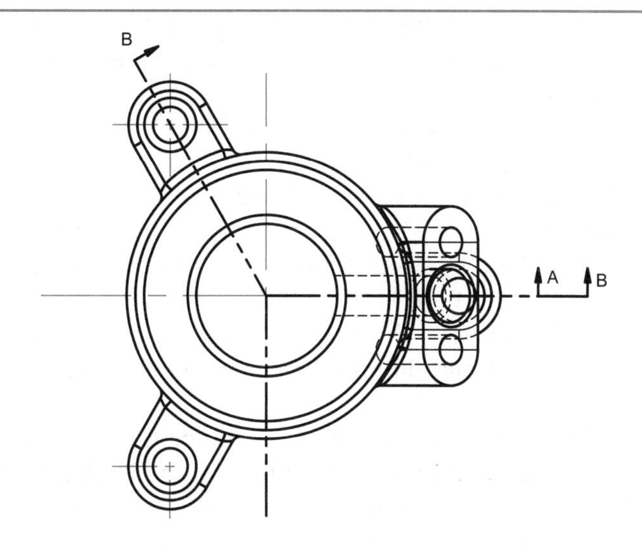

Figure 9-5.
The completed bolt circle. Note: The centerlines at the three o'clock and ten o'clock positions are under the cutting-plane lines.

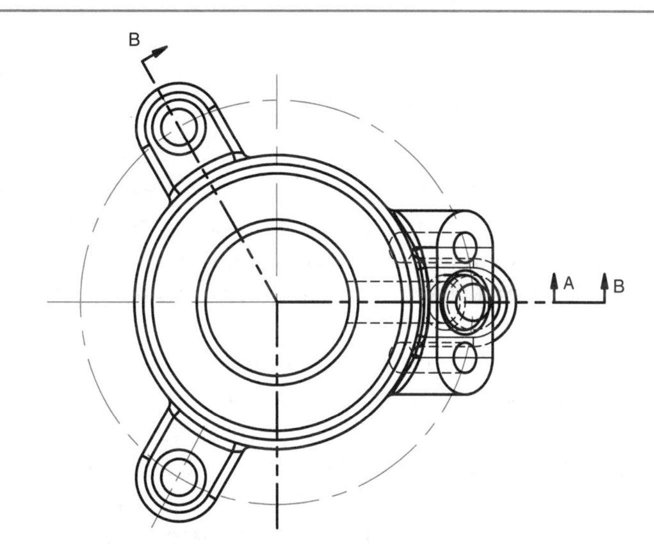

To make the centerline circle a full circle, pick the patterned centerline and drag the lower endpoint to meet the starting point (at the three o'clock position). This type of centerline is known as a ***bolt circle.*** Next, drag the endpoints of the center mark so that each line ends outside of the bolt circle. See **Figure 9-5.**

PROFESSIONAL TIP

If the values used to create center marks need to be edited, use the **Style and Standard Editor** dialog box. Expand the Center Mark branch, select Center Mark (ANSI), and change the values in the **Center Mark Style** area on the right-hand side of the dialog box.

Adding Centerlines to the Section Views

The section views you created for this part in Chapter 8 are good examples of circular features shown in a side view, or in elevation. Zoom in on SECTION B-B. The circular feature appears as a rectangle in these views. Typically, the **Centerline** or **Centerline Bisector** tool is used to create a centerline in this type of view. Please take note that both section views have an object line running down the center of the part in each view. This is due to the type of section views that were created. Change the visibility of the object lines so the view looks like **Figure 9-6.** There are several lines, so use a window to select them, then right-click and uncheck **Visibility** in the pop-up menu to hide them.

Either the **Centerline** or **Centerline Bisector** tool can be used to draw the large centerline running down the center of the part. For this example, pick the **Centerline** button in the **Drawing Annotation Panel**. Next, select the bottom center point of the large cylindrical feature that makes up the main body of the part and then the top center point. Right-click, select **Create** in the pop-up menu, and then press [Esc] to end the tool. Notice there is a centerline running through the part. See **Figure 9-6.** If needed, pick and drag the endpoints of the centerline to extend the line from the part. Now, use the **Centerline** tool to create centerlines for the two small bolt holes in SECTION B-B.

To finish up, use the **Centerline Bisector** tool to apply a centerline to the angled hole. Pick the **Centerline Bisector** button in the **Drawing Annotation Panel**. Next, pick one of the long, angled edges of the angled hole, then select the opposite edge. A centerline is created in the middle of the hole. Press [Esc] to end the tool. Then, drag the endpoints of the centerline as shown in **Figure 9-6.** This centerline will be used to dimension the angle.

Adding Centerlines to the Auxiliary View

Pan and zoom Example_09_01.idw so that the auxiliary view is centered in the graphics window. You will now place centerlines on this view. Because of the projection of this view, the circular edges are shown as elliptical edges. However, Inventor interprets the elliptical edges as circular so they can be used as a basis for centerlines.

Pick the **Center Mark** button in the **Drawing Annotation Panel**. Select the outer elliptical edge that represents the body of the part. Continue by selecting the circles that represent the bolt holes of the mounting feet. See **Figure 9-7.** These center marks can really help a view make more sense.

Using the **Centerline** tool, pick the leftmost dashed circle that represents the threads of the tapped hole in the angled feature. Next, select the largest circle representing the counterbored hole on the angled feature. Finish by selecting the threads on the rightmost tapped hole. Right-click and select **Create** in the pop-up menu. The center marks are aligned with the face's axis. See **Figure 9-8.** Press [Esc] to end the tool.

Zoom and pan to display the base view. Then, left-click on the center mark in the view (not on the bolt circle) to select it. Notice the two drop-down lists on the right-hand side of the **Inventor Standard** toolbar, **Figure 9-9.** The left-hand drop-down list displays the layer on which the annotation object resides. It reads By Standard (Center Mark (ANSI)) because the layer for this center mark is determined by the drafting standard in effect when it was created. You can change the annotation to another layer

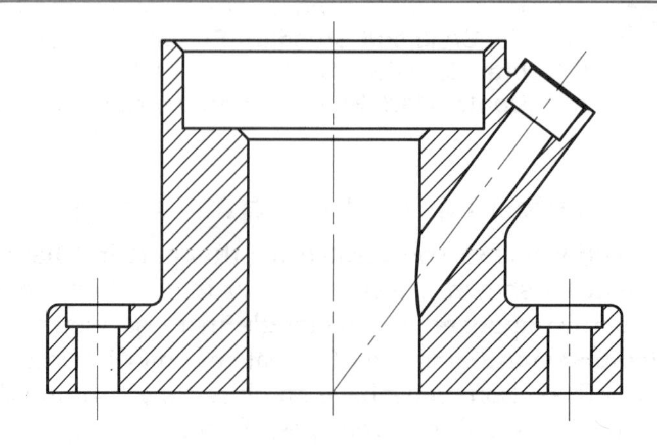

Figure 9-6.
Centerlines are added to SECTION B-B.

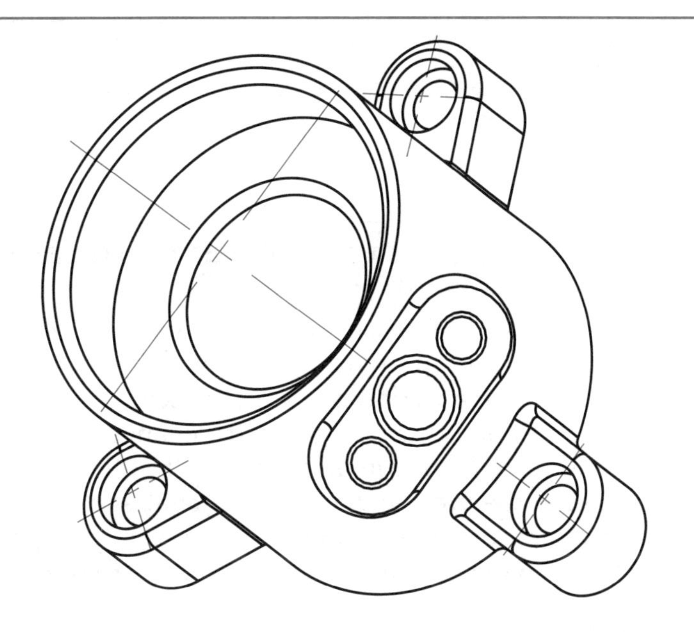

Figure 9-7.
Centerlines are added to the auxiliary view.

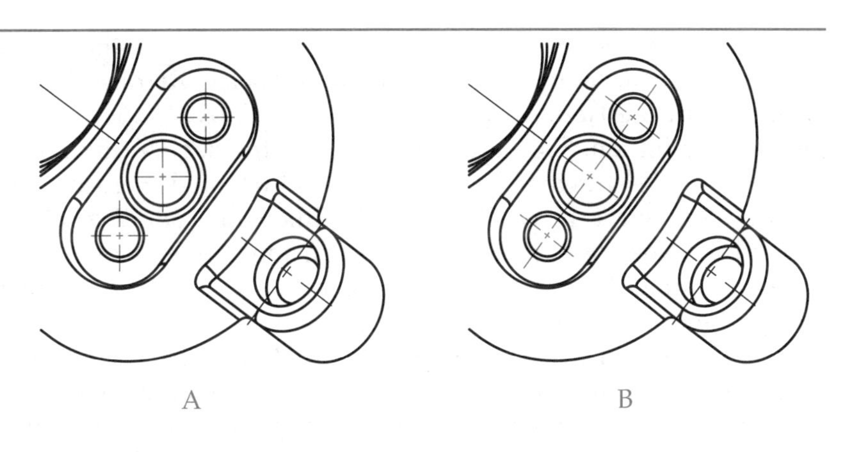

Figure 9-8.
A—The **Center Mark** tool is used to add center marks to the mounting foot hole. However, when used on the three holes in the angled feature, the center marks are not aligned with the face's axis. B—Center marks are aligned with the face's axis when the **Centerline** tool is used.

Figure 9-9.
The **Layer** and **Standard** drop-down lists on the **Inventor Standard** toolbar (shown floating).

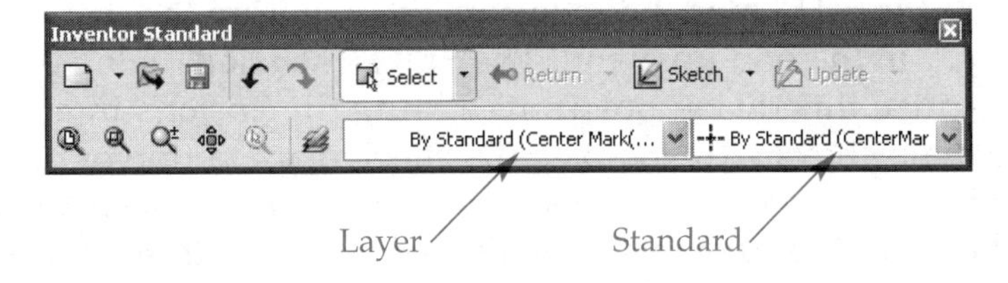

by selecting the layer in this drop-down list. You can also turn the Center Mark (ANSI) layer off by picking the lightbulb next to its name in the drop-down list (the list must be displayed). When the lightbulb is off, the layer is off.

The right-hand drop-down list indicates the standard used by the annotation. It reads By Standard (Center Mark (ANSI)). The ANSI drafting standard is defining the center mark. To select a different standard, pick it in the drop-down list. Try picking several different standards and notice how the linetype of the center mark changes. When done, select By Standard (Center Mark (ANSI)).

NOTE

Centerlines can be put on isometric views, but this is not a common practice.

Dimensioning in Inventor

The two types of dimensions used in Inventor are parametric and reference. ***Parametric dimensions*** are those placed on the sketches of the part. These dimensions can be displayed on a drawing view. If they are changed either on the part or the drawing, both part and drawing geometry change. In other words, this is a bidirectional relationship. ***Reference dimensions,*** on the other hand, are added to the drawing using the tools in the **Drawing Annotation Panel**, such as the **General Dimension** tool. They are driven by the part geometry. If the part changes the reference dimensions change in the drawing. However, they cannot be arbitrarily changed in the drawing.

Parametric Dimensions

Open Example_09_02.ipt and Example_09_02.idw. In the drawing, zoom in on the front view. Then, right-click on the view and select **Retrieve Dimensions...** in the pop-up menu. In the **Retrieve Dimensions** dialog box that is displayed, pick the **Select Dimensions** button. See **Figure 9-10.** All of the sketch dimensions for this view are displayed in the drawing. You need to select which dimensions are to be displayed on the drawing. Select the ∅1.5″ and ∅2.5″ dimensions and then pick the **OK** button in the dialog box. Now, only the selected dimensions are shown on the drawing. They can be moved around the drawing view as needed. Since these dimensions are from the sketch, they are parametric dimensions.

Right-click on the **Drawing Views Panel** and select **Drawing Annotation Panel** in the pop-up menu to display that panel. Pick the **General Dimension** tool and select the arc representing the ∅2.5″ feature; do not pick to place the dimension. Because the feature is represented as an arc, not a circle, the dimension defaults to a radius dimension. However, you can change this by right-clicking before picking the location for the dimension and selecting **Dimension Type>Diameter** in the pop-up menu. Then, pick a location for the dimension and end the tool. Since this dimension is not obtained from the sketch, it is a reference dimension.

Right-click on the ∅2.5″ model dimension (the one from the sketch) and pick **Edit Model Dimension...** in the pop-up menu. The **Edit Dimension** dialog box is displayed. You should recognize this dialog box from editing dimensions in a sketch. Change the value in the dialog box to 3.0 and pick the check mark button or press [Enter]. The part, drawing view, and reference dimension all change to match the new value. If you switch to the part file and display the dimensions for Extrusion1, you will see that the diameter has changed to 3.000″. You may need to pick the **Update** button on the **Inventor Standard** toolbar to see the change.

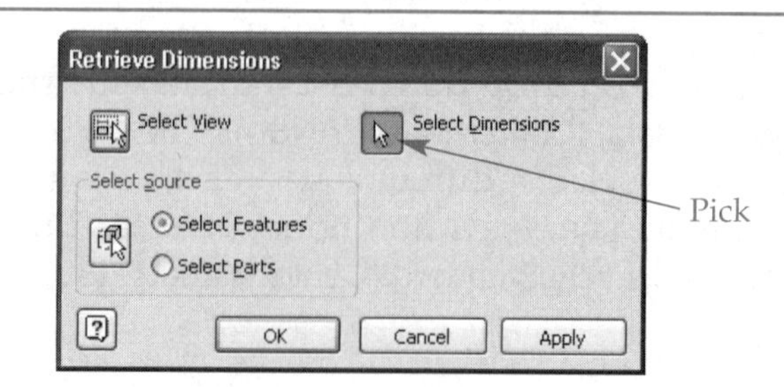

Figure 9-10.
Adding dimensions from the sketch to the drawing.

Now, in the drawing, right-click on the ∅3.0″ reference dimension. There are many things you can change about the dimension, such as style and precision, but you cannot change the value. Which dimensions you use in your drawings depends on your design intent, but most designers use reference dimensions and make model changes in the part file, not the drawing file. Close both Example_09_02.ipt and Example_09_02.idw without saving either file.

Dimension Styles

Inventor uses a ***dimension style*** to control the appearance of dimensions in the drawing. Open the **Style and Standard Editor** dialog box. Select Local Styles in the **Filter Styles** drop-down list. Then, expand the Dimension branch. Notice that there are a number of predefined dimension styles listed in the Dimension branch. Try to use the ANSI standard and its dimension styles, without edits, as much as possible. However, in some instances you may need to use a different standard, such as ISO or JIS.

Normally, changes should not need to be made to a dimension style. However, a company may find it necessary to make their own dimension style based on the national drafting standard for their current project. Keep in mind that as changes are made, the drawing's dimensions are updated. To make changes to a dimension style, select the style in the Dimensions branch. Its settings are displayed in the **Dimension Style** area on the right-side of the dialog box. See **Figure 9-11.**

If you want dimensions to appear in a different color, changing the color property of the Dimension layer will not work. This is because the drafting standard controls the color of the lines, arrows, and text that make up a dimension. Make the Example_09_01.idw file current, if it is not already. The **Style and Standard Editor** dialog box must be used to change the color of the dimension lines and text. Select Active Standard in the **Filter Styles** drop-down list in the dialog box. Then, expand the Dimension branch and select DEFAULT-ANSI (make sure the current standard is ANSI). The settings for this dimension style are displayed in the **Dimension Style** area on the right-hand side of the dialog box.

Figure 9-11.
Select a dimension style in the **Style and Standard Editor** dialog box and its settings are displayed on the right-hand side of the dialog box.

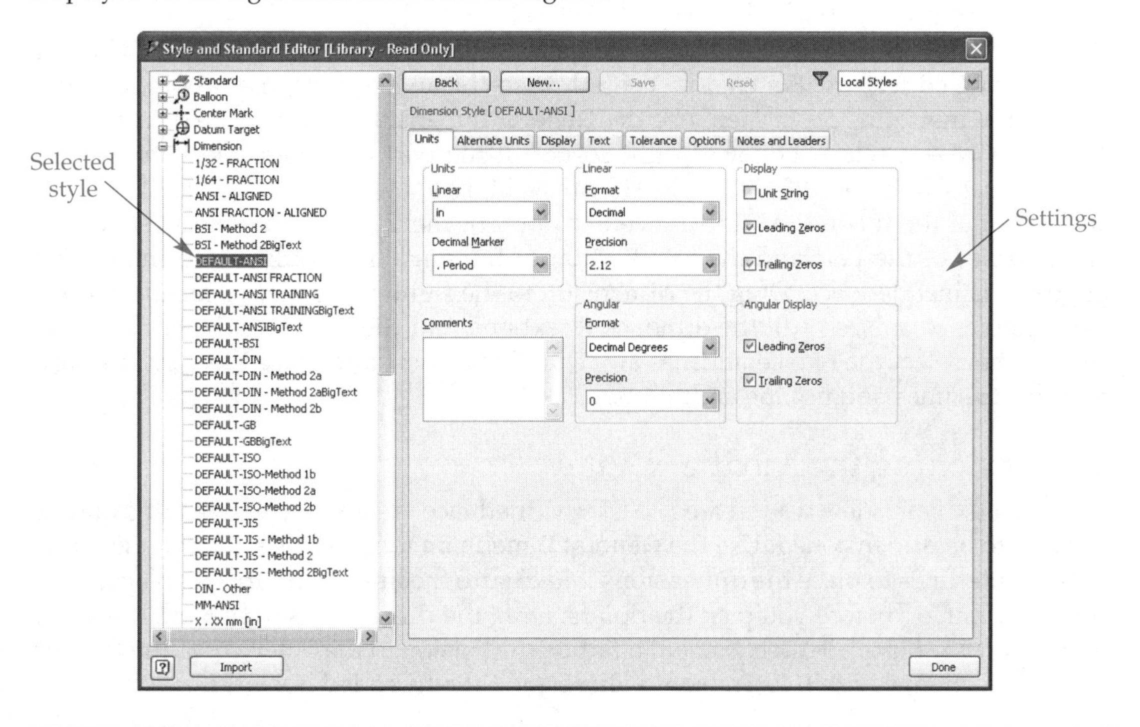

Pick the **Display** tab in the **Dimension Style** area. Then, pick the color swatch in the **Line** area of the tab. In the **Color** dialog box, select a dark blue and pick the **OK** button to exit the dialog box. Next, pick the **Text** tab. Notice that there is no color swatch here. However, you can see that DEFAULT-ANSI is selected in the **Primary Text Style** drop-down list. This indicates that the text is controlled by that style. Save the edits. Now, expand the Text branch and select DEFAULT-ANSI. Notice the color swatch in the **Text Styles** area on the right. Use it to select the same dark blue as selected for the dimension style. Save these changes and exit the dialog box. Now, you are ready to add dimensions to the drawing.

Dimensioning the Top View

As dimensions are added to this part, move the drawing views as needed to get more space. **Figure 9-12** shows how the views should look once they are dimensioned. Using the **General Dimension** tool, select the large bolt circle centerline and place the dimension text to the upper right. Do the same with the large circle for the body of the part. Also, dimension the outside arc of the upper-left mounting foot. Press [Esc] to end the tool. Dimensions can be moved quite easily by selecting them and then picking and dragging one of the green endpoints (grips). Try this with the dimensions that you have placed.

The standard associated with a particular dimension can be quickly changed. Select the dimension in the drawing and then select a different dimension style from the far-right drop-down list on the **Inventor Standard** toolbar. For example, select one of the international dimension styles. The inch dimension text is changed to its metric equivalent. Select By Standard (DEFAULT-ANSI) to reassociate the ANSI standard with the dimension.

Right-click on the last dimension you placed (on the foot) and select **Text...** from the pop-up menu. The **Format Text** dialog box is displayed. Text can be added to the default dimension using this dialog box. Place the cursor at the end of the chevrons (<<>>) in the text box. The chevrons indicate the default text and cannot be overwritten. Type a space, TYP, press [Enter] to start a new line, and then type 3 PLACES. Pick the **OK** button to exit the dialog box and update the dimension. Using the same method, add REF to the end of the default text for the bolt circle diameter.

Dimensioning the View SECTION A-A

A few dimensions are needed in the half-section view to define the part's height and the angled feature. First, add a centerline to the angled hole, main bore, and the hole in the mounting foot. Then, use the **General Dimension** tool to place dimensions. In SECTION A-A, select the lower-right corner of the part and then select the upper-right corner. Drag the 2.25 dimension to the right, as shown in **Figure 9-12.** To define the height of the theoretical starting center point of the angled hole, select the lower-right corner of the part and the intersection of the angled centerline and the profile edge of the inclined face. Drag the dimension to the right of the view. This dimension is for reference only, so edit the dimension text and add REF to the end. For the angle dimension, select the two centerlines and drag the dimension above the view. Be sure to select the lines and not the endpoints.

Dimensioning the Auxiliary View

The auxiliary view was created so the inclined face is shown true size. This allows the face to be dimensioned. Use the **General Dimension** tool and pick the intersections of the centerlines to place the dimensions. Make sure the linear cursor is displayed, not the angle cursor, before you pick the points. Drag the dimensions to the left, but at an angle. Refer to **Figure 9-12.** Next, add a radius dimension to the outer arc of the face. In the next section, you will learn how to dimension the threaded hole.

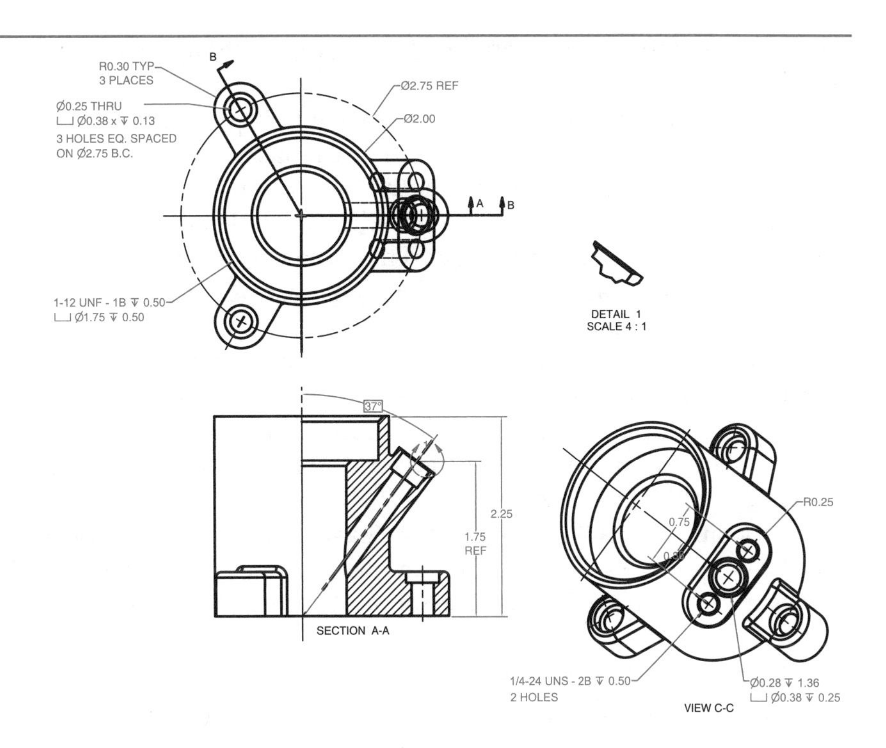

Figure 9-12.
The dimensioned views.

Adding Hole or Thread Notes

A ***hole note*** describes the properties of a given hole. For example, the depth of a drilled hole; whether or not the hole is countersunk, counterbored, or spotfaced; and the type of threads (if the hole is tapped) are indicated in a hole note. The part in Example_09_01.idw contains seven parametric hole features. Three of these are tapped. Hole notes are needed to help describe these features and complete the drawing views.

Zoom and pan to center the top view in the screen. Next, pick the **Hole/Thread Notes** button in the **Drawing Annotation Panel**. Select the hole on the upper-left mounting foot. Drag the hole note to the left of the view. When finished, press [Esc] to exit the tool. Then, right-click on the hole note text and select **Text...** from the pop-up menu. Add 3 HOLES EQ. SPACED ON ∅2.75 B.C., as shown in **Figure 9-13.** To add the diameter symbol (∅), pick the **Insert symbol** button above the text box and select the

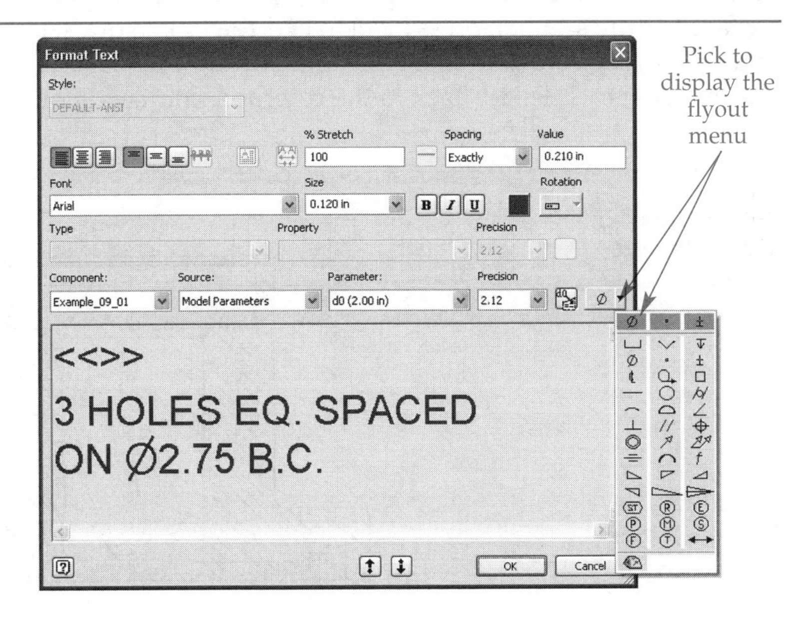

Figure 9-13.
Adding text to the default dimension. Special symbols are available by picking the button as shown.

symbol from the flyout menu. The hole note should appear in the color that was specified for text.

In this case, the hole note leader does not properly extend to the circle representing the hole. The arrowhead should touch the actual edge of the selected hole. This is because the hole is part of a parametric pattern and is beneath the feature level that the **Hole Note** tool can "see." To prove this, place a hole note on the three o'clock mounting hole on the right side of the view. This hole operation is farther up the part tree and the **Hole Note** tool can locate it. Unfortunately, this is not a good placement for the note. Delete this note, but keep the first hole note.

Place an additional hole note to describe the central bore. Select the small diameter. Notice that the thread and counterbore information is included in the note. In the auxiliary view, place hole notes describing the three small holes; only two notes are needed. Additional text is required for the hole note on the small tapped holes. Add 2 HOLES to this hole note.

Repositioning a hole note is quite easy. Pick on the note once to activate the green grips. Then, pick a grip on the leader (not the one at the arrow's end) and drag the hole note around the hole. Release the mouse button when the note is positioned.

Surface Texture Symbols

Surface texture symbols are typically used to describe the quality of the machined finish on a given face or surface. For example, the face of the angled feature and the inside faces of the central bore diameter are to be machined. The quality of the machined surface is indicated on the drawing by surface texture symbols. See **Figure 9-14.** A thorough study of surface textures and related terminology is beyond the scope of this textbook. However, you will learn how to place surface texture symbols on a drawing.

Zoom in on SECTION B-B. Pick the **Surface Texture Symbol** button in the **Drawing Annotation Panel**. Then, pick a point on the line that represents the angled face. This starts the leader callout portion of the symbol. Pick a second point for the leader, then right-click and select **Continue** in the pop-up menu. The **Surface Texture** dialog box is displayed. See **Figure 9-15.** If you do not want a leader, pick a start point, right-click, and select **Continue** to proceed directly to the **Surface Texture** dialog box. In the dialog box, enter 32 for the **A'** value, leave all of the other settings at their defaults, and pick the **OK** button to place the symbol.

Using the **Surface Texture Symbol** tool, pick one point on the central bore diameter, drag out the leader, and pick a second point. Then, right-click and pick **Continue.** Enter 63 for the **A'** value and pick the **OK** button. To have multiple leaders on this surface texture symbol, press [Esc] to end the tool. Now, select the symbol, right-click, and select **Add Vertex/Leader** from the pop-up menu. Pick two points to define each leader. Bring each leader back to the common point, as shown in **Figure 9-14.**

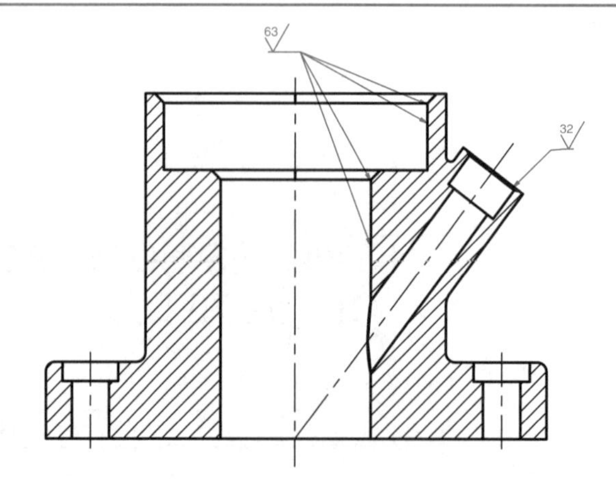

Figure 9-14. SECTION B-B with surface texture symbols.

Figure 9-15.
The **Surface Texture** dialog box is used to define a surface texture symbol.

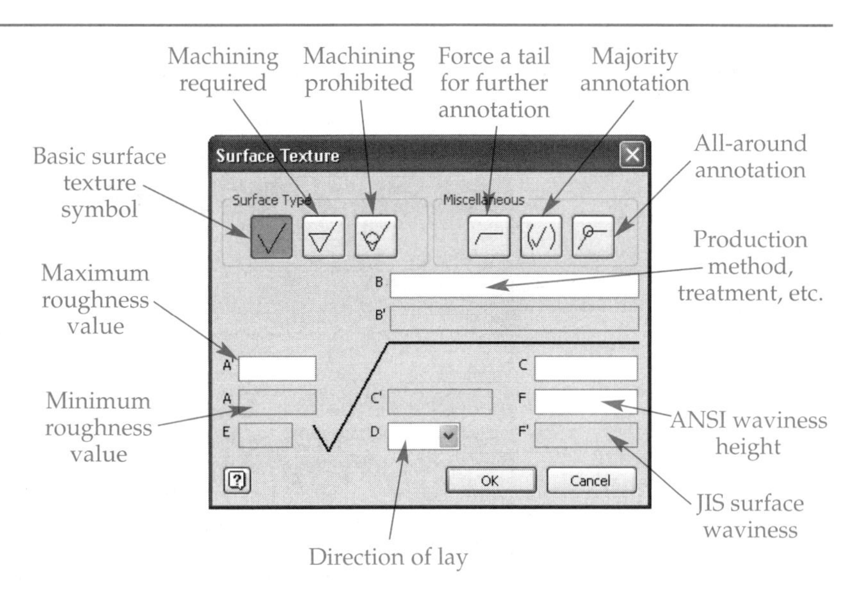

Feature Control Frames

A ***feature control frame*** is used in geometric dimensioning and tolerancing (GD&T) to define the geometric tolerancing characteristics of a feature. Symbols appear in the feature control frame to indicate which characteristics are being specified. The feature must fall within the tolerances specified in the feature control frames or the part is considered unacceptable. The appendix on the Student CD contains a chart of GD&T symbols.

With Example_09_01.idw open, zoom in on SECTION B-B. **Figure 9-16** shows the section with the feature control frames applied. To apply these feature control frames, pick the **Feature Control Frame** button in the **Drawing Annotation Panel**. To apply a feature control frame to the centerline of the angled feature, pick the end of the centerline. Move the cursor to the right and pick a second point to specify the length and direction of the leader. Then, right-click and select **Continue** in the pop-up menu. The **Feature Control Frame** dialog box is displayed. See **Figure 9-17.** Pick the **Geometric characteristic symbol** button in the **Sym** area to display the flyout menu. For this feature, select the angle symbol in the menu. In the top text box in the **Tolerance** area, type .01. Insert the maximum material condition symbol by picking the **Maximum**

Figure 9-16.
SECTION B-B with GD&T feature control frames and datums added.

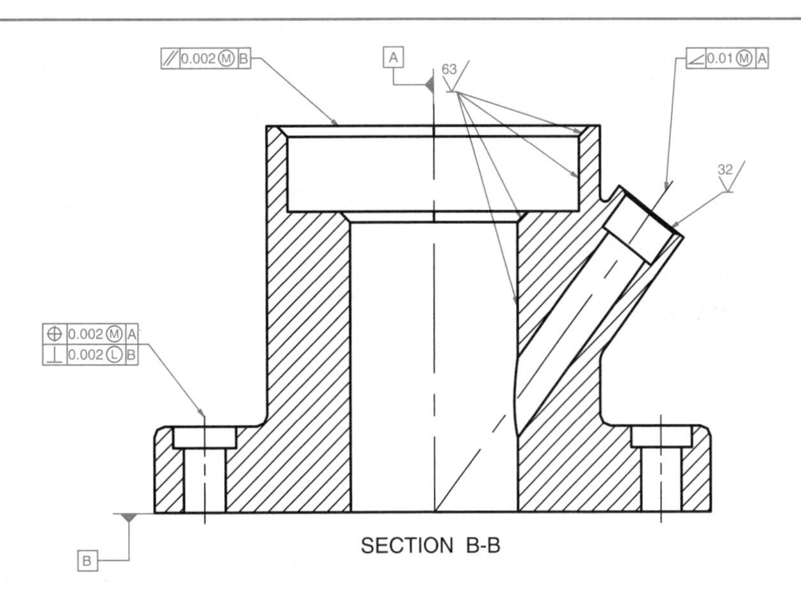

Figure 9-17.
The **Feature Control Frame** dialog box is used to define a feature control frame.

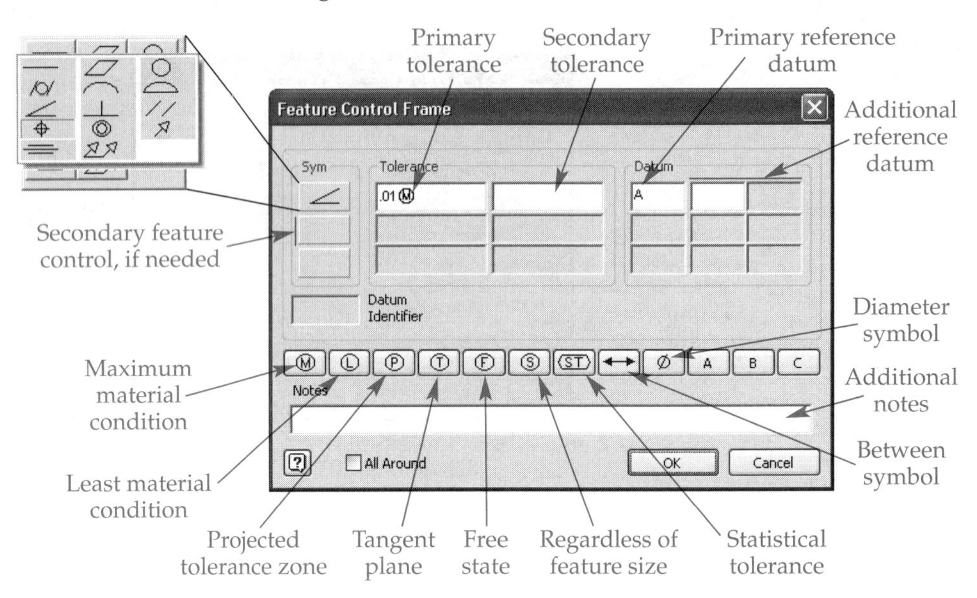

Material Condition button at the bottom of the dialog box. In the first text box in the **Datum** area, type A. Finally, pick the **OK** button to create the feature control frame.

Defining datums

The angle of the main bore's centerline and the feature that uses it are now controlled in reference to Datum A. However, Datum A has yet to be defined. Also, in order for this feature control frame to be valid, the angle dimension in SECTION A-A has to be changed to reflect that it is now a basic dimension. Right-click on the dimension and select **Edit...** from the pop-up menu. The **Edit Dimension** dialog box is displayed, **Figure 9-18.** In the Precision and Tolerance tab, select Basic in the **Tolerance Method** list and pick the **OK** button to exit the dialog box. This specifies that the dimension is a nominal value and the real value is controlled by the feature control frame, which indicates .01 at maximum material condition. This means the measured dimension on the actual part can vary from 37.01° to 36.99°. This is a simplified explanation, but will suffice for the scope of this section.

Now, Datum A needs to be specified. You will also add a second datum, Datum B. Typically, there are three datums that are perpendicular to one another. However, this example requires only two. The **Datum Identifier Symbol** tool is used to specify datums. Pick the **Datum Identifier Symbol** button in the **Drawing Annotation Panel**.

Figure 9-18.
The **Edit Dimension** dialog box is used to select the tolerancing method.

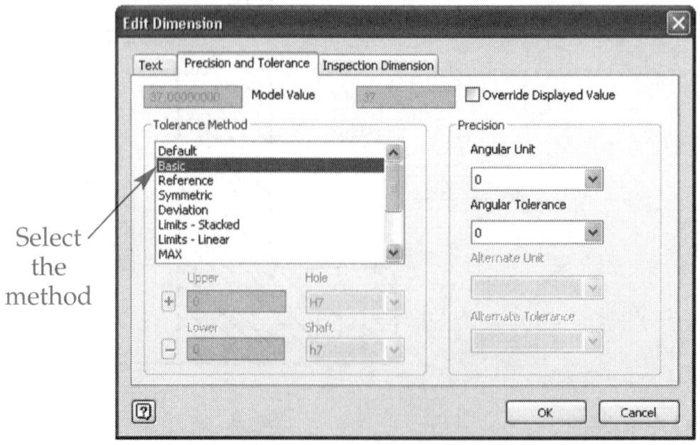

Select the method

Then, in SECTION B-B, pick the end of the centerline of the main bore diameter. Move the cursor to the left and pick another point. Move the cursor up and pick a third point. The **Format Text** dialog box is displayed. There should already be an A in the text box, so pick **OK** to exit the dialog box. Specify Datum B by picking the bottom surface of the part. Follow the same step for placing the datum, but change the letter to B in the text box.

Adding additional feature control frames

Now that the two datums are in place, the remaining feature control frames need to be added. To specify that the top surface of the part is to be parallel to Datum B, activate the **Feature Control Frame** tool. Next, pick the top surface of the part in SECTION B-B. Move the cursor to the left and pick a second point to specify length and direction of the leader. Right-click and select **Continue** in the pop-up menu. In the **Feature Control Frame** dialog box, specify the parallelism symbol and enter the values as shown in **Figure 9-19.** Pick the **OK** button to exit the dialog box.

With the **Feature Control Frame** tool active, specify the centerline of the left-hand mounting hole to be positioned in relation to Datum A and perpendicular to Datum B. Pick the end of the mounting hole centerline. Move the cursor to the left and pick a second point to specify length and direction of the leader. Then, right-click and select **Continue** in the pop-up menu. In the **Feature Control Frame** dialog box, specify the position symbol for the primary feature control and perpendicularity symbol for the secondary feature control. When a symbol is selected for the second feature control frame, the text boxes are enabled. Enter the values shown in **Figure 9-20.** Pick the **OK** button to exit the dialog box. Press [Esc] to exit the **Feature Control Frame** tool.

Take a good look at the last feature control frame you created. See **Figure 9-21.** Notice that the alignment is not correct on the right-hand end. To fix the alignment, select **Style and Standard Editor...** from the **Format** pull-down menu. Expand the Feature Control Frame branch and select FeatureControlFrame(ANSI). In the **Options** area of the **General** tab, pick the button below the **Cell Alignment** label to turn it on. Save the changes and exit the dialog box. The feature control frame appears as shown

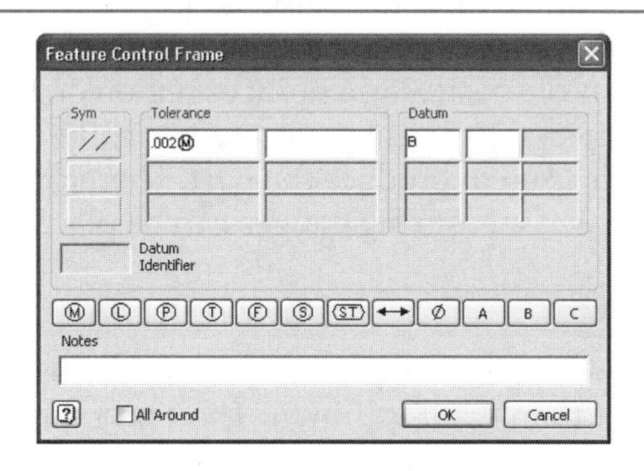

Figure 9-19.
Defining a feature control frame that specifies the top surface of the part is parallel to Datum B.

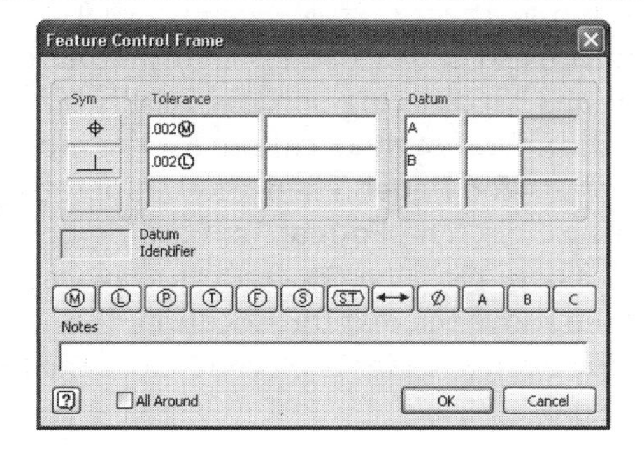

Figure 9-20.
Defining a feature control frame that specifies the center-line of the mounting hole is positioned in relation to Datum A and perpendicular to Datum B.

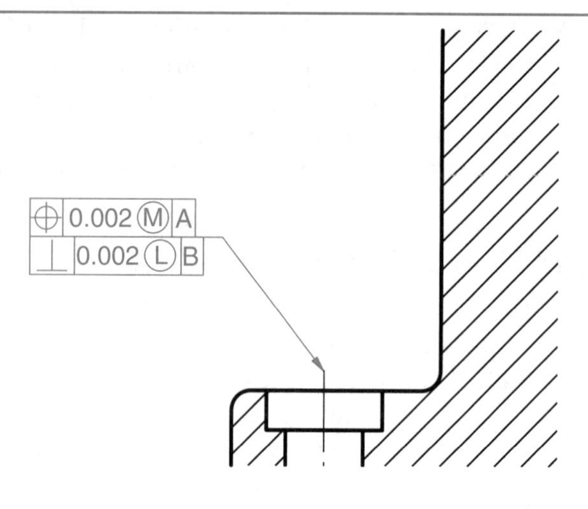

Figure 9-21.
The right-hand end of the feature control frame is misaligned.

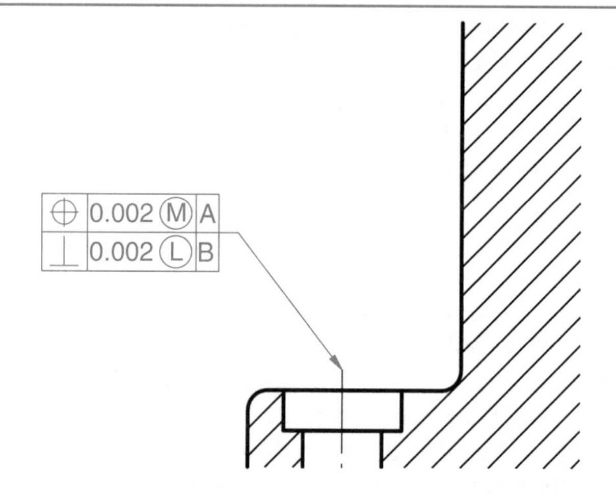

Figure 9-22.
The correct alignment for the right-hand end of the feature control frame.

in **Figure 9-22.** Notice how the right-hand edge is aligned. You may need to select the feature control frame for it to be updated.

The **Datum Target** tool is used to insert datum target symbols. These symbols provide a means of specifying a point or region on the part as a reference point or plane for tolerance checking. These areas are typically not altered by the various manufacturing methods used to produce the part. Therefore, they provide a consistent foundation against which measuring gauges can be placed.

Adding Text

Everyone working with prints eventually learns that what is written on a drawing is as important than what is drawn. The **Text** tool on the **Drawing Annotation Panel** is used to place text. Text in Inventor is based on a text style. Any changes made to the text style will be reflected in the text on the drawing. Use the **Style and Standard Editor** dialog box to change the text style. See **Figure 9-23.**

Keep in mind, text can be used in many situations—sketches in part modeling, for emboss operations, on drawing borders, in title blocks, etc. In this example, text will be used to finish the GD&T symbol callout for the mounting holes. Pick the **Text** button in the **Drawing Annotation Panel**. Then, pick a point below the dual feature control frame for the mounting hole. The **Format Text** dialog box is displayed. In the text box, type TYP. 3 HOLES. Then, pick the **OK** button to place the text. Finally, press [Esc] to end the tool and then move the text into location. Using the **Leader Text** tool is the same as adding text. However, a leader is placed before the **Format Text** dialog box is displayed and text is entered.

Figure 9-23.
Changing a text style in the **Style and Standard Editor** dialog box.

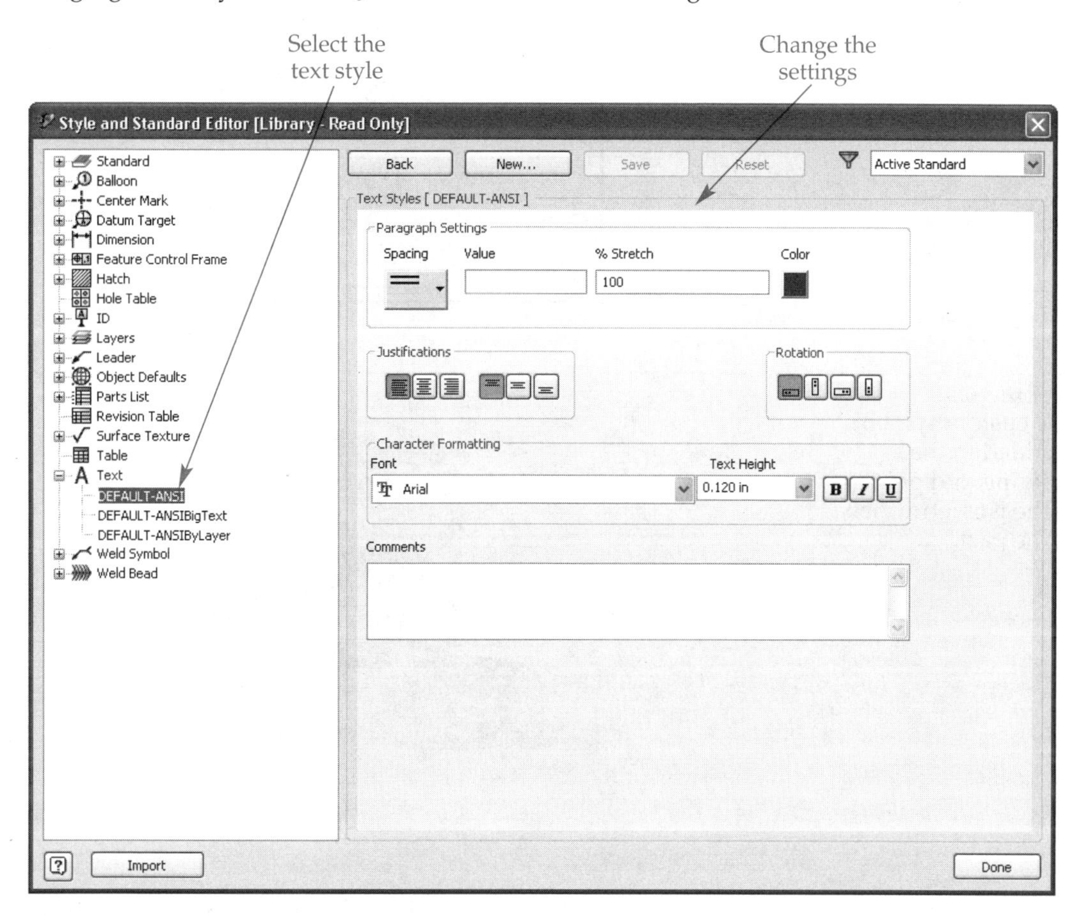

NOTE

The **Format Text** dialog box is only for the text object being created. Text styles govern the global appearance of text in your drawing.

Sketched Symbols

Sketched symbols are user-defined, saved graphics that can be inserted into the drawing at any time. Sketched symbols can be created by first expanding the Drawing Resources branch in the **Browser**. Then, right-click on Sketch Symbols and pick **Define New Symbol** in the pop-up menu. Inventor enters sketch mode and the **Drawing Sketch Panel** is displayed in the **Panel Bar**. Use the tools to create a symbol, right-click in the graphics window, and select **Save Sketched Symbol** from the pop-up menu. In the **Sketched Symbol** dialog box that appears, give the symbol a meaningful name. The new symbol can be found in the Sketch Symbols branch under the Drawing Resources branch in the **Browser**.

To insert a symbol, pick the **Symbols** button in the **Drawing Annotation Panel**. The **Symbols** dialog box is displayed, **Figure 9-24.** Here, you can select which symbol to insert, scale the symbol, and rotate the symbol. The symbol can be placed with a leader if the **Leader** check box is checked. In Example_09_01.idw there is a symbol named Mark Part. This is similar to a custom annotation required by internal drafting standards. Select Mark Part in the **Symbols** dialog box, enter a scale of 1.000, do not enter a rotation, and check the **Leader** check box. Then, pick the **OK** button to close the dialog box. The symbol appears in the graphics screen attached to the cursor. In the upper-right isometric view, pick a point on the vertical, cylindrical edge. Pick a second point to

Figure 9-24.
Custom symbols are added to the drawing using the **Symbols** dialog box.

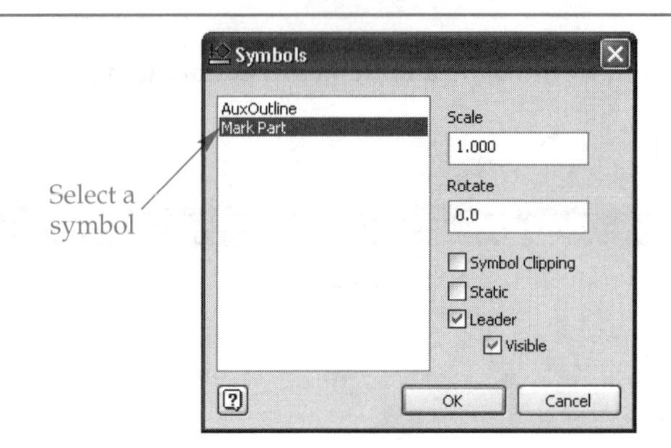

Figure 9-25.
The custom symbol is added to the drawing and points to the isometric view of the part.

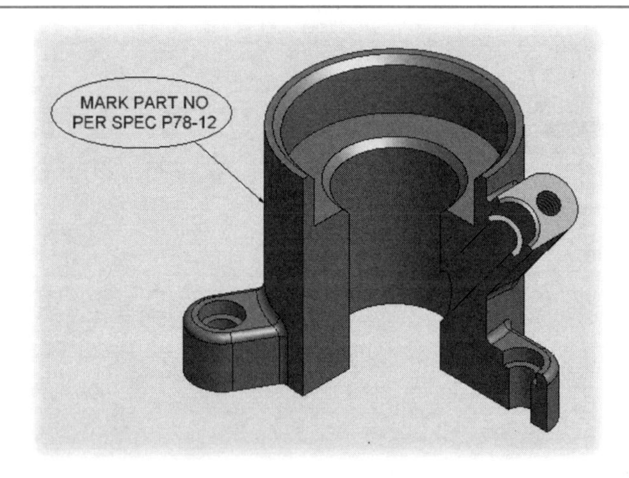

the upper-left of the view, right-click, and select **Continue** from the pop-up menu. The symbol is placed and points to the isometric view. See **Figure 9-25.** Press the [Esc] key to exit the tool.

PROFESSIONAL TIP

Another method for inserting a symbol without being scaled or rotated and without a leader is to right-click on the symbol in the Drawing Resources branch in the **Browser** and pick **Insert** from the pop-up menu. Then, pick the points for symbol placement, right-click, and select **Continue** in the pop-up menu. Finally, press the [Esc] key to exit the tool.

Revision Table and Revision Tag

The ***revision table*** is used to document the revisions that have been made to a drawing. It is typically located in the upper, right-hand corner of the drawing border, but it can be located in any of the corners.

For the part in Example_09_01.idw, the vertical height needs to be revised. If the 2.25 dimension on the height was a parametric dimension, then the change could be made there. However, it is driven, so the change has to be made to the extrusion distance. Open the part Example_09_01.ipt. If it is already open, switch to it using the **Window** pull-down menu. Edit Extrusion1, change the extrusion distance to 2.00, and save the part. Switch to Example_09_01.idw using the **Window** pull-down menu. Notice the drawing views have updated. Move any of the annotations or views as needed to properly space everything.

To create a revision table for the drawing, pick the **Revision Table** button in the **Drawing Annotation Panel**. The **Revision Table** dialog box is opened, **Figure 9-26.** Select the scope for the table and how the index will be created, then pick the **OK** button. A rectangle representing the table is attached to the cursor. Place the table in the upper-right corner of the drawing border. The edges of the table will "snap" against the border as you move the cursor. Once placed, the table is automatically filled out with default information.

The content in the revision table can be edited by right-clicking on the table and selecting **Edit** in the pop-up menu. Then, in the **Revision Table** dialog box, edit the table cells as needed and pick the **OK** button to update the table. For example, change the description text as shown in **Figure 9-27** and edit the approved text to your initials.

Before the revision is finished, a revision tag must be placed next to the new 2.00 dimension. This provides a visual locator to the change listed in the revision table. Pick the **Revision Tag** button in the **Drawing Annotation Panel**; it is located in the flyout with the **Revision Table** button. Then, pick a point next to the 2.00 dimension, right-click, and select **Continue** from the pop-up menu. Press [Esc] to exit the tool. See **Figure 9-28.** Save the drawing.

Figure 9-26.
When placing a revision table, this dialog box is used to select the scope and how the table is indexed.

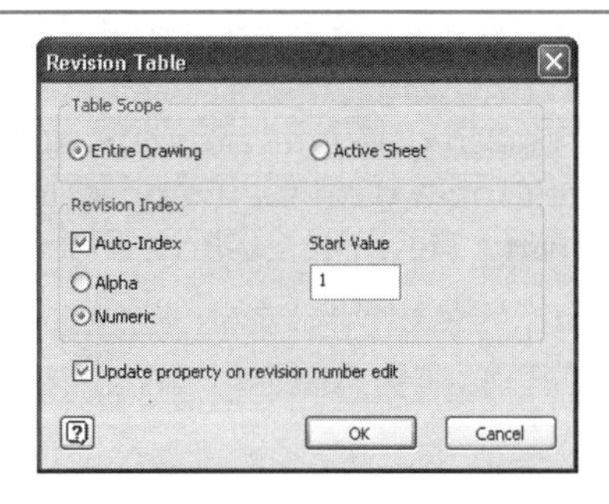

Figure 9-27.
The edited revision table.

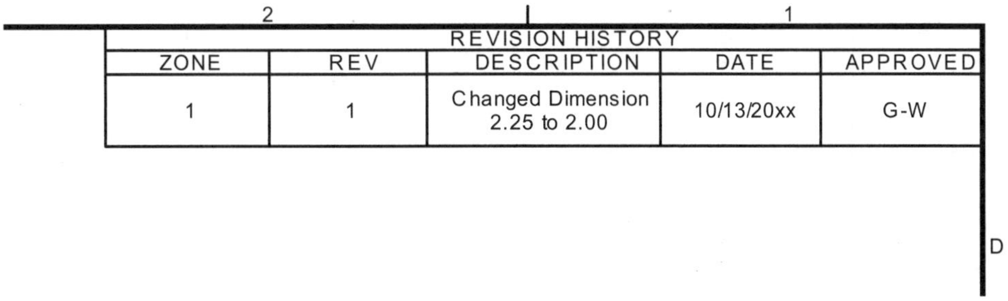

2 1

REVISION HISTORY				
ZONE	REV	DESCRIPTION	DATE	APPROVED
1	1	Changed Dimension 2.25 to 2.00	10/13/20xx	G-W

D

Figure 9-28.
Placement of the revision tag.

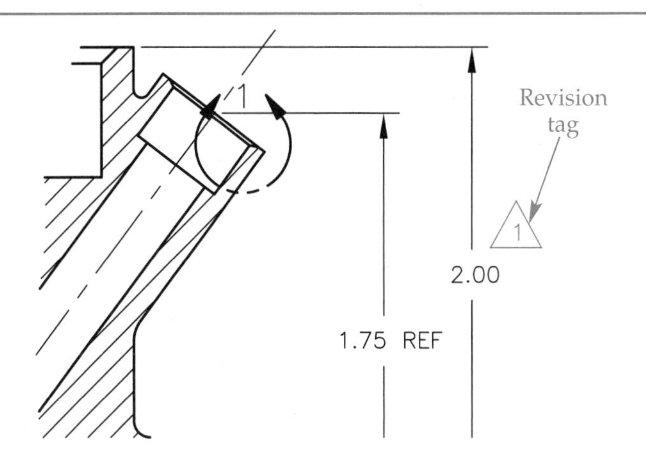

Weld Symbols

When two components of a part are to be joined by welding, the edges of the joint are typically prepared to receive the weld. Weld symbols are placed on the drawing to describe the type and size of the weld to be made at the joint. Open Example_09_03W.iam. This is an Inventor assembly file. Examine the weldment. Also, look at the weldment features in the **Browser**. This assembly file and another assembly file will be used to create a drawing layout and apply weld symbols. There is no association between these weldment features and the weld annotations that will be applied. Weld symbols are drafting conventions used on the drawing only.

Drawing setup

A finished drawing showing weldment features and annotations is shown in **Figure 9-29.** To create this drawing, begin a new drawing based on the ANSI (in).idw template and set the sheet size to C. Also, open Example_09_03.iam. Create a base view from the top view of Example_09_03.iam (*not* Example_09_03W.iam) at a scale of 1/2 and place it in the upper-left corner of the sheet. Create a projected view below the base view. From this projected view, create another view projected to the right. The drawing now contains three orthographic views, similar to **Figure 9-29,** and they are all based on Example_09_03.iam. Notice that there are no weldment features at the joint between the tee and the reducer.

To create the rendered isometric view, use Example_09_03W.iam. The isometric view is made by creating a base view and selecting the current view of Example_09_03W.iam in the **Drawing View** dialog box. Place the isometric view in the upper-right corner of the sheet. Also, shade the view. In the next section, you will see why two assemblies are used.

Figure 9-29.
Weldment features and annotations are added to the drawing.

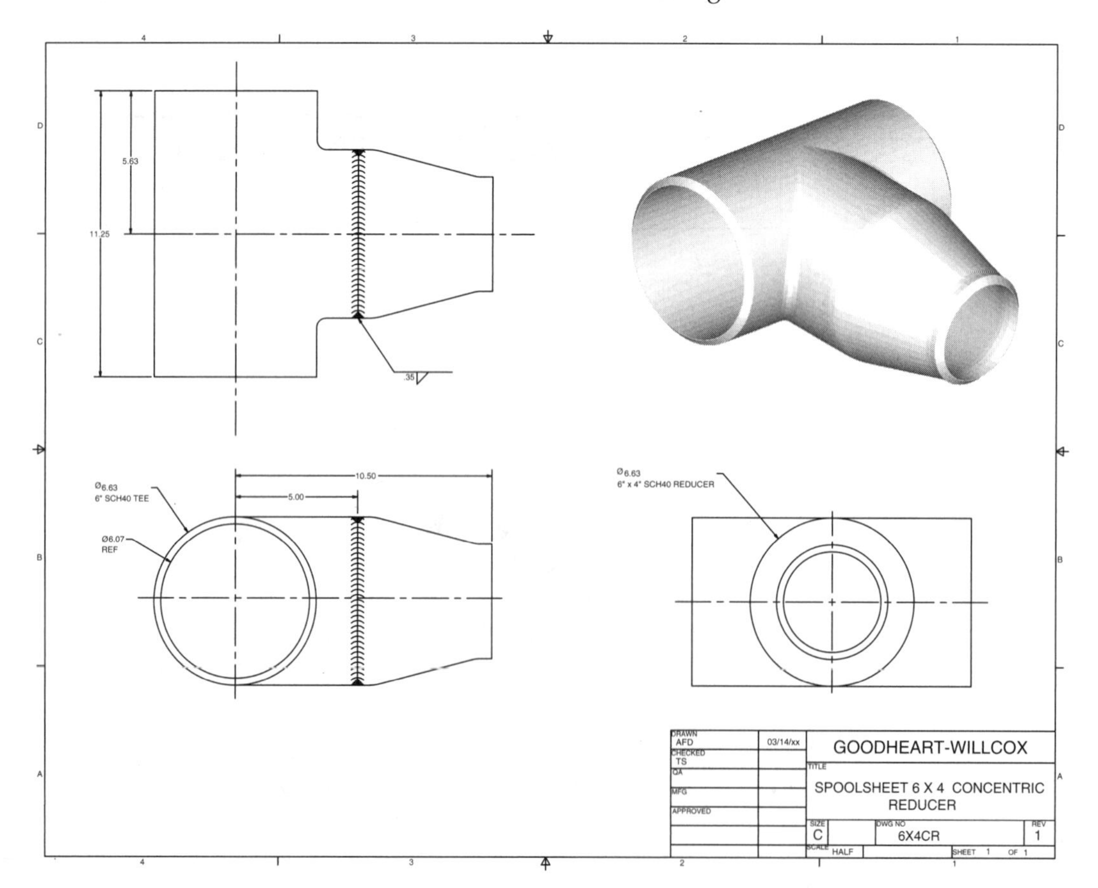

Applying weld annotations

To add the necessary weld symbol to the welded seam in the example drawing, pick the **Welding Symbol** button in the **Drawing Annotation Panel**. Then, pick a point on the joint in the top view, pick another point to determine leader line length and direction, right-click, and pick **Continue** from the pop-up menu. The **Welding Symbol** dialog box is displayed, **Figure 9-30.** Pick the weld symbol button in the middle of the dialog box and select the standard weld reference symbol, as shown in the figure. Enter .35 in the **Prefix** text box, also shown in the figure. Finally, pick the **OK** button to place the symbol. Press [Esc] to cancel the tool.

There are two other annotations used to depict a welded seam. These are produced using the **End Fill** and **Caterpillar** tools. The **End Fill** tool places filled-in triangles on the weld seam. Two are required in the top view and two are required in the front view. Refer to **Figure 9-29.** The **Caterpillar** tool creates a drafted symbol for the weld.

Pick the **End Fill** button in the **Drawing Annotation Panel**; it is located in the **Caterpillar** flyout. The **End Fill** dialog box is displayed, **Figure 9-31.** In the **Preset Shape** area of the **Type** tab, pick the **V-Type** button. Also, change the width to .500. The values in the **Options** tab control the size of the end treatment symbol. The default values are okay for this drawing. Next, pick the endpoint of the line representing the seam in the top view.

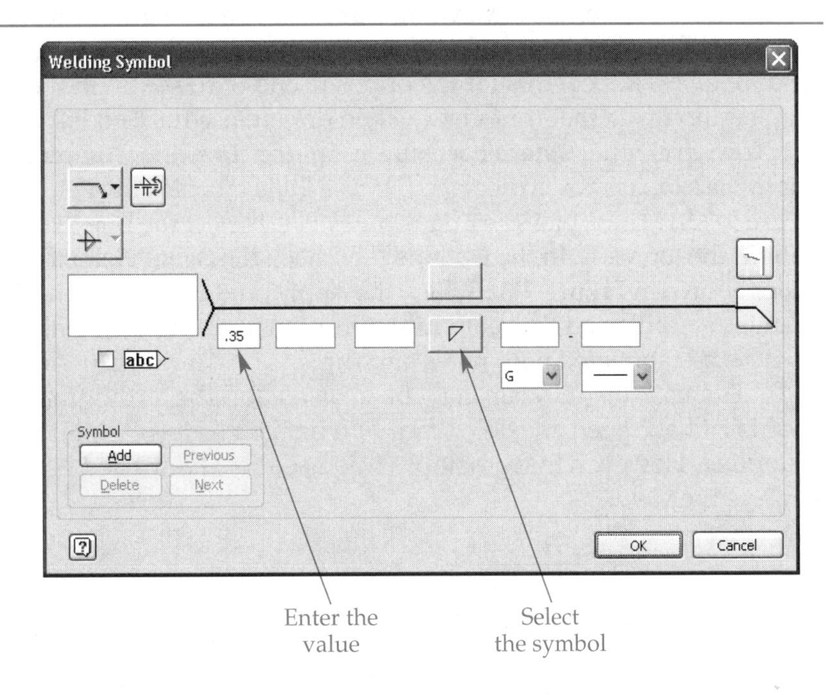

Figure 9-30.
Defining a weld symbol in the **Welding Symbol** dialog box.

Figure 9-31.
The **End Fill** dialog box.

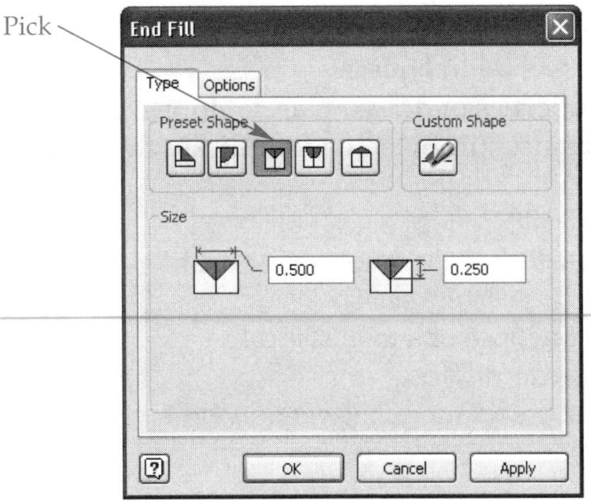

Figure 9-32.
A—The **Style** tab of the **Weld Caterpillars** dialog box. B—The **Options** tab of the **Weld Caterpillars** dialog box.

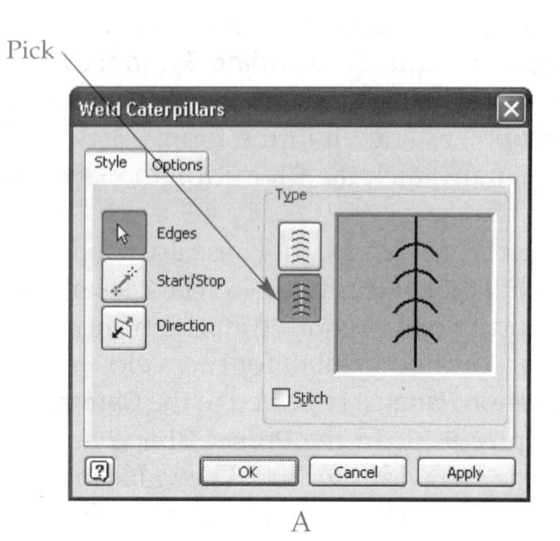

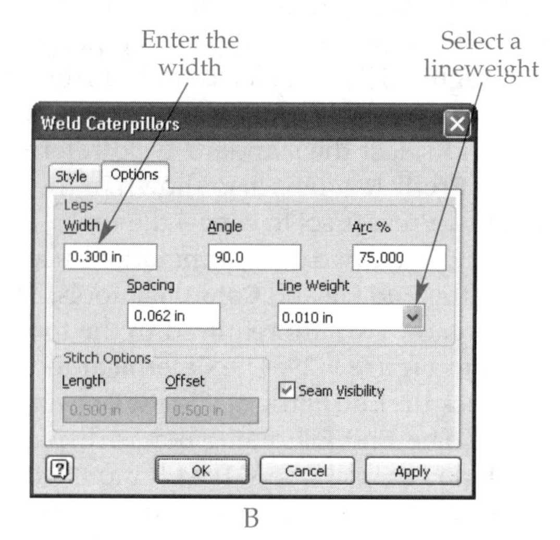

Either end can be selected. Move the cursor so the triangular preview points toward the other end of the seam and pick to determine the direction. Then, pick the **Apply** button in the dialog box. Repeat this for the opposite end of the seam in the top view and the two seam endpoints in the front view. When done, close the **End Fill** dialog box.

Now, pick the **Caterpillar** button in the **Drawing Annotation Panel**. The **Weld Caterpillars** dialog box is displayed. In the **Style** tab, pick the **Full** button in the **Type** area, **Figure 9-32A.** Then, pick the **Edges** button and select the line that represents the welded seam in the top view. In the **Options** tab, check the **Seam Visibility** check box and use the values shown in **Figure 9-32B.** Pick the **Apply** button to create the weld caterpillar. Now, create a caterpillar on the seam in the front view using the same settings. The drawing can now be completed with notes and dimensions and can be saved.

If Example_09_03W.iam had been used for the orthographic views, then there would not have been a visible seam to which the caterpillar and end treatments could be applied. However, that assembly was used for the shaded isometric view because it displays the weld.

Chapter Test

Answer the following questions on a separate sheet of paper or complete the electronic chapter test on the Student CD.

1. What is the difference between a *parametric dimension* and *reference dimension?*
2. What is a *feature control frame?*
3. How are properties of drawing annotations, such as color or linetype, controlled?
4. List the tools used in Inventor to indicate the center of circular and symmetrical features.
5. What is a *hole note?*
6. What is a *sketched symbol* and when would you use one?
7. What is a *surface texture symbol?*
8. What is the purpose of a *weld symbol?*
9. What is a *revision table?*
10. What is the association between weld symbols and the referenced Inventor weldment?

Chapter Exercises

Exercise 9-1. Support Beam.

Complete the exercise on the Student CD

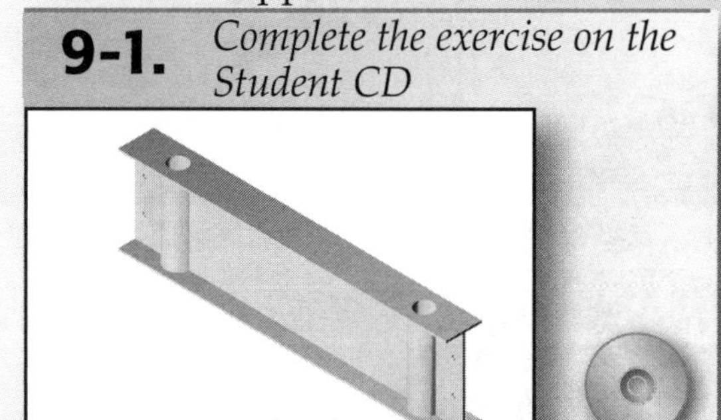

Exercise 9-2. Mass-Produced Part.

Complete the exercise on the Student CD

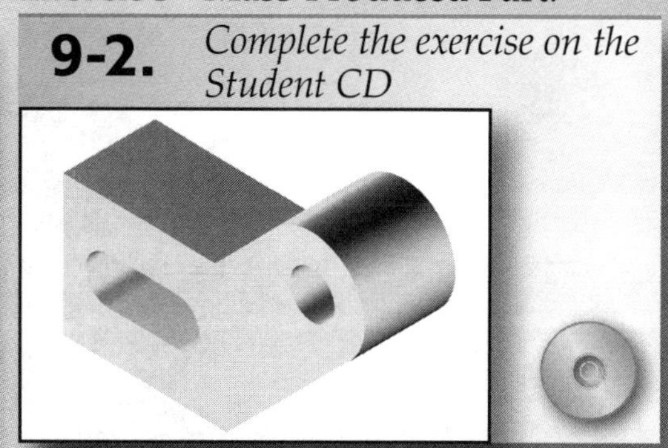

Exercise 9-3. Precision Part.

Complete the exercise on the Student CD

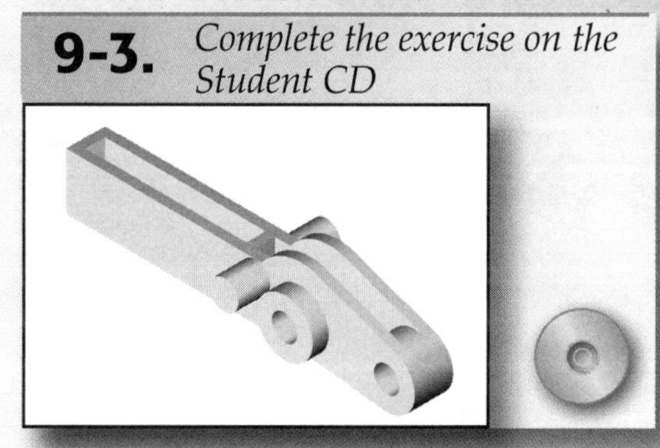

Exercise 9-4. 150 lb. Flange.

Complete the exercise on the Student CD

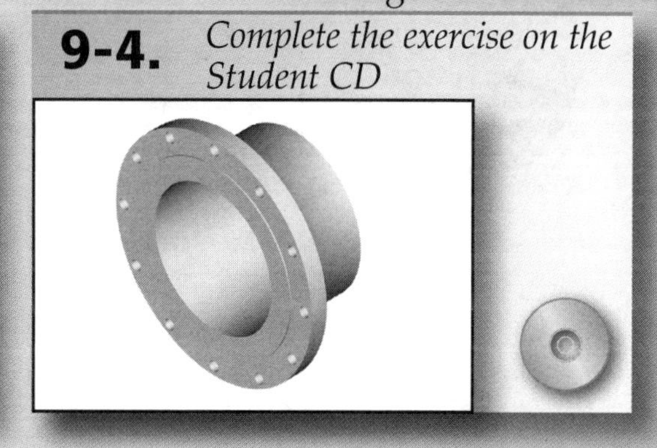

Exercise 9-5. Brace Shaft with Broken View.

Complete the exercise on the Student CD

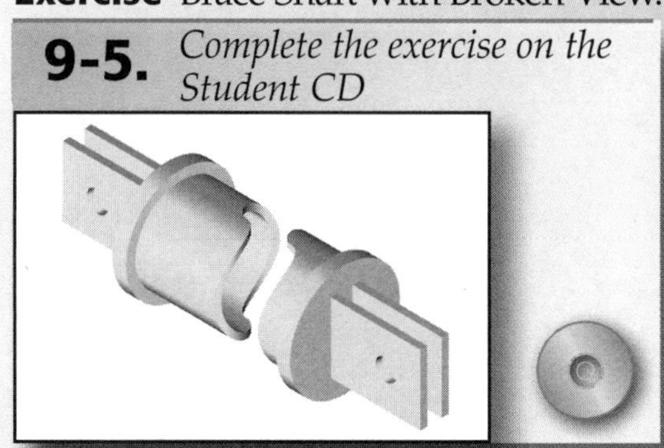

Exercise 9-6. *Handwheel.*

Complete the exercise on the Student CD

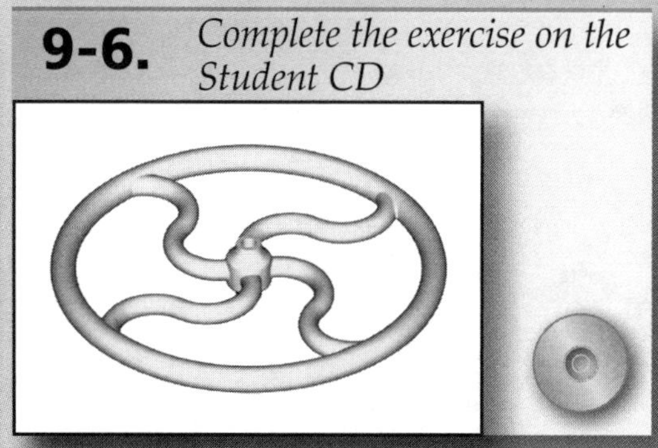

Sweeps and lofts are well suited for organic models. The overall shape of this shaver has few straight lines. Instead, the body is primarily formed from compound curves. This type of model can best be created from sweeps and lofts. (Model courtesy of Autodesk, Inc.)

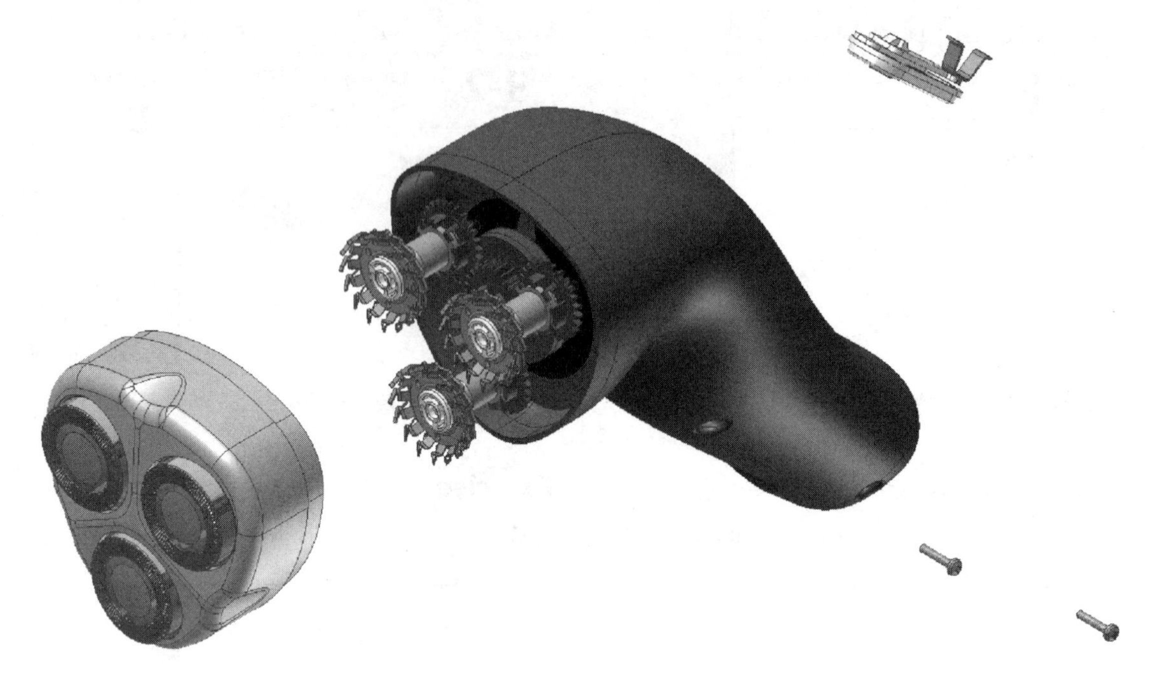

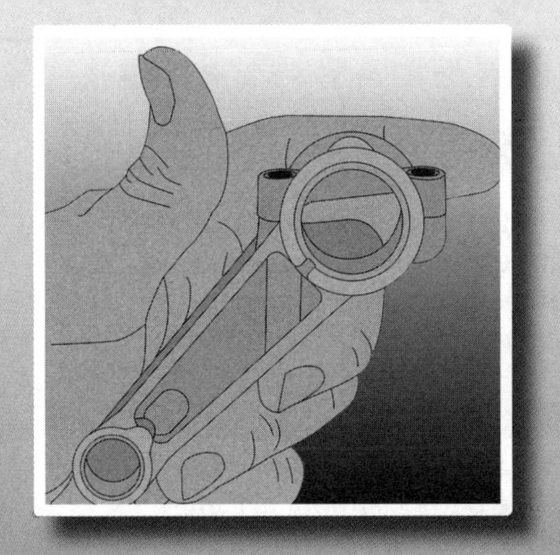

Chapter 10

Sweeps and Lofts

Objectives

After completing this chapter, you will be able to:

- Explain the difference between an extrusion, sweep, and loft.
- Explain the process for creating a sweep.
- Create 3D sketches.
- Create 2D and 3D sweeps.
- Define inline work features.
- Create inline work features.
- Create lofts.

User's Files

The Student CD included with this text contains several files required for this chapter. Refer to the file File List.txt *in the* \Ch10 *folder for the comprehensive list.*

Sweeps and Lofts

As you have seen, many complex parts can be created using extrusions and revolutions. However, Inventor has two other modeling techniques that are very powerful—sweeps and lofts. A ***sweep*** is a solid object created by extruding, or "sweeping," a profile along a path. With the **Extrude** tool, the path is an implied straight line the length of which is set by the extrusion terminator (Distance, To, or From To). A ***loft*** is similar to a sweep except that you can use multiple profiles to create a solid of varying cross sections. In lofts, the path is called a ***rail.*** There may be multiple rails or none at all.

Creating Sweep Features

Sweep features are typically created for piping or tubing, but are not limited to these uses. You may be able to quickly create a feature as a sweep that would take much longer to create as an extrusion. For example, a picture frame with mitered corners is best modeled as a sweep. Sweeps have a variety of applications in modeling piping, tubing, and similar parts.

Creating a Sweep Path

The first step in creating a sweep is to create the ***path.*** The path is simply a sketch that is dimensioned and constrained just as any other sketch. It may be open or closed, depending on the shape of the final part. For example, if the part is piping, then the path will be open. If the part is a picture frame, then the path is closed.

Open Example_10_01.ipt and examine the two sketches. Select Sketch for Path in the **Browser**. Notice that the path lies on a plane. Therefore, it is considered a ***2D path.*** The path was created on the XZ plane, but any sketch plane can be used. Later in this chapter, you will learn about 3D paths and the ways of creating them.

Example_10_01.ipt is an example of piping. In this type of application, it is best to plan the path so that it originates at a known point. In this case, the path originates at the CSI origin.

Creating a Cross-Sectional Profile

Like the path, the ***cross-sectional profile*** is a sketch that is dimensioned and constrained. However, the profile sketch must be separate from the path sketch. In order to use the **Sweep** tool, there must be two unconsumed sketches—the path and the profile. In this example, the profile was sketched on the XY plane. The sketch is of two concentric circles centered on the origin. Since the path is on the XZ plane and starts at the origin, this ensures that the cross section is at the start of the path and intersects the path. Inventor requires that the cross section *intersect* the path in some way. However, the cross section does not need to be at the start of the path. It may intersect the path somewhere along the path's length.

Creating the Sweep

With the path and the profile created, it is time to create the sweep using the **Sweep** tool. Pick the **Sweep** button in the **Part Features** panel. The **Sweep** dialog box is displayed. See **Figure 10-1.** With the **Profile** button on, pick the profile in the graphics window. Select the area between the two concentric circles. Since the cross section is an ambiguous sketch, you must select the profile. If there is a single, closed, unambiguous sketch, the profile is automatically selected. Next, pick the **Path** button in the dialog box to turn it on and select the path in the graphics window. The path is the S-shaped line.

Below the **Path** and **Profile** buttons, you can determine which type of part is constructed. Usually, you will be creating solids. Therefore, make sure the **Solid** button is on. However, if you want to create a surface, pick the **Surface** button.

The three buttons in the middle of the dialog box determine which type of operation is performed. The buttons are **Join**, **Cut**, and **Intersect**. These operations are the same as the corresponding operations in the **Extrude** tool. Since this sweep is the first feature of the part constructed, the only available option is **Join**.

You can set a taper angle in the **Taper** text box. This allows you to gradually increase or decrease the size of the profile as it is extruded along the path. A negative angle decreases the size and a positive angle increases the size.

You may have noticed when you selected the path and profile that a preview of the sweep appears along the path. Notice how the profile follows the path, adjusting

Figure 10-1.
The **Sweep** dialog box is used to make the settings for a sweep.

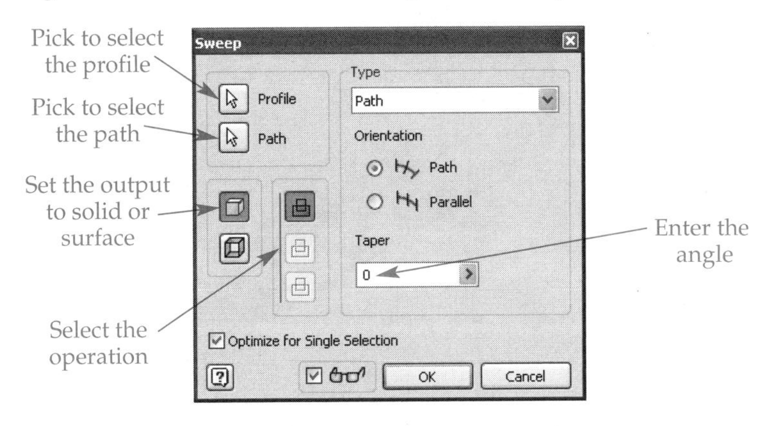

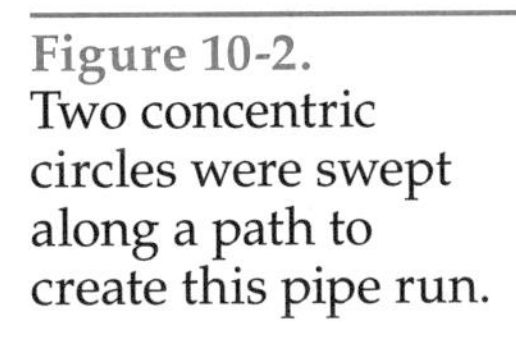
Figure 10-2.
Two concentric circles were swept along a path to create this pipe run.

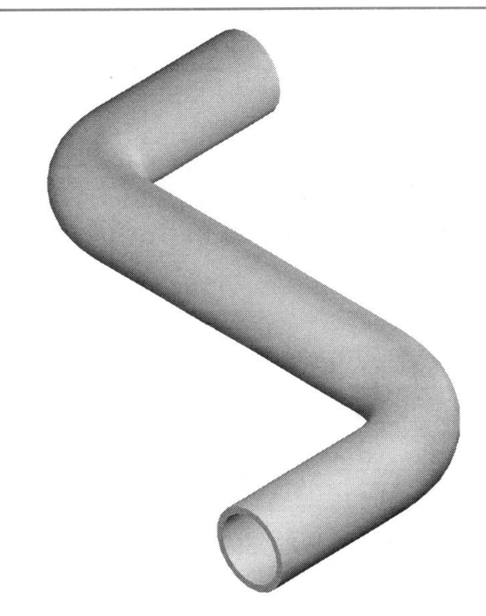

its orientation so it is always perpendicular to the path. Now, pick the **OK** button in the dialog box to create the sweep. See **Figure 10-2.**

Profile not Perpendicular to the Path

In the previous example, the cross section was sketched perpendicular to the path. To see the importance of this, open Example_10_02.ipt. The path is the same as in the previous example. The profile is also the same; however, notice that the sketch is *not* perpendicular to the path. The profile sketch is angled at 30° to the plane on which the path lies. Using the same procedure described above, create a sweep of the profile. Notice that the resulting cross section of the part is *elliptical,* **Figure 10-3A.** If you rotate the view, you can clearly see the end of the pipe is elliptical, **Figure 10-3B.** This is because the circular profile is projected onto a plane that is perpendicular to the path plane. When a circle is viewed at an angle, it appears as an ellipse. Since the circular profile is not perpendicular to the path plane, the projected profile is an ellipse, which is then extruded along the path. It is important to remember this as you sketch paths and profiles.

Open Example_10_03.ipt. Pick the **Sweep** button to open the **Sweep** dialog box. Pick the concentric circles as the profile and the sketched spline as the path. Set the output to a solid and notice the previewed sweep. Now, pick the **Parallel** radio button in the **Orientation** area. Notice how the preview has changed. Pick the **OK** button to create the

Figure 10-3.
A—If the circle profile is not perpendicular to the path, an elliptical cross section is created.
B—Rotating the view clearly shows the elliptical shape.

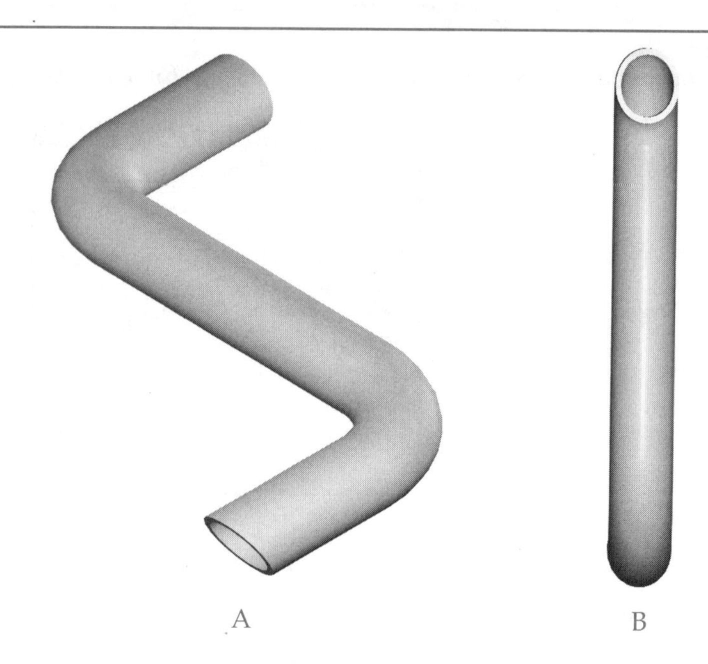

sweep. Rotate the view and examine the part. The cross section created by the concentric circles always remains parallel to the initial orientation of the circles. Notice how the tube changes thickness or diameter as the spline curves away from the XZ plane.

Closed Paths

In the previous examples, the paths were open. However, the path can be closed. Open Example_10_04.ipt. This file contains two unconsumed sketches—a profile and a path. Notice that the path is a rectangle; in other words, it is closed. Use the **Sweep** tool to create the part as a solid. The resulting part is picture frame molding. Pay particular attention to how the profile is extruded ("swept") around the corners. Sharp corners are created. Even if a circle is swept on this path, sharp corners are created because the path has sharp corners.

Practical Example of Sweeps

Look at **Figure 10-4.** This is a guard used in an industrial application to prevent large debris from entering a pipe inlet. Now, open Example_10_05.ipt. The two bottom rings are already created in the file. Also, the file contains a work plane and the Y axis and XY plane have been made visible. Work Plane1 is parallel to the rings and 8" above the top ring. You will now model one of the spokes as a sweep, create a circular pattern for the remaining spokes, and create the cap.

Creating the Sweep Path

1. Start a new sketch on the XY plane.
2. Project both the work plane and the Y axis onto the sketch plane.
3. Right-click and select **Slice Graphics** from the pop-up menu.
4. Zoom in on the left-hand end of the rings.
5. Project the outer radius of the top ring. See **Figure 10-5.** This results in an arc and a center point.
6. Zoom in on the right-hand end of the rings and project the same geometry.
7. Pick the **Three Point Arc** button in the **2D Sketch Panel**.

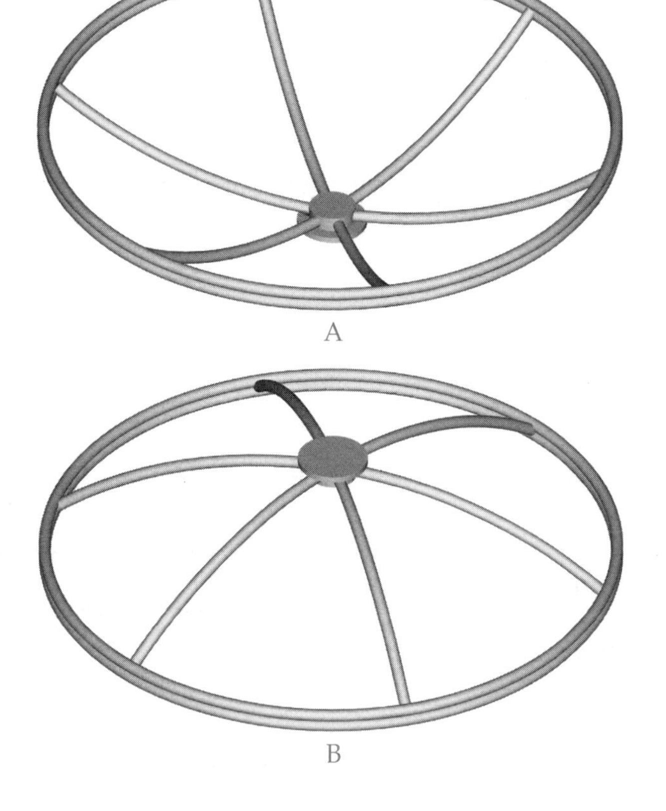

Figure 10-4.
This is a guard for a pipe inlet. A single spoke was created as a sweep, then a circular pattern was created. A—A top, isometric view. B—A bottom, isometric view.

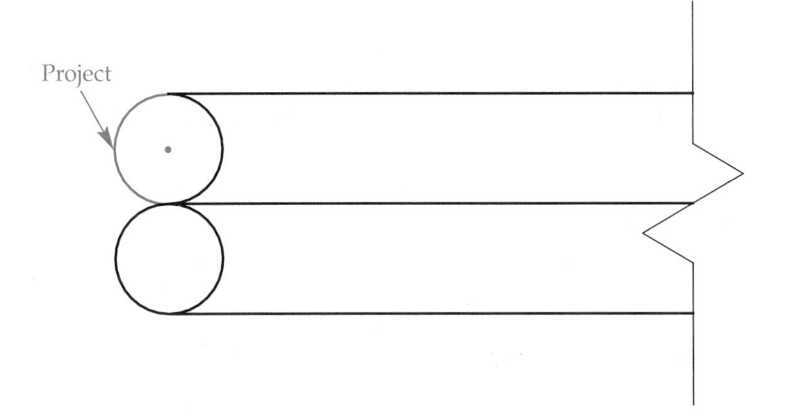

Figure 10-5.
To draw the path for one spoke, first project the geometry on the left-hand side. Then, project the same geometry on the right-hand side before drawing the arc path.

8. Select the start point of the arc as the left-hand center point.
9. Select the endpoint of the arc as the right-hand center point.
10. Select the third point—the point on the arc—as the intersection of the project axis and projected plane. To select this point, right-click and select **Intersection** from the pop-up menu. Then, pick the projected Y axis line and the projected work plane line.
11. This arc is the path for a sweep. See **Figure 10-6.** Finish the sketch.

Creating the Cross Section

The sketched profile to be swept does not have to be at the end of the path. In this example, it is easiest to create the cross-sectional profile at the midpoint of the path.

1. Display the isometric view.
2. Start a new sketch on the YZ plane by right-clicking on YZ Plane in the Origin branch of the **Browser** and selecting **New Sketch** in the pop-up menu.

Figure 10-6.
Drawing the profile for one spoke.

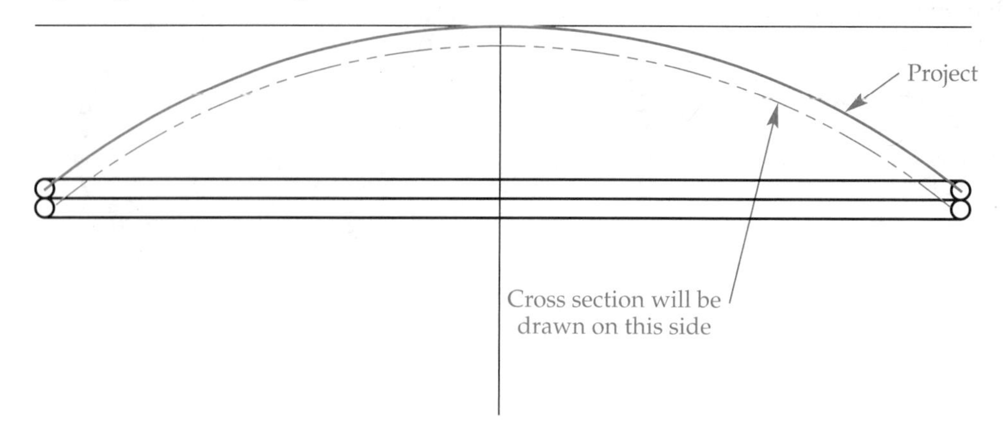

3. Project onto the sketch plane the arc created earlier as the path. This results in a line.
4. Draw a circle with its center on the projected line and passing through the top endpoint of the projected line.
5. Dimension the diameter of the circle as 1.0″. This is the cross-sectional profile.
6. Finish the sketch.

Completing the Part

1. Pick the **Sweep** button in the **Part Features** panel to open the **Sweep** dialog box.
2. Select the circle as the profile; it should be automatically selected since it is the only closed, unambiguous sketch.
3. Select the arc as the path.
4. Set the output to a solid, pick the **Path** radio button in the **Orientation** area, and pick the **OK** button to create the sweep.
5. Pick the **Circular Pattern** button in the **Part Features** panel to display the **Circular Pattern** dialog box.
6. Pick the sweep as the feature and the Y axis as the rotation axis. Enter 3 in the **Occurrence Count** text box and 360 in the **Occurrence Angle** text box. Then, pick the **OK** button to create the pattern.
7. To create the center cap, start a new sketch on Work Plane1.
8. Project the Y axis onto the sketch plane. This results in point.
9. Zoom in on the projected point.
10. Sketch a circle centered on the projected point.
11. Dimension the diameter of the circle as 4″ and finish the sketch.
12. Extrude the circle down 1.25″ with no taper.
13. Start another new sketch on Work Plane1.
14. Project the Y axis onto the sketch plane.
15. Sketch a circle centered on the projected axis point, dimension the diameter as 6″, and finish the sketch.
16. Extrude the circle up .5″ with no taper.

The part is now complete. Among other things, this example shows the reasons for building parts about the origin and inline with the basic work planes.

10-1 Complete the practice problem on the Student CD.

3D Sweeps

Two-dimensional planar sweeps have many applications, but there are also times when more-complex, nonplanar 3D sweeps are required. Examples of these situations are pipe or tubing runs and wire harnesses. In order to create a 3D sweep, the path must be created as a 3D sketch. To start a new 3D sketch, pick the arrow next to the **Sketch** button on the **Inventor Standard** toolbar to display a flyout. See **Figure 10-7.** Then, pick **3D Sketch**. A new 3D sketch is started and the **3D Sketch** panel is displayed in the **Panel Bar**.

A 3D sketch can be composed of a 3D line, 3D spline, or helical curve. Existing points or work points can be selected to create these curves. Coordinates can also be entered using the **Inventor Precise Input** toolbar.

First Example of a 3D Sketch

The next example is a pipe run. The most efficient way to construct the sketch is by starting with an existing point and then inputting the coordinates of subsequent points. Open Example_10_06.ipt.

In a 3D sketch, the **Line** command can automatically create fillets and rounds between line segments. The radius is set in the **Document Settings** dialog box. Open the dialog box by selecting **Tools>Document Settings...** and then display the **Sketch** tab. Enter the value in the **Auto-Bend Radius** text box; 6.0″ for this example.

Start a 2D sketch on the top of the ∅4″ stub. The two circles representing the pipe OD and ID are automatically projected into the sketch. These will be used as the cross section for a sweep. Also, the center point of the circles is projected into the sketch. This point will be used to start a 3D path. Finish this 2D sketch.

Now, start a 3D sketch. Pick the **Line** button in the **3D Sketch** panel. The **Inventor Precise Input** toolbar is automatically displayed. Notice that the **Z:** text box is not grayed out. Since this is a 3D sketch, you can enter X, Y, and Z coordinates. Pick the center point of the stub as the first point. The coordinate system triad moves to the point. Remember, red = X, green = Y, and blue = Z. Turn on the auto-bend option, by right-clicking in the graphics window and checking **Auto-Bend** in the pop-up menu. Now, input these coordinates for the pipe run using the **Inventor Precise Input** toolbar:

X	Y	Z	Comment
0	0	6	Press [Enter] to set the point. The triad moves to the new point.
0	–42	0	The line starts with the 6″ radius (in effect, replacing the first segment).
36	0	0	
0	0	54	
0	36	0	
30	0	0	Press [Enter] to set the point, then [Esc] to end the **Line** tool.

Note that you must input the zeros each time. When the last point has been entered, press [Esc] to end the **Line** tool. Now, right-click and select **Finish 3D Sketch** from the pop-up menu to finish the sketch and return to part mode.

Figure 10-7.
Starting a new 3D sketch.

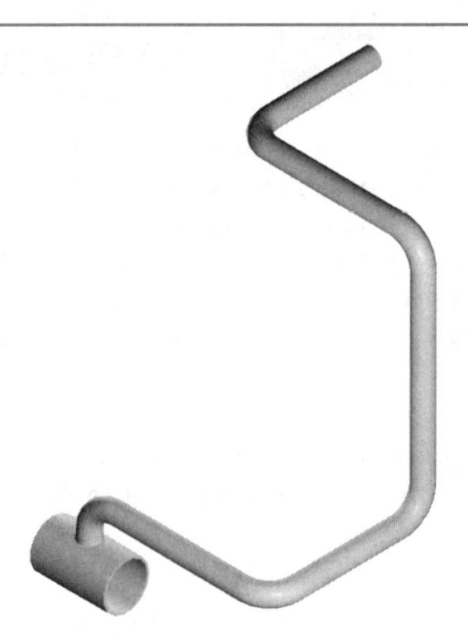

Figure 10-8.
The completed pipe run.

Pick the **Sweep** button in the **Part Features** panel to display the **Sweep** dialog box. Select the two concentric circles on the top of the stub as the profile. Select the 3D line as the path. Then, create the sweep. See **Figure 10-8.**

The path for this example is relatively easy to create, but the pipe run is difficult to edit. The next example creates the same pipe run. The method used to create the path is more complex, but the path is easy to edit.

Second Example of a 3D Sketch

Open Example_10_07.ipt. Notice the work planes in the **Browser**. Each is named according to the offset values used to create it. The finished version of the example is the same as the one shown in **Figure 10-8.** These work planes will be used to create the path for the 3D sweep. To create the path, you first need to create seven work points. Start a new 3D sketch and continue as follows.

1. Select the **Work Point** button in the **3D Sketch** panel. Any three intersecting planes that do not share the same edge view intersect at a single point. Therefore, pick the following three work planes in any order to define the first work point: XZ plane, YZ plane, and Work Plane Start. Remember, if the plane is not visible in the graphics window, it can be selected in the **Browser**. A work point, a diamond, appears at the intersection of the three work planes. See **Figure 10-9.**
2. Pick the **Work Point** button again. To create the second work point, pick the XZ plane, the YZ plane, and Work Plane 6in.
3. To create the third point, pick the **Work Point** button again and pick the YZ plane, Work Plane 6in, and Work Plane 42in.
4. Pick the **Work Point** button again. Then, pick Work Plane 6in, Work Plane 36in, and Work Plane 42in.
5. Pick the **Work Point** button again. Then, pick Work Plane 36in, Work Plane 42in, and Work Plane 54in.
6. Pick the **Work Point** button again. Then, pick the XZ plane, Work Plane 36in, and Work Plane 54in.
7. Pick the **Work Point** button again. Then, pick the XZ plane, Work Plane 30in, and Work Plane 54in.

As you can see, a good command of visualization in 3D space is required to create a 3D sketch. Now, to finish the 3D path you need to draw a line connecting the work points. First, set the auto-bend radius to 6″. Then, pick the **Line** button in the **3D Sketch** panel. The **Inventor Precise Input** toolbar is displayed. Since existing work points will be used to define the endpoints of the line, this toolbar can be closed. Right-click and

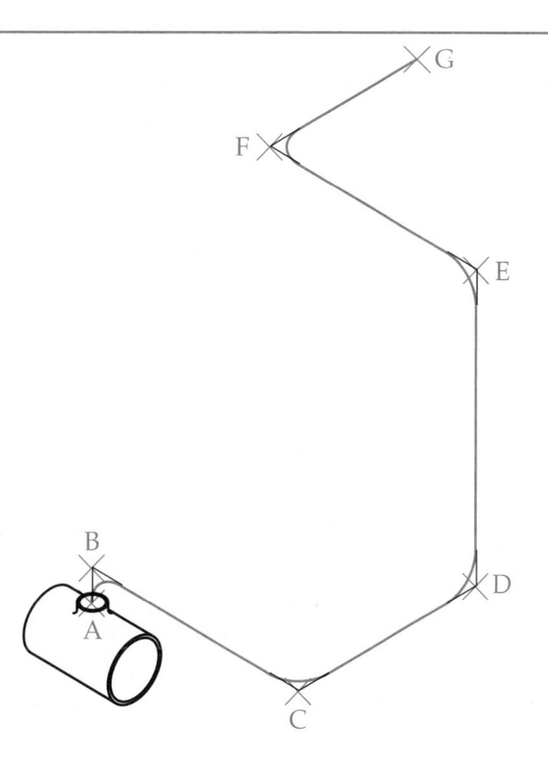

Figure 10-9.
Creating work points. The points are created in order from A to G. Notice the effect of the auto-bend feature.

make sure **Auto-Bend** is checked in the pop-up menu. Now, in the graphics window, pick the work points one by one in the order in which you created them. As you connect the work points, notice that Inventor automatically applies a 6″ radius fillet to each corner. After picking the seventh work point, press [Esc] to end the **Line** tool. Then, finish the 3D sketch.

Now, create the profile. Start a new 2D sketch on Work Plane Start. Zoom in on the pipe stub on top of the pipe tee. Using the **Project Geometry** tool, project the ID and OD circles of the pipe. Then, finish the 2D sketch.

Pick the **Sweep** button in the **Part Features** panel to display the **Sweep** dialog box. Select the area between the two concentric circles as the profile. Select the 3D line in the graphics window as the path. Then, create the 3D sweep. The part now appears as shown in **Figure 10-8.**

Expand the part tree for the sweep. Notice the structure. It contains two sketches, just as a 2D sweep would. Also, notice that the work points are in the part tree for the 3D sketch. See **Figure 10-10.** All of these items can be renamed to be more meaningful in the design. For example, the default name Work Point*x* quickly becomes meaningless in a complex piping layout. Instead, name the work points by location, pipe type, or other meaningful naming convention.

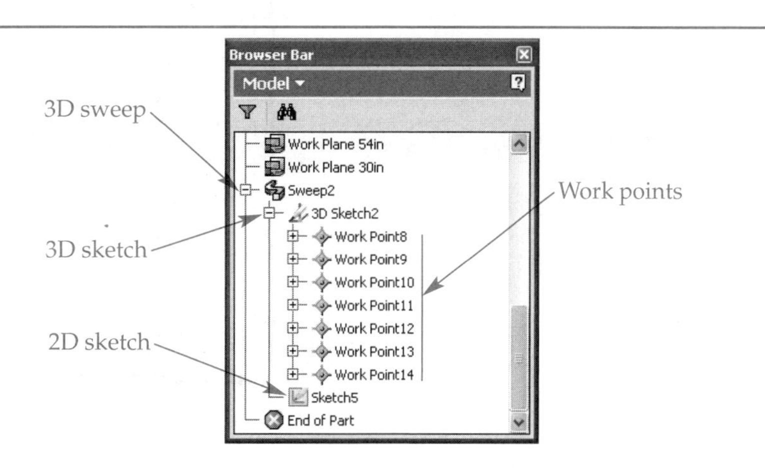

Figure 10-10.
The part tree for the completed pipe run.

This 3D sweep is easier to edited than the previous example because the work planes can be quickly altered. When the position of a work plane is changed, the work point based on it also moves. In this way, the path of the sweep can be edited.

PROFESSIONAL TIP

If you choose to draw the line with auto-bend off, the **Bend** tool can be used to manually apply a bend radius to the corners.

Inline Work Features

In the previous example, all of the work planes were created before the 3D sketch was started. In this section, you will create the work planes as you are creating the 3D sketch. Work features created on the fly, or within an active tool, are called ***inline work features.*** As you will see, the method used in the previous example is easier than the one you will use in this section. However, both methods are valid for creating 3D sweeps in Inventor. Why bother learning the more difficult method? The method presented in this section is more widely used in creating assemblies. Also, it is a more sophisticated process that is required in many real-world applications.

Open Example_10_08.ipt. This file is similar to the original file in the previous example. However, there is only one work plane—Work Plane Start. **Figure 10-11** shows the part tree for the completed 3D sweep in the **Browser**. Notice that the work points have work planes listed below them in the part tree.

Start a new 3D sketch. Then, use the following procedure to create the work points for the 3D sketch of the path.

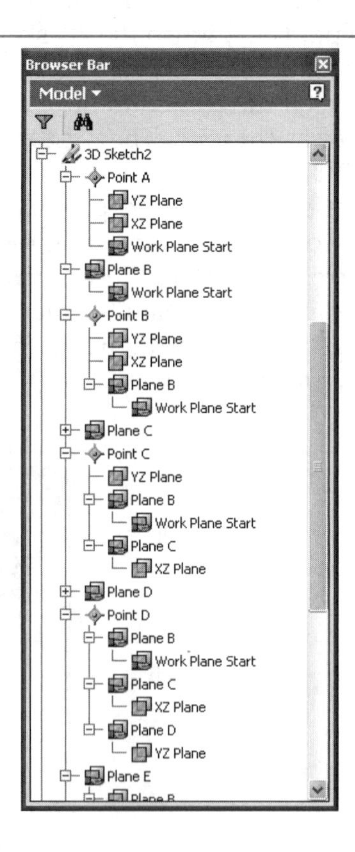

Figure 10-11.
The part tree for the completed pipe run created with inline work features. Notice that the work planes are below the work points in the part tree.

1. Pick the **Work Point** button in the **3D Sketch** panel. Then, select the XZ plane, YZ plane, and Work Plane Start. A work point appears at the intersection of the three planes. Note: This step is the same as in the previous example.
2. In the **Browser**, expand the part tree for the 3D sketch. Then, rename the new work point to Point A. This is done by picking the name and then picking it again; do not double-click. Then, type the new name in the text box that appears and press [Enter]. By renaming the work features as you create them, it is easier to keep track of items as you work through this procedure. Refer to **Figure 10-9.**
3. Pick the **Work Point** button in the **3D Sketch** panel. Then, pick the XZ plane and the YZ plane. For the third work plane, right-click in the graphics window and select **Create Plane** from the pop-up menu, **Figure 10-12.** Pick Work Plane Start in the graphics window and drag up. Release the mouse button, enter a value of 6 in the **Offset** dialog box that appears, and press [Enter].

A new work plane is created offset 6″ above Work Plane Start. At the same time, a work point appears at the intersection of the XZ plane, YZ plane, and the new work plane. Notice that the inline work plane was created last in Step 3. The new work plane is located under the new work point in the part tree. In the **Browser**, expand the part tree, rename the new work point to Point B, and rename the new work plane to Plane B. Also, turn on the visibility of Plane B. Continue as follows.

4. Make sure no work plane is selected. Pick the **Work Point** button in the **3D Sketch** panel. Pick the YZ plane and Plane B. Then, right-click in the graphics window and select **Create Plane** from the pop-up menu. Pick the XZ plane, drag to the right, release the mouse button, and enter –42 in the dialog box. A work point appears at the intersection of the two existing planes and the new work plane.
5. In the **Browser**, rename the new work point to Point C and the new work plane to Plane C. Also, turn on the visibility of Plane C.
6. Make sure no work plane is selected. Pick the **Work Point** button in the **3D Sketch** panel. Pick Plane B and Plane C. Then, right-click and select **Create Plane** from the pop-up menu. Move the cursor over the YZ plane name in the **Browser** until the plane is outlined in red in the graphics window. Then, pick the YZ plane in the graphics window, drag to the upper right, release the mouse button, and enter 36 in the dialog box.
7. In the **Browser**, rename the new work point to Point D and the new work plane to Plane D. Also, turn on the visibility of Plane D.
8. Make sure no work plane is selected. Pick the **Work Point** button in the **3D Sketch** panel. Pick Plane C and Plane D. Then, right-click and select **Create Plane** from the pop-up menu. Pick Plane B, drag up, release the mouse button, and enter 54 in the dialog box.
9. In the **Browser**, rename the new work point to Point E and the new work plane to Plane E. Also, turn on the visibility of Plane E.
10. Make sure no work plane is selected. Pick the **Work Point** button in the **3D Sketch** panel. Pick the XZ plane, Plane D, and Plane E. An inline work plane is not needed.

Figure 10-12.
Creating an inline work plane.

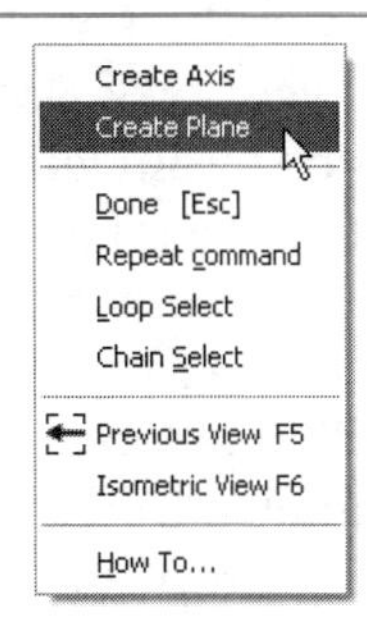

11. In the **Browser**, rename the new work point to Point F.
12. Pick the **Work Point** button in the **3D Sketch** panel. Pick the XZ plane and Plane E. Then, right-click and select **Create Plane** from the pop-up menu. Pick Plane D, drag to the right, release the mouse button, and enter 30 in the dialog box.
13. In the **Browser**, rename the new work point to Point G and the new work plane to Plane G. Also, turn on the visibility of Plane G.

Now, the sketch appears exactly as it did in the previous example after you created the seven work points. The rest of the procedure is the same as for the previous example. Set the auto-bend radius to 6 and draw a line between the work points. Then, finish the sketch, start a new 2D sketch on Work Plane Start, and project the ID and OD circles. Finish that sketch and sweep the circles along the path. The final part in this example is the same as in the previous examples. However, the process is a lot more tedious. Remember, this more complex process is useful when creating 3D sweeps in assemblies.

Editing a 3D Sweep

Editing a 3D sweep mainly involves editing the 3D sketches used for the sweep—the path and the profile. For example, you can change the offset values used to create the work planes. Since the work planes are used to create work points, and in turn the 3D path, this will alter the final part. You may also change the offset values to equations. In this manner, you can constrain the sweep to other features in the part. Additionally, if you right-click on the name of the sweep in the **Browser** and select **Edit Feature** from the pop-up menu, the **Sweep** dialog box is displayed. This allows you to select a different profile or path, change the output or creation method, or enter or change a taper angle.

Lofts

A ***loft*** is a solid or surface that is generated from a series of cross sections. The cross sections are created as sketches and do not need to be parallel to one another. The path for a loft is called a ***rail.*** The process of creating a loft is called ***lofting.*** Lofting is usually done to create organic shapes or parts with very complex, compound curves. Examples of where lofting is useful include a car body, ship hull, ergonomic mouse, or human hand. Other applications include modeling transitions between two known geometric shapes, such as a square-to-round transition found in HVAC ductwork.

Basic Loft

Open the file Example_10_09.ipt. There are three parallel work planes on which three sketches were created. The sketches were aligned using the Z axis, which is visible. Pick the **Loft** button in the **Part Features** panel to display the **Loft** dialog box. With the **Curves** tab displayed, pick Click to add in the **Sections** area. Then, pick the three sketched shapes in order from left to right (or right to left). As you pick sketches, they are added to the **Sections** area. The previewed part reflects those sketches selected. Finally, pick the **OK** button to create the lofted part. See **Figure 10-13.**

Right-click on the loft name in the **Browser** and select **Edit Feature** from the pop-up menu. Display the **Conditions** tab in the **Loft** dialog box. Notice that the sketches at each end of the loft are available in this tab. See **Figure 10-14.** Select Sketch1 and pick the arrow button to display a drop-down list. Select the Direction Condition entry in the drop-down list and notice how the preview changes. Pick in the **Weight** column for

Figure 10-13.
Creating a simple loft.

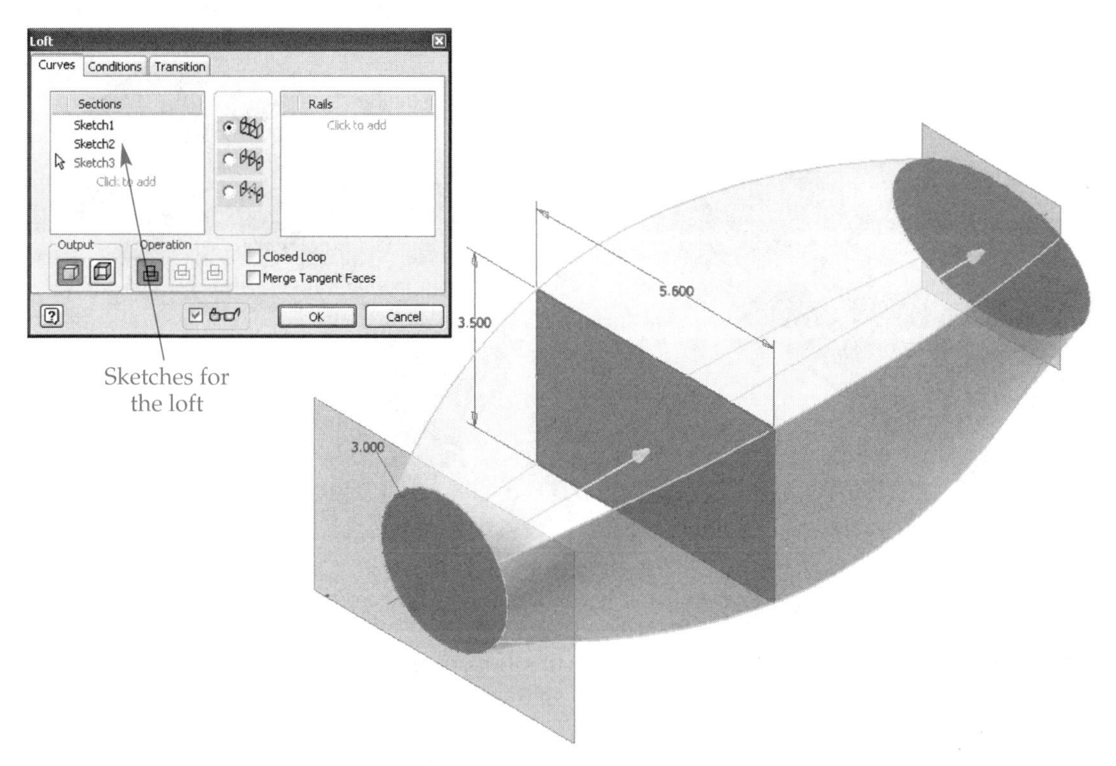

Figure 10-14.
The **Conditions** tab of the **Loft** dialog box.

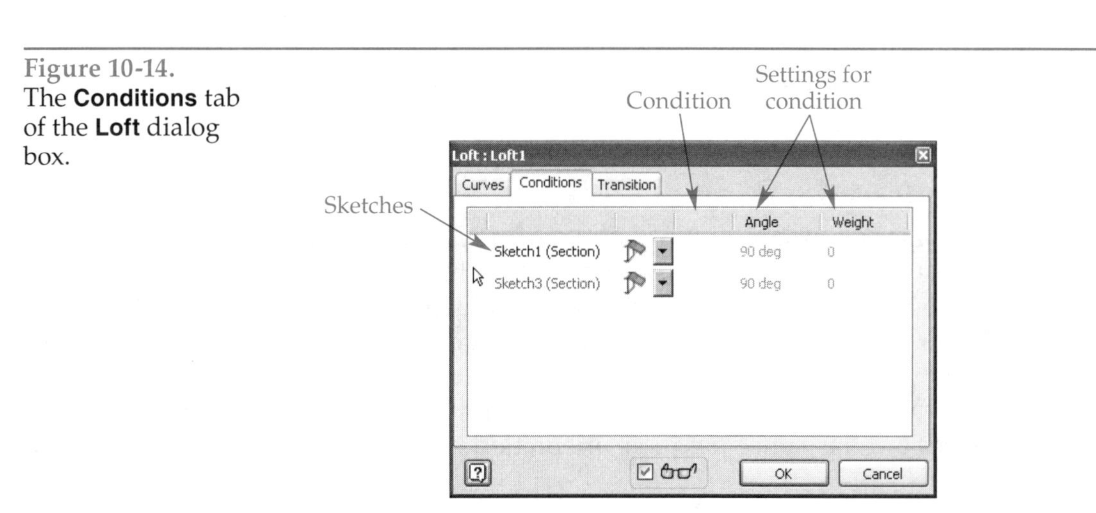

Sketch1 and enter a value of 3. This controls how far **Sketch1** maintains its shape as it lofts to the next sketch (Sketch2). A low value creates an abrupt change whereas a high value gives a more gradual change.

The sketches used for a loft do not have to be in parallel planes. Complex shapes can be created using rails to guide the shape from one sketch to another. Open Example_10_10.ipt. Shape 1 is the closed spline on the left and Shape 2 is on the right. Work Plane1 intersects both shapes. It was used to create the sketch named Rail, which is a spline. The spline must intersect each shape at one point order to be used as a rail.

Pick the **Loft** button in the **Part Features** panel to display the **Loft** dialog box. Pick the two closed shapes as sections. Notice how the feature preview goes straight from one shape to the other. In the **Curves** tab, pick Click to add in the **Rails** area. Then, pick the spline as the rail and note how it warps the lofted solid. See **Figure 10-15.**

Figure 10-15.
Using a rail to control a loft. A—Without a rail, the loft extends in a straight line between the two curves (profiles). B—By adding a rail that is curved, the loft is also curved.

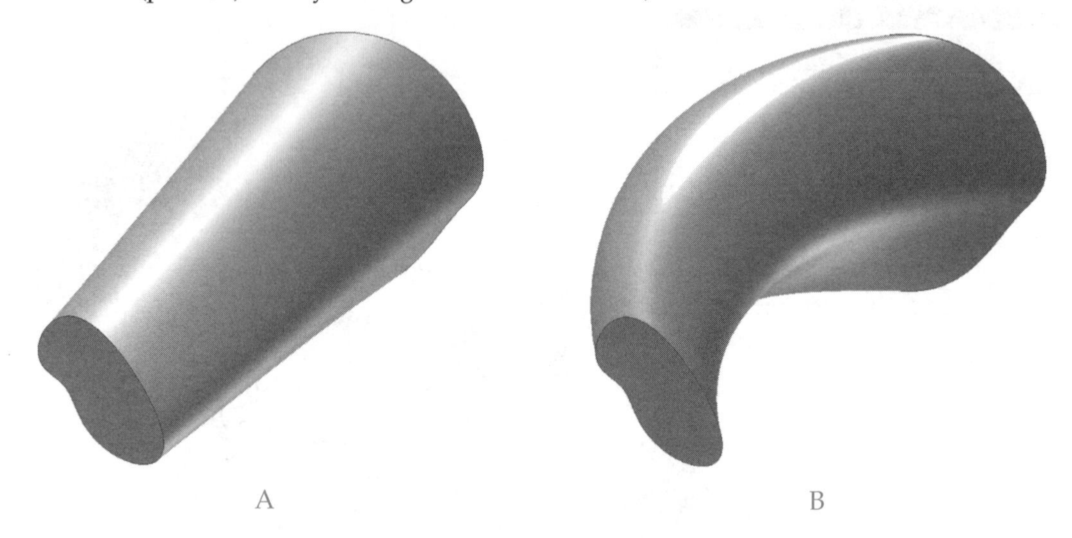

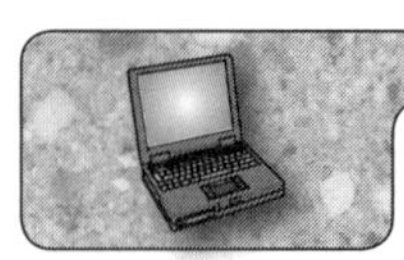

PROFESSIONAL TIP

Fillets and rounds can be applied to the edges of a loft.

PRACTICE 10-2 Complete the practice problem on the Student CD.

Closed-Loop Loft

The two previous examples are open-loop lofts. An ***open-loop loft*** does not begin and end at the same point. A ***closed-loop loft,*** which begins and ends at the same point, can also be created. Closed loops are typically used in sheet metal applications, such as the airplane engine cowling shown in **Figure 10-16.** This is a very complex, closed-loop loft.

Open Example_10_11.ipt. There are 16 sketches on 16 different work planes. All of the sketches are the same, just their orientation in 3D space is different. The work

Figure 10-16.
This is a closed-loop loft. It begins and ends at the same point.

planes and sketches are controlled by a shared sketch. Note: Each of the 16 sketches required several dimensions to locate and constrain the sketch. To improve visibility in the file, these dimensions have been deleted.

Pick the **Loft** button in the **Part Features** panel. In the **Curves** tab of the **Loft** dialog box, pick the Click to add label in the **Sections** area. Then, in the **Browser**, pick the first sketch (Sketch1). The sketches were created in order and, thus, appear in order in the **Browser**. Hold down the [Shift] key and pick the last sketch in the **Browser** (Sketch16). It may take a minute or so for Inventor to add the 16 sketches to the loft. Once all 16 sketches are listed in the **Sections** area of the dialog box, look at the preview. Notice the gap at the top. Now, check the **Closed Loop** check box in the **Curves** tab. This check box must be checked to connect the first and last cross sections. If it is not checked, this segment is not added and an open-loop loft is created. Also, you can check the **Merge Tangent Faces** check box. When checked, an edge is not created between segments whose faces are tangent. This eliminates an undesirable seam between the segments. However, in this example, do not check the **Merge Tangent Faces** check box. Finally, pick the **OK** button to create the loft. It may take Inventor some time to calculate the final part.

Complete the practice problem on the Student CD.

Editing a Loft Feature

A loft can be edited in basically the same way as a sweep. Editing the work planes on which cross-sectional sketches are created will alter the loft's shape. The individual sketches can also be edited, which may have the most effect on the final shape and appearance. If you right-click on the loft name in the **Browser** and select **Edit Feature** from the pop-up menu, the **Loft** dialog box is displayed. You can alter the settings in this dialog box. For example, if you forgot to check the **Closed Loop** check box, edit the feature and check it. You can also change the operation, from **Join** to **Cut**, for example. Editing the transitions by adjusting the point set is usually done after the cross sections are lofted. In this way, you can see the transitions that need adjustment.

Chapter Test

Answer the following questions on a separate sheet of paper or complete the electronic chapter test on the Student CD.

1. What is the difference between an *extrusion* and a *sweep?*
2. What is the difference between a *sweep* and a *loft?*
3. What is a *rail?*
4. What is the difference between a *2D path* and a *3D path?*
5. What happens if you sweep a circle along a path if the circle is *not* perpendicular to the path?
6. If a circle is swept around a square corner, what shape does the corner have on the finished part?
7. Describe the function of the auto-bend feature.
8. How do you turn on the auto-bend feature?
9. What is the difference between a *2D sweep* and a *3D sweep?*
10. Define *inline work feature.*
11. What is the process of creating a loft called?

12. When are lofts typically created?
13. List the three tabs in the **Loft** dialog box.
14. Explain the difference between an *open-loop loft* and a *closed-loop loft*.
15. When a curve is used to control the shape of a loft, what is the curve called?

Chapter Exercises

Exercise 10-1. Mating Part.
Complete the exercise on the Student CD

Exercise 10-2. Hitch Pin.
Complete the exercise on the Student CD

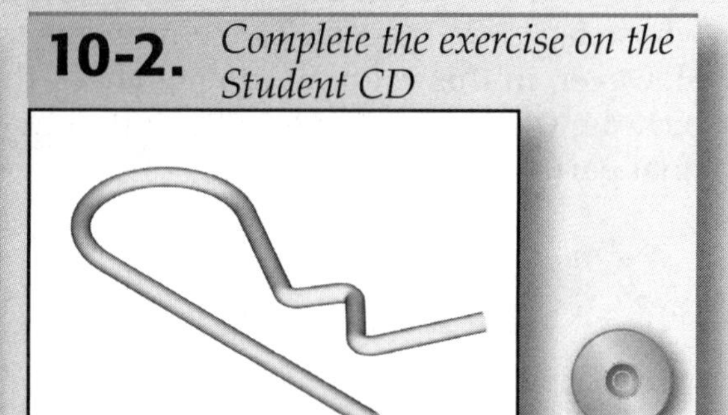

Exercise 10-3. Handwheel.
Complete the exercise on the Student CD

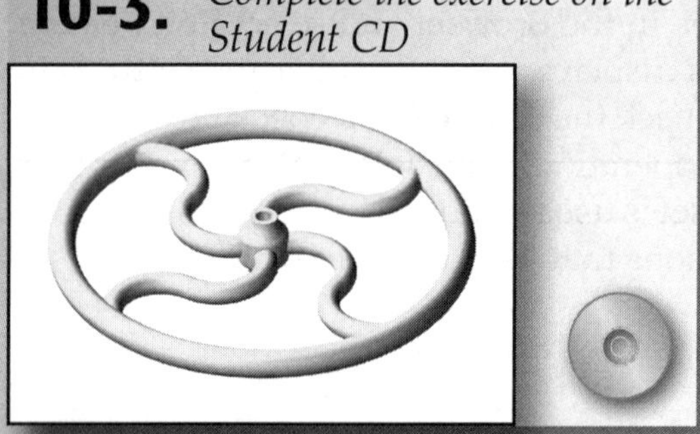

Exercise 10-4. Engine Cover.
Complete the exercise on the Student CD

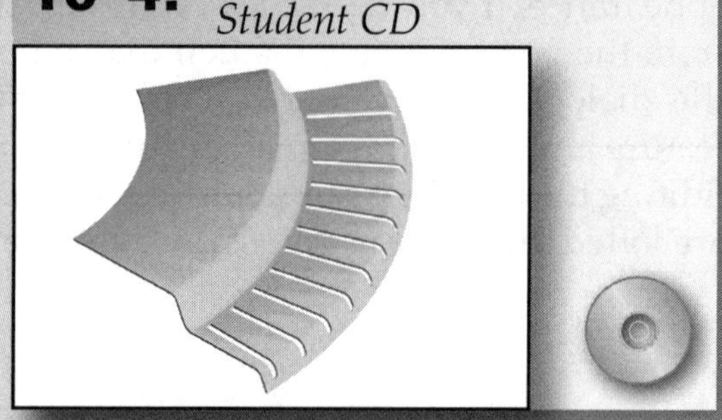

Exercise 10-5. Möbius Strip.
Complete the exercise on the Student CD

Exercise 10-6. Hook.
Complete the exercise on the Student CD

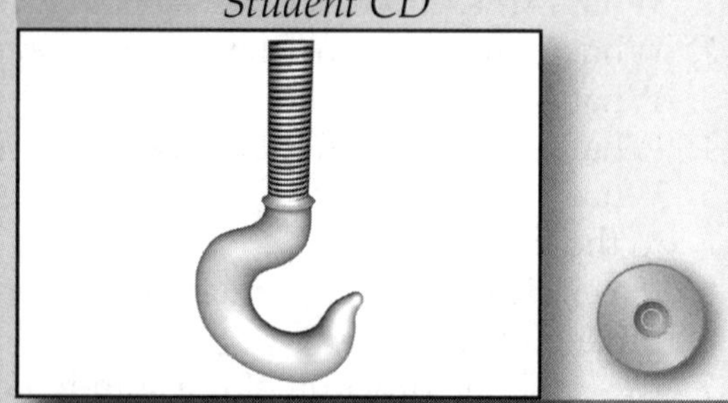

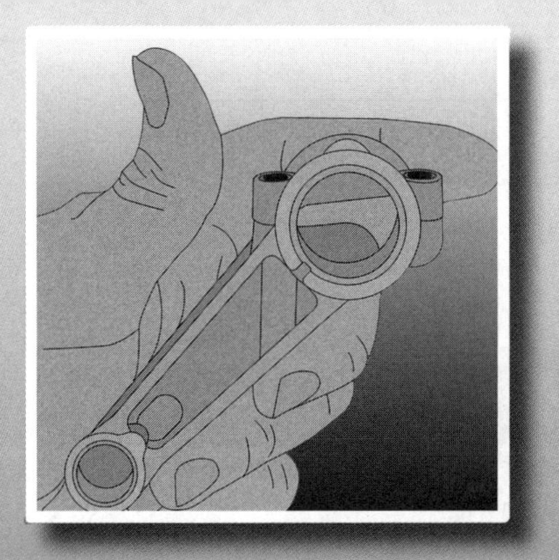

Chapter 11

Building Assemblies with Constraints

Objectives

After completing this chapter, you will be able to:

- Create and use projects.
- Create an assembly from existing parts.
- Apply mating constraints using various options.
- Apply insert constraints using various options.
- Apply tangent constraints using various options.
- Edit constraints placed in an assembly.
- Edit parts in place within an assembly.
- Place standard fasteners into an assembly.

User's Files

The Student CD included with this text contains several files required for this chapter. Refer to the file File List.txt in the \Ch11 folder for the comprehensive list.

Building an Assembly File

In this chapter, you will learn how to build an assembly file. An ***assembly*** shows how parts fit together to create the final product. For example, the assembly file shown in **Figure 11-1** contains six different parts—11E01_Lower_Base, 11E01_Base_Plate, 11E01_Bearing_Block, 11E01_Bushing, 11E01_CR_Locating Pin, and a standard cap screw selected from Inventor's fastener library. Some of these parts, such as the cap screw, are placed in the assembly file multiple times. When a part is placed in an assembly file, it is called an ***instance*** of the part, which is similar to an OLE link to the original part file. All parts are positioned in an assembly with respect to each other using assembly constraints. Also, in this chapter you will create an Inventor project. A ***project*** is a system for keeping track of the locations of all of the files in an assembly. It directs Inventor where to look for the files contained in an assembly.

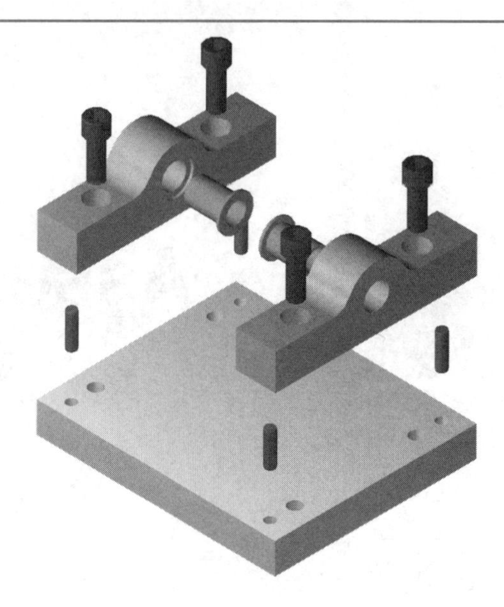

Figure 11-1.
This is an exploded view of the assembly you will create in this chapter. The lower base has been hidden.

Creating a Project with Folders

The assembly that you will build in this chapter, like most assemblies, requires a variety of files and file types:

- The assembly file (IAM), which shows all of the parts put together.
- The part files (IPT) of the modeled components.
- A project file (IPJ), which keeps track of all the part (IPT) and assembly (IAM) files.

To keep these files organized and to get access to the correct folders, Inventor uses projects to locate required folders. The reason for this seemingly complex process is that an Inventor assembly project may contain a large number of files, which may be stored on many computers, and may be worked on by many people. This is called a ***shared environment.*** Autodesk Vault, which is a separate installation from Inventor, is designed to be used in a shared environment. This software ensures that only one user can have a file open at any given time, thus eliminating the possibility of changes not being saved. In this chapter, you will use single-user projects. These are isolated environments and are suited for situations where only one user has access to the files, such as on a stand-alone workstation.

At least two projects were created when Inventor was installed, one called Default and the other tutorial_files. It is also likely that a third called samples was created. Depending on which options were selected during the installation, there may be more projects as well. The Default project can be thought of as an umbrella project that allows access to all user files. The single-user project you will create has a much more narrow focus by searching for files in only one or two folders.

In order to create a new project and make it active, there cannot be any Inventor files open. Make sure Inventor is open and that all files are closed. Now, select **Projects...** from the **File** pull-down menu to display the **Projects** dialog box. See **Figure 11-2.** The active project is indicated by a check mark. Pick the **New** button at the bottom of the dialog box. The **Inventor project wizard** is launched. In the first page of the wizard, pick the **New Single User Project** radio button and then the **Next** button to continue to the second page. In the second page, fill in the following information. See **Figure 11-3.**

Figure 11-2.
The available projects are listed at the top of the **Projects** dialog box. Pick the **New** button at the bottom of the dialog box to create a new project.

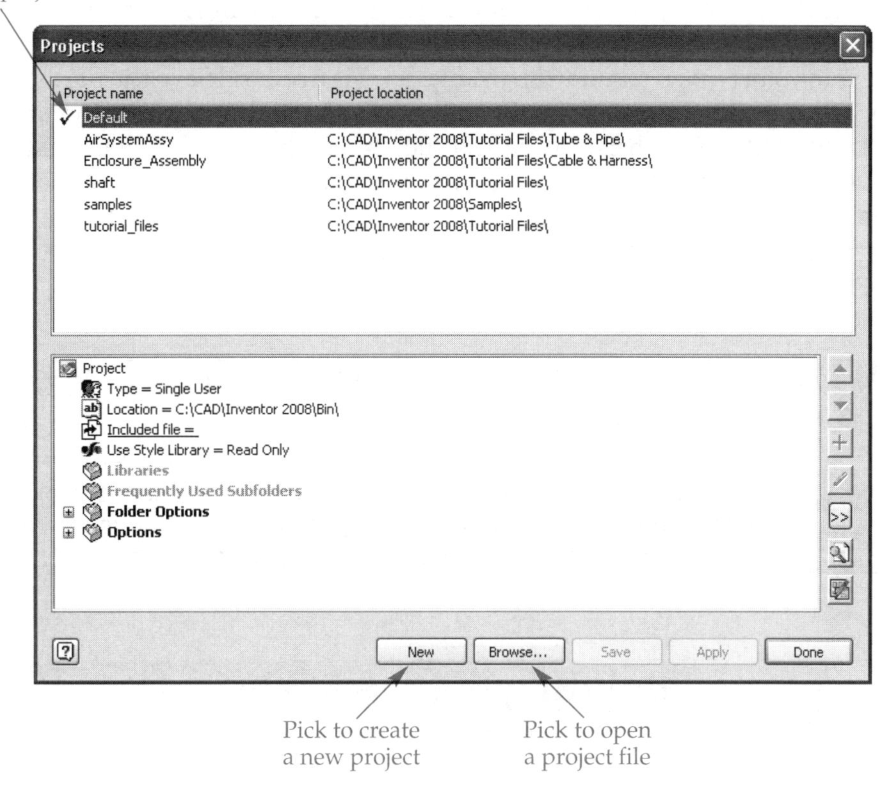

- In the **Name** text box, type Example_11_01.
- In the **Project (Workspace) Folder** text box, type c:\Program Files\G-W\Learning Autodesk Inventor 2008 Student CD\Ch11\Examples\Example_11_01. If the Student CD is installed in a different location on your computer, type the correct path. You can use the browse button (**...**) to locate the folder instead of typing the path.

Notice the **Project File to be created** field, which is grayed out. Inventor creates a project file (IPJ) with the same name as entered in the **Name** text box. Pick the **Finish** button to complete the project wizard. Now, double-click on the new project name or select

Figure 11-3.
Specifying the project name and the location of files.

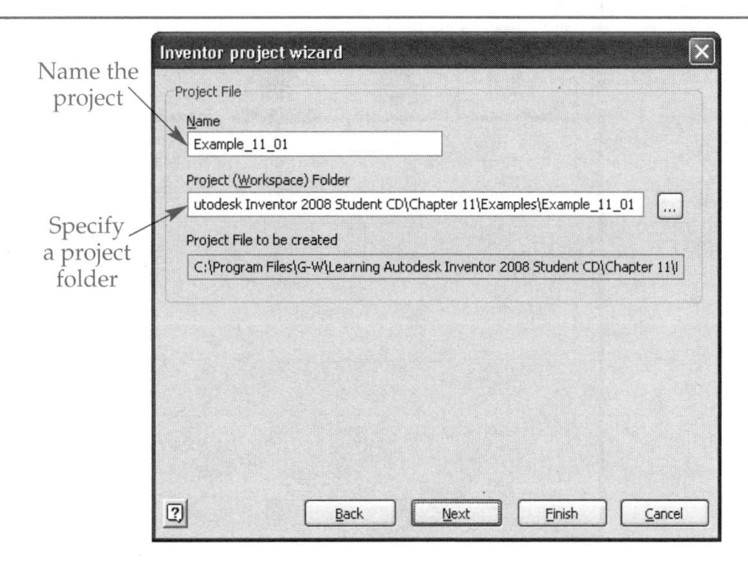

it and pick the **Apply** button to make the project active. Then, pick the **Done** button to close the **Projects** dialog box.

Now, the project file Example_11_01.ipj is created and placed in the \Example_11_01 folder. The IPJ file is a data file that points Inventor to the locations of all of the files. Do *not* double-click on the IPJ file in Windows Explorer as it will launch another session of Inventor.

In Inventor, start a new standard IAM file. See **Figure 11-4.** Save the file as Example_11_01.iam. The default location of the file in the **Save As** dialog box is automatically the Example_11_01 folder, which is where the file should be saved. Next, right-click on Example_11_01 in the **Browser** and select **iProperties...** from the pop-up menu. In the **Properties** dialog box, pick the **Project** tab. You can input additional information about this project similar to that shown in **Figure 11-5.** When done, pick the **OK** button to close the **Properties** dialog box.

NOTE

By default, you can only create a single-user project. Autodesk recommends that Vault be used in place of shared and semi-isolated projects. To create a shared or semi-isolated project, the **Enable creation of legacy project types** check box must be checked in the **General** tab of the **Application Options** dialog box.

Building the Assembly

Now, you are ready to build the assembly by placing the parts and adding assembly constraints. Pick the **Place Component...** button in the **Assembly Panel** or press the [P] key. The **Place Component** dialog box appears listing the part and assembly files in the folder defined in the project. The first part you will place is the file 11E01_Lower_Base.ipt.

Figure 11-4.
Starting a new assembly file for the Example_11_01 project.

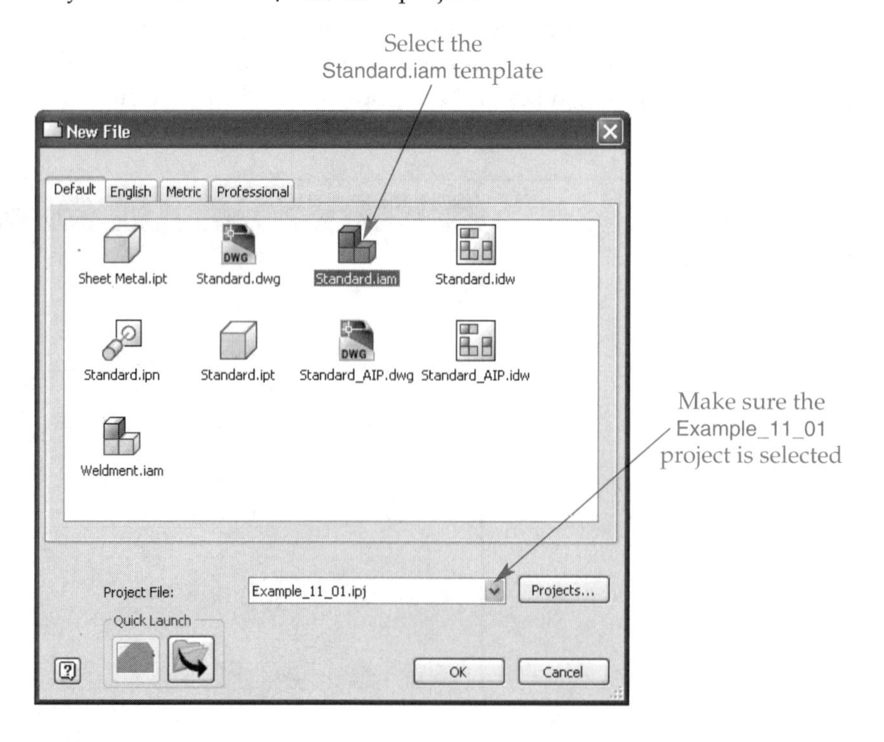

Figure 11-5.
Additional information about a project can be included using the **Properties** dialog box.

Example_11_01.iam Properties

General | Summary | Project | Status | Custom | Save | Physical

Location:	C:\Program Files\G-W\Learning Autod
File Subtype:	Assembly
Part Number:	Example_11_01
Stock Number:	
Description:	Base with bearing blocks
Revision Number:	R123
Project:	Example_11_01
Designer:	G. Camilli
Engineer:	T. Short
Authority:	F. Espino
Cost Center:	
Estimated Cost:	$148.50
Creation Date:	6/12/2007
Vendor:	
WEB Link:	

OK | Cancel | Apply

Highlight its name in the dialog box and pick the **Open** button. The part is placed in the assembly file. Notice how a copy of the part appears attached to the cursor. This allows you to place multiple instances of the part. Since only one instance of the base is needed, press [Esc] or right-click and select **Done** in the pop-up menu to end the command.

Look at the **Browser** and note the pushpin icon in front of the name 11E01_Lower_Base:1. This means that the part is ***grounded***—locked to the assembly's coordinate system—and cannot be moved. The first part in an assembly is automatically grounded and the following parts are typically constrained to it. In order to move or rotate the lower base, it is necessary to remove the ground. To do so, right-click on the part in the **Browser** and select **Grounded** in the pop-up menu to remove the check mark. However, in this example, leave the part grounded.

The isometric orientation of the part within the assembly is the same as its orientation in the part file, which in this case places the part on edge. However, for this assembly, it would be better if the lower base appears horizontal. See **Figure 11-6.** Use the **Common View** tool to change the viewpoint and then redefine the isometric view. Notice the colored faces on the part. Using the **Place Component** tool again, insert one instance of the file 11E01_Base_Plate.ipt somewhere above the lower base. Since this is

Figure 11-6.
The lower base is placed into the assembly. Later, this part will be hidden.

not the first part in the assembly, one instance is not automatically placed. Pick once to place the instance and then press [Esc] to end the tool.

When an ungrounded part is placed in an assembly, it has six degrees of freedom. It can rotate about each of the X, Y, and Z axes. It can also move along each of these axes. Putting it another way, it would take three coordinates (X, Y, and Z) and three angles to specify exactly where the part is in space and its orientation. As *constraints* are applied, degrees of freedom are removed. For example, a mating constraint applied face-to-face removes one linear and two angular degrees of freedom.

In the **View** pull-down menu, select **Degrees of Freedom** or press the [Ctrl][Shift][E] key combination. An icon appears above the base plate showing the six degrees of freedom (DOF)—three translational and three rotational. By applying assembly constraints, these degrees of freedom can be removed so the part is fully constrained.

Removing Degrees of Freedom

Note the green faces on the right-hand ends of both parts. These two faces will be aligned with a ***flush assembly constraint.*** Pick the **Constraint...** button in the **Assembly Panel** or press the [C] key. The **Place Constraint** dialog box is displayed, **Figure 11-7.** In the **Type** area of the **Assembly** tab, pick the **Mate** button. In the **Solution** area, pick the **Flush** button. Next, make sure the **First Selection** button is on in the **Selections** area and pick the green face on the lower base. A small arrow is placed on the face and the face is highlighted in color. The **Second Selection** button in the dialog is automatically turned on once the first face is selected. Pick the green face on the base plate. If the **Show Preview** check box is checked in the dialog box (which it is by default), the base plate moves into alignment so that both arrows point in the same direction and both faces are in the same plane. Pick the **Apply** button in the dialog box to accept the constraint and keep the dialog box open.

The Flush:1 constraint is listed in the **Browser** under both the 11E01_Lower_Base and 11E01_Base_Plate branches. This one constraint removes three degrees of freedom from the base plate—one translational (movement) and two rotational. There are still three more degrees of freedom to remove for the assembly to be fully constrained. This is reflected in the DOF icon on the base plate.

Now, a flush assembly constraint with an offset will be used to line up the red faces on the parts. In the **Place Constraint** dialog box, pick the **Mate** button in the **Type** area and the **Flush** button in the **Solution** area. With the **First Selection** button on, pick the red face on the base plate. With the **Second Selection** button on, pick the red face of the lower base. Then, type 1.0 in the **Offset:** text box in the dialog box and pick the **OK** button to place the constraint. This constraint is named Flush:2.

Notice in the **Browser** that the offset is listed at the end of the Flush:2 constraint name. If you select Flush:2 in the **Browser**, the corresponding faces on the parts are highlighted. Also, the offset value appears in a text box at the bottom of the **Browser**. See **Figure 11-8.** To change the offset, type a new value in this text box and press [Enter]; positive and negative numbers are accepted.

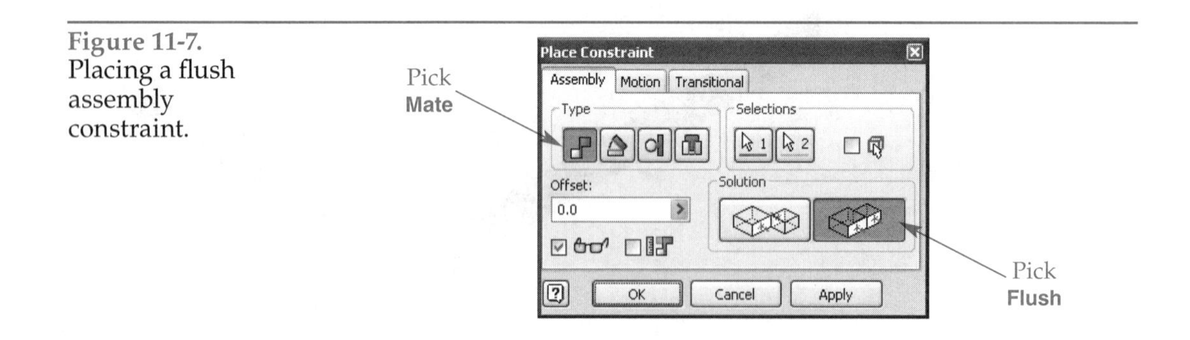

Figure 11-7.
Placing a flush assembly constraint.

Figure 11-8.
The constraints on a part are displayed in the part tree in the **Browser**. The offset can be changed, if needed.

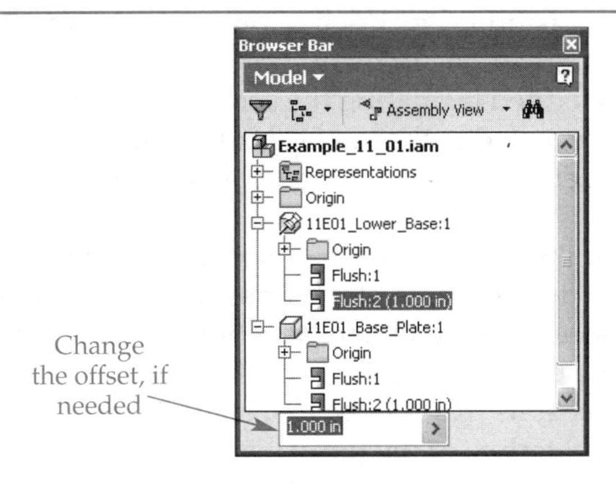

Figure 11-9.
The base plate is placed in the assembly and constrained to the lower base.

Now, the base plate can only be moved up and down; there is only one degree of freedom left. To fully constrain the part, a ***mate assembly constraint*** will be placed between the top face of the lower base and the bottom face of the base plate (the blue faces). Open the **Place Constraint** dialog box and pick the **Mate** button in the **Type** area. Also, pick the **Mate** button in the **Solution** area. With the **First Solution** button on, pick the blue face of the lower base; the arrow points upward. With the **Second Solution** button on, pick the blue face on the base plate. You may need to rotate the view to see the blue face. The preview shows the two parts in contact. Pick the **OK** button in the dialog box to place the constraint. See **Figure 11-9.** The base plate is now fully constrained with no degrees of freedom.

To review the constraints placed so far:

- The flush constraint aligns the preview arrows in the *same* direction and puts the faces in the same plane. A positive offset moves the faces apart. A negative offset moves the faces toward each other.
- The mate constraint aligns the preview arrows in the *opposite* direction and puts the faces in the same plane. A positive offset moves the faces apart. A negative offset moves the faces toward each other.

Constraining Centerlines

The next parts in the logical assembly order are the locating pins. The geometry in the part file complies with standards and has a flat on one side. With the **Place Component** tool, place the part 11E01_CR_Locating_Pin.ipt. Place four instances of the part into the assembly, one near each corner of the base plate. The exact location is not important because constraints will be used later to precisely locate the pins. To help with visibility, right-click on 11E01_Lower_Base in the **Browser** and uncheck **Visibility** in the pop-up menu.

The first pin is located by mating the centerlines of the pin and the hole. Then, the top of the pin will be mated to the top face of the base plate with an offset. Zoom in on the lower-right corner of the base so the hole and pin are visible. Change the display to wireframe.

Pick the **Constraint...** button in the **Assembly Panel**. The **Place Constraint** dialog box is displayed. In the **Type** area of the **Assembly** tab, pick the **Mate** button if it is not already on (depressed). Also, pick the **Mate** button in the **Solution** area. With the **First Selection** button on, move the cursor over the pin in the graphics window. There are several possible selections that will be displayed as you move the cursor around over the pin. Select the centerline of the pin. Then, with the **Second Selection** button on in the dialog box, move the cursor over the hole and select its centerline. A preview of the pin aligned with, and placed inside of, the hole is displayed. Pick the **OK** button in the dialog box to place the constraint. See **Figure 11-10.**

Four degrees of freedom are removed by the application of this constraint. The remaining degrees of freedom are movement along and rotation about the axis of the pin. If you pick the pin in the graphics window, you can drag it up and down along the centerline of the hole. However, you cannot move the pin around the face of the base plate.

Now, apply another mating constraint between the top of the pin and the top face of the base. In the **Constraint** dialog box, pick the **Mate** button in the **Type** area and the **Mate** button in the **Solution** area. However, since the blind hole in the base plate is .5″ deep, enter –.5 in the **Offset:** text box. This will make the pin extend .5″ into the hole and partially extend out of the hole. See **Figure 11-11.** There is still one rotational degree of freedom left. Removing it is not necessary as the orientation of the flat in the hole is not critical for this example.

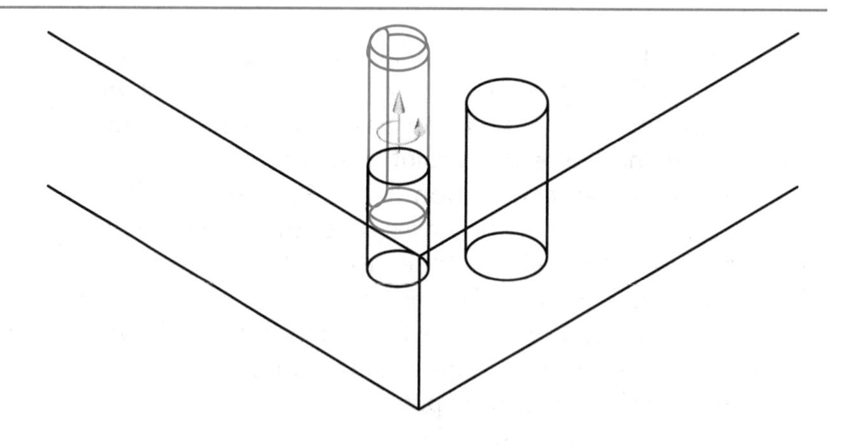

Figure 11-10.
The centerline of the pin has been aligned with the centerline of the hole using a mate assembly constraint. While the pin extends into the hole, this depth is not constrained. Notice the DOF icon.

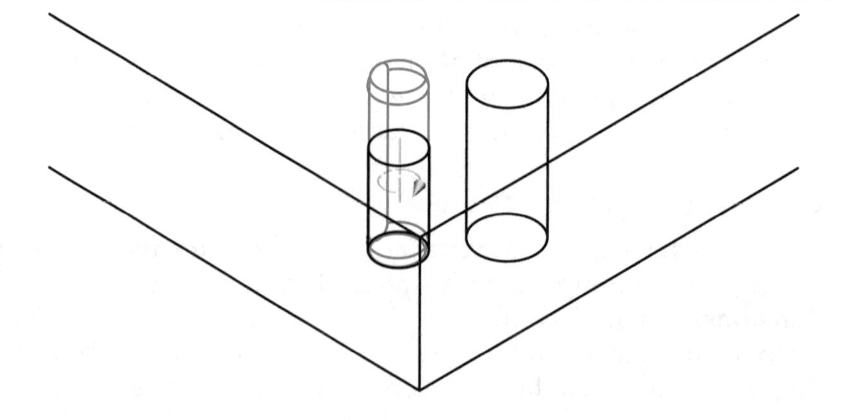

Figure 11-11.
Another mate assembly constraint is applied between the end of the pin and the top face of the base. An offset was specified so the pin extends into the hole. Notice the DOF icon; the pin can still rotate.

Figure 11-12.
Applying an insert assembly constraint.

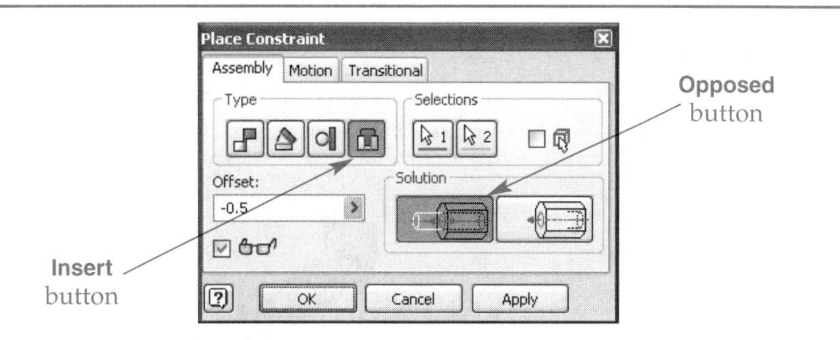

Using the Insert Constraint

The process of constraining fasteners in holes is so common in mechanical assemblies that Inventor has a special constraint that in one step applies both of the constraints discussed in the previous section. This is called an ***insert assembly constraint.*** Zoom in on the upper-right pin and hole. Then, continue as follows.

1. Pick the **Constraint...** button in the **Assembly Panel**.
2. Pick the **Insert** button in the **Type** area of the **Assembly** tab. Also, pick the **Opposed** button in the **Solution** area. See **Figure 11-12.**
3. In the graphics window, pick the end face of the pin. The direction arrow should be parallel to the centerline. Then, pick the top circle of the hole.
4. Enter –.5 in the **Offset:** text box in the dialog box.
5. Notice in the wireframe view that arrows and an axis are shown in color. This indicates that the constraint is applied both to the faces (arrows) and the centerlines.
6. Pick the **OK** button to apply the constraint and close the dialog box.

If you look at the **Browser**, you can see this one constraint takes the place of two separate mating constraints. However, it gets even better! Zoom in on the upper-left (or top) pin and hole. Open the **Place Constraint** dialog box, pick the **Insert** button in the **Type** area, pick the **Opposed** button in the **Solution** area, and enter –.5 in the **Offset:** text box. Now, pick and hold on the end of the pin. The direction arrow should be parallel to the centerline. Drag the pin over the hole. As the cursor touches the edge of the hole, the preview shows the pin inserted in the hole. Release the mouse button. Then, pick the **Apply** or **OK** button in the dialog box to apply the constraint.

The last pin will be pressed all of the way into the hole using an insert constraint without an offset. Zoom in on the lower-left pin and hole. Open the **Place Constraint** dialog box. In the **Type** area of the **Assembly** tab, pick the **Insert** button. Also, select the **Opposed** button in the **Solution** area. With the **First Solution** button on, pick the bottom edge of the pin. With the **Second Solution** button on, pick the bottom edge of the hole. Pick the **OK** button to close the dialog box and place the constraint.

Placing the Bearing Blocks

Zoom so that the entire assembly is visible. Leave the lower base hidden. Change to a shaded view. Then, turn off the display of the DOF icons by selecting **Degrees of Freedom** in the **View** pull-down menu; it is a toggle.

Now, place one instance of the part 11E01_Bearing_Block.ipt into the assembly. Remember, its exact location is not important at this point. The bearing block has one red and one green face that will eventually line up with the faces on the base plate of same color. You may need to rotate the view to see the colored faces. The way a technician or assembler would install the bearing block is to place it over the pins and then press it down. This is exactly how you will do it using constraints.

Open the **Place Constraint** dialog box. Pick the **Mate** button in the **Type** area and the **Mate** button in the **Solution** area. Next, refer to **Figure 11-13.** With the **First Solution** button on, pick the centerline of the locating pin, as shown in the figure. With the

Figure 11-13.
Constraining the bearing block on the pins.

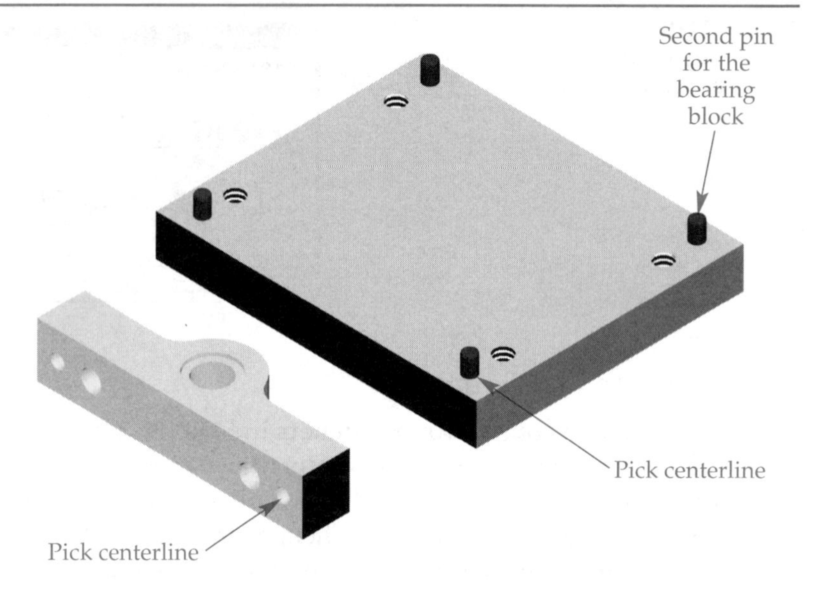

Second Solution button on, pick the centerline of blind hole in the bearing block, as shown in the figure. Make sure you select the *centerlines*. Pick the **Apply** button to place the constraint. Do not be concerned if the bearing block is out of position at this point. When the next mate constraint is applied, it will move to the correct position. Now, put the same type of constraint on the other pin and the hole in the bearing block. Refer to **Figure 11-13.** If needed, change to a wireframe display or rotate the view to see the other hole.

Finally, the bottom of the bearing block needs to be constrained to the top of the base plate. Open the **Place Constraint** dialog box. Pick the **Mate** button in the **Type** area and the **Mate** button in the **Solution** area. With the **First Selection** button on, pick the top face of the base plate. With the **Second Selection** button on, pick the bottom face of the bearing block. Then, pick the **OK** button to place the constraint. All degrees of freedom have been removed from the bearing block. Notice how the red and green faces align. See **Figure 11-14.**

Placing the Bushings

An insert constraint will be used to constrain a bushing into the hole in the bearing block. Place one instance of the part 11E01_Bushing.ipt into the assembly. Then, rotate the view so you can see the inside surface of the bearing block. Continue as follows.

Figure 11-14.
The bearing block is completely constrained. Notice that the colored faces line up.

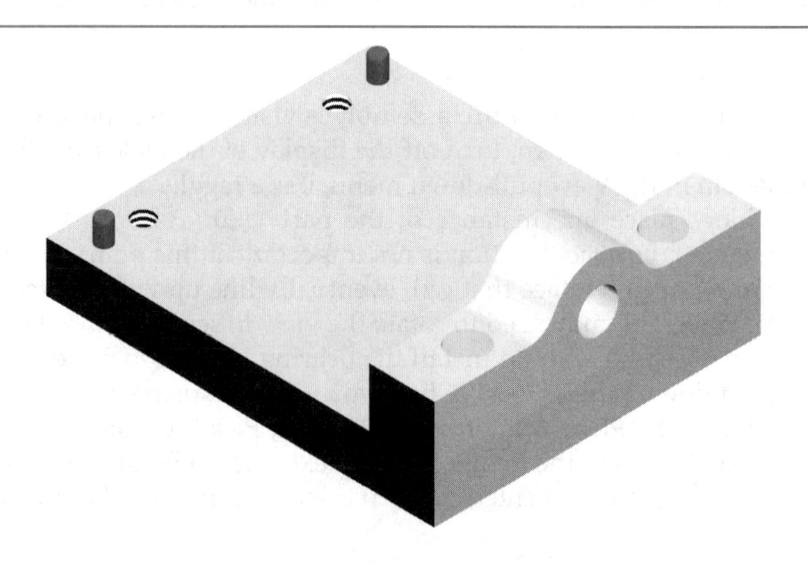

Figure 11-15.
Pick the outer edge of the back of the flange when constraining the bearing.

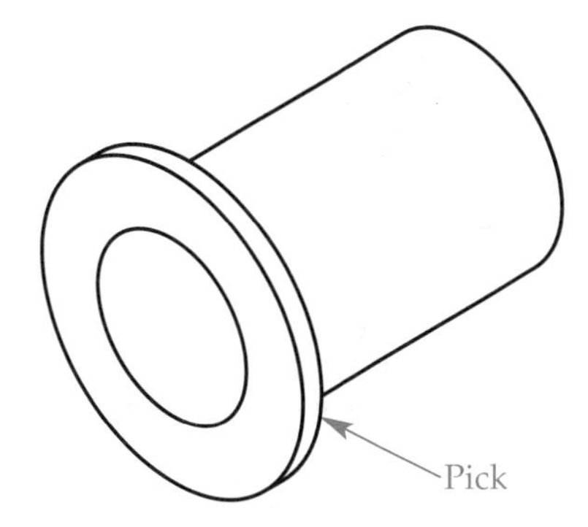

1. Open the **Place Constraint** dialog box. In the **Type** area of the **Assembly** tab, pick the **Insert** button. Also, pick the **Opposed** button in the **Solution** area.
2. Pick and hold on the outer edge of the back of the flange, as shown in **Figure 11-15.** Drag the bushing into the bearing block until the preview shows the bushing seated. The outer face of the bushing should be flush with the outer face of the bearing block.
3. Pick the **OK** button in the dialog box to apply the constraint and close the dialog box.
4. Redisplay the isometric view.

Determining the Cap Screw Specifications

Before putting in the cap screws that hold down the bearing block, the thread specifications need to be checked on the threaded holes in the base. There are two choices for accessing the base plate—open the part file or edit it in place within the assembly. To edit it in place, right-click on Base_Plate:1 in the **Browser**. The :1 part of the name indicates that this is the first instance placed, which in this case is the only instance. You can also right-click on the part in the graphics window. Then, select **Edit** in the pop-up menu. See **Figure 11-16A.**

All of the parts other than the base plate are displayed semitransparent in the graphics window and are grayed-out in the **Browser**. The part tree for the base plate is also displayed in the **Browser**. See **Figure 11-16B.** Right-click on Hole 3/8 Bolts in the **Browser** and pick **Edit Feature** in the pop-up menu. The **Hole** dialog box is displayed. In the **Threads** area, the pitch is indicated as 3/8-24 UNF, which is a fine thread with 24 threads per inch. This is correct for this example. Record this information; you will need it to select the correct cap screw. In the **Termination** area, verify that the termination is set to **Through All**. Pick the **OK** button to close the dialog box.

Now, you need to determine the proper length for the cap screw. Right-click on Extrusion1 in the **Browser** and select **Show Dimensions** in the pop-up menu. The extruded height is .75 inches; record this information. Now, pick anywhere in the graphics window, right-click, and select **Finish Edit** from the pop-up menu. This returns you to the assembly.

Using the edit in place method, edit Bearing_Block:1. Show the dimensions on Extrusion1. The distance from the bottom of the bearing block to the top of the holes for the cap screws is .875″. Record this information. Edit the feature Hole for Cap Screws. The depth of the counterbore is .375″. Record this information as well. Close the dialog box and return to the assembly.

Looking at the information you recorded, the cap screw is a 3/8-24 UNF thread. Also, the length must be longer than .5″ (.875 – counterbore depth of .375) but no longer than 1.25″ (.75 + .875 – .375). Generally, a thread engagement length equal to at least one diameter is sufficient. In this case, the diameter is 3/8 (.375), so the length should

Figure 11-16.
A—By selecting **Edit** from the pop-up menu, you can edit a part in place within the assembly.
B—When editing in place, only the features in the part being edited are available.

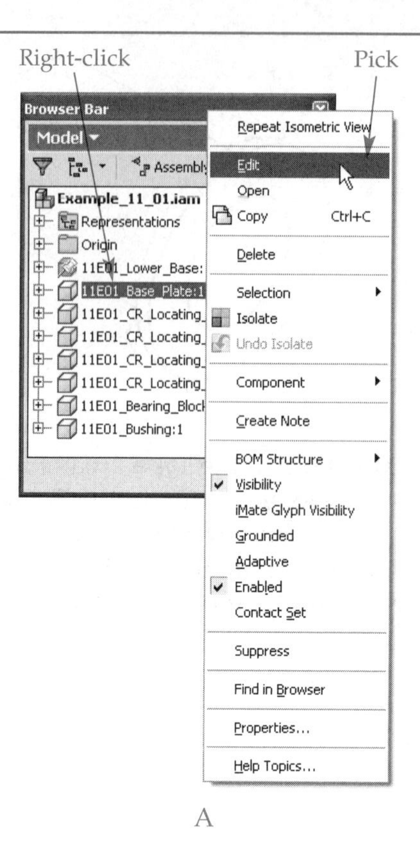

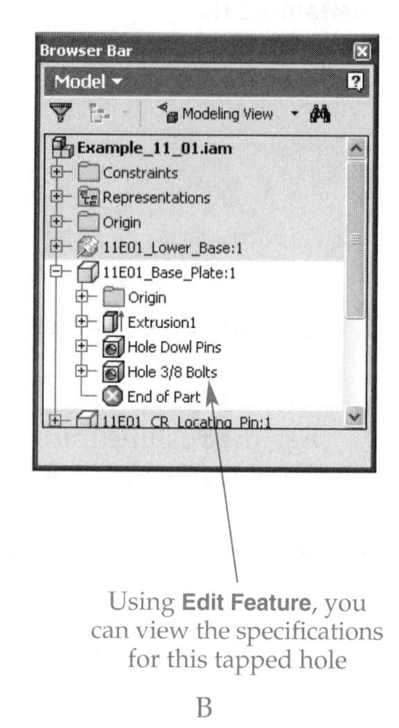

be between .875″ (minimum of .5 + one diameter of .375) and 1.25″ (maximum length). For this example, you will use 1″ long cap screws. They provide more thread engagement while not exceeding the maximum length.

PROFESSIONAL TIP

When editing in place, if you change the part(s) in some way, such as changing the thread size, the referenced files are updated when the assembly is saved.

Placing Standard Parts

Inventor has a library that contains catalogs of a wide variety of standard parts and components. ***Standard parts*** conform to industry standards for size and other specifications. Inventor calls this library the ***Content Center.*** Augmented by Internet resources, such as the Thomas Register, these can greatly reduce the time required to build parts for an assembly.

To display the Content Center library, pick the **Place from Content Center...** button in the **Assembly Panel**. The **Content Center** dialog box is displayed, **Figure 11-17.** There are seven categories of standard parts: Cable & Harness, Fasteners, Features, Other Parts, Shaft Parts, Steel Shapes, and Tube & Pipe. Each of these categories contains several subcategories.

Content Center includes parts for seven different standards and one manufacturer. Pick the arrow next to the **Filters** button on the toolbar to display the standards. Since ANSI standard cap screws are needed for this example, select ANSI in the flyout. Only those parts that meet the filter criteria are available in the tree. Expand the tree for Fasteners, then Bolts, and finally select Socket Head in the tree. The available part types conforming to ANSI standards are displayed on the right-hand side of the dialog box.

Figure 11-17.
The Content Center contains a library of standard parts. Expand the tree to see the categories available. Here, the content is filtered for ANSI.

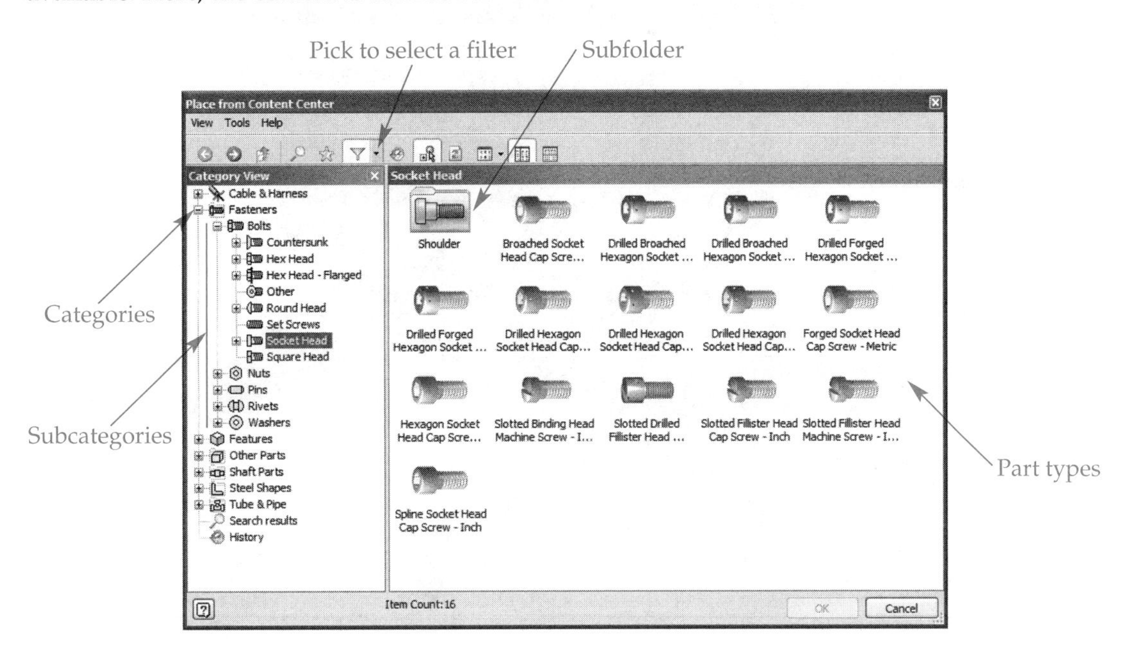

On the right-hand side of the dialog box, double-click on Hexagon Socket Hex Head Cap Screws—Inch. Inventor takes a moment to "build" the part, then Content Center dialog box is closed and a preview of the part appears attached to the cursor. Pick anywhere in the assembly and the specifications dialog box is displayed. See **Figure 11-18.** Select 3/8 in the **Thread Description** column, 1 in the **Nominal Length** column, and UNF in the **Thread Type** column. Then, pick the **OK** button to place the part.

Notice that the part is not placed where you picked. A preview of the "built" part appears attached to the cursor. Pick twice to place two instances of the cap screw into the assembly. Then, press [Esc] to end the tool.

Figure 11-18.
Entering specifications for the selected Content Center part.

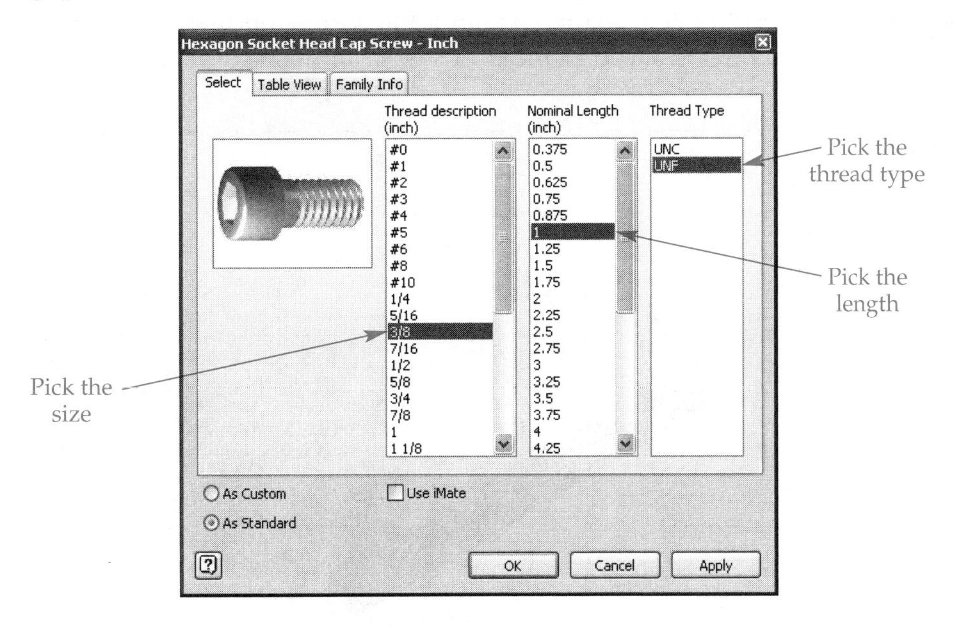

Figure 11-19.
Two instances of the standard part are placed into the assembly. Standard parts contain iMates.

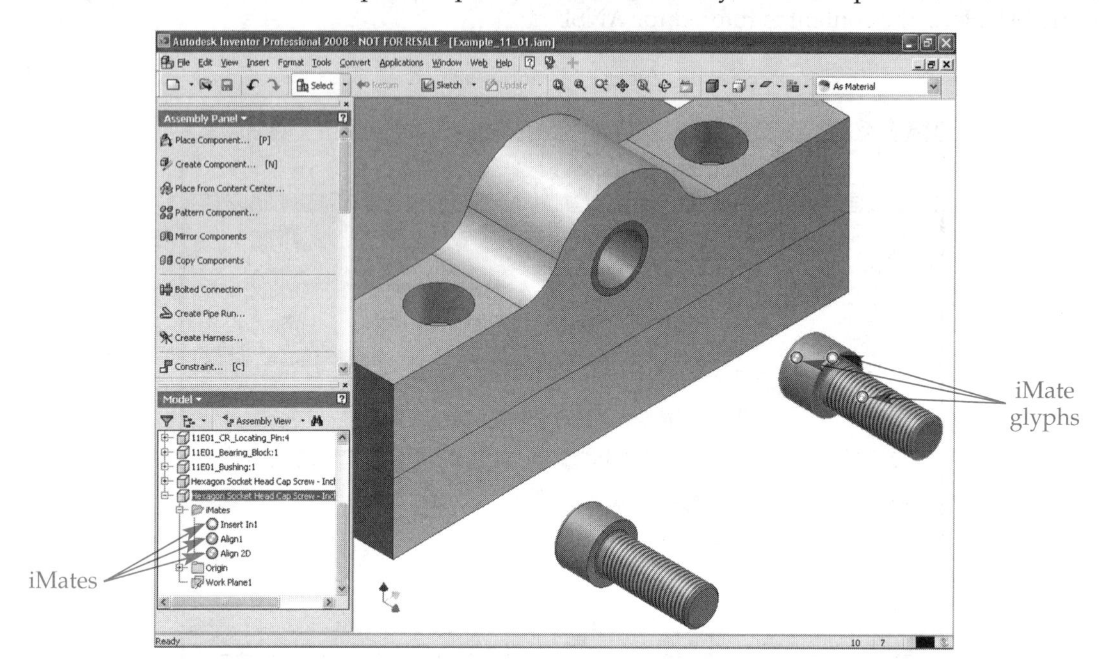

The standard parts are created with iMates, which are partial assembly constrains that make inserting the screws easier. The iMates are represented on the screw by an icon called an iMate glyph. See **Figure 11-19.** Chapter 14 covers iMates in detail.

To insert the first screw into the hole, select the screw in the graphics window. This displays the glyphs. Then, open the **Place Constraint** dialog box. Pick the **Insert** button in the **Type** area. Then, with the **First Selection** button on, pick the screw by the glyph on the bottom of the head. Drag the glyph (and screw) over the bottom of the counterbored hole and drop it. The top of the screw should be flush with the top of the bearing block. Pick the **OK** button in the dialog box to apply the constraint. Then, constrain the second screw in the other hole in the same manner. See **Figure 11-20.** Save the assembly file.

The two cap screws are listed in the **Browser** with very long names. You can change the names if you wish, but it is not necessary. Using Windows Explorer, look in the \Example_11_01 folder. Notice that there is not a file corresponding to the cap screws. The file for the cap screws is saved in the user's \My Documents\Inventor\Content Center

Figure 11-20.
The two cap screws are properly located and constrained in the assembly.

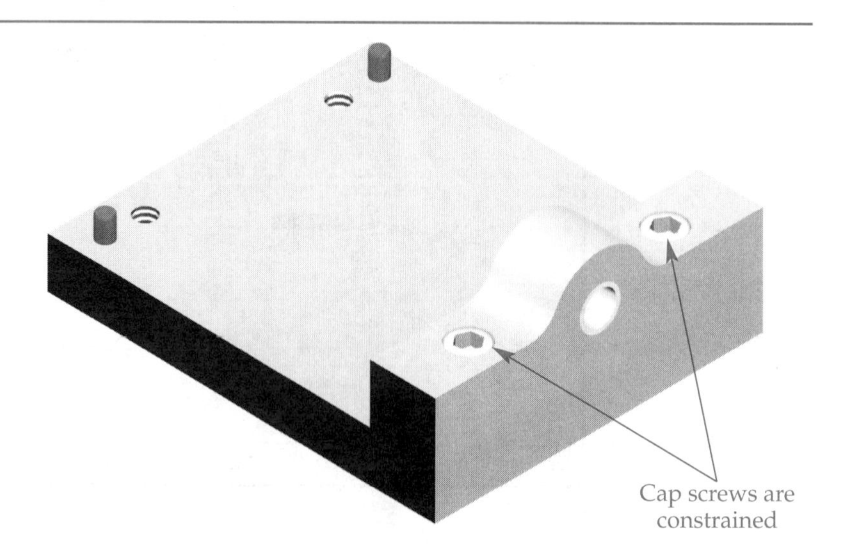

Files\R2008\en-US\ANSI B18.3(05) folder. The file name is ANSI B18.3-3_8-24-UNF-1(31).ipt. Both instances of the cap screw in the assembly file reference this part file.

PRACTICE 11-1 Complete the practice problem on the Student CD.

Constraining Edges of Parts

Close all files, but leave Inventor open. Now, select **Projects...** from the **File** pull-down menu. Pick the **New** button at the bottom of the dialog box. In the first page of the **Inventor project wizard**, select the **New Single User Project** radio button and then pick the **Next** button. On the second page of the wizard, fill in the following information.

- In the **Name** text box, type Example_11_02.
- In the **Project (Workspace) Folder** text box, type c:\Program Files\G-W\Learning Autodesk Inventor 2008 Student CD\Ch11\Examples\Example_11_02. If the Student CD is installed in a different location on your computer, type the correct path.

Finish creating the project. Then, make the Example_11_02 project current. Then, start a new assembly file and save it with the name Base_with_Molding in the \Example_11_02 folder.

Pick the **Place Component...** button in the **Assembly Panel**. Add one instance of the part 11E02_2nd_Base.ipt. The part is placed in an orientation that is acceptable for this assembly, as shown in **Figure 11-21.**

To prevent selecting unwanted features during operations, suppress the hole features. To do this, right-click on the name 11E02_2nd_Base:1 in the **Browser** and select **Edit** from the pop-up menu. Next, right-click on the name Hole Dowel Pins in the part tree for the base and select **Suppress Features** from the pop-up menu. The holes are no longer displayed on the part. Right-click in the graphics window and select **Finish Edit** from the pop-up menu. You are returned to the assembly.

Now, place one instance of the part 11E02_Molding.ipt into the assembly. The part is not properly aligned with the base. However, as constraints are applied, the part will be correctly oriented. Open the **Place Constraint** dialog box. Pick the **Mate** button in the **Type** area and the **Mate** button in the **Solution** area. With the **First Selection** button

Figure 11-21.
The base is placed into the assembly.

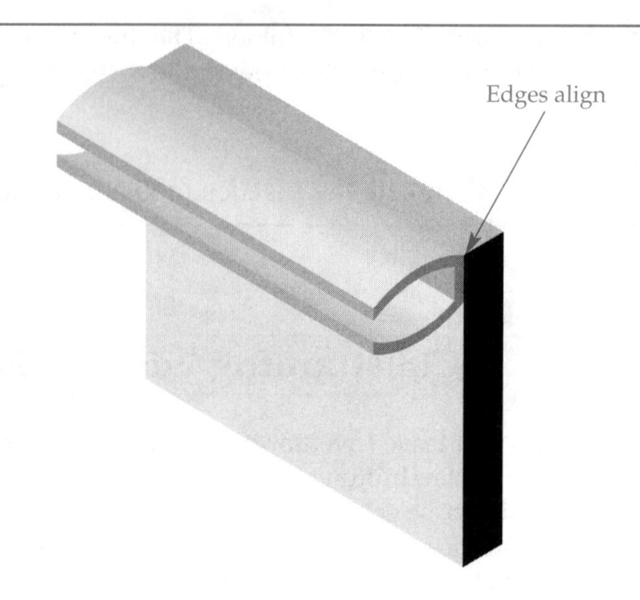

Figure 11-22.
The molding is placed into the assembly and properly constrained. Notice the location of the curved face in relation to the top face of the base.

on, pick the flat, bottom face of the molding. You may need to rotate the view or the part to see the proper face on the molding. With the **Second Selection** button on, pick the front face of the base. Pick the **Apply** button to place the constraint. Apply another mating constraint between the U-shaped end of the molding and the right-hand edge of the base. The faces should be flush. See **Figure 11-22.**

The top edge of the molding's rectangular face is located at the top of the base, as shown in **Figure 11-22.** To locate the molding at the top edge of the base, apply a mating constraint between the top edge of both parts. When applying a mating constraint, it can be applied between faces, centerlines, points, and edges when the **Mate** button in the **Solution** area is on. Remember, when the **Flush** button is on, you can only select faces. Save the assembly file and keep it open for the next section.

Tangent Constraint

Now, suppose the upper, curved face of the molding needs to be tangent to the top face of the base, instead of the edges being coincident. This can be achieved with the ***tangent constraint,*** which makes two curved faces or one curved and one planar face tangent to each other. Parts can be tangent to the inside or outside of each other. See **Figure 11-23.**

In the **Browser**, right-click on the last mating constraint applied and select **Delete** from the pop-up menu. If you do not delete this constraint, an error will be generated when the tangent constraint is applied. Open the **Place Constraint** dialog box and pick the **Tangent** button in the **Type** area. Also, pick the **Inside** button in the **Solution** area. Do not enter an offset. Then, with the **First Selection** button on, pick the upper, curved face of the molding. With the **Second Selection** button on, pick the top face of the base. Finally, pick the **OK** button in the dialog box to apply the constraint. See **Figure 11-24.**

Figure 11-23.
When using the tangent constraint, you have two choices as to how the parts are aligned. A—**Outside**. B—**Outside**. C—**Inside**. D—**Inside**.

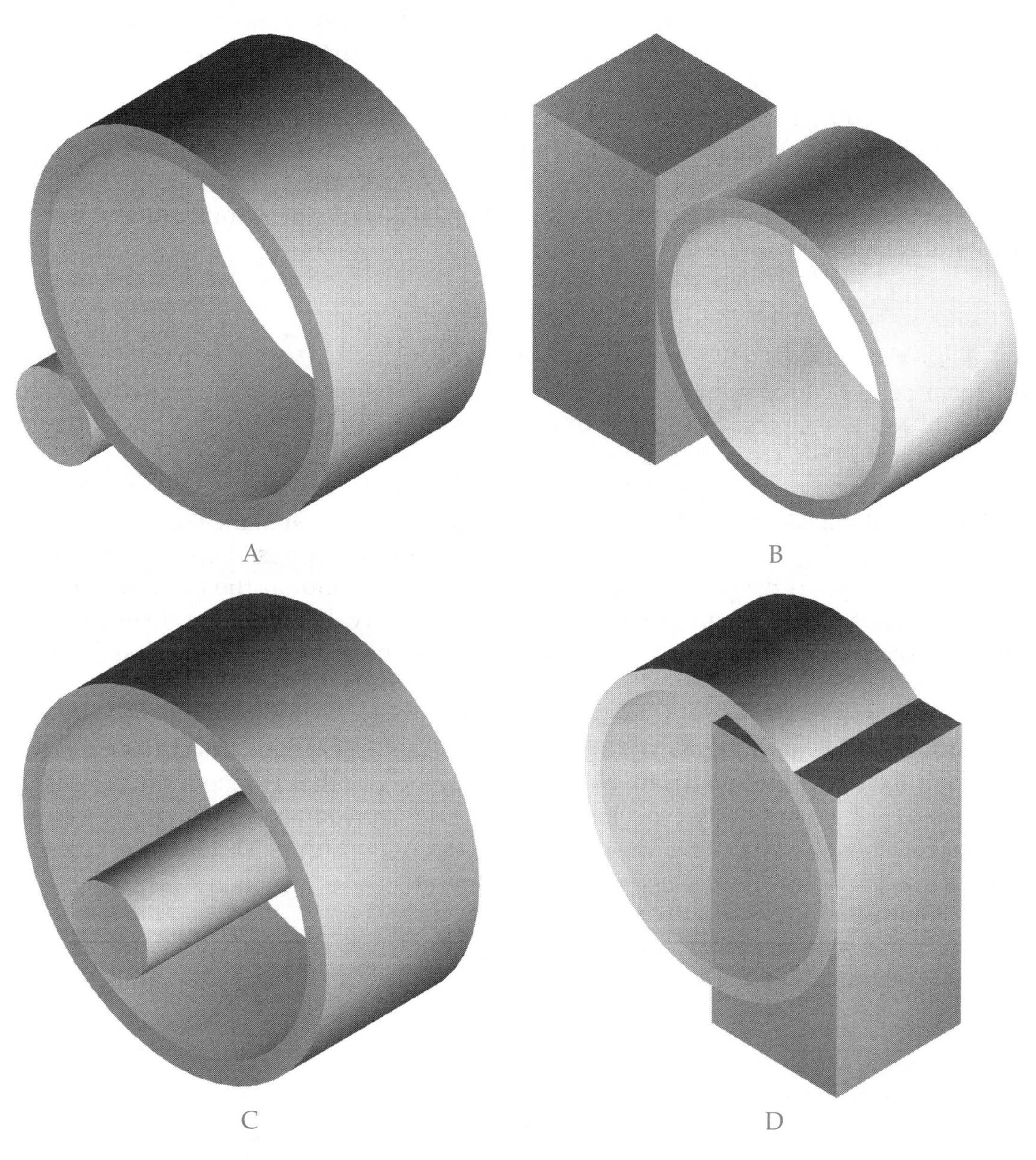

Figure 11-24.
The curved surface is constrained tangent to the top face of the base. Compare this to Figure 11-22.

Derived Parts

The word *derive* means to obtain from a specified source, so a ***derived part*** in Inventor is a part that is based on one or more existing parts. In this section, you will look at two examples of derived parts:

- Combining two gears into one part without redrawing or sharing sketches.
- Subtracting one part from another to create a mold base.

In Inventor, a derived part is linked to the original part. Any changes in the original are reflected in the derived part. This link can be broken if you want the derived part to be independent.

Derived Gears

Figure 11-25 shows two gears that need to be combined into one part. This type of component is common in automatic automobile transmissions. The procedure is to create an assembly, place both parts in it, constraint them in the correct position and alignment, and then create the derived part.

Start a new assembly file based on the Standard (in).iam template. Save the file as Example_11_03.iam in the \Example_11_03 folder. Place one instance of 11E03_Ring_Gear.ipt and 11E03_Pinion_Gear.ipt into the file. Place an opposed insert constraint between the hole in the pinion gear and the hole in the hub on the back of the ring gear. Then, put a mate assembly constraint between the two work planes to line up the teeth. To place a constraint on a work plane, pick the edge of the work plane. Finally, save the assembly file.

Now, start a new part file based on the Standard (in).ipt template. Finish the sketch without drawing anything to enter part mode. Next, pick the **Derived Component** button in the **Part Features** panel. In the **Open** dialog box that is displayed, select the Example_11_03 assembly file you just created. The **Derived Assembly** dialog box is displayed, since a part is being created from an assembly, **Figure 11-26.** The symbol in front of the part name indicates the status of the part:

- Plus sign means add the part
- Minus sign means subtract the part
- Slash means exclude the part

Figure 11-25.
Two views of the gears derived from an assembly.

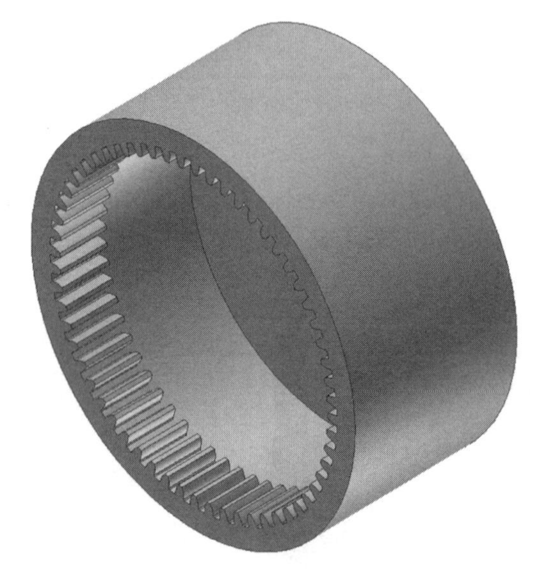

Figure 11-26.
Specifying how parts are to be treated when the derived part is created.

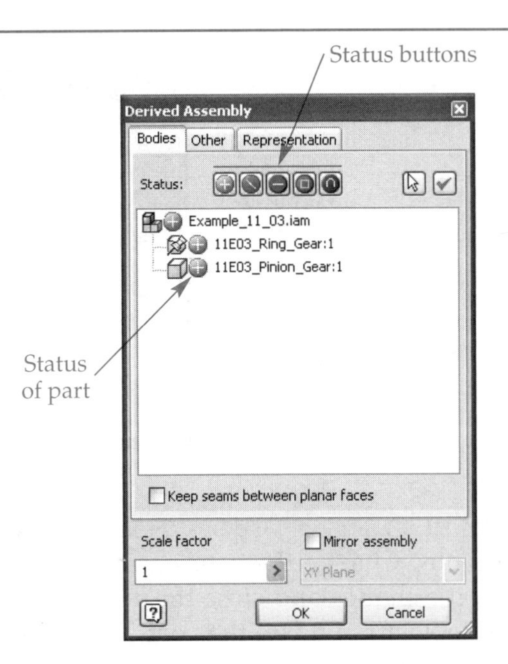

- Box means include the bounding box of the part.
- Upside-down U means the part will be intersected with the derived result.

Picking the symbol in front of the part name cycles through the choices. You can also select the part name and then pick the corresponding **Status:** button above the tree. Since both of these parts are to be added to the derived part, make sure a plus symbol is displayed in front of both part names.

Display the **Other** tab. This tab shows all of the details of the assembly and its parts. Notice that each is currently excluded, as indicated by the slash. For this example, the work plane for the pinion gear should be included. Expand the tree for the pinion gear and change the symbol in front of **Work Geometry** to a plus sign. See **Figure 11-27.**

Now, pick the **OK** button to create the derived part. Notice that the assembly file is listed in the **Browser**. If you expand the tree, the parts in the assembly are shown. Notice that the plus sign symbols are displayed in front of the two part names. The status cannot be changed here, but you can suppress either of these parts. Save this part as Example_11_03_Derived_Part.ipt and close both it and the assembly file.

Figure 11-27.
Changing the status of the work plane so it will be included in the derived part.

Figure 11-28.
An independent derived part has no link to the original parts.

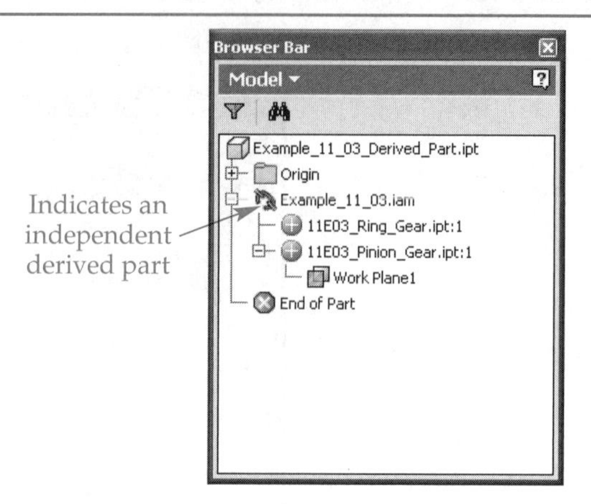

Open the file 11E03_Pinion_Gear.ipt. Select **Tools>Parameters...** to display the **Parameters** dialog box. Change the user parameter NT from 20 to 30. Then, pick the **Done** button to set the change and close the dialog box. Pick the **Update** button on the **Inventor Standard** toolbar to update the part. The number of teeth changes from 20 to 30 and the gear is larger.

Now, open the derived part, change the viewpoint so you can see the pinion, and update the part. This gear changes to match the number of teeth and size of 11E03_Pinion_Gear. This is because the derived part is linked to it sources. To break the link, right-click on the part in the derived part file and select **Break Link With Base Assembly** from the pop-up menu. You must right-click on the name (Example_11_03.iam) in the **Browser**. A broken chain icon appears in the **Browser** in front of the part name indicating that the part is an independent derived part, **Figure 11-28.**

Derived Mold

In this example, you will start the design for the rear cavity of a mold base for injection molding a plastic part. See **Figure 11-29.** This will be done by subtracting the part from a rectangular block. Of course a completed mold base has many more components and features than you will be creating, but it is a start.

Start a new assembly file based on the Standard (in).iam template. Save it as Example_11_04.iam in the \Example_11_04 folder. Using the **Place Component** tool, place one instance 11E04_Rear_Cavity.ipt and 11E04_Plastic_Link_S_H.ipt into the assembly.

Figure 11-29.
The injection mold half derived from an assembly.

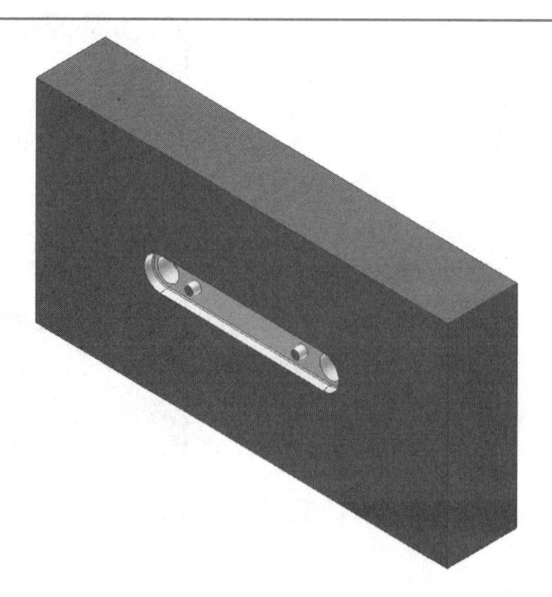

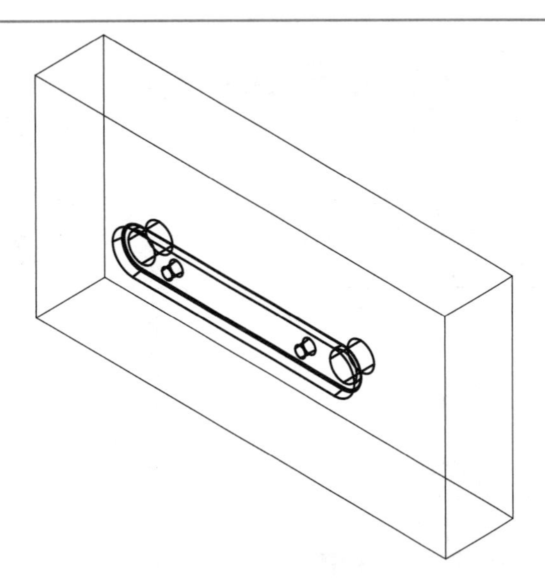

Figure 11-30.
The proper orientation of the two parts. Notice that the large pins extend into the block.

Using the **Place Constraints** tool, place a flush assembly constraint between the front face of the block and the face color green on the plastic part. You may need to rotate the view to see the green face. Next, place a mate assembly constraint between the vertical work planes. Finally, place a flush assembly constraint between the horizontal work planes. **Figure 11-30** shows the plastic part in its final, constrained position in a wireframe display. Save the assembly file.

Start a new part file based on the Standard (in).ipt template. Finish the sketch without drawing anything to enter part mode. Using the **Derived Component** tool, select the Example_11_04 assembly file. In the **Derived Assembly** dialog box, change the symbol in front of 11E04_Plastic_Link_S_H to a minus sign. This will subtract the part from the other part. In the **Other** tab, expand the branch for 11E04_Rear_Cavity and change the symbol in front of Work Geometry to a plus. Pick the **OK** button to create the derived part. The block should look like **Figure 11-29** with the addition of the work planes. Save the part as Example_11_04_Derived.ipt.

Chapter Test

Answer the following questions on a separate sheet of paper or complete the electronic chapter test on the Student CD.

1. Define *assembly.*
2. What is an *instance?*
3. What is an Inventor *project?*
4. Define *shared environment.*
5. Which type of Inventor files must be open to set a project current?
6. When a part is placed in an assembly, how many degrees of freedom does it have?
7. What is a *grounded* part?
8. Briefly describe *editing in place.*
9. What is a *standard part?*
10. How can you see the degrees of freedom on the parts in an assembly?

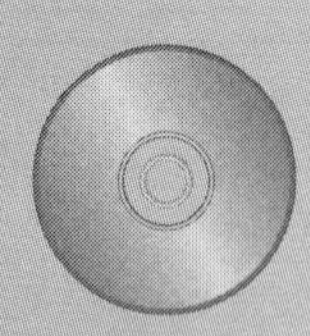

Chapter Exercises

Exercise 11-1. Wheel Assembly.
Complete the exercise on the Student CD

Exercise 11-2. Coin and Box.
Complete the exercise on the Student CD

Exercise 11-3. Ball Socket Clamp.
Complete the exercise on the Student CD

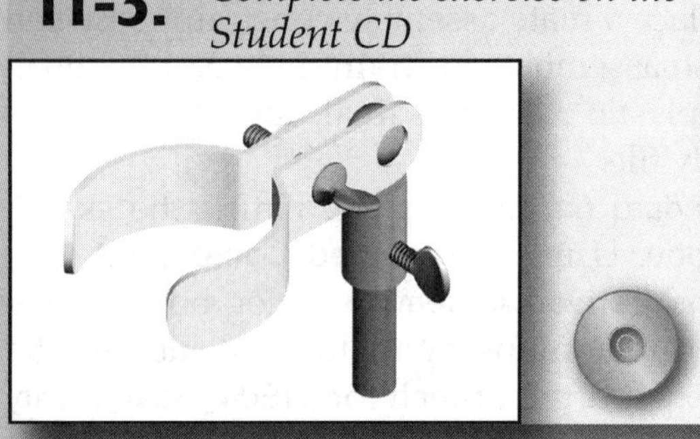

Exercise 11-4. Doorbell Remote Switch.
Complete the exercise on the Student CD

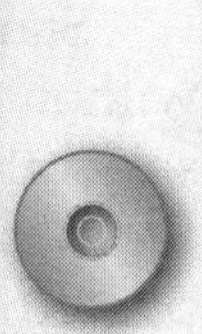

Exercise 11-5. Injection Mold.
Complete the exercise on the Student CD

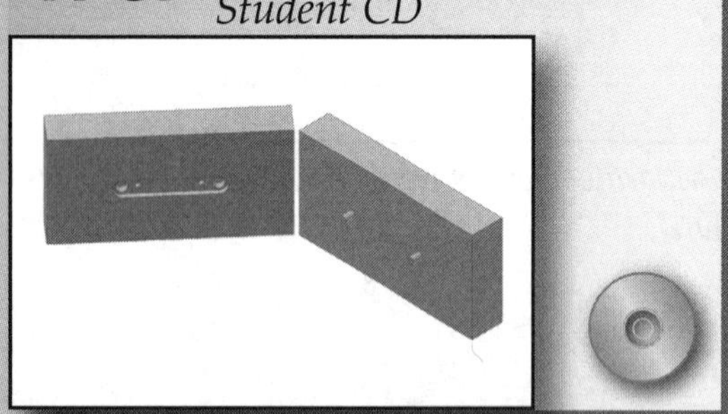

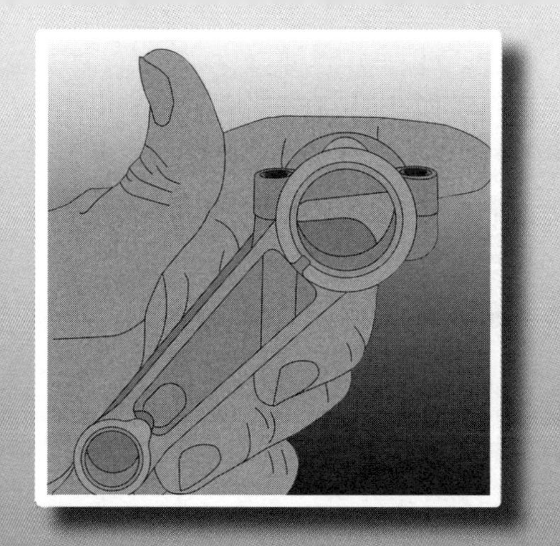

Chapter 12

Working with Assemblies

Objectives

After completing this chapter, you will be able to:

- Add paths to a project.
- Create a part from within the assembly.
- Apply the angle constraint.
- Drive constraints.
- Constrain work planes.
- Apply assembly constraints to sketches.
- Create adaptive parts.
- Set the visibility of parts in an assembly.
- Create and use design views.

User's Files

The Student CD included with this text contains several files required for this chapter. Refer to the file File List.txt in the \Ch12 folder for the comprehensive list.

Setting Additional Paths in Projects

In many projects, the part and assembly files may be in different folders or even on different computers. For example, the slider clamp assembly shown in **Figure 12-1** contains part files and a subassembly. The part files for the slider, coupler, and coupler link are located in the \Ch12\ Examples\ Example_12_01\Linkage folder. The component 12E01_Slider_Clamp_Base_Assembly is a subassembly, indicated by the icon in front of the name in the **Browser**, and it is in the \Ch12\ Examples\ Example_12_01\Base_ Assembly folder.

Remember from the previous chapter that a project is a system for keeping track of the locations of all of the files in an assembly. This folder structure is kept in a file called the project file, which has an IPJ file extension. The projects you created so far only had one folder to search. In some cases, it may be necessary to search multiple

Figure 12-1.
A—This is an Inventor assembly. B—The assembly tree in the **Browser**. Notice the subassembly.

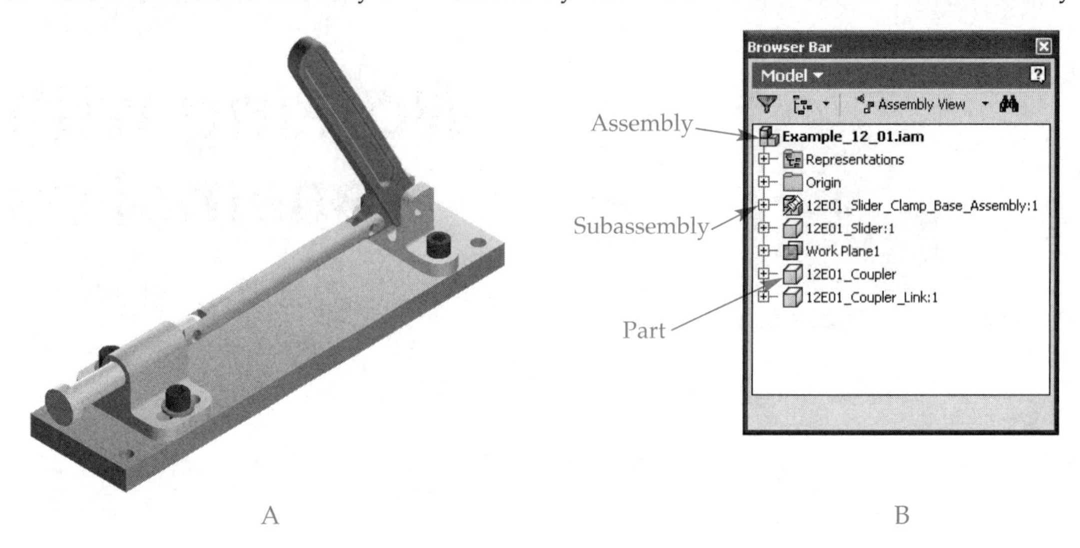

folders that may be located on other drives or server locations. For example, a company may have a library of unique, standard parts that rarely, if ever, change. This library needs to be added to each individual project so the parts in the library can be located by the assembly file.

Create a new project. In the Inventor project wizard, name the project Example_12_01. Also, specify the project (workspace) folder name as c:\Program Files\G-W\Learning Autodesk Inventor 2008 Student CD\Ch12\Examples\Example_12_01\Linkage. See **Figure 12-2.** If the Student CD is located in a different folder, specify the correct folder. Once the project is created, double-click on it in the **Projects** dialog box to make it current.

Now, Inventor will look in the \Linkage folder, and in any subfolders it contains, when attempting to locate the parts in an assembly. However, the 12E01_Slider_Clamp_Base_Assembly subassembly is not located in that folder structure. If you try to open the assembly file, Inventor will not be able to locate the subassembly and the parts it contains. Inventor is looking for the parts in the folder specified above and the **Resolve Link** dialog box is displayed, **Figure 12-3.** This is Inventor telling you it needs help finding the file.

Figure 12-2.
Creating a new project.

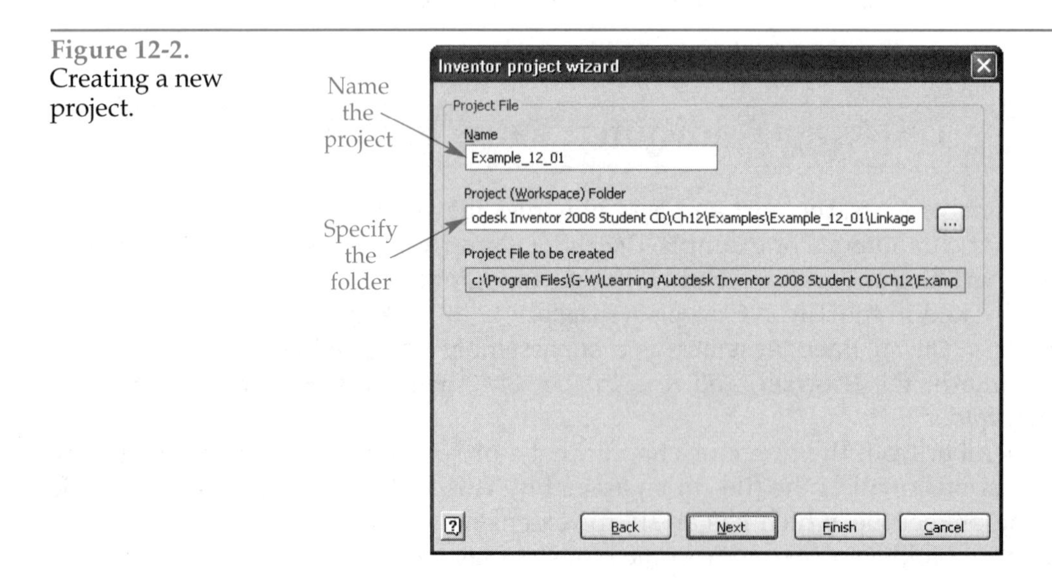

Figure 12-3.
The **Resolve Link** dialog box is displayed when Inventor cannot find a file in an assembly.

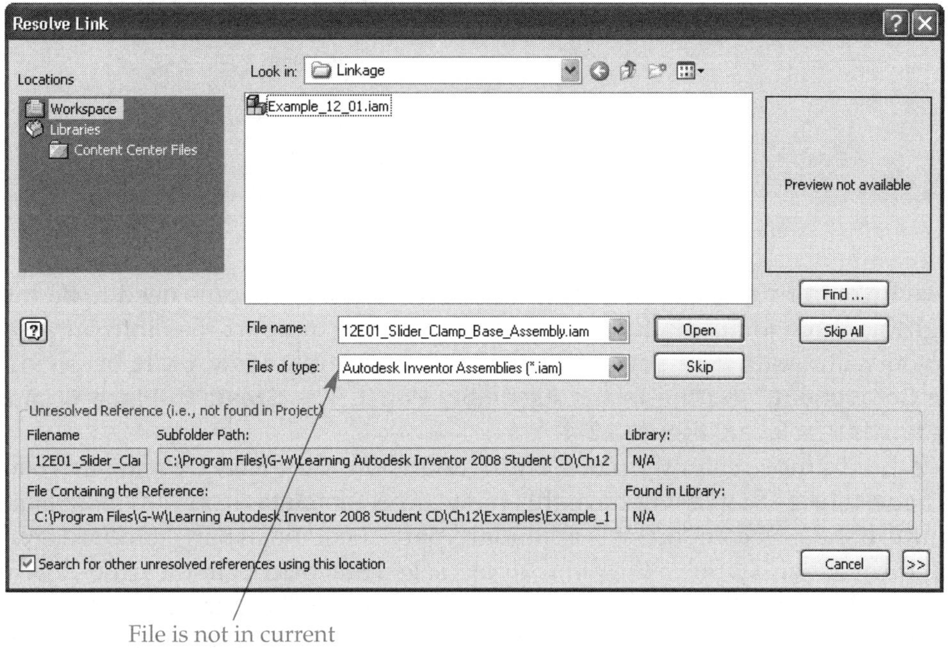

There are a couple of ways to add additional folders to the project search path. The first is using the Libraries entry and the second is using the Included file entry. Libraries are designed for parts and assemblies that never change, such as corporate/engineering standard parts or purchased parts. These parts are read only so they cannot be changed from within the project.

The Included file entry is for specifying another project file (IPJ) so the parts and assemblies within it can be used in the current project. If there was an IPJ file for the base assembly, it could be included in the Example_12_01 project by using the Included file entry. However, the base assembly does not have an IPJ file, so the best solution for this example is to use Windows Explorer to copy the \Base_Assembly folder into the \Linkage folder as a subfolder. Do this now.

NOTE

When using projects, be careful to assign unique file names for parts and assemblies. Duplicate file names may result in errors, as Inventor will search until it finds the first part with the required name.

Creating Parts in the Assembly View

Make sure the Example_12_01 project is set up as described in the previous section and set current. Then, open the assembly Example_12_01.iam, which is located in the \Linkage folder. Since the project file is located in this folder, it is displayed as the current folder in the **Open** dialog box. Also, since the \Base_Assembly folder is now a subfolder of \Linkage, that assembly can be found by Inventor.

Figure 12-4.
Creating a new component from within an assembly.

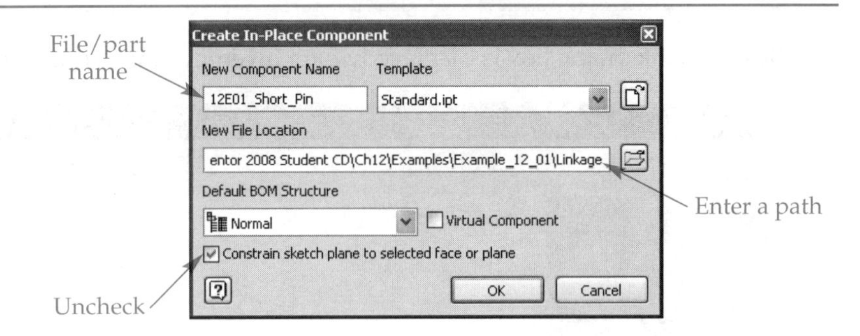

The links in this mechanism have been constrained, but pins need to be inserted through the three joints. So far, you have placed existing part files into an assembly. Now, you will create new parts directly in the assembly view. Start by picking the **Create Component...** button in the **Assembly Panel**. The **Create In-Place Component** dialog box is displayed, **Figure 12-4.**

For this example, enter the name 12E01_Short_Pin in the **New Component Name** text box. The resulting IPT file will have this name. The **Template** drop-down list is used to determine if the new component is a standard part, sheet metal part, standard assembly, or weldment assembly. Since the pin is a part, select Standard.ipt in the drop-down list.

The path where the new component will be saved is specified in the **New File Location** text box. By default, this is the folder for the current project. You can type a new path in the text box or pick the button to the right of the text box to locate the folder. By browsing for the folder, you can check current file names, thus helping to avoid duplicating a file name. For this example, leave the location as the \Linkage folder.

When the **Constrain sketch plane to select face or plane** check box is checked, a mate constraint is automatically placed between the new part and a face that you are asked to pick. For this example, uncheck the check box. Then, pick the **OK** button to start creating the part.

Now, you must select a sketch plane. Pick in a blank area of the graphics window. If you picked a face on one of the existing parts, the sketch plane would be placed on that face. Create a sketch as follows.

1. Draw a circle at the origin.
2. Dimension the circle to a diameter of .25″.
3. Finish the sketch.
4. Extrude the circle a distance of .5″.
5. Select **Active Standard...** from the **Format** pull-down menu.
6. In the **Document Settings** dialog box, select Bronze, Soft Tin in the **Active Material Style** drop-down list and then pick the **OK** button.
7. Right-click in the graphics window and select **Finish Edit** from the pop-up menu to return to the assembly.

A new part is created and added to the assembly. The pin is on the screen, but it is not yet saved. Save the assembly file and, when prompted, pick **OK** to save its dependents. Now, the pin needs to be constrained to the slider. Turn off the visibility of the coupler link to make it easier to see and select the slider. You may also want to turn off the display of the work planes. Then, continue as follows.

1. Place an insert constraint between the pin and the hole in the slider. Enter an offset of –.125″. Select the proper solution so the pin is centered. See **Figure 12-5.**
2. Pick the **Place Component...** button in the **Assembly Panel** and place another instance of 12E01_Short_Pin into the assembly. In order to do this, you must have saved the assembly and its dependents so that the part file for the pin is created.
3. Apply an insert constraint between the second pin and the hole in the input link. Use an offset of –.125″ and select the correct solution so the pin is centered.
4. Turn on the visibility of the coupler link.

Figure 12-5.
A—Applying an insert constraint between the pin and the hole. B—The pin is inserted into the hole.

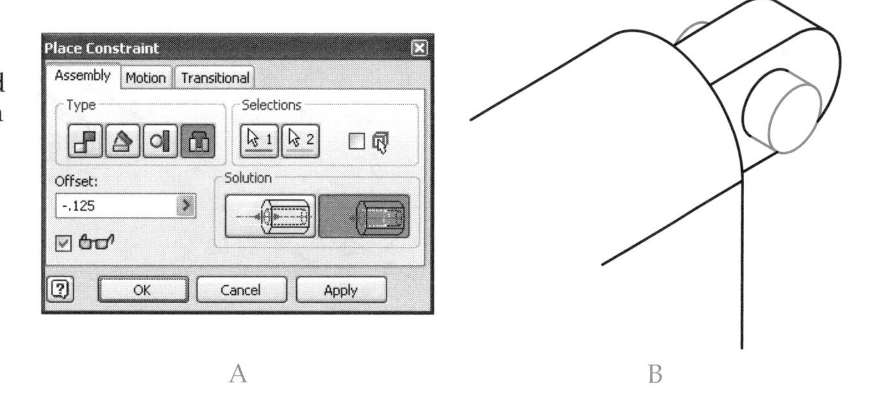

Figure 12-6.
The assembly is complete.

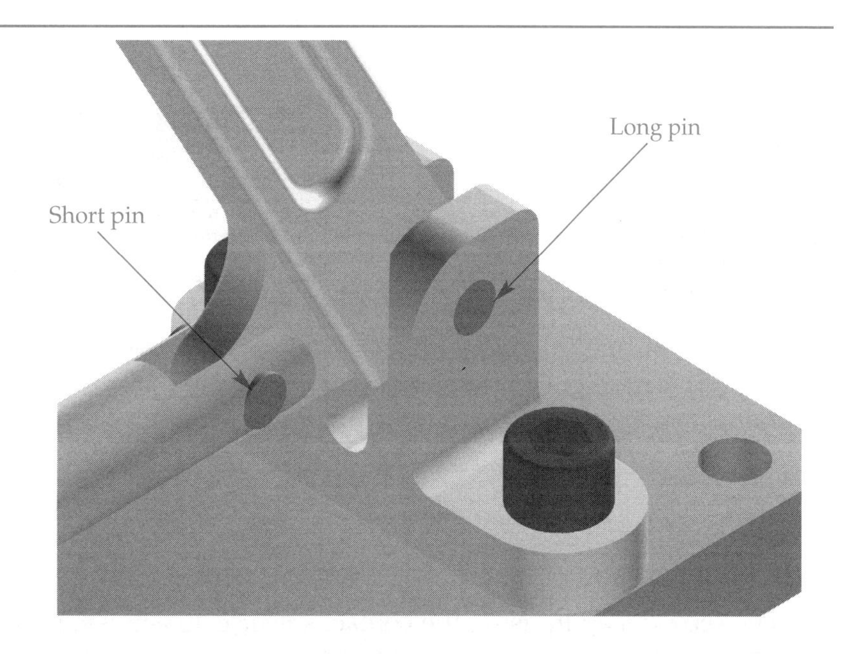

Using this same procedure, create another part called 12E01_Long_Pin. Make its diameter .25″ and its length .75″. Apply an insert constraint between the pin and the hole between the base assembly and the input. The end of the pin should be flush with the outer face. See **Figure 12-6.** Save the assembly file and its dependents.

Contact Solver

In the assembly Example_12_01, the input line is free to rotate about its pivot on the base. Try this by picking the input link and dragging. Notice how the coupler link, slider, and the two pins you created are constrained. It can also be moved to positions where the parts overlap or interfere with each other, positions that could never exist in reality. See **Figure 12-7.** Inventor has a tool that will prevent parts from moving into these positions. It does this by stopping motion when parts contact each other and, therefore, is called the ***contact solver.***

The first step in using this tool is to assign parts and assemblies to the *contact set*. This is done by selecting the parts in the **Browser**, right-clicking, and checking **Contact Set** in the pop-up menu. An icon then appears in front of the part in the **Browser** to indicate this status. Do this now. Select the base assembly, slider, and input link, right-click, and select **Contact Set**. The coupler link should not be part of the contact set. See **Figure 12-8.**

Figure 12-7.
Without the contact solver enabled, the assembly can be manipulated in a way that results in parts overlapping.

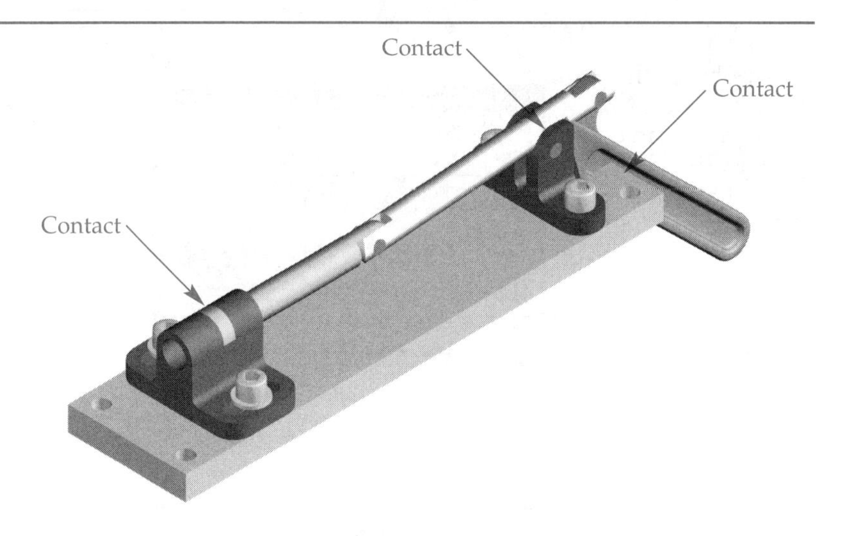

Figure 12-8.
To add parts to the contact set, select them in the **Browser**, right-click, and select **Contact Set** in the pop-up menu.

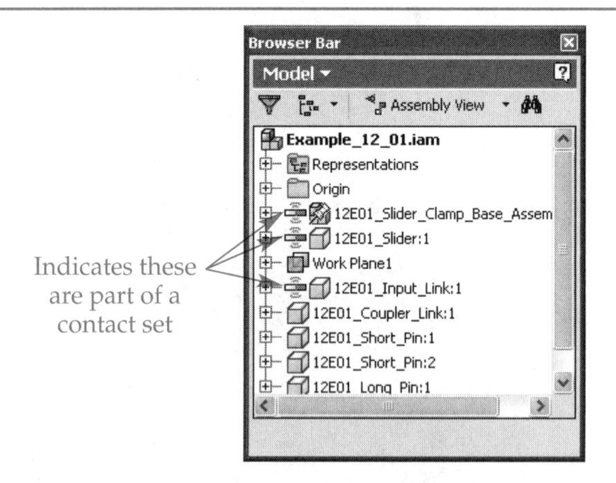

The second step in using the contact solver is to activate it. Select **Tools**>**Activate Contact Solver**. Notice the icon in front of the command name in the pull-down menu is the same one that appears in the **Browser**. The icon in the pull-down menu is depressed when the contact solver is on. Now, rotate the input link. It will stop whenever it or the slider contacts the base assembly. To turn off the contact solver, select **Tools**>**Activate Contact Solver** again so the icon is not depressed. Do this now.

PROFESSIONAL TIP

It is important to keep the contact solver off unless you are actually displaying motion, as it will interfere with or slow down other operations (such as applying constraints).

Angle Constraint

An ***angle assembly constraint*** works on part edges and faces, work planes, and work axes. It sets an angle between any two of these features on two different parts. You will now place an angle constraint between the input link and the base.

1. Pick the **Constraint...** button in the **Assembly Panel** to open the **Place Constraint** dialog box.

2. Pick the **Angle** button in the **Type** area of the **Assembly** tab.
3. Using the two buttons in the **Selections** area of the **Assembly** tab, pick the top face of the rectangular base in the base assembly as the first selection and the narrow, bottom face of the input link as the second selection.
4. In the dialog box, enter 120 in the **Angle:** text box. The angle is defined as the angle between the two arrows if you put them tail-to-tail.
5. In the **Solution** area, pick the **Directed Angle** button.
6. Pick the **OK** button to apply the constraint and close the dialog box.

The coupler is now fully constrained and cannot be moved.

Driving Constraints

To *drive* a constraint is to animate the constrained parts through motion allowed by the constraint. By driving the angle constraint applied in the previous section, the motion of the mechanism can be animated. In the **Browser**, expand the part tree for 12E01_Input_Link. Right-click on the angle constraint name and pick **Drive Constraint** from the pop-up menu. The **Drive Constraint** dialog box is displayed. See **Figure 12-9.** Expand the dialog box by picking the **>>** button.

1. In the **Start** text box, enter 120. This is the beginning angle value, or in this case, the maximum "closed" value for the link.
2. In the **End** text box, enter 30. This is the ending angle value. In this case, the link will rotate from 120° to 30°, or through 90° of motion.
3. The **Increment** area is used to set the angle for each step in the motion or the total number of steps. Pick the **amount of value** radio button and enter 2 in the text box so that each step is 2°.
4. The **Repetitions** area is used to set how the parts cycle through the motion. Since the link oscillates (travels back and forth), pick the **Start/End/Start** radio button. Also, enter 5 in the text box so the motion (either forward or backward) is repeated five times.
5. The parts are currently at the end of the motion. Therefore, pick the **Reverse** button in the controls to animate the motion. If the parts were at the beginning of the motion, you would pick the **Forward** button.
6. Pick the **Cancel** button to close the dialog box.
7. Save Example_12_01.iam and close all files.

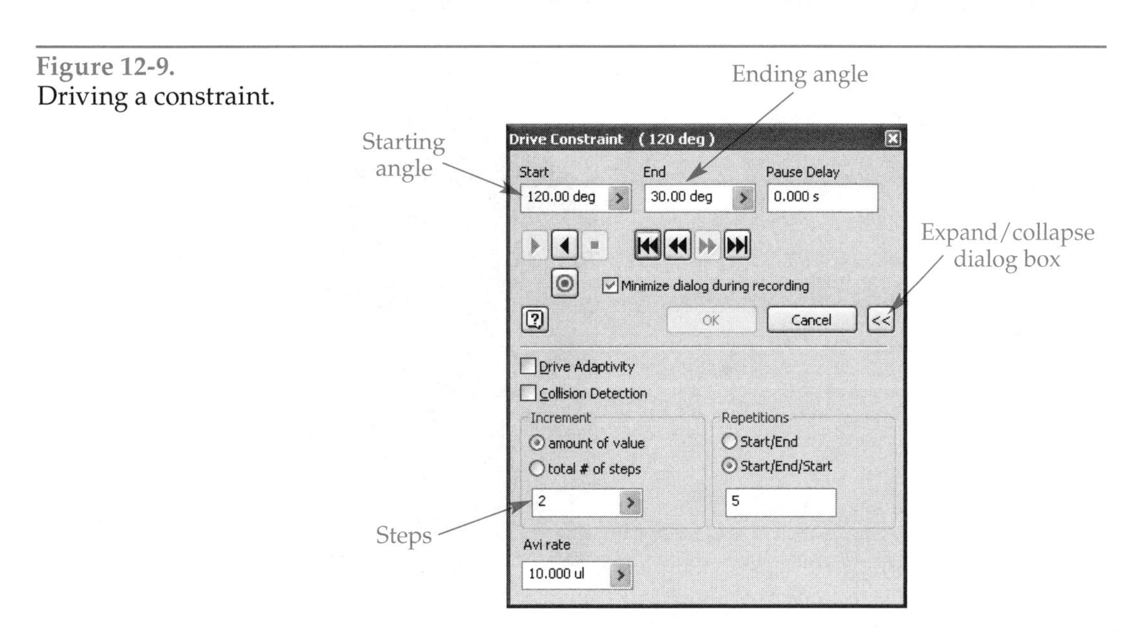

Figure 12-9.
Driving a constraint.

Any constraint that has an offset, such as mating, flush, angle, tangent, or insert, can be driven using this same procedure. This allows you to create all kinds of rotating and sliding animations. However, you can only drive one constraint at a time.

PROFESSIONAL TIP

By picking the **Record** button in the **Drive Constraint** dialog box, the motion can be saved to an AVI video file that can be played in Windows Media Player. In this way, you can share the animated part with others who do not have access to Inventor.

Constraining Work Planes and Axes

Constraints can be applied to work planes and axes, as well as part faces. This can simplify the assembly process. In the example in this section, you will be working with a simple model of a one-cylinder engine. All of the parts have work planes through their centers, which will be used to align the parts.

Create a new project named Ch12 with the project folder as c:\Program Files\G-W\Learning Autodesk Inventor 2008 Student CD\Ch12. This project will allow access to all of the files for Chapter 12. Set it current. Then, open the file Example_12_02.iam. This assembly currently includes the engine block and crankshaft (crank). The problem is how to align the center of the crank with the center of the bore without calculating offset values. Notice the two work planes. First, you will constrain the two planes to each other. Then, you will constrain the rotational centerline of the crankshaft with the work axis on the engine block. By doing this, the crankshaft is free to rotate and the resulting assembly is much more flexible. For example, if the block dimensions are changed, the crankshaft will remain centered within the bore.

1. Pick the **Constraint...** button in the **Assembly Panel**.
2. In the **Place Constraint** dialog box, pick the **Mate** button in the **Type** area and the **Flush** button in the **Solution** area. **Flush** is used because the flat on the crank is to be on the left side of the block.
3. With the **First Selection** button on, pick the work plane on the crank so the direction arrow points up. With the **Second Selection** button on, pick the work plane on the block so the direction arrow points to the left.

Figure 12-10.
The crankshaft is properly located and constrained.

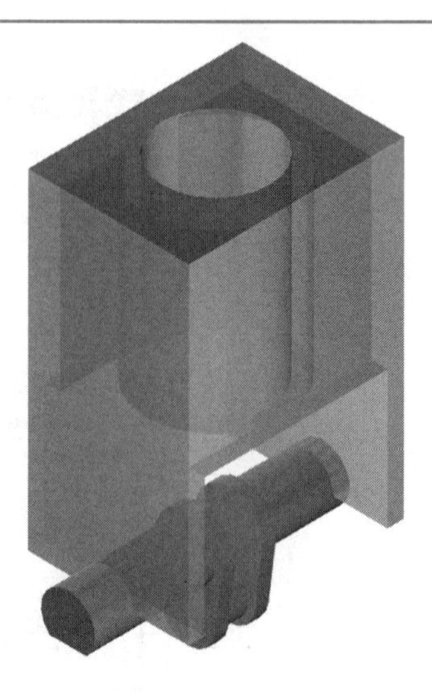

4. Pick the **Apply** button in the dialog box to apply the constraint and leave the dialog box open.
5. Pick the **Mate** button in the **Type** area and the **Mate** button in the **Solution** area.
6. With the **First Selection** button on, pick the centerline of the crankshaft. With the **Second Selection** button on, pick the work axis through the bearing housings on the block.
7. Pick the **OK** button in the dialog box to apply the constraint and close the dialog box.

The crankshaft is centered under the bore. Also, the rotational centerline of the crankshaft is inline with the centerline of the bearings. See **Figure 12-10.** The crankshaft is free to rotate. Pick it and drag to make it rotate.

Now, the connecting rod needs to be placed into the assembly and constrained to the crankshaft. This will be easier to do if you turn off the visibility of the block. Then, place one instance of 12E02_Con_Rod.ipt into the assembly. Constrain the large end of the connecting rod to the crankshaft using the process described above. The centerline of the large hole should be constrained to the centerline of the journal, not the centerline of the bearing surfaces. See **Figure 12-11.** Since the rod is symmetrical, it does not matter if you use a mating or flush solution on the work planes.

Using a similar process, place an instance of 12E02_Piston_Pin.ipt, constrain its work plane to the work plane on the connecting rod, and constrain its centerline to the centerline of the small hole in the connecting rod. Place an instance of 12E02_Piston.ipt, constrain its work plane to the work plane on the connecting rod, and constrain the centerline of its hole to the centerline of the pin. See **Figure 12-11.** Your assembly may look slightly different, depending on where you picked planes and centerlines. Continue as follows.

1. Turn on the visibility of the block.
2. Constrain the centerline of the piston to the centerline of the bore. If the piston is inside the block in your assembly, you may need to move the piston in order to acquire the centerline.
3. Place an instance of 12E02_Flywheel.ipt into the assembly.
4. Constrain the flat on the flywheel to the flat on the crank. Also, constrain the centerlines. Finally, constrain the end of the flywheel flush to the end of the crank. The green circle on the flywheel should face away from the engine block.
5. Drag and rotate the flywheel. The piston goes up and down.

Now, automate the motion. Put an angle constraint between the work plane on the flywheel and the top of the block. Both direction arrows should point up; use the buttons in the **Solution** area of the **Place Constraint** dialog box as needed. Also, set the angle to 0°. Next, right-click on the constraint name in the **Browser** and select **Drive Constraint** from the pop-up menu. Set the values as shown in **Figure 12-12** and "run the engine." Save the assembly and close all files.

Figure 12-11.
The piston, piston pin, and connecting rod are added to the assembly.

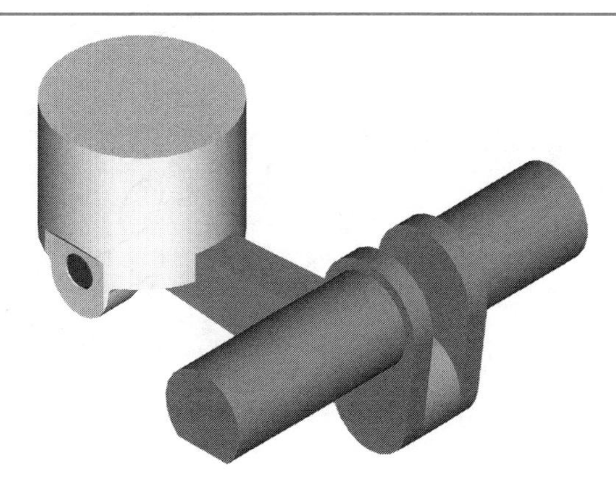

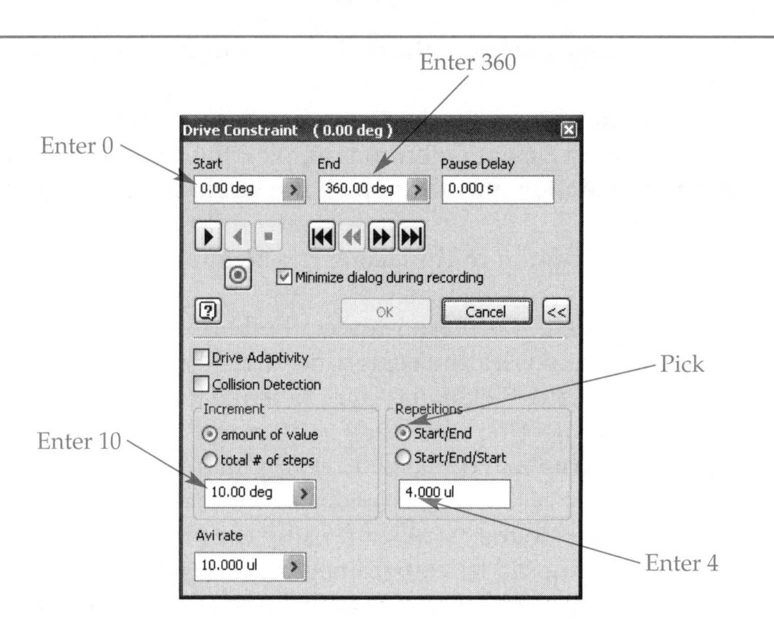

Figure 12-12. Driving the angle constraint to rotate the flywheel.

Assembly Constraints and Sketches

Assembly constraints can also be applied to circles, arcs, and lines in sketches. This is useful for conceptual designs where you are trying to develop the correct kinematic geometry, or geometry of motion, before constructing the detail parts. In this example, you will look at the kinematic layout of a scissors lift.

Make sure the Ch12 project is active. Then, open the assembly Example_12_03.iam. See **Figure 12-13.** This assembly contains eight parts: one base, one top, four instances of the link, one control, and one backboard.

Except for the backboard, each part is a simple sketch that has not been extruded; you can see the sketch dimensions. In the isometric view, you can see that only the backboard has depth. The circles on the ends of the links have mating constraints to

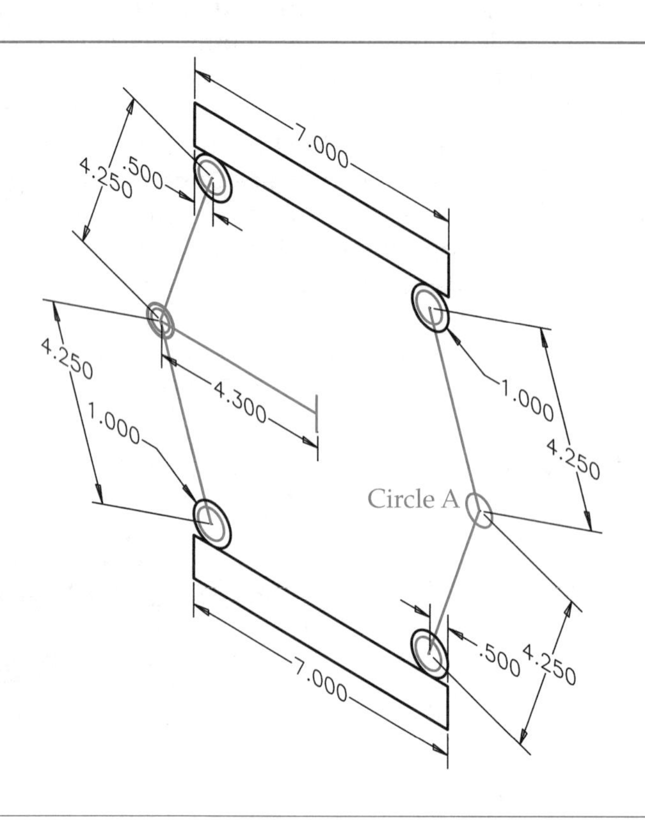

Figure 12-13. The sketch for a scissors lift. Note: The backboard, which is the extruded part, is hidden.

the circles on the base and the top, and to each other at the center. The left edges of the top and the base have mating constraints so the parts stay lined up and parallel.

When the sketched circles are mated, they are not constrained in the Z direction. As such, in an isometric view, their motion can be confusing when you try to move them in just the XY plane. This is the reason for the backboard. A line in each of the parts is constrained to the backboard with a mating constraint. This keeps all of the parts in the same plane and parallel to the backboard.

The base and backboard are both grounded. The control link has an angle constraint with the top that keeps it horizontal. There is a mating constraint between the top and the base with an offset of 9″, which has been suppressed. All of this means that you can drag the top up and down and the entire linkage will move. Be careful not to move the top up too far. The link motion may reverse. If this happens, undo the drag.

A second control link needs to be placed on the right side. It will be constrained to the existing control so they line up. Then, you will move the top down as if each of the controls is a hydraulic cylinder. Finally, you will measure the distance between the controls. The procedure is:

1. Place an instance of 12E03_EM1203_Control.ipt into the assembly.
2. Using the **Look At** tool, display a plan view of the control you just placed.
3. Pick the **Rotate Component** button in the **Assembly Panel**. Pick the inserted control. Then, rotate the control approximately 180° about the screen's Z axis by picking outside of the onscreen trackball and dragging. Finally, press [**Esc**] to end the tool.
4. Apply a mating constraint between the circle in the inserted part and circle A shown in **Figure 12-13.**
5. Apply a mating constraint between the long line in the inserted control and the face of the backboard.
6. Apply a mating constraint between the long line in the inserted control and the corresponding long line in the existing control. This mating constraint keeps the long lines in the two controls inline.
7. In the **Browser**, expand the assembly tree for the top. Right-click on the mating constraint with the 9″ offset and uncheck **Suppress** in the pop-up menu to unsuppress the constraint.
8. Edit the mating constraint on the top that currently has a 9″ offset and change the offset to 3″. See **Figure 12-14.**
9. Select **Measure Distance** from the **Tools** pull-down menu. Then, pick the short vertical line in each control. The distance between the two lines is displayed in the **Measure Distance** dialog box, **Figure 12-15.**
10. Save the assembly and close all files.

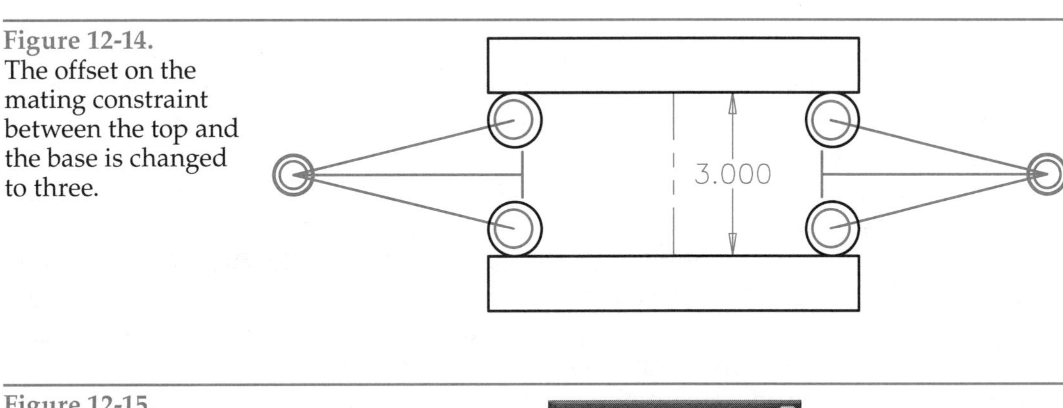

Figure 12-14.
The offset on the mating constraint between the top and the base is changed to three.

Figure 12-15.
Measuring the distance between the two controls.

Measure Distance
5.675 in
Length
0.83 in
Measured distance

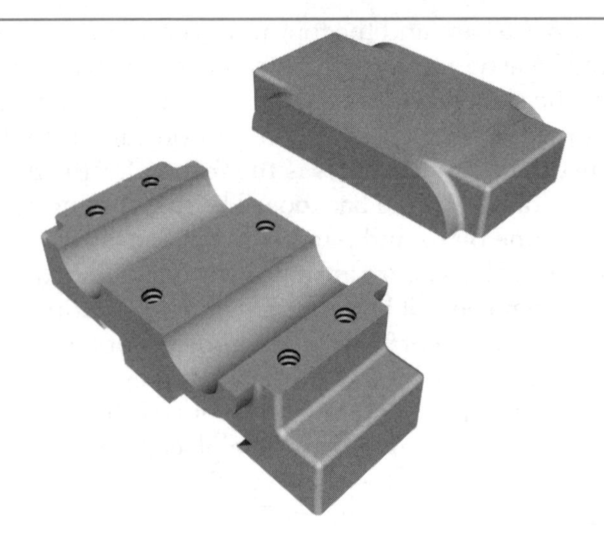

Figure 12-16. Adaptive clearance holes will be added to the cap. Note: The cap is shown out of place so the threaded holes in the base are visible.

Adaptive Parts

An ***adaptive part*** has unconstrained features, the sizes and locations of which are controlled by another part. For example, you can locate features on a part based on the geometry of a second part. Then, if the second part's features are changed, those on the first part change as well. For instance, the pins created in the Example_12_01 assembly could have been created as adaptive parts. In this way, if the hole size changes, the diameter of the pin changes to match. The assembly process can be simplified by working in a 2D front view of the parts.

Adaptive Location

Make sure the Ch12 project is active. Then, open the file Example_12_04.iam. This assembly is shown in **Figure 12-16** with the cap moved to expose the six threaded holes in the base. Matching clearance holes need to be placed in the cap. The holes in the base can be projected onto a sketch plane on the cap. Then, holes can be created in the cap, which will be adaptive. The procedure is:

1. Right-click on the cap name in the **Browser** and select **Edit** from the pop-up menu.
2. Display a wireframe view so you can see the holes in the base.
3. Start a new sketch with the top face of the cap as the sketch plane.
4. Project the circle tops of the six holes in the base onto the sketch plane.
5. Using the **Point, Center Point** tool, place hole centers at the center points of all six circles.
6. Finish the sketch.
7. Pick the **Hole** button in the **Part Features** panel. Using the **Holes** dialog box, place clearance holes for 1/2″ diameter cap screws at each of the points. Use the Through All setting. See **Figure 12-17.**
8. Right-click in the graphics window and select **Finish Edit** from the pop-up menu.
9. Display a shaded view.

A pair of circular arrows appears in the assembly tree in the **Browser** in front of the cap name, **Figure 12-18A.** These red and green arrows indicate that some feature of the part is adaptive. If you edit the cap, the adaptive symbol appears in front of the hole feature and the sketch for the holes as well. See **Figure 12-18B.**

Now, right-click on the base name in the **Browser** and select **Edit** from the pop-up menu. Next, edit the sketch for Hole1 and change the 3″ dimension to 1.5″. Finish the sketch and then finish the edit. The corresponding hole in the cap has moved to remain inline with the threaded hole in the base. See **Figure 12-19.** Save the assembly and close all files.

Figure 12-17.
Adding clearance holes to the top part.

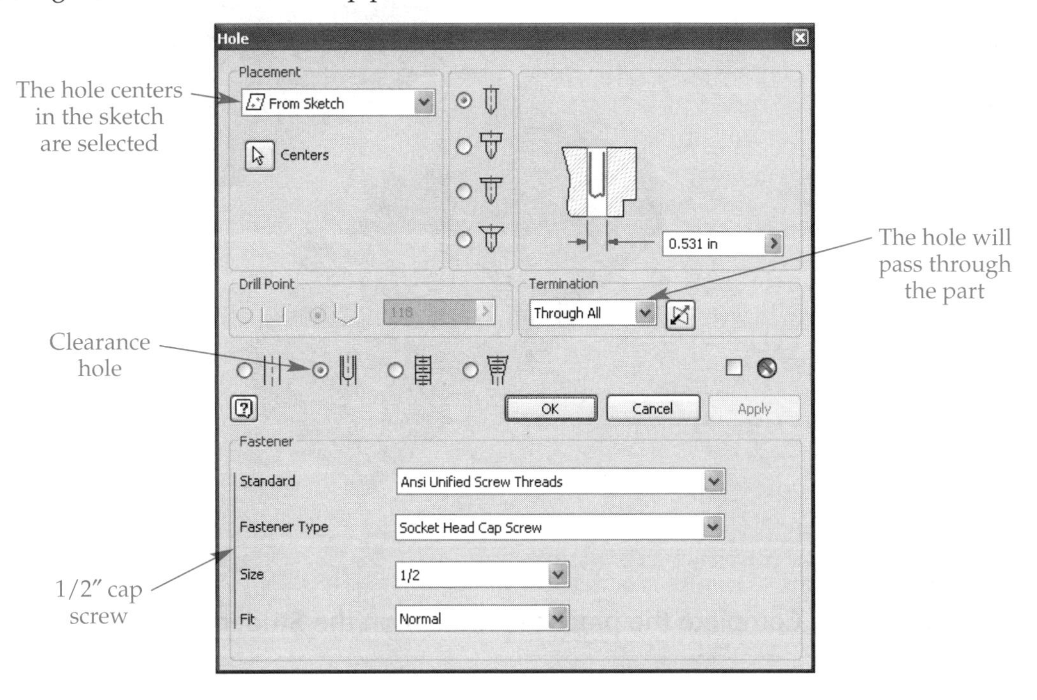

Figure 12-18.
A—The part has the adaptive symbol next to its name in the assembly tree. B—The feature of the part and its sketch have the adaptive symbol next to their names in the part tree when the part is edited.

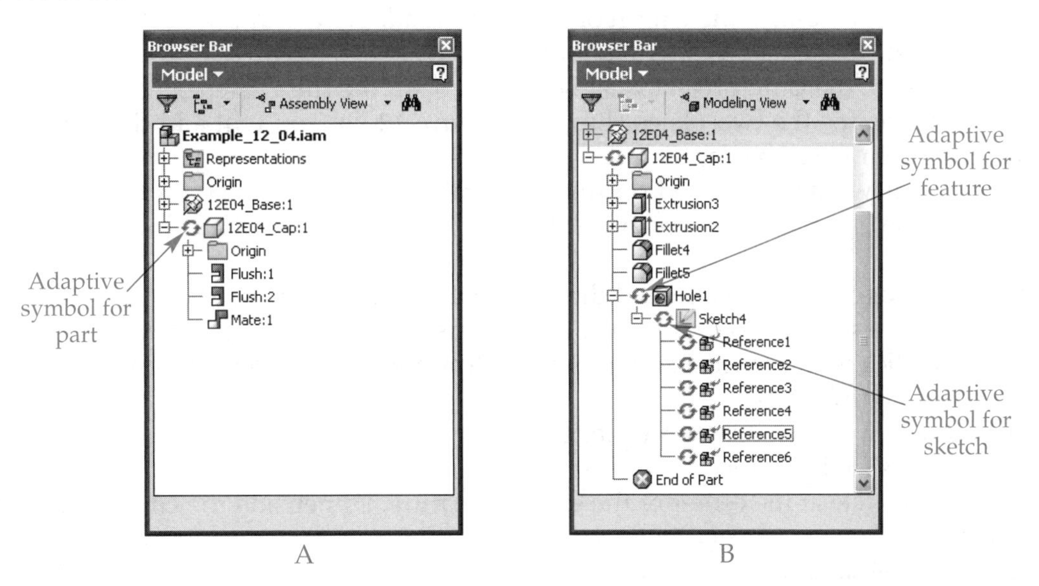

PROFESSIONAL TIP

Experienced users suggest that you leave only a few parts adaptive. When you get the assembly the way you want it, remove adaptivity by right-clicking on the component and unchecking **Adaptivity** in the pop-up menu; a check mark appears next to the menu item when adaptivity is on. Do this for both the part and the sketch.

Figure 12-19.
When a threaded hole in the base is moved, the corresponding clearance hole in the cap moves as well.

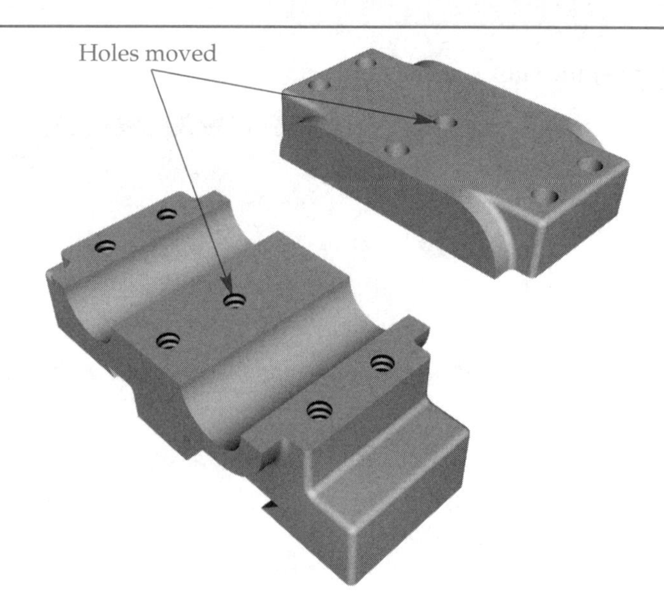

PRACTICE 12-1 Complete the practice problem on the Student CD.

Adaptive Size

The second application of adaptivity ties the size of features in a part to the size of features in another part in the assembly. For example, the two plates shown in **Figure 12-20** are constrained with a 4″ offset. Each post is constrained with an insert constraint to each plate. The length of each post is adaptive. If the 4″ offset is changed to a different value, the posts will automatically change length.

Make sure the Ch12 project is active. Then, open the file Example_12_05.iam. This assembly contains the two plates shown in **Figure 12-20.** You will add the four posts and make their lengths adaptive.

1. Start a new part file based on the Standard (in).ipt template. (You do not need to close the assembly file.) Turn on the visibility of the XY plane and display the isometric view.
2. Draw a circle at the origin and dimension it to a diameter of .75″.
3. Finish the sketch and extrude the circle 1″ using the midplane option.
4. Right-click on the extrusion name in the **Browser**. Select **Adaptive** from the pop-up menu, **Figure 12-21.**
5. Start a new sketch on the XY plane.
6. Display sliced graphics.
7. Draw a circle at the center of the existing feature. Dimension the circle to a diameter of .5″.
8. Finish the sketch.
9. Select **Parameters...** from the **Tools** pull-down menu.

Notice in the **Parameters** dialog box that d1 is the extruded length of the first extrusion. The .5″ diameter circle will be extruded so it protrudes out of each end of the .75″ diameter shaft. The parameter d1 will be used in an equation for the extruded length so that the .5″ diameter shaft is always longer than the .75″ diameter shaft. Close the **Parameters** dialog box and continue as follows.

10. Extrude the circle using the midplane option. For the extents, select the Distance option and enter the equation d1 + 1.0 in the **Depth** text box. Pick the **OK** button. See **Figure 12-22.**
11. In the material drop-down list on the **Inventor Standard** toolbar, select Metal-Brass.

Figure 12-20.
The four posts are adaptive so their length will change as the offset between the two plates is changed.

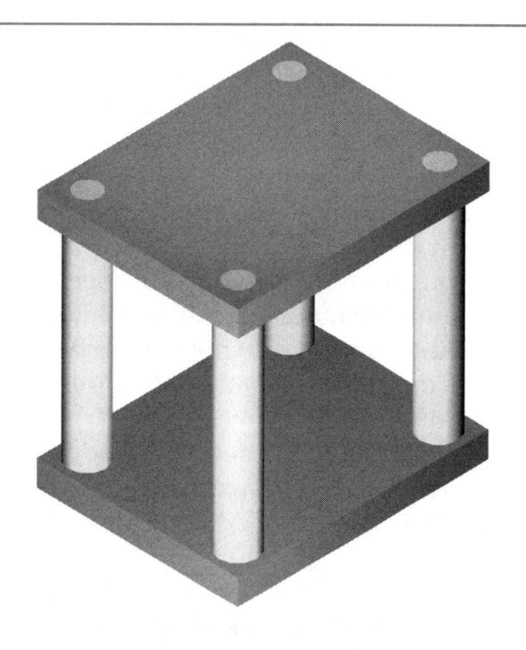

Figure 12-21.
Setting a feature to be adaptive.

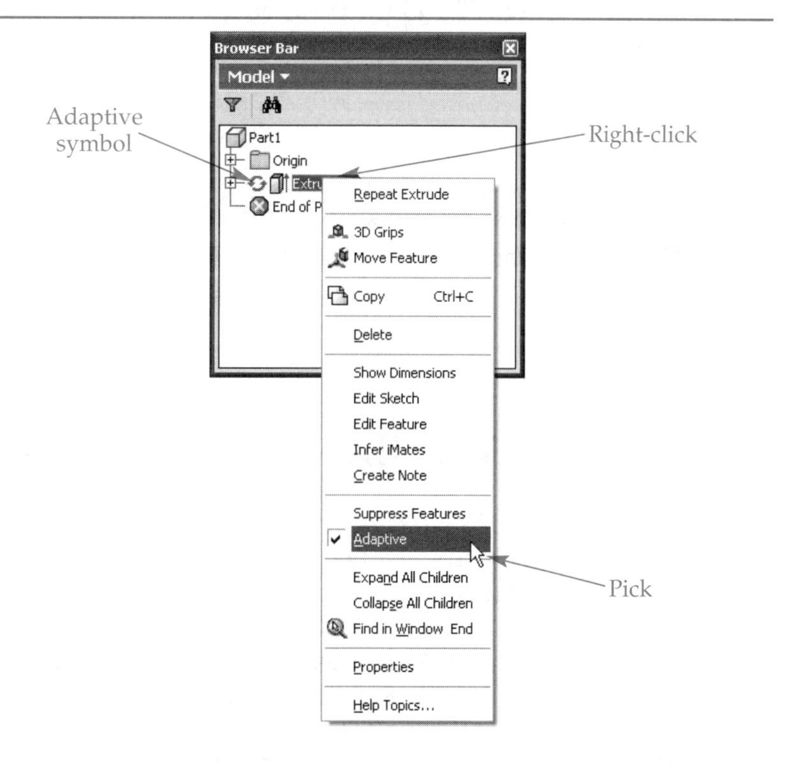

Figure 12-22.
The post is created as a part file. A—Entering an equation ties the extents of the two shafts together. B—The final part.

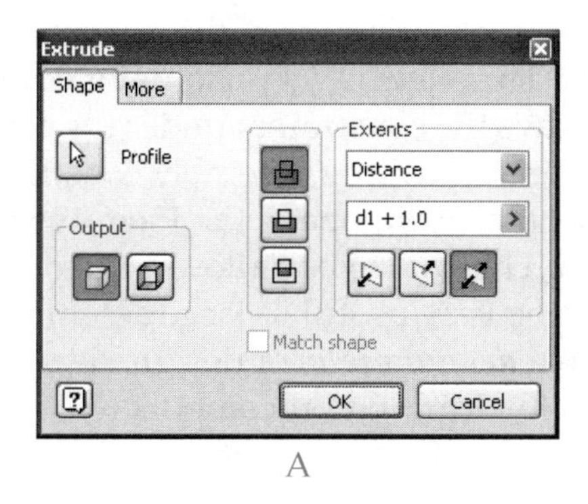

A

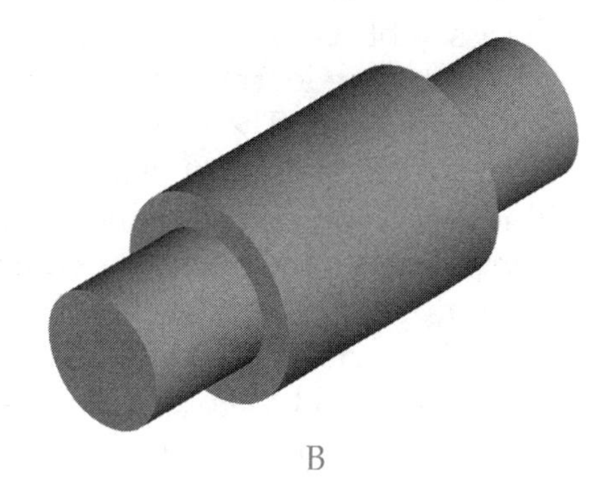

B

12. Save the file as 12E05_Post.ipt and close the file.
13. In the assembly file, place one instance of 12E05_Post.ipt at the lower-right corner.
14. Right-click on the post name in the **Browser** and select **Adaptive** from the pop-up menu.

Adaptivity was turned on in the part file for the .75″ diameter shaft. However, the part also must be made adaptive in the assembly. Continue as follows.

15. Place an insert constraint between the .75″ diameter shaft and the hole in the plate. Make sure you pick the large diameter. See **Figure 12-23.**
16. Place three more instances of the post into the assembly. Constrain each in the same way. These instances do not have to be set to be adaptive. In fact, **Adaptive** is grayed out in the pop-up menu for the three additional instances; they will match the behavior of the first instance.
17. Rotate the view to show the bottom of the assembly.
18. Place an insert constraint between the first post and the hole in the top plate, again selecting the large diameter. All four posts change length to fit the space between the two plates.
19. Display the isometric view. The assembly should now look like **Figure 12-20.**

To see the adaptability of the shafts, edit the mating constraint for the plates. Change the 4″ offset to 5″ and the posts adapt to the new value. You can also drive the offset in the mating constraint. In order to do this, you must check the **Drive Adaptivity** check box in the **Drive Constraint** dialog box. See **Figure 12-24.** Save the assembly and close all files.

PROFESSIONAL TIP

Adaptive sizing needs to be used carefully; users have reported problems with large assemblies containing multiple adaptive parts.

Visibility and Design Views

Make sure the Ch12 project is active. Then, open the file Example_12_06.iam. This is the completed assembly from earlier in this chapter. Look at the **Browser**. There are four tools or menus. See **Figure 12-25.** These are:

- **Browser filters.** See **Figure 12-25A.** Controls the display of items in the **Browser**. For example, if **Hide Work Features** has a check mark next to it, the Origin branch is not displayed in the **Browser**. Note that this does not affect the display of the model.
- **Representations.** See **Figure 12-25B.** Lists recently restored design views. Picking **Other...** opens the **Design View Representations** dialog box.
- **Browser views.** See **Figure 12-25C.** The modeling view displays the features that create each part and any features that are added to the assembly. All of the assembly constraints are located in a branch at the top of the assembly tree, rather than under the corresponding part. This view is very useful for editing a feature of a part. The assembly view displays the constraints under the component itself. This is most useful for applying and editing assembly constraints.
- **Find.** The **Find** button, which has a binocular icon, opens the **Find Assembly Components** dialog box. See **Figure 12-26.** This dialog box allows you to search for components in the assembly. This can be very useful in large assemblies.

The most important thing you can control in a ***design view***, or design view representation, is component visibility. This is useful for working on components that are obscured by others. For example, in the engine assembly you might want to work on

Figure 12-23.
The post part file is added to the assembly and inserted into a hole.

Figure 12-24.
Driving a constraint and the adaptability of parts.

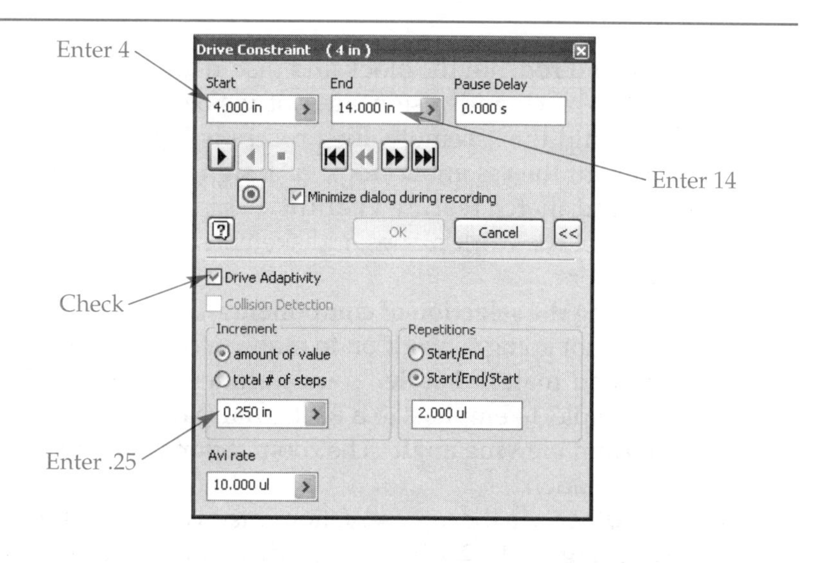

Figure 12-25.
A—The browser filters menu. B—The representations menu. C—The browser views menu.

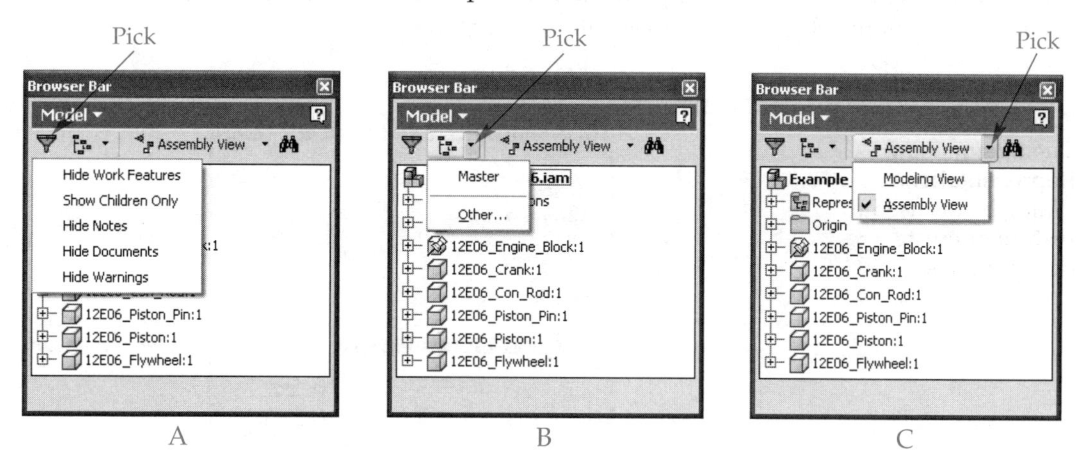

Figure 12-26.
Searching for components in the **Browser**.

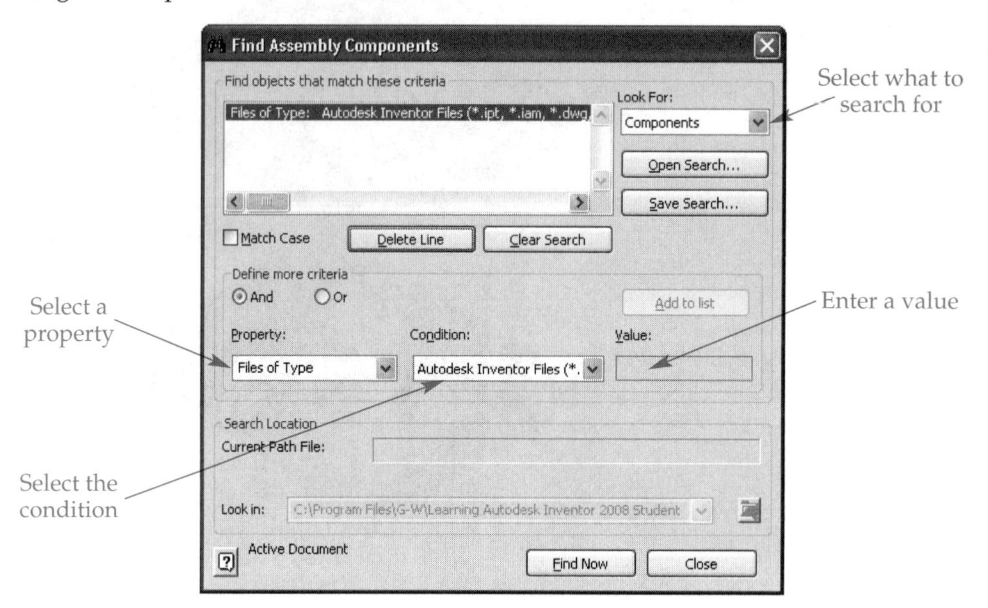

the piston pin and rod, but the block and piston are in the way. You can create a design view where the block and piston are not visible. Also, if multiple design views are saved as you build the assembly, the process is "recorded." In this way, views can be used to document the assembly steps. A design view can also save settings of:

- **Work and sketch feature visibility.** This is an easy way to display the assembly with all of the work planes and sketches hidden, without individually turning off each one.
- **Disabling the selection of components.** Disable the selection of components that you are not going to work on to make selection of relevant components easier.
- **Color and material style.** You can hide or override these for any component; for example, to emphasize a particular part.
- **Zoom and viewing angle.** The zoom factor and viewing angle are saved with the representation.

Selecting **Other...** in the representations menu opens the **Design View Representations** dialog box. See **Figure 12-27.** To restore a design view, select its name and then pick the **Activate** button. To create a new design view of the current display and settings, type the name in the text box to the left of the **New** button. You can make display changes with the dialog box open. Then, pick the **New** button to create the design view. There are two formats in which new representations can be saved:

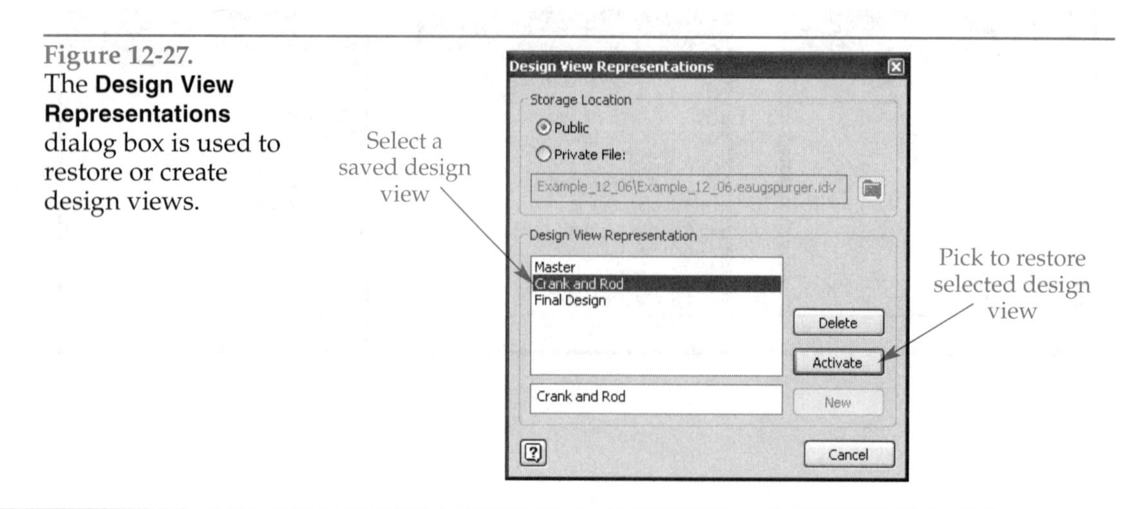

Figure 12-27.
The **Design View Representations** dialog box is used to restore or create design views.

- **Public.** The information is saved in the design file. This format is associative and changes are displayed in the drawing files.
- **Private.** The information is saved within a separate file with an IDV extension. This file is located in the folder shown in the path. Design view representation files are not associative to the drawing file.

Now, set up the assembly for another new view:

1. Turn off the visibility of the engine block.
2. Right-click on 12E06_Flywheel in the **Browser** and uncheck **Enabled** from the pop-up menu. See **Figure 12-28.** This disables the flywheel so it cannot be selected. A check mark appears next to **Enabled** when a part can be selected. Also, disable the crank.
3. Turn off the work planes by selecting **View**>**Object Visibility**>**Origin Planes** to uncheck the item.
4. Change the color of the connecting rod to Red (Bright).
5. Change the viewpoint and zoom in on the connecting rod and piston.
6. Open the **Design View Representations** dialog box.
7. Enter the name Piston and Rod and pick the **New** button. Then, close the **Design View Representations** dialog box.

Now, in the **Browser**, select **Master** from the representations menu. Also, display the isometric view. Notice how the parts appear in the graphics window. Next, select **Piston and Rod** from the representations menu. Notice how the parts appear in the graphics window. Open the **Design View Representations** dialog box and restore the design view Final Design. Save the assembly.

NOTE

Any design view representations that are created in an assembly can be used in the drawing for that assembly. If you anticipate making an assembly drawing, you may want to build a representation that can be used both in the assembly creation and as a view to be placed in the assembly drawing.

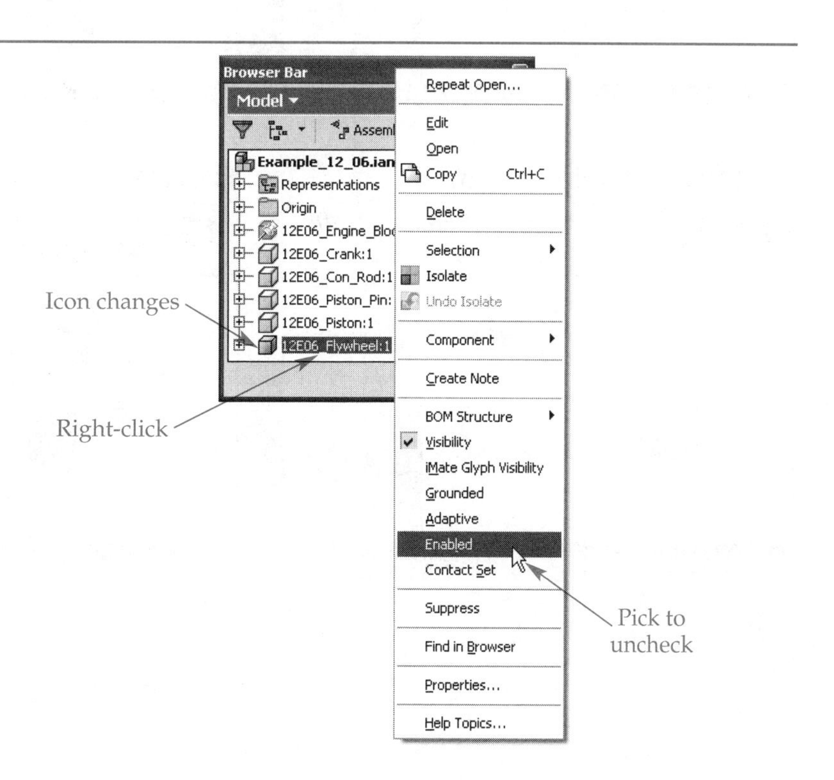

Figure 12-28.
Disabling a part so it cannot be selected.

Chapter Test

Answer the following questions on a separate sheet of paper or complete the electronic chapter test on the Student CD.

1. Why would you need to add additional paths to a project?
2. Which tool allows you to create new parts from within an assembly?
3. What does the *contact solver* do?
4. What does the *angle constraint* do?
5. What does it mean to *drive* a constraint?
6. How do you open the **Drive Constraint** dialog box?
7. Why would you apply a constraint between two work planes, as opposed to the corresponding parts?
8. What is an *adaptive part?*
9. What are the two types of adaptive parts?
10. Briefly describe how to save a new design view in Inventor.

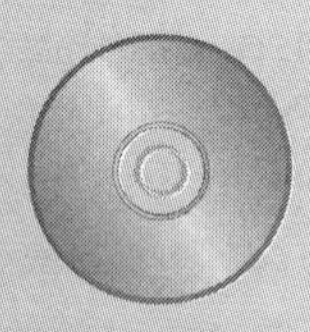

Chapter Exercises

Exercise 12-1. Linkage.

Complete the exercise on the Student CD

Exercise 12-2. Pivot.

Complete the exercise on the Student CD

Exercise 12-3. Holding Fixture.

Complete the exercise on the Student CD

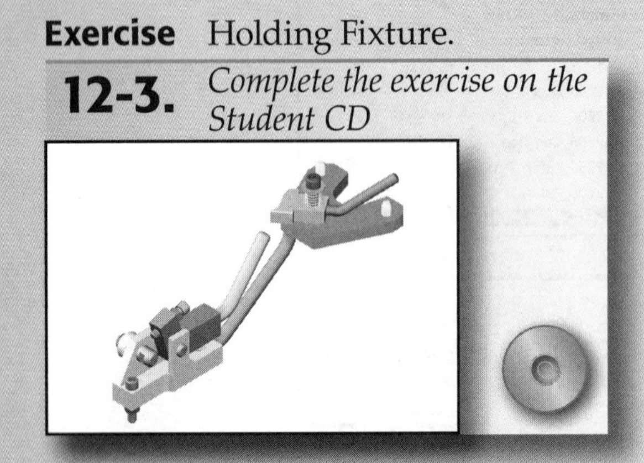

Exercise 12-4. Base and Cap.

Complete the exercise on the Student CD

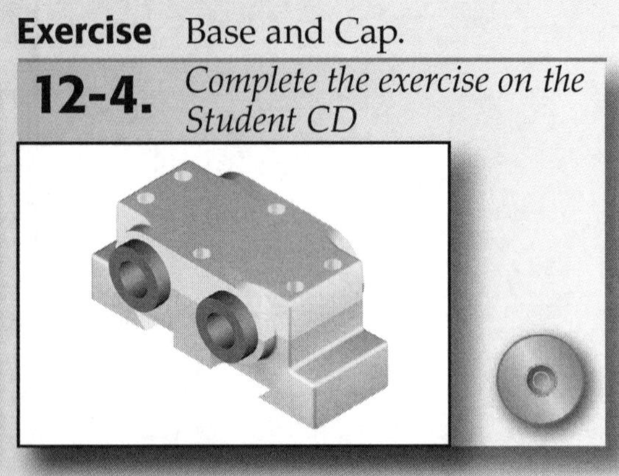

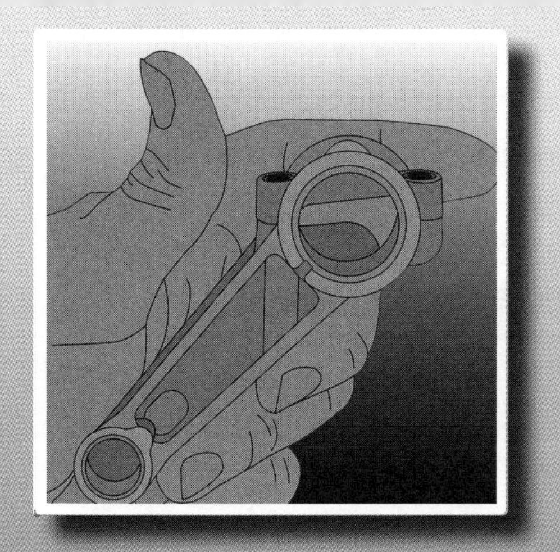

Chapter 13

Motion Constraints and Assemblies

Objectives

After completing this chapter, you will be able to:

- Use the rotation motion constraint.
- Use the rotation-translation motion constraint.
- Use the collision detector feature of the **Drive Constraint** dialog box.
- Use the transitional motion constraint.

User's Files

The Student CD included with this text contains several files required for this chapter. Refer to the file File List.txt in the \Ch13 folder for the comprehensive list.

Motion Constraint

Besides the assembly constraints for positioning components introduced in the last two chapters, there are three motion constraints that can relate the relative motion of one component to another. The tools for applying these constraints are found in the **Motion** and **Transitional** tabs of the **Place Constraint** dialog box. See **Figure 13-1.**

- **Rotation.** Used for rotating part to rotating part, such as a pair of gears. There are two possible solutions when this constrain is applied.
- **Rotation-Translation.** Used for rotating part to translating part, such as a gear and rack. There are two possible solutions when this constraint is applied.
- **Transitional.** Used for surface contact, such as a cam and follower. There is only one possible solution when this constraint is applied.

Figure 13-1.
A—The **Motion** tab of the **Place Constraint** dialog box. B—The **Transitional** tab of the **Place Constraint** dialog box.

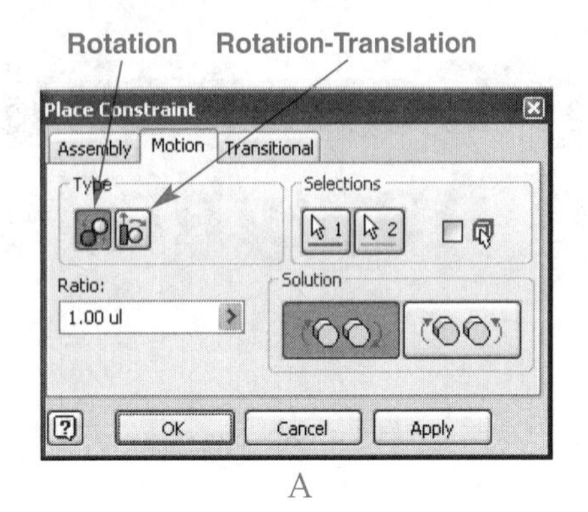

A

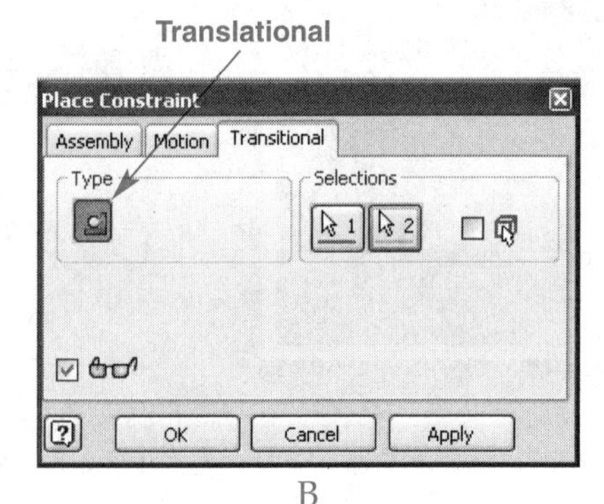

B

Rotation Constraint

A ***rotation constraint*** transfers the rotation of one part to another part. Make sure the Default project is active. Then, open Example_13_01.iam. This assembly has two friction wheels that turn together without slipping. This concept is the basis for gear design; the diameters of the wheels represent the pitch diameters of the gears. The small wheel has a 2″ diameter and has an angle constraint named DRIVE ME Angle. This can be seen if the tree for 13E01_Small_Wheel is expanded in the **Browser**. See **Figure 13-2.** This constraint controls the angle between the work plane on the wheel and the top face of the base. Therefore, if the constraint is driven from 0° to 360° the wheel rotates once.

Opposite Direction of Rotation

The diameter of the large wheel is 4″. Since it is twice the size of the small wheel, it should rotate once for every two rotations of the small wheel. To create a constraint that will do this:

1. Pick the **Constraint...** button in the **Assembly Panel** to display the **Place Constraint** dialog box.
2. Pick the **Motion** tab and pick the **Rotation** button in the **Type** area of the tab.
3. With the **First Selection** button on, pick anywhere on the *curved* face of the small wheel. With the **Second Selection** button on, pick anywhere on the *curved* face of the large wheel. Make sure you pick in this order because it makes a difference when the ratio is input.
4. Since these wheels turn in opposite directions, pick the **Reverse** button in the **Solution** area.

Figure 13-2.
The angle constraint used to drive the wheel.

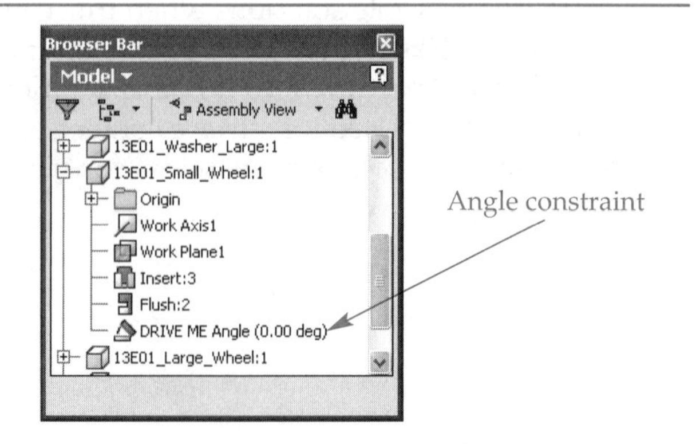

Figure 13-3.
Placing a rotation constraint on the two wheels.

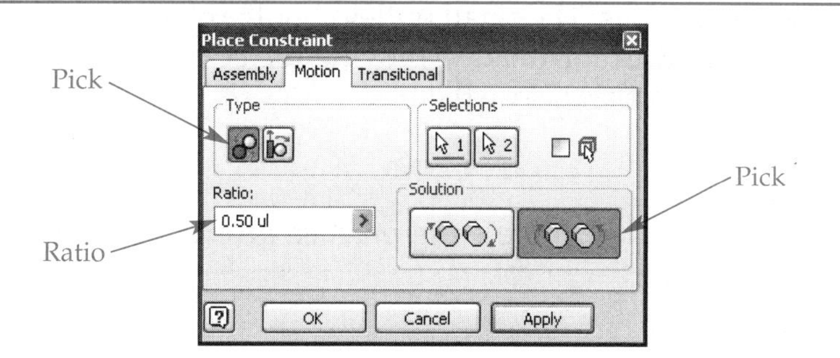

Notice the rotational vectors on the face of the wheels do not necessarily show the correct directions. Also, notice the ratio that was automatically calculated. See **Figure 13-3.** If the flat faces of the wheels were picked, the rotational constraint would work but the ratio would be 1:1. The ratio would have to be manually corrected.

The value of the ratio may seem counterintuitive. You might think the ratio is the angular velocity of the first pick over the second—it is not. It is actually the ratio of the diameters, the first pick (2″) over the second (4″).

Pick the **OK** button to place the constraint. If the **Browser** is set for Modeling, the constraint appears in the Constraints branch. If the **Browser** is set for Assembly, it appears in the 13E01_Small_Wheel branch. Above the Rotation:1 constraint is DRIVE ME Angle. Use this constraint to drive the angle constraint on the small wheel from 0° to 360° with two repetitions. The small wheel rotates twice and the large wheel rotates once. Save and close Example_13_01.iam.

Same Direction of Rotation

Make sure the Default project is active. Then, open Example_13_02.iam. Apply a rotation motion constraint between the curved faces of the two pulleys, picking the small pulley first. However, pick the **Forward** button in the **Solution** area of the **Motion** tab in the **Place Constraint** dialog box. The ratio is automatically calculated as .25, or 1/4. Pick the **OK** button to place the constraint. In the **Browser**, expand the tree for 13E02_Pulley_Pinion. See **Figure 13-4.** Now, drive the angle constraint from 0° to 360°

Figure 13-4.
The belt and pulley assembly.

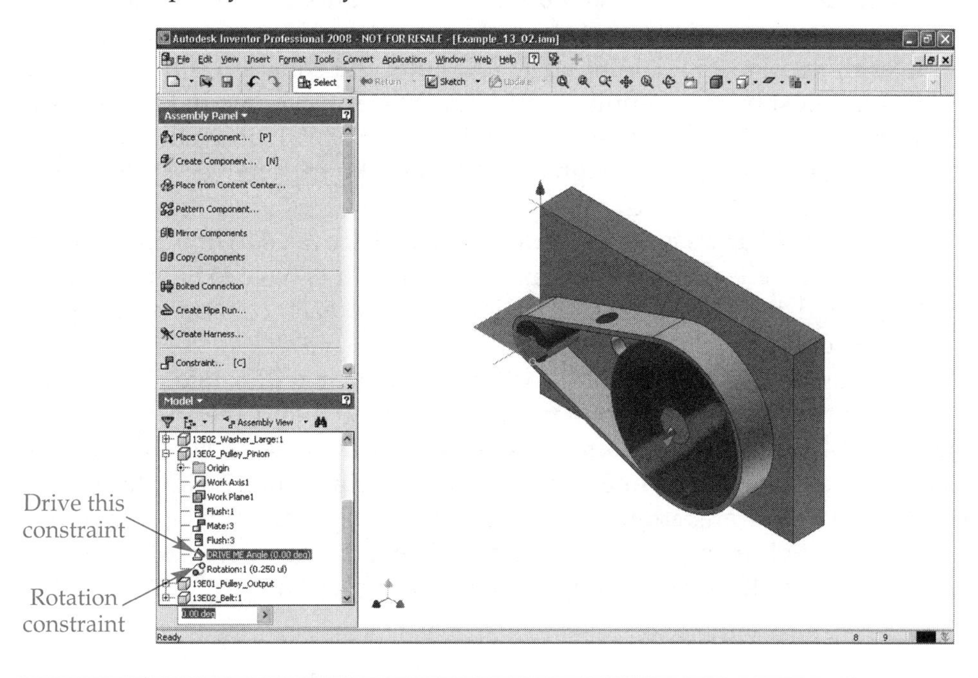

with four repetitions. The small pulley should rotate four times, while the large pulley rotates one time. Unfortunately, moving the belt is something to be figured out. Save Example_13_02.iam and close the file.

Complete the practice problem on the Student CD.

Rotation-Translation Motion Constraint

Make sure the Default project is active. Then, open Example_13_03.iam. The pinion gear is inserted on the shaft. The rack is constrained to the slide on the frame with two mates. Three things need to be done to finish this example. You must constrain the work planes to line up the teeth, apply a motion constraint between the pinion and rack, and apply a mate constraint between the rack and the frame to drive the motion. A ***rotation-translation constraint*** changes rotational motion into linear motion, or linear motion into rotational motion.

Aligning the teeth is first. Apply a mate constraint between the work plane on the pinion and the one at the center of the frame. See **Figure 13-5A.** Rename this constraint Suppress Me 1. Apply a mate constraint between the work plane on the rack and the one on the frame. See **Figure 13-5B.** The rack will slide into position. Rename this constraint Suppress Me 2.

Next, apply a rotation-translation motion constraint between the pinion and the rack. See **Figure 13-6.** Open the **Place Constraint** dialog box and display the **Motion** tab. Pick the **Rotation-Translation** button in the **Type** area. Also, pick the **Forward** button in the **Solution** area. Then, with the **First Selection** button on, pick the pinion on its front face. With this motion constraint, the first part rotates. The other part translates, which is the rack. With the **Second Selection** button on, pick the rack. Select a face or work plane that is perpendicular to the direction of motion, such as the end face of the rack. The vector (arrow) preview should point in the direction of the translation, which is along the length of the frame. As the gear turns, the rack will slide on the green rail.

The value for the relationship between the rotation and translation motions is entered in the **Distance:** text box in the **Motion** tab. The distance is how for the second part (the rack) moves for one rotation of the first part (the pinion). This is equal to the pitch diameter times pi (3.75 × 3.1416), or 11.781. Enter the equation (PI)*3.75 in the **Distance:** text box (pi must be in parentheses and uppercase), as shown in **Figure 13-6.** The value is bidirectional. This means if the rack is driven 11.781 units, the gear will rotate once, or if the gear is driven through one rotation, the rack moves 11.781 units. Pick the **OK** button to apply the constraint.

The constraints called Suppress Me 1 and Suppress Me 2 are used to line up the teeth and prevent any motion of the rack and pinion. Both constraints have to be suppressed before the rack can be driven back and forth. Right-click on each constraint and pick **Suppress** in the pop-up menu. See **Figure 13-7.**

Now, the rack can be moved back and forth and the gear will rotate. To do this, put a mate constraint between the right-hand end of the rack and the inside face of the right-hand end of the frame. Rename this constraint Drive Me. Then, drive this constraint using the settings shown in **Figure 13-8.** If the gear is rotating in the wrong direction, edit the rotation-translation constraint by right-clicking on its name in the **Browser** and selecting **Edit** from the pop-up menu. Then, change the solution in the **Edit Constraint** dialog box. Keep this file open for the next section.

Figure 13-5.
A—Applying a flush mating constraint between the pinion and the frame.
B—Applying a mating constraint between the rack and the frame.

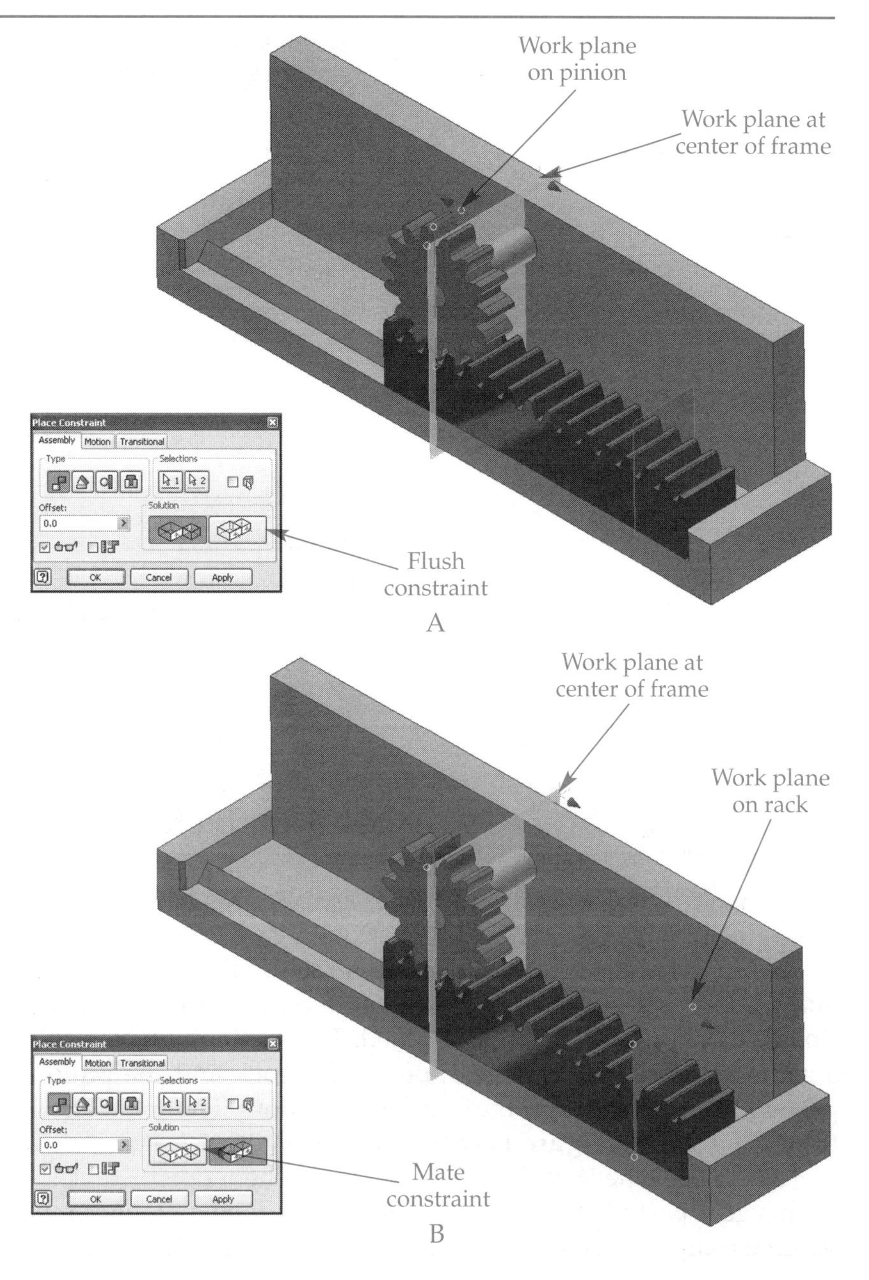

Figure 13-6.
Applying the rotation-translation constraint. The distance value can be input as an equation.

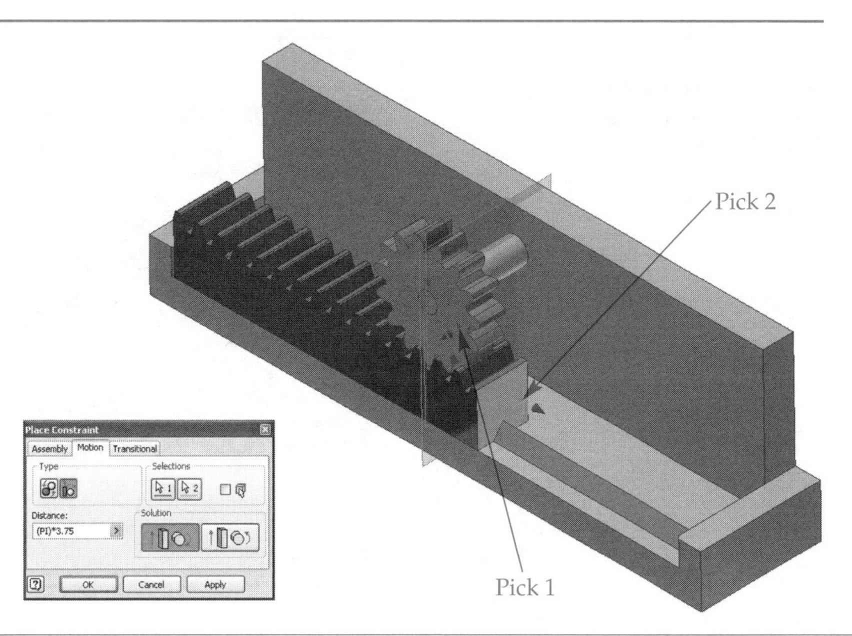

Figure 13-7.
Two constraints need to be suppressed in order to drive the mating constraint between the rack and the frame.

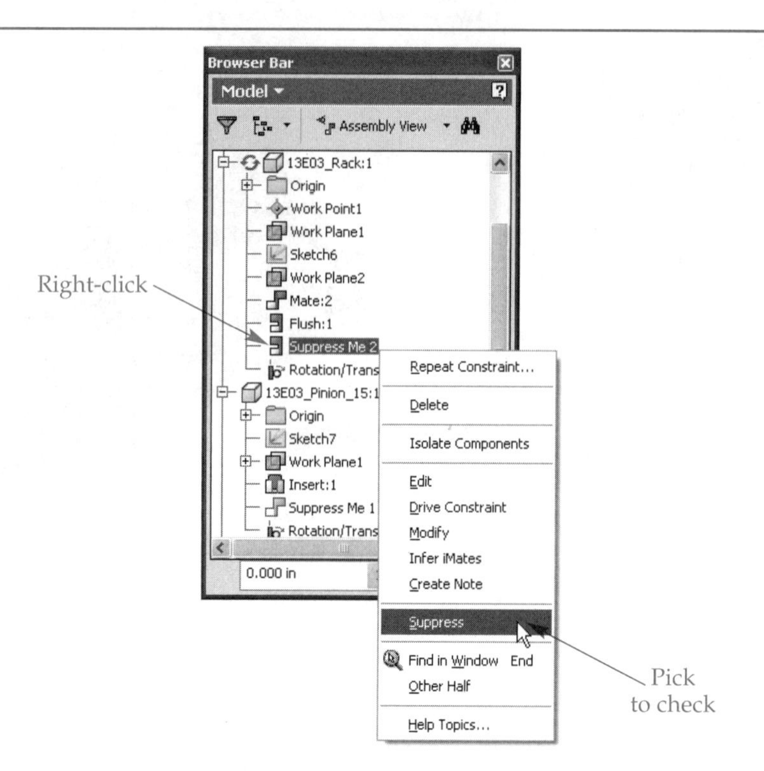

Collision Detector

When driving a constraint, the assembly does not know if the rack collides with the frame or if the gear teeth interfere with each other unless the **Collision Detection** check box is checked in the **Drive Constraint** dialog box. See **Figure 13-8.** In the rack and pinion example, the teeth do interfere. Drive the Drive Me constraint with the **Collision Detection** check box checked. The motion will stop after one increment and a collision detection message appears. The rack and the pinion (the two interfering parts) are highlighted in the graphics window. The interference can be seen in the front view shown, **Figure 13-9.** Save Example_13_03.iam and close the file.

Make sure the Default project is active. Then, open Example_13_04.iam. The teeth in this gear part file have been redesigned so they no longer interfere. Drive the Drive Me constraint. Notice that the end is set at 10″ and collision detection has been turned

Figure 13-8.
The **Drive Constraint** dialog box is used to drive the constraint.

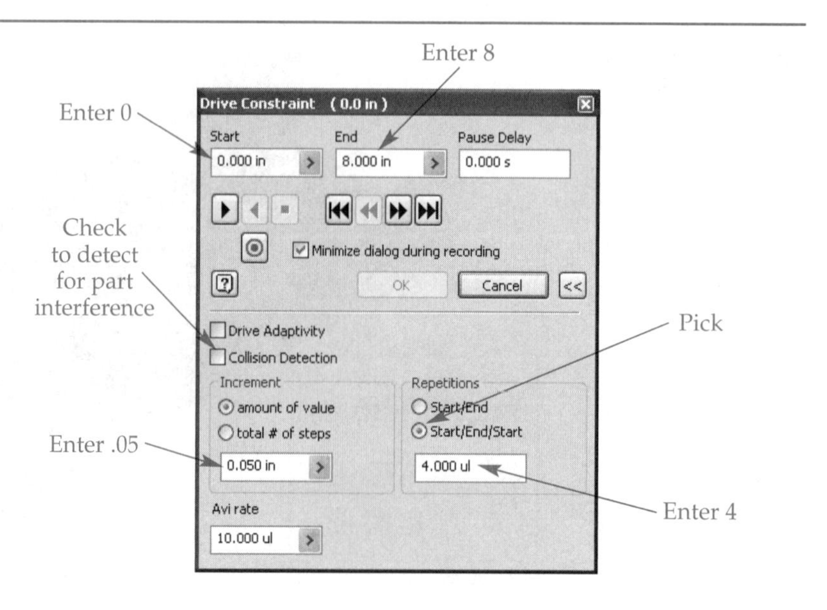

Figure 13-9.
The part interference is evident when a wireframe of the front view is displayed.

on. The collision detector will stop the motion when the rack runs into the left end of the frame. If you set the end to 8″, the interference is corrected and the motion can be previewed with collision detection on.

Rotation-Translation Constraint—Second Solution

Make sure the Default project is active. Then, open Example_13_05.iam. The threaded shaft has right-hand threads and is supported in the frame. See **Figure 13-10.** When the shaft is rotated, the threaded nut moves. If the gear is rotated in a counterclockwise direction (as viewed from the right-hand end of the assembly), the nut moves toward the gear.

Apply a rotation-translation constraint to the assembly, picking the gear as the first selection and then the block as the second selection. Refer to **Figure 13-10.** In the

Figure 13-10.
An assembly with a threaded shaft and threaded nut.

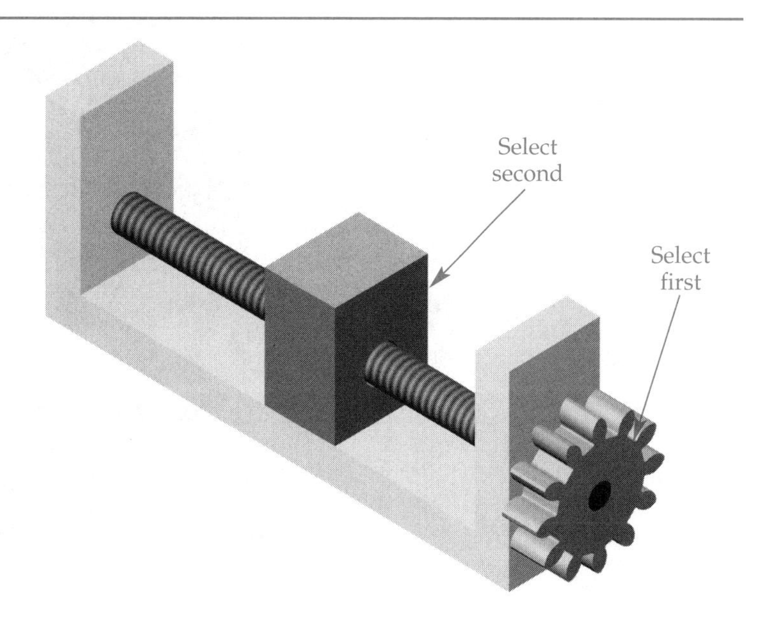

Place Constraint dialog box, pick the **Forward** button in the Solution area of the **Motion** tab. The shaft has 12 threads per inch, so set the **Distance** to 1/12. One rotation of the shaft results in 1/12″ of translation (linear movement). Remember, the direction of the rotation vector on the gear is meaningless. It is just an icon.

Drive the DRIVE ME angular constraint that is on the shaft using the values saved in the file. The block slowly moves toward the gear when the **Forward** button is selected. Edit the Rotation/Translation constraint and pick the **Reverse** button in the **Solution** area of the **Motion** tab in the **Place Constraint** dialog box. This is the correct rotation for left-hand threads. Now, if the gear is driven in the same direction, the block moves away from the gear. Save Example_13_05.iam and close the file.

Transitional Motion Constraint

The ***transitional constraint*** relates the motion of contacting faces on two parts, such as the cam and follower shown in **Figure 13-11.** Make sure the Default project is active. Then, open Example_13_06.iam. This simple cam is a circle with an offset center. The cam rotates on the shaft, pushing on the follower that slides up and down in the frame. Since there is no return spring, this is a low-speed device. It relies on the weight of the follower to maintain contact between the two surfaces. Rotate the view so the bottom cam surface and the bottom face of the follower are visible. See **Figure 13-12.**

Open the **Place Constraint** dialog box and display the **Transitional** tab. The only button in the **Type** area is the **Transitional** button, which should be on. With the **Moving Face** button (labeled **1**) in the **Selection** area on, pick the contact face of the cam. With the **Transitional Face** button (labeled **2**) in the **Selection** area on, pick the bottom of the follower. Once this is done, the follower moves down to contact the cam. Pick the **OK** button to apply the constraint.

Figure 13-11.
A cam and follower assembly.

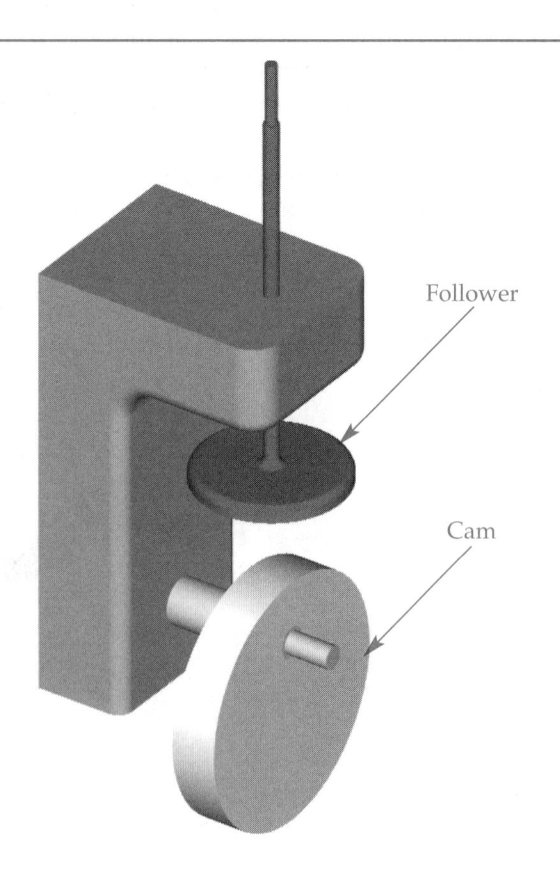

Figure 13-12.
Pick the cam surface as the first selection and the bottom of the follower as the second selection.

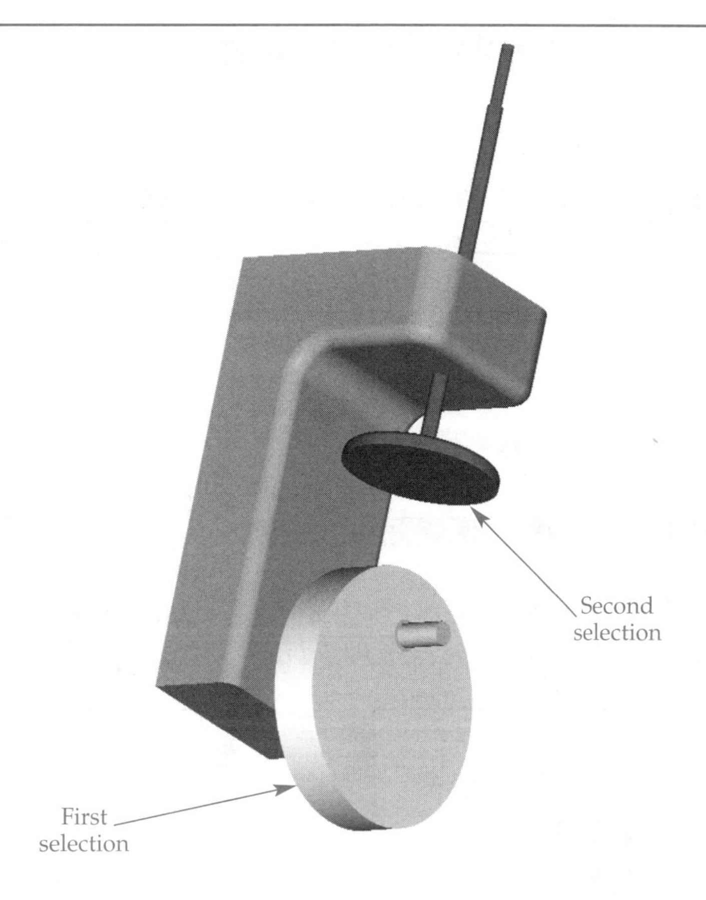

In the **Browser**, expand the tree for 13E06_Shaft. Drive the DRIVE ME Angle constraint. The follower moves up and down, remaining in contact with the face on the cam. Any pair of contacting faces on two parts can be used, as long as the face for the second choice is one face or contiguous multiple faces. The exercises at the end of this chapter provide examples of this. Save Example_13_06.iam and close the file.

Chapter Test

Answer the following questions on a separate sheet of paper or complete the electronic chapter test on the Student CD.

1. What are the three motion constraints that relate relative motion of one component to another?
2. The ______ motion constraint is used to transfer the rotation of one part to another part, such as one gear driving a second gear.
3. The ______ motion constraint is used to translate the rotation of one part into linear movement of another part, such as a gear driving a rack.
4. The ______ motion constraint is used to maintain contact between two surfaces, such as between a cam and a follower.
5. What can be used to determine if components in motion interfere with each other?

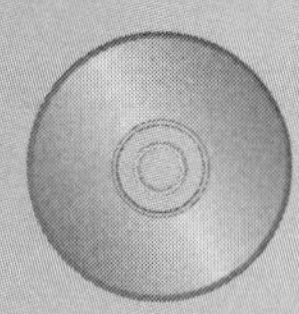

Chapter Exercises

Exercise 13-1. Three Gears.

Complete the exercise on the Student CD

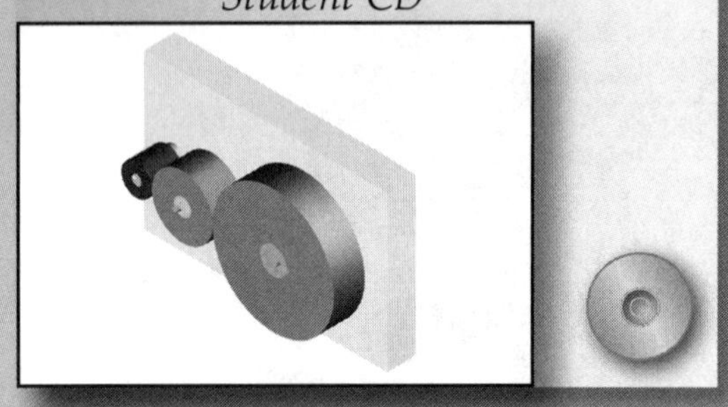

Exercise 13-2. Four Gears.

Complete the exercise on the Student CD

Exercise 13-3. Planetary Gear Set One.

Complete the exercise on the Student CD

Exercise 13-4. Planetary Gear Set Two.

Complete the exercise on the Student CD

Exercise 13-5. Sliding Blocks.

Complete the exercise on the Student CD

Exercise 13-6. Latch Bolt.

Complete the exercise on the Student CD

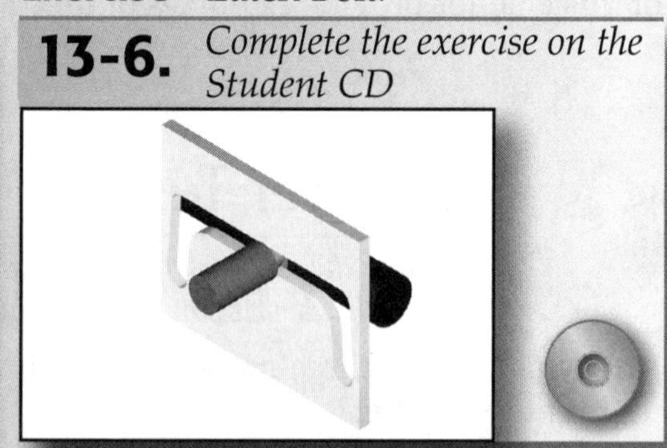

Exercise 13-7. Scissors Lift.

Complete the exercise on the Student CD

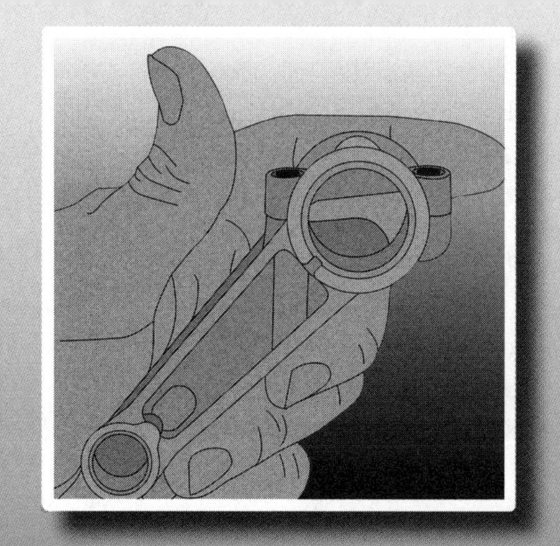

Chapter 14

iParts and Factories

Objectives

After completing this chapter, you will be able to:

- Explain the process of creating a part from an iPart using a factory.
- Create parts from iParts using a factory.
- Explain iMates.
- Create and use iMates.
- Create an iPart from scratch.

User's Files

The Student CD included with this text contains several files required for this chapter. Refer to the file File List.txt *in the* \Ch14 *folder for the comprehensive list.*

Inventor allows you to create and use iParts. An ***iPart*** is an "intelligent" part with parameters that are tied to a spreadsheet, allowing multiple sizes and configurations from a single part. All of the possible variations produce a group of similar parts called a ***family of parts.*** For example, a basic socket-head cap screw can be created. Then, its dimensions, such as length and diameter, can be tied to a spreadsheet that provides data for all variations in the group of similar parts. By using iParts, you can:

- Create parts with versions.
- Develop libraries of standardized parts.
- Create a basic design and use it in many different assemblies by varying data.
- Adhere to corporate specifications for design consistency in assemblies.

The process of placing an iPart into an assembly is called *publishing* in a ***factory*** because you assemble the specifications for, or "manufacture," the part being inserted from the iPart "raw material." There are two types of iParts—standard iParts and custom iParts. A ***standard iPart*** cannot be altered from a set list of parameters as it is inserted into an assembly. A ***custom iPart*** has certain aspects of its definition that can be manually entered when the part is inserted into an assembly. For example, an iPart bolt may allow the designer to enter a custom value for the bolt length, rather than select from a list, as the part is inserted. This would be considered a custom iPart.

An *iMate* is a named, partial constraint definition that allows a part to be automatically constrained after it has been inserted into an assembly. The constraint within an iMate will have an "i" added to the constraint name, such as iInsert, iAngle, and so on. In order to use an iMate, there must be a matching iMate in the assembly when the part is inserted. Multiple constraints can be defined in an iMate and applied as a group. This is called a ***composite iMate.***

In this chapter, you will insert custom parts into an assembly by manufacturing them in a factory from iParts. You will also create your own iParts from scratch. Finally, you will create iMates, apply them to parts, and insert the parts into an assembly using the iMate feature.

Assembling iParts

In this example, you will build a piping system off a storage tank. See **Figure 14-1.** Most of the various fittings you will add are iParts that have been created for you. Open Example_14_01.iam and zoom in on the flanged outlet on the side of the tank.

Placing an iPart

All iParts for this example are located in the \Ch14\ Examples\ Example_14_01 folder. The first step is to put a 10″ long, 12″ nominal diameter pipe on the end of the 12″ OD flange. A custom iPart called Pipe.ipt will be used for this. Using the **Place Component...** tool, place an instance of the iPart Pipe.ipt into the assembly. In the **Place Component** dialog box, do *not* check the **Use iMate** check box. See **Figure 14-2.**

Normally, when you pick the **Open** button in the **Place Component** dialog box, the dialog box is closed and a preview of the part appears attached to the cursor. However, since this is an iPart, an additional dialog box is displayed in which you can specify values for the part. See **Figure 14-3.** You can change the values by picking in the **Value** column for the appropriate property. Pick the value next to PipeNominal; a list box appears. See **Figure 14-4.** Select 12 from the list. Next, pick the value next to PipeLength. This time, a text box appears in place of the value. Type 10 and press [Enter]. Since you can manually enter a property, this is a custom iPart.

As you make settings in the dialog box, you are actually defining a part for which there is no part file (IPT). You need to specify where to save the file. By default, the part file will be saved in the same folder where the assembly file (IAM) is located. However, a generic name, such as Part1, is specified. To change the name and/or location, pick the **Browse...** button. The **Save As** dialog box appears in which the name and location can be specified.

Figure 14-1.
This is the completed assembly. You will add the components from the flange outward.

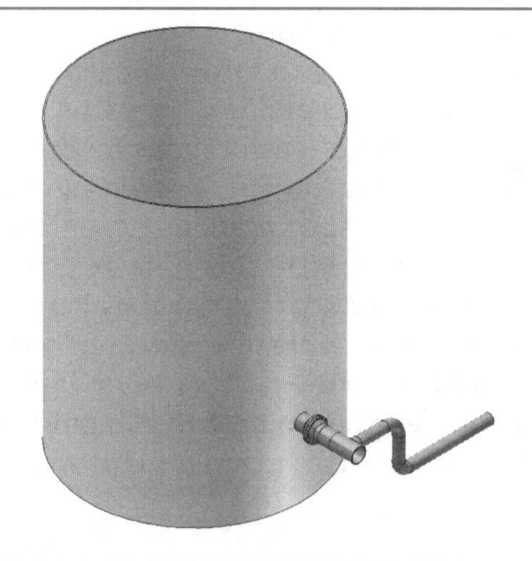

Figure 14-2.
Selecting an iPart for placement in an assembly.

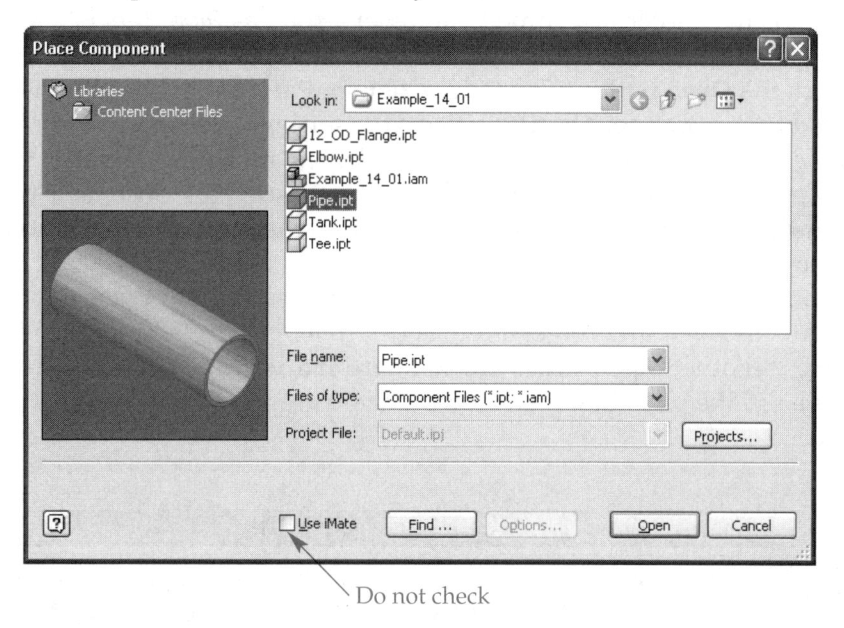

Figure 14-3.
Specifying the part parameters for a custom iPart.

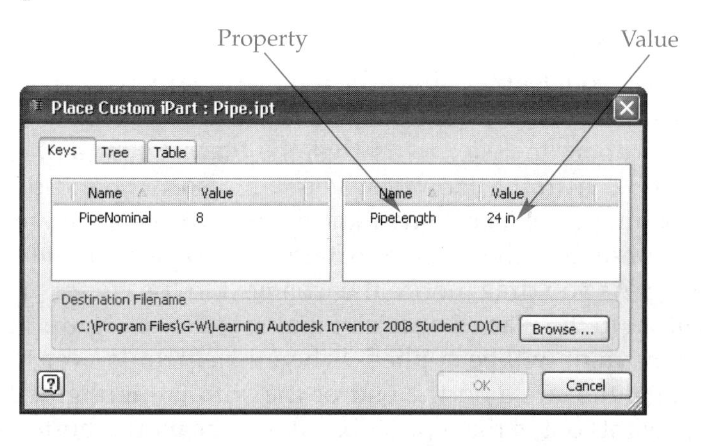

Figure 14-4.
Setting a parameter by selecting a value from a set list of options.

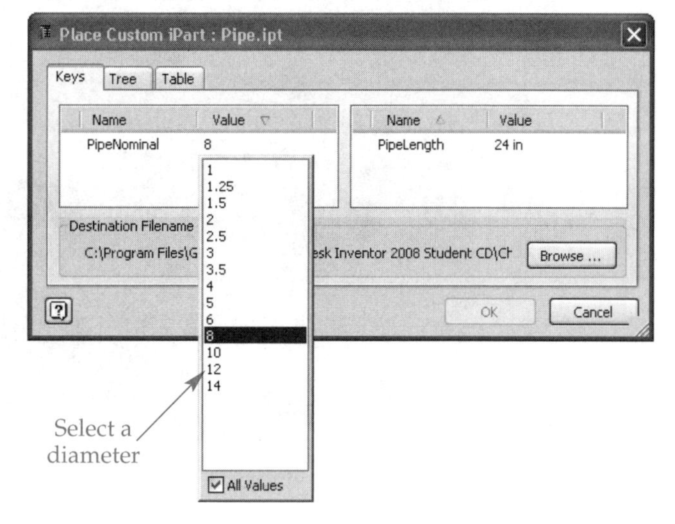

For this example, pick the **Create New Folder** button in the **Save As** dialog box and create a new subfolder named Pipe. In a complex assembly with many parts, it is a good idea to create logically named subfolders, such as Pipe, Tee, Elbow, and so on. Then, navigate to the new subfolder, enter the name Pipe12x10 in the **File name:** text box, and pick the **Open** button. The **Place Custom iPart** dialog box is redisplayed and the path next to the **Browse...** button reflects the setting.

With the **Place Custom iPart** dialog box still open, move the cursor in the graphics window. A preview of the part is attached to the cursor. Pick once near the flange to place one instance. Then, pick the **Dismiss** button in the **Place Custom iPart** dialog box or press [Esc].

The pipe is different from all of the other fittings you will place in this example because it is a custom iPart. You entered a custom length for the part. This length can be any value. All other pipe fittings are standard iParts. You will select sizes from list boxes, as you did the pipe diameter. However, you are limited to the sizes provided in the list box.

PROFESSIONAL TIP

The process of saving a custom iPart as it is placed or creating an iPart in a factory is called "publishing" the part. Keep this in mind when using the Inventor documentation. The term is used often without much explanation.

Constraining the iPart

In Chapter 11, you learned how to remove degrees of freedom by applying constraints to parts. One of the methods you learned was to "drag-and-drop" the part into the correct location. In order to do this, the **Place Constraint** dialog box is open and set up for the constraint. However, you can also drag-and-drop to constrain a part without opening the **Place Constraint** dialog box. Hold down the [Alt] key and pick the circle representing the outside surface of the pipe near one end. Drag the pipe toward the flange. As you drag, notice the icon next to the cursor. It appears the same as the **Insert** button in the **Place Constraint** dialog box. See **Figure 14-5.** This indicates that an insert constraint will be applied. When the cursor is over the end of the pipe stub on the flange, the circle on the end of the stub is highlighted red and the red direction arrow points out of the pipe stub. Release the mouse button to place an insert constraint between the pipe and the flange.

Figure 14-5.
Press the [Alt] key and then constrain the pipe using iMates.

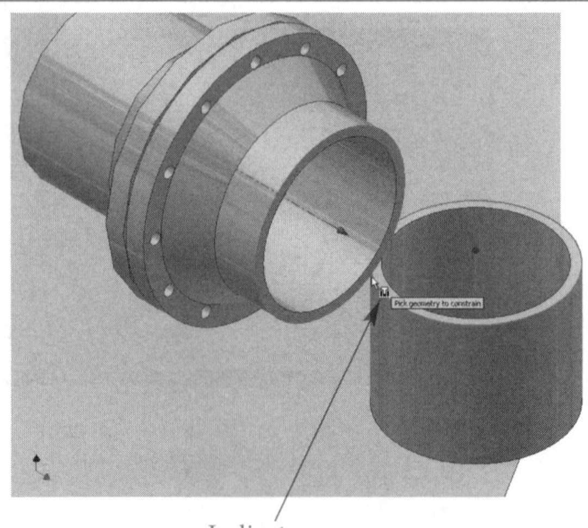

The pipe is now properly aligned with the pipe stub on the flange. However, the pipe is still free to rotate about the centerline of the flange. This could cause alignment problems for subsequent parts. To prevent the pipe from rotating, apply a flush mating constraint between the XZ plane of the pipe and XZ plane of the flange. To do this, expand the Origin branch in the **Browser** for both the flange and the pipe. Then, open the **Place Constraint** dialog box by picking the **Constraint...** button in the **Assembly Panel**. Using the two buttons in the **Selections** area, pick the two planes in the **Browser**. Finally, pick the **Flush** button in the **Solution** area of the dialog box and apply the constraint. The pipe is now completely constrained and has zero degrees of freedom.

Placed iPart

Now, take a look at the assembly tree in the **Browser**. See **Figure 14-6A.** The part is listed with the name you specified when placing the iPart, which is Pipe12x10. The :1 indicates that it is the first instance of this part file. Since this version of the part is now saved using the file name and path specified earlier, additional instances of the part can be inserted by placing the IPT file into the assembly. Each additional instance will have a sequential number added to its name.

The Table branch below the Pipe12x10 branch displays the possible selections that appeared in the list box for the diameter when the part was placed. Right-click on the Table branch and select **Change Component** from the pop-up menu. The **Place Custom iPart** dialog box is displayed. On the **Table** tab, the size you selected is highlighted, as shown in **Figure 14-6B.** The size or length of the pipe can be changed. However, if you do so, the part file will no longer reflect the logical file name you provided.

Other branches below Pipe12x10:1 include Origin, iMates, and the two constraints that were applied. The Origin branch contains the three standard planes and axes for the part. The constraints can be edited in the same way as for other parts. The iMates branch contains the automated constraints defined by the iMate. Constraining with iMates is discussed in the next section.

Constraining iMates

Now, you will place a reducing tee into the assembly. Refer to **Figure 14-1.** Pick the **Place Component...** button in the **Assembly Panel**. In the **Place Component** dialog box,

Figure 14-6.
A—The pipe in the assembly tree. B—The available part variations in the family of parts.

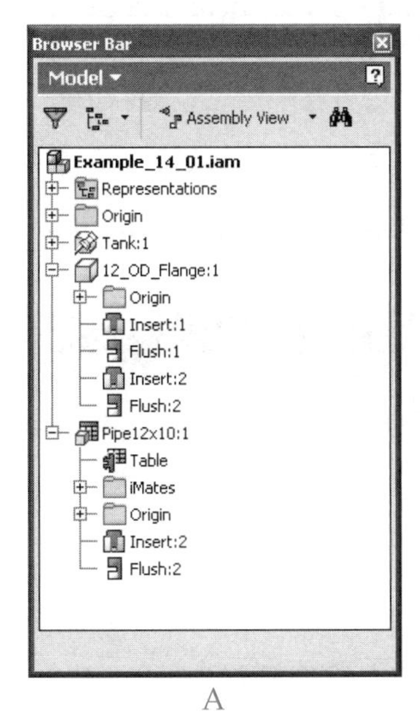

A

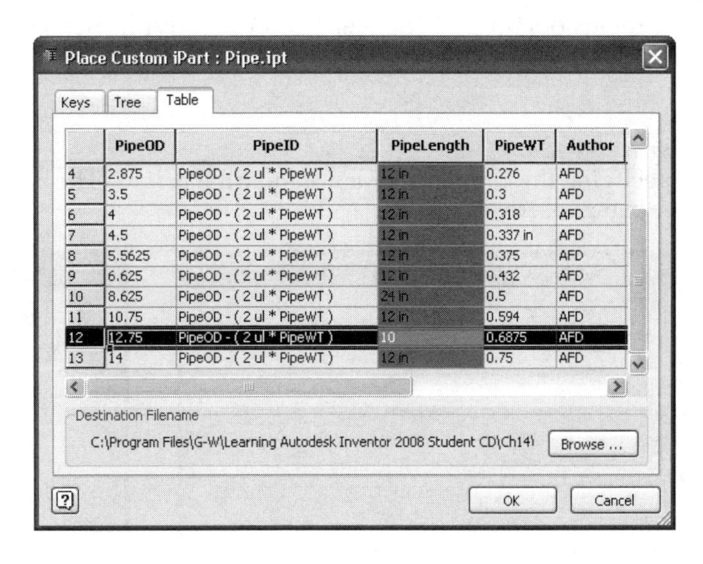

	PipeOD	PipeID	PipeLength	PipeWT	Author
4	2.875	PipeOD - (2 ul * PipeWT)	12 in	0.276	AFD
5	3.5	PipeOD - (2 ul * PipeWT)	12 in	0.3	AFD
6	4	PipeOD - (2 ul * PipeWT)	12 in	0.318	AFD
7	4.5	PipeOD - (2 ul * PipeWT)	12 in	0.337 in	AFD
8	5.5625	PipeOD - (2 ul * PipeWT)	12 in	0.375	AFD
9	6.625	PipeOD - (2 ul * PipeWT)	12 in	0.432	AFD
10	8.625	PipeOD - (2 ul * PipeWT)	24 in	0.5	AFD
11	10.75	PipeOD - (2 ul * PipeWT)	12 in	0.594	AFD
12	12.75	PipeOD - (2 ul * PipeWT)	10	0.6875	AFD
13	14	PipeOD - (2 ul * PipeWT)	12 in	0.75	AFD

B

navigate to the \Ch14\ Examples\ Example_14_01 folder and select Tee.ipt. Even though you will be using iMates, do not check the **Use iMate** check box in the **Place Component** dialog box. When you pick the **Open** button, the **Place Standard iPart** dialog box is displayed, **Figure 14-7.**

Pick the value next to BasePipeNom to display a list box. Select 12 from the list box. This is the diameter of the pipe you inserted and constrained to the flange. Then, pick the value next to StubPipeNom to display a list box. Select 8 from the list box. This is the reduced diameter of the tee.

Notice how both of the settings provide list boxes, not text boxes. Since the tee is a standard iPart, there are no user-defined settings. All settings must be selected from available values or options. Also, notice that you do not need to specify a file name and location for the part. However, Inventor will create a subfolder named \Tee and save this version of the part, a 12″ × 8″ tee, in the folder under a descriptive name. Finally, pick once near the pipe in the graphics window to place one instance of the tee. Press [Esc] to close the **Place Standard iPart** dialog box.

The tee has defined iMates. You can verify this by looking at the assembly tree in the **Browser**. Notice that the Tee branch of the assembly tree indicates the selected values in its name. Also, there is an iMates branch below it. This indicates that iMates have been defined for the part. The pipe you inserted earlier also had an iMates branch because iMates were defined for it.

If you expand the iMates branch for the Tee, you will see that three iMates have been defined. See **Figure 14-8.** These three iMates happen to be composite iMates, which are discussed later. If you expand the individual iMate branches, you can see which constraints are contained within the composite iMate, as shown in the figure.

When a part that has iMates is selected, icons appear on the part in the graphics window indicating where the iMate constraints are located. See **Figure 14-9.** Inventor calls these icons ***iMate glyphs.***

By default, the iMate glyphs are only displayed for a part when it is selected. However, they can be displayed for a part when it is not selected. Right-click on the tee in the graphics window to display the pop-up menu. See **Figure 14-10.** Select **iMate Glyph Visibility** to display the iMate glyphs on the part, even when the part is not selected. A check mark appears next to the menu item when the display is turned on. Pick the menu item again to turn off the display of the iMate glyphs except when the part is selected. For this example, turn on the iMate glyph display for both the tee and the pipe.

Now, you are ready to constrain the tee to the pipe using the iMates set up for each part. Hold down the [Alt] key and pick and hold on the iMate glyph near the large-diameter opening of the tee. The iMate glyph turns green. You may need to move the cursor slightly for the iMate glyph to be displayed in green. Drag the tee by the iMate

Figure 14-7.
Specifying the parameters for a standard iPart. A—The default values. B—The values for the tee being inserted.

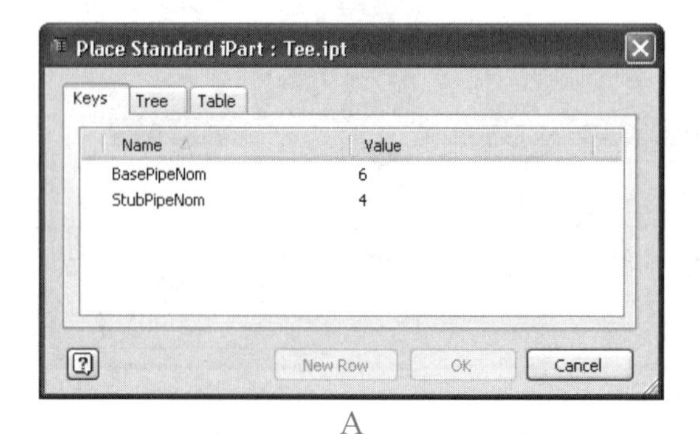

A

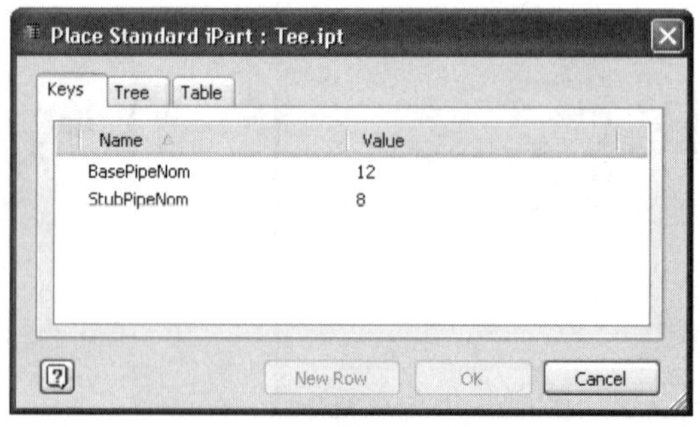

B

Figure 14-8.
The iMates for the tee. These are composite iMates. The constraints contained within a composite iMate are displayed below it in the tree.

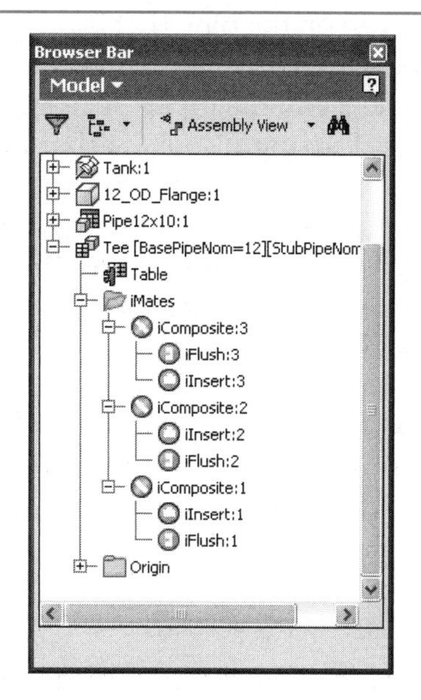

Figure 14-9.
The iMate glyphs indicate where iMates are located.

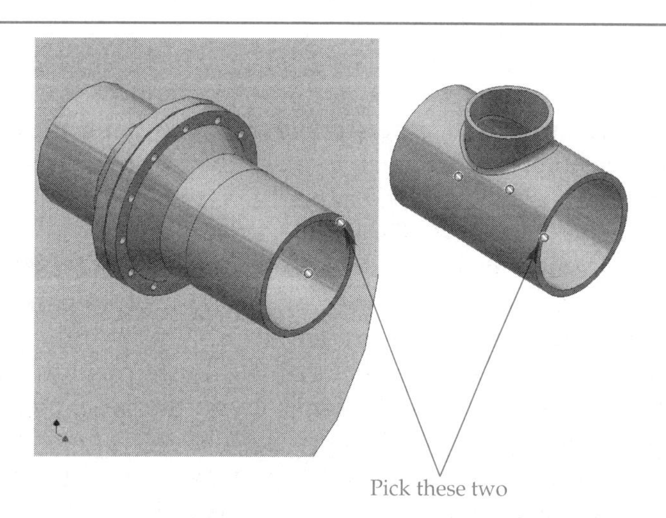

Figure 14-10.
Displaying iMate glyphs for a part.

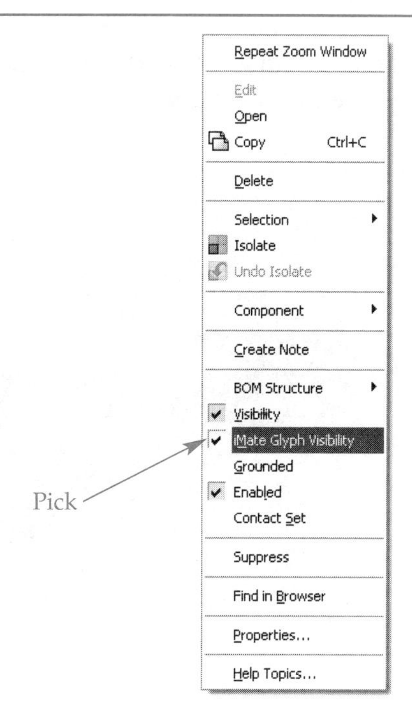

glyph to the iMate glyph on the pipe that is near the edge. Refer to **Figure 14-9.** When the two iMate glyphs meet, they turn red and the tee is properly aligned. Release the mouse button to constrain the tee to the pipe. Once the tee is in place, release the [Alt] key.

A big advantage of using iMates is that all six of the degrees of freedom can be eliminated on a part in a single step. Because of the composite iMates set up for the tee and the pipe, the tee is now fully constrained. All six degrees of freedom were eliminated simply by dragging the tee into the correct location using the iMates. If you try to move the tee now, you will not be able to do so.

PROFESSIONAL TIP

As you drag using iMates, only the valid iMates on the mating part are displayed.

Placing the Remaining Parts

There are several more parts to be added to the assembly. These include a series of elbows and pipes added to the reduced-diameter pipe run from the tee. Zoom and pan as needed as you work.

Place one instance of the part Pipe.ipt, which is the same part file you used earlier. This time, set PipeNominal to 8 and PipeLength to 24. Specify the filename as Pipe8x24 and the location as the \Pipe subfolder. If you get a message asking to save the file and its dependents as you pick to place the instance, pick the **OK** button. Then, press [Esc] to close the dialog box. Using iMates, constrain the pipe to the small diameter connection on the tee. See **Figure 14-11.**

Next, place an instance of the part Elbow.ipt. Set PipeNominal to 8 and ElbowRad to 12. Since this is a standard iPart, you do not need to name it. However, keep in mind that Inventor will create a subfolder named \Elbow and save this version of the part in it. Using iMates, constrain either end of the elbow to the pipe you just placed. There will be only one iMate glyph displayed on the pipe. Pick it as the target even if the iMate glyph is displayed at the opposite end of the pipe. The elbow will automatically point down when constrained. If the preview points up, drag the elbow away from the pipe and release the mouse button. Then, repeat the process, but select the iMate glyph on the opposite end of the elbow. Continue as follows.

1. Place another instance of the part Pipe.ipt. Set PipeNominal to 8 and PipeLength to 20. Specify the filename as Pipe8x20 and the location as the \Pipe folder. Using iMates, constrain the pipe to the elbow.

Figure 14-11.
Constraining the pipe to the small-diameter connection on the tee.

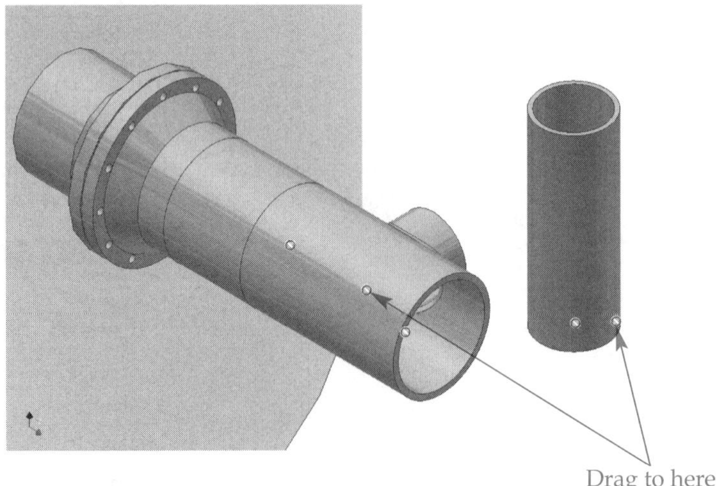

Drag to here

2. Place another instance of the part Elbow.ipt. Set **PipeNominal** to 8 and **ElbowRad** to 8. Using iMates, constrain the elbow to the pipe you just placed. Refer to **Figure 14-1** for the orientation.
3. Place another instance of the part Pipe.ipt. Set **PipeNominal** to 8 and **PipeLength** to 72. Specify the filename as **Pipe8x72** and the location as the **\Pipe** folder. Using iMates, constrain the pipe to the elbow.

The reduced-diameter pipe run is complete.

Creating an iPart from Scratch

So far, you have used existing iParts. However, you can create your own iParts from scratch. For example, you will now create a gate valve symbol as an iPart for use in a piping application. The basic process is:

1. Create the part.
2. Constrain the part.
3. Select the iParts authoring tool.
4. Build the spreadsheet for the family of parts.
5. Publish the iPart.

When creating the part, pay close attention to the relationship between part features. Also, fully constrain everything. Try to have all dimensions relate back to one or two key features. For example, the thread diameter on a bolt may be a "base" dimension for a family of bolts. A pipe OD may be used as the "base" dimension for a family of pipe fittings. Once the part is modeled and constrained, the spreadsheet built, and the part published, you have a factory for that iPart.

Modeling the Part

Open Example_14_02.ipt. This is a symbolic model of a valve that is fully dimensioned and constrained. Otherwise, you would start with a design, create sketches, and then turn the sketches into features. However, you will need to complete the iPart definition.

The first thing you need to do is change a couple of dimension equations and add comments to several dimensions. Select **Tools>Parameters** to display the **Parameters** dialog box. To rename a dimension parameter, simply pick on its name to display a text box. Then, type the new name and press [Enter]. Several dimensions have been renamed for you and comments provided. Comments are added by picking in the **Comment** column to display the text box. Alter equations and add comments as indicated in **Figure 14-12.** Also, add a user parameter named PipeNominal with a nominal value of 1. Then, close the **Parameters** dialog box. Now, these parameters will make sense to anyone using this iPart.

Creating iMates

The next step is to create iMates on the valve. It is not required for an iPart to have iMates, but iMates can greatly simplify the assembly process, as you saw earlier. Pick the **Create iMate** button in the **Part Features** panel. The **Create iMate** dialog box is opened, **Figure 14-13.** This dialog box is similar to the **Place Constraints** dialog box. However, there are two major exceptions. The **Transitional** tab is not included. Also, in the **Selections** area there is only one button. Remember, an iMate is one-half of a constraint definition.

Pick the **Insert** button in the **Type** area of the **Assembly** tab. Then, with the button in the **Selections** area on, pick the inside cylindrical edge of the valve body. Refer to **Figure 14-13.** The centerline should be highlighted in red. Pick the **Apply** button in the dialog box to apply the constraint and leave the dialog box open. An iMate glyph now appears at the end of the valve.

Figure 14-12.
Change the dimension parameters as indicated in this chart. The highlighted cells are those that need to be edited.

Parameter Name	Unit	Equation	Nominal Value	Tol.	Model Value	Comment
PipeOD	in	4.5 in	4.500000		4.500000	Outside diameter of the pipe.
PipeWT	in	0.337 in	0.337000		0.337000	Pipe wall thickness—SCH80.
d2	deg	35 deg	35.000000		35.000000	Slope of valve body.
Half_length	in	4 in	4.000000		4.000000	Half of the length of the valve body.
d5	in	PipeWT	0.337000		0.337000	Wall thickness on slope.
d7	deg	30 deg	30.000000		30.000000	Angle for the stem's work plane.
d8	in	0.4 in	0.400000		0.400000	Stem diameter.
d9	in	PipeOD*2 ul	9.000000		9.000000	Stem extrusion height.
d10	deg	0 deg	0.000000		0.000000	Stem extrusion angle.
d11	in	PipeOD	4.500000		4.500000	Handle diameter.
d12	in	0.25 in	0.250000		0.250000	Handle extrusion height.
d13	deg	0 deg	0.000000		0.000000	Handle extrusion angle.
PipeNominal	in	1.0 in	1.000000		1.000000	Nominal pipe diameter.

Rotate the view to the opposite end of the valve. Then, apply an insert constraint to the other end in the same manner. Pick the **OK** button to apply the constraint and close the dialog box. An iMate glyph is displayed on this end of the valve as well. Expand the iMates branch of the part tree in the **Browser**. There should be two constraints listed, each named iInsert and a sequential number.

You will now add a mating constraint to fully constrain the valve when the iMates are used to constrain it within an assembly. Select the **Create iMate** button again. Pick the **Mate** button in the **Type** area of the **Create iMate** dialog box. Also, pick the **Flush** button in the **Solution** area. Select the XZ plane in the Origin branch in the **Browser** and then pick the **Apply** button. Now, exactly repeat this step and then close the dialog box. There should be two constraints named iFlush in the **Browser**. The flush constraint was applied twice so one can be used on each end of the valve.

Now, you must create the composite iMates. Select iInsert:1 in the **Browser**, hold down the [Ctrl] key, and select iFlush:1. The two glyphs on the part turn green. Right-click in the **Browser** or graphics window to display the pop-up menu and select **Create Composite**. The two separate constraints are replaced by one named iComposite. If you expand this branch, the two constraints are listed below it. Create another composite constraint from iInsert:2 and iFlush:2. Now, there should be two iComposite constraints in the iMates branch, one for each end of the valve.

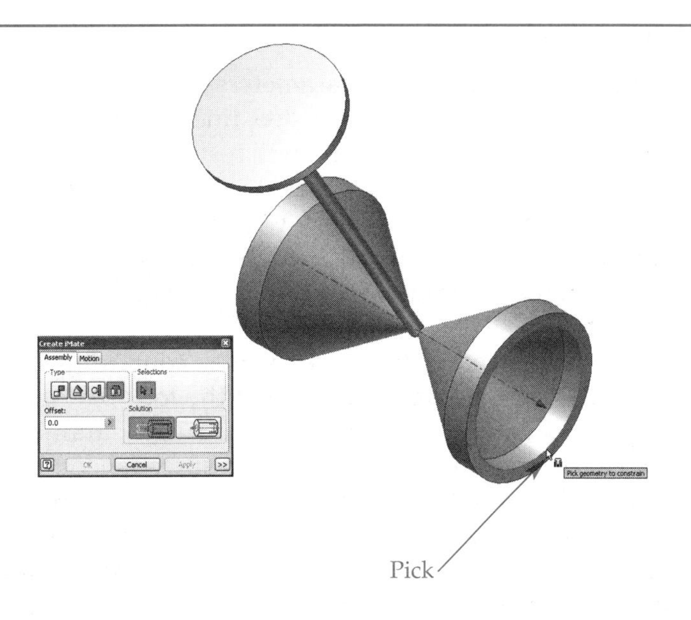

Figure 14-13.
Creating an iMate constraint.

PROFESSIONAL TIP

You can have Inventor create iMates (composite or not) from the assembly constraints on a part already in the assembly. After the part is inserted and constrained, right-click on the name of the part in the assembly tree and select **Infer iMates...** from the pop-up menu. Then, make appropriate settings in the **Infer iMates** dialog box.

iPart Authoring

So far, you have prepared the part to be an iPart. However, you have not yet *created* an iPart. If you save the file now, the valve can be placed into an assembly. However, you cannot "manufacture" a series of valves in a factory. In order to manufacture different valve sizes for use in assemblies, you must author (publish) the iPart. Select **Create iPart...** from the **Tools** pull-down menu. The **iPart Author** dialog box is displayed, **Figure 14-14.**

Figure 14-14.
The **Parameters** tab of the **iPart Author** dialog box.

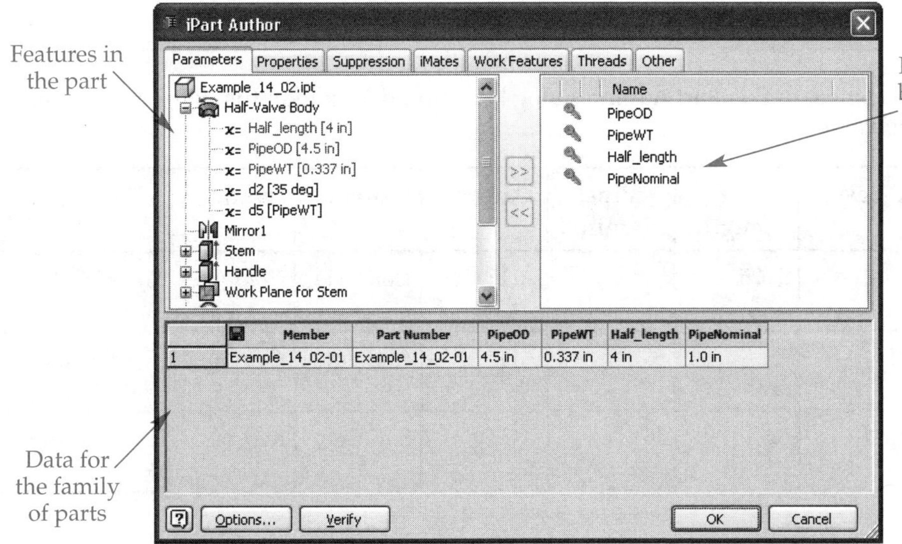

Parameters tab

The upper-left pane of the **Parameters** tab lists all of the features that make up the part. Dimension parameters are listed underneath each feature's name. The upper-right pane lists the parameters that will be used to manufacture the part in the factory. Currently, the four parameters that do not have default names are listed. Inventor assumes that renamed parameters will be used to manufacture the part. To add a parameter to the upper-right pane, highlight it in the upper-left pane and pick the **>>** button. To remove a parameter from the upper-right pane, highlight it on the right and pick the **<<** button.

The bottom pane is the spreadsheet containing all of the values needed to generate the family of parts. Notice that the four parameters listed in the right-hand pane appear as column headings. As you add parameters to the right-hand pane, they appear as column headings in the spreadsheet. Each row in the spreadsheet represents a different version of the iPart. If there are five different versions, there will be five rows. Since there is currently only one version of the part, there is only one row. An intersection between a row and a column is called a cell. The cell contains the data corresponding to the column. A cell can contain a value, Inventor equation, or Microsoft Excel formula.

Entering Data for the Family of Parts. For this family of valves, there will be three possible versions. Therefore, you need to have a total of three rows. However, there is currently only one row. Right-click anywhere on the row, except in the current cell, and pick Insert Row from the pop-up menu. A new row is added. Do this one more time so there are three rows; notice the sequential numbers on the left of the spreadsheet.

Notice how the data in the cells are identical for each row. The data now need to be edited to reflect the family of parts. Pipe and related components, such as valves, are specified by the nominal diameter. The actual dimensions of a pipe—the inside diameter (ID) and outside diameter (OD)—are based on the nominal size. For example, a 4″ nominal pipe fitting has a 4.5″ OD. Refer to the chart in **Figure 14-15.** Enter the information shown in the chart for the PipeOD, PipeWT, Half_length, and PipeNominal columns. You will add the other columns shown in the figure later.

PROFESSIONAL TIP

When editing the cells within a row, start with the left-hand cell. Then, using the number pad on your keyboard, type the number and press the [Enter] key on the number pad. The next cell to the right is automatically made current for editing.

Figure 14-15.
The final spreadsheet in the **iPart Author** dialog box should look like this. Refer to this chart when entering values.

PipeOD	PipeWT	Half_ Length	Pipe Nominal	Author.	Stem	Handle	Description
6.625	0.432	5.625	6	ADF	Compute	Compute	6″ OD Gate Valve
8.625	0.5	7	8	ADF	Compute	Compute	8″ OD Gate Valve
10.75	0.594	8.5	10	ADF	Supress	Supress	10″ OD Gate Valve

Setting the Key. Notice the key icons next to the parameter names in the right-hand pane of the **Parameters** tab. Left-click on the key icon next to PipeNominal. The key turns blue and the number 1 appears next to the key. The column in the spreadsheet for the PipeNominal parameter is now the key column. A ***key column*** is used to select which version of the part will be "manufactured." Since the PipeNominal column is the key column, you will select the nominal diameter of the pipe when placing the valve into an assembly. The pipe fittings you placed into the tank assembly earlier in this chapter had the nominal pipe diameter set up as the key.

There may be multiple keys (or key columns), one primary and up to eight secondaries. For example, the tee you inserted into the tank assembly had a primary key for the base nominal OD and a secondary key for the allowable sizes based on the primary key. The primary key *filters* the available selections for the secondary keys.

Setting the Default Version. A default version can be set up for the iPart. For example, if most of the gate valves you will be using are for 8" pipe, you can have the settings for this version displayed in the "place iPart" dialog box. Right-click on the row that contains the value 8 in the PipeNominal column. Then, select **Set as Default Row** from the pop-up menu. The background of the row is changed to a light green, indicating this row defines the default version.

Setting Custom Parameters. There are two basic types of iParts—custom and standard. After a custom iPart is placed into an assembly, additional features may be added. Additional features cannot be added to a standard iPart. To create a custom iPart, right-click on a column and select **Custom Parameter Column** from the pop-up menu. The background for the column changes to dark blue to indicate that the value for this column can be changed as the part is placed into an assembly. For example, the pipe iPart from the tank assembly allowed you to enter a pipe length. The pipe iPart has a length column set up as a custom parameter column. A custom iPart can have multiple custom parameter columns. To disable the custom feature for a column, right-click on the column and select **Custom Parameter Column** in the pop-up menu to uncheck it. Since you are creating a standard iPart, no columns should be set up as custom parameter columns.

Properties tab

The properties listed in the **Summary**, **Project**, and **Physical** tabs of the **Properties** dialog box can be added to an iPart definition. These properties appear in a tree in the left-hand panel of the **Properties** tab of the **iPart Author** dialog box. If you add any of these properties to the iPart definition, they will be available when a drawing (IDW) is created.

Highlight Author in the Summary branch in the left-hand pane. Then, pick the **>>** button to place it in the right-hand pane. Notice that a column has been added to the spreadsheet. See **Figure 14-16.** The column is named Author and each cell in the column displays the author of the IPT file you have open. These cells can be edited, if needed.

Suppression tab

The **Suppression** tab is used to set up certain features to be excluded from certain versions of the part. For example, the 10" valve does not have a stem and handle. Highlight Stem in the left-hand pane of the **Suppression** tab and pick the **>>** button. Do the same for Handle. The two features are now listed in the right-hand pane and two columns have been added to the spreadsheet. See **Figure 14-17.**

The default text in the cells of the new columns is Compute, which means to generate the feature. Valid entries for Compute include Compute, U, u, C, c, ON, On, on, and 1. To suppress a feature, type Suppress in the cell. Valid entries for Suppress include Suppress, S, s, OFF, Off, off, and 0. To suppress the stem and handle for the 10" valve, type Suppress in the cells in the Stem and Handle columns for the 10" nominal size. Refer to the chart in **Figure 14-15.**

Figure 14-16.
The **Properties** tab of the **iPart Author** dialog box.

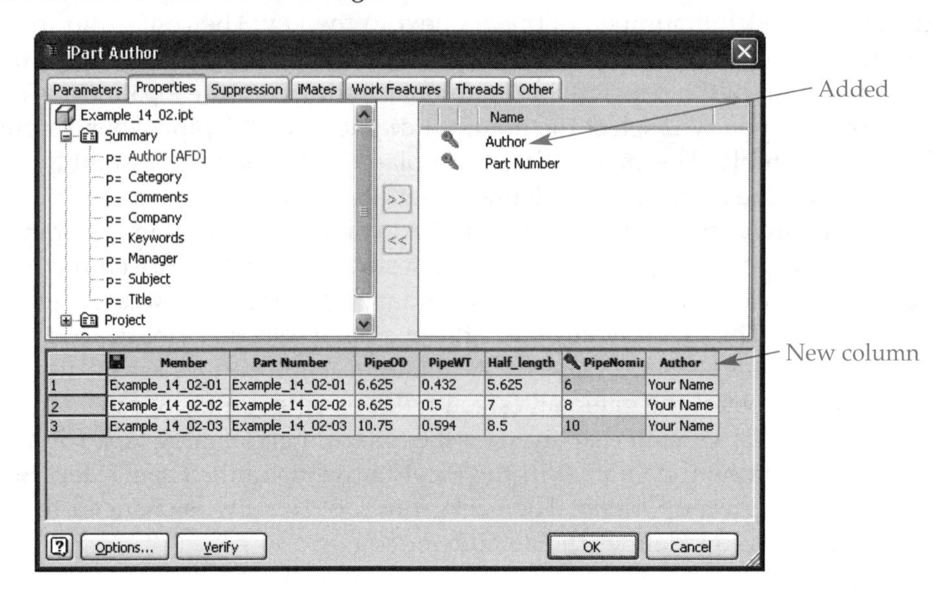

Figure 14-17.
The **Suppression** tab of the **iPart Author** dialog box.

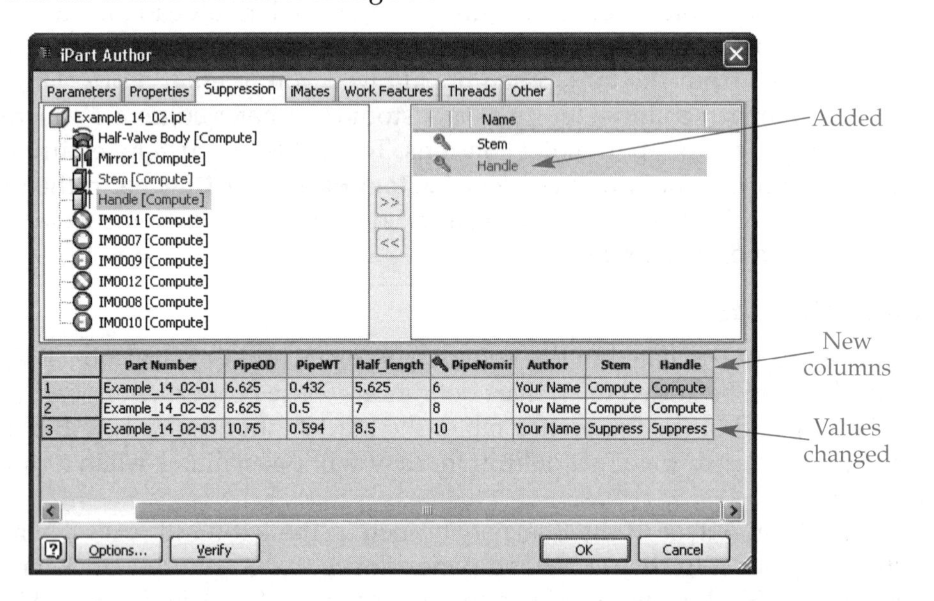

iMates tab

The **iMates** tab is used to fine-tune the iMates set up for the part. You can include or exclude individual iMates for different part versions. In addition, you can specify the offset values for the iMates, matching names, and sequence numbers. The iMates set up for the part are listed in the left-hand pane. Highlight a name in the left-hand pane and pick the **>>** button to display it in the right-hand pane and create a column for it in the spreadsheet. The values in the cells can then be edited. However, for the gate valve, you do not need to fine-tune the iMates. Therefore, do not add any to the spreadsheet.

Work features tab

Work features are useful in iParts to constrain parts in assemblies and to create pins in electrical parts. Work features are created in a part before it is transformed into an iPart factory. The **Work Features** tab is used to add work features to the iPart. You can then determine which work features to include or exclude in iPart instances.

Work features have default Include or Exclude settings. You can override the setting by selecting work features to include or exclude in the iPart table. Each row, which represents a version of the iPart, can have work features included or excluded. Default settings are: Included for work features constrained with iMates, Included for pins (work points) in electrical parts, and Excluded for all other work features (except those constrained with iMates). The setting for each row in the table can be edited as needed. However, for the gate valve, no settings need to be made in this tab.

Threads tab

If the part has any thread features, the **Threads** tab is used to add the feature to the spreadsheet and provide parametric control over the thread specifications. For example, for a family of threaded valves, the threads per inch, class, and orientation can be specified for each version of the part in the family. However, this is a family of welded valves, so no settings need to be made in this tab.

Other tab

The **Other** tab is used to add custom columns to the spreadsheet. For the gate valve, add a Description column. Refer to **Figure 14-18.** First, pick the entry Click to add value in the pane at the top of the tab. A text box appears in the **Name** column of the pane. Type the name Description and press [Enter]. A new custom column named Description is added to the spreadsheet. Next, edit each cell in the column to the values shown in **Figure 14-15.**

A default file name can be set up for the part that is saved when the iPart is placed into an assembly. The default file name is based on a column. For the gate valve, the Description column will be used. Right-click on that column heading and select **File Name Column** from the pop-up menu. Now, the part that is placed into an assembly will be saved to an IPT file with a name based on the description and the file name of the iPart. Since this is a standard part, the file name is automatically created by Inventor. For a custom part, you need to specify the file name as you place the part.

Figure 14-18.
The **Other** tab of the **iPart Author** dialog box.

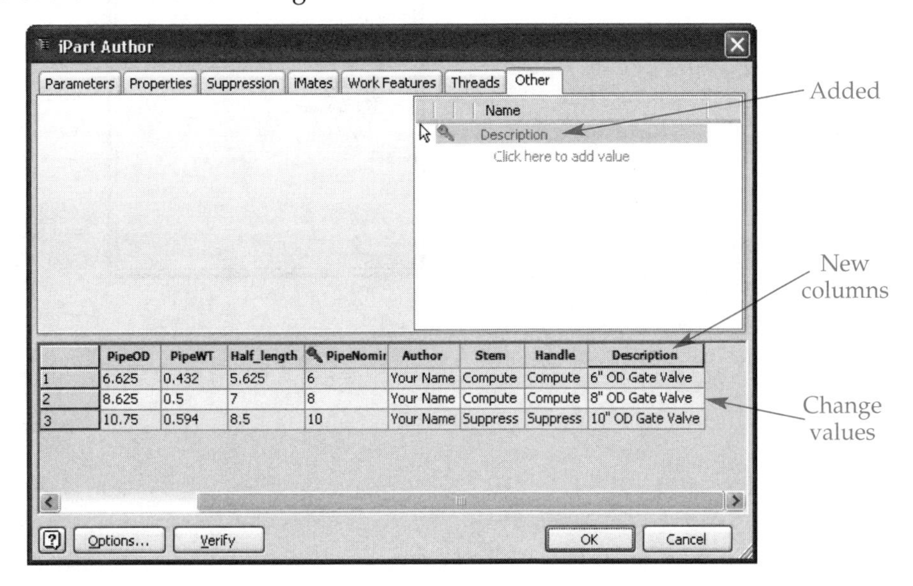

Testing the iPart in the Factory

The iPart is now completely set up. Close the **iPart Author** dialog box. Then, select **Save As...** in the **File** pull-down menu and save the part as Valve.ipt. Now, you need to "test build" the part to simulate the factory.

First, expand the Table branch of the part tree in the **Browser**. There should be three listings of PipeNominal = and a number corresponding to the nominal size. See **Figure 14-19.** Also, notice that the PipeNominal = 8 entry has a check mark next to it. This indicates that the size is the default part.

To simulate the factory, double-click on the first entry (PipeNominal = 6). The part should update in the graphics window to reflect the settings in the spreadsheet. Also, the name in the **Browser** has a check mark next to it. Repeat this procedure for the 10" valve. Check that the handle and stem are suppressed. The version of the part that is "manufactured" when the file is saved is set as the default part. Therefore, since the 8" valve is supposed to be the default, double-click on its entry in the Table branch before saving the file.

If you receive an error message while testing the parts, right-click on Table in the **Browser** and choose **Edit Table...** from the pop-up menu. The **iPart Author** dialog box is displayed. Verify the data against the chart in **Figure 14-15.** Look for typos or missing data. Edit the spreadsheet as needed and pick the **OK** button to close the **iPart Author** dialog box. If you have Microsoft Excel installed and prefer to correct the spreadsheet in that software, select **Edit via Spread Sheet...** from the pop-up menu when you right-click on Table. The spreadsheet is opened in Excel. When you save and exit Excel, the spreadsheet is updated as if you had edited it in the **iPart Author** dialog box.

Once you have tested all versions of the part by simulating a factory, and all versions are correct, the iPart is complete. Save and close the IPT file. Be sure that the correct version is set up as the default before you save the file. The iPart is ready to be placed into an assembly.

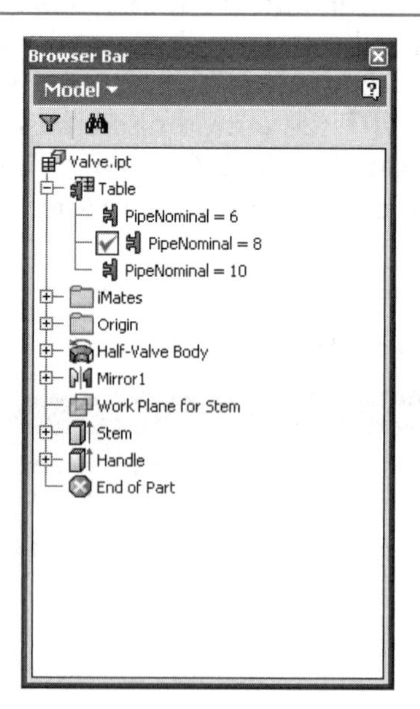

Figure 14-19.
The Table branch of the part tree contains all of the part variations set up in the spreadsheet.

Chapter Test

Answer the following questions on a separate sheet of paper or complete the electronic chapter test on the Student CD.

1. What is an *iPart?*
2. Define *family of parts.*
3. What is a *factory?*
4. What is the difference between a *standard iPart* and a *custom iPart?*
5. What is an *iMate?*
6. What is a *composite iMate?*
7. List three advantages of using iParts.
8. What are the icons called that are used with iMates?
9. What is the basic process for constraining parts using iMates?
10. How can you use iMates on a subassembly?
11. What are the five basic steps in creating an iPart from scratch?
12. What is a *key column?*
13. How many key columns, or keys, can an iPart have?
14. How can you test the iPart in a simulated factory?
15. How do you specify that the iPart you are creating is a custom iPart?

Chapter Exercises

Exercise 14-1. Raised Face Weld Neck Flange (RFWN), 300 lb., SCH80.

Complete the exercise on the Student CD

Exercise 14-2. Mounting Bracket for Pressure Vessel.

Complete the exercise on the Student CD

This model is of a motorcycle fork assembly. Using parameters in this type of assembly allows for multiple configurations within a single file. (Model courtesy of Autodesk, Inc.)

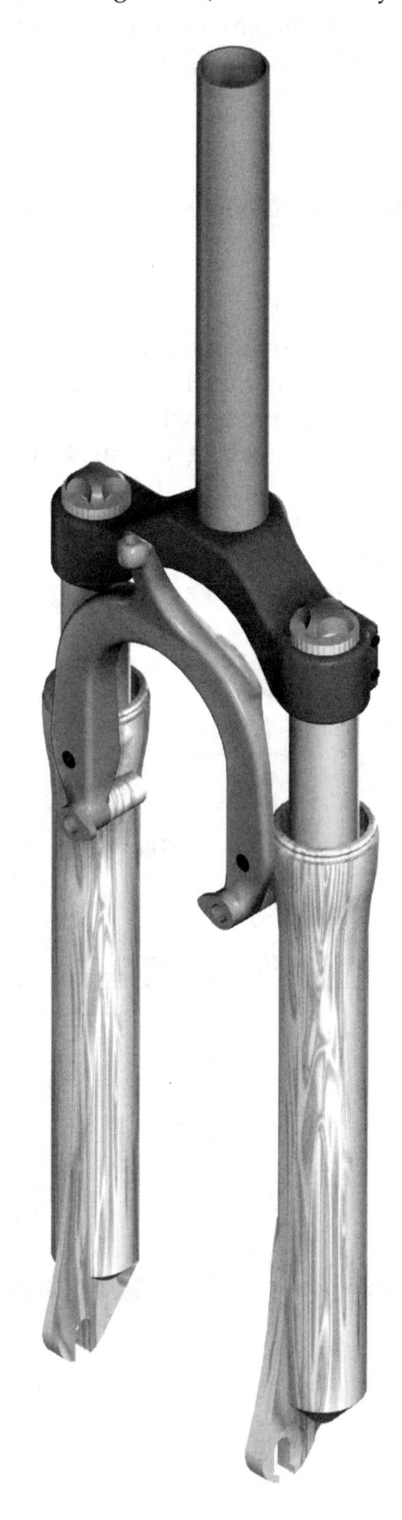

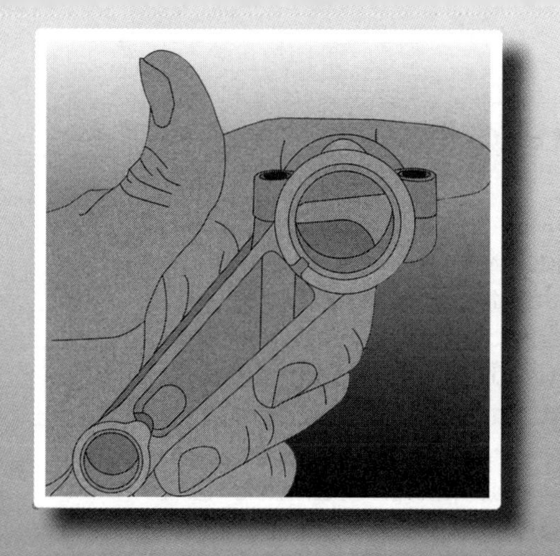

Chapter 15

Parameters in Assemblies

Objectives

After completing this chapter, you will be able to:

- Explain how parameters are used to control an assembly.
- Create a spreadsheet to control the parameters in an assembly.
- Use parameters to control an assembly.

User's Files

The Student CD included with this text contains several files required for this chapter. Refer to the file File List.txt in the \Ch15 folder for the comprehensive list.

Process of Working with Parameters

In this chapter, a part will be modeled with special attention paid to the parametric relationships needed to provide control over the part. When working with parameters, the dimension parameters are typically renamed to meaningful names in the **Parameters** dialog box. The equation relationships between the dimensions are also set up in this dialog box. However, to have a spreadsheet control the dimensions, the parameter names must be specified in the spreadsheet. User parameters can also be added as necessary, such as Force or Wall Thickness. Microsoft Excel's math functions can also be used to create relationships between the parameters.

PROFESSIONAL TIP

User parameters are typically only used for a variable that is not dimensioned on the part. If no dimension parameter exists and another variable is needed, then a user parameter must be added.

Controlling a Part with a Spreadsheet

Create a new project named Ch15 with the project folder as c:\Program Files\G-W\Learning Autodesk Inventor 2008 Student CD\Ch15. This project will allow access to all of the files for Chapter 15 and can be used throughout this chapter. Set it current.

The first example you will look at is a simple block. An existing spreadsheet that contains the parameter names, values, and units will be used. In a new part file, you will link the spreadsheet to the file, providing access to the parameter names.

In Microsoft Excel, open the spreadsheet Example_15_01.xls. See **Figure 15-1.** The parameter names are in column A, starting in row 3. The values are in column B. Cell B4 contains the equation =B2/2 so the WIDE parameter is always one-half of the value of the LONG parameter. The unit of measure is specified in column C. Units must be in the Inventor format for units: in, mm, deg, and so on.

In Inventor, start a new part file based on the Standard (in).ipt template. The file must be saved before a spreadsheet can be linked to it. Therefore, finish the sketch. Then, save the file as Example_15_01.ipt.

Pick the **Parameters...** button in the **Part Features** panel to display the **Parameters** dialog box. You can also select **Tools>Parameters...** to display the dialog box. Next, pick the **Link** button at the bottom of the **Parameters** dialog box. In the **Open** dialog box that is displayed, select the Example_15_01.xls file. Also, enter A3 in the **Start Cell** text box as this is where the parameter names start. Make sure the **Link** radio button is selected. Refer to **Figure 15-2.** Then, pick the **Open** button. The parameters are now listed in the **Parameters** dialog box. They are ready to use in the sketches. Pick the **Done** button to close the **Parameters** dialog box. Notice the 3rd Party branch has been added to the tree in the **Browser**. If you expand this branch, the spreadsheet file is listed below it.

Right-click on Sketch1 in the **Browser** and select **Edit Sketch** from the pop-up menu. Then, draw a rectangle. Place dimensions on the bottom and left side of the rectangle. Edit the dimension on bottom of the rectangle and enter LONG. Remember, parameters are case sensitive. To enter LONG, you can type it or right-click, select **List Parameters** from the pop-up menu, and select the name in the dialog box that is displayed. Edit the dimension on the left side and enter WIDE.

Figure 15-1.
Notice how the value of cell B4 is related to cell B3.

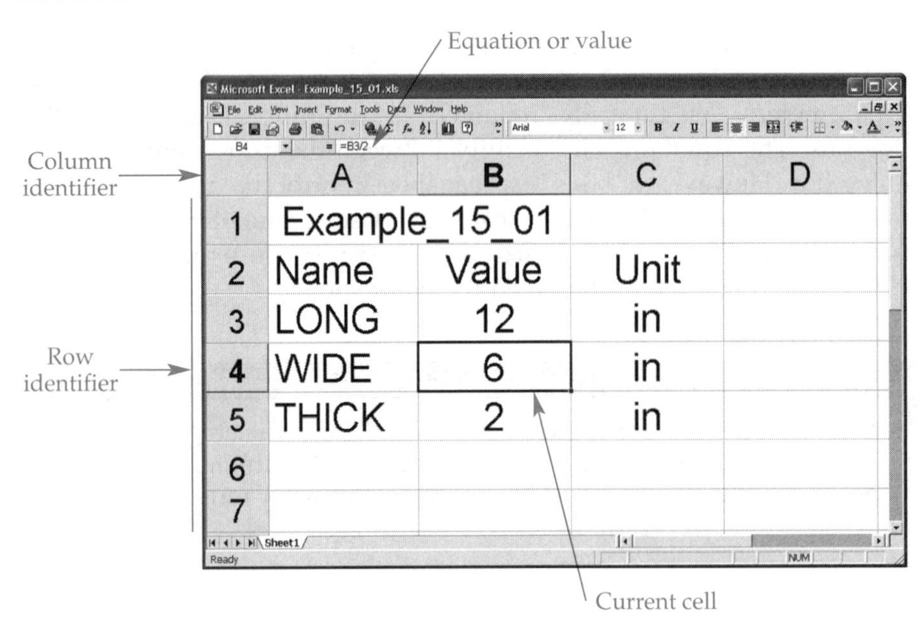

Figure 15-2.
Opening a spreadsheet to be linked.

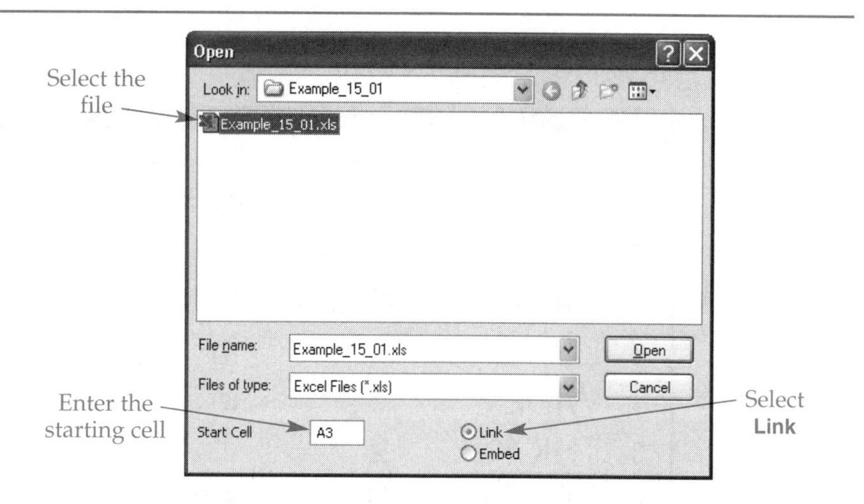

Finish the sketch and display the isometric view. Then, pick the **Extrude** button in the **Part Features** panel. The sketch is automatically selected, since it is the only unconsumed, unambiguous sketch. For the extrusion distance, enter THICK. You can type the value or right-click and select **List Parameters** from the pop-up menu. See **Figure 15-3.** Finally, extrude the part and save the file.

To show that the part is controlled by the spreadsheet, expand the 3rd Party in the **Browser**, right-click on Example_15_01.xls, and select **Edit** from the pop-up menu. The spreadsheet is opened in Excel. Change the value of LONG to 6 and save the file. Return to Inventor and pick the **Update** button on the **Inventor Standard** toolbar to update the part. The length and width of the part are decreased.

Open the **Parameters** dialog box. Note in the Model Parameters branch that d0 is set to LONG, d1 is set to WIDE, and d2 is set to THICK. Also notice that d3 is set to a value of 0, not to a parameter. The units for this value is degrees. You will now add this parameter to the spreadsheet. Close the **Parameters** dialog box and open the

Figure 15-3.
One of the available parameters can be entered as a dimension value.

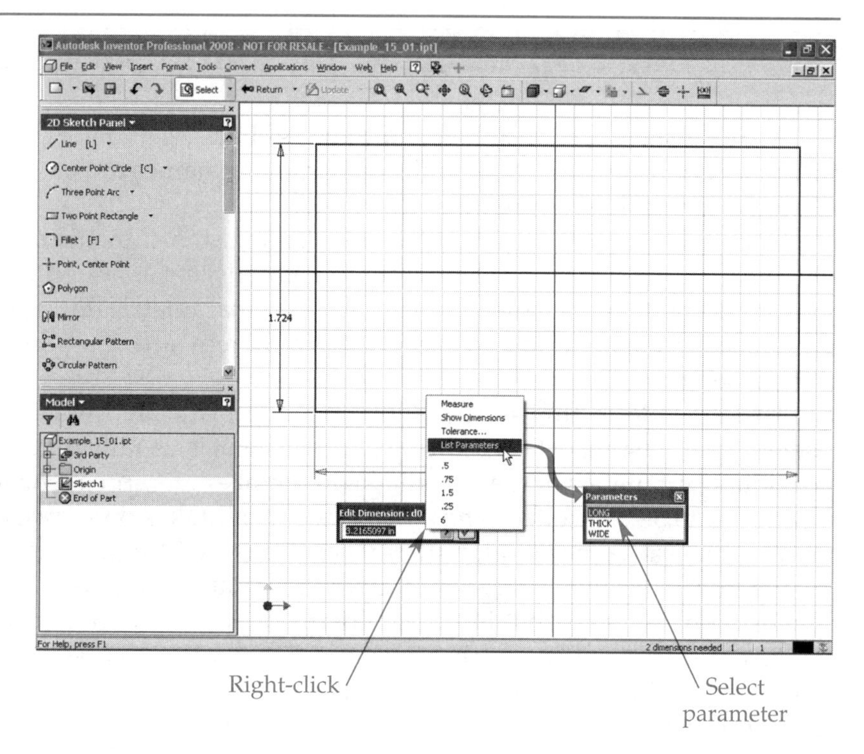

Figure 15-4.
Adding another parameter to the spreadsheet.

	A	B	C	D
1	Example_15_01			
2	Name	Value	Unit	
3	LONG	6	in	
4	WIDE	3	in	
5	THICK	2	in	
6	DRAFT	6	deg	
7				

New parameter added

spreadsheet for editing (if it is not already open). In row 6, type DRAFT in column A. Type 6 in column B and deg in column C. Refer to **Figure 15-4.** Save the spreadsheet, return to Inventor, and update the part.

The draft angle did not take effect because the parameter in the spreadsheet is not yet associated with the draft angle in Inventor. Open the **Parameters** dialog box. In the Equation column for d3, pick in the cell and type DRAFT. Pick the **Done** button to close the dialog box. Then, update the part. The parameter in the spreadsheet is now linked to the part.

This example shows how easy it is to control a part with a spreadsheet. In the next section, you will look at how assemblies can be controlled by a spreadsheet.

Using a Spreadsheet to Control an Assembly

A spreadsheet can also be used to control an assembly. Be careful to avoid any duplicate parameter names. If duplicate parameter names are used, an error will occur when the spreadsheet is linked to the model.

Once the spreadsheet is linked and loaded, the dimension parameters can be edited and the named parameters from the spreadsheet assigned to them. This must be done for all parts in the assembly that are to be parametrically controlled. Finally, the spreadsheet must be linked to the assembly, thereby affecting the parametric control of the assembly.

The example in this section illustrates the principle involved in controlling an assembly with a spreadsheet. Editing a spreadsheet is usually much easier than editing individual parameters in the part file(s). Remember, the spreadsheet is linked to every part that is to be controlled. The spreadsheet itself takes on a special status because it has all of the values needed for all parts contained within it.

Open Example_15_02a.ipt and open the **Parameters** dialog box. Review the dimension parameters that have been defined. At the bottom of the dialog box, pick the **Link** button. Navigate to the \Example_15_02 folder and select Example_15_02.xls. Make sure the **Link** radio button is selected in the **Open** dialog box. Also, enter A3 in the **Start Cell** text box so Inventor will start reading in values from the first cell in the third row. Pick the **Open** button and the spreadsheet is linked to the part. Review the named parameters that have been imported.

The final step is to assign the named parameters in the spreadsheet to the dimension parameters defined in the part. The dimension parameter d0 controls the length of the part. Pick once in the **Equation** field for d0. Then, pick the arrow and select **List Parameters** from the flyout menu. In the list box that appears, select BaseLength. This assigns the value of BaseLength to control the dimension parameter d0. You can also type the parameter name. Using the same procedure, assign the parameter values listed in **Figure 15-5.** Any skipped dimension parameter, such as d8, will not be driven by the spreadsheet. When finished, pick the **Done** button. Update the part, save your work, and leave the file open.

Open Example_15_02b.ipt and link the same spreadsheet (Example_15_02.xls) to this part. Specify the starting cell as E3. There are two sets of values defined within

Figure 15-5.
The dimension parameters used for Example_15_02a.

Parameter Name	Unit	Equation	Nominal Value	Tol.	Model Value	Comment
d0	in	BaseLength	6.000000		6.000000	Length of base.
d1	in	BaseWidth	4.000000		4.000000	Width of base.
d2	in	BaseHeight	1.000000		1.000000	Height of base.
d3	deg	0 deg	0.000000		0.000000	Extrusion angle.
d4	in	BasePostDia	0.500000		0.500000	Diameter of post.
d5	in	BasePostOffset	0.750000		0.750000	Corner offset to locate post.
d6	in	BasePostOffset	0.750000		0.750000	Corner offset to locate post.
d7	in	BasePostDepth	0.750000		0.750000	Height of post.
d8	deg	0 deg	0.000000		0.000000	Extrusion angle on post.
d9	in	0.06 in	0.060000		0.060000	Chamfer.
d11	ul	BasePostCountX	2.000000		2.000000	Number of posts in X direction.
d13	in	BasePostDistX	4.500000		4.500000	Distance between posts in X direction.
d14	ul	BasePostCountY	2.000000		2.000000	Number of posts in Y direction.
d16	in	BasePostDistY	2.500000		2.500000	Distance between posts in Y direction.

Figure 15-6.
The dimension parameters used for Example_15_02b.

Parameter Name	Unit	Equation	Nominal Value	Tol.	Model Value	Comment
d0	in	TopLength	6.000000		6.000000	Length of top.
d1	in	TopWidth	4.000000		4.000000	Width of top.
d2	in	TopHeight	1.000000		1.000000	Height of top.
d3	deg	0 deg	0.000000		0.000000	Extrusion angle of top.
d4	in	TopHoleOffset	0.750000		0.750000	Corner offset to locate hole.
d5	in	TopHoleOffset	0.750000		0.750000	Corner offset to locate hole.
d6	in	TopHoleDia	0.500000		0.500000	Diameter of hole.
d13	ul	TopHoleCountX	2.000000		2.000000	Number of holes in X direction.
d15	in	TopHoleDistX	4.500000		4.500000	Distance between holes in X direction.
d16	ul	TopHoleCountY	2.000000		2.000000	Number of holes in Y direction.
d18	in	TopHoleDistY	2.500000		2.500000	Distance between holes in Y direction.

the spreadsheet—one for each part. Refer to **Figure 15-6** for the dimension parameters with the named parameters from the spreadsheet. When finished, pick the **Done** button. Update the part, save your work, and leave the file open.

Editing the Part

The dimension parameters are now using the named parameters from the spreadsheet. So, to change the shape of the features that make up the part, the spreadsheet must be edited. First, switch to Example_15_02a.ipt. In the **Browser**, right-click on the spreadsheet in the 3rd Party branch and select **Edit** from the pop-up menu. The spreadsheet is opened in Microsoft Excel.

Notice the two sets of values for Example_15_02a and Example_15_02b. This spreadsheet will be used to drive the assembly of both parts. The two parts are "linked" in the spreadsheet using Excel's capability of referencing values in other cells. Double-click on cell F3, which is the TopLength dimension in Example_15_01b.ipt. Notice that the value is = B3, which means that the value of F3 is controlled by cell B3. See **Figure 15-7A.** The value in B3 is the BaseLength dimension in Example_15_01a.ipt; change it to 8.00. Notice that the value in cell F3 also changes to 8.00, thereby affecting the length of both parts. See **Figure 15-7B.** Both parts will then update accordingly in the assembly.

Take a moment to study the format of the cells for each part. *Each* part in the assembly that is to be controlled needs its *own* entry in the spreadsheet. There are three columns for each part—Name, Value, and Unit. Name is for the name of the parameter,

Value is for the numeric value, and Unit is for the type of units specified. The unit column is optional. Distances use inches (in) or millimeters (mm). The number of items in a pattern is unitless and denoted by ul. If the unit column is not included, then all values will use the default unit value of the part file, such as inches or millimeters.

After changing the value in B3, save the spreadsheet and exit Excel. In Inventor, update both parts to see the changes take effect. Then, save and close both part files.

Open Example_15_02.iam. If you look at the **Browser**, you will notice that there is no spreadsheet linked to the file. Using the **Assembly Panel**, open the **Parameters** dialog box and pick the **Link** button (the command is not located in the **Tools** pull-down menu in an assembly file). Locate the spreadsheet Example_15_02.xls and specify A3 as the starting cell. Open the file and then pick the **Done** button in the **Parameters** dialog box. In the **Browser**, right-click on the spreadsheet in the 3rd Party branch and select **Edit** from the pop-up menu. Change the value for BaseWidth to be the same as for BaseLength, which is 8.00. Save the spreadsheet, exit Excel, and update the assembly to

Figure 15-7.
A—The relationship between cells B3 and F3. B—Changing the value in cell B3 changes the value in cell F3 as well.

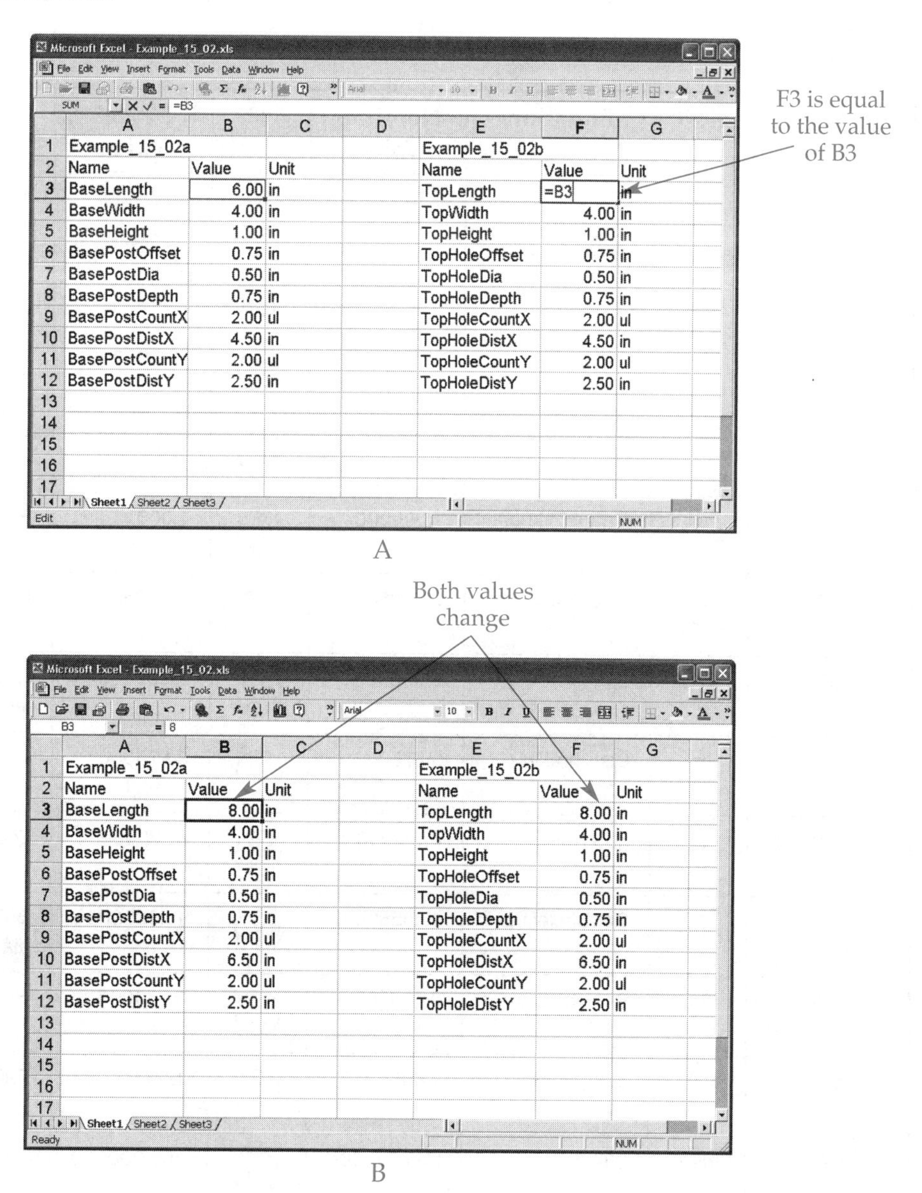

see the revisions. Take a moment to review the **Browser**. Notice that the assembly and both parts have a 3rd Party branch that contains a link to the spreadsheet. Save your work and close the file.

Flanged Pipe Run—An Advanced Example

Open Example_15_03.iam. Examine the assembly structure in the **Browser**. In this example, you will use named parameters to control parametric values in an assembly.

Right-click on Pipe in the **Browser** and choose **Open** from the pop-up menu. Then, in the part file, open the **Parameters** dialog box and link the spreadsheet Example_15_03.xls. Specify the starting cell to be A3 since the values for the pipe start there. Next, pick once on the equation for d0 and change it to the named parameter PipeOD. Change the equation for d2 to Length. Then, pick the **Done** button. Update the part and the length should change from 24″ to 36″. Save the part and then close the file.

In the assembly file, open the Flange_300lb part (either instance) using the **Browser**. Open the **Parameters** dialog box and link the same spreadsheet starting at cell E3. Change the equations as shown in **Figure 15-8.** Update, save, and close the part.

Figure 15-8.
The dimension parameters used for Flange_300lb in Example_15_03.iam.

Parameter Name	Unit	Equation	Nominal Value	Tol.	Model Value	Comment
d0	in	d1 – (2 in * 0.432 ul)	5.761000	●	5.761000	ID
d1	in	PipeOD	6.625000	●	6.625000	OD
d2	in	FlangeOD	12.500000	●	12.500000	Flange OD
d3	in	d12 – 1.5 in	9.125000	●	9.125000	Face OD
d4	in	FlangeThkness	1.000000	●	1.000000	Flange thickness.
d5	in	0.0625 in	0.062500	●	0.062500	Raised face.
d7	in	2 in	2.000000	●	2.000000	Angled body length.
d8	in	DistThruHub	3.875000	●	3.875000	Length through hub (Y).
d10	in	d3 – 1 in	8.125000	●	8.125000	Locate angle.
d12	in	BoltCircleDia	10.625000	●	10.625000	Bolt circle diameter.
d13	in	BoltHoleDia	0.750000	●	0.750000	Hole diameter.
d20	ul	BoltCount	12.000000	●	12.000000	Pattern count.
d22	deg	360 deg	360.000000	●	360.000000	Pattern fit angle.

In the assembly file, open the Elbow part using the **Browser**. Open the **Parameters** dialog box and link the same spreadsheet starting at cell I3. Change the equation for d0 to PipeOD. Change the equation for d2 to Radius. Update, save, and close the part.

In the assembly file, open the Hexhead_Bolt part using the **Browser**. This is located in the Component Pattern branch. Open the **Parameters** dialog box and link the same spreadsheet starting at cell M3. Change the equation for d0 to HeadDia, d7 to BoltDia, and d8 to BoltLength. Update, save, and close the part.

In the assembly file, open the Hexhead_Nut part using the **Browser**. This is also in the Component Pattern branch. Open the **Parameters** dialog box and link the same spreadsheet starting at cell Q3. Change the equation for d0 to BodyDia and d7 to BoltDia. Update, save, and close the part.

The final step is to link the spreadsheet to the assembly. This allows for full control of the assembly. Open the **Parameters** dialog box in the assembly file. Notice that there is an assembly parameter (d37) for the count of the bolts used in the pattern. Link the same spreadsheet starting at cell E3. Why start at cell E3? If you recall, this is where the values for the flange begin. One of the parameters used in the flange controls the number of bolt holes in the bolt circle. If the flange's parameters are imported into the assembly, then they can be used to control the assembly's parameters, such as the count of hex head bolts used in the assembly pattern.

Once the spreadsheet is linked, change the equation for d37 to BoltCount and edit the equation for d36 to read 360/BoltCount. In Inventor, assembly pattern angles are specified explicitly, as opposed to being specified by an included angle—usually 360°—as in part feature patterns. Update and save the assembly file.

Right-click on the linked spreadsheet in the 3rd Party branch in the **Browser** and select **Edit** from the pop-up menu. In the spreadsheet, change the values as shown below.

- Change cell B3 to 4.50. This is the pipe OD.
- Change cell B4 to 24.00. This is the length of the straight section of pipe.
- Change cell F4 to 10.00. This is the flange OD.
- Change cell F5 to .75. This is the flange thickness.
- Change cell F6 to 3.375. This is the depth of the flange.
- Change cell F7 to 7.875. This is the bolt circle diameter.
- Change cell F9 to 8. This is the count of bolts in the pattern.
- Change cell J4 to 6.00. This is the long radius for the elbow.

Save the spreadsheet, close the file, and update the assembly. The assembly should change from a 6″ pipe run to a 4″ pipe run with the correct number of bolts. If there are any problems, go into the individual part files and investigate the spreadsheet links.

Why are not more of the values parametrically related to the pipe OD so you would only change the pipe OD to change everything? All of the values are taken from standard tables for Schedule 80 Pipe. You must change each of the values because they are not functions of the pipe OD. For example, the flange OD is not always 1.5 times the pipe OD. There is no relationship between the dimensions controlling the parts.

NOTE

Alternately, the pattern of bolts in this assembly could be controlled by using the Inventor capability of controlling an assembly pattern. This can be done by linking it to a feature pattern on one of the assembly's parts (for example, the pattern of bolt holes on one of the flanges).

PRACTICE 15-1 Complete the practice problem on the Student CD.

Summary

You have seen how important the proper use of parameters in an assembly can be for providing control and making revisions a simple process. In essence, this ability provides a central storage of named parameters and a means of importing them for each part in the assembly. Once the links are established, simply change the spreadsheet to revise the assembly. Without this capability, a designer would have to use the tedious method of changing each part's dimensions to revise the assembly.

Chapter Test

Answer the following questions on a separate sheet of paper or complete the electronic chapter test on the Student CD.

1. How can parameters and a spreadsheet be used to control an assembly?
2. When linking a spreadsheet to a part or assembly file, what is the purpose of the **Start Cell** text box?
3. What happens if a spreadsheet contains duplicate parameter names?
4. Once a spreadsheet is linked to a part or assembly file, how are dimension parameters linked to the spreadsheet?
5. Which parts in an assembly must be linked to a spreadsheet?
6. Where is the spreadsheet located in the **Browser** tree?
7. After a spreadsheet has been edited, what must be done in Inventor so the changes to the parameters are reflected in the part or assembly?
8. When data for multiple parts are contained within a single spreadsheet, what must be true of the data?
9. Once a spreadsheet is linked to a part or assembly file, how can it be displayed for editing from within Inventor?
10. Within a spreadsheet in Excel, how can you "link" parameters between parts?

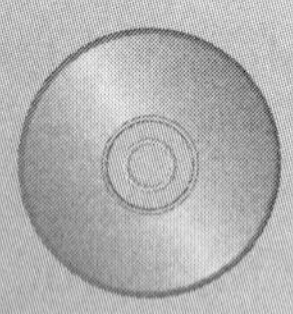

Chapter Exercises

Exercise 15-1. Flanged Pump Head.
Complete the exercise on the Student CD

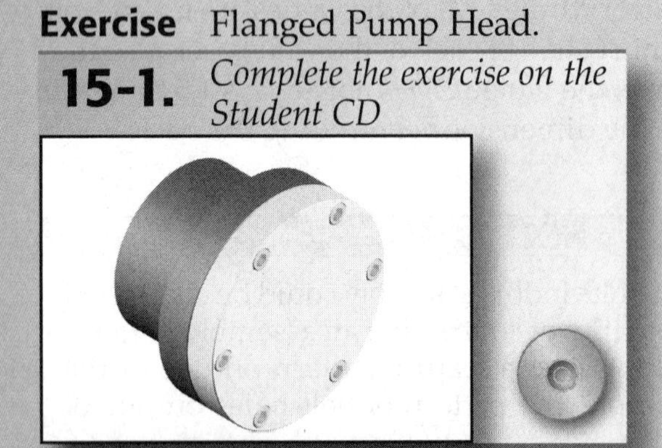

Exercise 15-2. 10" Pipe Run.
Complete the exercise on the Student CD

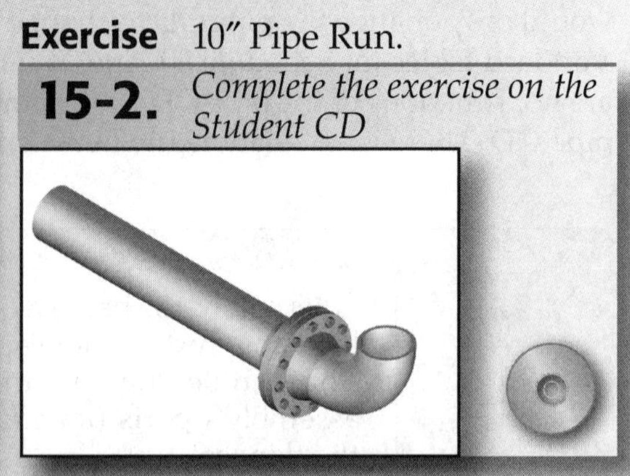

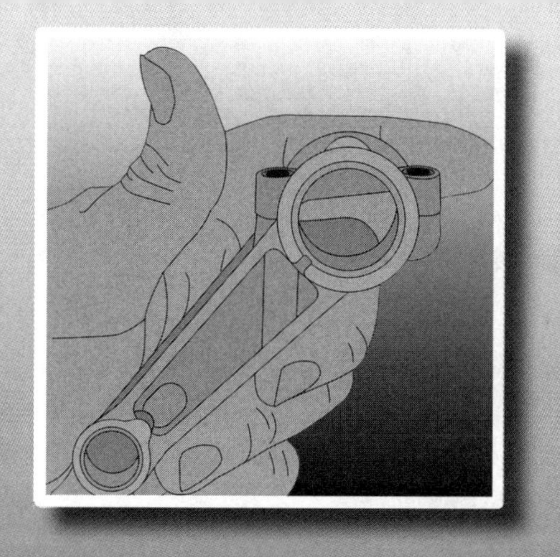

Chapter 16

Surfaces

Objectives

After completing this chapter, you will be able to:

- Explain the differences between surfaces and solids.
- Adjust the display of a surface.
- Explain the basic process for creating a surface.
- Offset a surface from a solid or surface.
- Replace a face on a solid with a face that matches the shape of a surface.
- Use surfaces as construction geometry.
- Stitch surfaces into a quilt.
- Visually analyze the faces of a part for draft angle or continuity of surface topology.

User's Files

The Student CD included with this text contains several files required for this chapter. Refer to the file File List.txt *in the* \Ch16 *folder for the comprehensive list.*

A ***surface*** defines the form and shape of an object, but does not have volume. A solid also defines the form and shape of an object, but it has volume. Inventor is primarily used to create solid objects. As you will learn in Chapter 19, even thin, sheet metal parts are created as solids. However, you can create surfaces to use as construction geometry. You used this method in Exercise 10-4 in Chapter 10. See **Figure 16-1.** In addition, there are some instances where you may want to create the final product as a surface model, such as when sharing a model with software that does not support solid modeling.

The process for creating a surface is basically the same as for creating a solid. First, sketch and constrain the geometry. The sketch for a surface does not need to be closed. Then, finish the sketch and select the feature creation tool you wish to use, such as **Extrude** or **Revolve**. In the feature creation dialog box, such as the **Extrude** dialog box, pick the **Surface** button in the **Output** area. See **Figure 16-2.** If the sketch is open, the **Solid** button will be disabled and the **Surface** button will be on. Then, select the profile, if needed, and create the feature.

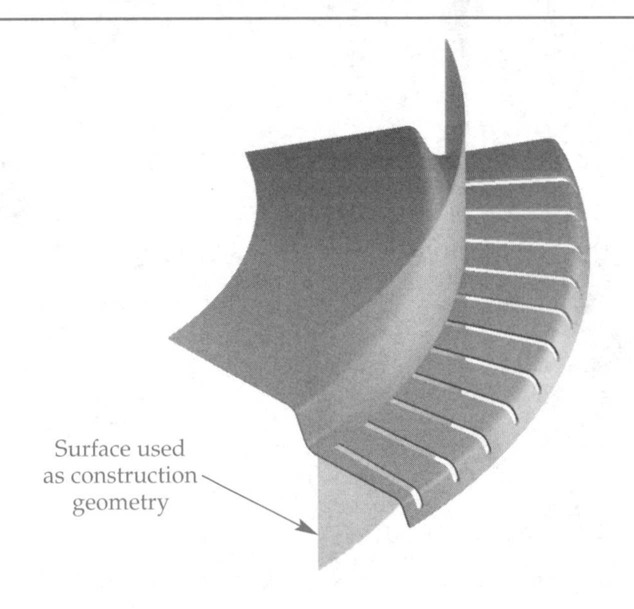

Figure 16-1.
In Exercise 10-4, you used a construction surface to terminate the extrusion when creating the slots.

Surface Display

By default, surfaces are displayed translucent in Inventor. This can be changed so that they are displayed opaque. Select **Application Options...** in the **Tools** pull-down menu. The **Options** dialog box is displayed. In the **Construction** area of the **Part** tab, check the **Opaque surfaces** check box. Then, pick the **OK** button to save the setting and close the dialog box. This setting is only applied to new surfaces you create. Existing surfaces are unaffected.

You can also turn translucency on and off for an individual surface feature. Open Example_16_01.ipt, which contains an opaque surface. Then, right-click on the surface feature name (LoftSrf2) in the **Browser** and select **Translucent** from the pop-up menu. A check mark appears next to the menu item when translucency is on. Notice how the part is now translucent, similar to a work plane.

Surfaces are normally displayed in orange, either translucent or opaque. You can change the material for a surface feature, as you can a solid feature, but the material set up for the part is not applied to surfaces. You can change the face color style of the surface. In the Example_16_01 file, right-click on the surface in the graphics window and select **Properties** in the pop-up menu. In the **Face Properties** dialog box, select Black Chrome in the Face Color Style drop-down list. Then, pick the **OK** button to apply the face color style.

Pick to create a surface

Figure 16-2.
Creating a surface extrusion.

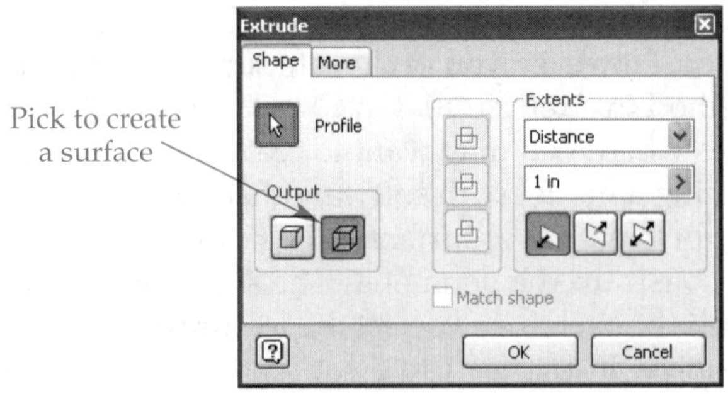

Extruded Surfaces

Open Example_16_02.ipt. This file contains an open, fully constrained sketch composed of a line and an arc. Pick the **Extrude** button in the **Part Features** panel. In the **Extrude** dialog box, select the **Profile** button in the **Shape** tab. Since there is only one unconsumed, unambiguous profile, it is automatically selected. If you need to select the profile, pick it in the graphics window with the **Profile** button on.

Notice in the dialog box that the **Surface** button in the **Output** area is automatically selected. This is because the sketch is open. A solid cannot be created from an open sketch. If the **Solid** button is selected, the **Examine Profile Problems** button (a red cross) is displayed at the bottom of the dialog box. Also, notice that the operation buttons—**Join**, **Cut**, and **Intersect**—are grayed out when the **Surface** button is selected. In this case, there are no existing solid features so the **Cut** and **Intersect** buttons would normally be grayed out. However, these buttons are always grayed out when the **Surface** button is selected.

Select Distance in the **Extents** drop-down list. Then, enter 1 in the text box below the drop-down list. Finally, pick the **OK** button to extrude the surface. Notice in the **Browser** that the feature is named ExtrusionSrf*x*. The Srf suffix indicates that the feature is a surface, not a solid.

Revolved Surfaces

Open Example_16_03.ipt. This file contains a sketch that you will revolve into a surface. Like the previous example, this is an open sketch. Pick the **Revolve** button in the **Part Features** panel. Since there is only one unconsumed, unambiguous profile and the axis of revolution was drawn as a centerline, these are automatically selected. Otherwise, pick the **Profile** button and select the profile in the graphics window, and then pick the **Axis** button and select the axis of revolution. Notice that the **Surface** button in the **Output** area of the dialog box is automatically turned on. Select Angle in the **Extents** drop-down list. Then, enter 180 in the text box below the drop-down list. Finally, pick the **OK** button to create the surface.

Lofted Surfaces

Open Example_16_04.ipt. This file contains two closed sketches that will be used to create a lofted surface. Pick the **Loft** button in the **Part Features** panel. In the **Curves** tab of the **Loft** dialog box, pick the **Surface** button in the **Output** area. Next, pick Click to add in the **Sections** column. Then, select both sketches in the graphics window. Pick the **OK** button to create the loft. A surface is generated between the two sketches. However, notice that the ends—the area of each closed sketch—are open. This is more apparent if translucency is turned off for the surface or if the part is rotated to an end view.

Swept Surfaces

Open Example_16_05.ipt. This file contains two sketches. The curved line will be used as the path for a swept feature. The circle sketch will be the cross section. Pick the **Sweep** button in the **Part Features** panel. In the **Sweep** dialog box, pick the **Surface** button in the **Output** area of the **Shape** tab. Then, pick the **Profile** button and select the circle in the graphics window. Pick the **Path** button in the dialog box and select the curved line in the graphics window. Finally, pick the **OK** button to create the swept surface. If you turn off the translucency for the surface and rotate the view, it is apparent that this is a tube, not a solid part.

Thicken/Offset Tool

The **Thicken/Offset** tool allows you to offset a surface, thereby creating a second, parallel surface at a distance. You can also thicken a surface with this tool, turning it into a solid. Open Example_16_06.ipt; it contains a single lofted surface.

1. Pick the **Thicken/Offset** button in the **Part Features** panel. The **Thicken/Offset** dialog box is displayed, **Figure 16-3.**
2. In the **Output** area of the **Thicken/Offset** tab of the dialog box, pick the **Surface** button.
3. Since this is a single surface, pick the **Face** radio button.
4. Enter .1 in the **Distance** text box.
5. Pick the **Select** button and then select the surface.
6. On the right-hand side of the dialog box, pick the direction button so the preview shows the offset to the left.
7. Pick the **OK** button to create the new surface. Notice the name of the surface in the **Browser** is OffsetSrft1.

Undo the operation. Then, using the **Thicken/Offset** tool again, pick the **Solid** button in the **Output** area. Select the surface and enter the same distance. When the **OK** button is selected, a solid is created. Note that the part takes on the assigned material of Aluminum (Polished). In the **Browser**, the name of the feature is Thicken1.

Open Example_16_07.ipt. This file has two lofted surfaces that share a common edge. Open the **Thicken/Offset** dialog box. Notice that you can only select one of the surfaces for the operation. Only one surface can be offset at a time. To be able to select both, you need to stitch the two surfaces together. Close the dialog box and pick the **Stitch Surface** button in the **Part Features** panel. Select both surfaces and pick the **OK** button in the **Stitch Surface** dialog box. This turns the two surfaces into a single quilt. A ***quilt*** is a single continuous surface consisting of two or more faces. Now, you can use the **Thicken/Offset** tool. With the **Face** radio button on in the **Thicken/Offset** dialog box, you can pick each surface. However, if the **Quilt** radio button is on, selecting either surface selects the entire quilt (both surfaces).

Figure 16-3.
Creating an offset surface. By setting the output to a solid, you can thicken a solid using this dialog box.

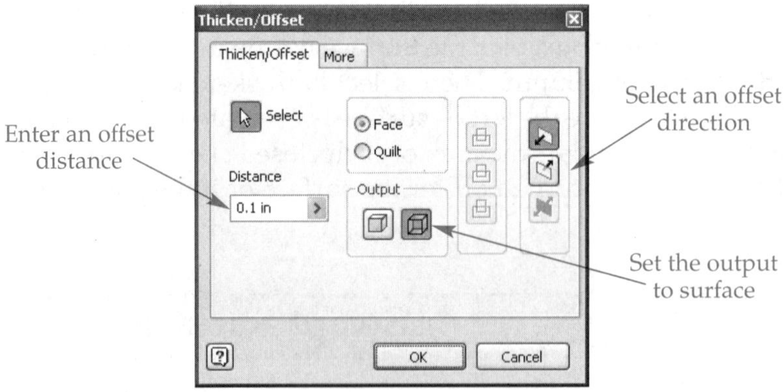

Thicken/Offset
Thicken/Offset More
Select
Face
Quilt
Distance
0.1 in
Output
OK Cancel

Enter an offset distance

Select an offset direction

Set the output to surface

Figure 16-4.
All faces in a chain can be selected with one pick when this check box is checked.

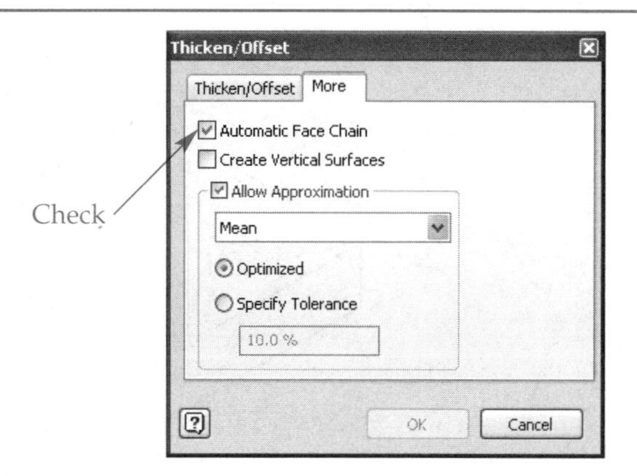

A surface can also be offset from a solid part. Open Example_16_08.ipt, which is a solid part. Next, open the **Thicken/Offset** dialog box. In the **Output** area, select the **Surface** button. Then, pick the **Select** button and the **Face** radio button. Then, pick any face on the part in the graphics window. A single face is selected and previewed offset. You can continue to pick individual faces to offset as a surface. However, the selected faces must be contiguous and form a single area. To create the offset surface, pick the **OK** button.

Suppose you want to offset all faces on the solid that are contiguous. You can individually select all faces. However, there is an easier method:

1. Undo the offset operation you just performed.
2. Open the **Thicken/Offset** dialog box.
3. Pick the **Surface** button and the **Face** radio button in the **Thicken/Offset** tab.
4. In the **More** tab, check the **Automatic Face Chain** check box, **Figure 16-4.**
5. Pick the **Select** button in the **Thicken/Offset** tab and select any face on the part other than the top or bottom face.

All of the faces around the perimeter of the part are selected because they are a chain. The top and bottom faces are not part of the chain because of the sharp edge. Next, pick the top face to add it to the selection set and then create the offset surface. A surface "shell" is generated at the specified offset distance from the solid.

Now, suppose you want to offset the surface you just created. The **Face** and **Automatic Face Chain** options work the same when selecting a surface as they do on a solid. However, you can select *all* faces on a contiguous surface with one pick using the **Quilt** option. Open the **Thicken/Offset** dialog box and pick the **Quilt** radio button. Then, with the **Select** button on, pick anywhere on the surface in the graphics window. The entire surface is selected and displayed in blue. Do not try to pick on the solid. You cannot select faces on a solid with the **Quilt** option.

Replace Face

In this example, you will create a loft surface and use it to replace the planar face on top of a rectangular solid. Open Example_16_09.ipt. This file contains a rectangular solid box and four unconsumed sketches. Two of the sketches will be used as the cross section. The other two sketches will be used as rails, or guides. The sketch names in the **Browser** reflect this.

Pick the **Loft** button in the **Part Features** panel. In the **Curves** tab of the **Loft** dialog box, pick the **Surface** button in the **Output** area. It is important to select this button before selecting cross sections or rails. Next, pick Click to Add in the **Sections** column. Then, pick the first cross section sketch. See **Figure 16-5.** Select the second cross section as well. Once the second cross section is selected, a preview appears in the graphics

Figure 16-5.
Creating a loft surface.

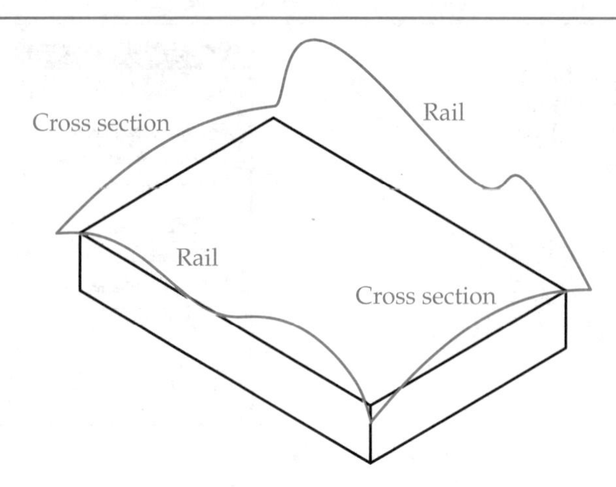

window. Then, pick Click to Add in the **Rails** column and select one rail. Select the second rail as well. Finally, pick the **OK** button to create the surface. See **Figure 16-6.** Notice that this surface overlaps the solid on all four sides, as viewed from the top.

Next, pick the **Replace Face** button in the **Part Features** panel. The **Replace Face** dialog box is displayed, **Figure 16-7.** With the **Existing Faces** button in the dialog box on, pick the top planar face on the solid in the graphics window. Then, pick the **New Faces** button in the dialog box. In the graphics window, select the loft surface you just created. Pick the **OK** button to replace the face, **Figure 16-8.**

Since the top face was replaced, the solid still has six faces. If the loft had been created as a solid that was joined to the existing solid, additional faces would have been created. Keeping the face count down is one of the benefits of using this procedure.

Figure 16-6.
The completed loft surface.

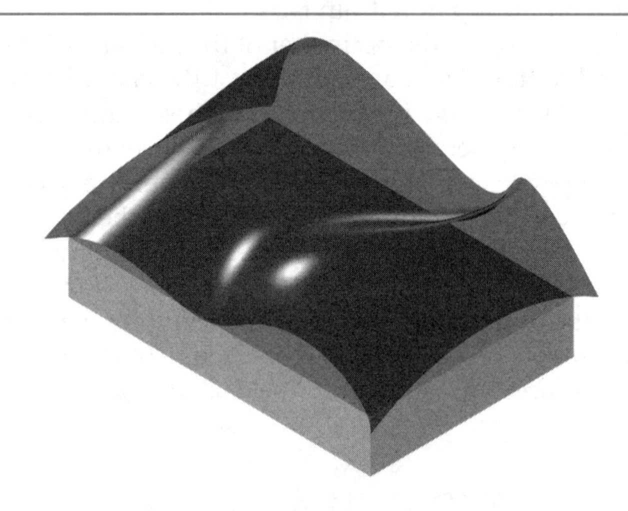

Figure 16-7.
Replacing a face.

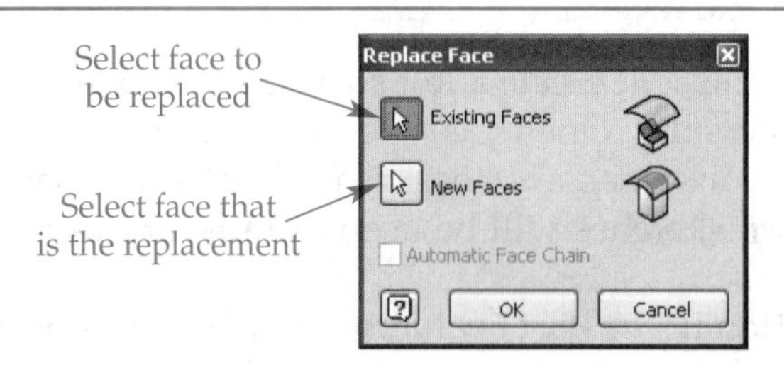

Figure 16-8.
The original flat face is replaced by a face that matches the shape of the loft surface.

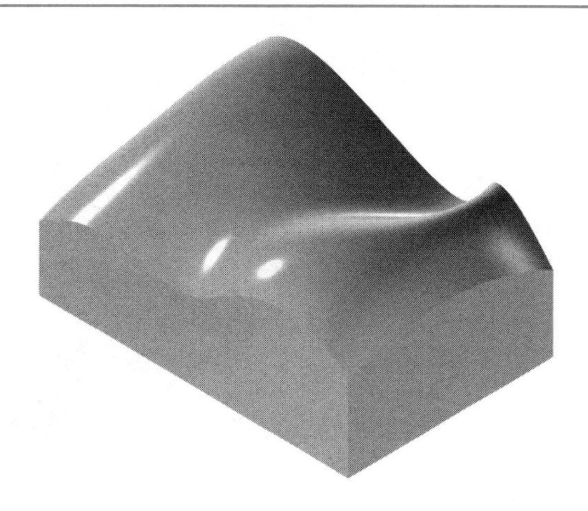

Surfaces as Construction Geometry

As mentioned earlier, surfaces are often used as construction objects. For example, a surface can be used to terminate an extrusion or revolution. A surface can also be used to split a part. The **Split** tool is discussed in Chapter 7. In the example in this section, a surface will be used to terminate an extrusion. Open Example_16_10.ipt. This file contains a rectangular part and two unconsumed sketches. One sketch will be used to create the construction surface. The other sketch will be extruded as a solid.

Using the **Extrude** tool, extrude the sketch named Sketch for Construction Surface a distance of 4.00″. See **Figure 16-9A.** Select the **Surface** button in the **Output** area of the **Extrude** dialog box and then, with the **Profile** button on, select the sketch. Pick the **Midplane** option and then pick the **OK** button to create the surface. The surface should project above and below the top face of the solid part. Notice that the surface is slightly curved.

Next, extrude the sketch named Sketch for Subtraction Solid. Pick the **Solid** button in the **Output** area of the **Extrude** dialog box. Then, select the profile in the graphics window. Select To in the **Extents** drop-down list and, with the "select" button below the drop-down list on, pick the surface in the graphics window. Finally, pick the **Cut** button and then the **OK** button to create the feature. See **Figure 16-9B.** Notice how the back of the subtracted feature matches the curvature of the surface.

Stitched Surfaces

As described earlier, a quilt is a single surface consisting of two or more surfaces that are connected as a single unit. The **Stitch Surface** tool allows you to "sew" separate surfaces into a quilt. This tool was introduced earlier. In order for the tool to work, the mating edges of the surfaces to be stitched need to be identical.

Open Example_16_11.ipt. Using the **Revolve** tool, revolve Sketch1 90° about the X axis. See **Figure 16-10.** Next, use the **Extrude** tool to extrude Sketch2 2″ extending away from the revolution. One edge on each surface touches the other surface. Additionally, those two edges are identical.

Pick the **Stitch Surface** button in the **Part Features** panel. The **Stitch** dialog box is displayed, **Figure 16-11.** With the **Surfaces** button in the dialog box on, pick both surfaces in the graphics window. Then, pick the **OK** button in the dialog box to stitch the surfaces into a quilt.

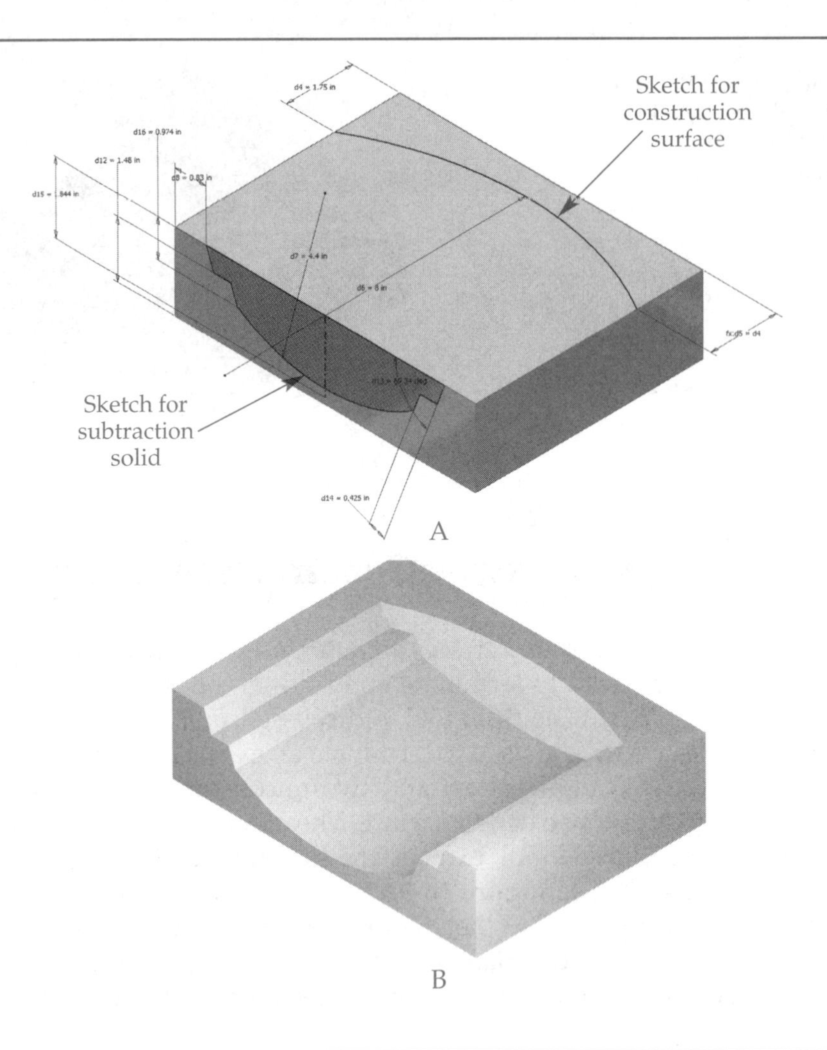

Figure 16-9.
A—The part showing the two unconsumed sketches. B—The curved back of the cutout feature was used as the terminator for the extrusion.

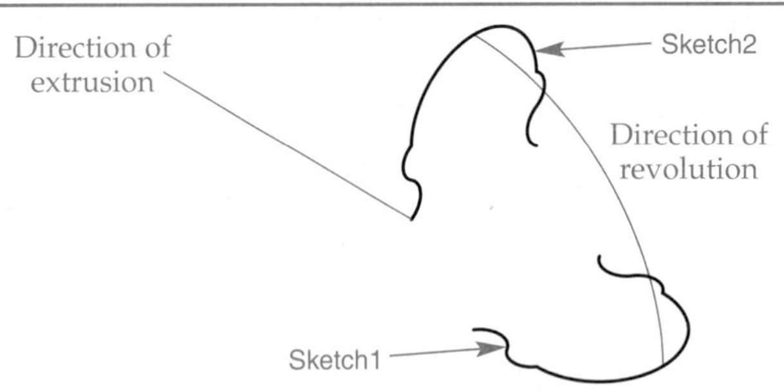

Figure 16-10.
These two sketches will be used to create two surfaces, which will then be stitched into a quilt.

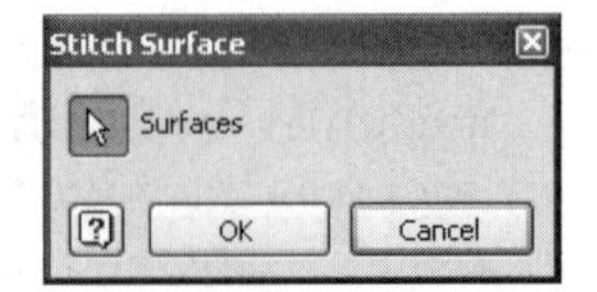

Figure 16-11.
Stitching surfaces into a quilt.

Boundary Patch

At times, you will need a flat surface to complete a model. A boundary patch is created from an enclosed area. This type of surface can be applied to any 2D sketch as long as it defines a closed region. It can also be created from a 3D surface model as long as there is a change of edges that will serve to define a closed region. A boundary patch is often used to put a flat top or bottom on a surface model.

Figure 16-12.
Select the area highlighted here when creating the first boundary patch.

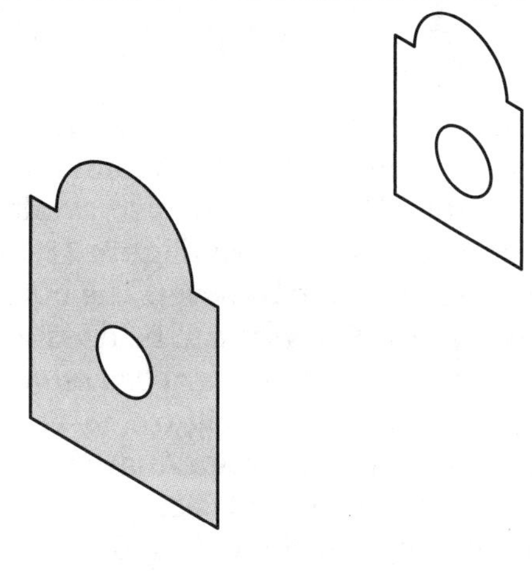

Open Example_16_12.ipt. Pick the **Boundary Patch** button in the **Part Features** panel. The **Boundary Patch** dialog box is displayed. On Sketch1, select the region indicated in **Figure 16-12.** Then, pick the **OK** button in the dialog box to create the boundary patch. Next, create a similar boundary patch on Sketch5. Then, loft a surface between the two boundary patches. Notice how a surface edge is not created between the circular hole in each boundary patch.

Surfaces in 2D Drawings

Surfaces can be included in a 2D drawing (IDW). If the part file contains only surfaces, simply place the base view and all projected views as you would for a solid part. If the part file contains both solids and surfaces, you can choose to either include or exclude the surfaces from the drawing. By default, only the solid features are included; surfaces are excluded.

To include both surfaces and solids in a drawing view, place the base view normally. Then, expand the tree for the view in the **Browser** until the features of the part are displayed. Finally, right-click on the name of the surface(s) to include and select **Include** from the pop-up menu. The surface(s) will be visible in the base view and all orthographic views derived from it. To hide a surface in a derived view, expand the tree for the view, right-click on the surface to hide, and uncheck **Include** in the pop-up menu.

PROFESSIONAL TIP

When certain files are imported, such as IGES files, they may come in as surfaces. In order to be used as a solid feature or part, these surfaces must be promoted to the part environment. By default, this is automatically done when importing the file. Once imported, a Construction branch is added to the part tree. To manually promote the surfaces, right-click on the branch and select **Edit Construction** from the pop-up menu. Then, stitch all surfaces into a quilt using the **Stitch** tool in the **Construction Panel**. Finally, use the **Copy Object** tool in the **Construction Panel** to copy the stitched surface to a surface in the part environment.

Analyzing Faces and Surfaces

Inventor provides tools for visually analyzing faces on parts and surfaces and the surface topology. There are five tools found in the **Analysis** cascading menu of the **Tools** pull-down menu. The tools are also available in the **Analysis Visibility** flyout on the **Inventor Standard** toolbar. See **Figure 16-13.** After using any of these tools, an Analysis branch is added to the **Browser**. The button on the toolbar or the check mark in front of the name in the **Browser** can be used to turn the analysis on or off.

The first tool in the list is called a ***zebra analysis*** because black and white stripes are displayed on the surface. See **Figure 16-14.** Open Example_16_13.ipt. To display the zebra analysis, pick the **New Zebra Analysis** button in the **Analysis Visibility** flyout or **Tools>Analysis>New Zebra Analysis**. The **Zebra Analysis** dialog box is displayed, **Figure 16-15.** The settings in the dialog box reflect the zebra settings. Pick the **OK** button to display the analysis. The zebra stripes are parallel on areas of flatness on the part. The zebra stripes will converge where there are areas of radical departure from

Figure 16-13.
Selecting an analysis mode.

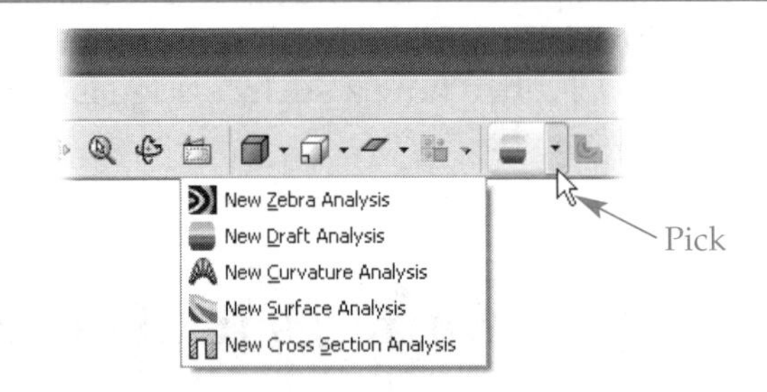

Figure 16-14.
Surface topology can be analyzed using zebra stripes.

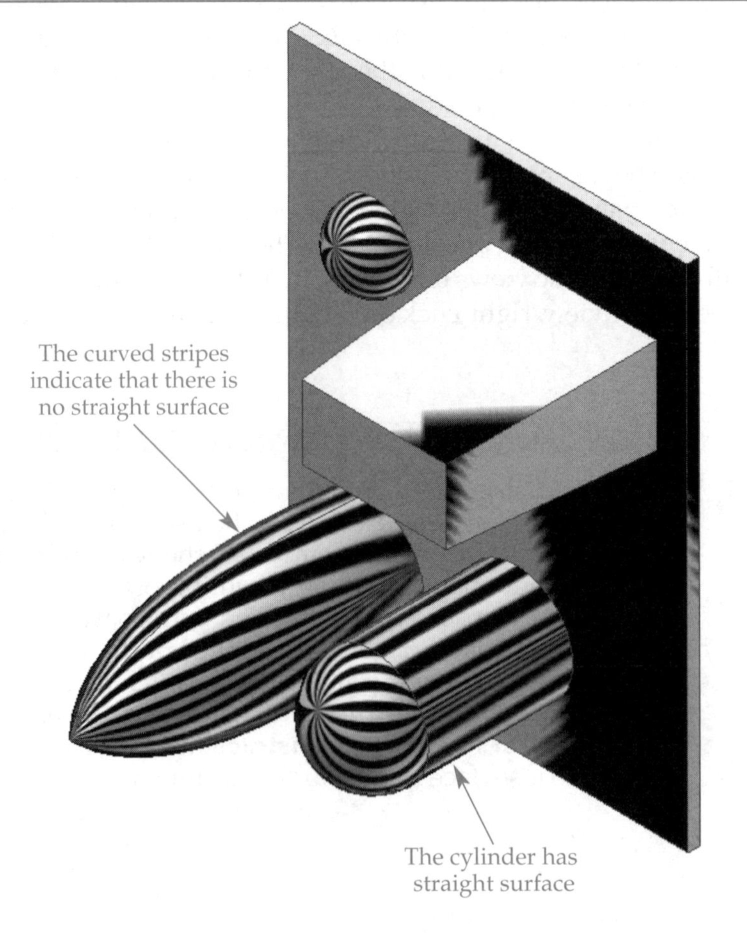

Figure 16-15.
Making settings for a zebra analysis.

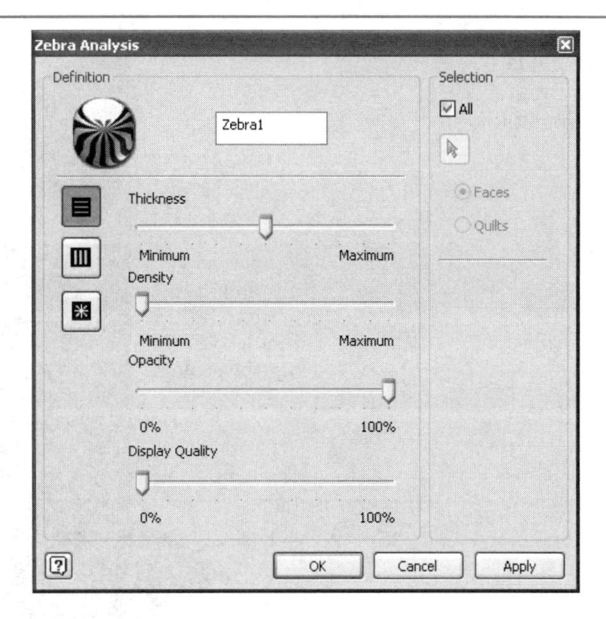

the curvature. One thing to keep in mind when displaying a zebra analysis is that the stripes are not "fixed" to the part. As you rotate the view, the lines will "move around" on the part.

The second tool in the list is a ***draft analysis.*** This helps the designer to determine whether or not the part can be removed from the mold. If not, then the draft angle will have to be changed and a new analysis run. To display the draft analysis, pick the **New Draft Analysis** button in the **Analysis Visibility** flyout or **Tools**>**Analysis**>**New Draft Analysis**. The **Draft Analysis** dialog box is displayed, **Figure 16-16.** The settings in the dialog box reflect the draft settings. In the middle of the dialog box are two text boxes in which the minimum and maximum draft angles are entered. The color spectrum shown below these text boxes is applied to the surface of the part. In the **Selection** area of the dialog box, you can choose to have the analysis applied to the entire part or to selected faces. To display the analysis, pick the **OK** button.

The part in Example_16_13.ipt will need to be pulled from a cavity of a mold. Because the draft angles vary, each shape will present a different degree of difficulty when pulling the part out of the mold. Set the start angle (left-hand text box) to –2° and the end angle (right-hand text box) to 10°. In the **Selection** area, check the **All** check box. Pick **OK** to see the draft analysis applied to the part. See **Figure 16-17.** The cylinder was extruded without any taper. Therefore, any face that is the same color as the cylinder (purple) has 0° of draft. Any area displayed in a color that is left of the purple on the

Figure 16-16.
Making settings for a draft analysis.

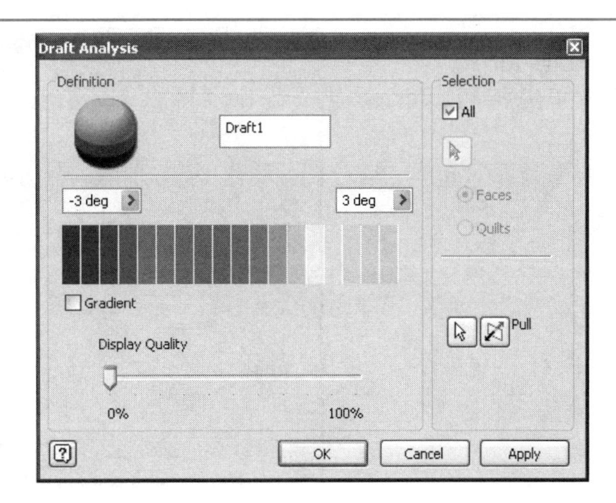

Figure 16-17.
The draft angle is represented by a corresponding color.

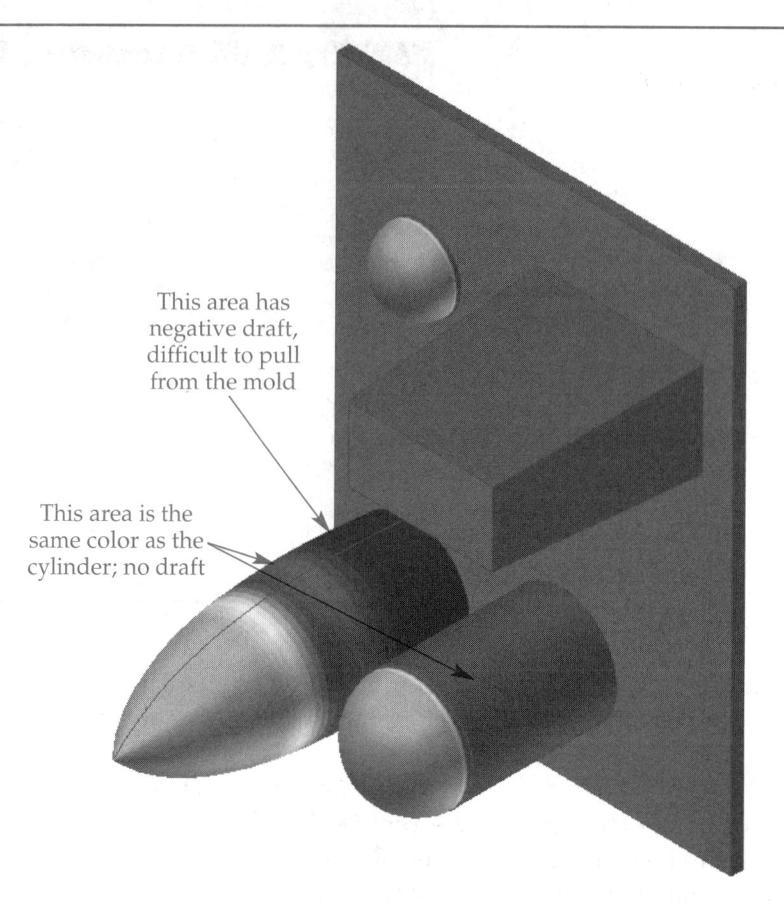

scale has a negative taper and could be difficult to remove from the mold. The bottom of the bullet-shaped feature is blue, which indicates an undercut situation that may make the part difficult to pull out of the mold.

The third tool in the list is a ***curvature analysis.*** This provides a visual analysis of the curvature and overall smoothness of the surface. Open Example_16_14.ipt. This part consists of two surfaces stitched into a quilt. To display the curvature analysis, pick the **New Curvature Analysis** button in the **Analysis Visibility** flyout or **Tools**>**Analysis**>**New Curvature Analysis**. The **Curvature Analysis** dialog box is displayed, **Figure 16-18.** Pick the **Faces** radio button in the **Selection** area and then pick both surfaces. Finally, pick the **OK** button to display the analysis. The resulting lines and curves look like a comb, **Figure 16-19.** This analysis can also be applied to model faces, sketch curves, and edges.

Figure 16-18.
Making settings for a curvature analysis.

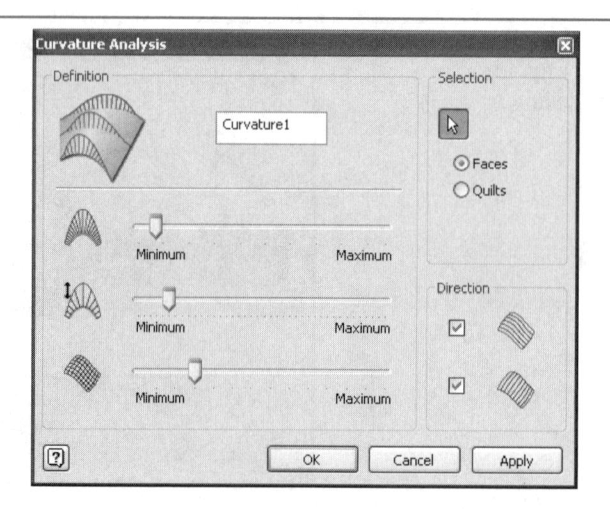

Figure 16-19.
A curvature analysis appears like a comb.

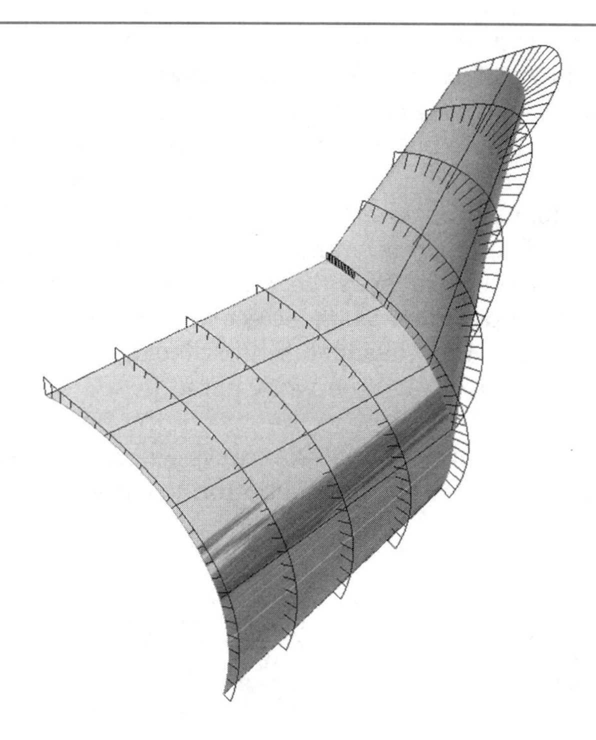

The fourth tool in the list is a ***surface analysis.*** It displays areas of high and low surface curvature using a color gradient. Open Example_16_15.ipt. To display the surface analysis, pick the **New Surface Analysis** button in the **Analysis Visibility** flyout or **Tools>Analysis>New Surface Analysis**. The **Surface Analysis** dialog box is displayed, **Figure 16-20.** In the **Selection** area, uncheck the **All** check box and pick the **Faces** radio button. Then, select the surface. Next, set the minimum curvature (left-hand text box) to –2° and the maximum curvature (right-hand text box) to 2°. In the drop-down list at the top of the dialog box, pick Max Curvature. Finally, pick the **OK** button to apply the analysis. The areas of maximum curvature, shown in blue and yellow, indicate where there may be problems manufacturing the part.

The last tool in the list is a ***cross section analysis.*** This is for the interior of solid parts. It can provide a simple analysis, which is a graphic view of the part cross section. It can also provide an advanced analysis, which gives detailed information, such as the area, and corresponding graphic for the section. The tool also analyzes whether or not the part adheres to the minimum and maximum wall thickness. Open Example_16_16.ipt.

Figure 16-20.
Making settings for a surface analysis.

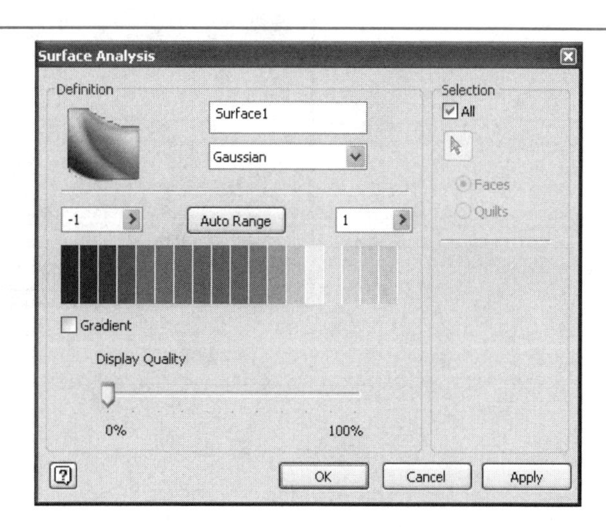

To display the cross section analysis, pick the **New Cross Section Analysis** button in the **Analysis Visibility** flyout or **Tools>Analysis>New Cross Section Analysis**. The **Cross Section Analysis** dialog box is displayed, **Figure 16-21.**

Pick the **Simple** button on the left side of the dialog box (this is on by default). Then, with the **Plane** button on, select the work plane in the graphics window. This is the cutting plane from which the cross section is created. Pick the **OK** button to display the analysis. The result is similar to the **Slice Graphics** option when in sketch mode. The portion of the part above the work plane is not visible.

Undo the operation. Then, select the tool again. In the **Cross Section Analysis** dialog box, select the **Advanced** button. Next, with the **Select** radio button on in the **Section Planes** area, pick the work plane in the graphics window. Finally, pick the **Calculate** button. This displays the data in the **Results** area of the dialog box. Also, in the graphics window, the analysis shows where the thickness exceeds the maximum. See **Figure 16-22.** Pick the **Cancel** button to close the dialog box. The graphic analysis remains until you turn it off.

Figure 16-21.
A—Making settings for a simple cross section analysis.
B—Making settings for an advanced cross section analysis.

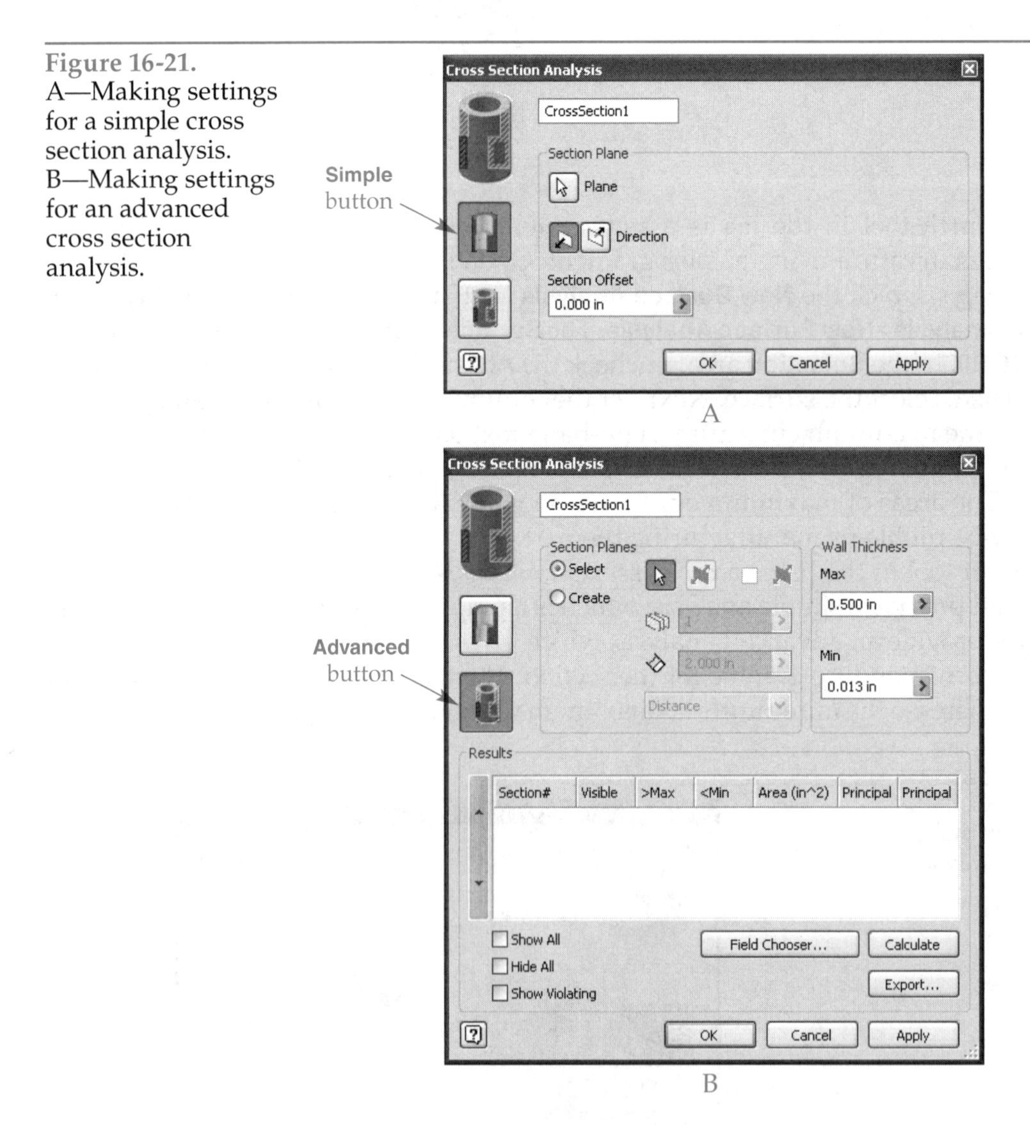

Figure 16-22.
An advanced cross section analysis displayed on the part.

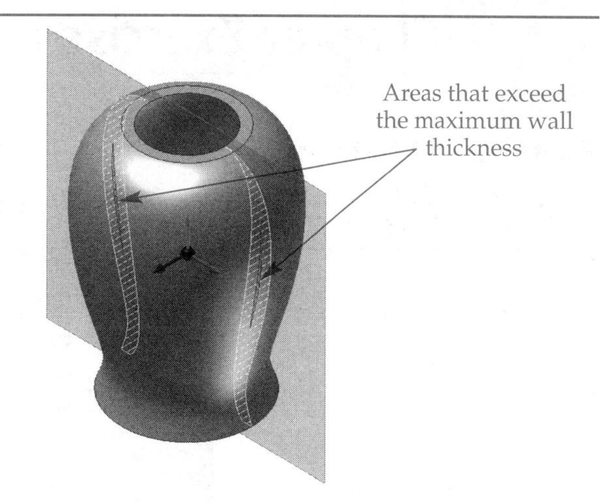

Chapter Test

Answer the following questions on a separate sheet of paper or complete the electronic chapter test on the Student CD.

1. Explain the differences between a surface and a solid.
2. How does the process for creating a surface differ from the process used to create a solid?
3. Briefly describe how to change the display of an existing surface from translucent to opaque.
4. Which tool is used to create a new surface that is offset from the original surface?
5. Define *quilt.*
6. Which tool is used to create a quilt from separate surfaces?
7. What must be true of two surfaces that are going to be joined using the tool in question 6?
8. Give two examples of how a surface can be used as construction geometry.
9. List two ways to initiate an analysis of the faces of a part.
10. What are the modes available for analyzing the faces of a part?

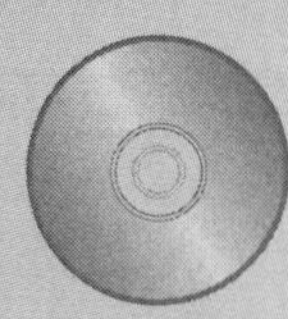

Chapter Exercises

Exercise 16-1. Offset Surface.

Complete the exercise on the Student CD

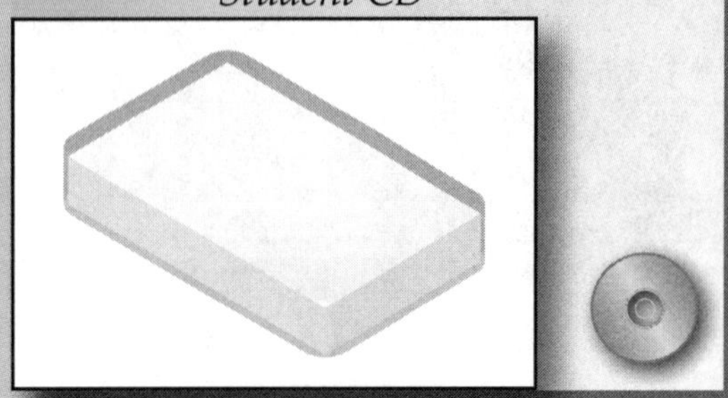

Exercise 16-2. Revolved Surface.

Complete the exercise on the Student CD

Exercise 16-3. Extruded Surface.

Complete the exercise on the Student CD

Exercise 16-4. Loft Surface.

Complete the exercise on the Student CD

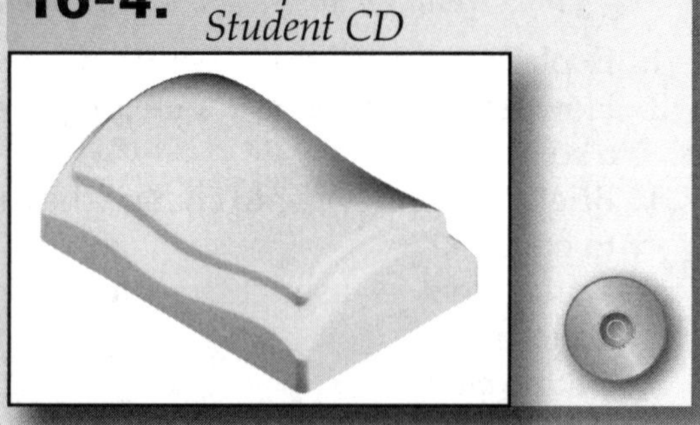

Exercise 16-5. Welding Fixture.

Complete the exercise on the Student CD

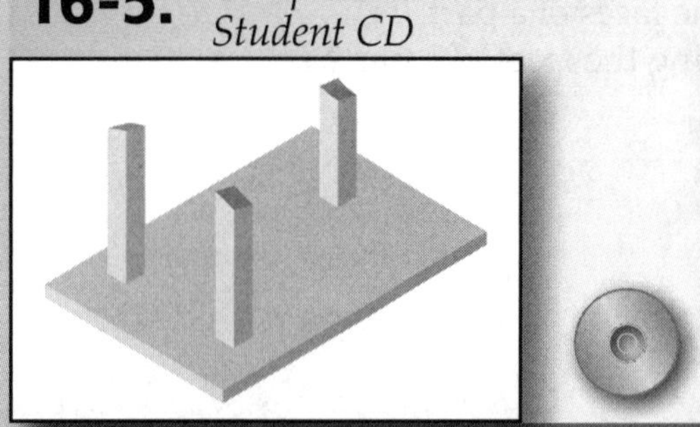

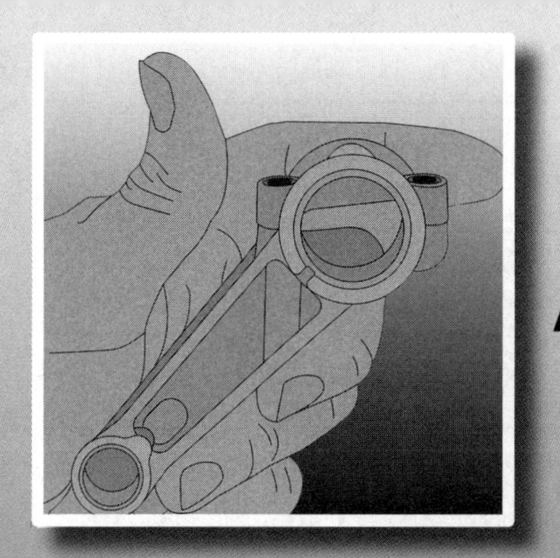

Chapter 17

Assembly Drawings

Objectives

After completing this chapter, you will be able to:

- Create 2D orthographic, section, and break out views from assemblies.
- Annotate assembly drawings.
- View, modify, and insert parts lists.
- Create drawing views using design view representations.

User's Files

The Student CD included with this text contains several files required for this chapter. Refer to the file File List.txt *in the* \Ch17 *folder for the comprehensive list.*

Views

Open Example_17_01.iam and examine the model to familiarize yourself with the various parts and subassemblies used in the assembly. See **Figure 17-1.** Notice that several of the work planes are visible—these will not appear in the drawing views. Pay attention to the orientation of the coordinate system. This plays a crucial part in specifying the viewing direction for the creation of the drawing views.

Create a new drawing (IDW) based on the ANSI (in).idw template and save it as Example_17_01.idw. Set the sheet size to E. Using the **Base View** tool, create a base view from the bottom orientation of the assembly. Pick in the lower-left corner of the drawing to place the base view. Refer to **Figure 17-2** for the correct orientation and location. Even though the bottom view orientation of the assembly was selected, it will be the front view for the drawing. You cannot always count on an assembly being properly oriented for the creation of a given view, so you must become comfortable with using the various options available.

Now, create three additional views projected from the front (base) view. Right-click on the base view and select **Create View>Projected** from the pop-up menu. See **Figure 17-3.** When the cursor is moved around the drawing, different views appear.

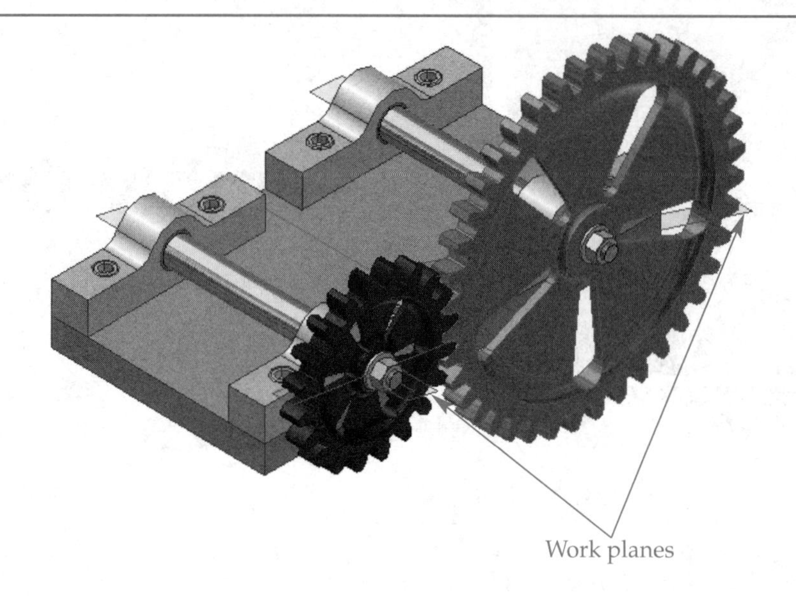

Figure 17-1.
An assembly drawing will be created from this assembly.

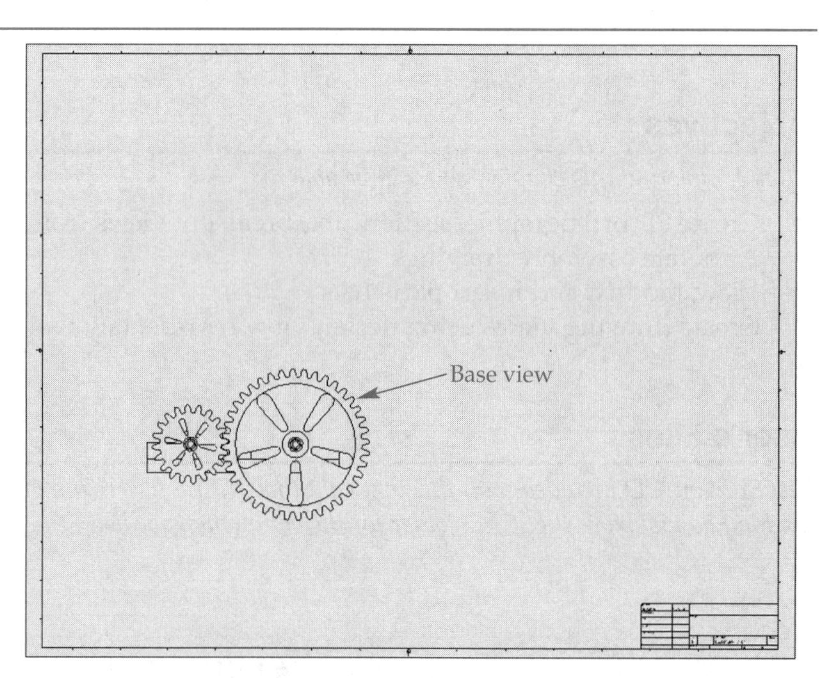

Figure 17-2.
The correct orientation of the base view.

Pick a location above the base view, to the right of the base view, and diagonally to the upper-right of the base view. When all three points have been specified, right-click and select **Create** from the pop-up menu. Inventor produces the three projected views.

The gears in the front view appear to be missing some geometry. There are some fillets that are not shown. Double-click on the view to display the **Drawing View** dialog box and then pick the **Display Options** tab. Check the **Tangent Edges** check box and pick the **OK** button. Notice the effect on the gears in the front view.

In the **Browser**, the views are not referred to as Front or Top, but instead have numbers assigned to them. Also, dependent views are listed underneath their parent views. For example, VIEW2 and VIEW3 are dependent on VIEW1 (the base view). See **Figure 17-4.**

Currently, all of the orthographic views inherit their display style from the front view, the view on which they are based. If you change the display style for the front view, all other views follow suit. Try it now—open the **Drawing View** dialog box for the front view and change the display style by picking the **Shaded** button. Pick the

Figure 17-3.
Creating projected views.

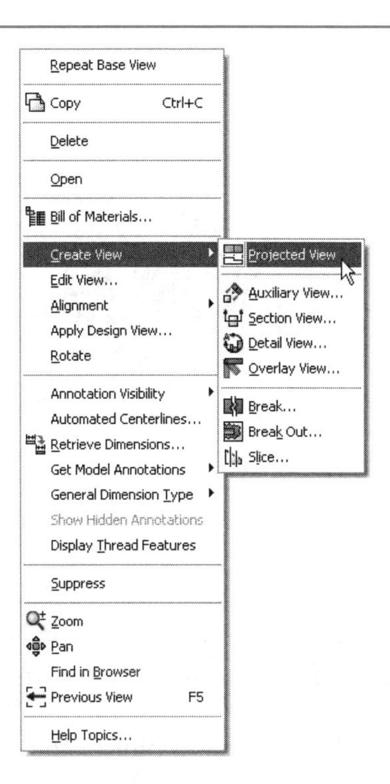

Figure 17-4.
View names and the layout of dependent views.

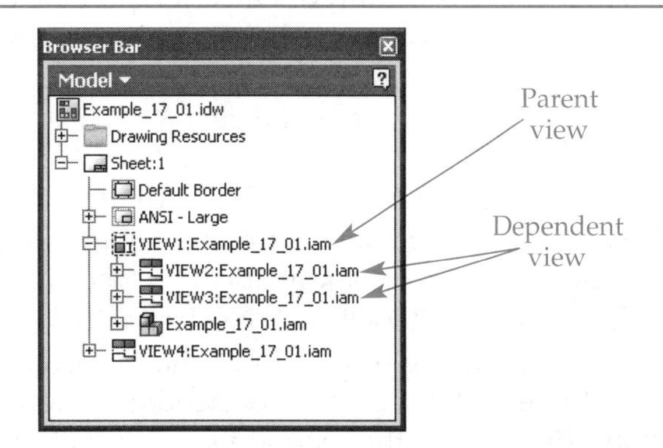

OK button to apply the change. Notice that the isometric view is unaltered. Open the **Drawing View** dialog box again for the front view, change the display style by picking the **Hidden Line** button, turning the **Shaded** button off, and applying the change. The hidden line and hidden line views can both be shaded or not. Next, change the display style of the isometric view to shaded. Save the drawing.

Section Views

The side view that was created does not convey many details of the assembly. A section view would reveal the interior details by cutting away some of the assembly. To properly describe the large gear, an aligned section view needs to be used. This type of section view is discussed in Chapter 8. Delete the side view by right-clicking its name in the **Browser** and picking **Delete** in the pop-up menu. Then, right-click on the front view and select **Section View...** from the **Create View** cascading menu in the pop-up menu. Continue as follows.

1. Hover over the center of the large gear and acquire the center point—do not pick. Refer to **Figure 17-5.**

Figure 17-5.
Creating an aligned section view.

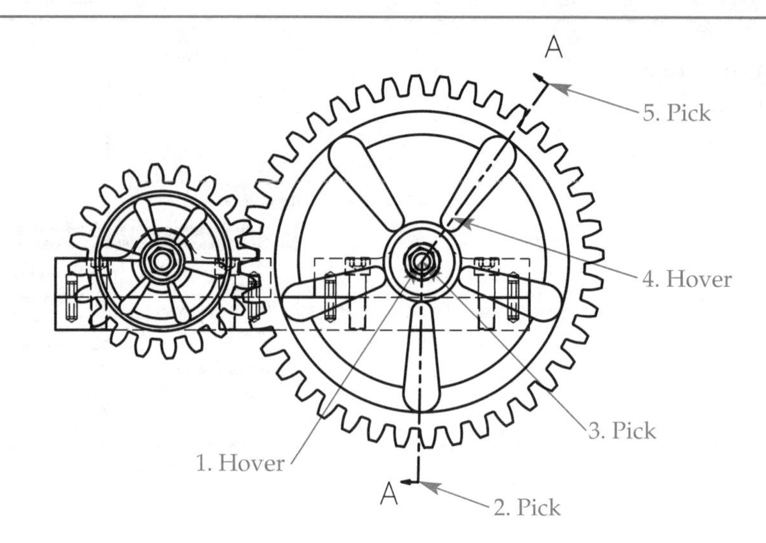

2. Slowly move the cursor straight down so a dotted line is being projected from the center of the gear. Pick a point that is outside of the gear and inline with the gear center.
3. Move the cursor back to the center of the gear and pick that point.
4. Slowly move the cursor toward the upper-right until the center of the web cutout's smaller arc is acquired. Once the point shows up green, move the cursor up and to the right so a line is being projected through the point and the point remains green.
5. Pick a point outside of the gear.
6. Right-click and select **Continue** from the pop-up menu. The **Section View** dialog box is displayed. Review the settings.
7. Drag the section view to the right of the front view and pick to place it. The **Section View** dialog box is automatically closed.

Zoom in on SECTION A-A. Notice that all of the parts in the assembly are sectioned. In accepted drafting conventions, shafts, fasteners, and most standard components are not sectioned. The section lines need to be removed from these parts. In the **Browser**, expand the tree for the section view until all of the parts that make up this assembly are shown. Hover over the second instance of the shaft. Notice how the sectioned shaft in the drawing is highlighted. Right-click on the shaft name and select **Section Participation>None** in the pop-up menu. The section view is updated and the shaft is no longer sectioned. In a similar manner, remove the section lining from the second instance of the hex nut and washer. Now, SECTION A-A follows accepted drafting conventions. Save the drawing and close all files.

Sliced and Offset Section Views

Open Example_17_02.idw. There are two section views and a base view. Right-click on the view SECTION B-B and pick **Edit Section Properties...** from the pop-up menu. The **Edit Section Properties** dialog box is displayed. In the **Slice** area of the dialog box, check the **Include Slice** check box and then check the **Slice All Part** check box. Pick the **OK** button to update the section view. The view is now a very thin slice of the assembly. See **Figure 17-6.** Only the features on the cutting plane are shown. This type of view is called a ***sliced section view.***

Right-click on the view SECTION A-A and pick **Edit Section Properties...** from the pop-up menu. In the **Section Depth** drop-down list in the **Edit Section Properties**

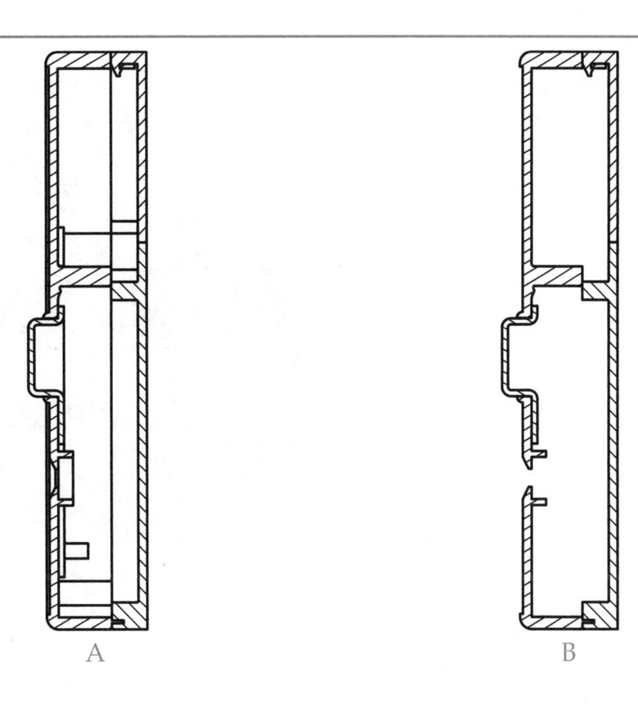

Figure 17-6.
A—A normal section view.
B—A sliced section view. Notice how many of the lines in A are not shown.

dialog box, select Distance. Then, in the text box below the drop-down list, change the value to 0.5 mm. This defines the thickness of the sliced section. A line appears on the main view offset from the section line. Only the material between the section line and the offset line is displayed in the section view. Pick the **OK** button to update the section view. This type of view is called an ***offset section view.*** You can think of an offset section view as a "fat" slice, whereas a sliced section view is a "thin" slice.

PRACTICE 17-1 Complete the practice problem on the Student CD.

Break Out Views

Inventor has the capability of removing a portion of an assembly and revealing what is underneath. For example, part of an outer cabinet can be removed to reveal the machinery inside. This is called a ***break out view.*** The first step in creating a break out view is to create a sketch inside of the view. The sketch determines the boundary of the break out. **Figure 17-7** shows the final view you will create in this section.

1. Open Example_17_03.iam.
2. Create a new metric drawing based on the ANSI (in).idw template and save it as Example_17_03.idw.
3. Change the sheet size to D.
4. Right-click on the sheet in the **Browser** and select **Base View...** from the pop-up menu.
5. In the **Drawing View** dialog box, set the scale to .25 and select the front orientation. Make sure the assembly file is selected in the **File** drop-down list.
6. Pick a point in the lower-left corner of the drawing to place the view.
7. Create projected views above and to the right of the base view.

You will see empty squares because the outer sheet metal cabinet is hiding the internal parts. In order to see the parts inside of the cabinet, double-click on the front view to open the **Drawing View** dialog box. Select the **Hidden Line** button in the **Style**

Figure 17-7.
The final projected view with a break out.

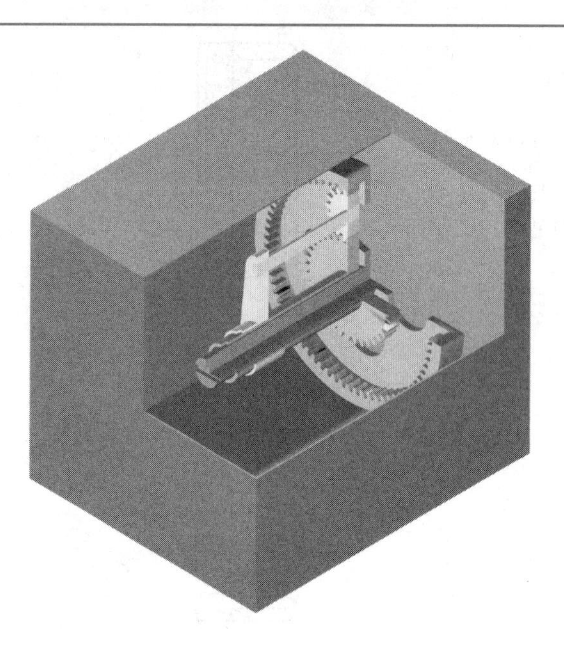

area of the dialog box. Make sure the **Shaded** button is off and then pick the **OK** button. All three views now display the hidden lines. See **Figure 17-8.** If the dependent views do not change to display hidden lines, double-click on a dependent view to open the **Drawing View** dialog box. Then, check the **Style From Base** check box in the **Style** area and pick the **OK** button. Repeat this for any dependent view that does not display hidden lines.

Figure 17-8.
Hidden lines are shown.

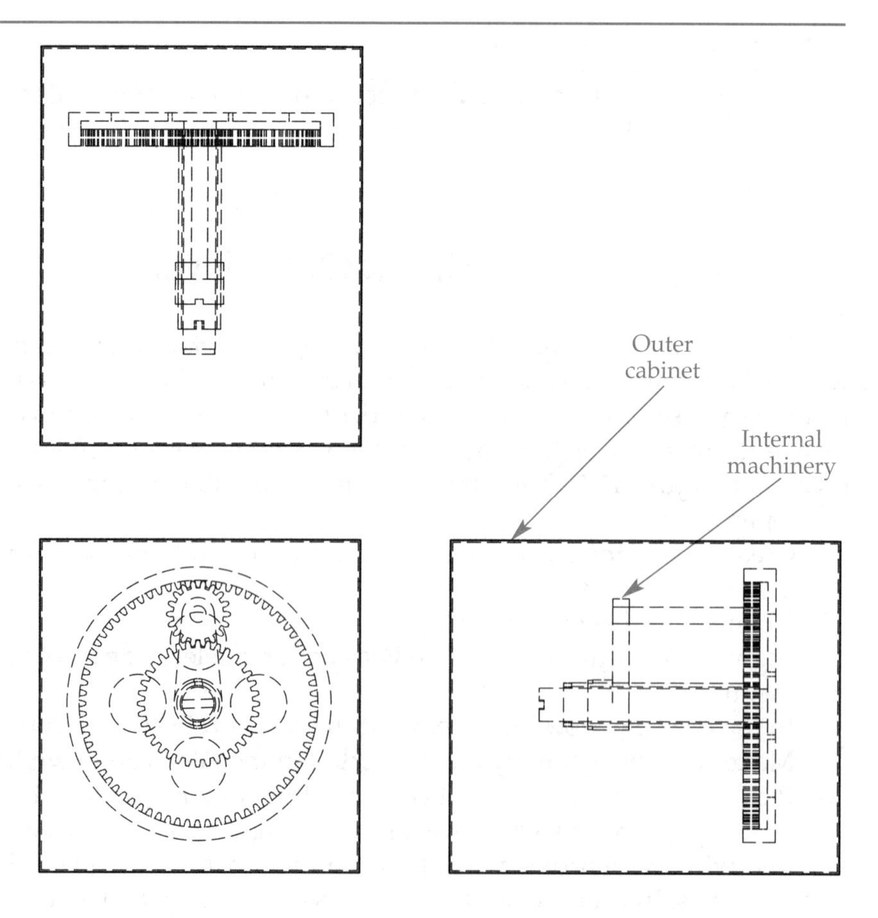

Next, pick the side view to activate it. Then, pick the **Sketch** button on the **Inventor Standard** toolbar to create a sketch that is considered a part of the view. If the view is not highlighted when the sketch is started, the sketch is part of the drawing, not the view. Draw a rectangle as shown in **Figure 17-9.** No need to dimension it, just finish the sketch. Finally, right-click on the side view and select **Create View>Break Out...** in the pop-up menu. The **Break Out View** dialog box is displayed, **Figure 17-10.**

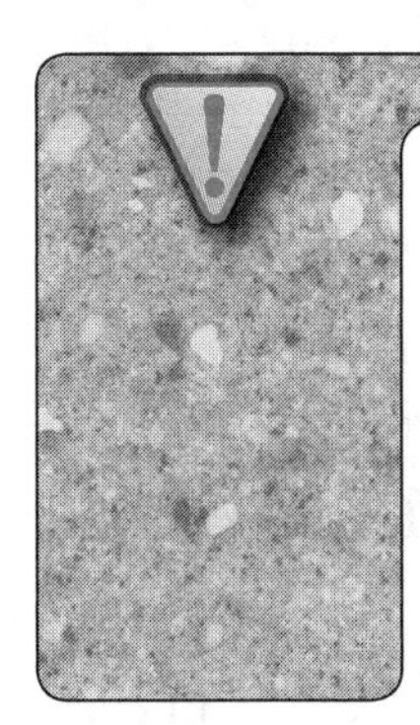

CAUTION

The following error message appears if the sketch is not part of the view. Use the **Undo** tool until the sketch is gone, and then select the view *before* selecting the **Sketch** tool.

Figure 17-9.
The rectangle defines the break out.

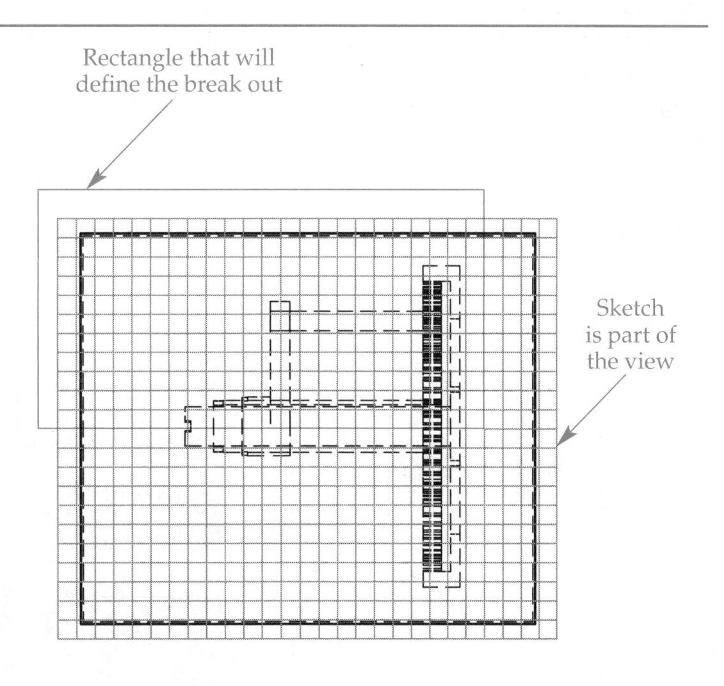

Figure 17-10.
The **Break Out View** dialog box with the **Depth** button selected.

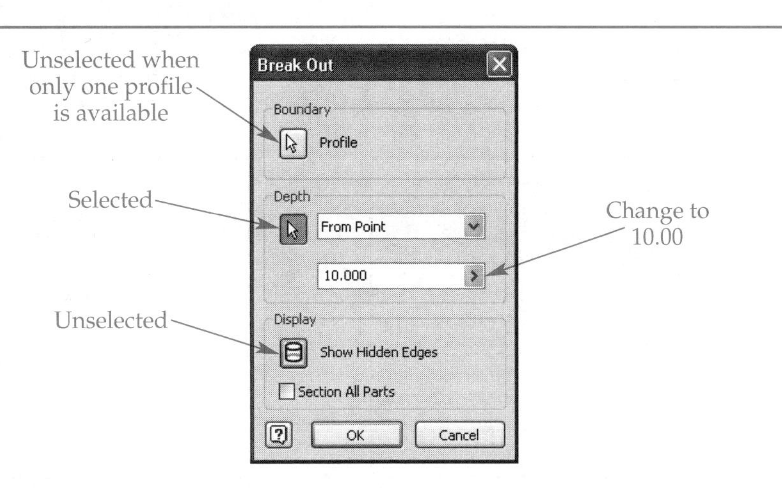

Inventor automatically selects the sketch profile because there is only one in this case. Next, in order to create a break out view, the depth of the break out must be specified. Make sure that the method of specifying the depth is set to From Point and the **Show Hidden Edges** button is not selected. In the **Depth** area, change the depth value to 10.00. There are other methods of specifying the depth of the cut:

- **To Sketch.** Uses an existing sketch in another view to specify depth.
- **To Hole.** Uses an existing hole feature in the assembly or part to specify depth.
- **Through.** The break out cuts all the way through the assembly or part.

With the **Selector** button on in the **Depth** area, pick the start point of the break out view in the *top* view. In other words, from where do you want Inventor to measure the 10.00″? In the top view, pick the lower-right corner of the rectangle representing the sheet metal cabinet. Then, pick the **OK** button to exit the **Break Out View** dialog box and the break out is created in the side view.

The break out view does not convey much information. An isometric view needs to be projected from the break out view to better represent the internal features. Right-click on the side view, select **Create View>Projected View** in the pop-up menu, and pick a point to the upper left. Then, right-click and select **Create** from the pop-up menu. Change the display style for the isometric view to shaded and the drawing should look like **Figure 17-11.**

The last view you created is not technically a break out view—only an isometric view of the break out view. The break out view is the side view. Always remember, the break out view is generated *into* the view that is perpendicular to the "paper."

Figure 17-11.
The completed views with a shaded isometric break out view.

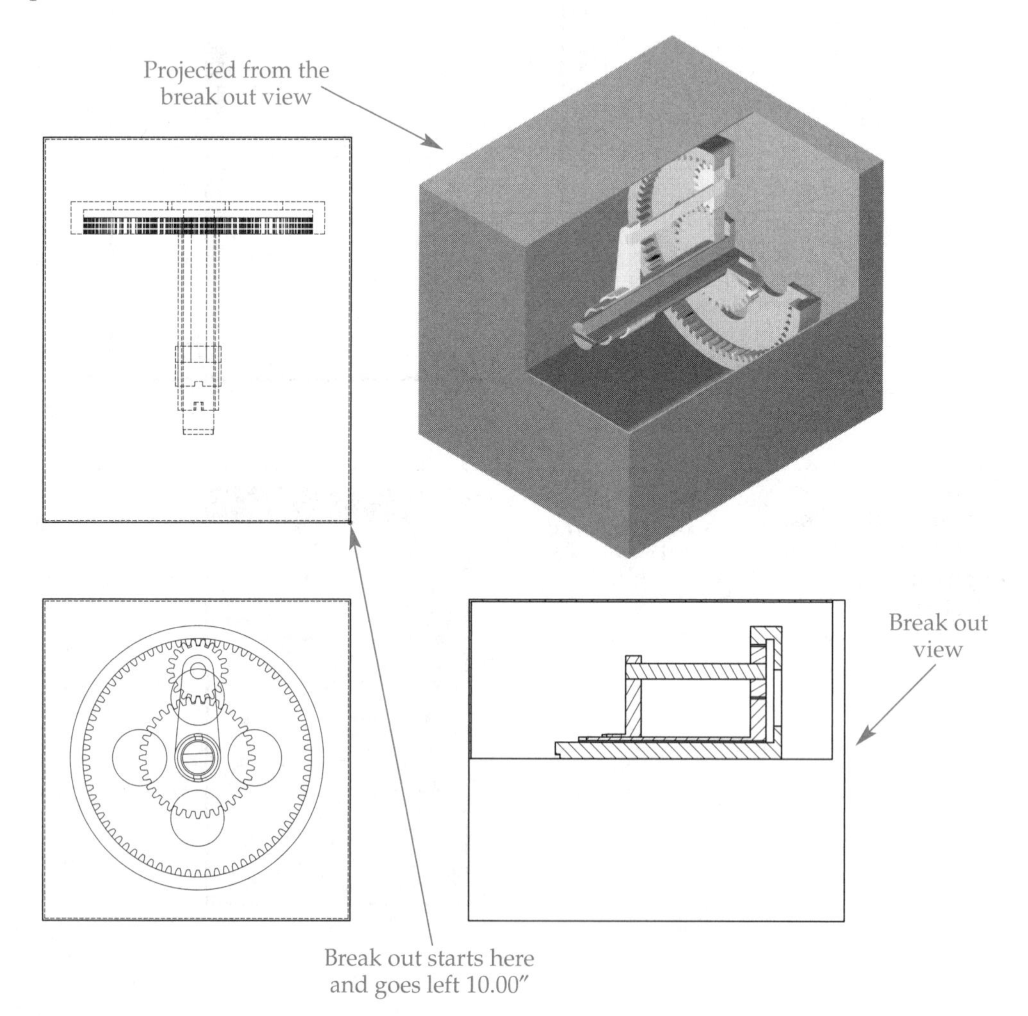

Annotations

As you learned in Chapter 9, various types of annotations can be applied to 2D drawings. All of the annotations that can be applied to a part drawing are available for an assembly drawing. This section discusses some of these annotations.

Centerlines

Open Example_17_04.idw. Centerlines can be added to this assembly drawing in the same way as they would be added to part drawings. For this example, however, the **Automated Centerlines** tool will be used.

Zoom in on the front view. Then, right-click on the view and select **Automated Centerlines...** from the pop-up menu. The **Automated Centerlines** dialog box is displayed, **Figure 17-12.** In the **Apply To** area of the dialog box, pick the **Hole Features** and **Cylindrical Features** buttons so they are on. This determines which features have centerlines applied to them. In the **Projection** area, make sure the **Objects in View, Axis Normal** button is on. The second button in this area is **Axis Parallel**. This button would be selected when placing centerlines on the side views of this assembly. In the **Radius Threshold** area, you can set minimum dimensions so small fillets and edges that should not have centerlines are ignored. Pick the **OK** button and the centerlines are applied to the view, **Figure 17-13.**

Some clean up is necessary with method. However, this cleanup is easier than manually placing each centerline. Experiment by placing centerlines on the three other views.

Dimensions

Dimensioning an assembly is similar to dimensioning a part. After activating the appropriate dimensioning tool, select the part edges or centerlines. Once the proper geometry is selected, the dimension can be placed on the drawing. Open Example_17_05.idw. Refer to **Figure 17-14** as you place the dimensions on the top view. On the front view, place a horizontal dimension between the two gears. This documents the distance between the two gear shafts. Use the endpoints of the center marks. On the section view, place the dimensions shown in **Figure 17-15.**

After placing a dimension, you may need to change its value or alter the appearance of the dimension text. There are several techniques available, as described in the next sections. Choose the appropriate one for the situation at hand.

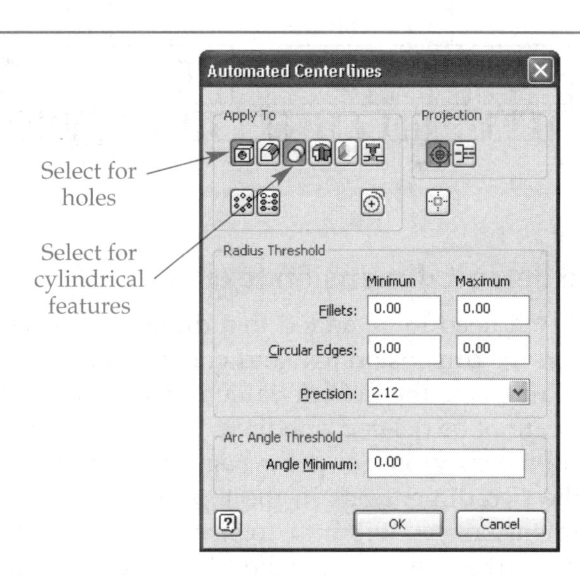

Figure 17-12. The **Automated Centerlines** dialog box is used to automatically apply centerlines to a view.

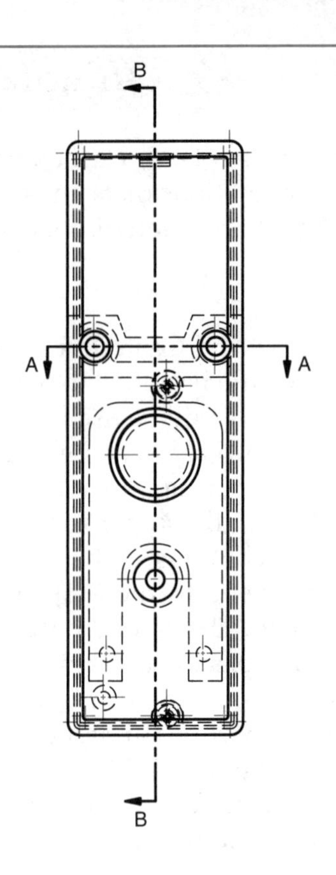

Figure 17-13.
The view after centerlines were automatically applied and some clean-up work performed. The centerlines are shown in color.

Figure 17-14.
The top view with dimensions.

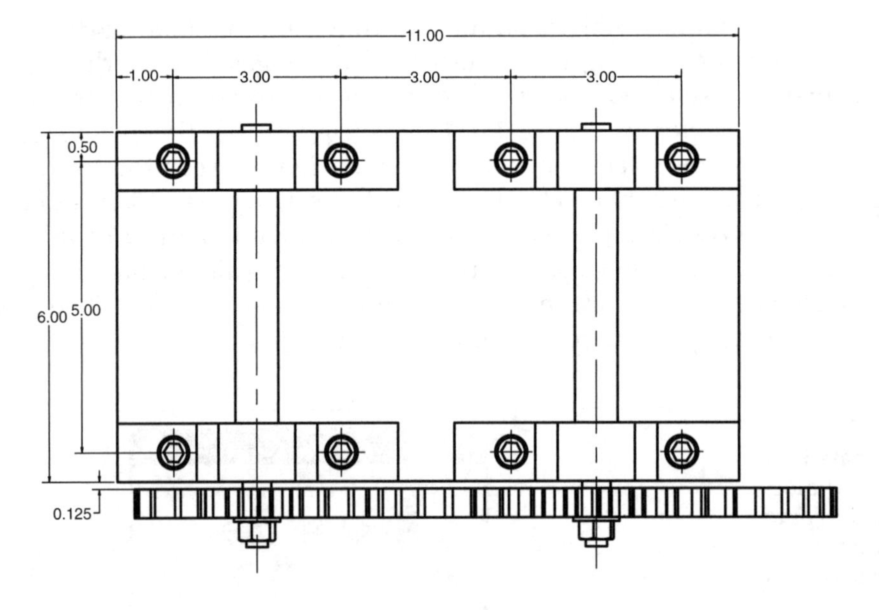

Add text to the default dimension text

At times text will need to be added to a dimension, such as TYP, REF, or 6X. To do so, right-click on the dimension text and choose **Text...** from the pop-up menu. In the **Format Text** dialog box, the default value is represented by the chevrons (<<>>). The default value cannot be deleted or edited. To add text to the dimension, place the cursor at the end of the chevrons (or at the beginning) and type the necessary text.

Do this for the two dimensions in the top view shown in **Figure 17-16.** To add TYP to the 1.00 dimension, you will have to place the cursor at the end of the chevrons and press [Enter] to start a new line. For the 0.50 dimension, just start typing after the

Figure 17-15.
The section view with dimensions.

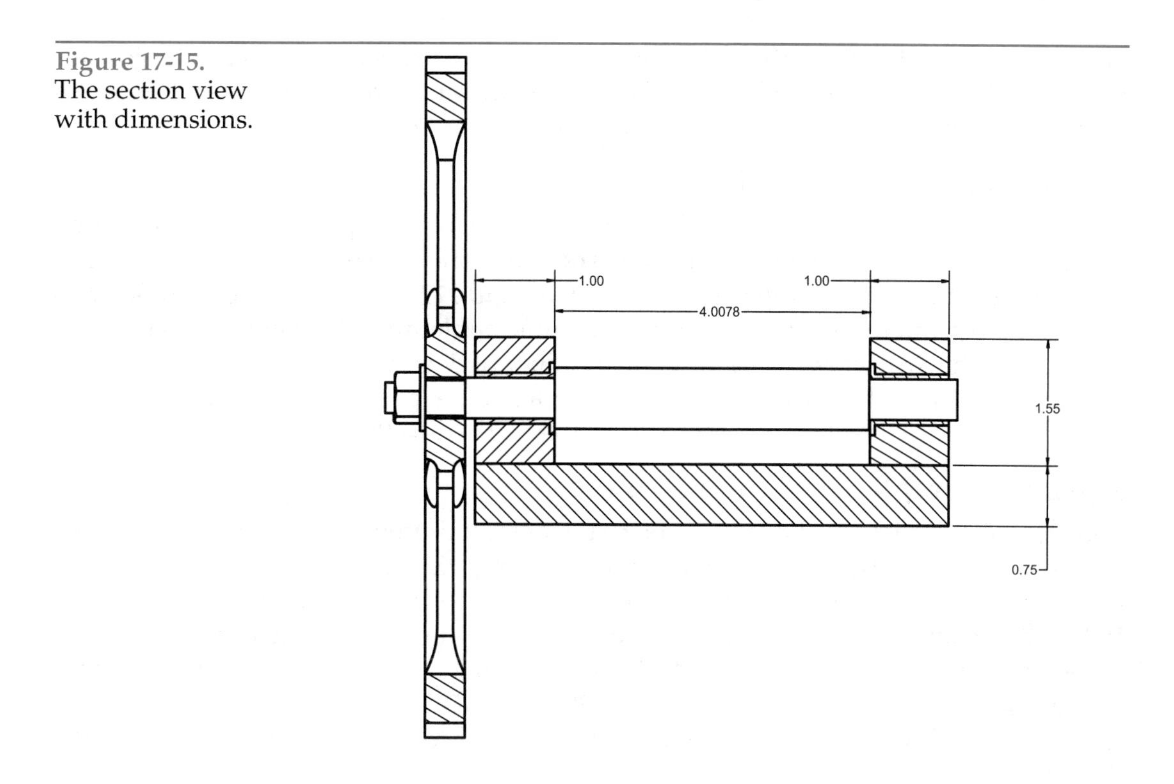

Figure 17-16.
Adding text to dimensions.

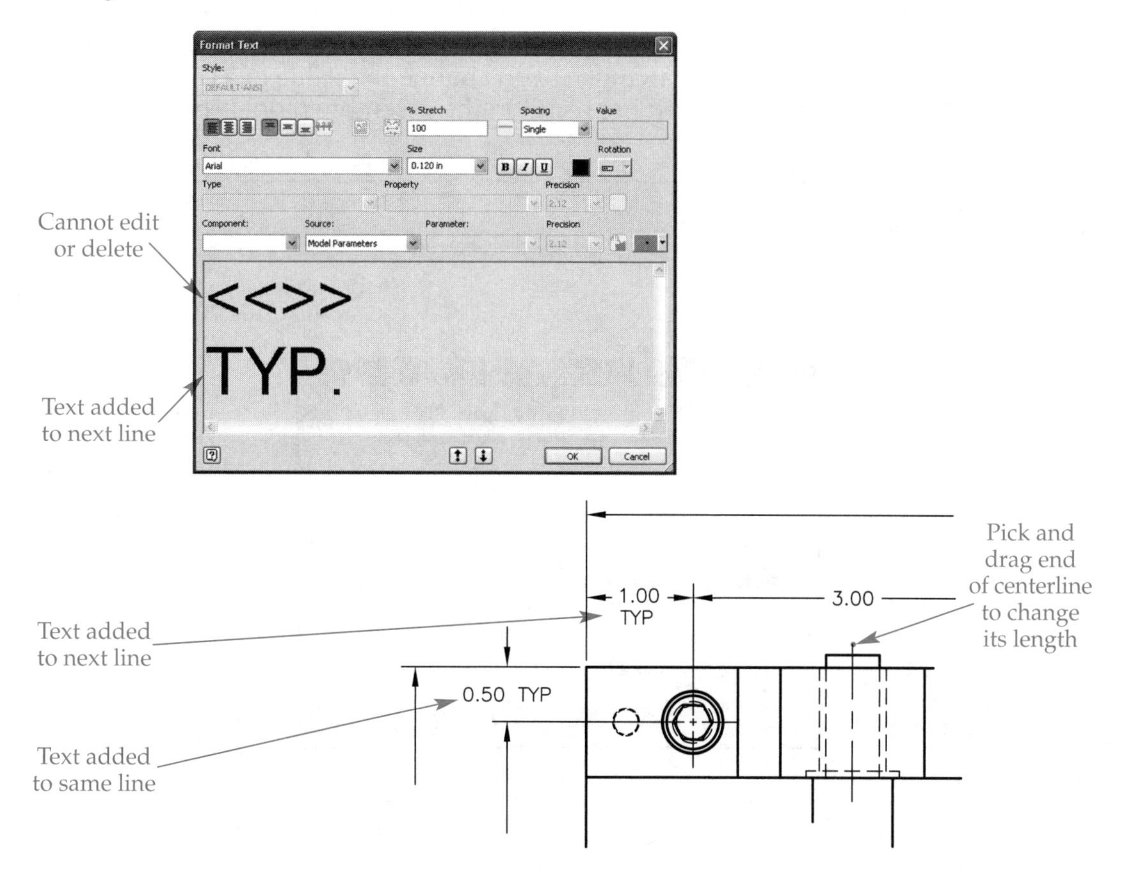

chevrons. The TYP designation signifies that those dimension values are typical for similar features in the assembly. In this case, the distance from the corner of the base plate to the location of the hole is typical for all similar holes.

Add tolerancing information

To add a tolerance, double-click on the dimension to display the **Edit Dimension** dialog box. Select the **Precision and Tolerance** tab. Then, select the type of tolerance desired in the **Tolerance Method** list. As shown in **Figure 17-17,** change the 6.00 dimension between the gears in the front view to a maximum tolerance. In the **Precision** area of the tab, use the drop-down lists to set the number of decimal places used in the value. To enter a value that is different from the model value, check the **Override Displayed Value** check box. Then, type a new value in the text box to the right of the check box.

Completely replace the dimension text

Occasionally, you will need to replace the default dimension text. Add a thickness dimension to the large gear shown in the top view, **Figure 17-18.** Then, double-click on the dimension to display the **Edit Dimension** dialog box. In the **Text** tab, check the **Hide Dimension Value** check box and then close the dialog box. The dimension's text is replaced by <TEXT>. Open the **Edit Dimension** dialog box for the dimension. Then, in the **Text** tab, type the new text, such as SEE VENDOR PRINT. Pick the **OK** button to exit the dialog box. This dimension now displays the text you entered in place of the default value.

PROFESSIONAL TIP

Assembly drawings are not bidirectionally associative. This means, unlike in part drawings, if you change the value of a dimension in the drawing it will not result in the assembly model changing.

Figure 17-17.
Adding a tolerance to a dimension.

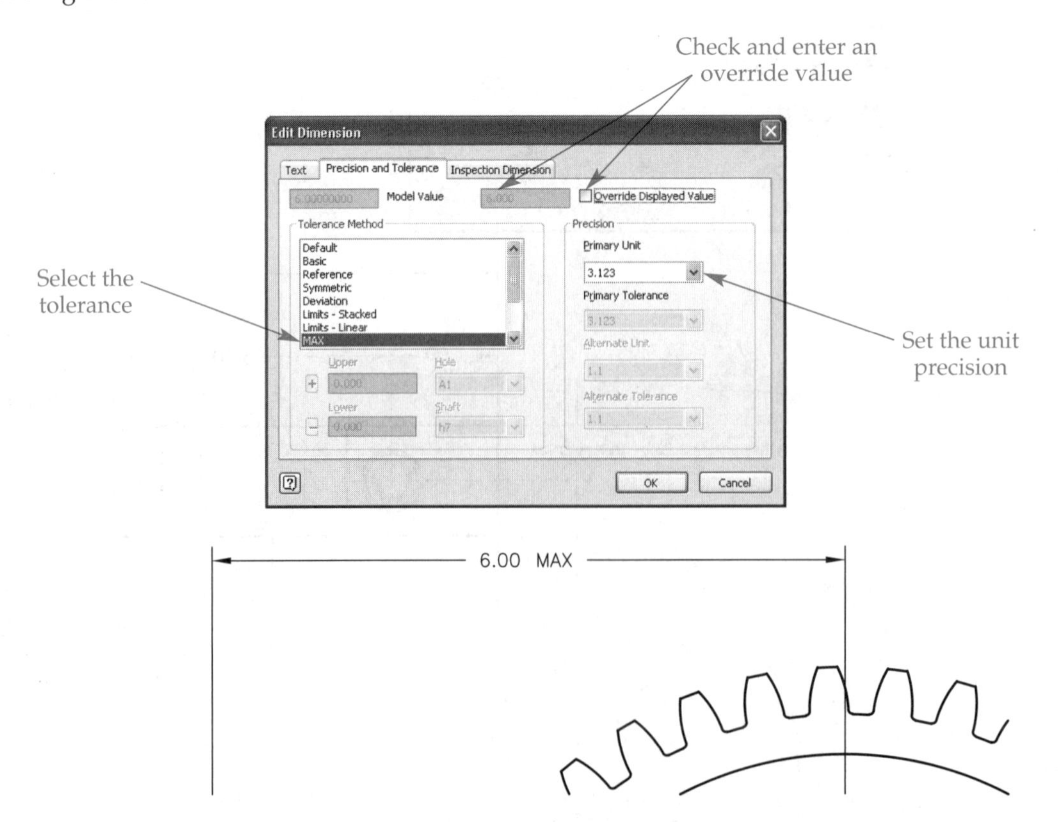

Figure 17-18.
Replacing the dimension with a note.

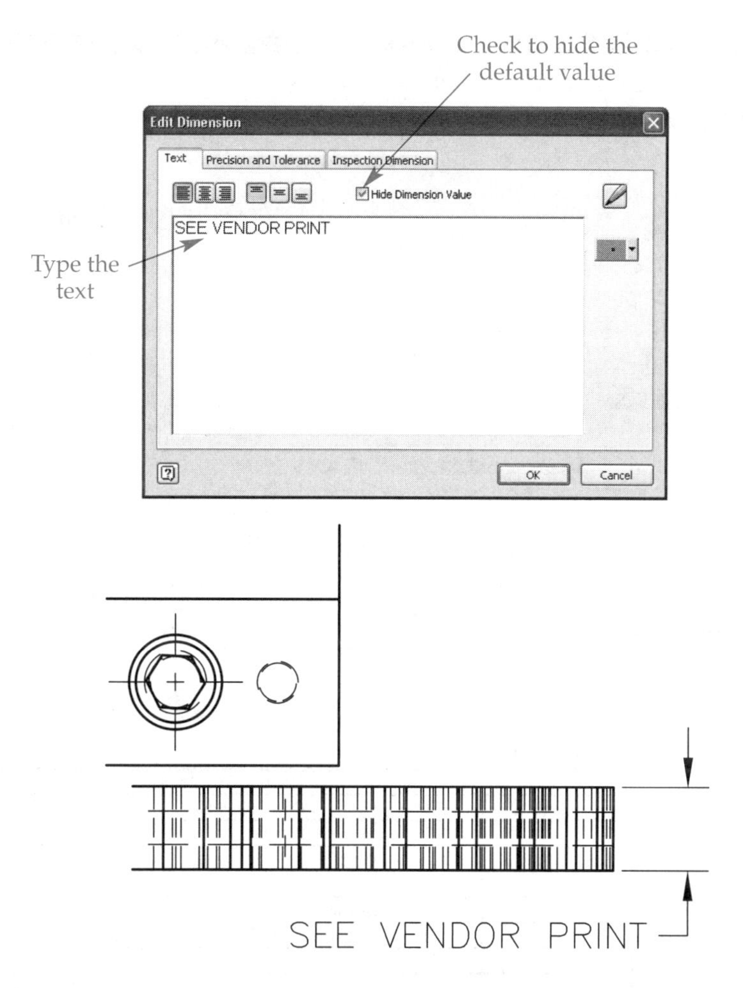

Leader Text

Leader text is added to an assembly drawing in the same manner as it is added to a part drawing. See **Figure 17-19.** Pick the **Leader Text** button in the **Drawing Annotation Panel**. Then, pick a point on a tooth of the large gear, pick in the drawing to locate the text, right-click, and select **Continue** in the pop-up menu. In the **Format Text** dialog box, type the text shown in the figure and then pick the **OK** button. Add a leader on the smaller gear with the text 20T PINION GEAR.

Balloons

Balloons are numbered callouts that identify the components of an assembly. The numbers are used to correlate the items in a parts list to the drawing. In this section, you will add balloons to a drawing. First, however, you will add a parts list to the drawing so you can see from where the balloon numbers come. Parts lists are described in detail later in this chapter.

Open Example_17_06.idw. Display the **Drawing Annotation Panel**. Then, select the **Parts List** button. The **Parts List** dialog box is displayed. With the **Select View** button on in the dialog box, pick the isometric view. Then, pick the **OK** button. A rectangle representing the parts list appears attached to the cursor. Place the parts list in the upper-right corner of the drawing. The rectangle snaps to the border as you move the cursor. Pick to place the parts list. See **Figure 17-20.**

Figure 17-19.
Adding leader text.

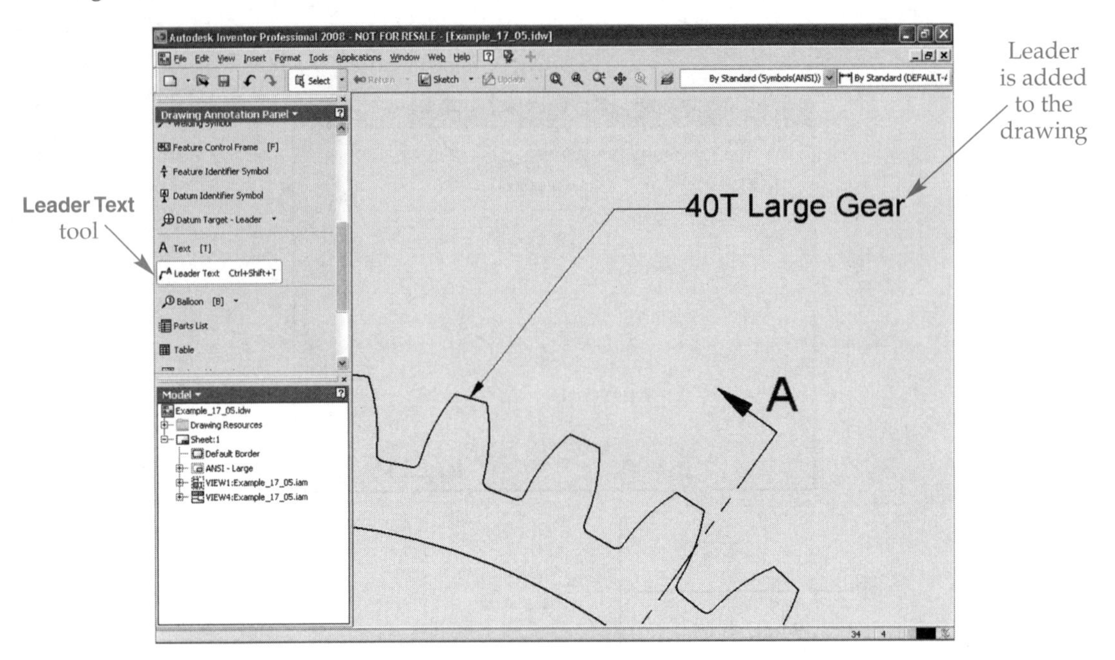

Figure 17-20.
A parts list has been added to the upper-right corner of this drawing.

Parts List			
ITEM	QTY	PART NUMBER	DESCRIPTION
1	1	Bearing Assembly	
2	2	Shaft	7 1/4" x 3/4" DIA ROUND BAR
3	1	Pinion_Gear	20 Tooth Pinion Gear
4	2	Washer .375	Washer A
5	2	Hex Nut	Hex Nut
6	1	Large_Gear	40 Tooth Spur Gear

Placing balloons

There are two tools used to place balloons: **Balloon** and **Auto Balloon**. The **Balloon** tool is used to place balloons one-at-a-time for each part in the view. The **Auto Balloon** tool puts balloons on all the parts in a selected view.

Pick the **Balloon** button in the **Drawing Annotation Panel**. Then, in the isometric view, select the large gear. Pick points for the leader and then right-click and select **Continue** from the pop-up menu. The balloon is drawn with the number 6, the item number for the large gear in the parts list. Undo the operation to remove the balloon. Then, pick the **Auto Balloon** button in the flyout. The **Auto Balloon** dialog box is displayed, **Figure 17-21.** Continue as follows.

1. With the **Select View Set** button on in the dialog box, select the isometric view.
2. With the **Add or Remove Components** button on, drag a window around all of the parts in the isometric view.
3. Make sure the **Ignore Multiple Instances** check box is checked. When checked, only one instance of each part will have a balloon.
4. In the **Styles Overrides** area of the dialog box, check the **Balloon Shape** check box and then pick the left-hand button (**Circular—1 Entry**). This creates a balloon with only the item number within a circle.

Figure 17-21.
This dialog box is used to automatically add balloons to a drawing.

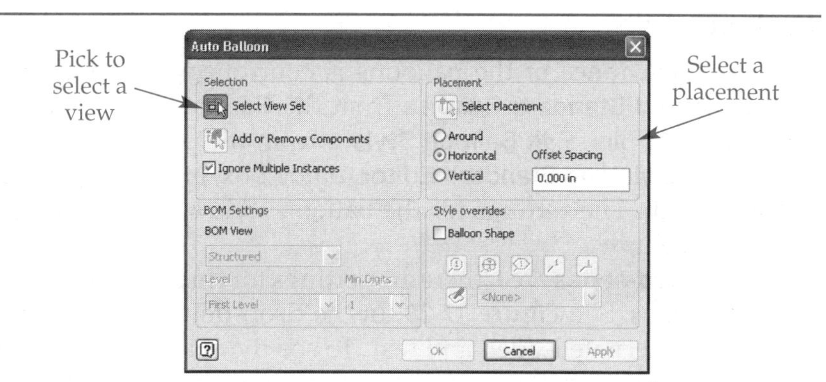

5. In the **Placement** area of the dialog box, pick the **Around** radio button. This places the balloons around the view in a circular arrangement.
6. Pick the **Select Placement** button and then move the cursor in the view. The location of the balloon previews changes as you move the cursor. Pick a point to set the location.
7. The dialog box remains open so you can change the selection, placement, or style. Pick the **OK** button to accept the settings.

You can now move the balloons to better positions, if needed. For example, the balloon leaders should not pass over parts whenever possible. See **Figure 17-22.** This is discussed in detail later in the chapter.

Undo the balloon operation. Using the **Auto Balloon** tool again, place balloons on the front view. Use a horizontal placement and set the balloons in a location of your choice. Also, enter 2 in the **Offset Spacing** text box in the **Placement** area. This is the distance between balloons. Finally, override the style and select the **Circular—2 Entries** button. This button creates balloons with the item number on the top and the quantity at the bottom. Pick the **OK** button to create the balloons.

Figure 17-22.
The isometric view after the **Auto Balloon** tool was used. Note: The balloon size has been increased here.

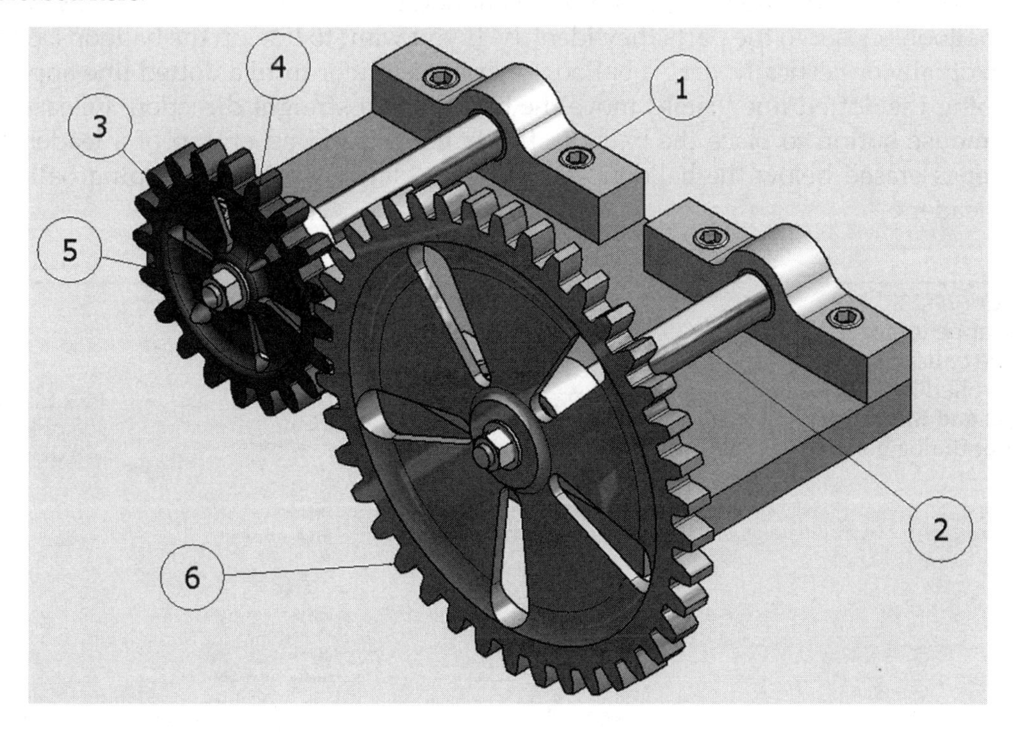

Balloon style

The appearance of the balloons is controlled by the **Style and Standard Editor**. Pick **Style and Standard Editor...** from the **Format** pull-down menu or right-click on a balloon and pick **Edit Balloon Style...** from the pop-up menu. In the tree on the left side of the **Style and Standard Editor** dialog box, expand the Balloon branch and select Balloon (ANSI). The settings for the balloon style are displayed on the right-hand side of the dialog box.

In the **Sub-styles** area, there are settings for three styles that control the appearance of the balloons. See **Figure 17-23.** In the drop-down lists in this area, pick a style for the leader, alternate leader, and text. To see the settings for the selected style, pick the pencil button next to the drop-down list. To return to the balloon style, pick the **Back** button along the top of the dialog box. When the **Save** button is picked, the changes are written to the style and the balloons are updated in the drawing.

Below the **Sub-styles** area is the **Default Offset** text box. This value controls the distance between balloon edges when the **Auto Balloon** tool is used. The value can be overridden in the **Auto Balloon** dialog box.

In the **Balloon Formatting** area, the shape of the balloon and which property is to be shown in the balloon can be changed. What is displayed in the balloon is controlled by the **Property Display** text box. The property is usually Item, but it can be changed by picking the **Property Chooser** button. The **Symbol Size** area controls the balloon size to accommodate any changes in text size. Uncheck the **Scale to Text Height** check box and the diameter of the balloon can be changed using the **Size** text box. The **Stretch Balloon to Text** check box allows the balloon shape to become an oval in order to accommodate long text items. The **Comments** text box is where comments can be entered, such as indicating the ANSI standard on which the balloons are based.

Now, select Label Text (ANSI) in the **Text Style** drop-down list. Then, pick the **Save** button. If the new text style is larger, the balloons are resized in the drawing to accommodate the size of the text. Finally, pick the **Done** button to close the **Style and Standard Editor** dialog box.

Editing balloon placement

Often, the balloons must be moved to create a legible drawing. A balloon is relocated by picking on it and dragging to a new location. Open Example_17_07.idw. Move all of the balloons closer to the parts they identify. If you want to line up the balloons, either horizontally or vertically, drag a balloon over its neighbor until a dotted line appears. Keeping the dotted line visible, move the balloon in a straight direction. Release the left mouse button to place the balloon. If a balloon is placed on top of a leader, the leader is erased below the balloon. For drawing clarity, avoid overlapping balloons and leaders.

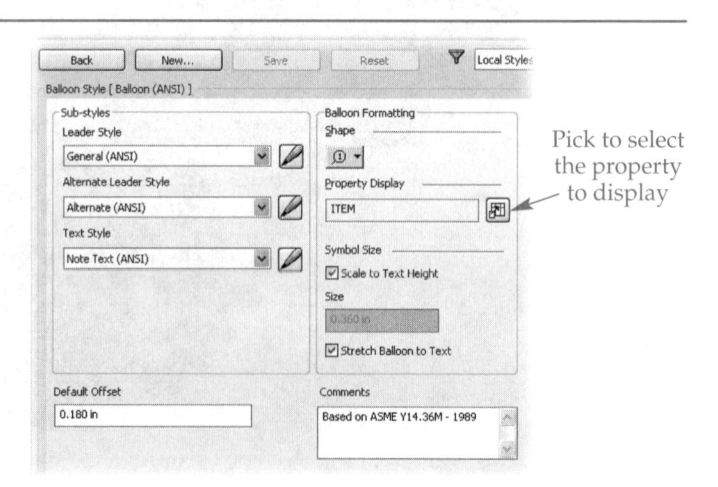

Figure 17-23.
The appearance of the balloons is controlled by the **Style and Standard Editor** dialog box.

The drawing is looking good so far, but there are too many balloons. Ideally, there should be only one or two balloons per item number. To accomplish this, some of the balloons will have multiple leaders. Refer to **Figure 17-24** and look at item 4 with four leaders pointing to four fasteners. The numbers in your drawing may differ. To create multiple leaders, delete all of the item 4 balloons except for one—select the balloons and press the [Del] key. Right-click on the remaining balloon and choose **Add Vertex/ Leader** from the pop-up menu. Pick another instance of the fastener for the origin of the leader. Lead it to the balloon and pick the center of the balloon. Continue doing this for the other two fasteners visible in the isometric view.

Continue to refer to **Figure 17-24** as you clean up the balloons. Delete as many as necessary to match the figure. It is not necessary to assign a balloon to every part visible in the view. For parts that are logically grouped together, such as nuts, bolts, and washers, the balloons are often shown together with only one leader pointing to the group of parts. Notice that this has been done for items 8 and 9 in the figure. Right-click on the item 8 balloon and select **Attach Balloon** from the pop-up menu. Pick the part that you want to group with item 8, in this case the hex nut—item 9. Finally, pick to place the location of the balloon. Repeat these steps for the other set of the washer and nut.

To edit a balloon and override the item number, right-click on the balloon, select **Edit Balloon...** from the pop-up menu, and enter a different number in the **Edit Balloon** dialog box. If the number is entered in the **Item** cell, all balloons identifying this part and their reference in the parts list change to the new item number. If the number is entered in the **Override** cell, only the selected balloon is changed. Pick the **OK** button in the **Edit Balloon** dialog box to apply the change.

Right-click on one of the balloons for one instance of item 2, select **Edit Balloon...** in the pop-up menu, and change the number in the **Override** cell to 1234. What happened to the balloon circle? It expanded into an oval to accommodate the longer number. Also notice that the other balloon for the other instance of item 2 is unchanged.

Figure 17-24.
The balloons have been relocated and modified.

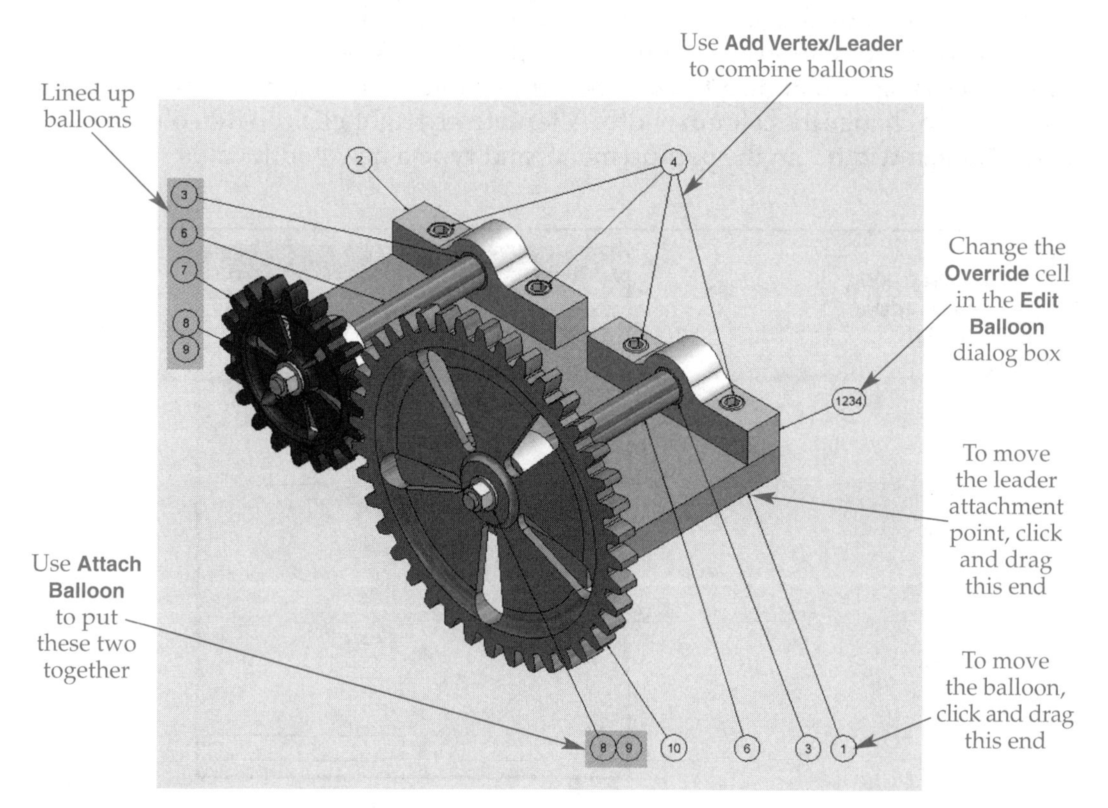

Parts Lists

A ***parts list*** is a table of information used to identify and describe the parts required for an assembly. Do not confuse the parts list with the bill of materials (BOM). The bill of materials is the iProperties database in the assembly. The parts list is the annotation of the information kept in the database and is a subset of the BOM. The parts list never contains *all* of the information in the BOM. A single drawing can contain several different parts lists for different subassemblies. You will continue to use Example_17_07.idw for this section. Pick the **Parts List** button on the **Drawing Annotation Panel**. The **Parts List** dialog box is displayed. See **Figure 17-25.**

The **Select View** button is active. You must first specify the assembly file from which the parts list will be derived. This can be done by selecting a drawing view. However, most parts list drawings do not have any graphic views, just parts lists containing information related to an assembly. If the drawing contains no views, then you may specify an assembly file by picking the **Browse for file** button and locating the file. Once you select a view or an assembly file, the text box below the **Select View** button displays the name of the assembly file and its path.

The BOM settings section is used to specify how you would like the BOM to be viewed. It can be displayed in "classic" BOM structure (reference parts, phantom parts, etc.) or parts only (ignoring reference parts and assembly structure). If parts only is disabled in the assembly file that is specified, this area is grayed out.

The settings in the **Table Wrapping** area control how a large parts list is fit onto the drawing sheet. When the **Enable Automatic Wrap** check box is checked, you can choose between a maximum number of rows and a specified number of sections. The value is entered in the text box next to the radio button. More than one section may be required if the parts list is too tall for the given drawing sheet. Specifying **Left** or **Right** instructs the parts list to continue in that direction if additional rows require another section. In this example, the parts list is small so leave the check box unchecked.

Pick the **OK** button to exit the dialog box. Then, pick a point on the drawing to locate the upper-right corner of the parts list. If you slowly move the cursor near the border or title block, the parts list will snap into position on it. Once placed, the parts list can be moved by dragging one of the green corner grips.

To widen the columns in the parts list, double-click on the parts list to open the **Parts List** dialog box, **Figure 17-26.** Pick the separator between two column headings and drag it to change the column width. Alternatively, right-click on the column header, select **Column Width...** in the pop-up menu, and type a new width value.

Figure 17-25.
The **Parts List** dialog box is used to add a parts list to the drawing.

Figure 17-26.
The **Parts List** dialog box is used to modify a parts list.

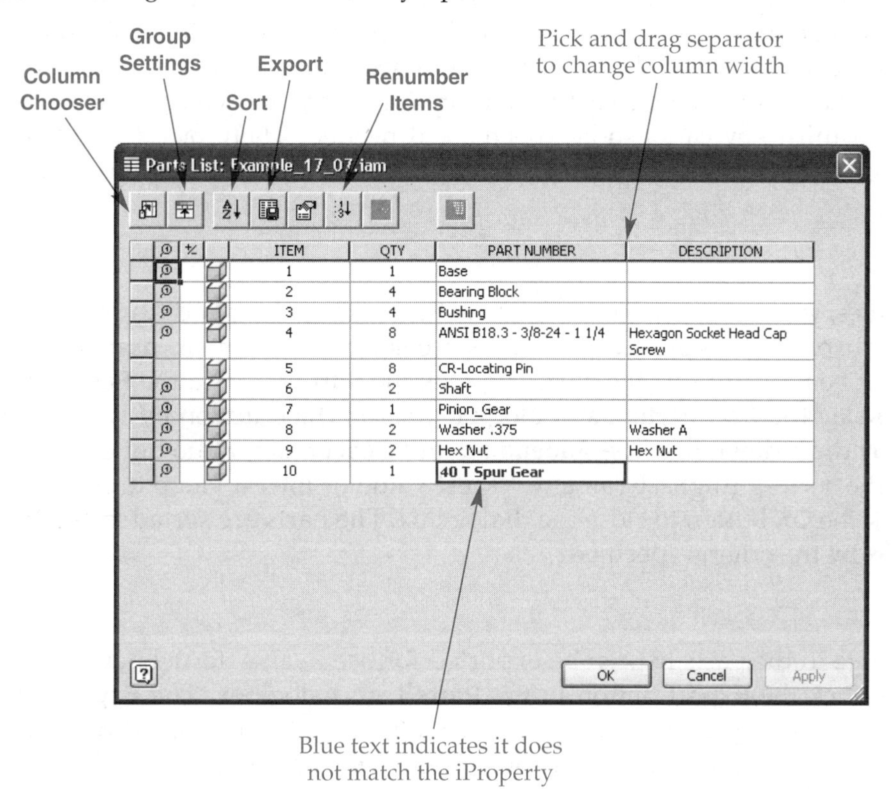

The cells can be edited in the **Parts List** dialog box. However, certain cells should not be edited, such as those in the Quantity column. When information is added or changed in this dialog box, the text in the cell turns blue. This indicates that the information does not match the part file's iProperty or the actual number of items, if the quantity is changed.

Many of the parts do not have descriptions in this example. Parts lists are actually populated, in part, by each part file's iProperties. To add descriptions for the gears, open each part file, edit its iProperties, and save the file. Use the descriptions shown below.

- **Pinion_Gear.ipt.** 20 Tooth Pinion Gear
- **Large_Gear.ipt.** 40 Tooth Spur Gear

When you are done, review the parts list in the assembly file. Only the top-level part descriptions have been updated. The parts contained within the Subassembly Bearing Assembly are not updated. Delete the parts list and place a new one to see the new descriptions.

The **Parts List** dialog box offers many ways to control the appearance of the parts list. These are discussed in the next sections.

Column chooser

You may want to add more columns than are currently in the parts list. To do so, pick the **Column Chooser** button in the **Parts List** dialog box to display the **Parts List Column Chooser** dialog box. In the **Available Properties:** list, select the column heading that you want to add. Then, pick the **Add->** button to include it in the list in the right-hand pane. For example, select Material in the **Available Properties:** list and then pick the **Add->** button. Press the **OK** button to close the dialog box. A new column labeled **Material** appears that lists the material used in each part. This information was specified when the part was created.

Row merge settings

By default, duplicate parts are grouped together in the parts list. To merge rows based on other criteria, such as material or description, pick the **Group Settings** button in the **Parts List** dialog box. In the **Group Settings** dialog box that is displayed, check the **Group** check box. Then, select a first key (property) by which to group parts. A second and third key can also be specified, if needed. Then, pick the **OK** button to close the dialog box. The parts are grouped in the **Parts List** dialog box by the key(s) specified.

Sort

At times, you may want to change how the parts list is sorted (the order in which parts are displayed). Typically, parts lists are sorted by the item number, but other criteria can be used. To sort the parts list, pick the **Sort** button in the **Parts List** dialog box to display the **Sort Parts List** dialog box. Select the criterion by which to sort in the **Sort by** drop-down list. The **Ascending** and **Descending** radio buttons control the order of the sorting (highest value to lowest value or lowest value to highest value). Then, pick the **OK** button to close the dialog box. The parts are sorted in the **Parts List** dialog box by the criteria specified.

Export

The data in the parts list can be exported for use in an external database or word processor. Pick the **Export** button in the **Parts List** dialog box. The **Export Parts List** dialog box is displayed. This is a standard Windows "save" dialog box. Select the format in which to export the data using the **Save as type:** drop-down list. Then, name the file and pick the **Save** button.

Table layout

To change the name of the parts list from the default Parts List, pick the **Table Layout** button in the **Parts List** dialog box. The **Parts List Table Layout** dialog box is displayed. In the **Heading and Table Settings** area, enter a new name in the **Title** text box. You can also suppress the title by unchecking the check box above the text box. The line spacing for the table can be changed in the **Line Spacing** drop-down list. The spacing can be set to single, double, or triple spacing by selecting the corresponding setting in the drop-down list. When done making settings, pick the **OK** button to redisplay the **Parts List** dialog box.

Renumber

If the parts list is sorted by any property other than item, such as quantity, the item numbers will not be sequential. To assign new, sequential numbers to the parts list, pick the **Renumber Items** button in the **Edit Parts List** dialog box. The order of the parts is not changed, but each part has a new item number assigned to it. The item numbers in the parts list now appear in order. When you pick **OK** to close the **Edit Parts List** dialog box, the balloons on the drawing are also renumbered. If you need to revert to the previous numbering, use the **Undo** tool.

Changing Standards

Standards related to the parts list are controlled in the **Style and Standard Editor** dialog box. Expand the Parts List tree and select Parts List (ANSI). The properties of this style are displayed on the right-hand side of the dialog box. See **Figure 17-27.** Many of the properties listed here are the same as those available when editing the parts list in the **Parts List** dialog box.

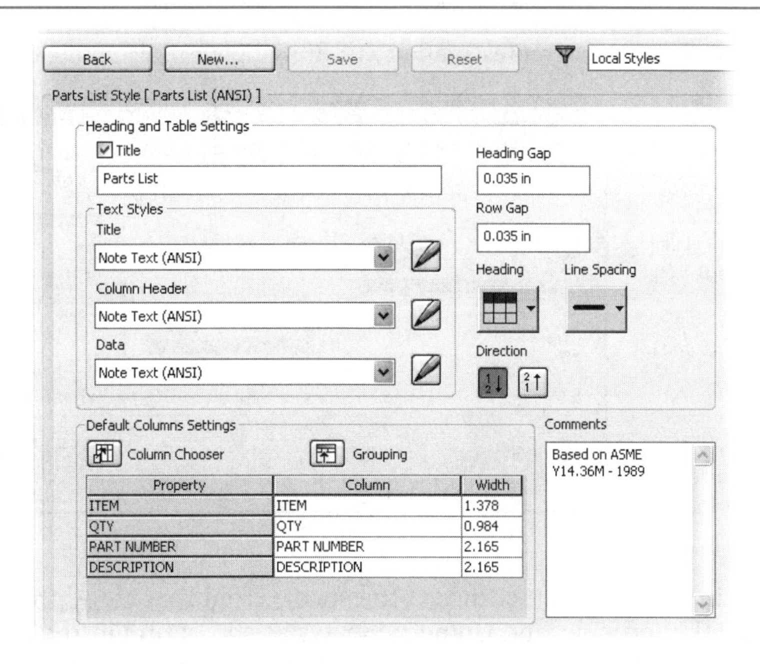

Figure 17-27.
The style of the parts list is set in the **Style and Standard Editor** dialog box.

Printing the Drawing

Many of the details of printing a drawing are controlled by Inventor. For example, lineweights are specified by the drafting standards and rarely have to be changed. Printing an assembly drawing is very similar to the procedure used to print a part drawing. Select the output device, paper size, and scale (if any) and print the drawing. For the most part, what you see on the graphics screen is what you will get on the paper. Try it by printing Example_17_07.idw.

Creating Drawing Views Using Design View Representations

You may recall that Inventor saves design views in each assembly. You can save many design views within an assembly file. These design views can also be used to make drawing views. If the design view is modified and saved, the drawing view automatically updates to show the current design view.

Open Example_17_08.iam and examine the assembly. It uses the contact solver, a feature that stops part movement when contact is made. Enable the contact solver by selecting **Tools>Activate Contact Solver** so the icon next to the menu item is depressed. Now, pick one of the Input_Link parts and drag it to animate the mechanism.

Now, take a look at the design views that have been saved. Expand the Representation branch in the **Browser**, then expand the View branch below it. Double-click on a view name to restore the view. You will use some of these design views to create drawing views: NE Isometric, SE Isometric, Locked down position, Open position, No fasteners, and Bottom view. The following three default design views cannot be used to establish associativity between the drawing views and the design view: system.nothing visible, system.all visible, and User.default.

Start a new drawing using a C-size sheet. Create a base view generated from the bottom orientation of the assembly, set the scale to 1:1, select the Locked down position design view, and check the **Associative** check box, as shown in **Figure 17-28.** Place the

Figure 17-28.
Using a design view representation as a drawing view.

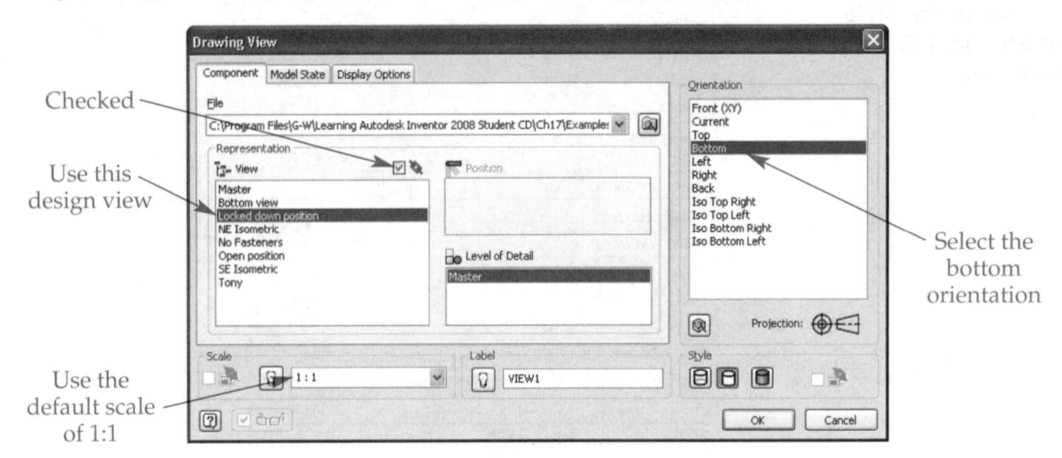

view in the lower-left corner of the sheet. From that view, project three more views to create the top, side, and isometric views as shown in **Figure 17-29.**

All of the drawing views are associative in relation to the linked design view. Display the drawing and assembly at the same time by picking **Arrange All** from the **Window** pull-down menu. See **Figure 17-30.** Within the assembly, drag one of the red input links to a new position. Now, pick in the drawing window to make it active. The views in the drawing update to reflect the changes in the assembly file.

It is possible to change the design view associated with an existing drawing view, but with mixed results. Maximize the drawing window. Then, double-click on the top view and select a different design view in the **Drawing View** dialog box. Any one that you choose will not affect the drawing view's orientation, but will affect the visibility of components in the drawing view. Try it with the design view No Fasteners. The

Figure 17-29.
The drawing with views projected from the design view representation.

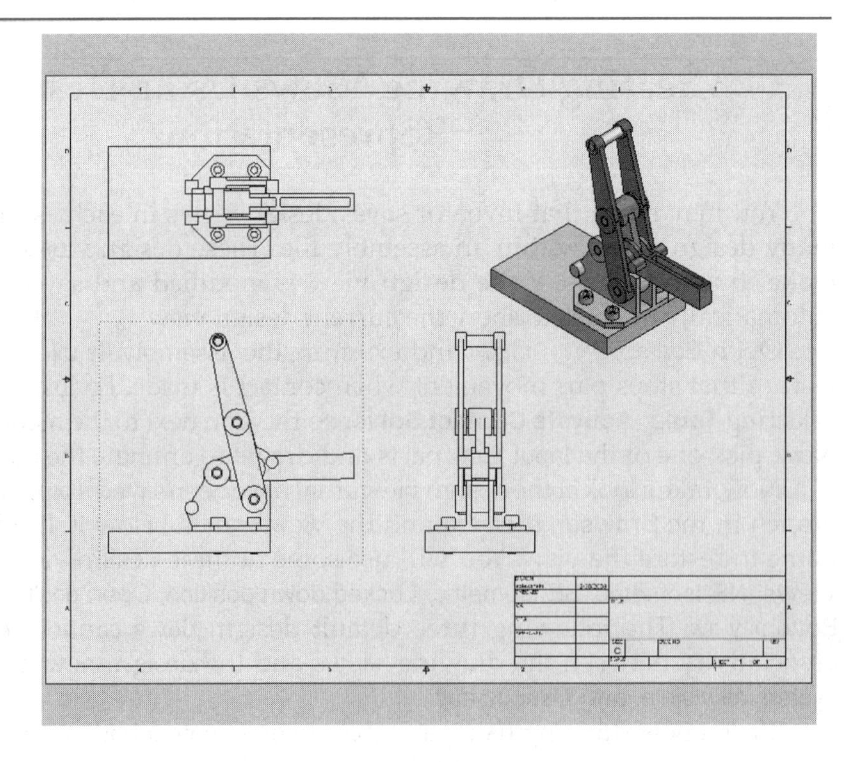

Figure 17-30.
When the assembly is moved, the drawing views are automatically updated since they are based on a design view representation.

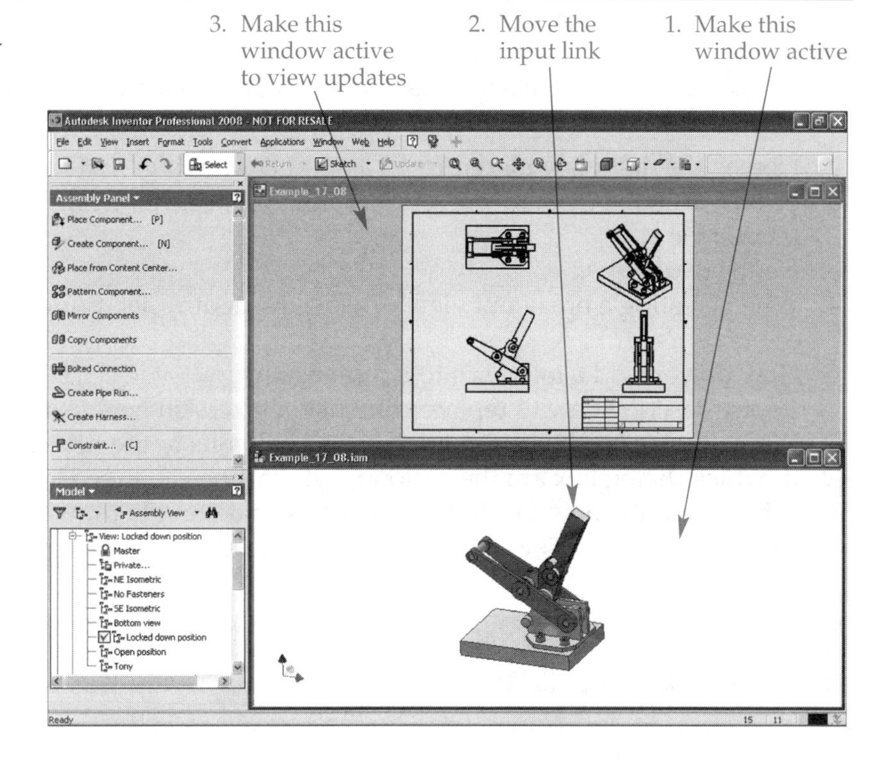

fasteners disappear in the drawing view; so far so good. Now, change the design view to Bottom View. Notice that there is no change. However, since the orientation of the drawing view is changed, other drawing views projected from it might be affected.

PROFESSIONAL TIP

Only public design views are associative. If you are experiencing problems getting drawing views to update, make sure that each is associated to the design view. To do so, double-click on each and ensure that the **Associative** check box is checked in the **Drawing View** dialog box.

Overlay Views

Overlay drawing views are used to document the multiple positions available in a work envelope for a given assembly. They are not actually drawing views, rather additions that are overlaid on an existing view. The existing view can be an unbroken base view, orthographic projected view, or auxiliary view. Overlay drawing views are based on positional representations and each can reference a different design view independent of the parent view. They can be annotated with dimensions to indicate the distance from one work position to another. A thorough study of overlay drawing views is beyond the scope of this book.

Chapter Test

Answer the following questions on a separate sheet of paper or complete the electronic chapter test on the Student CD.

1. How is the first view placed in an assembly drawing?
2. Briefly describe how to prevent certain components from being sectioned in a section view.
3. Define *break out view.*
4. When creating a break out view, what is the depth specifying when the From Point option is used?
5. How do you add a tolerance to a dimension?
6. Briefly describe how to replace dimension text with new text.
7. Which tool is used to automatically add balloons to the drawing?
8. In which dialog box are the balloon style settings changed?
9. Briefly describe how to add a parts list to an assembly drawing.
10. How can a design view representation be used as a drawing view?

Chapter Exercises

Exercise 17-1. Single Cylinder Engine.
Complete the exercise on the Student CD

Exercise 17-2. Adaptive Assembly.
Complete the exercise on the Student CD

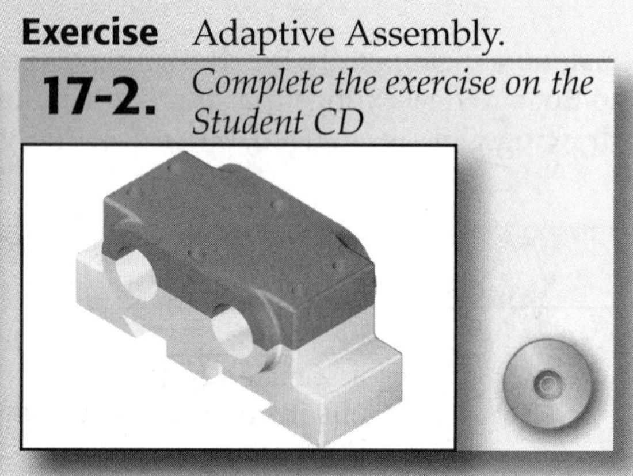

Exercise 17-3. Sliding Clamp.
Complete the exercise on the Student CD

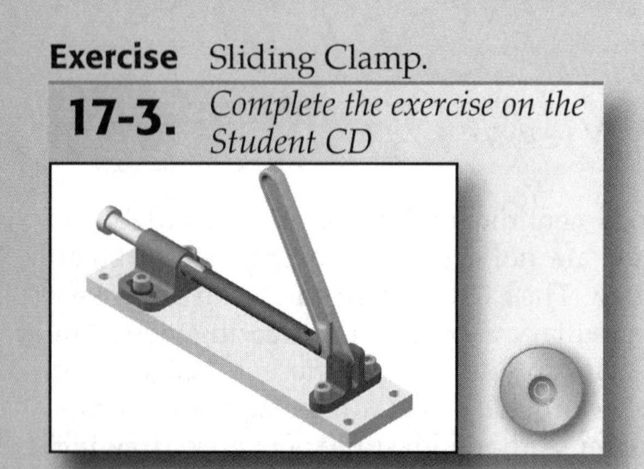

Exercise 17-4. Scissors Lift.
Complete the exercise on the Student CD

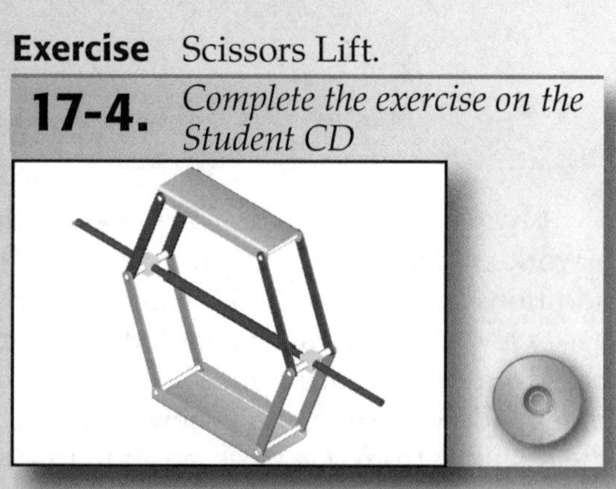

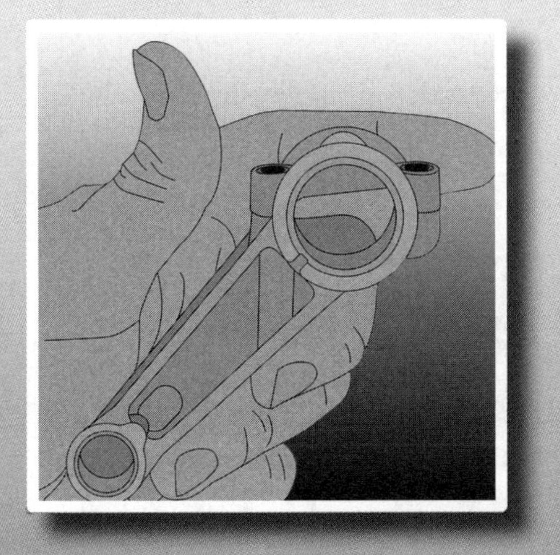

Chapter 18

Presentation Files

Objectives

After completing this chapter, you will be able to:

- Create automatic and manual explosions.
- Add linear and rotational tweaks to create and modify exploded views.
- Modify animations to make the components move at the correct time and speed.
- Use design views to create multiple exploded views.
- Set different camera angles to emphasize specific areas of an assembly.
- Rearrange the sequences of an animation.
- Apply colors, styles, textures, and lights to create a refined presentation.

User's Files

The Student CD included with this text contains several files required for this chapter. Refer to the file File List.txt *in the* \Ch18 *folder for the comprehensive list.*

Presentation Files

Exploded views of an assembly are created in a separate file called a ***presentation file.*** A presentation file has an IPN extension and is related to a specific assembly file. A single presentation file can contain many exploded views of the assembly. Each exploded view is based on a saved design view. Once exploded, the view can be animated to reassemble the components. Video files in AVI format can be created from the animation and then viewed outside of Inventor.

A wooden valet will be used as an example to study the features, principles, and steps in creating presentation files. The valet is designed to be built with simple hand tools, such as a saw, drill, and screwdriver. It has four subassemblies: the drawer, drawer front, drawer box, and drawer box top. The final assembly of the components is shown in **Figure 18-1.**

Figure 18-1.
The valet assembly and components.

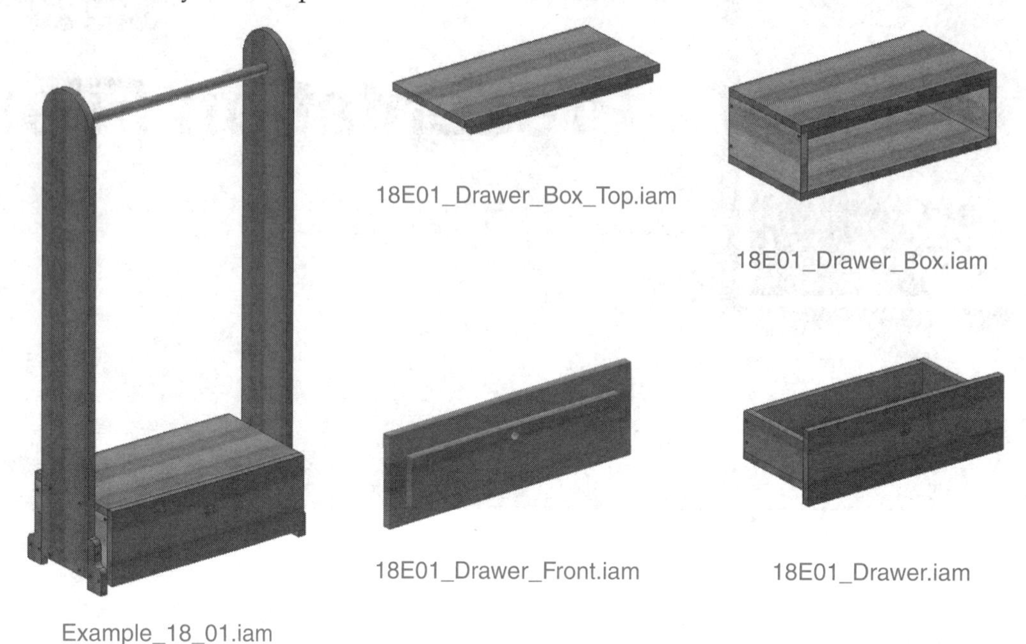

The first subassembly to work on is the drawer box top. Make sure the **Default** project is active. Then, open the assembly file 18E01_Drawer_Box_Top.iam located in the \Example_18_01 folder. The file opens with the BOTTOM design view representation active, but there is also a TOP design view saved. See **Figure 18-2.**

To create an exploded view, start a new presentation file using the English Standard (in).ipn template. Then, pick the **Create View...** button in the **Presentation** panel to display the **Select Assembly** dialog box. See **Figure 18-3A.** The default assembly file is 18E01_Drawer_Box_Top.iam since it is the only assembly file open. The assembly file does not have to be open to create a presentation file. Continue as follows.

1. Pick the **Options...** button to open the **File Open Options** dialog box, **Figure 18-3B.** Select BOTTOM in the drop-down list in the **Design View Representation:** area. Pick the **OK** button to return to the **Select Assembly** dialog box.
2. In the **Explosion Method** area, pick the **Automatic** radio button and enter 2 in the **Distance:** text box. This automatically explodes the assembly.

Figure 18-2.
The file opens with the BOTTOM design view representation active.

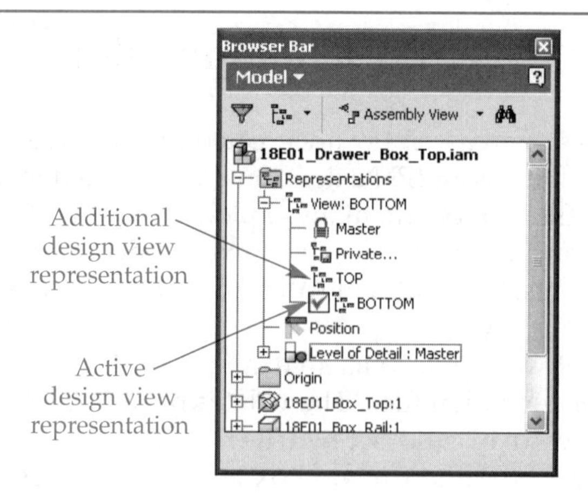

Figure 18-3.
A—Choose an assembly file and an explosion method. B—Choose a design view.

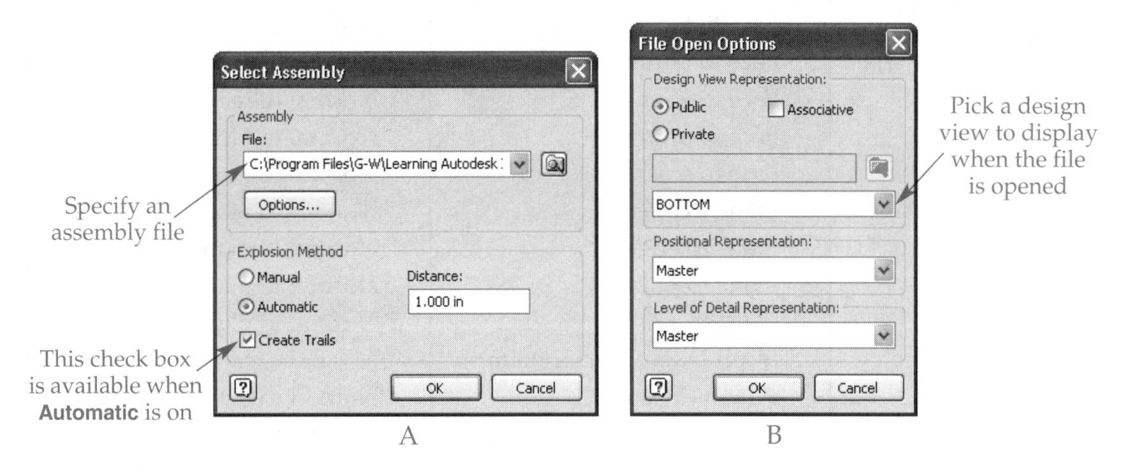

Figure 18-4.
An exploded view automatically created.

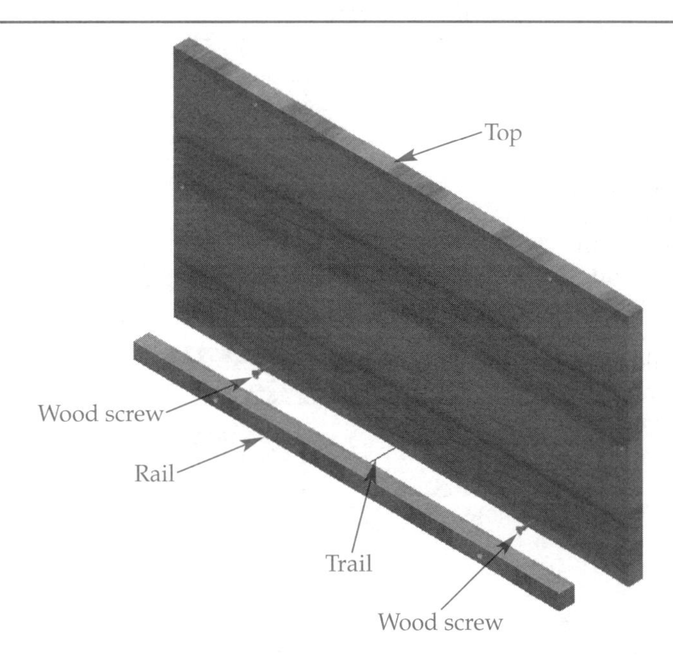

3. Check the **Create Trails** check box, which becomes available when the **Automatic** radio button is on. ***Trails*** are thin lines relating the assembled position to the exploded position.
4. Pick **OK** to create the first exploded view. See **Figure 18-4.**

Automatic explosion applies only to mate constraints that involve a face or a plane, to the tangent constraint, to some types of insert constraints, and to some iMate constraints. See **Figure 18-5.** It does not apply to angle constraints, or angle or plane iMates.

In the example, the top is the grounded part so it does not move. The rail was mated face to face with the top. During the explosion, the rail moved 2″ perpendicular (normal) to the mate constraint. This is called a ***tweak*** in the **Browser**. The screws will not move automatically because they were put in with an insert constraint.

Right-click on 18E01_Drawer_Box_Top.iam in the **Browser** (not the very top level) and select **Auto Explode**. See **Figure 18-6.** The distance entered in the **Auto Explode** dialog box will be added to the existing tweaks. With 1.000 entered in the text box, pick the **OK** button. The rail now has two tweaks on it, as seen in the **Browser**.

Figure 18-5.
These are the only mating conditions that allow a move during an automatic explosion.

Constrained Geometry	Movement
Face to face	Normal to face
Face to plane	Normal to face
Face to edge	Normal to face
Face to point	Normal to face
Face to axis	Normal to face
Plane to plane	Normal to plane
Plane to edge	Normal to plane
Plane to point	Normal to plane
Plane to axis	Normal to plane
Insert (axis and face mate)	Normal to face along axis
iMate—insert	Normal to face along axis
Tangent	Normal to planar face
iMate—tangent	Normal to planar face
iMate to face	Normal to face

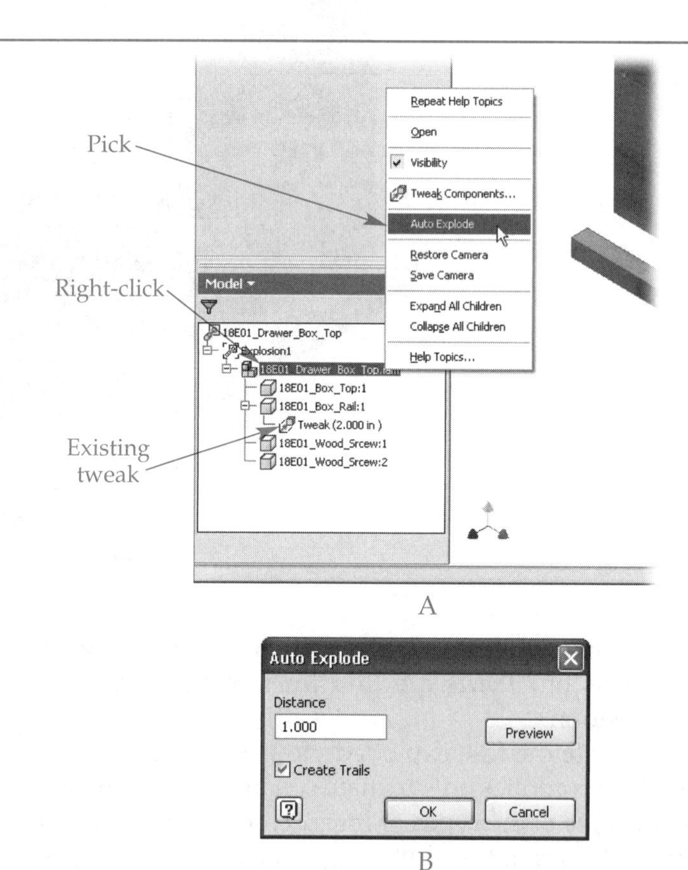

Figure 18-6.
A—Right-click on the assembly in the **Browser** and select **Auto Explode** from the pop-up menu.
B—The distance entered in the **Auto Explode** dialog box is added to any existing tweaks.

Adding Tweaks

To move the wood screws, the **Tweak Components** tool must be used. This tool adds translational (movement) or rotational tweaks. It is also used to edit existing trails. To translate, pick a direction, select one or more components, and drag-and-drop on the screen or enter a value in the current units.

1. Pick the **Tweak Components...** button in the **Presentation Panel** or press the [T] key.

2. In the **Tweak Component** dialog box, make sure the **Display Trails** check box is checked. Also make sure the **Direction** button is on in the **Create Tweak** area and the **Z** button is on in the **Transformations** area.
3. Move the cursor around the assembly in the graphics window. When the cursor is on an edge, the Z axis aligns with that edge. When the cursor is on a face, the Z axis extends perpendicular to that face. Pick the front face to set the direction of the Z axis, as shown in **Figure 18-7.**
4. The **Components** button is now on in the **Create Tweak** area of the dialog box and the cursor has changed. Press the [Ctrl] key and select the two wood screws, either in the assembly or in the **Browser**.
5. Pick and hold on the screws, drag them about 6″ away from the assembly, and release the mouse button. The distance is displayed in the text box in the **Transformations** area of the dialog box. Also, notice that the XYZ tripod moves with the screws.
6. Pick the **Close** button in the dialog box to accept the tweak and end the tool. Picking the **Clear** button accepts the tweak, but does not end the tool.

You will practice using the **Tweak Components** tool and learn its other features in later examples. For now, save the file as Example_18_01.ipn in the \Example_18_01 folder and leave it open. Then, start a new ANSI (in).idw file. Pick the **Base View** button in the **Drawing Views** panel to open the **Drawing View** dialog box. In the **File** drop-down list, select Example_18_01.ipn. In the **Orientation** list, select **Current**. In the **Scale** area, enter 0.5 in the text box. Then, pick a location in the drawing to place the view. The drawing with balloons and a parts list is shown in **Figure 18-8.**

Sometimes, some of the trails are not necessary for a drawing. For example, the screw shown in **Figure 18-9** has a two-segment trail. You can remove an entire trail or a segment of the trail. To remove the entire trail, right-click on the trail in the presentation file and pick **Hide Trails** in the pop-up menu. To remove a segment of a trail, right-click on the segment in the presentation file and uncheck **Visibility** in the pop-up menu.

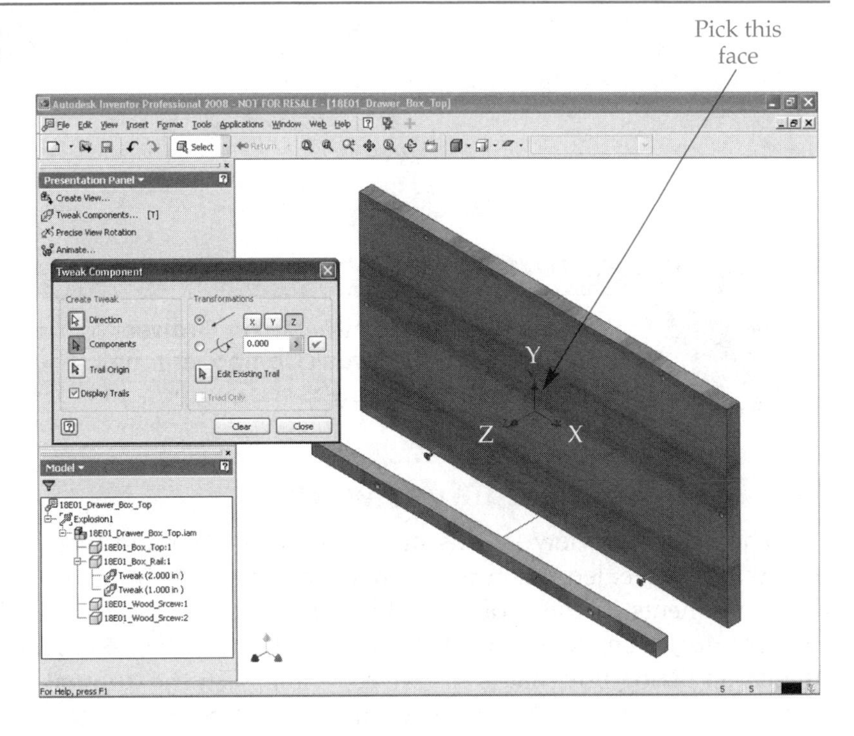

Figure 18-7.
Tweaking the screws to move them.

Figure 18-8.
A drawing with an exploded view, balloons, and a parts list.

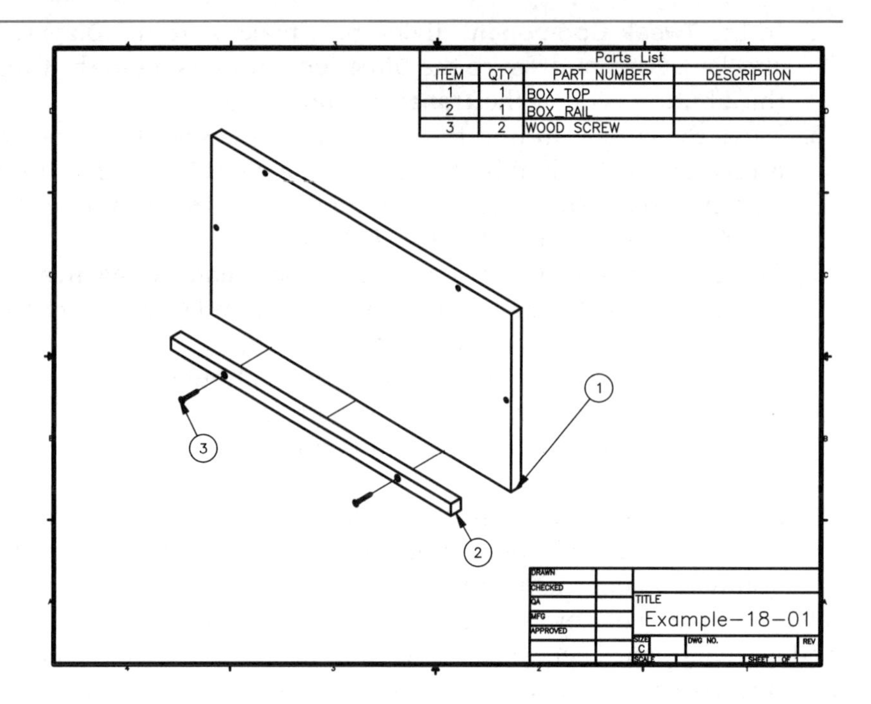

Figure 18-9.
You can hide the trails on a tweak or segments of the trail.

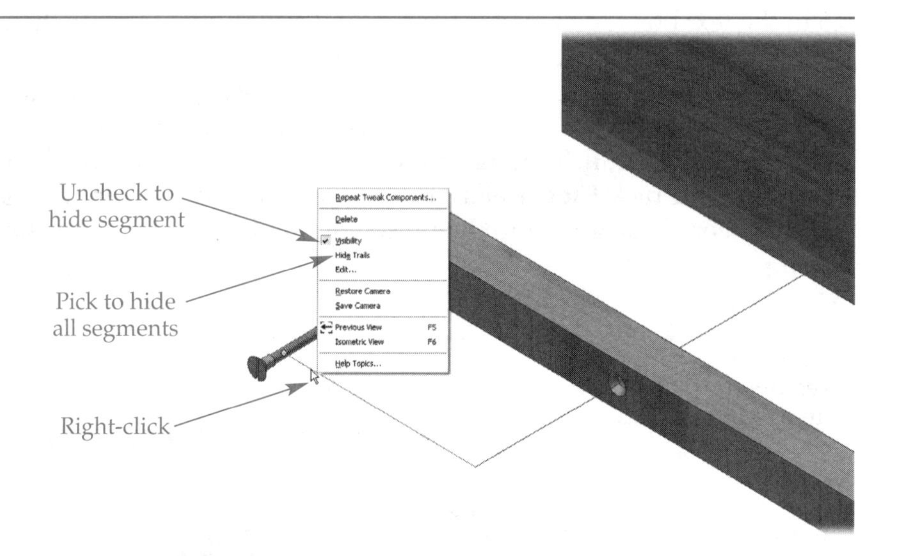

PROFESSIONAL TIP

You can select parts, either in the **Browser** or in the graphics window, and then open the **Tweak Component** dialog box. The selected parts are automatically used as the components for the tweak.

Animate Tool and Editing Tweaks

The exploded view in this presentation file worked well in the drawing, but a flaw will be revealed when the view is animated. Animation temporarily assembles the components in the order in which they were exploded. In the presentation file Example_18_01.ipn, pick the **Animate...** button in the **Presentation** panel. In the **Animation** dialog box that is displayed, enter 15 in the **Interval** text box. The ***interval*** is the number of steps for each tweak. In the **Repetitions** text box, enter 1. Pick the **Apply** button and then pick the **Play Forward** button. See **Figure 18-10.** Note: This is an extremely calculation-intensive operation and usually requires a high-end graphics

Figure 18-10. These are the controls to animate an explosion.

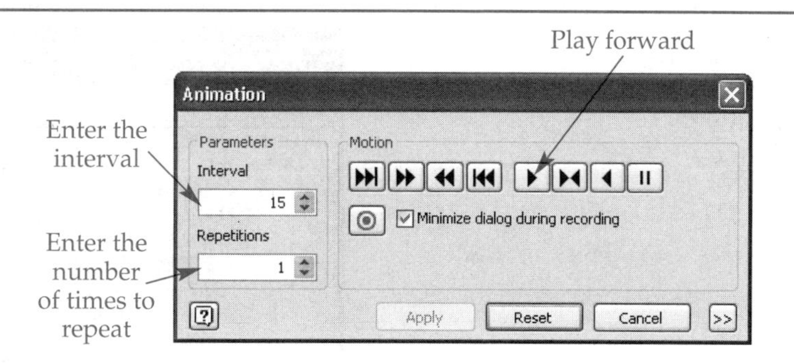

card. If parts "drop out" of the animation or do not appear when they should, it is likely that your graphics card is not up to the task.

The speed of the screws and the rail during the animation is controlled by the distance each has to travel and the number of steps to complete the movement. The screws move quickly since their tweak distance is 6". The screws have to move 6" in 15 steps. This means that each step will move the screws 6/15". Compare this with the distance the rail must travel during each step. The rail has two tweaks: the first is 1", therefore 1/15" per step, and the second at 2/15" per step. Watch carefully as you play this and you can see the rail speed up during the second tweak.

The screws should be assembled after the rail moves into place. However, notice how they move through the rail, and then the rail moves over the screws. Pick the **Reset** button in the **Animation** dialog box to return the parts to their original positions. Then, pick the **>>** button to expand the dialog box. Select 18E01_Wood_Screw:1 in the **Animation Sequence** list. Notice that the **Move Up** and **Move Down** buttons are now active. See **Figure 18-11.** Also note that since you picked both screws earlier for one tweak, the **Ungroup** button is available. Pick the **Ungroup** button. Then, use the **Move Down** button to move 18E01_Wood_Screw:1 to the bottom of the list. The part moves one step at a time, so you need to pick the button multiple times. Also, move 18E01_Wood_Screw:2 to the bottom of the list. Then, pick the **Apply** button. Now, when

Figure 18-11. The **Animation** dialog box is expanded to reveal the sequence of tweaks.

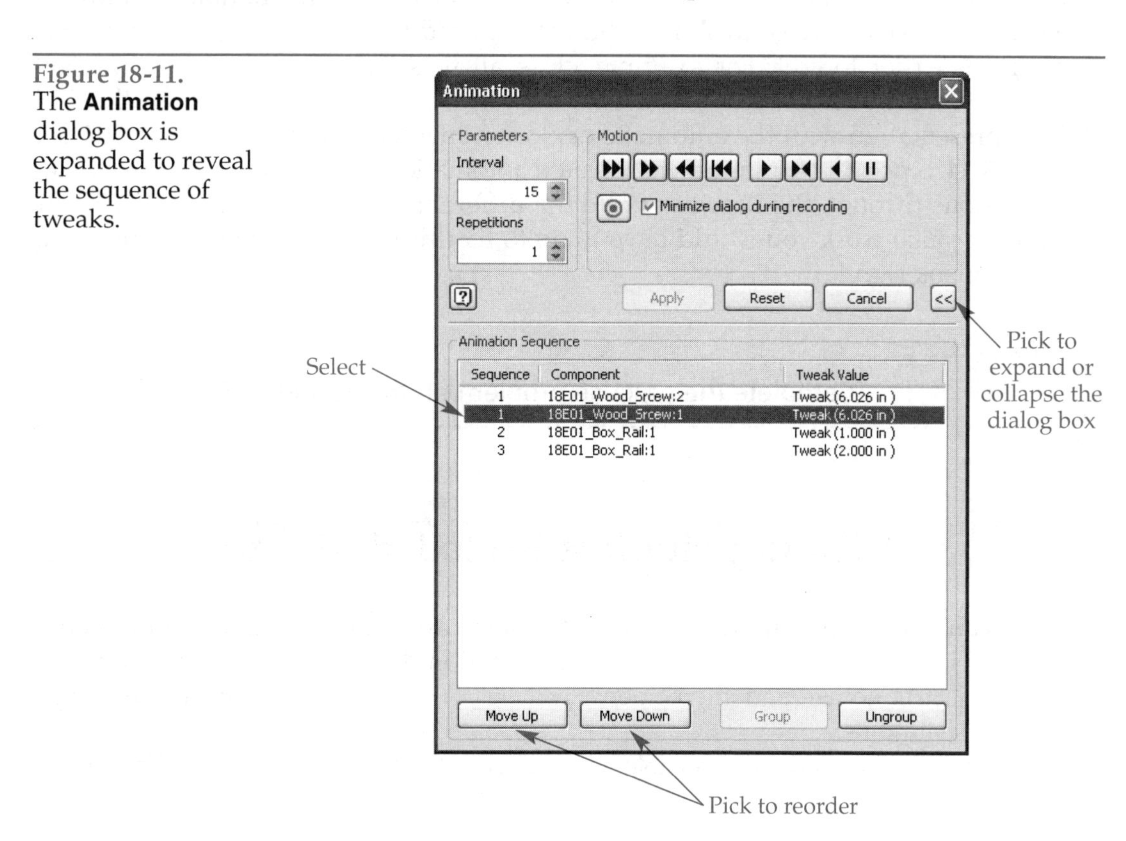

Figure 18-12.
The distance of an auto-exploded tweak can only be changed in the **Browser**.

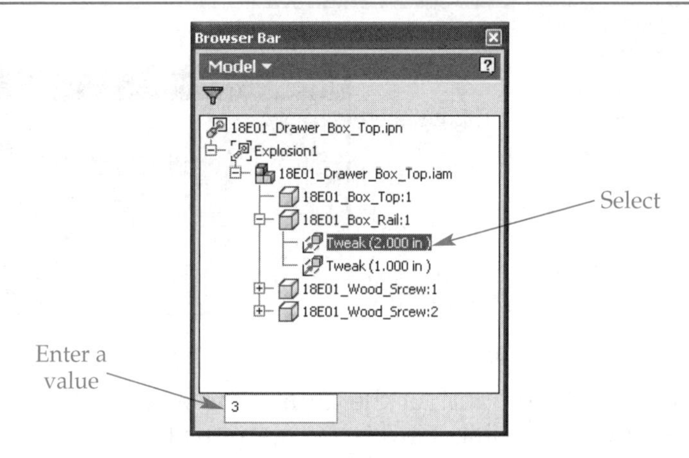

the **Play Forward** button is picked, the rail is assembled first and then the screws are assembled one at a time. After the animation plays, pick the **Cancel** button to close the dialog box and return the parts to their original locations.

The names and values of the tweaks *cannot* be edited in the **Animation** dialog box. To change the value of a tweak, select the tweak in the **Browser**. The tweak distance appears in a text box at the bottom of the **Browser** and can be edited there. See **Figure 18-12.** Select the 2″ tweak for the rail, type 3 in the text box, and press [Enter] to change the tweak value for the rail to 3″. Now, right-click on the 1″ tweak for the rail and select **Delete** from the pop-up menu.

The value of a manually applied tweak can also be changed by picking the trail in the graphics window and dragging the green grip. See **Figure 18-13A.** You can also right-click on a trail and select **Edit...** in the pop-up menu to display the **Tweak Component** dialog box. See **Figure 18-13B.** Enter an exact value in the text box in the **Transformations** area or drag the component back and forth in the graphics window.

In more complex assemblies, an automatic explosion may generate a view that requires many manual tweaks. A more efficient process is to manually apply all tweaks. This gives more control over the values, produces fewer steps in the animation, and is easier to edit. For example, close all files, open 18E01_Drawer_Box.iam, and create a new presentation file. Create a view of the assembly based on the TOP VIEW representation with the automatic explosion method and a distance of 4.0″. See **Figure 18-14.** Since this explosion was created automatically, any editing of the tweaks must be done through the **Browser**. Note the different directions of the trails. Think about how much work you would have to do to fix this exploded view and make the animation look right!

Complete the practice problem on the Student CD.

Creating Multiple Exploded Views

You can have many different views of the same assembly in one presentation file. Each view is based on a design view representation. Close all files, make sure the Default project is active, and then open Example_18_02.iam. Check the three design view representations:

- TOP. All components are visible, viewed from the top-left isometric viewpoint.

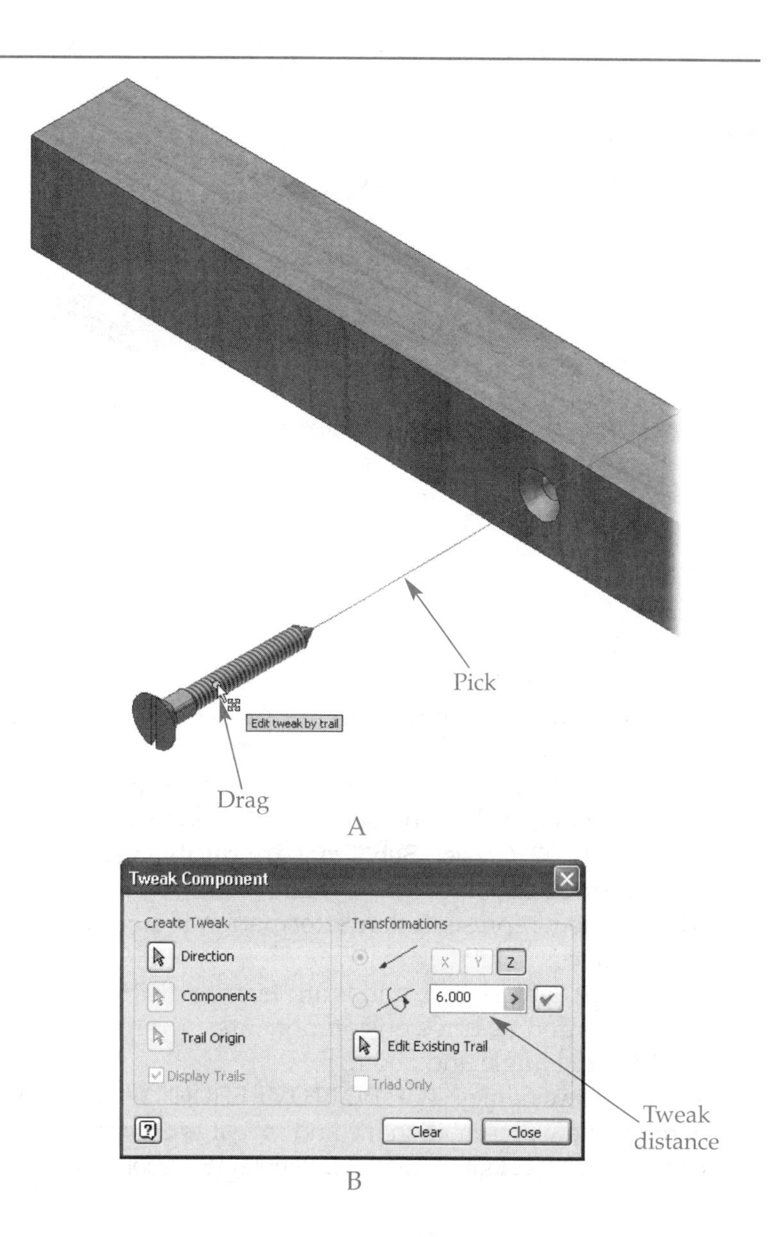

Figure 18-13. Two methods of editing a manually applied tweak. A—Drag the green grip on the trail. B—Change the distance in the **Tweak Component** dialog box.

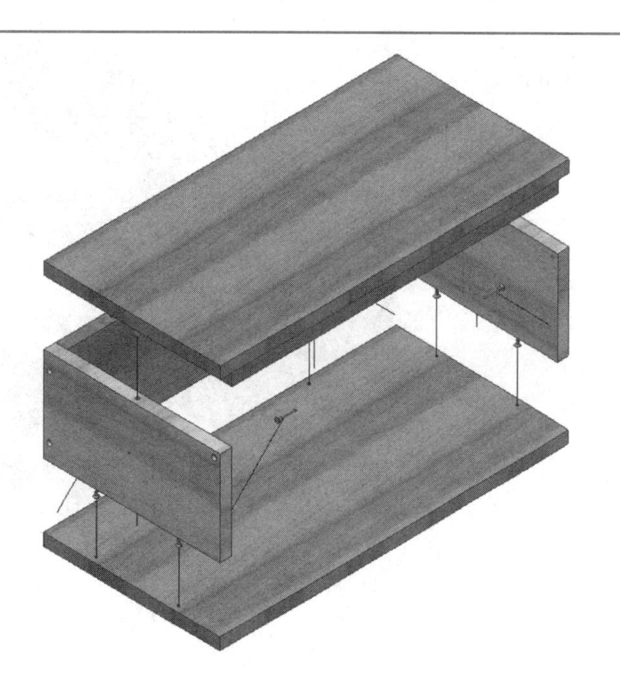

Figure 18-14. Automatic explosions can sometimes create confusing views, such as this one.

Figure 18-15.
Components with **Enabled** unchecked appear transparent.

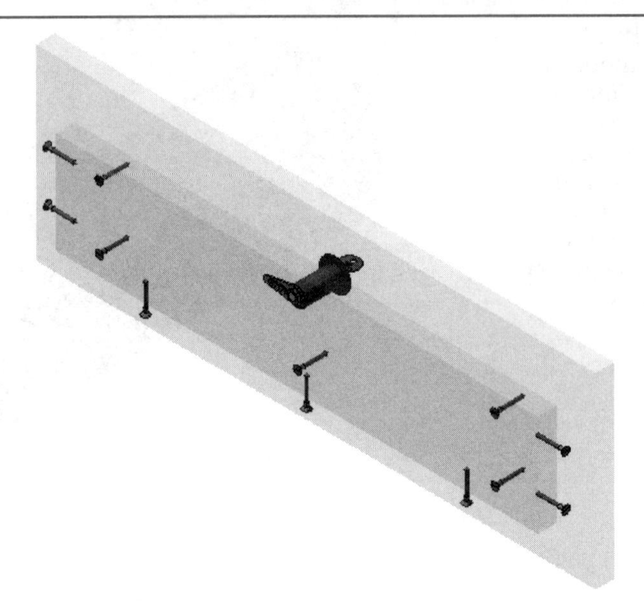

- BOTTOM. All components are visible, viewed from the bottom-left isometric viewpoint.
- FRONT WITH LOCK. Only the front subassembly and lock components are visible, viewed from the top-back isometric viewpoint. The parts 18E02_Drawer_Front and 18E02_Drawer_Sub_Front are disabled so you can see inside of them. See **Figure 18-15.**

The design view representation information is stored in the assembly file. This information includes:

- Component, sketch, and work feature visibility.
- Colors and textures on components.
- Viewing angle and zoom.
- The view names: TOP, BOTTOM, FRONT WITH LOCK, and the default Master.

Create a new presentation file and save it as Example_18_02.ipn in the \Example_18_02 folder. Create an exploded view of Example_18_02.iam using the TOP design view with:

Figure 18-16.
An exploded view created with manual tweaks.

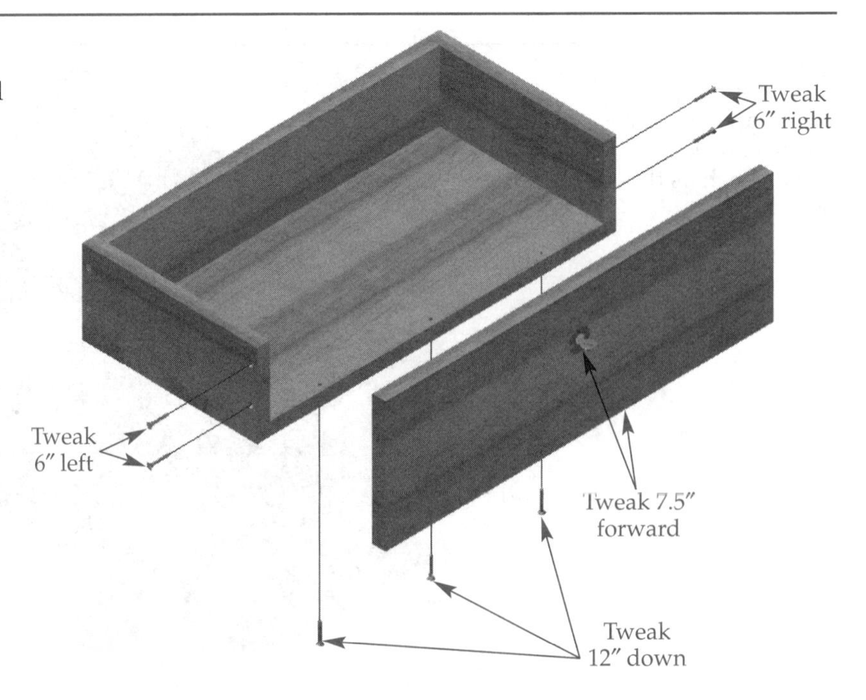

- The 18E02_Drawer_Front subassembly and the five lock components moved out 7.5″ with no trails displayed. See **Figure 18-16.**
- The four screws in the sides of 18E02_Drawer_Front moved 6″ with trails.
- The three screws in the bottom of 18E02_Drawer_Front moved down 12″ with trails.

Create a new IDW file and place the current exploded view as a base view at a scale of .25. Save the file as Example_18_02.idw.

In the presentation file, pick the **Create View...** button in the **Presentation Panel**. In the **Select Assembly** dialog box, pick the **Manual** radio button in the **Explosion Method** area and select the FRONT WITH LOCK design view representation. Then, create the view. A new view called Explosion2 is listed in the **Browser**. Note that wooden parts are enabled even though disabled in the assembly file. Create an exploded view similar to that shown in **Figure 18-17.** Place this view as another base view in the Example_18_02.idw drawing with a scale of .5. See **Figure 18-18.** Save and close all files.

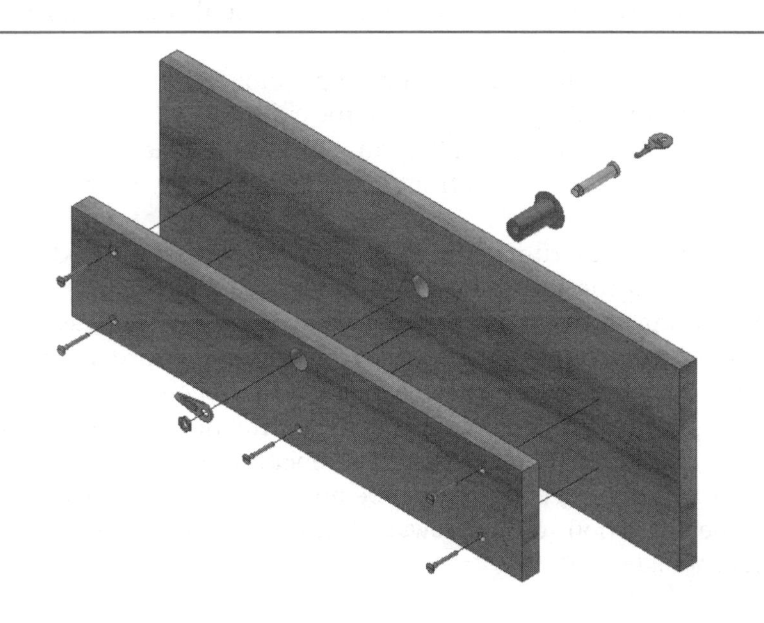

Figure 18-17.
Some of the tweaks in this manually exploded view do not have trails.

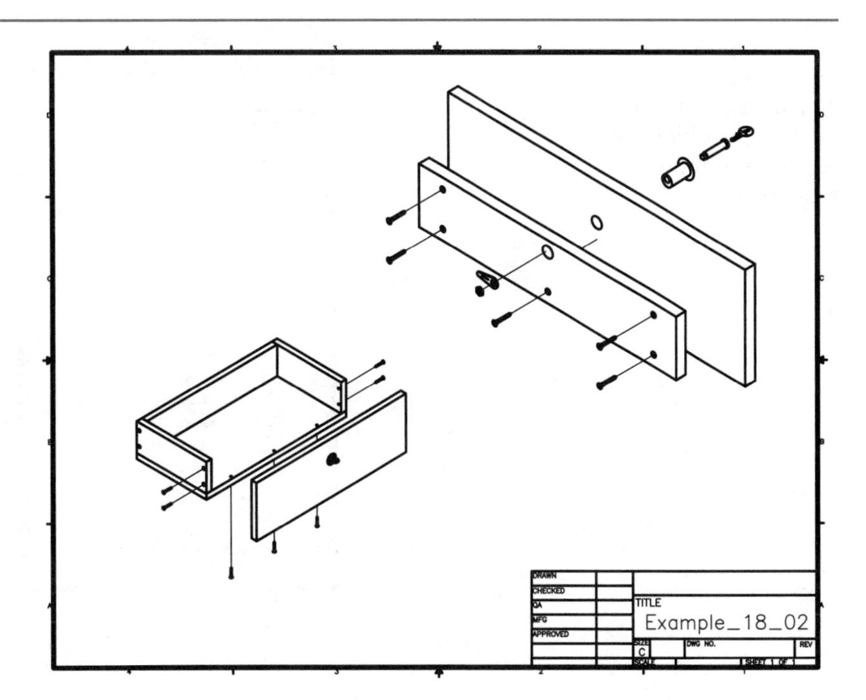

Figure 18-18.
The exploded views have been placed in a drawing.

Complex and Rotational Tweaks

Open Example_18_03_Done.ipn and run the animation. In a complex animation like this one, it is best to start out with a storyboard of the steps you want to see. A ***storyboard*** is a list of the steps in the animation and may contain some rough hand sketches. The storyboard for the drawer front is:

1. In the isometric view, put the drawer subfront on the drawer front from an original position about 3″ away. All lock parts are hidden.
2. Put four of the screws in at one time from original positions about 7″ away.
3. Put the fifth screw in from an original position about 7″ away. Rotate the screw as if it is being driven in.
4. Rotate the view point, make the lock parts visible, and install the lock housing from an original position about 7″ away.
5. Install the lock cylinder from an original position about 8″ away.
6. In the isometric view, zoom in and install the washer from an original position about 7″ away.
7. Install the flag from an original position about 7.5″ away.
8. Bring the nut up to the end of the cylinder.
9. Rotate the nut 360°. At the same time, move the nut .1″ onto the cylinder.
10. Change the view to a front view. Bring the key in from an original position about 9″ away.
11. Turn the key, tumbler, flag, and nut 90° with the wooden parts invisible.

To create the animation, all of these tweaks need to be applied in the reverse order. Each animated tweak is listed as Sequence*x* in the **Browser**. Under Explosion1, expand the Task1 branch. The eleven steps from the storyboard are listed as Sequence1 through Sequence11. See **Figure 18-19.** Select the Sequence1 branch in the **Browser**. The part is highlighted and an arrow indicates the direction of movement. When a new tweak is added, it becomes Sequence1 and all existing tweaks are renumbered. This is why the tweaks on the storyboard must be applied in reverse order. A tweak can be dragged to a new position in the **Browser**, but the tweak in the top position is always named Sequence1.

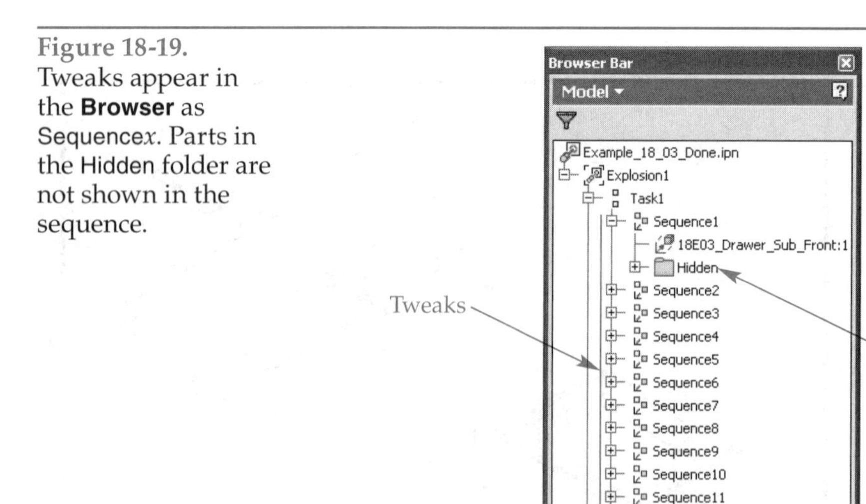

Figure 18-19. Tweaks appear in the **Browser** as Sequence*x*. Parts in the Hidden folder are not shown in the sequence.

After the tweaks are added, the view in each sequence can be adjusted. The **Edit Task & Sequences** dialog box is used for this, which is discussed later. Also, you can determine which parts are visible in the sequence. If you expand Sequence1 in the **Browser**, the branch contains a Hidden folder. The parts listed within this folder are not visible in the sequence.

Tweaks

Now, you will create this animation. Close Example_18_03_Done.ipn without saving any changes and open Example_18_03_Original.ipn. If the rotation tweak and final movement tweak on the nut are grouped, it will look like the nut is being threaded onto the cylinder. Open the **Animation** dialog box by picking the **Animate...** button in the **Presentation Panel**. Expand the dialog box by picking the **>>** button. Holding down the [Ctrl] key, select Sequence8 and Sequence9 in the **Animation Sequence** area of the dialog box, **Figure 18-20.** Next, pick the **Group** button and then the **Apply** button. Close the dialog box. The sequences in the **Browser** are changed; there are only ten sequences. This operation could also have been done by dragging the .1" tweak from the original Sequence11 and dropping it into the original Sequence10 branch.

Currently, all five wood screws are assembled at the same time. However, in the final animation (according to the storyboard) 18E03_Wood_Screw5 is to be assembled by itself. Also, the screw should rotate as if it is being driven in. Open the **Animation** dialog box and expand it. Then, select 18E03_Wood_Screw5 in the **Animation Sequence** list. Next, pick the **Ungroup** button and then the **Apply** button. Close the dialog box. This wood screw is now in new Sequence3. Also, nothing is hidden in this new sequence. The visibility will be addressed later.

Next, the screw needs to rotate. Apply a new tweak:

1. Pick the **Tweak Components...** button in the **Presentation Panel** to display the **Tweak Component** dialog box.
2. With the **Direction** button on in the **Create Tweak** area and the **Z** button on in the **Transformations** area, zoom in on the head of the screw and pick its circumference. Make sure the Z axis is along the long axis of the screw.
3. Pick the rotational radio button in the **Transformations** area of the dialog box (it is the lower radio button).

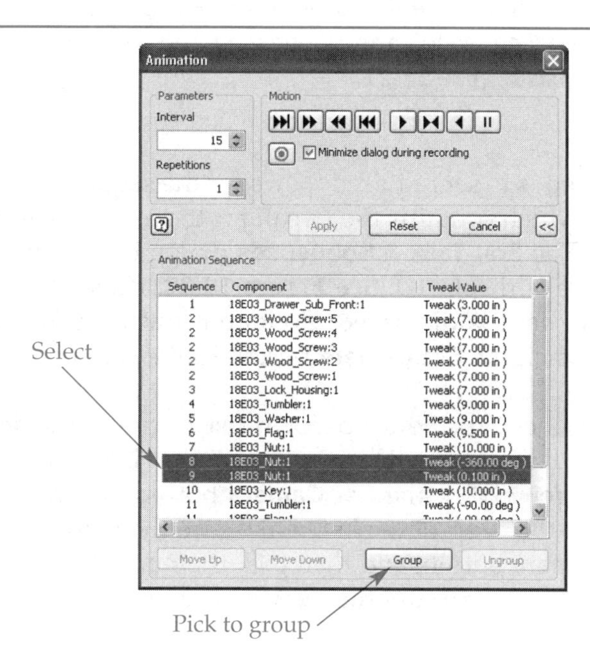

Figure 18-20.
Grouping the translation and rotation for the nut.

Figure 18-21.
Tweaks are added and adjusted for 18E03_Wood_Screw:5.

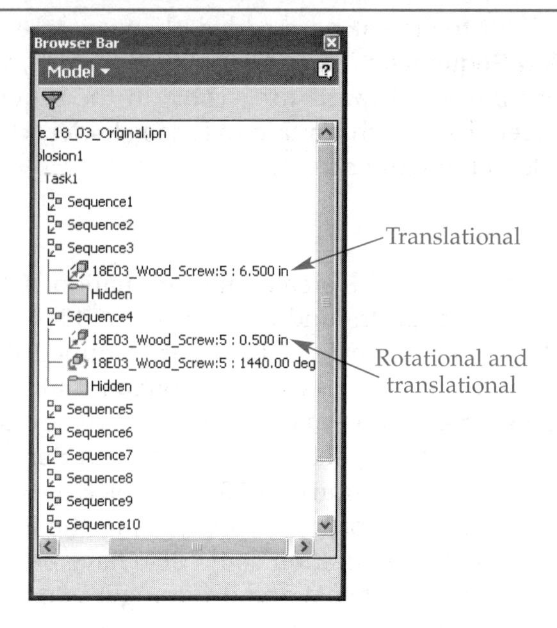

4. With the **Components** button on in the **Create Tweak** area of the dialog box, pick 18E03_Wood_Screw5. It is safest to select the part in the **Browser**.
5. Enter 4*360 in the text box in the **Transformations** area of the dialog box and then pick the check mark button.
6. Close the dialog box.

This rotation tweak is now Sequence1. If you expand the Sequence1 branch, it contains the rotation tweak with a value of 1440° (4 * 360 = 1440). The rotation should take place as the screw moves the final .5″ into the drawer subfront. Therefore, edit the translational tweak for 18E03_Wood_Screw5 (it should be in Sequence4) and change the value from 7″ to .5″. Then, drag the rotational tweak in Sequence1 and drop it into Sequence4. This groups the two tweaks and renumbers the sequences.

Next, the screw needs to be moved out 6.5″ (.5″ + 6.5″ = 7″) to place it inline with the other four screws. Apply a translational tweak to 18E03_Wood_Screw5 moving it 6.5″. This becomes the new Sequence1. Now, the 6.5″ movement should occur directly before the rotation/movement combination. Drag Sequence1 (*not* the tweak within it) down in between Sequence3 and Sequence4 and drop it. The first three sequences are reordered. Refer to **Figure 18-21.**

Views

The next step is to set required views for the sequences. In the **Browser**, right-click on Sequence5, which contains the tweak for the lock housing, and select **Edit...** in the pop-up menu. The **Edit Task & Sequences** dialog box is opened and the current view for the sequence is displayed. See **Figure 18-22.** Zoom and rotate the view until the exact view you want is displayed. Zoom out until the key is visible in the graphics window. Next, pick the **Set Camera** button in the dialog box and then pick the **Apply** button.

Select Sequence3 in the drop-down list in the **Sequences** area of the **Edit Task & Sequences** dialog box. Then, display the isometric view and zoom in on 18E03_Wood_Screw5. Pick the **Set Camera** button and then the **Apply** button. Select Sequence4 from the drop-down list, make sure the view is the same as for Sequence3, pick the **Set Camera** button, and then pick the **Apply** button. Pick Sequence5 from the drop-down list and make sure the view is correct (as you set it earlier).

Figure 18-22.
The **Edit Task & Sequences** dialog box is used to set the view for sequences.

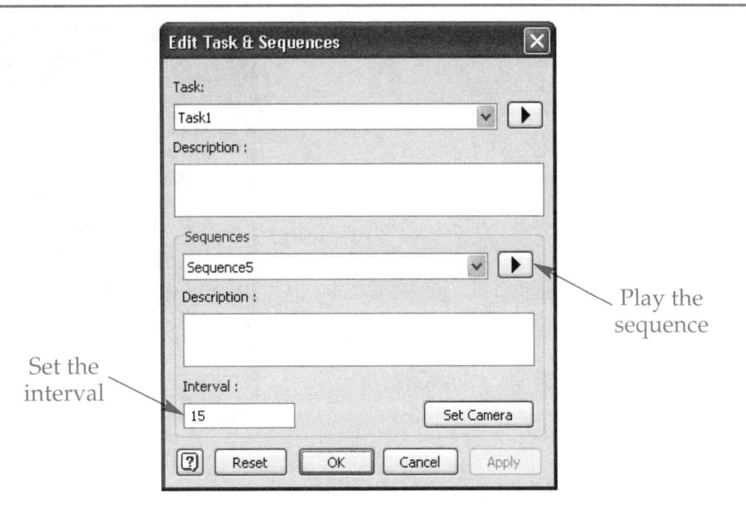

To see the result of a sequence, pick the **Play Forward** button in the **Sequences** area. The **Play Forward** button next to the **Task:** drop-down list plays the entire set of sequences. Any sequence can be selected in the drop-down list in the **Sequences** area, so you can walk through them all, setting the camera as you go.

The interval in the lower-left corner defaults to the setting in the **Animate** dialog box. However, this value can be changed for each individual sequence if needed. Note that object visibility is not reflected as you step through the sequences. Pick the **OK** button to close the dialog box.

Visibility

Finally, the visibility of the components needs to be set for each sequence. Expand Sequence1 in the **Browser** and notice the Hidden branch; expand it. The parts listed in this branch are not visible in the sequence. To hide parts in a sequence, drag and drop the part from the assembly in the **Browser** or from another sequence. To remove a part from the Hidden branch so it is visible in the sequence, right-click on the part name and select **Delete** from the pop-up menu. Visibility must be set for every sequence.

Currently, all parts are visible in Sequence3 (the screw moving 6.5″) and Sequence4 (the screw moving .5″ and rotating). The lock parts need to be hidden in these two sequences. Expand Sequence2 and Sequence3, select all of the parts in the Hidden branch, and drag-and-drop them into the Hidden branch for Sequence3. You can also select the parts in the Hidden branch of Sequence2, press the [Ctrl][C] key combination, select the Hidden branch in Sequence3, and press the [Ctrl][V] key combination. Similarly, hide the lock parts in Sequence4.

Play the animation to check your work. Pick the **Animate...** button in the **Presentation Panel**. Then, pick the **Play Forward** button in the **Animation** dialog box. If the animation is correct, save and close the file. If not, make corrections as needed.

Styles

In previous chapters, you learned how to create new styles, modify styles, and apply styles to parts. Styles can be used to enhance presentation files. In this section, you will apply styles to highlight specific part locations and functions. Next, you will create custom lighting to expose details that would otherwise be hidden in a shadow. One benefit of performing these tasks within a presentation file is that the assembly and part files remain unchanged.

Figure 18-23.
Select the six parts in the assembly and look at the color selector drop-down list on the **Inventor Standard** toolbar.

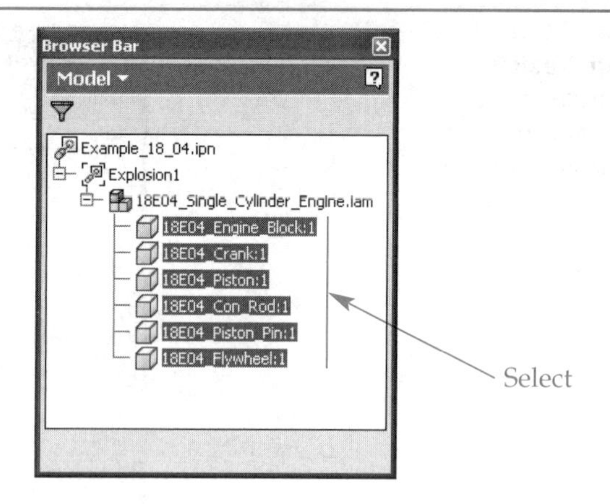

Part Colors in Assemblies

In Chapter 12, you created a design view representation of a single-cylinder engine. The engine block was made invisible in this representation so that the internal components could be viewed and highlighted. To highlight various parts of an assembly without creating multiple design view representations, contrasting color styles can be applied to the parts that require highlighting. This method works well in a meeting, where the discussion is too unpredictable to preplan multiple design view representations.

Open Example_18_04.ipn. In the **Browser**, expand the branches for the explosion and the assembly. Then, select the six parts in the assembly. See **Figure 18-23.** Now, look at the color selection drop-down list at the right-hand end of the **Inventor Standard** toolbar. It currently displays As Material. If the drop-down list is grayed-out, then something other than a part is selected in the **Browser**. Pick Polycarbonate (Clear) from the drop-down list. If you pick the arrow to display the list, you can then press the [P] key to jump to the beginning of the names that start with P. After selecting the material, all of the selected parts should be transparent. Now, select Crank:1, either in the **Browser** or the graphics window, and change the color to Lime. Undo the color change and practice highlighting other parts of the assembly. Then, close the file without saving any changes.

Lighting

In a presentation file, lighting is the only type style available in the **Document Settings** dialog box. This dialog box is displayed by picking **Active Standard...** from the **Format** pull-down menu. Lighting can be applied to the entire presentation, whereas each part must be selected to change its color in the presentation file. In Chapter 2, you learned what each lighting effect does and how to change it. In this section, you will learn how to apply lighting to a presentation.

Open Example_18_05.ipt. This is a molded plastic part with its color set as black. Since this part is made of black plastic, some of the inner details are difficult to see. Now, open Example_18_05.ipn. This is a presentation file that contains an assembly of the molded plastic part. Presentation files can only contain assemblies. To bring in the molded part into an assembly, the part first had to be placed in an assembly file. As you can see in the presentation file, the details are easier to understand than in the part file because of the lighting style that has been applied.

Open the **Document Settings** dialog box by selecting **Format>Active Standard...** in the pull-down menu. Pick the **Active Lighting Style** drop-down list to see the available styles. See **Figure 18-24.** Pick the **Close** button to close the dialog box. Then, select **Style and Standard Editor...** from the **Format** pull-down menu to display the **Style and**

Figure 18-24.
Lighting is the only style that changes the entire presentation.

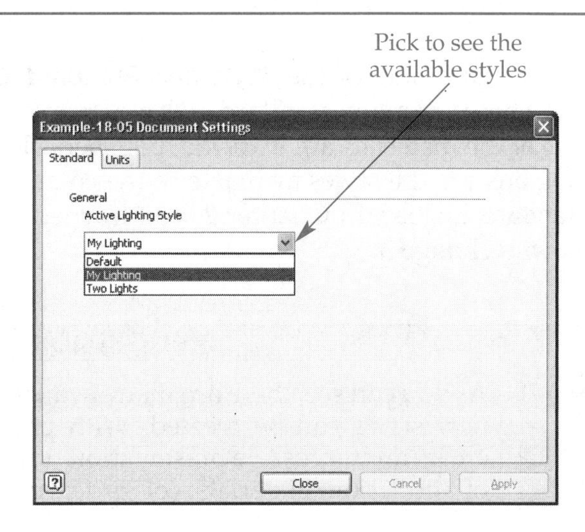

Standard Editor dialog box. Expand the Lighting branch on the left-hand side of the dialog box. Then, select the My Lighting style. See **Figure 18-25.** Notice that this style has three lights out of a possible four. Pick each radio button in the **Settings** area on the right-hand side of the dialog box to view the location and color of each light. Notice the color swatch when the third light is selected. This light is colored gray so that it is not as bright as the other lights. Changing the color is the only way to control the brightness of individual lights. Leave the **Style and Standard Editor** dialog box open and continue to the next section.

Figure 18-25.
The settings for the My Lighting style are displayed on the right-hand side of the dialog box.

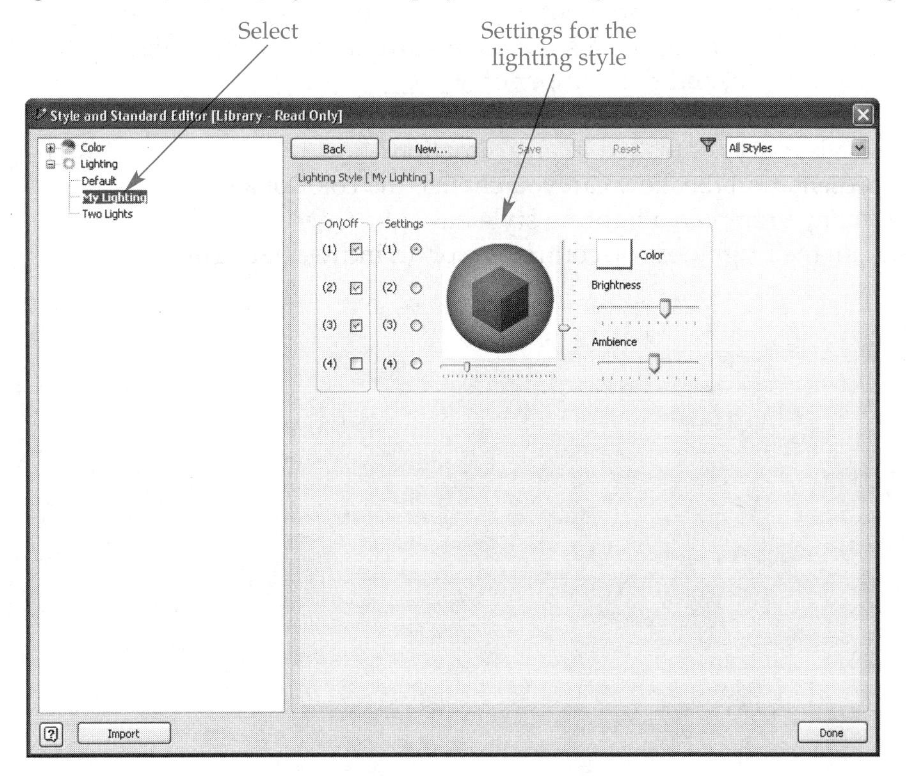

Colors

On the left-hand side of the **Style and Standard Editor** dialog box, expand the Color branch. There are many standard styles available, including color, materials, and Default. Some of the materials are textured, polished, clear, or chrome. The styles listed in this dialog box are the styles available in the color selection drop-down list on the **Inventor Standard** toolbar. In Chapter 2, you learned what each material component does and how to change it.

PROFESSIONAL TIP

As you can see, there are many available standard styles. If needed, new styles can be created. With practice, you should be able to create an impressive presentation utilizing the proper combination of animation, materials, colors, textures, and lighting.

Chapter Test

Answer the following questions on a separate sheet of paper or complete the electronic chapter test on the Student CD.

1. Exploded views of an assembly are created in a separate file called a(n) _______ file.
2. The type of animation file that Inventor can create for an exploded assembly is _______.
3. The two methods of creating an explosion are _______ and _______.
4. Each movement of a component during an explosion is called a(n) _______.
5. The speed of the components during an animation is controlled by the distance each has to travel and the number of _______.
6. Briefly describe how to set a view for a given sequence.
7. How is visibility controlled in an animation?
8. In a presentation file, how can you change the color of a part?
9. How many lights can a lighting style have?
10. How can the brightness be controlled for an individual light?

Chapter Exercises

Exercise 18-1. Slider Clamp.
Complete the exercise on the Student CD

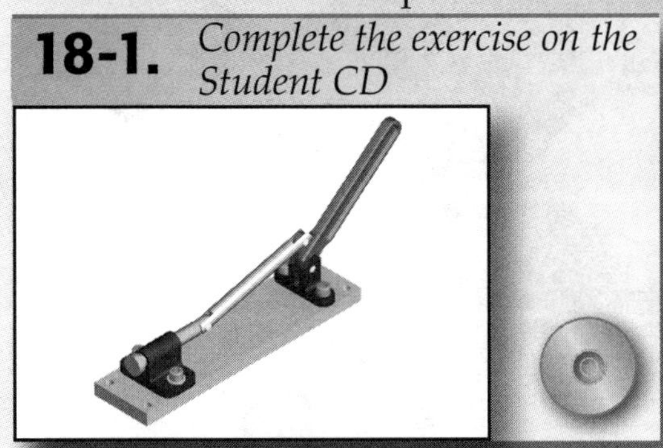

Exercise 18-2. Exploded View Drawing.
Complete the exercise on the Student CD

Exercise 18-3. Clamp Linkage.
Complete the exercise on the Student CD

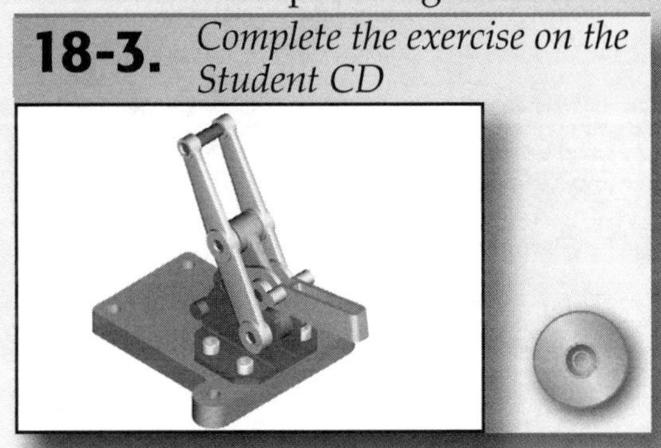

Exercise 18-4. Latch Assembly.
Complete the exercise on the Student CD

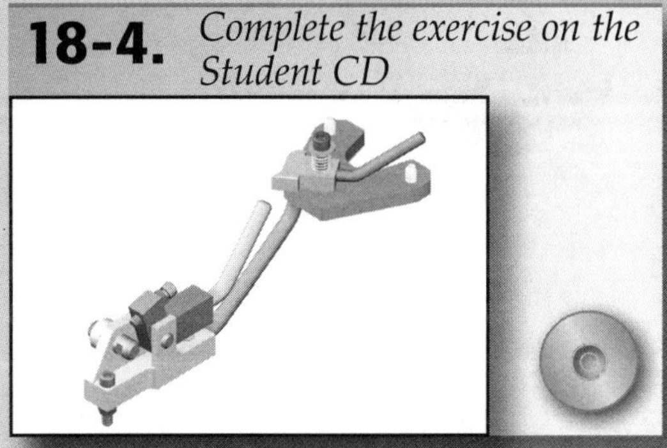

Exercise 18-5. Valet.
Complete the exercise on the Student CD

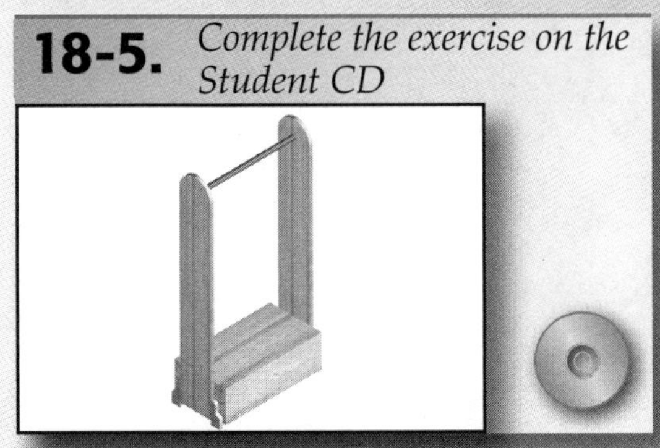

This sheet metal part was created from a combination of faces, contour flanges, folds, corner rounds, and holes. (Model courtesy of Autodesk, Inc.)

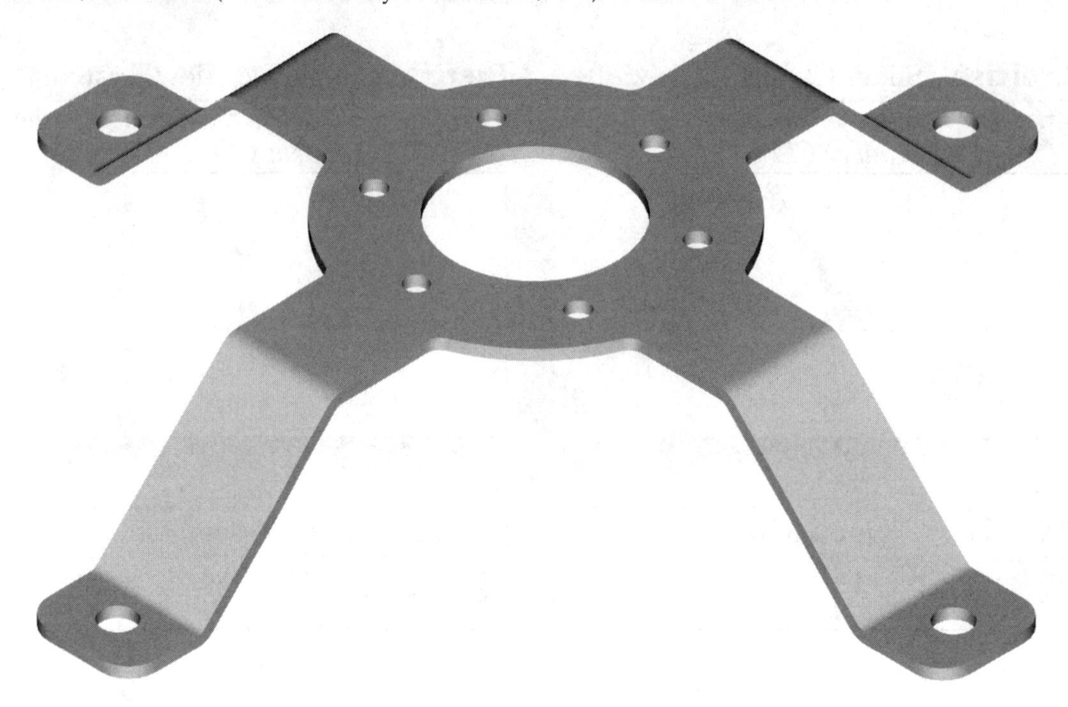

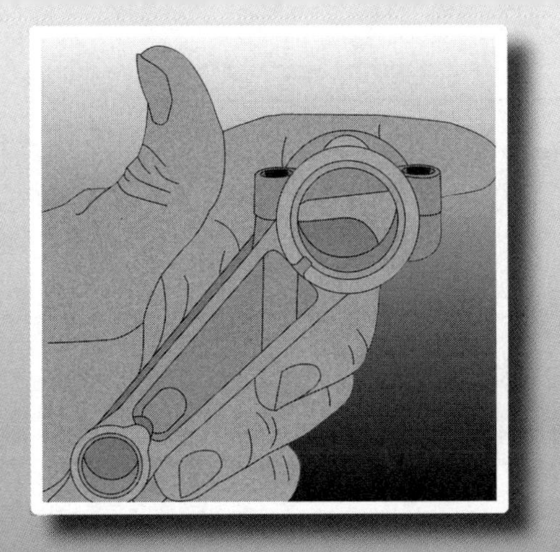

Chapter 19

Sheet Metal Parts

Objectives

After completing this chapter, you will be able to:

- Define terms related to sheet metal parts.
- Create sheet metal styles.
- Add features to a sheet metal part with the **Face** tool.
- Specify relief settings.
- Override sheet metal styles.
- Create a flat pattern from a folded part.
- Create cutouts in sheet metal parts.
- Add features to a sheet metal part with the **Flange** tool.
- Create rounds and chamfers on the corners of sheet metal parts.
- Create hems on sheet metal parts.
- Create cutouts across bends.
- Create contour flanges.
- Create a folded sheet metal part from a developed pattern.
- Control the shape of the seam between two edges using the **Corner Seam** tool.
- Use punch tools to create holes or emboss sheet metal parts.

User's Files

The Student CD included with this text contains several files required for this chapter. Refer to the file File List.txt *in the* \Ch19 *folder for the comprehensive list.*

Sheet Metal Parts

The term ***sheet metal part*** generally refers to a part formed from relatively thin, flat sheet metal stock, which is typically between 0.01″ and 0.18″ thick. In the United States the thickness of sheet metal is often specified by a gage number; the larger the number, the thinner the material. For example, 12 gage cold rolled steel is 0.1046″ thick. A table of the gages for several materials is included in the appendix. There are many manufacturing processes for turning flat stock into parts. See **Figure 19-1.**

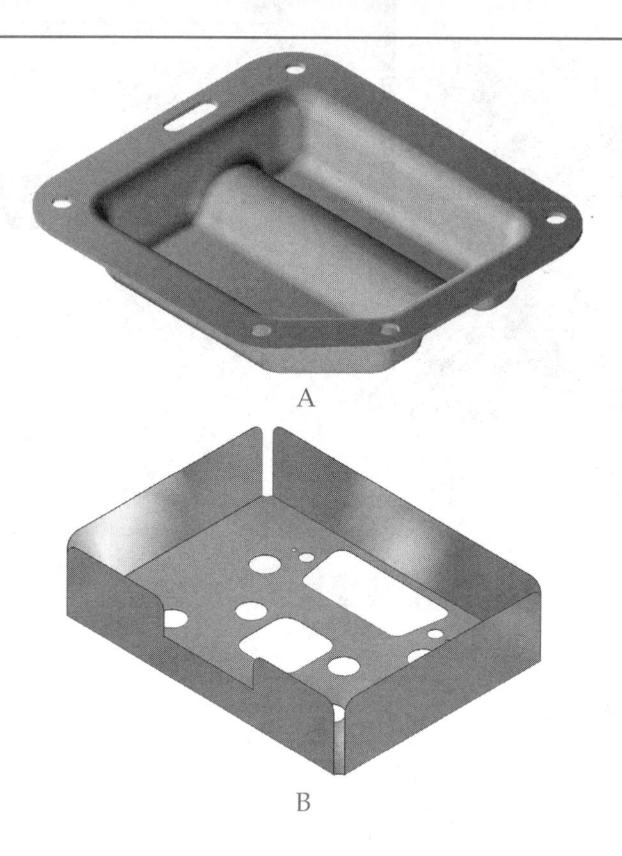

Figure 19-1.
A—This Inventor part was created using the **Part Feature** tools. The actual part would be fabricated in a press with matching pairs of dies. B—This Inventor part was created using the **Sheet Metal Features** tools. The actual part would be fabricated by punching the features into a metal blank and then performing four bends.

These processes include spinning, sheet hydro-forming, and roll forming, but two of the most common are:

- Drawing.
- Punching and bending.

Drawing uses matching pairs of dies in a press to stretch and flow the metal into complex curves about two axes. The metal is generally forced between two mating dies that match the contours of the formed part. Solid models of drawn parts can be created in Inventor using sketches and the standard **Part Feature** tools, such as the **Extrude** and **Revolve** tools.

Punching and ***bending*** use tools in a press to cut the profile of the part, punch holes, form ribs and other small features, and create bends. The bends are each about one axis, much like a piece of cardboard is bent to form a box. Solid models of these parts are created using the tools in the Inventor **Sheet Metal Features** panel. The basic steps in fabricating this type of part are:

1. Shear (cut) the sheet into a flat, rectangular shape called the ***blank.*** This is the basic shape from which the sheet metal part is produced.
2. Punch, drill, or broach various shaped holes, notches, and slots into the blank.
3. Bend the blank to form the part into its final shape.

The problem is that the designer creates a part with a final shape that meets the design criteria. However, the dimensions of the flat pattern (the blank) must be calculated. These dimensions are called the ***developed length*** or layout dimensions. This is not simply a process of adding the dimensions to arrive at a total. The metal dimensionally changes as it is bent. The outside surface of the bend gets longer, the inside surface gets shorter, and the neutral axis moves away from the center of the bend.

For example, the simple U-shaped bracket shown in **Figure 19-2** has to be 2″ deep and 2.5″ wide on the inside. The sheet metal is 11 gage (.12″) mild steel. The inside radius of the bend is .12″, which is equal to the metal thickness. The minimum recommended bend radius is usually the same as the material thickness. Using Inventor's sheet metal

Figure 19-2.
A—Two orthographic views of the folded part on the left and the flat pattern on the right.
B—The completed part.

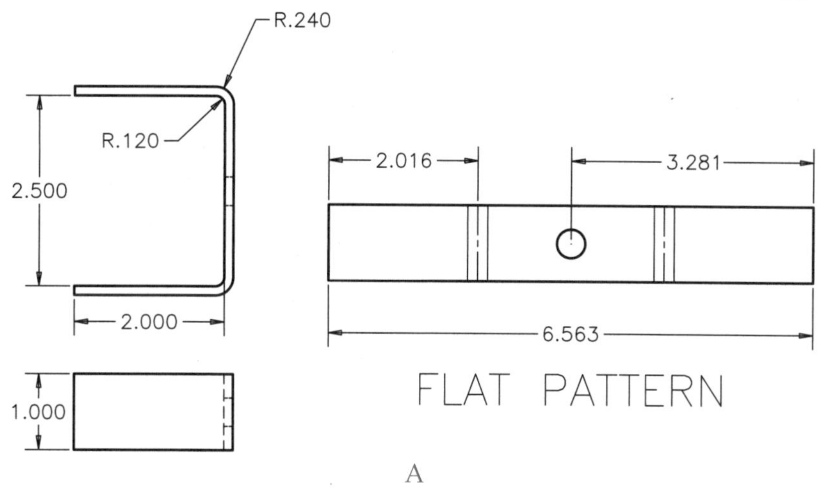

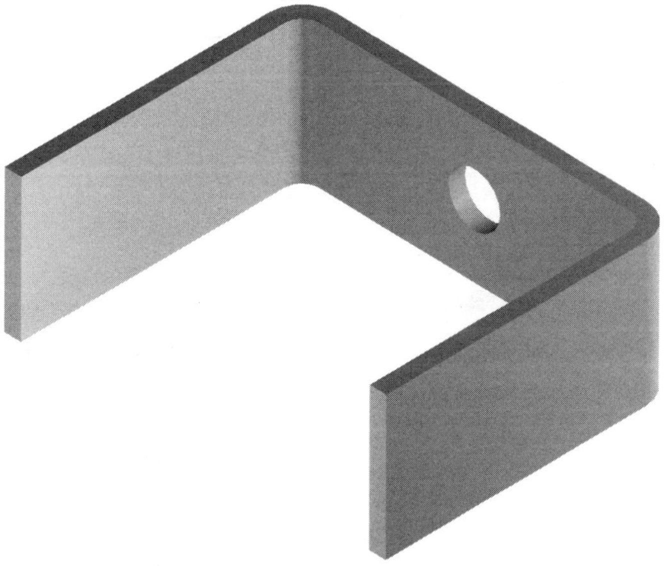

function, a flat pattern is automatically generated from the solid model and the length of the blank is calculated as 6.563″. The center of each bend is also calculated. The flat pattern can then be inserted into a drawing file and dimensioned.

Style Tool

In a standard part file, a profile is sketched and extruded a specified distance to give the part thickness. In a sheet metal part file, the thickness of the material is first specified using the **Sheet Metal Style** tool. The sketch is then extruded to the specified thickness.

Open Example_19_01.ipt, which is a sheet metal part file. See **Figure 19-3A.** You will create the finished sheet metal part shown in **Figure 19-3B.** First, a style needs to be set for the part. The style for a sheet metal part includes information such as the material, material thickness, and several default values. To set the style, pick the **Sheet Metal Styles** button in the **Sheet Metal Features** panel. The **Sheet Metal Styles** dialog box is opened, **Figure 19-4.**

Figure 19-3.
A—The sketch for creating a face on a sheet metal part.
B—The completed part.

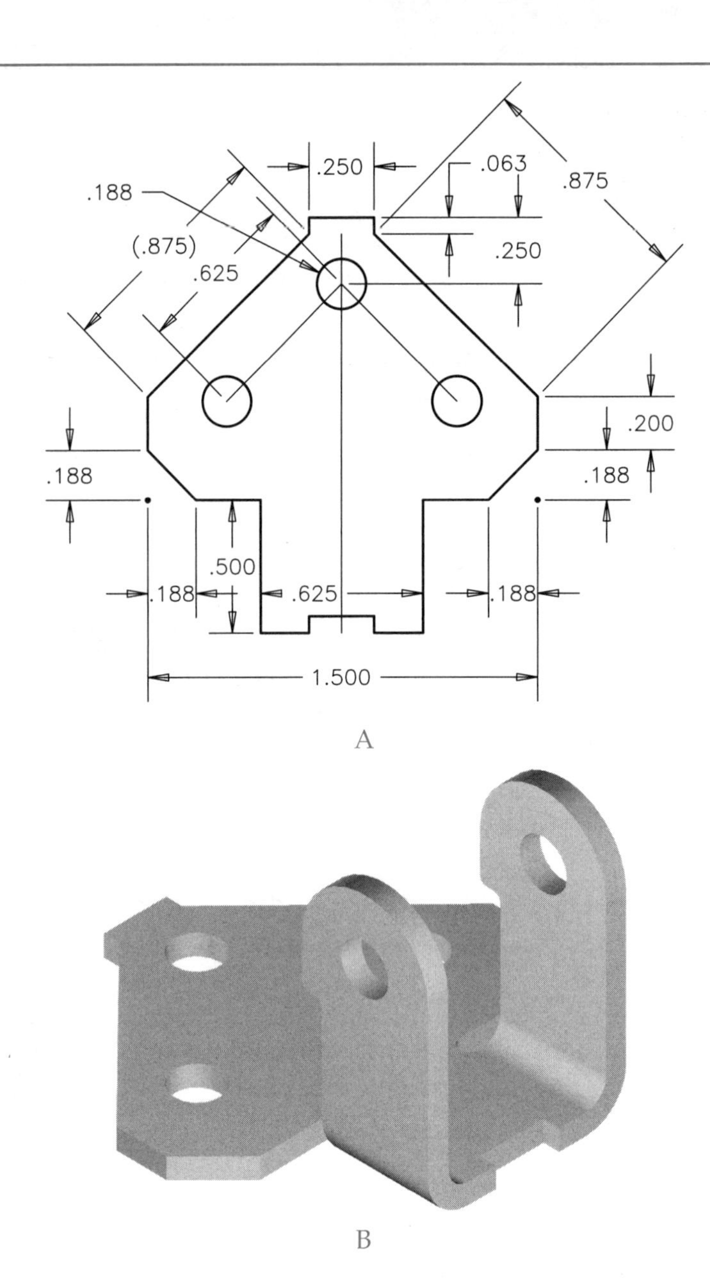

The saved styles appear in the list at the left of the dialog box with the style Default. The active style appears in the drop-down list below the list of saved styles. To create a new style based on the current style, pick the **New** button. The new style is initially given the name Copy of *style,* which in this case is Copy of Default. To rename it, type a new name in the text box above the styles list. For this example, name the new style Example_19_01.

The type of material is set in the **Sheet** tab of the **Sheet Metal Styles** dialog box. The type of material is only important for the material shading color and the physical properties analysis. Pick the **Material** drop-down list in the **Sheet** area of the tab and select Steel, Mild. Also in this area, set the material thickness to .08″ by typing the value in the **Thickness** text box.

The other options in the **Sheet Metal Styles** dialog box are discussed later. For now, pick the **Save** button to save the new style. Then, select Example_19_01 from the **Active Style** drop-down list. Finally, pick the **Done** button to close the dialog box.

Figure 19-4.
The **Sheet Metal Styles** dialog box.

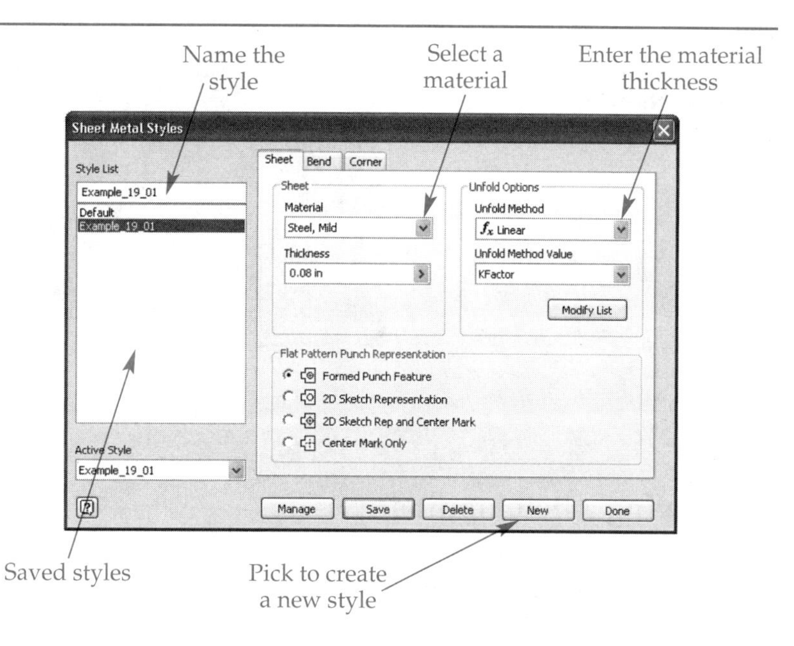

Face Tool

The **Face** tool is used to create sheet metal faces on the part. The *face* for this sheet metal part you are developing is a base feature. It is generated from a sketch containing holes and cutouts that is assigned the material thickness. As the extruded faces of the part are created, they will be joined with seams or bends to create the final sheet metal part. If you think of a folded sheet metal box that is open on one side, there are five faces to that part. The five faces are joined by four bends and four seams.

To create the first extruded face on the sheet metal part in this example, pick the **Face** button in the **Sheet Metal Features** panel. The **Face** dialog box is displayed, **Figure 19-5.** With the **Profile** button in the **Shape** tab on, select the profile. The area inside of the circles should not be part of the profile. A preview of the extruded profile appears in green in the graphics window, just as it does when using the **Extrude** tool on a standard part. Picking the **Offset** button in the **Shape** tab changes the direction of the extrusion. Since this is the first face on the part, the direction is not important. Pick the **OK** button to create the face, **Figure 19-6.**

Adding a Second Face

Now, you will add another extruded face to the sheet metal part. The new face will be at 90° to the face you just created. Open Example_19_02.ipt, in which a sketch of the new face has already been created. The side face shown in red was selected as the sketch plane. A vertical construction line is used in the sketch to align the arc with the base part.

Figure 19-5.
The **Face** dialog box.

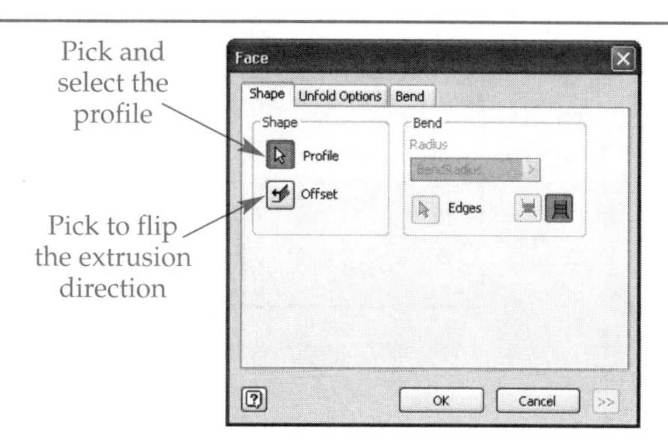

Figure 19-6.
The face is created based on the sketch in Figure 19-3A.

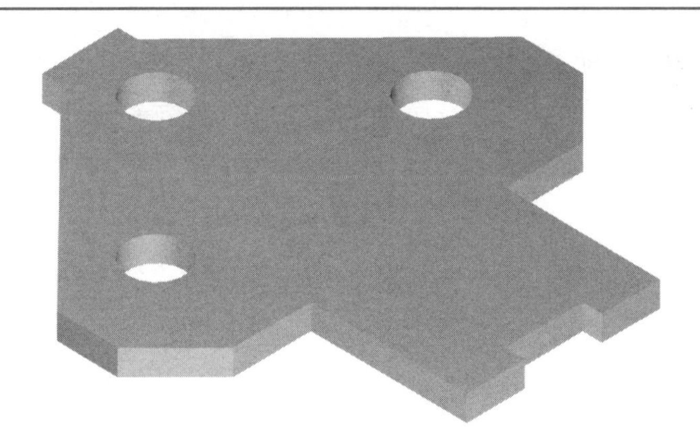

Specifying a bend radius

Before creating the face, a bend radius specification needs to be added to the style. Open the **Sheet Metal Styles** dialog box and select the **Bend** tab. See **Figure 19-7.** The entry in the **Bend Radius** text box is the inside radius of the bend. It is a good idea to express the radius as a function of the material thickness. In this way, if the material thickness is changed, the bend radius is automatically updated. Type Thickness in the **Radius** text box, if it is not already displayed. In other applications, you may type a formula here.

Specifying a relief

A ***relief*** is a notch punched, sawed, or laser cut in the blank that allows the material to bend without tearing. There are three choices for the shape of the relief—Round, Straight, and Tear. A round relief has a circular end or radiused corners. A straight relief has square corners. When Tear is selected, no relief is created and the material is torn. The preview below the **Relief Shape** drop-down list reflects the selected shape.

If the relief is close to the edge of a part, a small "prong" or leftover piece of material may be created. This unwanted leftover material is called a ***remnant.*** Refer to C in the preview in the **Bend** tab. The minimum allowable remnant size is set in the **Minimum Remnant** text box in the tab. This is often expressed as a formula, as is the case with the default. If the remnant is less than this value, the width of the relief is increased to eliminate it.

Figure 19-7.
Creating a sheet metal style.

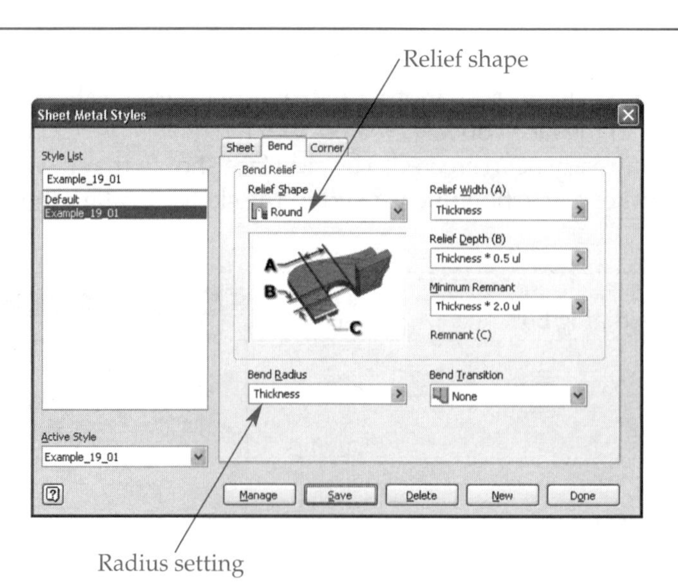

The **Relief Width** setting determines the width of the relief. Refer to A in the preview in the tab. As with the bend radius, the relief width is often related to the material thickness. This is because a thicker material requires a wider relief to prevent distortion.

The **Relief Depth** setting determines how far the relief extends into the part beyond the start of the bend. Refer to B in the preview in the tab. The minimum value is half of the material thickness.

For this example, set the **Relief Shape** to Round. Leave all other relief settings at their defaults.

Saving the style

Once you have made changes to the style, it must be saved. Pick the **Save** button in the **Sheet Metal Styles** dialog box to save the style. Then, pick the **Done** button to close the dialog box.

Creating the face

Open the **Face** dialog box. Then, select the profile, **Figure 19-8.** The offset is important here as it determines the width of the part. The extruded face should extend out from the existing extruded face, not into the existing face. If needed, pick the **Offset** button in the **Shape** area of the **Face** dialog box to reverse the extrusion direction. A new bend radius can be entered in the **Radius** text box of the **Shape** tab to override the value set in the style. Also, the settings in the **Bend** tab override the relief settings for the style. Overriding the style is discussed later. For now, pick the **OK** button to accept the default settings and create the new face.

Notice that a bend is automatically applied between the new face and the existing face. See **Figure 19-9.** The bend is generated based on the settings in the style and any overrides. Also, notice how the relief is created in relation to the bend.

Adding a Third Face

Now, a third face needs to be added that is the mirror image of the face you just created. The **Mirror Feature** tool will work in this situation, but here you will practice the sketch technique. A new 2D sketch needs to be created and the geometry from the second extruded face projected onto the sketch plane. Then, the symmetrical feature can be created.

Figure 19-8.
The preview of the new face. The offset is important and should appear as shown here.

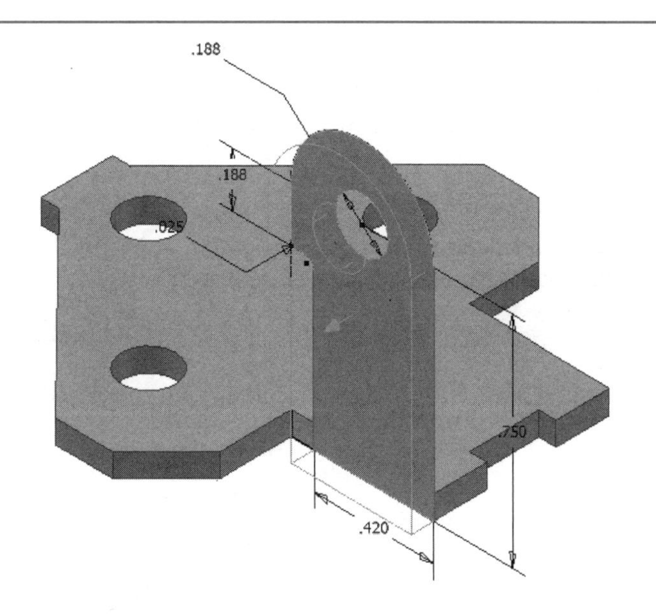

Figure 19-9.
A—The new face is created. The bend is automatically applied. B—The top view clearly shows the relief that was created.

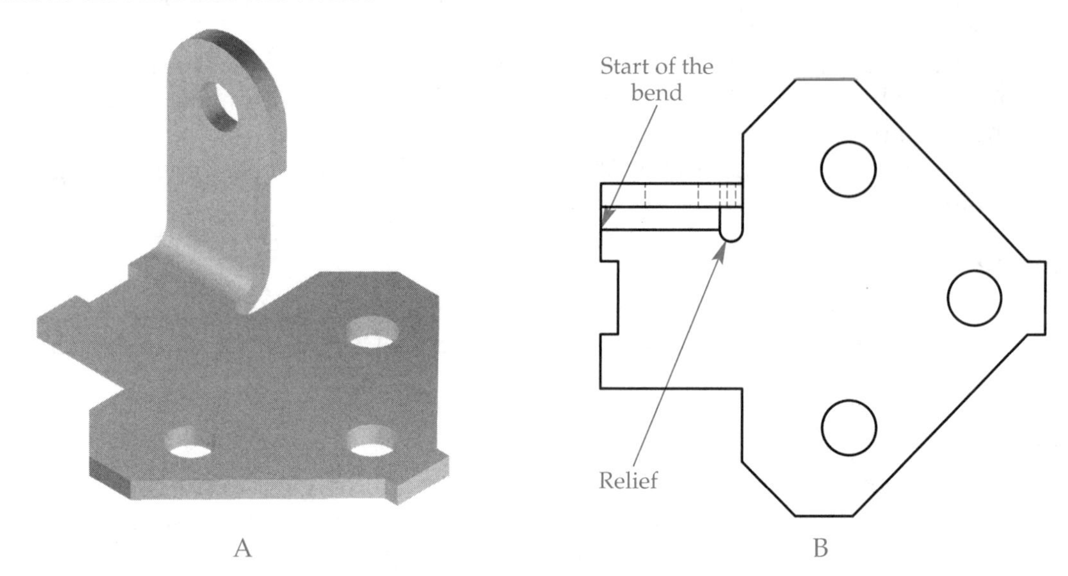

Start by selecting as the sketch plane the face on the first extruded face that is opposite of the original red face. This face is colored purple in the file. Then, use the **Project Geometry** tool to project the geometry from the second extruded face. If you pick within the area of the surface, as shown in **Figure 19-10,** the outline is projected. Notice how there is a gap between the projected geometry and the existing extruded face. This is okay because the bend will fill in this area. Finish the sketch.

Next, open the **Face** dialog box and select the profile. Again, the extrusion should extend out from the existing extruded face. Now, pick the **Edges** button in the **Bend** area of the **Shape** tab. Pick the top edge of the existing extruded face, as indicated in **Figure 19-10.** Finally, pick the **OK** button in the dialog box to create the new face. Notice how Inventor fills in the bend and creates the relief.

Overriding Styles

The settings for the style in the **Sheet Metal Styles** dialog box control the entire part. However, some of these settings can be overridden when creating features. For example, some of the settings in the **Bend** tab that control the relief can be set when

Figure 19-10.
Creating a third face by projecting the second face.

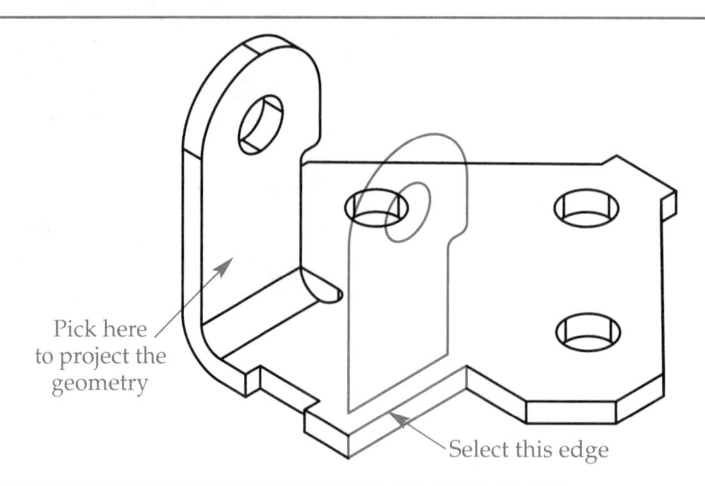

creating a new face. The bend radius can also be changed for an individual bend. However, some settings, such as material thickness, cannot be overridden. Settings that can be overridden are:

- Radius.
- Relief shape.
- Minimum remnant.
- Relief width.
- Relief depth.

For example, you can set the default choice in the **Bend** tab of the **Sheet Metal Styles** dialog box to Round, as in the previous example. Then, for each individual bend, you can choose Default, Tear, Round, or Straight in the **Relief Shape** drop-down list in the **Bend** tab of the **Face** dialog box. This override only applies to the bend on the face you are currently creating. To use the style setting, choose Default.

PROFESSIONAL TIP

If you want a unique relief for a particular bend, construct the reliefs as sketches and cut them from the part using the **Extrude** tool. Make sure the sketched relief is deep enough so that it goes past the start of the bend. Then, when creating the face, override the style by selecting Tear in the **Relief Shape** drop-down list in the **Bend** tab.

Creating a Flat Pattern

A ***flat pattern,*** or ***developed blank,*** results when the sheet metal part is unfolded into a flat sheet. Inventor can automatically develop and unfold a sheet metal part into a flat pattern. This pattern can be inserted into a drawing layout. Once in the drawing, the flat pattern can be dimensioned. The **Flat Pattern** tool is used to create a flat pattern.

Open Example_19_03.ipt. You will create a flat pattern of this folded sheet metal part. Pick the **Flat Pattern** button in the **Sheet Metal Features** panel. A new branch called Flat Pattern is added to the tree in the **Browser** and the pattern is displayed on the screen. The actual sheet metal part is called Folded Model in the **Browser**. See **Figure 19-11.** To switch between a display of the part and the folded pattern, double-click on the name in the **Browser**.

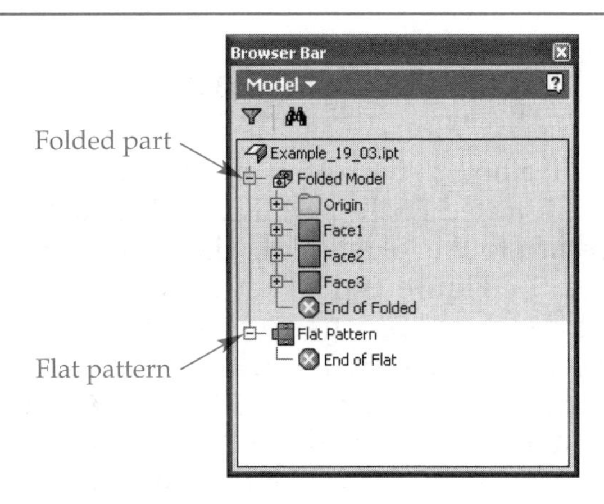

Figure 19-11.
The folded part and the flat pattern are contained in separate branches in the **Browser**.

Figure 19-12.
A drawing shows the dimensions of the folded part and the overall dimension of the flat pattern.

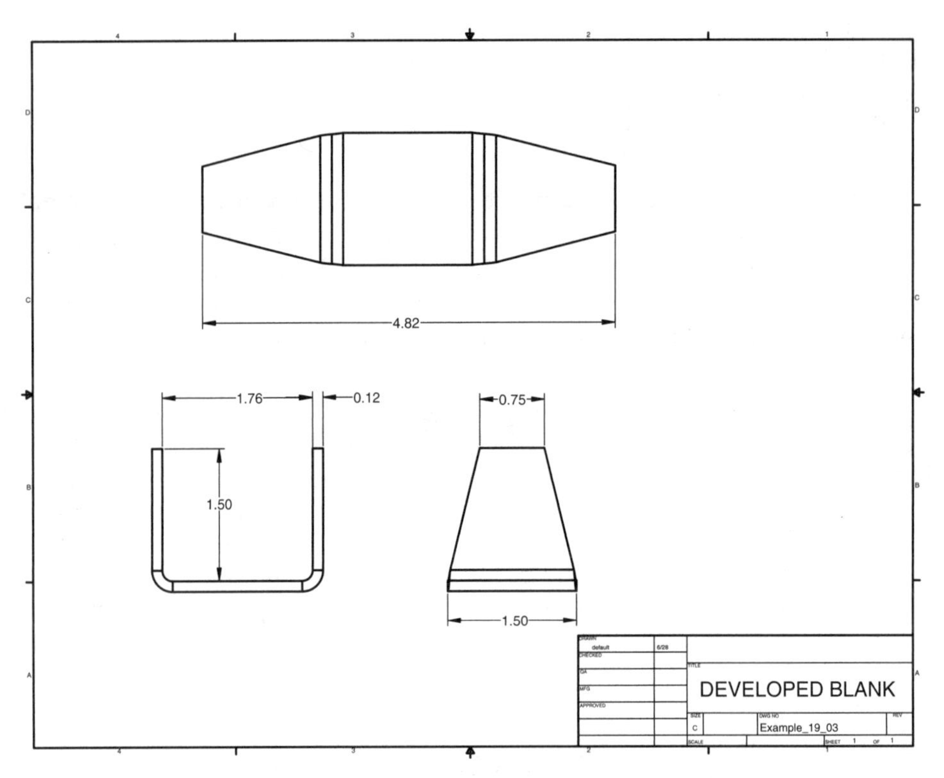

Now, open Example_19_03.idw, which is a drawing set up with two orthographic views of the folded part. Pick the **Base View...** button in the **Drawing Views Panel**. In the **Component** tab of the **Drawing View** dialog box, make sure Example_19_03.ipt is selected in the **File** drop-down list. Then, select the **Flat Pattern** radio button in the **Sheet Metal View** area. In the **Scale** text box, enter 2 for the scale. Then, place the view in an open area of the drawing.

Finally, using the **General Dimension** tool in the **Drawing Annotation Panel**, place an overall dimension on the drawing, as shown in **Figure 19-12.** Notice how only one dimension is required on the flat pattern. This is because the two fold lines are perpendicular to the dimensioned length. If the part had another fold not perpendicular to this length, the height would need to be dimensioned as well.

Cut Tool

Now, you will add a hole to the folded sheet metal part Example_19_03.ipt and see how the flat pattern is affected. In the part file, double-click on the Folded Model branch in the **Browser** to return to the folded part. Then, start a new 2D sketch on the right-hand vertical surface. See **Figure 19-13.** Sketch a circle and dimension it as shown in the figure. Finish the sketch and pick the **Cut** button in the **Sheet Metal Features** panel. The **Cut** dialog box is opened, **Figure 19-14.**

The **Cut** tool is equivalent to the **Cut** option of the **Extrude** tool used to create part features. With the **Profile** button on in the **Shape** area of the dialog box, select the interior of the circle as the profile. Then, in the **Extents** area of the dialog box, pick All in the upper drop-down list. Finally, pick the **OK** button to create the hole. Notice how the hole passes through both upright extruded faces on the part.

Figure 19-13.
Sketching a circle to be used with the **Cut** tool.

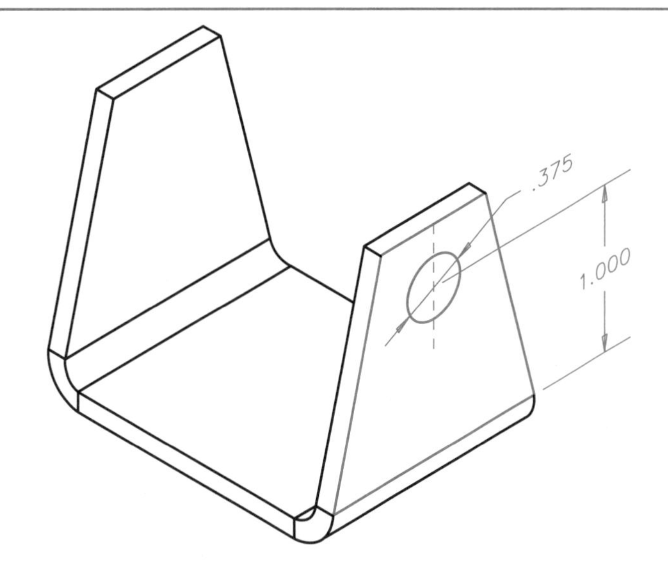

Figure 19-14.
The **Cut** dialog box.

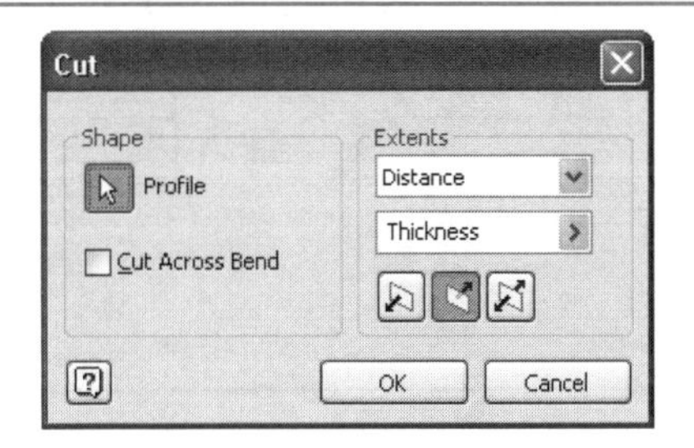

Return to the drawing layout. Notice how the flat pattern view is automatically updated to reflect the addition of the hole. Dimension the hole as shown in **Figure 19-15.** Save and close all files.

Projecting a Flat Pattern into a Sketch

You can project a tracing of the flat pattern into a sketch. Open Example_19_04.ipt. Edit Sketch3, which is a 2D sketch on the top surface. Refer to **Figure 19-16.** Then, pick the arrow next to the **Project Geometry** button in the **2D Sketch Panel**. From the flyout, select the **Project Flat Pattern** button and then pick the vertical face. The outline of the unfolded part is projected onto the sketch plane.

Now, the projected flat pattern can be used to create a hole. Construct and dimension the circle shown in **Figure 19-17.** Finish the sketch and pick the **Cut** button in the **Sheet Metal Features** panel to open the **Cut** dialog box. Select the interior of the circle as the profile. Then, check the **Cut Across Bend** check box. This will project the circle onto the surface that was projected to create the sketch. If you do not check this check box, an error will result. Also, when the check box is checked, the **Extents** options are grayed out. You will examine these options later in this chapter. Finally, pick the **OK** button to create the hole. See **Figure 19-18.** Save and close the file.

Complete the practice problem on the Student CD.

Figure 19-15.
The drawing of the part is automatically updated when the hole is cut in the part. Add the dimensions related to the hole.

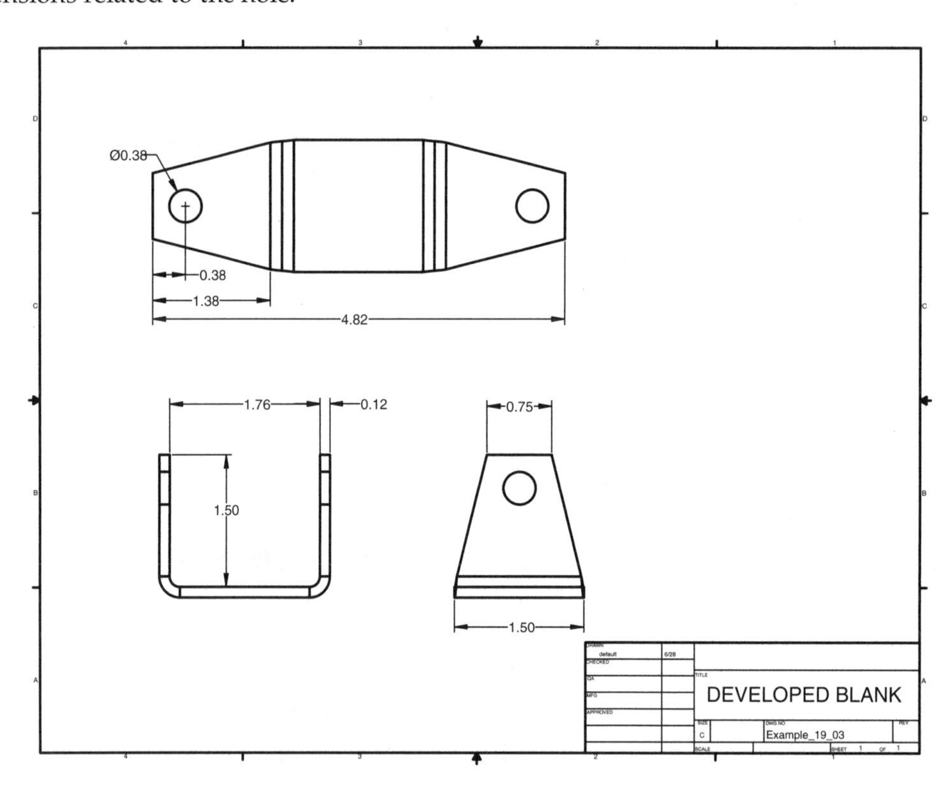

Figure 19-16.
Projecting a flat pattern into a sketch.

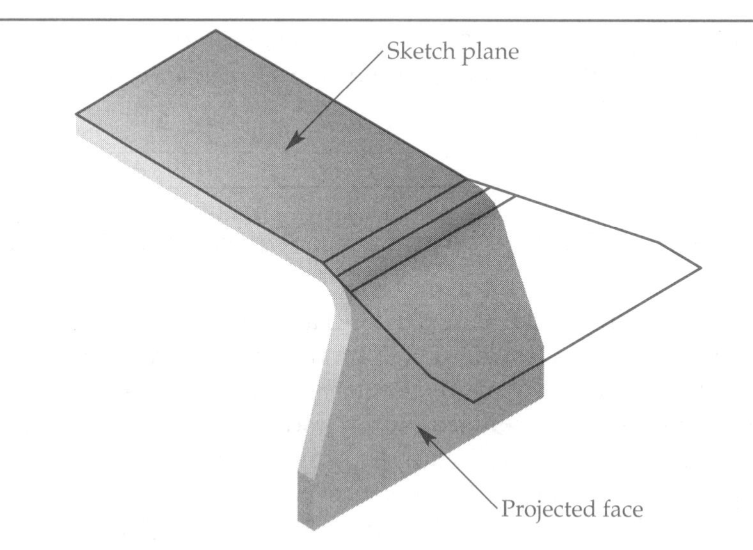

Flanges

A *flange* is a rectangular face on a sheet metal part that is at an angle to an existing face. The **Flange** tool is used to add a flange to a sheet metal part. The advantage of this tool is you can easily specify the bend angle. The entire edge or a portion of an edge can be flanged with or without reliefs. However, the disadvantage of the **Flange** tool is that you can only create a rectangular face.

Open Example_19_05.ipt, which is a sheet metal part with a base face already created. Then, pick the **Flange** button in the **Sheet Metal Features** panel. The **Flange** dialog box is displayed, **Figure 19-19.** In the **Shape** tab, pick the 0 Selected entry in the

Figure 19-17.
Sketching a circle to use with the **Cut** tool.

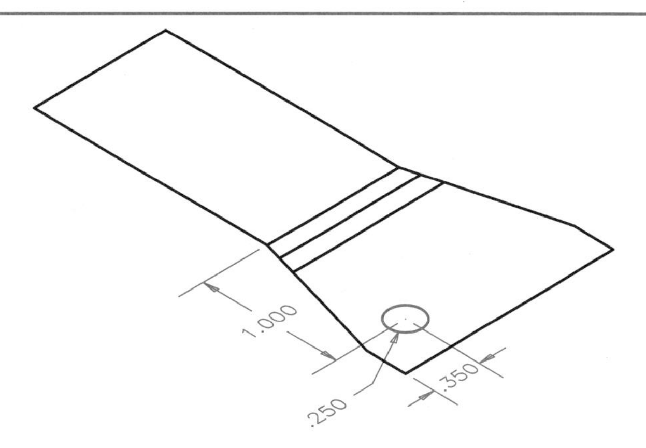

Figure 19-18.
The hole is created using the **Cut Across Bend** option of the **Cut** tool.

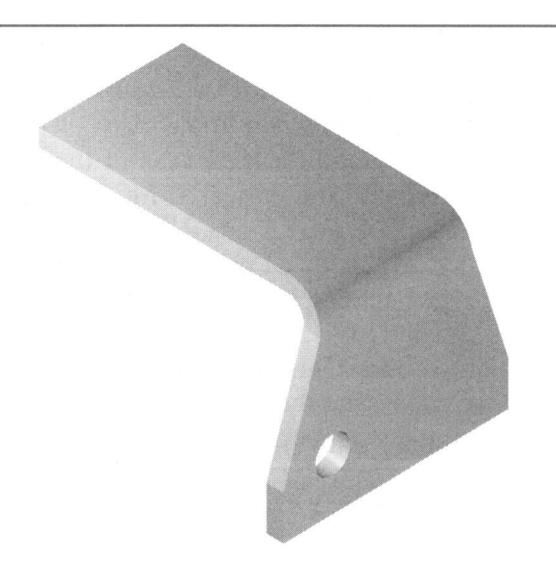

Figure 19-19.
The **Flange** dialog box.

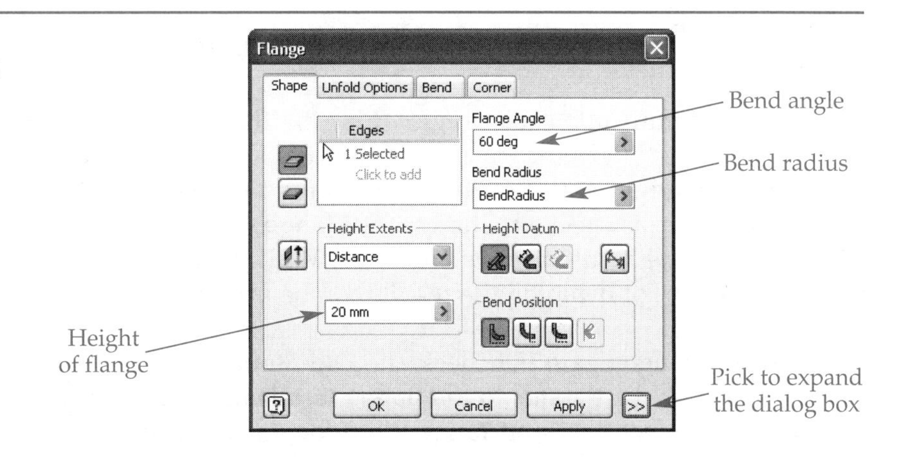

Edges column and then pick edge A shown in **Figure 19-20.** A preview of the flange appears in the graphics window.

In the **Height Extents** area, the drop-down list sets how the height is determined. The options in the drop-down list are Distance and To. Select Distance. Then, enter 20 in the bottom text box below the drop-down list. This is the height of the flange. The distance must be a positive value. However, you can change the direction of the flange by picking the **Flip Direction** button next to the **Height Extents** area.

Also, enter 60 in the **Flange Angle** text box. Notice how the angle is measured from the selected edge. The angle can be negative to bend the flange in the opposite

Figure 19-20.
Selecting the edge for creating the flange.

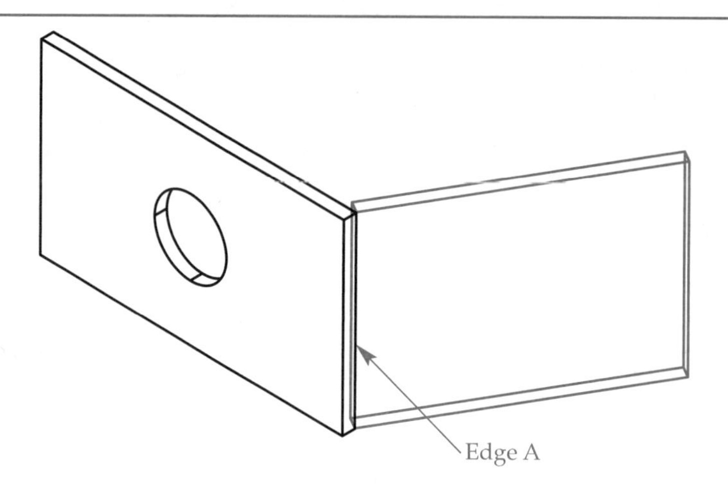

direction. A **Flange Angle** value of 90 creates a right angle flange. Realistically, a flange would not be created with a **Flange Angle** value of zero.

The buttons in the **Bend Position** area of the **Shape** tab determine if the inside, middle, or outside edge of the flange is inline with the selected edge. In effect, this determines if the width of the existing extruded face remains the same or is increased. For this example, select the **Inside of bend face extents** button (the left-hand button).

Once all settings have been made, pick the **OK** button to apply the flange and close the dialog box. You can now modify the flange by sketching on one of its faces and using the **Face** tool. You can also create cuts by putting a sketch plane on the original extruded face and projecting the flange.

You can edit the flange feature by right-clicking on its name in the **Browser** and selecting **Edit Feature** from the pop-up menu. The **Flange** dialog box is displayed. Make the necessary changes and pick the **OK** button to update the feature.

More on Flanges

The maximum allowable angle value in the flange dialog box is 180°. Open Example_19_06.ipt and edit Flange1. If the dialog box is not expanded, pick the **>>** button to expand it. Notice in the **Design** area in the expanded section that the **Old Method** check box is checked. Change the **Flange Angle** to 168, pick the **OK** button, and notice the strange result. See **Figure 19-21.** With the old method, which was used in previous releases of Inventor, the results can be unpredictable if an angle greater than 90° is entered.

Edit the flange again. Uncheck the **Old Method** check box. Also, in the **Height Datum** area, pick the **Parallel to the flange termination detail face** button. In the **Bend Position** area, select the **Bend Tangent to Side Face** button. Then, pick the **OK** button to apply the change. This bend is shown in **Figure 19-21C.** Notice how the flange is correctly applied to the right-hand end of the first feature.

More than one edge can be selected for flanges. Open Example_19_07.ipt. Pick the **Flange** button to display the **Flange** dialog box. Then, select the three inside, straight edges of the green face. Do not select the curved edges. Then, in the **Height Extents** area of the dialog box, select Distance and enter 1.12. Then, pick the **OK** button to create the flange. See **Figure 19-22.** Notice that three flanges are actually created because you selected three edges. However, they are all considered one operation in the **Browser**. In this example, if the height is too great, the flanges will intersect and the operation will fail. The maximum height for this example is about 1.12, but this depends on which buttons are selected in the **Height Datum** and **Bend Position** areas.

Figure 19-21.
The difference between old and new methods of calculating a flange. A—The flange is 90° with the old method. B—Changing the angle to 168° with the old method. C—The angle is 168°, but the method is new.

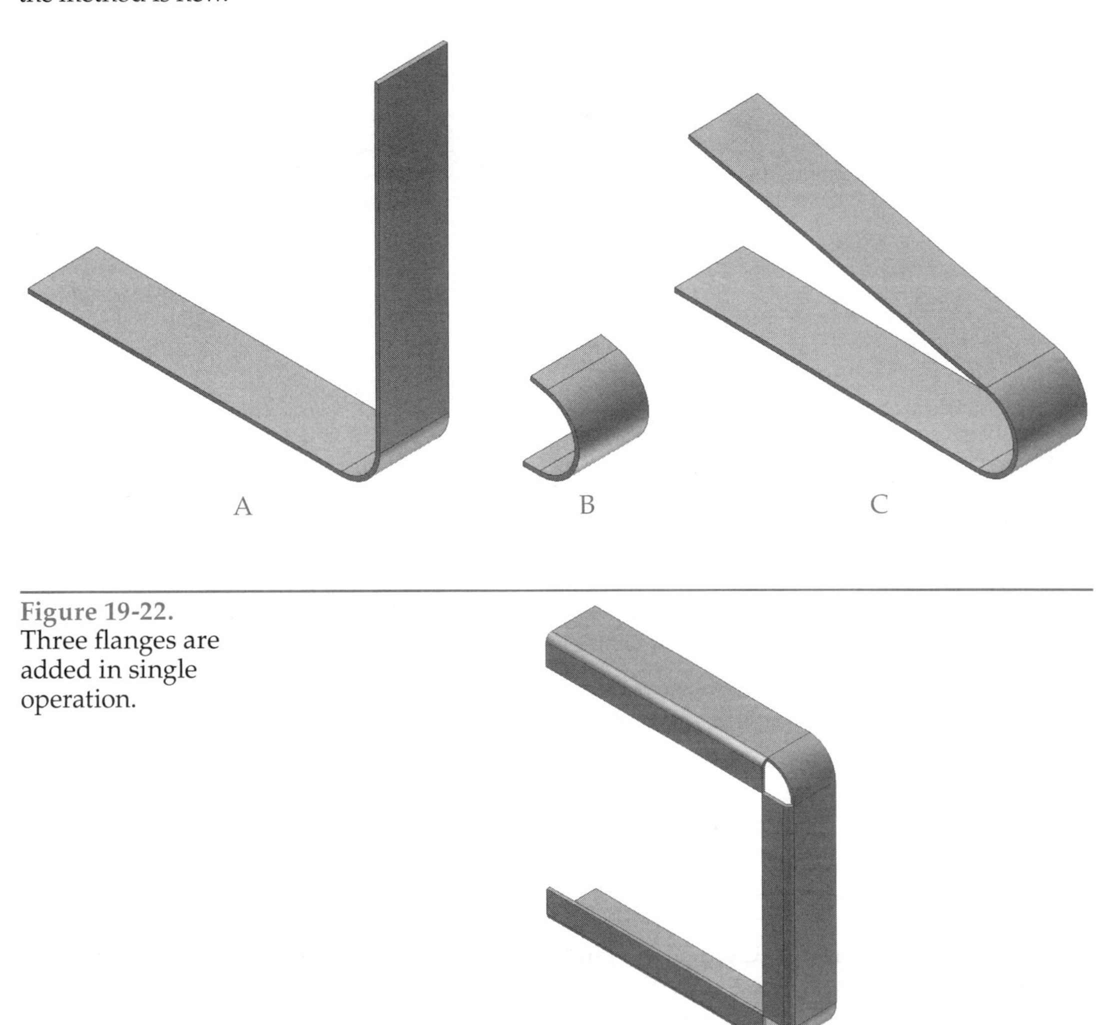

Figure 19-22.
Three flanges are added in single operation.

When creating a flange, you can select all of the straight edges in a face with one pick. Open Example_19_08.ipt. Pick the **Flange** button to open the **Flange** dialog box. Notice the two buttons to the left of the **Edges** column. By default the **Edge Select Mode** button is on, which allows you to pick individual edges. Pick the **Loop Select Mode** button to activate it. Then, pick any edge on the front face of the part. All straight edges are selected. In the **Height Extents** area of the dialog box, select Distance and enter 0.5 in the text box. Then, pick the **OK** button to create the flange. See **Figure 19-23.** Notice that flanges are created on all of the straight edges, but not on the curved edges. Also, notice the reliefs at the curved edge. These flanges are all considered one operation in the **Browser**.

Applying a Flange to a Portion of an Edge

The flanges you have worked with so far have extended across the edge selected in the operation. However, you can create a flange that is only applied to a portion of the edge. Open Example_19_09.ipt. This file contains a sheet metal face; the material

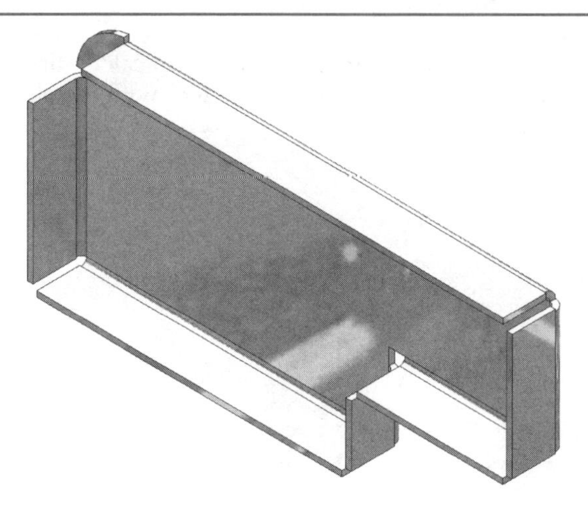

Figure 19-23.
All of these flanges were added by selecting a single edge.

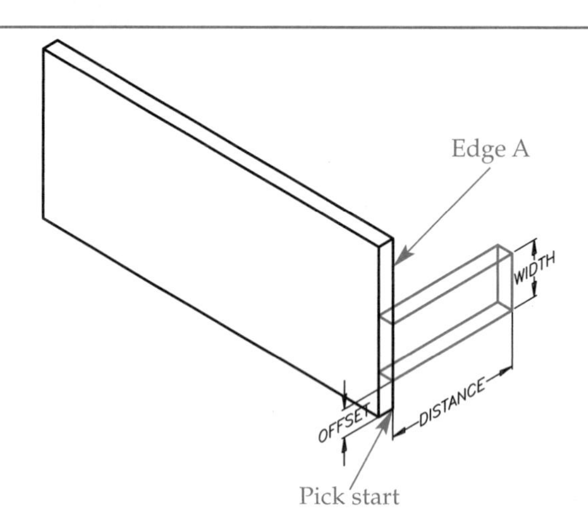

Figure 19-24.
A flange can be applied to a portion of the selected edge.

is 3 mm thick aluminum. Open the **Flange** dialog box and pick edge A indicated in **Figure 19-24.** In the **Height Extents** area, select Distance and enter 25 in the text box. Also, enter 90 in the **Flange Angle** text box.

Next, expand the dialog box by picking the **>>** button, **Figure 19-25.** There are four choices in the **Type** drop-down list in the **Extents** area—Edge, Width, Offset, and From To. The Edge option creates the flange across the entire edge, which is the default option. The Width option allows you to specify a portion of the edge on which the flange is created. The Offset option allows you to create a flange offset a distance between two points. The From To option allows you to pick the starting and ending points for the flange.

Select the Width option and then pick the **Offset** radio button that appears. Then, pick the **Offset** button (with the arrow) to select a point from which the offset will be measured. Pick the point indicated in **Figure 19-24.** Notice the preview in the graphics window. If necessary, pick the **Offset Flip** button to move the flange onto the part. In the **Bend Position** area of the dialog box, pick the **Bend from the adjacent face** button. Finally, enter 5 in the **Offset** text box, 10 in the **Width** text box, and pick the **OK** button to create the flange. Notice that a relief is generated based on the settings in the sheet metal style. Leave the file open for the next section.

Figure 19-25.
The expanded **Flange** dialog box.

Pick to expand or collapse the dialog box

Select a type

Select to create an offset flange

Pick to select the offset origin

Enter an offset

Enter a width

Rounds and Chamfers

You can apply rounds and chamfers to sheet metal parts. Fillets, which are internal arcs, are generally created on sheet metal parts by the bending and folding operations. The bend radius is the fillet radius. Rounds and chamfers are applied to the corners of sheet metal parts. Corners are parallel to the thickness of the material. The tools for placing rounds and chamfers on sheet metal parts are **Corner Round** and **Corner Chamfer** located in the **Sheet Metal Features** panel. The tools work in much the same way as the **Fillet** and **Chamfer** tools found in the **Part Features** panel for standard parts.

Using the **Corner Round** tool, place a 6 mm round on the two corners opposite the flange you added in the last section. See **Figure 19-26.** Notice as you select the corners that you cannot select the edges of the large face. Save and close the file.

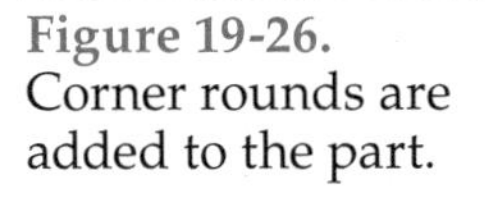

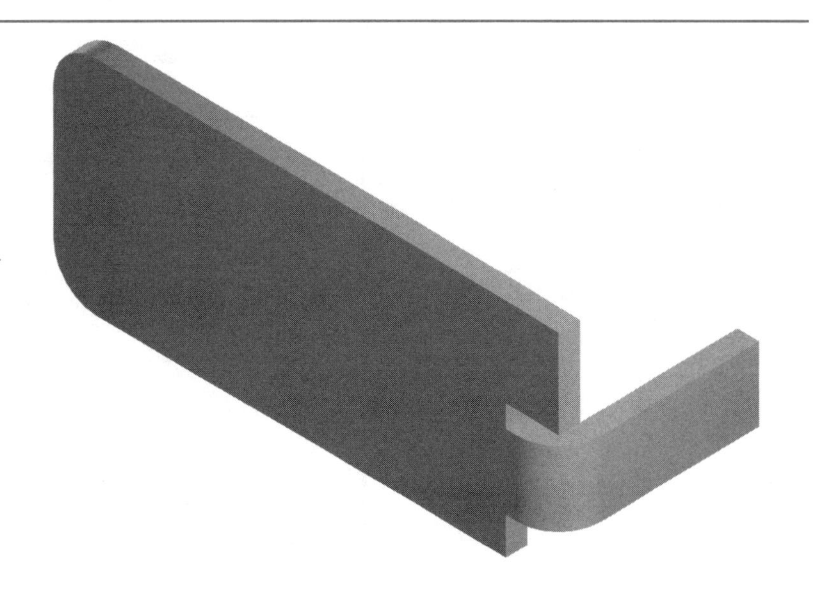

Figure 19-26.
Corner rounds are added to the part.

PROFESSIONAL TIP

You may need to add a round to create a safer edge or add a chamfer for welding purposes. These are applied using the **Fillet** and **Chamfer** tools listed in the **Part Features** panel.

Hems

A ***hem*** is a short distance of metal at the edge of a sheet metal part that is folded back onto itself. Hems are often added to strengthen a long span of unfolded sheet metal. Open Example_19_10.ipt and refer to **Figure 19-27.** This part has a single hem placed on the right-hand edge. Right-click on Hem1 in the **Browser** and select **Edit Feature** from the pop-up menu. The **Hem** dialog box is displayed, **Figure 19-28.** As with other features, the dialog box used to edit the feature is the same as the one used to create the feature.

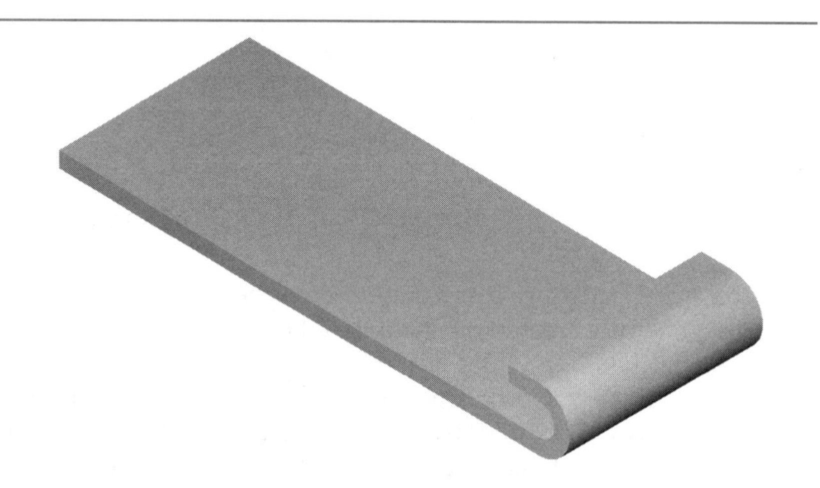

Figure 19-27.
A hem is added to the right edge of this part.

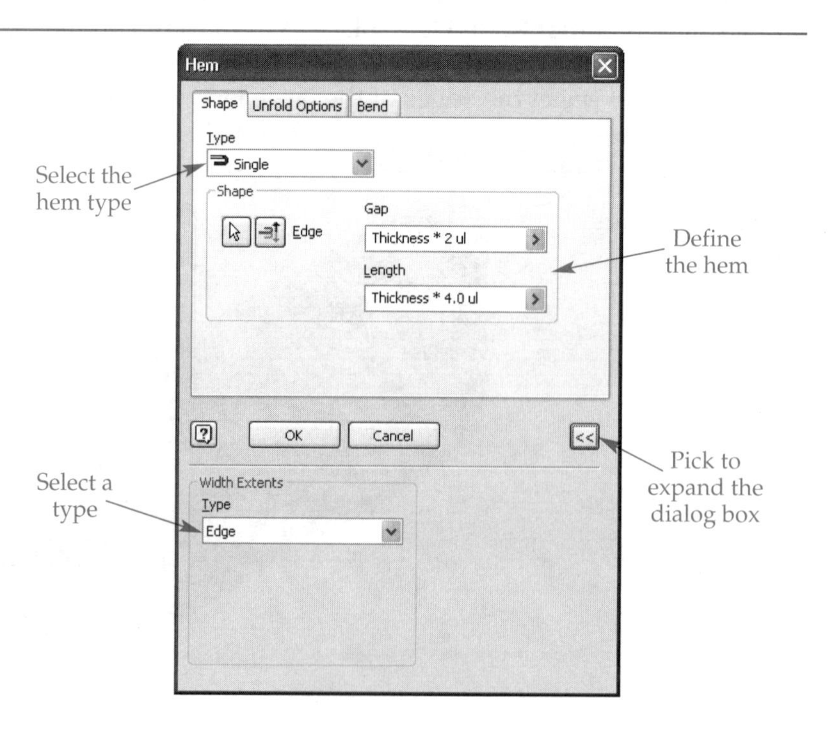

Figure 19-28.
The **Hem** dialog box, shown expanded.

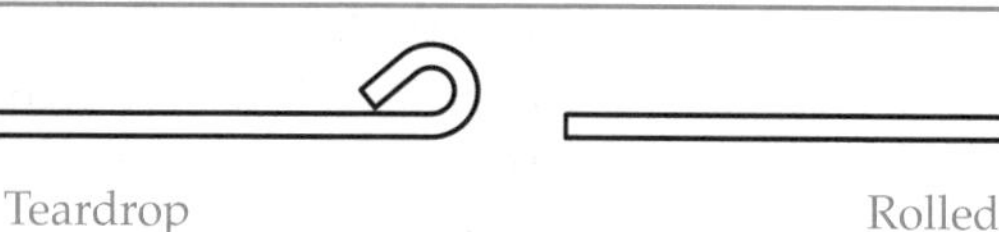

Figure 19-29.
These two hems were created with the same radius and angle settings. Notice the difference between the two.

There are four types of hems available in the **Type** drop-down list in the **Shape** tab. The type is currently set to Single. This type of hem is defined with a gap and length. The **Gap** setting is currently twice the material thickness. The **Length** setting is currently four times the thickness. The overall length of the part will remain unchanged, which is 4″ in this case. The angle for a single hem is fixed at 180°. If you expand the dialog box, the four choices available for the width of the hem are the same as those offered with the flange.

Select Teardrop in the **Type** drop-down list in the **Shape** tab. The settings for defining the hem change to **Radius** and **Angle**. Also, notice the preview in the graphics window. The value entered in the **Angle** text box must be greater than 180° and less than 360°.

Select Rolled in the **Type** drop-down list. The settings for defining the hem are **Radius** and **Angle**. The value entered in the **Angle** text box must be greater than 0° and less than 360°. Notice the difference between the previous teardrop hem and the rolled hem. With the same settings, the hem appears differently. See **Figure 19-29.**

Now, select Double in the **Type** drop-down list. The settings for defining the hem are **Gap** and **Length**. For a double hem, the **Length** value must be greater than the **Gap** value plus twice the material thickness. Also, notice the preview of the hem. This type of hem is common for strengthening an edge of the part.

Close the dialog box. Then, close the file without saving.

Holes and Cuts across Bends

Holes and slots that are punched or laser cut on or near the bend line in the flat pattern change shape when the part is bent (folded). If you create a sketch on a folded face of the part and cut a hole using the sketch, the hole will not be distorted. However, these features should display in their true, distorted shape in the folded part. To do this, the folded faces need to be projected onto a sketch plane, as you did earlier in this chapter. Open Example_19_11.ipt and refer to **Figure 19-30.** The

Figure 19-30.
The folded part before cuts across bends are added.

left-hand, rectangular face was created as a flange and the right-hand, trapezoidal face was created as a face, both with a bend radius of .5″.

First, you will add a circular cutout through the bend between Face1 and Flange1. Face1 is the original face. Start a new 2D sketch on the top face of Face1. Refer to **Figure 19-31.** Next, pick the **Project Flat Pattern** button in the **2D Sketch Panel.** Pick the flat portion of the inside face of Flange1 to project it onto the sketch plane. You must pick the inside face because the sketch plane is on the inside face of Face1. If the sketch plane is on the outside face of Face1, you would need to pick the outside face of Flange1.

At this point, create a sliced graphics display or change to a wireframe view. Next, construct a .75″ diameter circle centered at the midpoint of the bend centerline. Finish the sketch and open the **Cut** dialog box. Select the interior of the circle as the profile. In the **Cut** dialog box, check **Cut Across Bend** check box. When you check this check box, all of the options in the **Extents** area are grayed out. Pick the **OK** button to create the circular cutout. Notice how the circle is distorted as the cutout is created. Then, use the **Flat Pattern** tool to see the developed pattern. See **Figure 19-32.**

Now, you will add a slot cutout to the other bend. Switch to the folded part. Then, rotate the view to see the inside of Face2, which is the trapezoidal face. Start a new 2D sketch on the top of Face1. Using the **Project Flat Pattern** tool, project the inside face of Face2. Remember, select the flat part of the face, not the bend. Sketch the slot as shown in **Figure 19-33.** Then, finish the sketch and, using the **Cut** tool, create the slot. The distortion may not be as apparent in this feature, but it is there. Finally, double-click on FlatPattern in the **Browser** to see the developed pattern, **Figure 19-34.** Save and close the file.

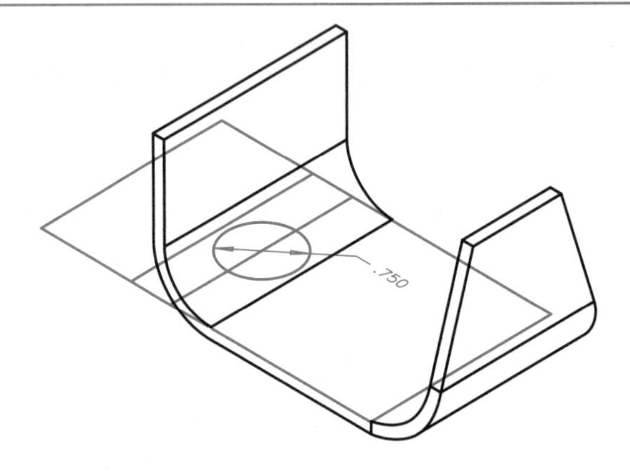

Figure 19-31.
The flat pattern is projected into the sketch and a circle is created.

Figure 19-32.
A—The circle is distorted as the hole is created. This is what actually happens when a bend is created through a hole. B—The hole is correctly shown in its true shape and size in the flat pattern.

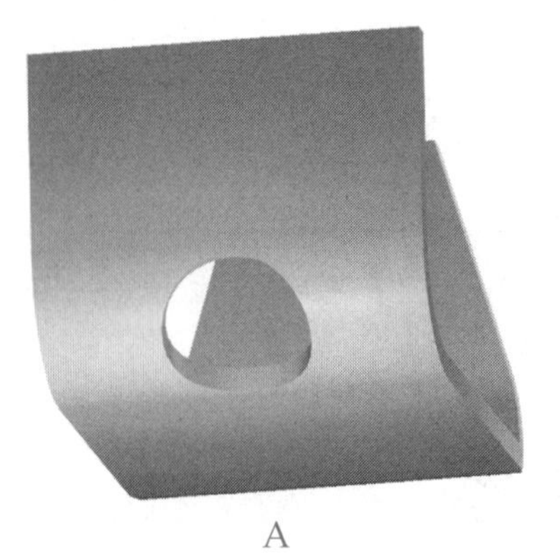

A

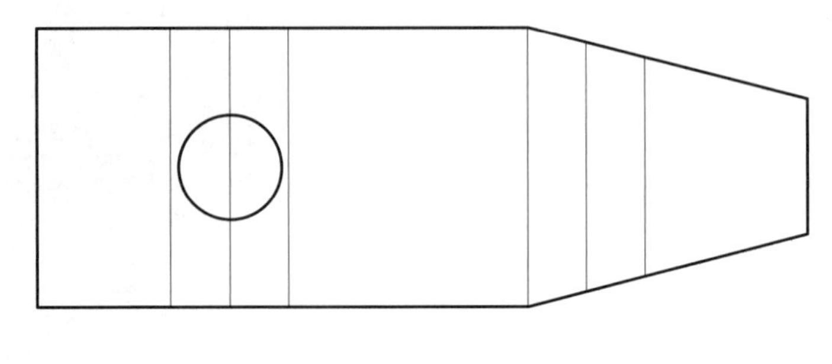

B

Figure 19-33.
In a second sketch, the slot is drawn.

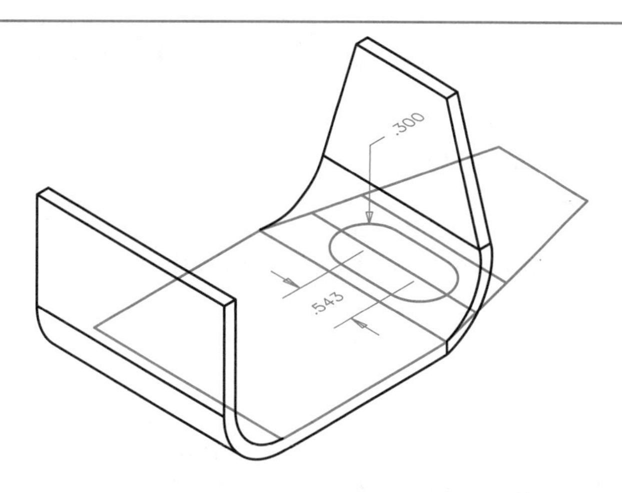

Figure 19-34.
A—The slot is distorted as it is created. B—In the flat pattern, the slot is shown in its true shape and size.

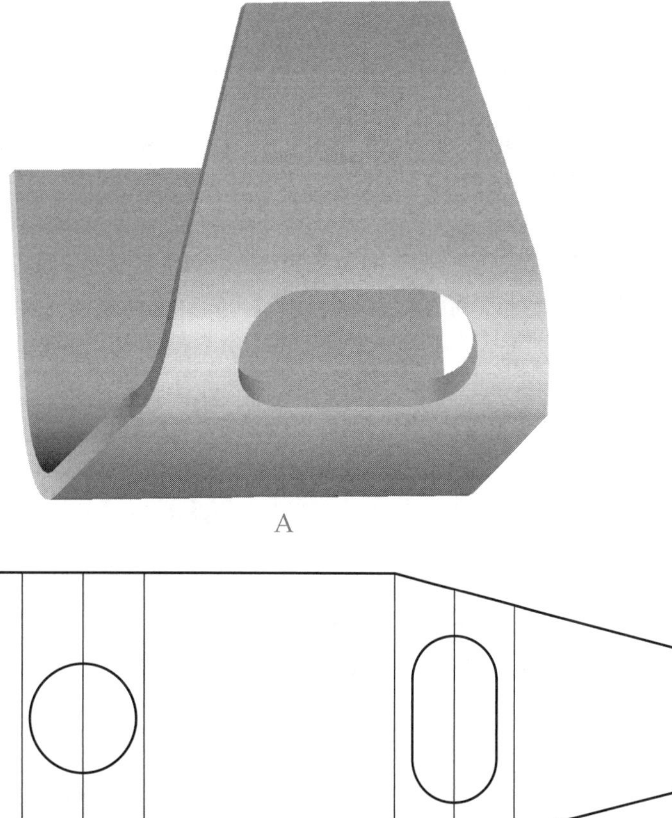

A

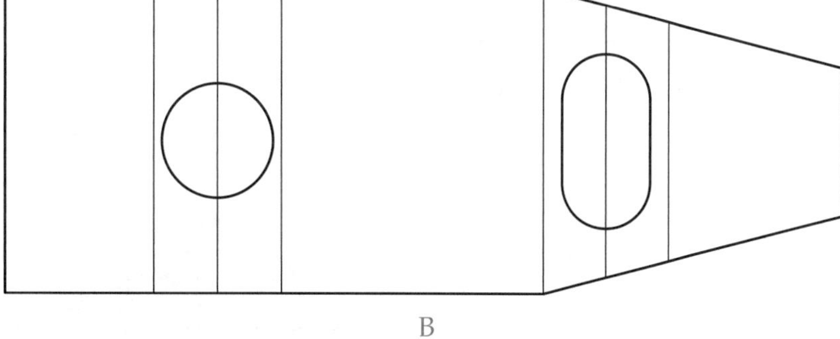

B

PROFESSIONAL TIP

One of the important things to note about projecting holes across bends is that you must have a flat portion of a face or flange to project onto the sketch plane.

Contour Flanges

If a part has a uniform cross section, it can be created from an open sketch with the **Contour Flange** tool. The sketch can contain lines, arcs, and splines, but it cannot have any sharp internal corners. A valid sketch and the resulting part are shown in **Figure 19-35.** Notice that a fillet has been applied to eliminate the square corner. Also, tangent constraints are required between the vertical lines and the large arc.

Creating a Contour Flange

Open the file Example_19_12.ipt, which contains the sketch shown in **Figure 19-35.** Pick the **Contour Flange** button in the **Sheet Metal Features** panel to display the **Contour Flange** dialog box. See **Figure 19-36.** With the **Profile** button on in the **Shape** tab, pick any part of the open sketch. Since there are no sharp corners, the entire sketch is selected. The **Flip Offset** button determines to which side of the profile the material thickness is applied.

Since this is the first feature to be created, the **Contour Flange** dialog box is expanded and Distance is automatically set in the **Type** drop-down list in the **Width Extents** area. The drop-down list is disabled, so you cannot select a different option. In the **Distance** text box, type 15. Then, pick the **OK** button to create the contour flange. Once the flange is created, you can cut slots and add rounds and chamfers to the part, as shown in **Figure 19-37.** Save and close the file.

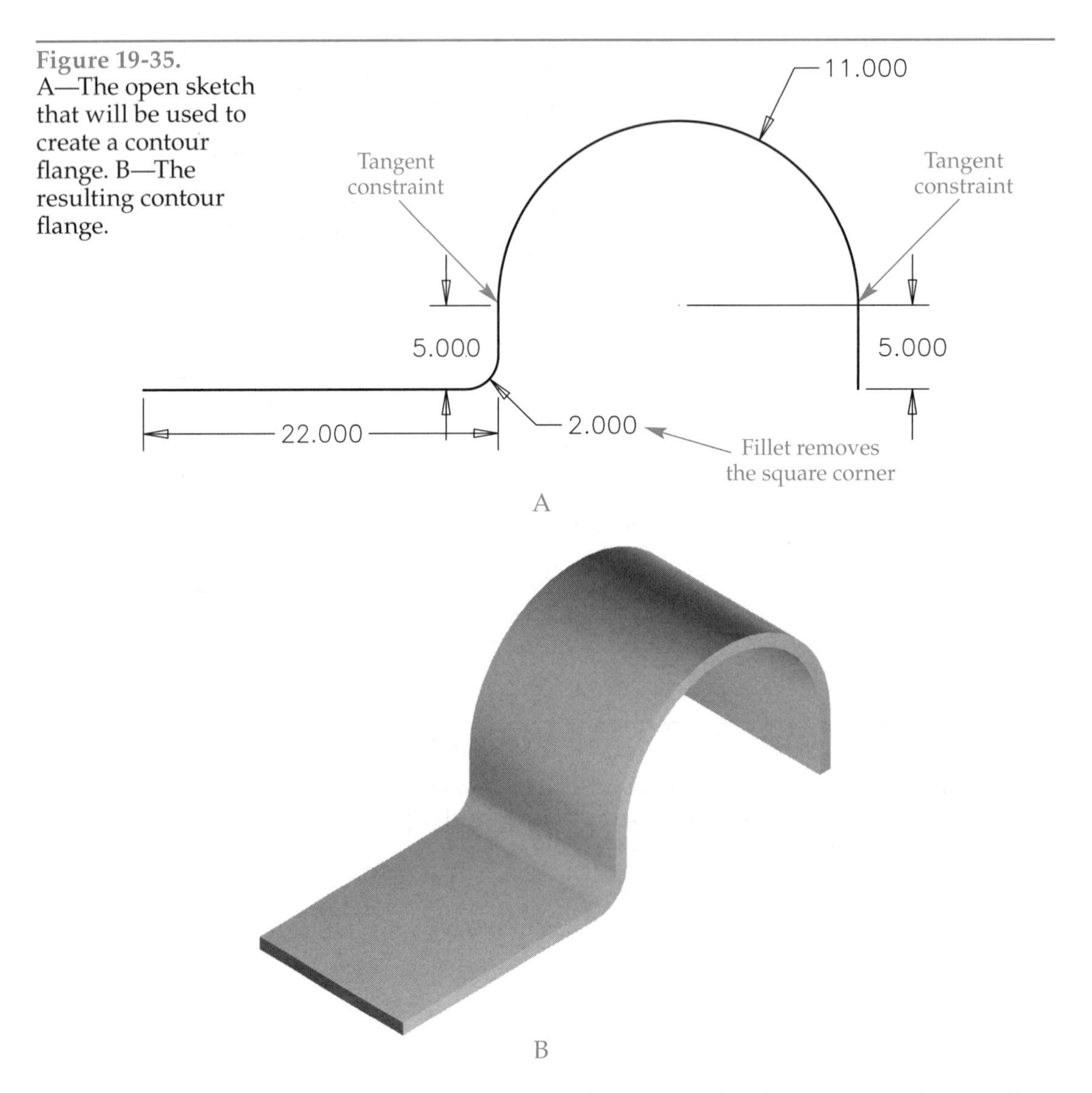

Figure 19-35.
A—The open sketch that will be used to create a contour flange. B—The resulting contour flange.

Figure 19-36.
The **Contour Flange** dialog box, shown expanded.

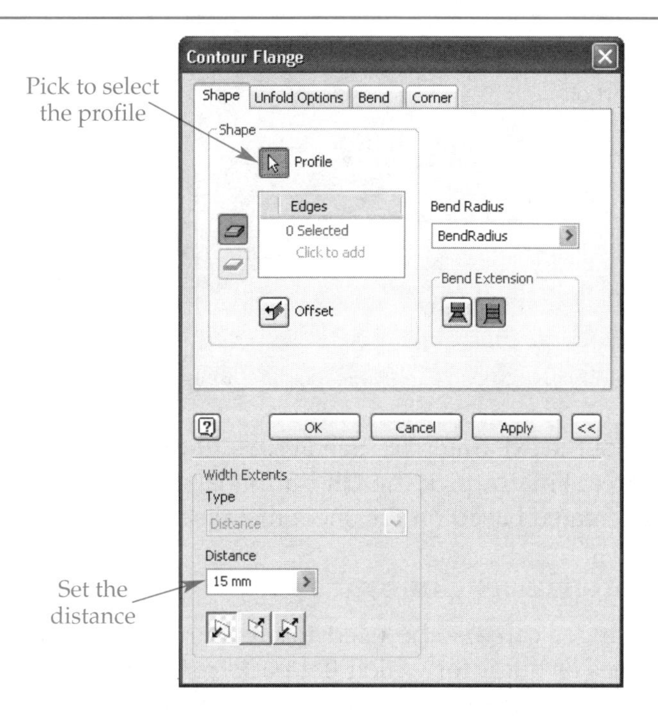

Figure 19-37.
Two corner chamfers and a cutout (slot) are added to the contour flange. The rounded end was created using the **Cut** tool, not the **Corner Round** tool.

Extending a Contour Flange from an Edge

A contour flange can be created from the edge of an existing sheet metal part based on the sketched profile. To do so, open the **Contour Flange** dialog box and first select the profile. Then, pick the **Edge** button in the **Shape** tab and select an edge from which the contour flange should extend. Expand the dialog box. The same options are available in the **Width Extents** area as for the **Flange** tool.

Open the file Example_19_13.ipt. Open the **Contour Flange** dialog box and select the sketch as the profile. Then, with the 0 Selected entry highlighted in the **Edge** column, select the bottom edge of the part. Now, expand the dialog box and select Width from the **Type** drop-down list in the **Width Extents** area. Then, pick the **Offset** radio button. With the **Offset** button (with the arrow) on, pick the left endpoint of the edge as the start. Then, enter .5 in the text box next to the **Offset** button. Also, enter 1.5 in the **Width** text box.

Notice the preview in the graphics window. If necessary, pick the **Offset Flip** button in the **Width Extents** area. The top surface of the contour flange should be flush with

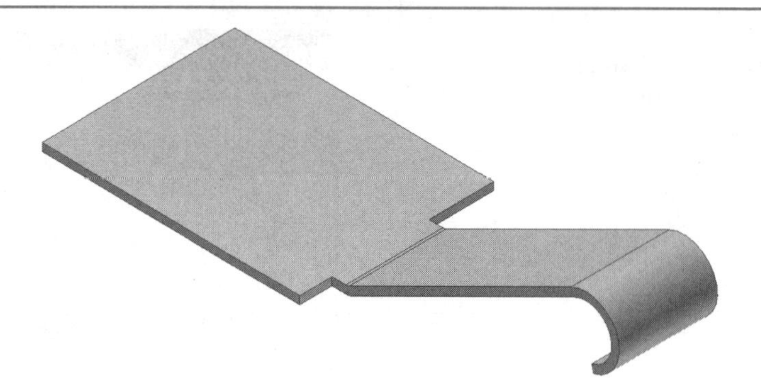

Figure 19-38.
Adding a contour flange to a portion of an existing edge.

the top surface of the existing part. See **Figure 19-38.** If not, pick the **Offset Flip** button in the **Shape** area. Finally, pick the **OK** button to create the contour flange. Notice the reliefs that are created based on the sheet metal style. Save and close the file.

Complex Patterns as Contour Flanges

Contour flanges can also be used to create round, elliptical, or complex-shaped sheet metal tubes or ducts for which flat patterns are required. The sketch, of course, must be open and the trick is to have a small straight line at one end. Open the file Example_19_14.ipt. This file contains three sketches, two of which you will use to create a round sheet metal duct with an angled end.

Open the **Contour Flange** dialog box and select the circular sketch as the profile. If you zoom in on the circle, you can see that this is not a circle, rather an arc and a straight line with a small gap between its ends. If you zoom in close, you will see the small straight line on the left side of the gap. This straight line is necessary for creating the flat pattern. The **Contour Flange** tool does not work with closed shapes. However, the gap can be made small enough to be within the allowable tolerances for sheet metal ducts. Next, enter 12 in the **Distance** text box and pick the **OK** button to create the contour flange. The part is now a round duct with two square ends. See **Figure 19-39.** This operation did not consume the sketch. Turn off the visibility of Sketch for Contour Flange.

Pick the **Cut** button in the **Sheet Metal Features** panel. The trapezoidal sketch should be automatically selected as the profile. If not, select it. Set the extents to All and pick the midplane button. Then, pick **OK** to create the cut. One end of the duct is angled. See **Figure 19-40.** Now, use the **Flat Pattern** tool to create the developed pattern, **Figure 19-41.** Save and close the file.

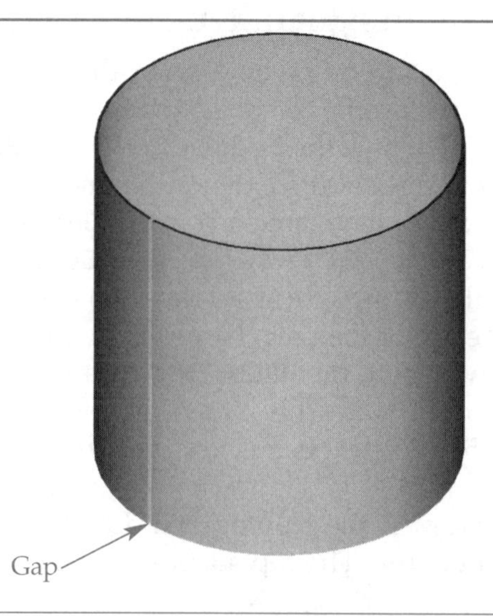

Figure 19-39.
A revolution is added to the small flange to create a round duct with square ends.

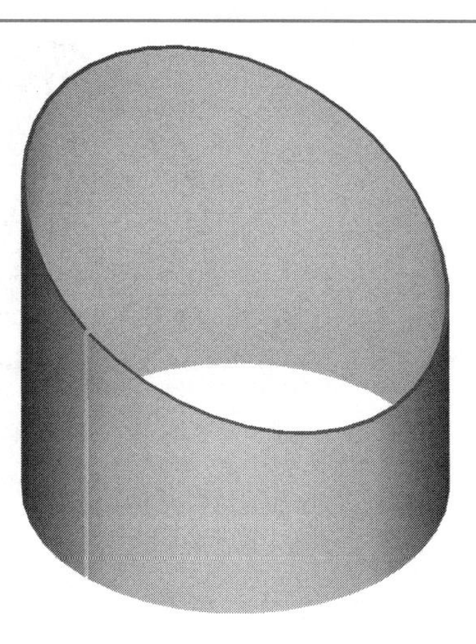

Figure 19-40.
The angled end is created using the **Cut** tool.

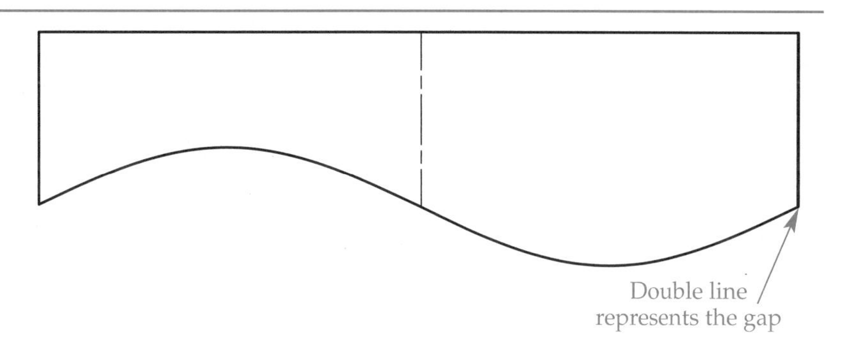

Figure 19-41.
The flat pattern for the duct with an angled end.

Cylindrical and conical ducts can also be created with the **Revolve** tool using an angle of revolution of just less than 360°. Open Example_19_15.ipt. This part has a long, narrow face extruded with the **Face** tool. It is named Face1 in the **Browser**. There is also an unconsumed sketch that was created by projecting Face1 onto a sketch plane. A centerline was drawn in the sketch to use as the axis of revolution.

Pick the **Revolve** button in the **Part Features** panel to display the **Revolve** dialog box. If this panel is not displayed, pick the arrow next to the current panel name and select **Part Features** from the drop-down list. The sketch should be automatically selected as the profile when the **Revolve** tool is selected. Also, since the axis is a centerline linetype, it is automatically selected as the axis. In the **Extents** drop-down list in the dialog box, select Angle. Then, enter 358 in the text box. Finally, pick **OK** to create the revolution. See **Figure 19-42.** Now, display the **Sheet Metal Features** panel. Then, pick the **Flat Pattern** button to create the developed pattern. See **Figure 19-43.** Save and close the file.

You can use the **Project Flat Pattern** tool in a sketch if the small flat face is selected as the sketch plane. The revolved part can be projected. However, a sketched circle will not generate a hole in the revolved part using the sheet metal **Cut** tool.

Fold Tool

The process you have used so far to create sheet metal parts is to create the folded part and then produce the unfolded, developed pattern. However, you can start with the developed pattern and then create the folded part. First, sketch the developed blank (flat pattern) and then use the **Face** tool to create a flat part. Next, sketch bend lines on

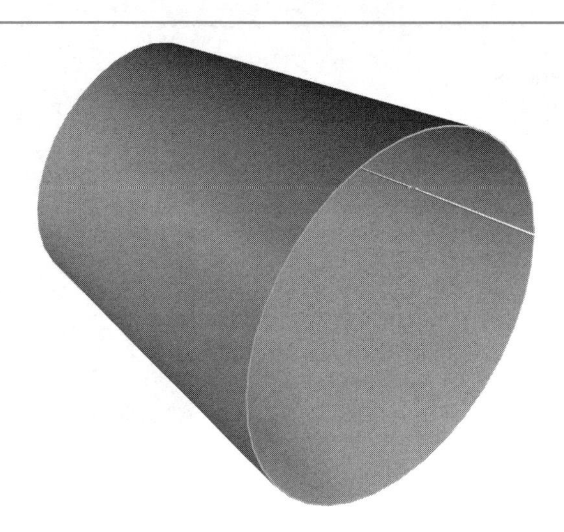

Figure 19-42.
A conical duct created with the **Revolve** tool.

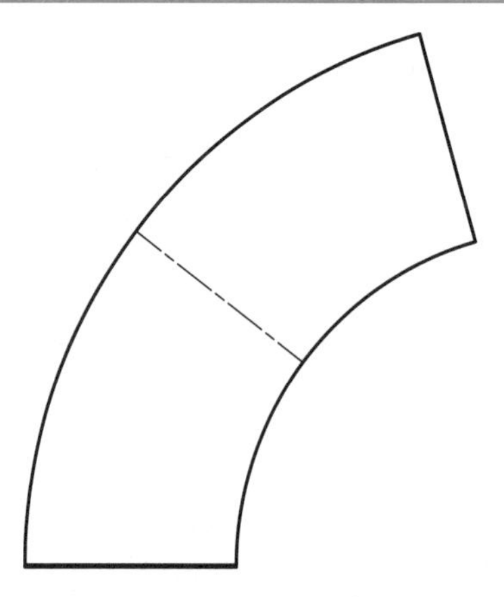

Figure 19-43.
The flat pattern for the conical duct shown in Figure 19-42.

the part using any linetype; however, the lines must be straight. Finally, use the **Fold** tool to create the folded part.

Open Example_19_16.ipt. There are two unconsumed sketches in this part. Each sketch is of a line that will be used as a bend line. Refer to **Figure 19-44.** A sketch is consumed as the **Fold** tool is used. Therefore, since there are to be two folds on the part, two sketches are necessary.

Pick the **Fold** button in the **Sheet Metal Features** panel to open the **Fold** dialog box, **Figure 19-45.** With the **Bend Line** button in the **Shape** tab on, pick the short vertical line in Sketch4 as the bend line. Refer to **Figure 19-44A.** Only one bend line can be selected. Once you select a bend line, you cannot select a different line without deselecting the first line. If you need to do so, pick the **Bend Line** button, hold down the [Shift] key and pick the first line again. Then, select the new line.

Now, look at the preview in the graphics window. See **Figure 19-46.** The bend line is displayed in blue. The area of the bend is outlined by a red box, as is the area of the relief. There are also two green arrows displayed, one straight and one curved. The straight arrow indicates which side of the line is going to be bent. The portion of the part to which the arrow points will be bent about the line. The curved arrow shows the direction of the bend. In the **Shape** tab of the dialog box, there are two buttons in the **Flip Controls** area. The **Flip Side** button is used to change the direction of the straight arrow. The **Flip Direction** button is used to change the direction of the curved arrow. Using these two buttons, there are four possible bends. Pick the buttons as needed so the arrows point as shown in **Figure 19-46.**

Figure 19-44.
A—This part was created from the flat pattern drawing. The two bend lines are shown here in color. B—The part after the two bends are created.

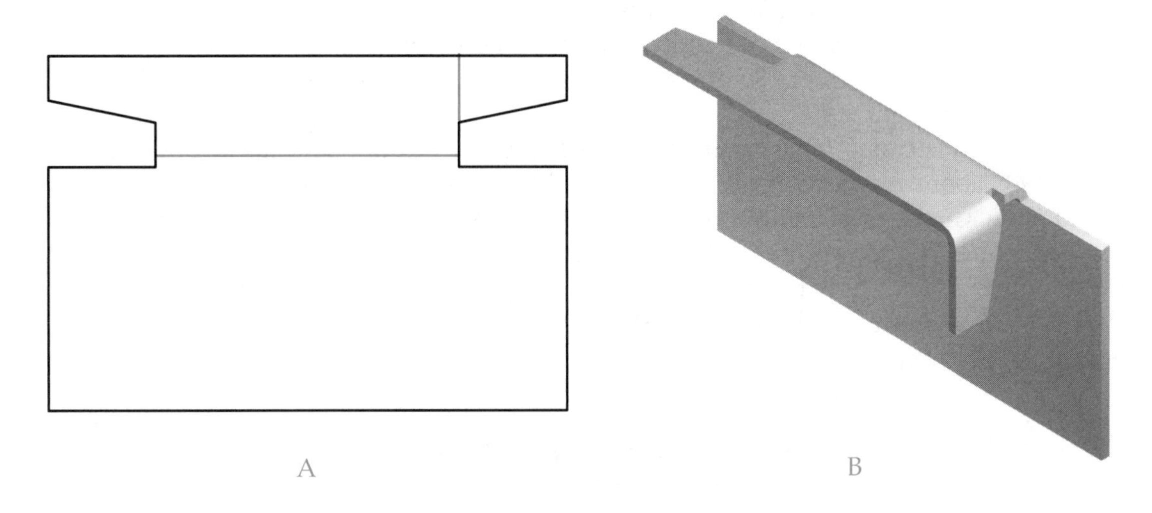

Figure 19-45.
The **Fold** dialog box.

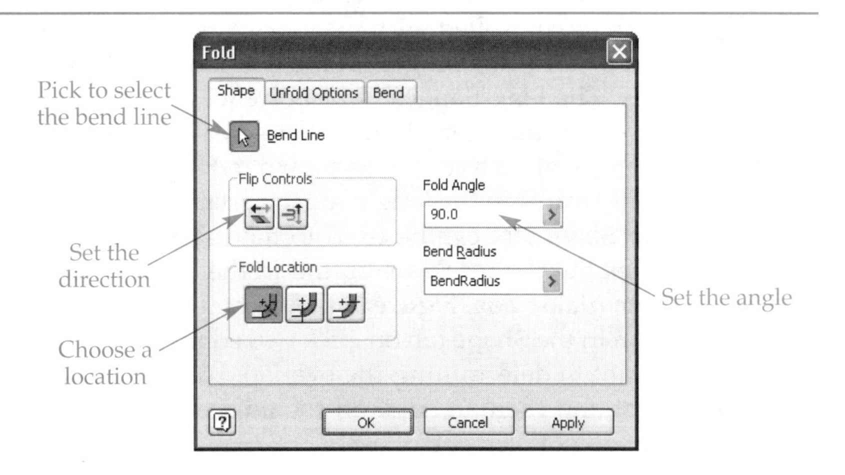

Figure 19-46.
The preview arrows show which side of the bend will be folded and the direction. The red lines represent the extent of the bend and relief.

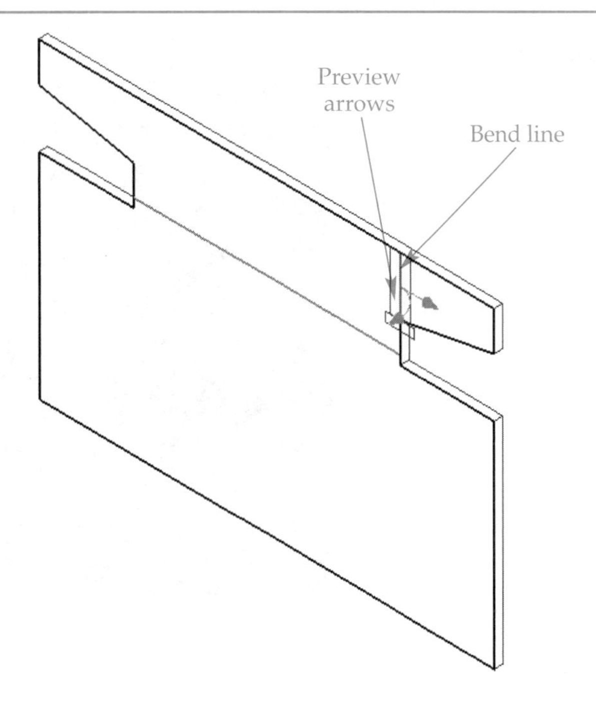

There are three buttons in the **Fold Location** area—**Centerline of Bend, Start of Bend**, and **End of Bend**. These buttons are used to determine how the bend is applied in relation to the bend line sketch. Pick the **Centerline of Bend** button so the bend is equally applied on each side of the bend line.

The bend can be from 0° and 180°. To set the bend angle, type a value in the **Fold Angle** text box. If you enter 180, the result is similar to that produced by the **Hem** tool. For this example, enter 90.

When all settings have been made, pick the **OK** button. The tab on the part is folded 90° about the bend line. A relief is added based on the settings in the sheet metal style. Create another fold using the horizontal line shown in **Figure 19-44A** as the bend line. The top of the part should bend 90° down. This time, pick the **Start of Bend** button. Notice how the red box representing the bend shifts. The final part with the two folds is shown in **Figure 19-44B.** Save and close the file.

Corner Seam Tool

The **Corner Seam** tool controls the shape of the seam between two edges. The gap and the relief can be controlled with the tool. Open Example_19_17.ipt. The sheet metal box was formed by creating flanges on the two long edges of the face and then on the two short edges. The long flanges have no relief as they were the full length of the edge. However, when the short flanges were created, reliefs were applied. This relief forms the hole or gap at each corner, as shown in **Figure 19-47.** The seam created by the flange operations results in the long flange overlapping the short flange.

The **Corner Seam** tool can be used to modify the seams between the short and long flanges. Pick the **Corner Seam** button in the **Sheet Metal Features** panel to open the **Corner Seam** dialog box, **Figure 19-48.** First, you must select the two edges. With the **Edges** button in the **Shape** tab on, pick two edges of one of the seams. The selection order is important in determining the overlap. For this example, pick an edge on the short flange first and then an edge on the adjacent long flange. Pick either inner or outer edges, but do not pick one of each.

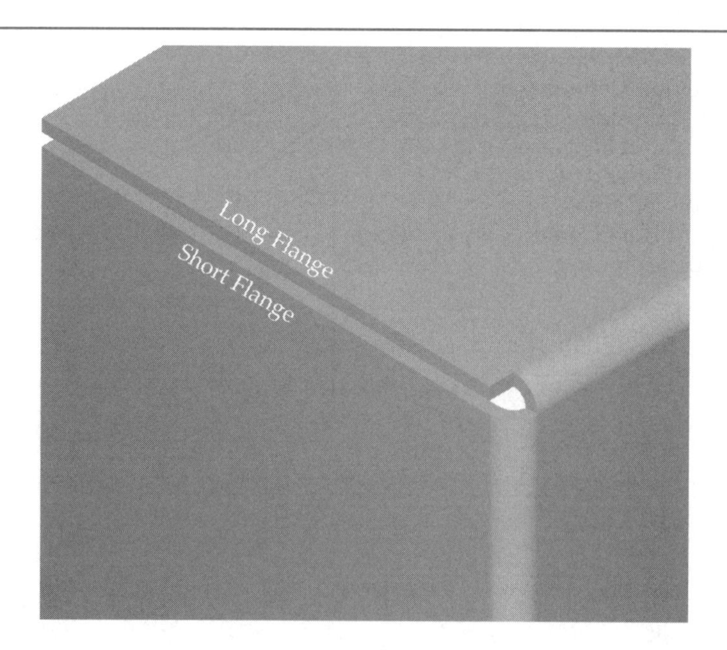

Figure 19-47.
The **Corner Seam** tool can be used to control the seam between two edges.

Figure 19-48.
The **Corner Seam** dialog box.

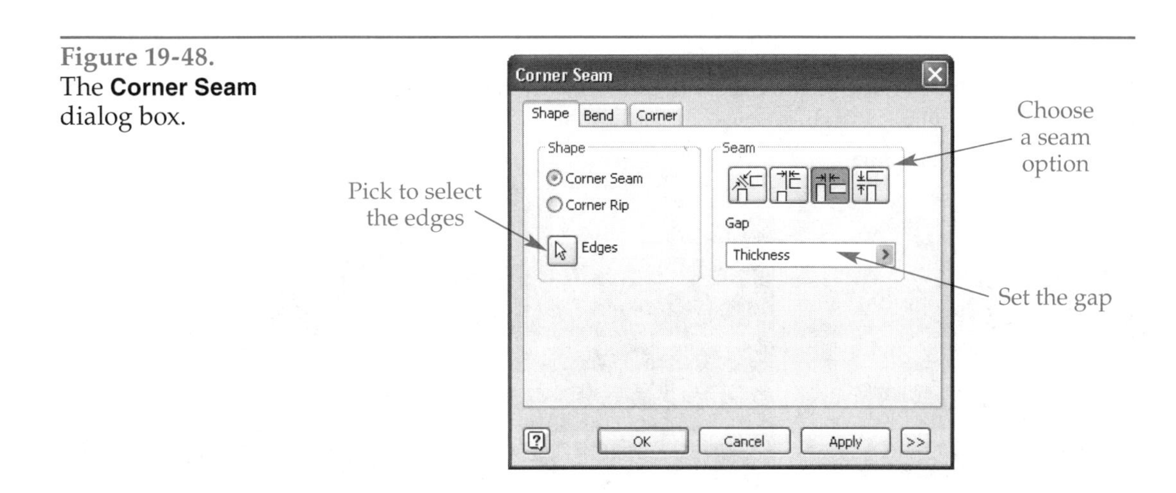

Figure 19-49.
There are three options for the type of seam. A—**Reverse Overlap**. B—**No Overlap**. C—**Overlap**.

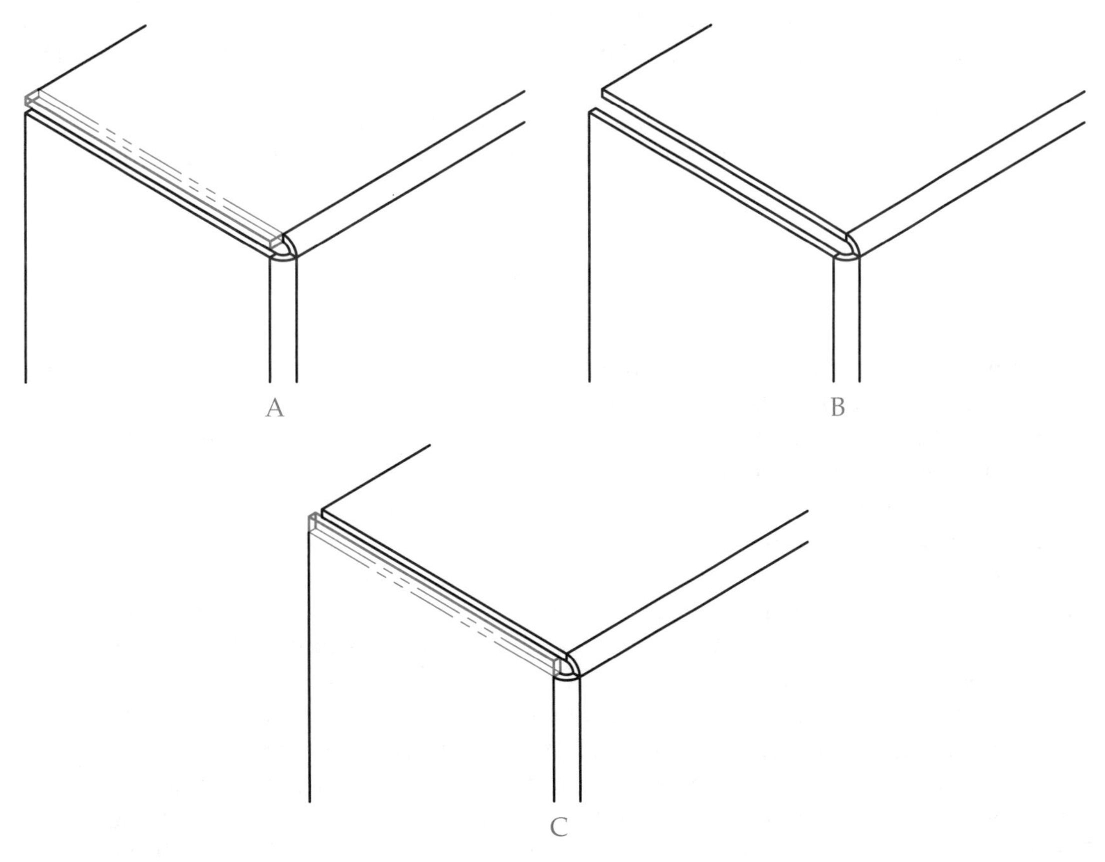

After selecting the edges, the type of seam can be set. There are four choices. See **Figure 19-49.** The **Symmetric Gap** and **No Overlap** buttons open the seam. For the **No Overlap** setting, the end of each face (flange) stops at the inner edge of the adjacent face (flange). The **Overlap** button lengthens the first selected face to overlap the second face. The gap is applied between the extension of the first face and the second face. The **Reverse Overlap** button lengthens the second selected face to overlap the first face. In this case, there is no change because as the second choice (long side) already overlaps the first choice (short face). For this example, pick the **Overlap** button.

The shape and size of the relief at the corner is controlled by the settings in the **Corner** tab. There are six options for the shape of the relief. These are selected in the

Figure 19-50.
You can control the shape of the relief created with the **Corner Seam** tool. A—Round. B—Square. C—Tear.

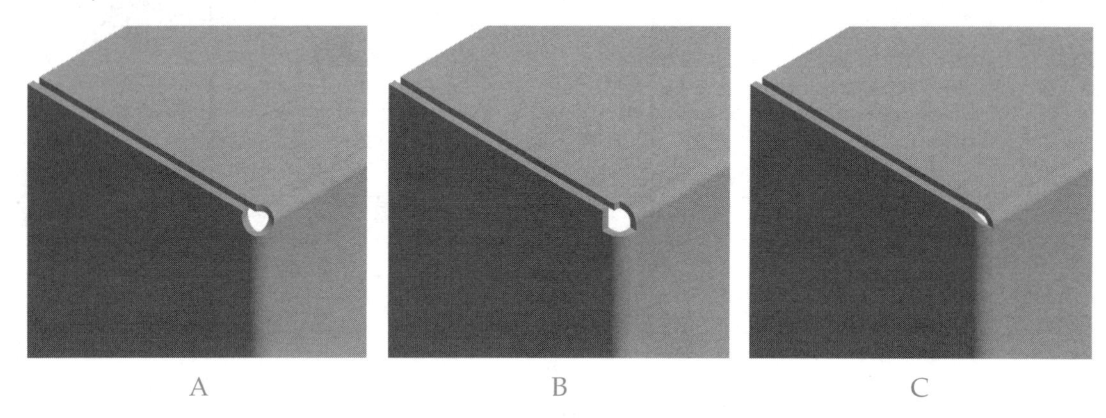

Relief Shape drop-down list in the **Corner** tab. The Default option creates the relief based on the settings in the sheet metal style. The Trim to Bend option produces no corner relief. The Round, Square, and Tear options are illustrated in **Figure 19-50.** The Linear Weld option creates a V-shaped relief.

Once all settings have been made, pick the **OK** button to create the corner seam. You can edit the feature as you would any other feature. Save and close the file.

Punch Tools

Punching is the process of creating holes in sheet metal. ***Stamping*** is similar, but the tool deforms or embosses the metal instead of cutting through it. Inventor has a catalog of standard punch tools that can be used to punch or stamp sheet metal parts. Nine of the punch tools provided with Inventor create holes and two emboss shapes into the part. **Figure 19-51** shows the shapes of the available punches. They all can be sized and positioned on the face of the part.

Open Example_19_18.ipt, which is a simple sheet metal part. In order to use the **Punch Tool**, the part must have an unconsumed sketch containing a center point. Start a new sketch on the large rectangular face. Then, use the **Point, Center Point** tool to place a point at the center of the face. Finish the sketch.

Pick the **Punch Tool** button in the **Sheet Metal Features** panel. The **Punch Tool Directory** dialog box is displayed, which is a standard Windows open dialog box. The folder displayed is the one set up as the punch tool location in the **Application Settings** dialog box. By default, this is located in the Inventor installation folder under \Catalog\Punches. Each punch tool is based on a definition contained in an IDE file. If you move up one folder to \Catalog, there are folders for geometric shapes, pockets, bosses, and slots. Select the Curved Slot.ide file and pick the **Open** button.

Once you pick the **Open** button, the **Punch Tool** dialog box is displayed. See **Figure 19-52.** The punch geometry is also shown on the sketch in relation to the center point. In the **Preview** tab of the dialog box, you can select a different tool, if needed. A preview of the tool is also displayed.

Since there was only one sketch containing a center point, that was automatically selected. However, the **Geometry** tab allows you to select additional center points, **Figure 19-53.** Pick the **Centers** button in the tab and then manually select the center points.

Figure 19-51.
The eleven punch tools that come with Inventor. Tools A through I cut, while tools J and K emboss. A—Curved slot. B—D-sub connector. 2. C—D-sub connector. 3. D—D-sub connector. 4. E—D-sub connector. 5. F—D-sub connector. G—Keyhole. H—Keyway. I—Obround. J—Round emboss. K—Square emboss.

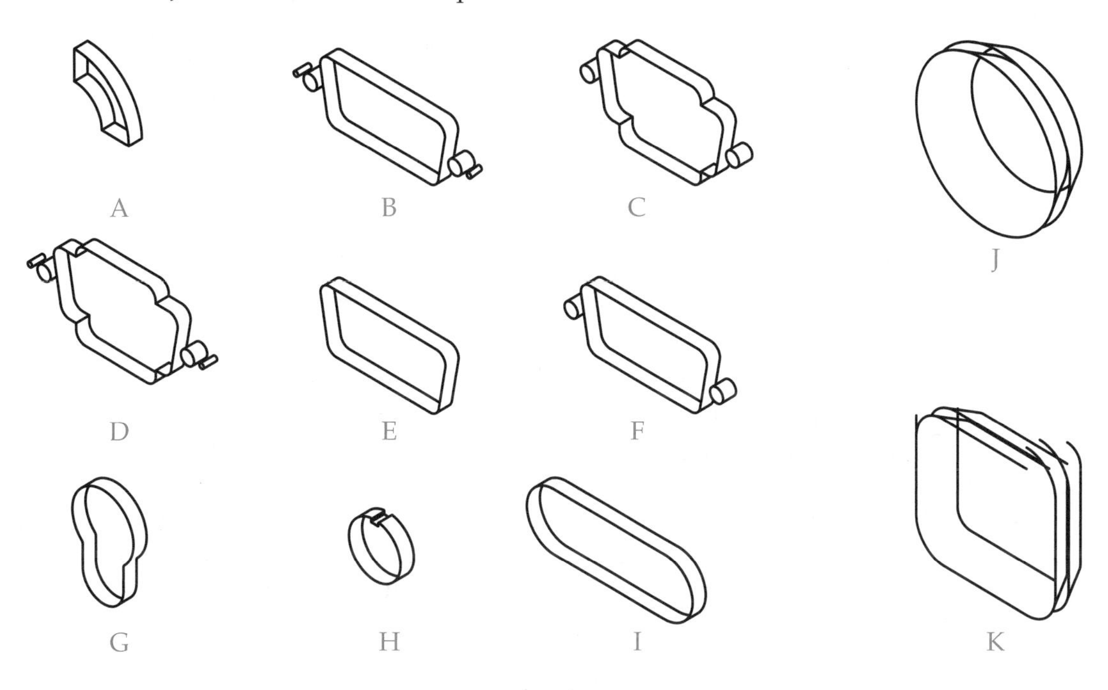

Figure 19-52.
The **Preview** tab of the **Punch Tool** dialog box is where you select a tool.

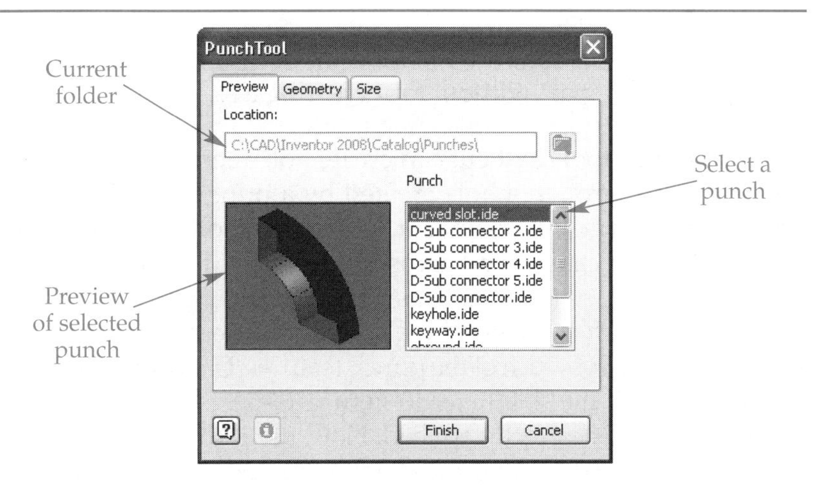

Figure 19-53.
The **Geometry** tab of the **Punch Tool** dialog box is where you select the center of the feature.

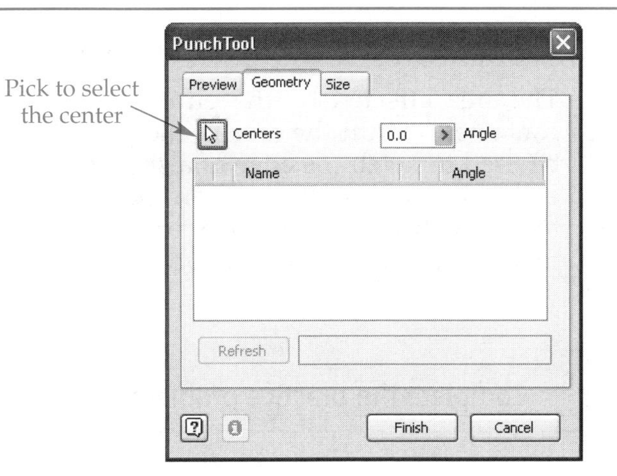

Figure 19-54.
The **Size** tab of the **Punch Tool** dialog box is where you change the dimensions of the tool to meet your requirements.

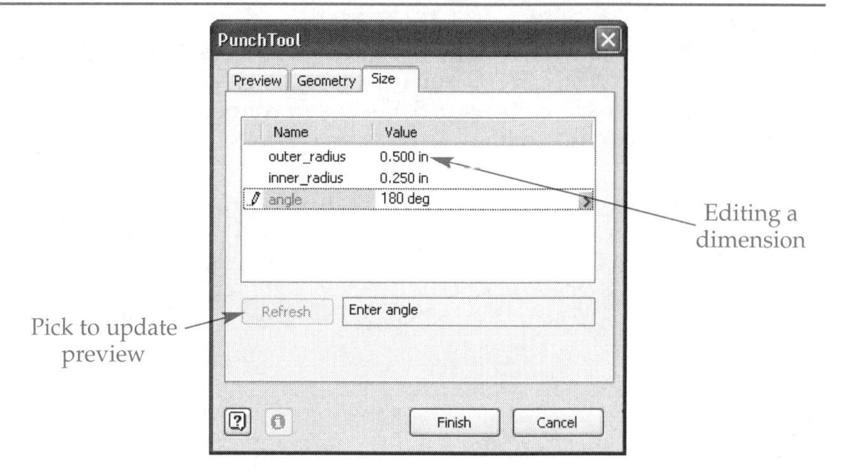

To change the dimensions of the punch, display the **Size** tab, **Figure 19-54.** The dimensions that are available to change depend on how the punch definition is set up. For this example, change the angle to 180°. Note that the preview on the part does not change. Pick the **Refresh** button to update the preview.

Pick the **Finish** button to create the punch feature. The 180° curved slot is punched through the part. In the **Browser** the feature is called iFeature*x*. Even though this feature is called a "punch," during the actual manufacturing process, the feature may be punched, laser cut, plasma cut, or waterjet cut.

PROFESSIONAL TIP

There is a tool in the **Sheet Metal Features** panel called **Insert iFeature**. This tool can be used to insert punches on any face of a part without first creating a sketch.

The punch tools that cut can create holes across bend lines. Also, the **Fold** tool can create a bend through a hole created by a punch tool. However, the punch tools that emboss or deform, but do not cut, cannot be applied across bend lines. The **Fold** tool is also unable to create a bend across an embossed punch feature.

Open Example_19_19.ipt. This part has two punch features, a slot created with the Obround.ide punch and a boss created with the Square Emboss.ide punch. A bend line has been sketched across each of the punch features. Using the **Fold** tool, make a 90° bend about the line through the slot. Refer to **Figure 19-55.** Since the punch feature is a cut feature (hole), the bend is correctly applied. Now, use the **Fold** tool to make a 90° bend about the line through the boss. An error is not generated. However, since the punch feature is an embossed feature, the bend is not correctly applied and the result makes no sense.

PROFESSIONAL TIP

The area affected by the punch tool does not need to be fully contained within the area of the part. A punch tool that cuts can be used to notch the edge of a sheet metal part. A punch tool that embosses can also be used on the edge of the part. However, the tool actually adds material to compensate for the area of the emboss tool that is not within the boundary of the part.

PRACTICE 19-2 Complete the practice problem on the Student CD.

Figure 19-55.
Bends can be created across punched features that are cut. A bend *cannot* be created across an embossed feature.

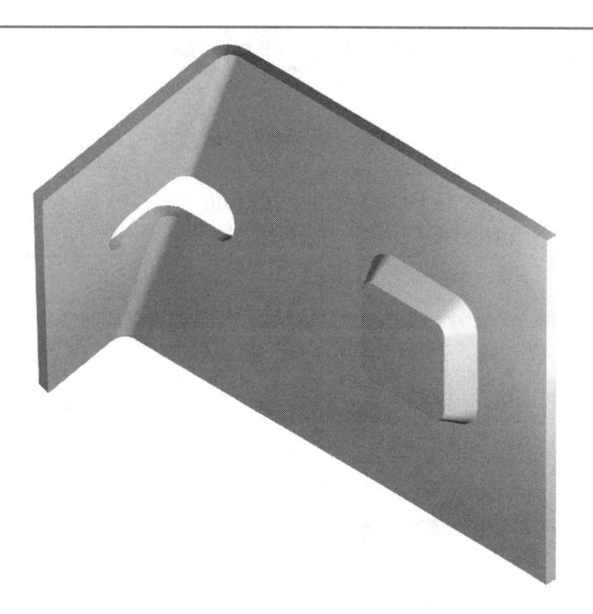

Chapter Test

Answer the following questions on a separate sheet of paper or complete the electronic chapter test on the Student CD.

1. What is a *sheet metal part?*
2. Define *drawing,* as related to sheet metal.
3. Define *bending.*
4. What is a *blank?*
5. What is the *developed length?*
6. Briefly describe a sheet metal style in Inventor.
7. What determines the thickness of a face created with the **Face** tool?
8. What is a *relief* and what purpose does it serve?
9. Define *remnant.*
10. List the five settings of a sheet metal style that can be overridden when creating a feature.
11. Define *developed blank.*
12. Which tool is used to create a developed blank?
13. What is the purpose of the **Cut** tool?
14. Using the **Project Flat Pattern** tool, you can project the flat pattern of a folded feature into a sketch. Why would you need to do this?
15. What is an advantage of using the **Flange** tool over using the **Face** tool?
16. Which tools are used to apply rounds and chamfers to the corners on sheet metal parts?
17. What is a *hem?*
18. Why is a hem created?
19. How many types of hems can be created?
20. From which type of geometry is a contour flange created?
21. What is the purpose of the **Fold** tool?
22. What does the **Corner Seam** tool do?
23. What are the five options for the relief shape when creating a corner seam?
24. What is the default location of the punch tools that come with Inventor?
25. Which type of punch tool can be used across a bend line, or through which a bend line can pass?

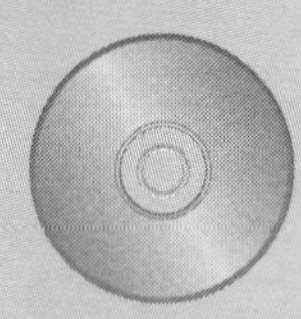

Chapter Exercises

Exercise 19-1. Tee Bracket.

Complete the exercise on the Student CD

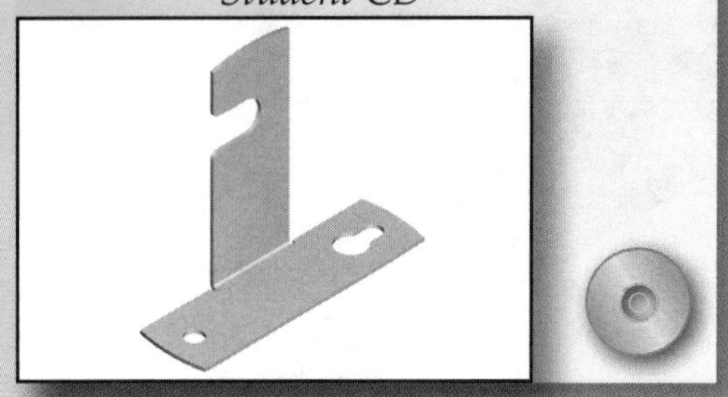

Exercise 19-2. Shelf.

Complete the exercise on the Student CD

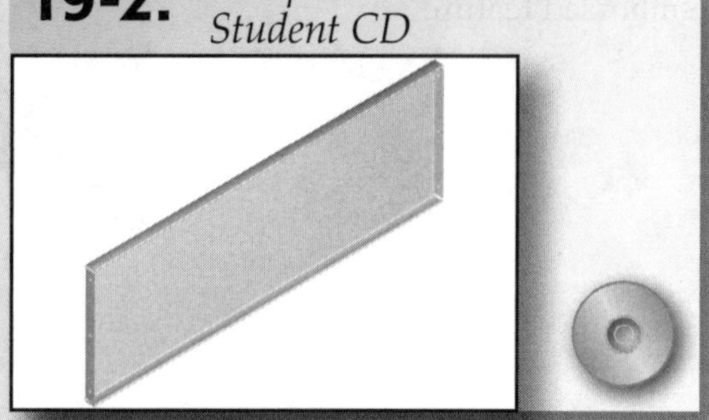

Exercise 19-3. Hinged Cover.

Complete the exercise on the Student CD

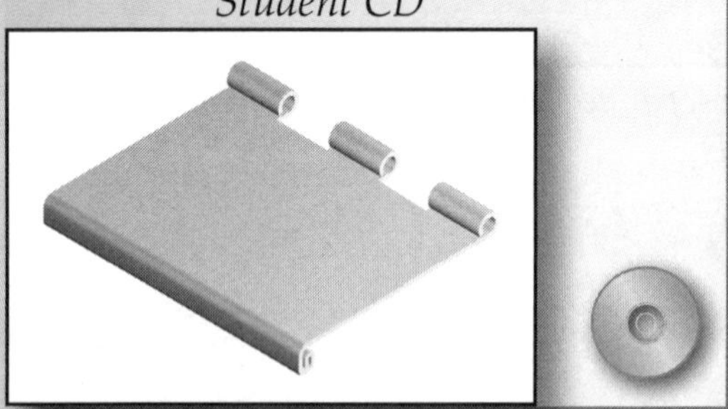

Exercise 19-4. Cover.

Complete the exercise on the Student CD

Exercise 19-5. Pan and Brackets.

Complete the exercise on the Student CD

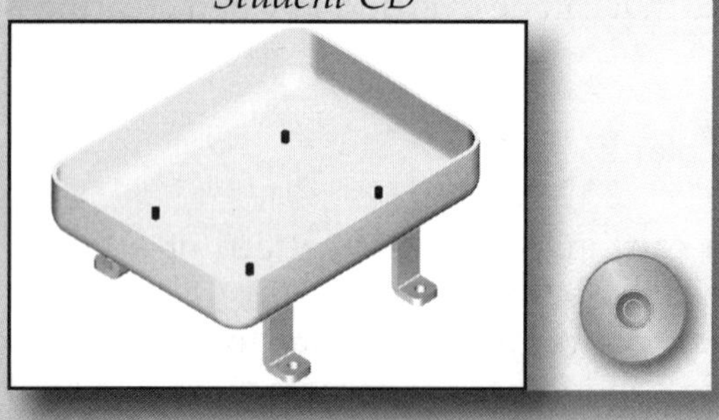

Exercise 19-6. Round Ducts.

Complete the exercise on the Student CD

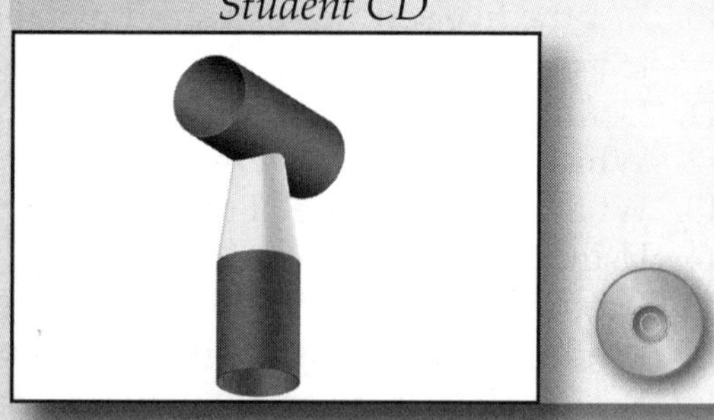

Exercise 19-7. Connection Panel.

Complete the exercise on the Student CD

Exercise 19-8. Flanged Bracket.

Complete the exercise on the Student CD

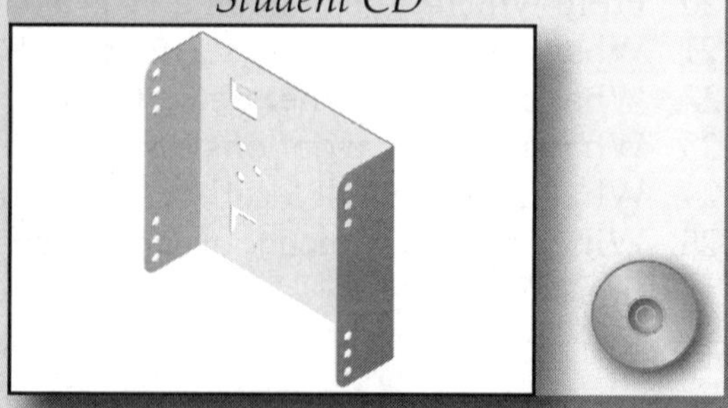

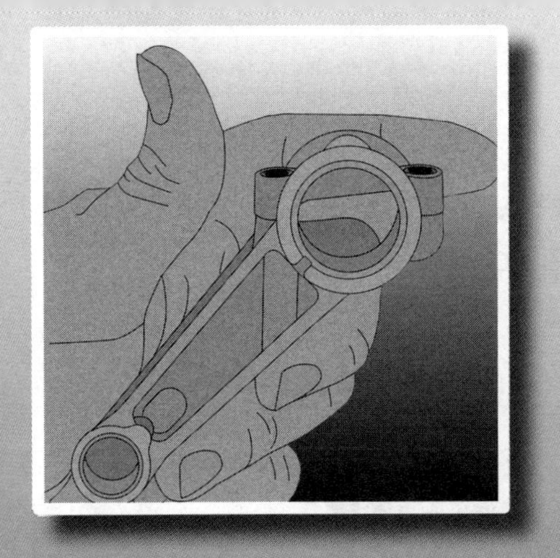

Chapter 20

Inventor Studio

Objectives

After completing this chapter, you will be able to:

- Describe the process of rendering in Inventor Studio.
- Describe the process of animating in Inventor Studio.
- Create a camera in an Inventor Studio scene.
- Add lighting to an Inventor Studio scene.
- Render an Inventor Studio scene.
- Animate components.
- Animate constraints.
- Animate a camera.
- Explain fade.
- Animate fade.
- Animate parameters.
- Render an animation in Inventor Studio.

User's Files

The Student CD included with this text contains several files required for this chapter. Refer to the file File List.txt in the \Ch20 folder for the comprehensive list.

Inventor Studio is an environment within Inventor used to create realistic images and animations. An animation is a series of still images, each with small differences, played in sequence to simulate movement. This is similar to a presentation file.

To enter Inventor Studio, with a assembly file or part (in part mode) open, select **Inventor Studio** from the **Applications** pull-down menu. The **Inventor Studio** panel is displayed in the **Panel Bar** and the **Browser** is labeled **Inventor Studio**. See **Figure 20-1.** Notice that there are Animations, Lighting, and Cameras branches displayed in the **Browser** along with the part tree.

Figure 20-1.
The Inventor Studio interface.

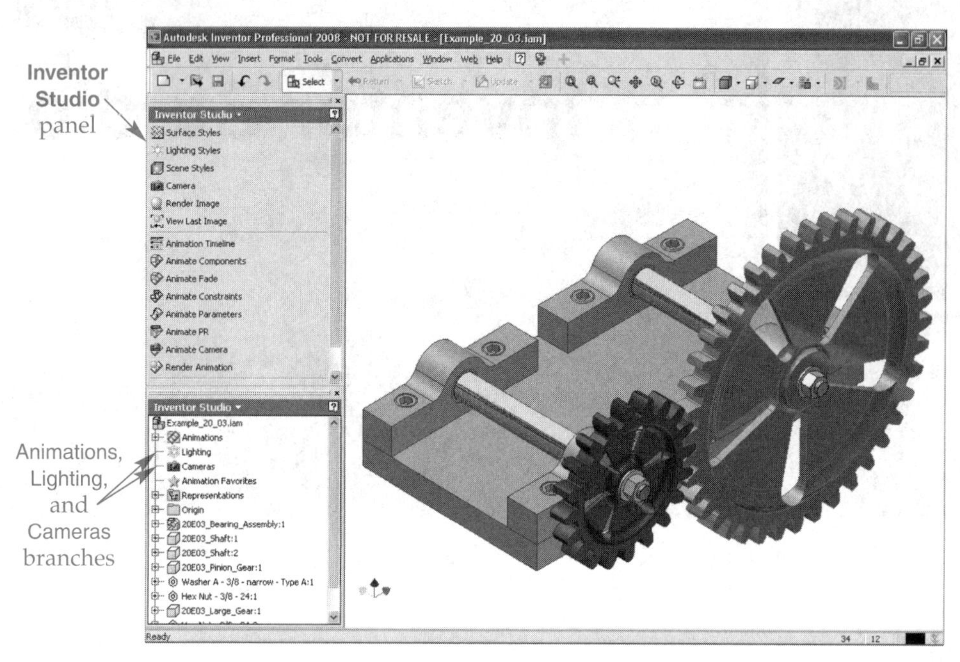

Rendering Process

A ***rendering*** is a still image of the scene. Whether creating a still image or an animation, the basic process is to define a scene and then render it. A ***scene*** consists of the geometry, surface styles ("materials"), lighting, background image, and viewpoint (camera).

The ***surface style*** is the "material" for the geometry. Do not confuse a surface style with the material defined for the geometry in the **Properties** dialog box. The surface style determines the basic appearance, such as color and texture, of the geometry when rendered.

A ***lighting style*** is used to provide the illumination in a scene and consists of one or more lights. Often, one of the lighting styles supplied with Inventor will work. However, many times you will need to create your own lighting style. Creating a lighting style is covered in Chapter 2. Creating new lights is covered later in this chapter.

The background image is not included in the part or assembly file. It is added when the scene is rendered. ***Scene styles*** are used to define the background image.

The viewpoint is the direction from which the scene is viewed. A ***camera*** is used to create a viewpoint by specifying a target for the camera and the camera's position. A zoom level is also set for a camera. The advantage of using a camera over a manually-generated viewpoint is that the camera and its target can be animated.

Animation Process

An animated scene needs to have a surface style, lighting style, scene style, and camera, just as with a still image. Remember, an ***animation*** is just a series of renderings (still images) strung together into a video. In addition, you need to add motion to the scene. Determine what will be animated:

- **Components.** The position and orientation of components can be changed.

- **Constraints.** The parameters of applied assembly constraints can be modified.
- **Camera.** The camera can be panned and its zoom level changed.
- **Fade.** The opacity/translucency of components can be altered.
- **Parameters.** Numeric user parameters can be changed.

Once you have determined what needs to be animated, create the animation. Then, specify any animation options. Finally, render the animation.

Rendering Components

Rendering is not only important in its own right, but an important part of creating animations. You need a realistic rendering for a good animation. Many of the tools are common to both processes .

Look at **Figure 20-2.** The image in **Figure 20-2A** is a photograph of a coil spring on a 10 cm grid. The image in **Figure 20-2B** is a rendering of an Inventor part, also on a 10 cm grid. As you can see, Inventor Studio allows you to create photorealistic images of parts and assemblies.

Open Example_20_01.iam. This is the assembly file used to create **Figure 20-2B.** In this assembly, the part 20E01_White_Paper was inserted ungrounded. A flush constraint was placed between it and the assembly's XZ plane. The part was positioned so the center point is in the correct position and then the part was grounded. The part 20E01_Spring was positioned on top of the "paper" using the **Move** and **Rotate Component** tools. Then, it was grounded.

There are four unconsumed sketches listed in the **Browser**. Two of these—Target and Camera—will be used to create the camera. The other two—Light1 and Light2—will be used to create lights. Each of the sketches contains a center point that will be selected as a location. Locations can be specified by entering coordinates or selecting a location with the mouse, but the latter approach provides a visual cue to where the locations are in relation to the rest of the assembly.

Now, you need to create a camera and set the surface, lighting, and scene styles. In this example, the current background will be used as the scene style. Using the **Applications** pull-down menu, switch to Inventor Studio. Then, continue to the next section.

Figure 20-2.
A—A photograph of a spring on a 10 cm grid. B—A spring created in Inventor and rendered in Inventor Studio.

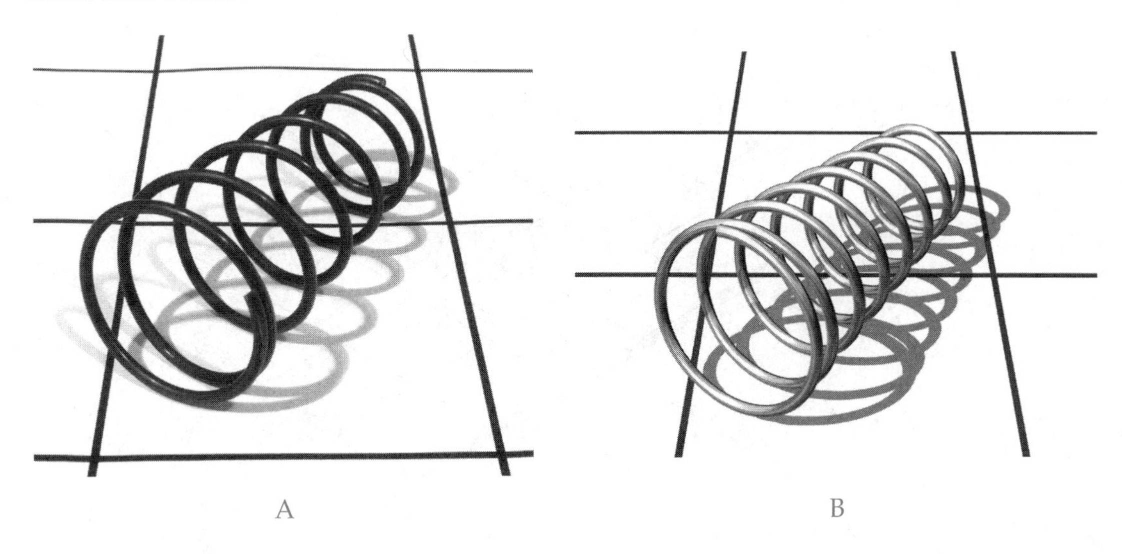

Figure 20-3.
Adding a camera to a scene in Inventor Studio.

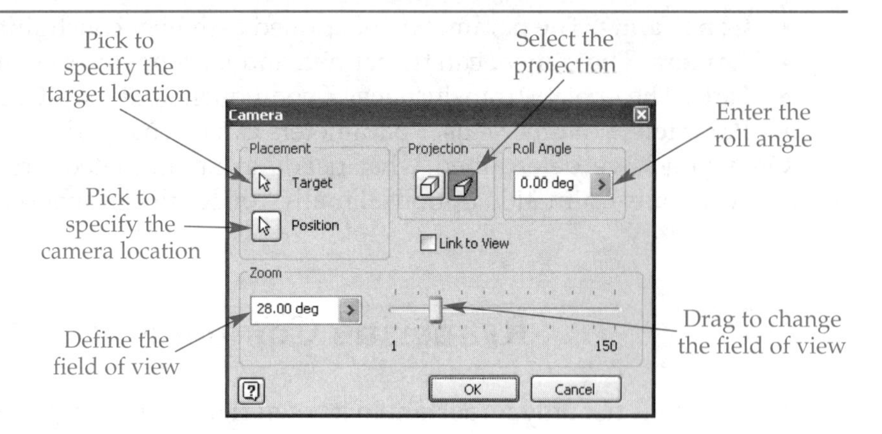

Creating a Camera

To create a camera, pick the **Camera** button in the **Inventor Studio** panel. The **Camera** dialog box is displayed. See **Figure 20-3.** A new camera is defined based on its position, target location, projection type (perspective or orthographic), roll angle, and viewing angle.

With the **Target** button on in the dialog box, pick the center point that is on the spring. Next, with the **Position** button on in the dialog box, pick the center point that is 10″ above the "paper." In the dialog box, set the zoom to about 28° by entering a value in the **Zoom** text box or dragging the slider. As the zoom is changed, the field of view dynamically changes in the graphics window just like a zoom lens on a camera. The roll angle is the rotation about the axis between the camera and target. Leave this set to 0.

When all settings have been made, pick the **OK** button to create the camera. See **Figure 20-4.** A new camera appears in the Cameras branch in the **Browser**. Cameras are sequentially named as Camera*x*, but they can be renamed just like any part or geometry.

Figure 20-4.
The camera has been added to the scene.

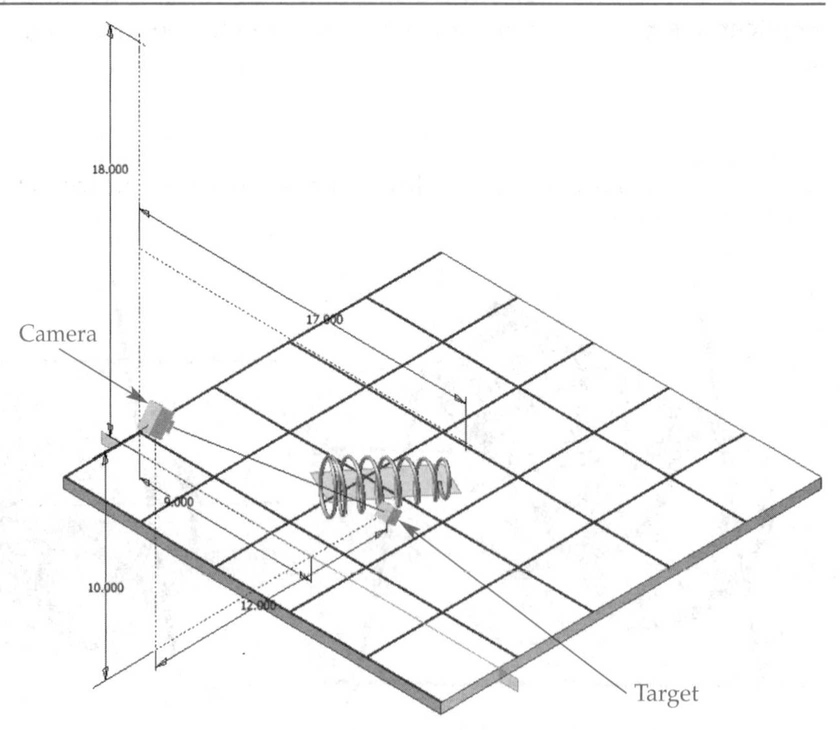

To display the view seen by the camera, right-click on the camera name in the **Browser** and pick **Set View to Camera** in the pop-up menu. This is the view that will be used for the rendering no matter what the current view is in the display. When setting the view to the camera, be careful not to select **Set Camera to View** in the pop-up menu. This will change the camera settings to match the current view, which in this case you do not want.

PROFESSIONAL TIP

An easy way to create a new camera is to pan and zoom until the view you want is displayed, right-click on Cameras in the **Browser**, and select **Create Camera from View** in the pop-up menu.

Creating a Lighting Style

Next, a lighting style needs to be created that approximates the illumination in the actual photograph of the spring. A lighting style consists of one or more lights. Like a camera, each light has a position and a target. A light can be one of three types:

- **Directional.** The light rays are parallel, similar to sunlight.
- **Point.** The light rays go in all directions from a point source, such as an incandescent lightbulb.
- **Spot.** The light rays project from the source in a cone, like a theatrical spotlight.

Most studio lighting setups have at least two lights—key light and fill light. See **Figure 20-5.** A third light, known as a background light, is sometimes included in the setup. The ***key light*** provides most of the illumination and can be to the left or right of the camera. The ***fill light*** is placed on the opposite side of the camera and is used to soften the shadows created by the key light. When used, the ***background light*** provides a visual separation between the subject (the part) and the background.

A fourth type of illumination is called ***ambient light.*** This is the natural light that shines in all directions and onto all faces. Ambient light is not represented by a light object in the graphics window. It is merely a setting in the lighting style.

Figure 20-5.
Most scenes should have at least a key light and a fill light. A—A basic setup of key and fill lights. B—Only the fill light is illuminating this bust. C—Both the key and fill lights are illuminating the bust. (Jack Klasey)

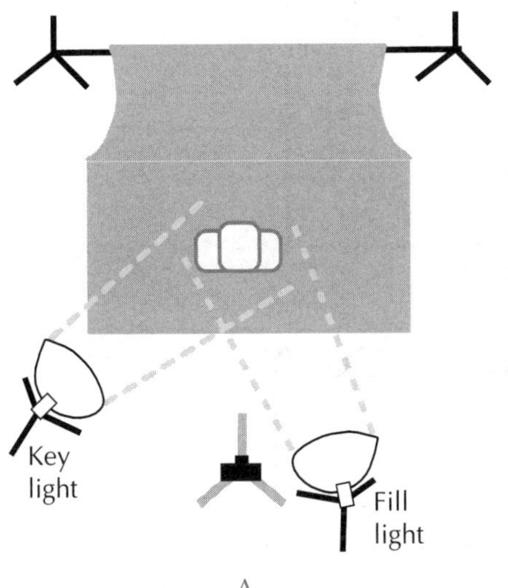

A

B

C

Inventor is supplied with many preset lighting styles for Inventor Studio. However, none of these will replicate the lighting in the spring photo. To create a new lighting style, first open the **Lighting Styles** dialog box by picking the **Lighting Styles** button in the **Inventor Studio** panel. See **Figure 20-6.** Then, continue as follows.

1. Pick the **New Lighting Style** button at the top, left of the dialog box. A new style named Default 1 is added to the list.
2. Right-click on Default 1 and select **Rename Lighting Style** from the pop-up menu. In the **Rename Lighting Style** dialog box that is displayed, enter Spring Lighting and pick the **OK** button, **Figure 20-7.**
3. Expand the Spring Lighting branch in the **Lighting Styles** dialog box. Delete the three default lights in the style by selecting each one and pressing the [Delete] key.
4. With the Spring Lighting branch selected, pick the **New Light** button at the top of the **Lighting Styles** dialog box. The **Light** dialog box is displayed, **Figure 20-8.**
5. In the **Type** area of the **General** tab, pick the **Directional** button to create a directional light.
6. With the **Target** button in the placement area on, select the center point that is on the spring.
7. Right-click in the graphics window and select **Previous View** in the pop-up menu to display all objects.
8. With the **Position** button on in the dialog box, pick the center point to the left and above the camera.
9. In the **Illumination** tab of the dialog box, enter 60 in the **Intensity** text box.
10. In the **Shadows** tab, uncheck the **Use Style Settings** check box. Then, pick the **Soft Shadows** button.

Figure 20-6.
The **Lighting Style** dialog box is used to manage lighting styles.

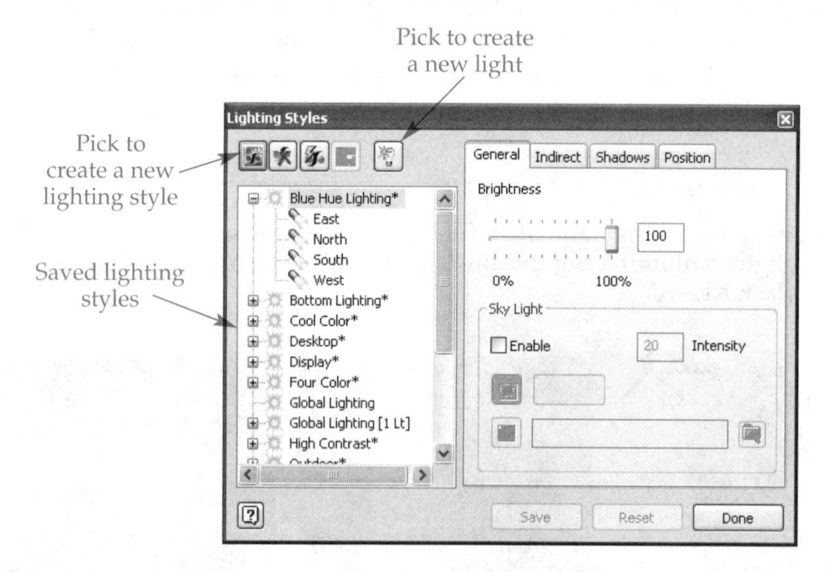

Figure 20-7.
Naming a new lighting style.

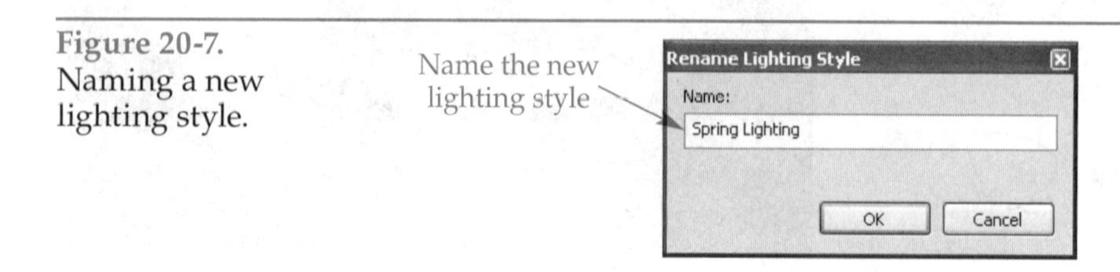

Figure 20-8.
Creating a new light.

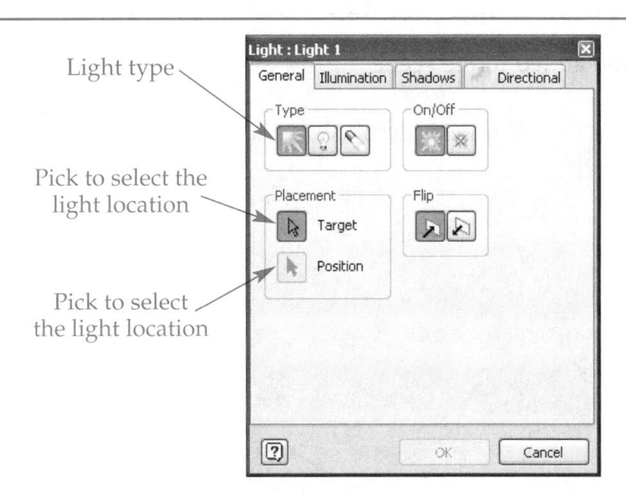

11. Pick the **OK** button to create the light and return to the **Lighting Styles** dialog box.
12. In the **Lighting Styles** dialog box, right-click on the new light under Spring Lighting and select **Rename Light** from the pop-up menu.
13. In the **Rename Light** dialog box, enter Key Light and pick the **OK** button.

The light you just created is the key light. For the fill light, create a second light with the same target and the hole center to the right of the camera as the location. Pick the **Spot** button in the **Type** area of the **General** tab of the **Light** dialog box. Also, in the **Illumination** tab, set the intensity to 40. In the **Shadows** tab, uncheck the **Use Style Settings** check box and pick the **No Shadows** button. This is equivalent to a floodlight with a diffuser.

In the **Lighting Styles** dialog box, rename the light to Fill Light. Also, right-click on the lighting style name and select **Active** from the pop-up menu. Then, pick the **Done** button to close the **Lighting Styles** dialog box. See **Figure 20-9.** Redisplay the camera view.

When creating a new light, the fourth tab in the **Light** dialog box changes for each type of light because of their different characteristics. For example, a directional light represents sunlight and, therefore, has latitude and longitude settings. A point light has a decay setting. A spotlight also has a decay setting along with settings for hotspot and falloff.

PROFESSIONAL TIP

If you select the first lighting style in the **Lighting Style** dialog box and use the down arrow key to scroll through the list, the effect of each lighting style is previewed in the graphics window.

Moving Lights and Cameras

Lights and cameras can be moved after they are created. For example, right-click on the camera name in the **Browser** and pick **Edit** from the pop-up menu. The **Camera** dialog box is opened. Display the previous view in the graphics window to see all objects. Next, pick the camera in the graphics window. The **3D Move/Rotate** dialog box is displayed and a triad is displayed on the camera. If you pick and drag the end of one of the triad vectors, the camera moves along that axis. If you pick the center of the triad, the camera can be moved in any direction. The camera target, light, and light target can be moved in the same manner. For this example, cancel the changes to retain your original settings.

Figure 20-9.
Two lights have been added to the scene.

Key light (directional light)
18.000
17.000
Camera
9.000
12.000
10.000
Fill light (spotlight)

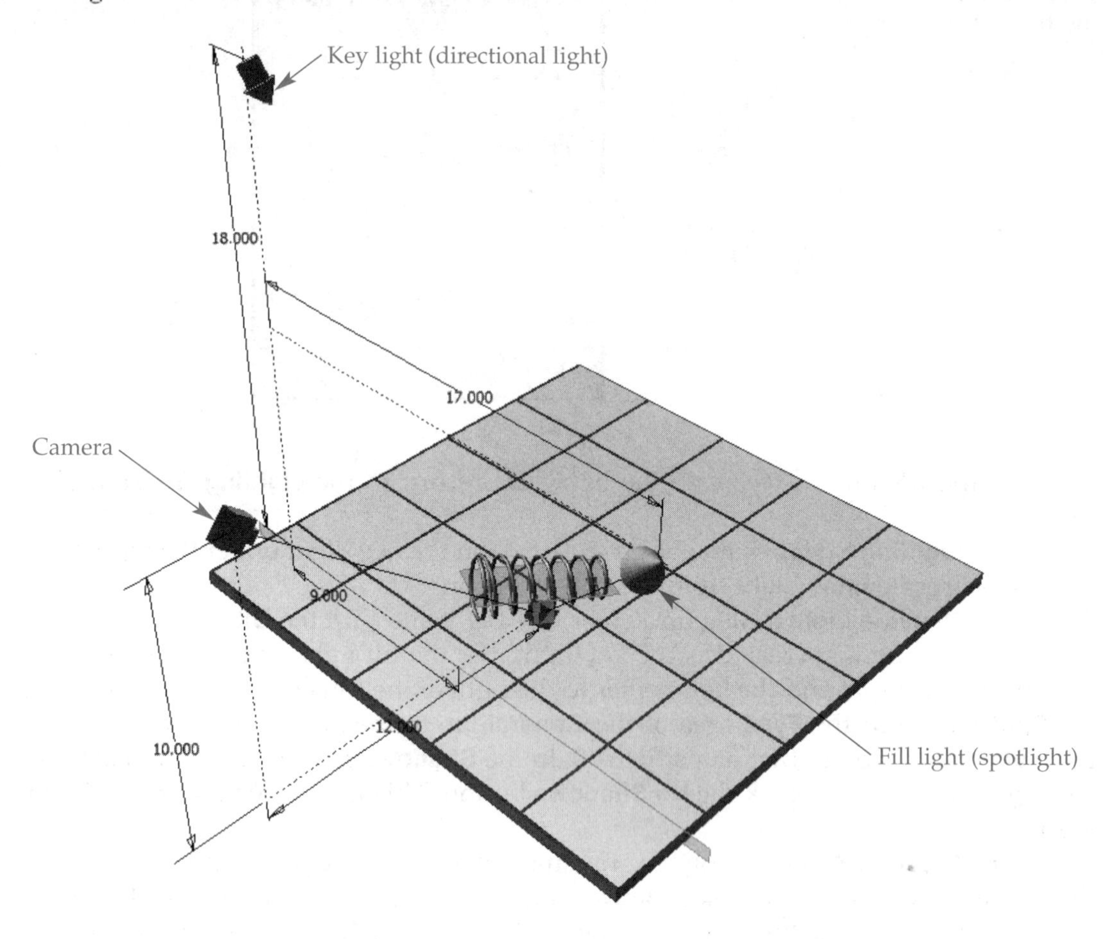

Rendering the Scene

Now, it is time to render the first image. Pick the **Render Image** button in the **Inventor Studio** panel. The **Render Image** dialog box is displayed, **Figure 20-10.**

In the **General** tab, set the pixel size, or resolution, of the image. Images displayed on a computer screen are measured in pixels, not inches. A low-resolution image, such as 320 × 240, renders quickly and has a small file size when saved. It is useful for

Figure 20-10.
Rendering an image.

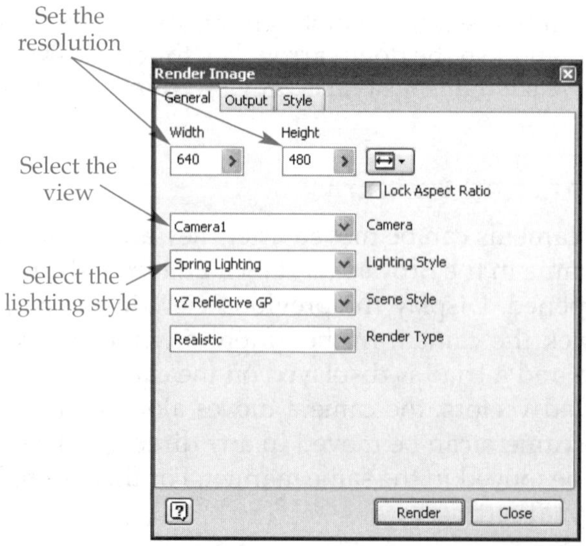

Set the resolution
Select the view
Select the lighting style
Render Image
General
Output
Style
Width
Height
640
480
Lock Aspect Ratio
Camera1
Camera
Spring Lighting
Lighting Style
YZ Reflective GP
Scene Style
Realistic
Render Type
Render
Close

Figure 20-11.
Increasing the resolution also increases the file size and the amount of time required to render the scene.

Resolution (pixels)	Render Time (seconds)	File Size (kilobytes)	Physical Size (inches)
320 × 240	3	11	3.3 × 2.5
640 × 480	9	30	6.6 × 5.0
800 × 600	13	40	8.3 × 6.2
1024 × 768	19	57	10.6 × 8.0
1600 × 1200	40	109	16.6 × 12.5

checking the image or for publishing on a web page. A high-resolution image, such as 1200 × 900, renders slowly and has a larger file size. High-resolution images are needed for presentations or printed copies. The table in **Figure 20-11** shows the resolution (pixel size), rendering time, file size, and physical size for a sample scene.

For this example, enter 800 in the **Width** text box and 600 in the **Height** text box. A preset output size can be selected using the **Select Output Size** button next to the **Height** text box. The **Lock Aspect Ratio** check box is used to keep the sizes at the same ratio as when the check box is checked. For example, if you set the size to 800 × 600, check the **Lock Aspect Ratio** check box, and then change the width from 800 to 900, the height automatically changes from 600 to 675 to maintain the same width-to-height ratio.

After setting the size in the **General** tab, you need to select which view will be rendered. In the **Camera** drop-down list, you can select the current view or any of the cameras created in the scene. For this example, select Camera1, which you created earlier.

Next, select the lighting style to use for the rendering in the **Lighting Style** drop-down list. You can select any of the saved styles. You can also choose to use the current lighting. Since you have created a lighting style to use for the rendering, select Spring Lighting in the drop-down list.

The **Scene Style** drop-down list is used to set the background. For this example, select Current Background. If the scene does not have a "floor" or "back wall," as this example does, choose one of the available backgrounds in this drop-down list.

There are two types of renderings that can be created. A realistic rendering is like a photograph. An illustration rendering is like a drawing or hand-rendered image. The type is selected in the **Render Type** drop-down list in the **General** tab.

In the final rendered image, lines at an angle can look stepped or jagged. The **Output** tab in the **Render Image** dialog box contains settings to control this. The buttons in the **Antialiasing** area of the tab are **No Antialiasing**, **Low Antialiasing**, and **High Antialiasing**. ***Antialiasing*** is a the softening of the edges in the rendered image to reduce the jagged effect. However, this may also result in a blurry image and will increase the rendering time. For this example, pick the **High Antialiasing** button.

The **Style** tab has only one setting: **True Reflection**. If this check box is checked, then the reflections on any shiny surfaces are calculated. This has little effect on the file size, but dramatically increases the rendering time. Since there are no shiny surfaces in this scene, the **True Reflection** setting has no effect whether or not it is checked.

To render the scene, pick the **Render** button at the bottom of the **Render Image** dialog box. The **Render Output** window is displayed, **Figure 20-12.** The scene is rendered in this window. When the rendering is complete, the image can be saved by picking the **Save Rendered Image** button at the top-right corner of the window. In the **Save** dialog box that is displayed, name the file and select a type. You have a choice of several image formats. After rendering the spring scene, save it as a JPG file, which is a compressed file format.

Figure 20-12.
A rendered image is displayed in the **Render Output** window. A—A photorealistic rendering. B—An illustration rendering.

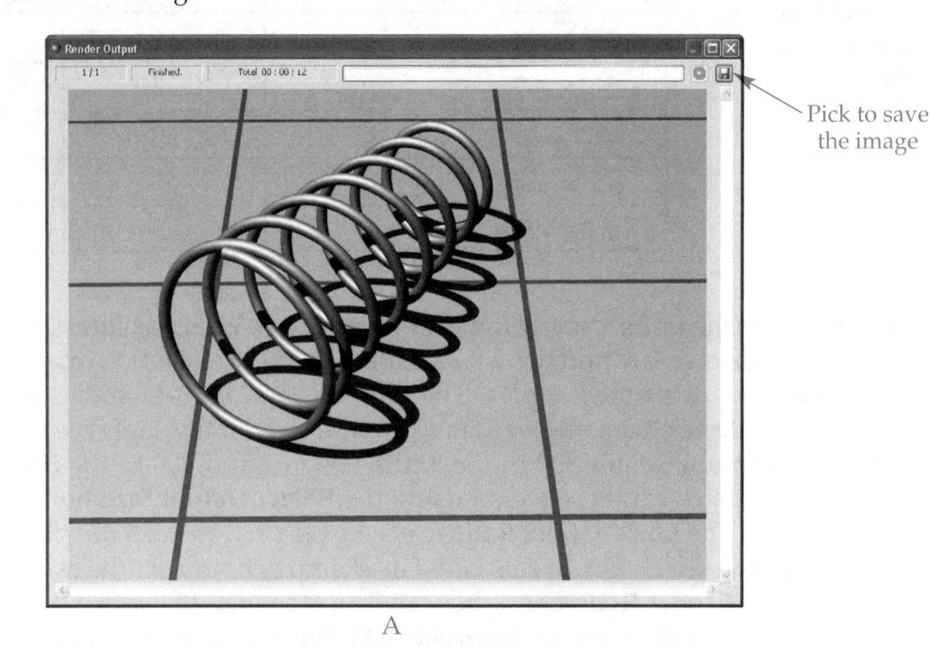

A

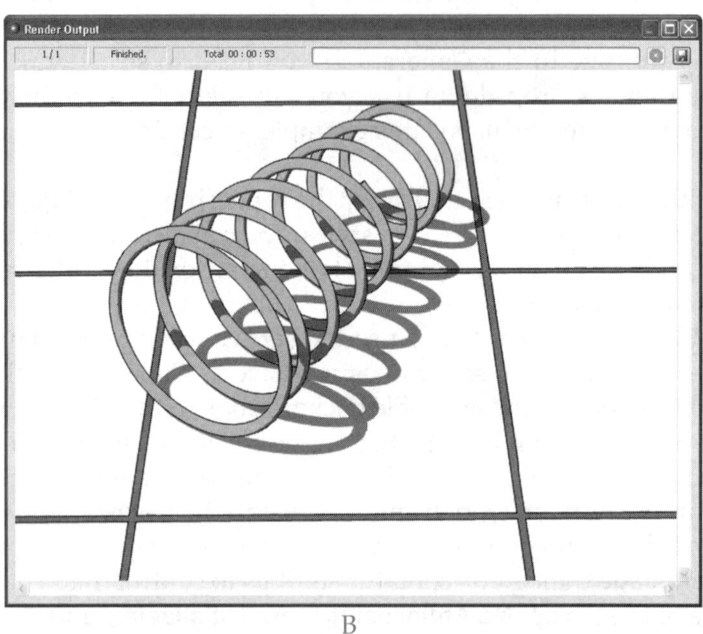

B

Creating a Surface Style

Now, it is time to create a surface style for the spring. Pick the **Surface Styles** button in the **Inventor Studio** panel to open the **Surface Styles** dialog box. See **Figure 20-13.** In the list at the top, left-hand side of the dialog box, there are categories for materials and several uncategorized surface styles. Expand the Metals branch to display the predefined metal surface styles. Then, select the Black (Casting) surface style. A preview of the surface style appears at the bottom of the dialog box and the color settings for the style appear on the right in the **Basic** tab. The four color settings are:

- **Ambient.** The color of the material in indirect light. Refer to **Figure 20-14.**
- **Diffuse.** The basic color of the material in direct light. For example, a red apple has a red diffuse color.
- **Specular.** The color of highlights on curved surfaces.

Figure 20-13.
The **Surface Styles** dialog box is used to manage surface styles.

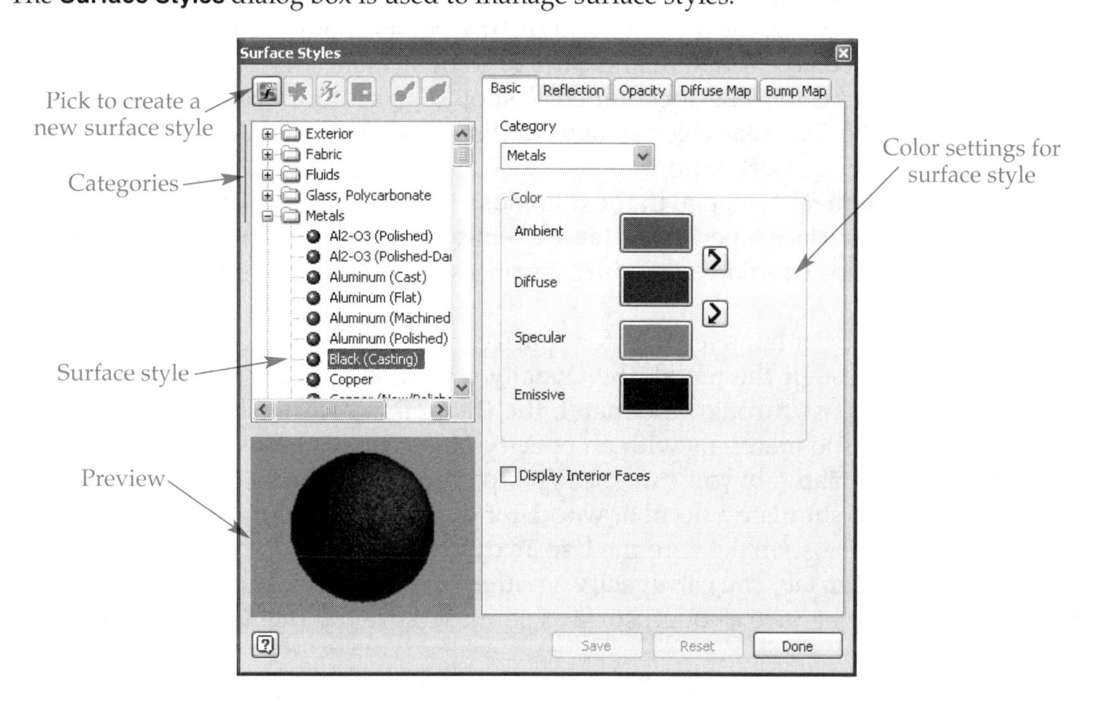

Figure 20-14.
The three color components of a surface style are diffuse, specular, and ambient. Also, notice how the surface roughness affects the specular highlight. A color version of this photo is located in the \Ch20 folder on the Student CD.

- **Emissive.** Color generated from the material itself, such as red from a hot steel part.

Pick the **New Style** button to create a new surface style. The style is added to the list of uncategorized styles below the folders and automatically selected. Right-click on the name of the new style in the tree (Default 1) and select Rename Surface Style in the pop-up menu. In the **Rename Surface Style** dialog box, name the new style For My Spring.

In the **Basic** tab, the colors for the surface style are set. Pick the **Diffuse** color swatch to display the color dialog box. Then, select the dark gray color and pick the **OK** button. For the spring, the ambient and diffuse colors should be the same. Pick the arrow button between the **Ambient** and **Diffuse** color swatches. This copies the diffuse color to the ambient color. The specular color should be set to white to give the bright highlights. Pick the **Specular** color swatch to display the color dialog box. Select the white color and pick the **OK** button.

In the **Reflection** tab, enter 40 in the **Shininess** text box. If you want to add a reflection image to the surface, check the **Use Reflection Image** check box and specify an image file. This helps simulate very shiny surfaces. For the spring surface style, leave this unchecked.

The **Opacity** tab contains settings for opacity (transparency) and diffraction. Since you cannot see through the metal, the **Opacity** setting should be left at 100%. Also, since no light can pass through the metal, the **Refraction** setting should be 0%. This setting only applies to materials with an opacity of less than 100%.

In the **Diffuse Map** tab, you can specify a pattern to be rendered on the surface. This can be used to simulate a decal or wood, for example. For the spring, there should be no surface texture, so make sure the **Use Texture Image** check box is unchecked.

In the **Bump Map** tab, you can specify a pattern to simulate a rough surface. Again, the spring should not have a rough surface, so make sure the **Use Bump Image** check box is unchecked.

Pick the **Save** button to save the surface style and then pick the **Done** button to close the dialog box. Next, return to the assembly view by selecting **Assembly** from the **Applications** pull-down menu. Select the spring in the graphics window and pick **For My Spring** in the color drop-down list at the right-hand end of the **Inventor Standard** toolbar. Then, return to Inventor Studio and render the camera view again. Notice the change in the spring. Save and close the assembly file.

Rendering a Scene Containing Threads

Open Example_20_02.iam and switch to Inventor Studio. This is a brass pad on a diamond plate. There is no camera, but the lighting style is set to the predefined Desktop style. Set the intensity of each light to 90%. Set the ambient light to 20%. This is done with the **Ambience** text box in the **Indirect** tab of the **Lighting Styles** dialog box, which is available when the style name is selected in the tree.

Zoom in on the brass pad, if needed, and render the current view. Notice that the vertical holes, which are threaded with the standard Inventor thread bitmap, are rendered as threaded. The horizontal hole, which has an actual thread cut as geometry, is also rendered as threaded. Close the file.

PRACTICE 20-1 Complete the practice problem on the Student CD.

Animating Components

Open Example_20_03.iam and examine the assembly. This is a good example to start with as the animation is similar to the presentation files you created in Chapter 18. Basically, components are moved in 3D space and those movements are recorded for playback. In this animation, the assembly will be disassembled in the same way the physical assembly would be. These types of animations are ideal for showing the staff on the factory floor how to assemble or disassemble a given machine.

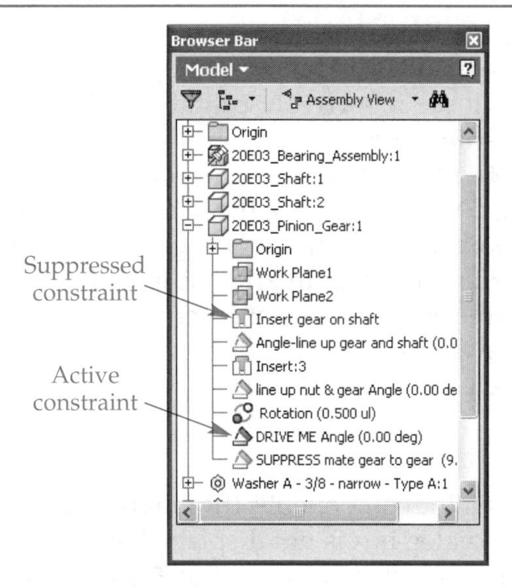

Figure 20-15.
Some of the constraints must be suppressed to animate the assembly.

Expand the tree in the **Browser** to reveal all of the assembly constraints on all of the parts. See **Figure 20-15.** Motion cannot be animated in Inventor Studio beyond what is allowed by constraints. That is why many of the constraints have been suppressed. Constraints should be suppressed before entering Inventor Studio. Next, enter Inventor Studio. To keep things simple, the default lighting is used in this example.

Creating a Camera

Every animation should have a camera or two. Pick the **Camera** button in the **Inventor Studio** panel to open the **Camera** dialog box. It can be very difficult to specify a point in 3D space for the camera's position. This is why the center points were used in the earlier example. In this example, two center points have been included. See **Figure 20-16.** Create a camera based on the points, as shown in the figure. Next, check the **Link to View** check box in the **Camera** dialog box. The screen display radically changes, but you can now use the familiar **Pan** and **Zoom** controls to obtain the view you want. As you zoom and pan, the values in the dialog box update to reflect your actions. Make sure you zoom out enough so there is plenty of space around the assembly for removing parts. When the view is correct, pick the **OK** button in the **Camera** dialog box to create the camera.

After the camera is created, rotate the view. The two objects that represent the camera and target can be hidden, if needed. Right-click on the camera name in the **Browser** and uncheck **Visibility** in the pop-up menu. The camera target is automatically hidden as well.

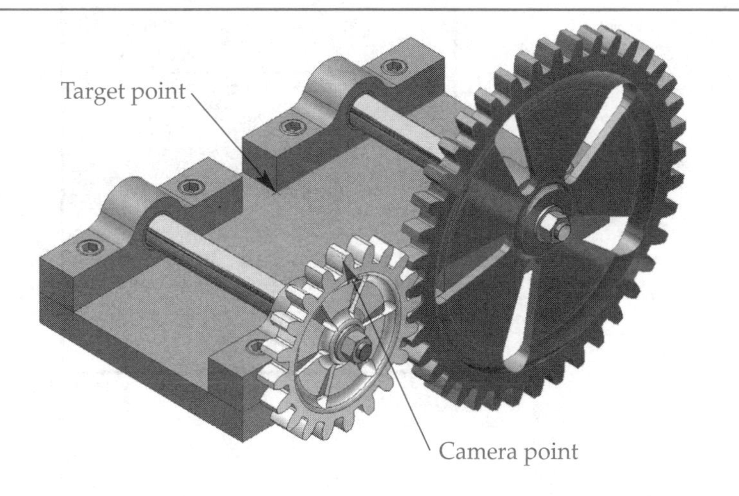

Figure 20-16.
Use the two center points in the assembly to initially locate the camera.

Entering Animation Mode

Look at the tree in the **Browser**. At the top is the Animations branch. If you expand this branch, you will see it currently contains only the Model State branch. The first step in creating an animation is to enter animation mode. Right-click on the Animations branch and select **New Animation** from the pop-up menu. A new branch is added with the name Animation*x*. You can rename the branch by left-clicking on it two single times to display the text box. To enter animation mode, simply double-click on the name of the animation. The current mode, animation or model, is indicated by a check mark next to its name in the tree. When you first enter animation mode, the **Animation Timeline** dialog box is displayed. This dialog box is used to control the timing of events in the animation. It is discussed in detail later.

Setting the Animation Time

Before animating the components, the length of the animation should be set. If the **Animation Timeline** dialog box is not displayed, pick the **Animation Timeline** button in the **Inventor Studio** panel. Pick the **Animation Options** button in the **Animation Timeline** dialog box, **Figure 20-17A.** The **Animation Options** dialog box is displayed. It is used to set the length of the animation, among other things, **Figure 20-17B.** Enter 16 in the **Seconds** text box to create an animation that is 16 seconds long.

Figure 20-17.
A—The **Animation Timeline** dialog box. B—The animation length is set in the **Animation Options** dialog box.

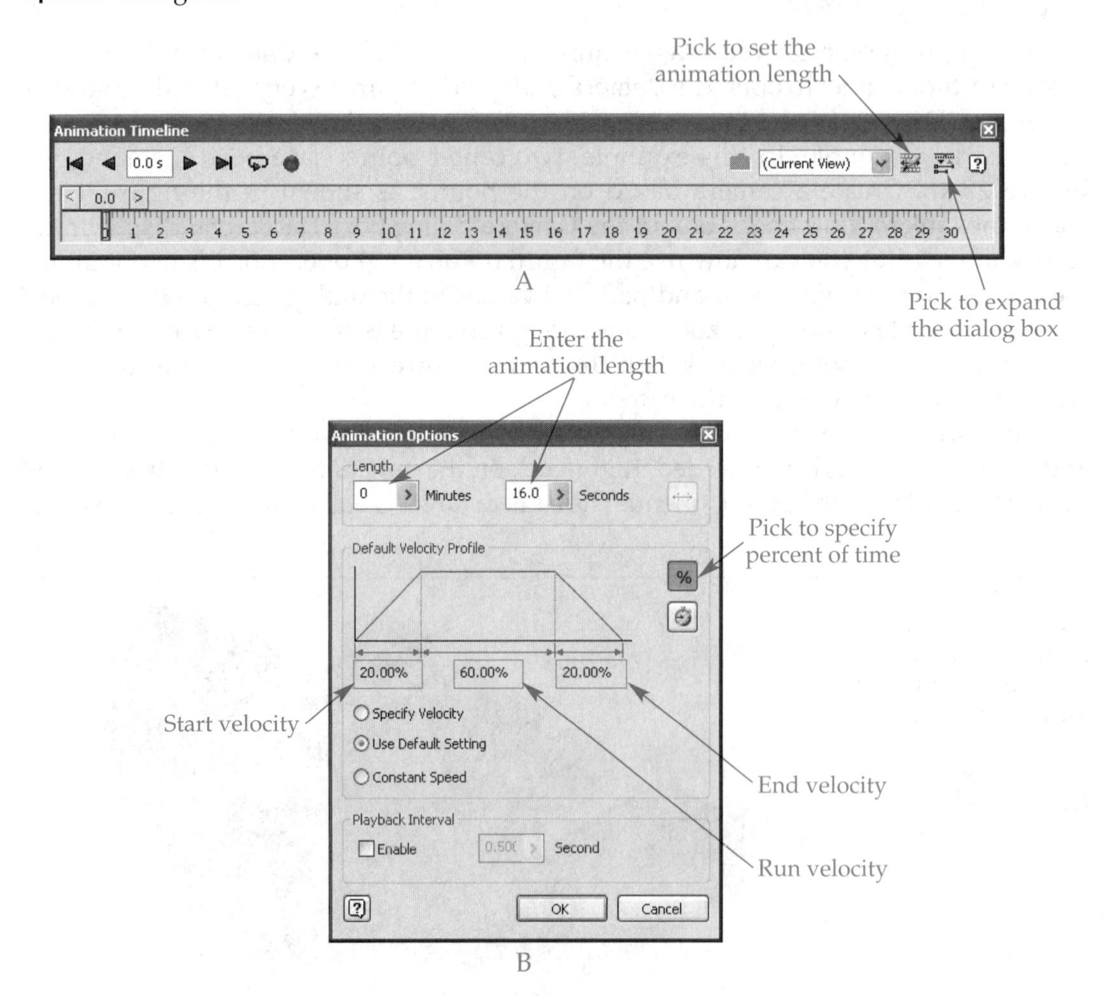

The default velocity profile determines how an animated component reaches its velocity. You can display this as a percentage or as time. If the **Specify Velocity** radio button is selected, you can change the settings. The three text boxes in the tab correspond to the start, run, and end velocity. The value in the left-hand text box is the amount of time the component takes to reach velocity. The value in the middle text box is the amount of time the component is at velocity. The value in the right-hand text box is the amount of time the component takes to decrease from velocity to zero. For this example, pick the **Use Default Setting** radio button.

Pick the **OK** button to close the dialog box. Note that the animation timeline is now set to 16 seconds. You can leave the **Animation Timeline** dialog box open.

Animating Components

Now, you are ready to add motion to the assembly. Pick the **Animate Components** button in the **Inventor Studio** panel opening. The **Animate Components** dialog box is displayed, **Figure 20-18.**

You are going to rotate and move the nut in three seconds to simulate unscrewing it from the shaft. With the **Components** button on, select the hex nut on the small gear. You can select it in the graphics window, but picking it in the **Browser** may be easier. In the **End** text box, enter 3. The start time and duration are automatically calculated.

Specifying the movement

Next, pick the **Position** button. The **Animate Components** and **Animation Timeline** dialog boxes are hidden and the **3D Move/Rotate** dialog box is displayed, **Figure 20-19.** Also, a triad is displayed on the nut in the graphics window. The ***triad*** represents the local X, Y, and Z axes and is used to apply movement or rotation on these axes. You saw this earlier when moving lights and cameras.

Figure 20-18.
The **Animate** tab of the **Animate Components** dialog box.

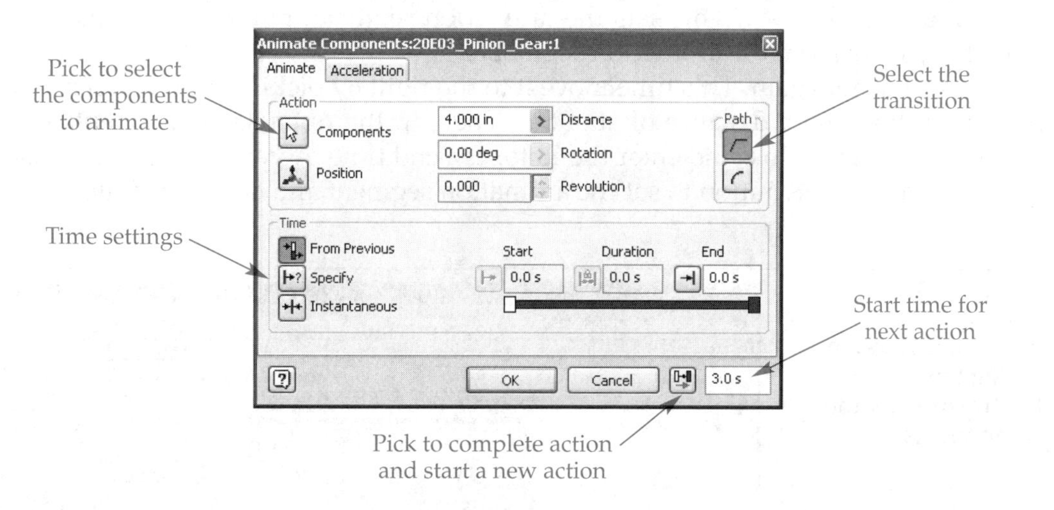

Figure 20-19.
The **3D Move/Rotate** dialog box.

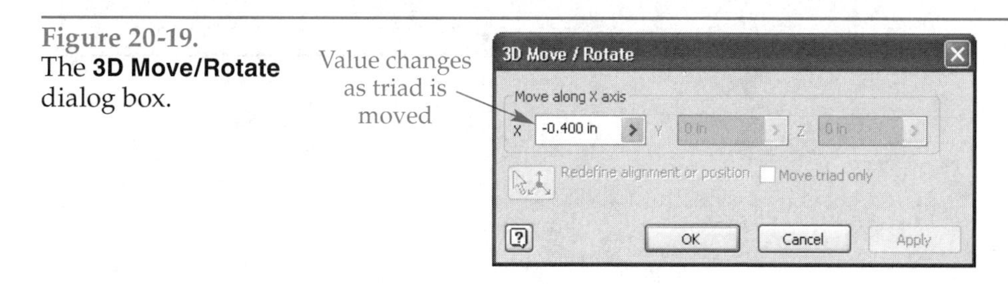

The red arrow on the triad corresponds to the local X axis. It is pointing into the assembly along the shaft's axis. Place the cursor over the shaft of the red arrow. Notice the rotational icon that appears next to the cursor. Picking and dragging the shaft of the arrow rotates the part. Place the cursor over the red arrowhead. Notice the translational (movement) icon that appears next to the cursor. Picking and dragging on the arrowhead moves the part along the local X axis.

Pick the red arrowhead and drag the nut away from the assembly about .4". Notice the value in the dialog box changes. Release the mouse button and the value is grayed out. This is okay because it can be edited later. Now, rotate the nut 90° about the red axis. This sets the directions of motion for this three-second animation segment. Pick the **OK** button to redisplay the **Animate Components** and **Animation Timeline** dialog boxes. Look at the **Distance** and **Rotation** text boxes in the **Animate** tab of the **Animate Components** dialog box. The values in these text boxes reflect how the nut was moved and rotated. Enter .4 in the **Distance** text box. Also, enter 5 in the **Revolutions** text box. This changes the value in the **Rotation** text box to 1800.

Display the **Acceleration** tab. After the nut is unscrewed, it should keep moving away from the assembly for another 5". Therefore, pick the **Specify Velocity** radio button and change the deceleration at the end to zero. The nut will accelerate for the first 20% of the time and be at full velocity for the remaining 80%.

Pick the **OK** button in the **Animate Component** dialog box to set this first animation segment. In the **Animation Timeline** dialog box, pick the **Expand Action Editor** button to expand the dialog box. Notice the first action is represented by a bar. See **Figure 20-20.**

Make sure the time slider at the top of the dialog box is at 3.0 seconds. Then, pick the **Animate Components** button again. Select the same nut and move it away from the assembly about 5" along the X axis. In the **Animate Components** dialog box, set the distance to 5. Also, be sure the **From Previous** button is on in the **Time** area. Then, enter 6 into the **End** text box. This results in another three-second animation segment. In the **Acceleration** tab, pick the **Specify Velocity** radio button and set the starting acceleration to zero.

Pick the **Complete action and start a new action** button at the bottom of the **Animate Components** dialog box. This sets the animation segment, but leaves the dialog box open for you to add more animation segments.

Move the nut again. This time move it to the right by picking on the green arrowhead and dragging a distance of about 6". Then, in the **Animate Components** dialog box, set the distance to 6 and enter the 10 for the end time. Accept the default acceleration and pick the **OK** button to set the animation segment and close the dialog box.

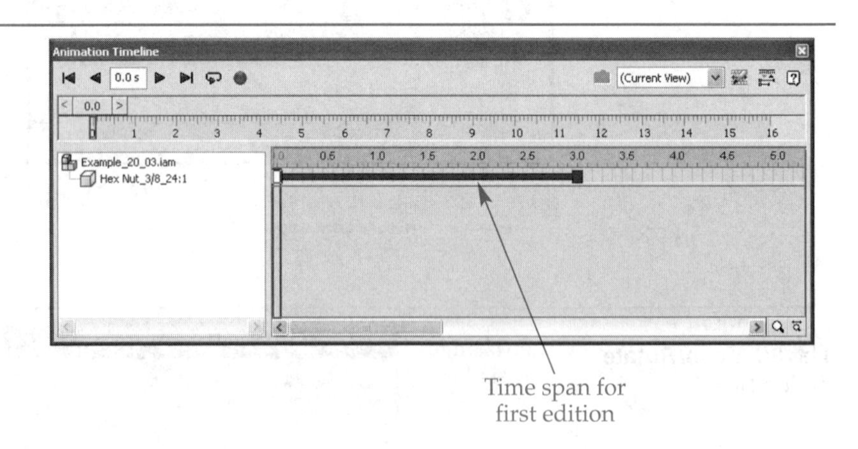

Figure 20-20.
The expanded **Animation Timeline** dialog box shows the timeline for the components.

Testing the animation

Now is a good time to test the animation to see if everything went as planned. Expand the **Animation Timeline** dialog box, if it is not already. Also, resize the dialog box as needed to see the timeline from zero to 10 seconds. If needed, use the **Zoom** button at the bottom of the dialog box to zoom the timeline.

The blue line indicates the time span over which the part is animated. For this example, the blue line should extend from zero to 10. Notice the three white nodes, one at the zero second mark, one at the three second mark, and one at the six second mark. There is also a blue node at the 10 second mark. See **Figure 20-21.** These nodes are called keys. A ***key*** contains the values for the state of the part at a given time. The time, or frame, at which a key appears is called a ***keyframe.*** Keys and keyframes are the basis for animation.

Pick the camera name from the drop-down list at the top, right of the **Animation Timeline** dialog box. Then, collapse the dialog box and move it toward the top of the screen. Pick the **Go to Start** (rewind) button in the navigation controls at the top, left of the **Animation Timeline** dialog box to set the timeline to zero. Then, pick the **Play Animation** button. The nut moves in the graphics window as you defined: away from the part and then to the left. Pick the **Go to Start** button to return the animation to zero on the timeline. This is a good time to save the file.

Editing animation segments

Expand the **Animation Timeline** dialog box. Hover the cursor over the blue timeline. Information about that operation is displayed as help text. If you right-click on the blue line between to keys and select **Edit** from the pop-up menu, the **Animate Components** dialog box is displayed with the values for that segment. You can edit the operation and then pick the **OK** button to update the animation.

You can also drag-and-drop the keys to change the times. When you previewed the animation, you should have noticed that the rotation of the nut is too fast. Drag the last key to the 12 second mark. Also, drag the second key to the 8 second mark and the first key to the 5 second mark. There are actually four keys in the middle of the timeline, not just the two you can see. For each of the two you see, there is another key beneath it. Be sure to move both keys. Next, edit the first segment and change the number of revolutions to 3 and pick the **Constant Speed** radio button in the **Acceleration** tab. Play the animation again.

Now, animate the washer on the small gear. Move it 1″ away from the assembly in one second. Then, move it 4″ to the right in three seconds. In the **Animation Timeline**

Figure 20-21.
The keys and timeline for the completed hex nut animation.

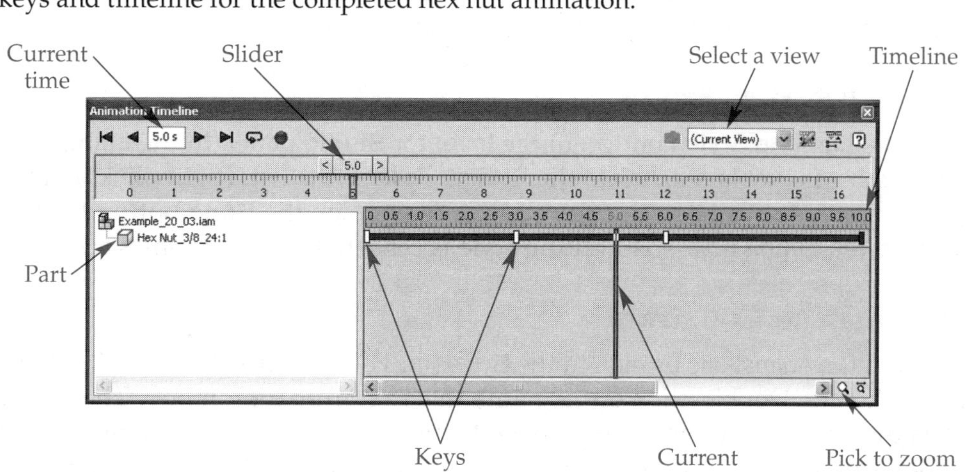

dialog box, notice that a blue timeline has been added for the washer. Drag the keys for the washer so the two animation segments occur from the 5 second mark to the 9 second mark. Pick the **Go to Start** button. Then, pick the **Play Animation** button to test the animation. The nut moves as before and the washer follows five seconds later. Save the file.

PROFESSIONAL TIP

When animating a scene, it is a good practice to save the file before adding the next animated segment and after the tested animation is correct. This can save hours of time if an error is encountered while animating the next segment.

PRACTICE 20-2 Complete the practice problem on the Student CD.

Animating Constraints

Open Example_20_04.iam and examine the assembly and its components. This is the rack and pinion assembly from earlier in the book. Expand the 20E04_Rack branch in the **Browser** and notice the DRIVE ME constraint. This is a mating constraint used to move the rack across the frame. Right-click on the constraint name in the **Browser** and select **Drive Constraint** from the pop-up menu. In the **Drive Constraint** dialog box that appears, pick the **Forward** (play) button and watch the movement. This same movement will be produced in the animation for this example. Pick the **Minimum** (reverse) button in the **Drive Constraint** dialog box to reset the motion and then close the dialog box.

Creating a Camera

Switch to Inventor Studio. Using the **Pan** and **Zoom** tools, obtain a good view of the assembly. The isometric view can be used, if you like. Then, right-click on the Cameras branch in the **Browser** and select **Create Camera from View** in the pop-up menu. A camera is created and appears in the graphics window. Right-click on the camera name in the Cameras branch in the **Browser** and select **Set View to Camera** in the pop-up menu. Then, right-click on the name again and uncheck **Visibility** in the pop-up menu. This hides the camera in the graphics window.

Adding Lighting

Pick the **Lighting Styles** button in the **Inventor Studio** panel. In the **Lighting Styles** dialog box that is displayed, locate the Display style in the list. Right-click on the style name and select **Active** from the pop-up list. Then, pick the **Done** button to close the **Lighting Styles** dialog box. The lighting style is reflected in the graphics window.

Animating the Constraint

Expand the Animations branch in the **Browser**. There is already one animation state added to the file. Double-click on its name to enter animation mode. The **Animation Timeline** dialog box is automatically opened when you enter animation mode. Pick the **Animation Options** button in the dialog box and, in the dialog box that is displayed, set the length of the animation to 24 seconds. The DRIVE ME constraint will be animated over a length of eight inches in six seconds and the movement will occur four times. Thus, the total animation time is 24 seconds.

In the **Browser**, right-click on the DRIVE ME constraint for the 20E04_Rack component and select **Animate Constraints** in the pop-up menu. The **Animate Constraints** dialog box is displayed. See **Figure 20-22.** This dialog box can also be displayed by picking the **Animate Constraints** button in the **Inventor Studio** panel, but the constraint will not automatically be selected. Notice that the time controls at the bottom side of the **Animate Constraints** dialog box are the same as those found in the **Animate Components** dialog box used earlier. The **Acceleration** tab also has the same settings as in the **Animate Components** dialog box.

In the **Animate** tab, notice the three buttons in the middle of the **Animate** area. The **Constraint** button is for specifying a numeric distance for the assembly constraint. This button will be used to animate the constraint in this example. The other two buttons are **Suppress** and **Enable**. They can be used in the middle of an animation to suppress or enable an assembly constraint. When doing so, use the **Instantaneous** button in the **Time** area for an instantaneous effect on the constraint.

The first movement is from 0.00″ to 8.00″. Pick the **Constraint** button so it is on and enter 8 in the **End** text box in the **Action** area. This movement spans six seconds. Pick the **From Previous** button in the **Time** area and enter 6 in the **End** text box. Finally, in the **Acceleration** tab, pick the **Constant Speed** radio button. The first action is defined. Since the next action also spans six seconds, enter 6 in the **New Action Increment** text box at the bottom-right corner of the dialog box and then pick the **Complete action and start new action** button. The first action is recorded and the dialog box is readied for a second action spanning six seconds.

Make sure 6 is entered in the **Start** text box in the **Time** area and 12 is entered in the **End** text box. Then, make sure the **Constraint** button is on and enter 0 in the **End** text box. This returns the rack to its starting position. Make sure 6 is entered in the **New Action Increment** text box and pick the **Complete action and start new action** button.

For the third action, move the rack 8″ from time 12 seconds to time 18 seconds. For the fourth action, move the rack back to the 0″ position from time 18 seconds to time 24 seconds. After the fourth action is defined, pick the **Done** button to close the **Animate Constraints** dialog box.

Finally, test the animation. Make sure the camera is selected in the drop-down list at the top, right of the **Animation Timeline** dialog box. If this dialog box is not open, pick the **Animation Timeline** button in the **Inventor Studio** panel. Then, pick the **Go to Start** button in the dialog box followed by the **Play Animation** button. The rack travels back and forth through two complete cycles. Save and close the file.

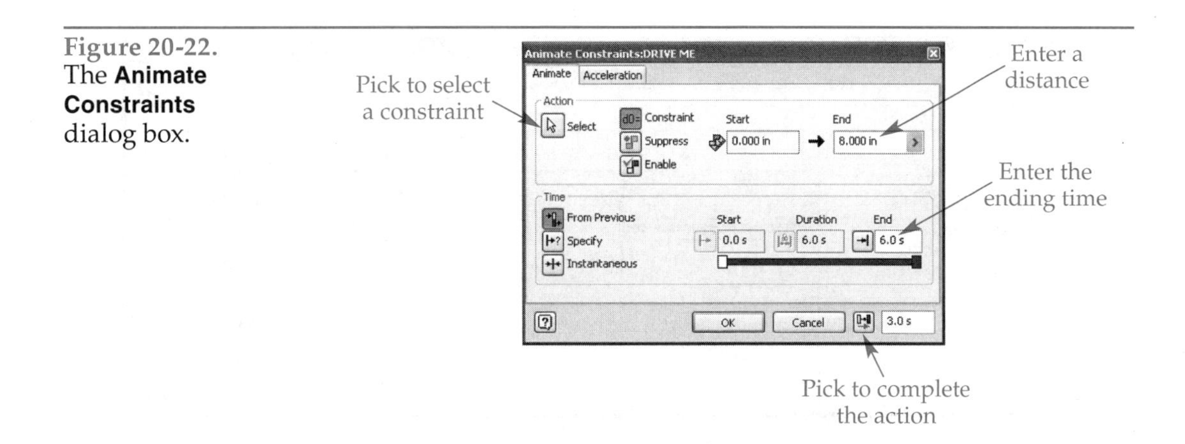

Figure 20-22. The **Animate Constraints** dialog box.

Animating a Camera

A camera can be animated to produce a panning or zooming effect in the animation. Open Example_20_05.iam and examine the assembly and its components. This is the planetary gear set from earlier in the book. Switch to Inventor Studio. The gear set has already been animated and a camera has been added to the scene. A lighting style has also been set up. Now, the camera needs to be animated.

There is the **Animate Camera** tool, but the simplest way to animate a camera is using the **Animation Timeline** dialog box. Double-click on the animation name in the Animations branch in the **Browser** to enter animation mode and display the **Animation Timeline** dialog box. Expand the dialog box. Select Camera1 in the view drop-down list at the upper, right-hand corner of the dialog box. Then, drag the time slider to the 1.5 second mark. Using the **Rotate** tool on the **Inventor Standard** toolbar, rotate the view in the graphics window so that the left side of the assembly is shown. Finally, pick the **Add Camera Action** button in the **Animation Timeline** dialog box. This records the camera's movement to keys. Notice that the camera now appears in the **Animation Timeline** dialog box and two keys have been created. See **Figure 20-23.** The key at the 0 second mark contains the camera's original position. The key at the 1.5 second mark contains the camera's current position.

Now, drag the time slider to the 7 second mark. Rotate the view in the graphics window so that the right side of the assembly is shown, viewed slightly from below. Pick the **Add Camera Action** button to record the movement. Another key is added for the camera. Drag the time slider to the end of the animation (15 second mark). Using the **Zoom** tool on the **Inventor Standard** toolbar, zoom in close on the top of the assembly. Also, rotate the view so you are looking slightly down on the assembly. Pick the **Add Camera Action** button to add the final key.

Test the animation. Notice that the camera is always in motion, which can be confusing. Pauses can be added to the camera's animation so that the view is not always in motion. In the expanded **Animation Timeline** dialog box, notice that there are three segments for the camera: one between the first and second keys, one between the second and third keys, and one between the third and fourth keys. Pick the third segment to select it. Then, drag the first key on that segment (7 second mark) to the

Figure 20-23.
Keys are added for the camera.

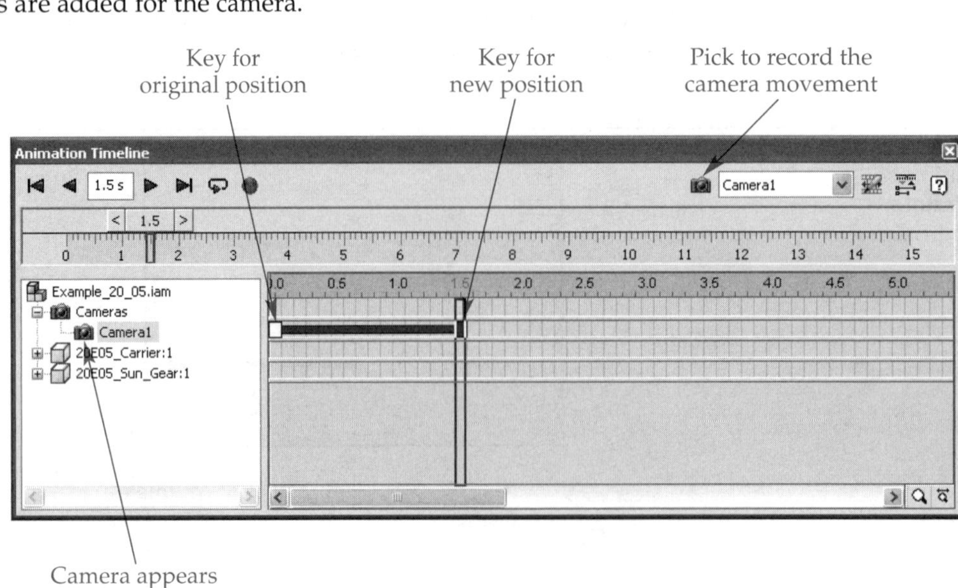

12 second mark. Remember, there are actually two keys at this point, one for the second segment and one for the third segment. Next, select the second segment. Then, drag the first key for that segment from the 1.5 second mark to the 4 second mark. Now, test the animation again. The camera pauses twice during the animation. Save and close the file.

PROFESSIONAL TIP

There is a more advanced method of animating a camera, which involves using a dialog box similar to the other "animate" dialog boxes you have used to this point. To open the **Animate Camera** dialog box, pick the **Animate Camera** button in the **Inventor Studio** panel or right-click on the camera name in the **Browser** and select **Animate Camera** from the pop-up list.

Animating Fade

The visibility of a part can be animated so that it appears to gradually appear or disappear. This is called ***fade.*** Open Example_20_06.iam. This is the simple engine from earlier in the book. Switch to Inventor Studio. The components have been animated, a camera has been created, and a lighting style has been set up. Now, you need to animate the fade of the engine block.

Switch to Inventor Studio and double-click on the name of the animation in the Animations branch in the **Browser** to enter animation mode. Then, pick the **Animate Fade** button in the **Inventor Studio** panel to open the **Animate Fade** dialog box. See **Figure 20-24.** Pick the **Components** button in the **Action** area and select the engine block. Next, in the **Time** area of the **Animate** tab, pick the **Specify** button. Then, enter 3 in the **Start** text box and 8 in the **End** text box. In the **Action** area of the tab, enter 0 in the **End** text box. Make sure 100 is displayed in the **Start** text box in the **Action** area. This means that from the 3 second mark to the 8 second mark the engine block will gradually disappear. Finally, pick the **OK** button to create the action.

Test the animation. Notice how the engine block gradually fades from the 3 second mark to the 8 second mark. The outline of the engine block visible in the preview after the fade has reached 0% will not be visible in the final animation rendered to a file. Rendering to a file is discussed in the next section.

Figure 20-24.
Animating fade.

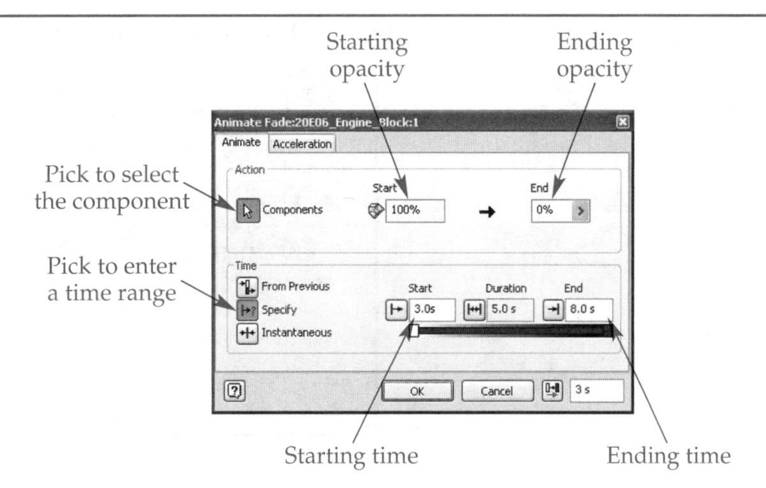

Rendering an Animation

Once an animation is set up, it can be rendered to a file. The file can then be played to view the animation. In the **Animation Timeline** dialog box, pick the **Record Animation** button. The **Render Animation** dialog box is displayed. See **Figure 20-25.** The settings in the **General** and **Style** tabs are similar to the corresponding tabs in the **Render Image** dialog box. Refer to the section Rendering the Scene earlier in this chapter. In the **General** tab, set the width and height to 640 × 480. In the **Style** tab, uncheck the **True Reflection** check box.

In the **Output** tab, pick the **Low Antialiasing** button. Also, make sure the time range is from 0 to 25 seconds. Make sure the **Video Format** button is on in the **Video** area. Enter 15 in the **Frame Rate** text box. Next, pick the "file browse" button at the top, right corner of the tab. In the **Save** dialog box that is displayed, name the file Example_20_06 and select the AVI format in the **Save as type:** drop-down list. Pick the **Save** button to return to the **Render Animation** dialog box. Finally, pick the **Render** button in the **Render Animation** dialog box. In the **Video Compression** dialog box that is displayed, select an appropriate codec, such as Microsoft MPEG-4, and pick the **OK** button to start the rendering. If you select the WMV file type, a different codec dialog box is displayed.

Rendering the animation will take some time. As each frame is rendered, it is displayed in the **Render Output** window. See **Figure 20-26.** When the animation is rendered, save and close the assembly file. Then, locate the animation file Example_20_06.avi using Windows Explorer and double-click on it to play the animation in the associated program (most likely Windows Media Player).

PROFESSIONAL TIP

When rendering, keep in mind only one camera can be recorded at a time. To create an animation that switches cameras during the sequence requires post-processing separate animation files in an editor. Record the animation once for each camera over the particular time span for that view. Then, combine the results in a video editing program. There are many shareware and freeware programs available on the Internet.

Figure 20-25. Rendering an animation.

Set the time span

Pick to save a file

Set the frame rate

Pick to create a video file

Figure 20-26.
As each frame of an animation is rendered, it is displayed in the **Render Output** window.

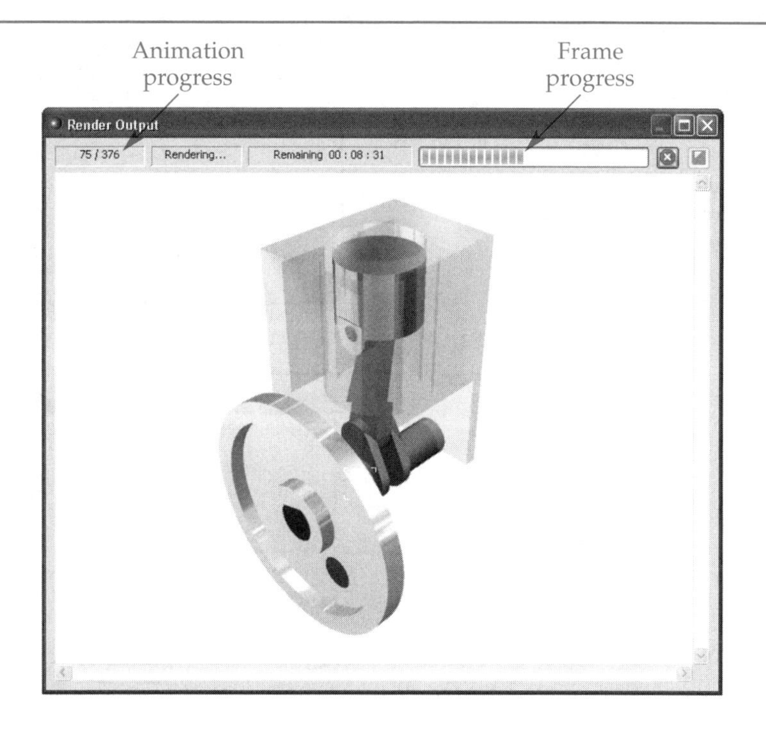

Animating Parameters

Animating parameters that have been set up in an assembly or part file is a very powerful feature of Inventor Studio. For example, this can remove the necessity of animating each and every assembly constraint. Inventor's adaptivity can also be modeled into an animation, as demonstrated by the example in the next section.

Adaptive Spring

Open Example_20_07.iam and examine the assembly and its components. Take a look at the constraint that will be used to give motion to the parts. In the **Browser**, expand tree for 20E07_Part:1 and notice the Drive This constraint. Right-click on the constraint name in the **Browser** and pick **Drive Constraint** from the pop-up menu. In the **Drive Constraint** dialog box, pick the **Forward** (play) button. The elasticity of the spring adapts to the distance between the plates so the spring remains in contact with both plates.

Switch to Inventor Studio. A camera has already been created. Set the view to the camera. Also, turn off the visibility of the camera so that the target is not visible in the graphics window. The animation length has also been set to 15 seconds and a lighting style has been set up. Double-click on the animation name in the **Browser** to enter animation mode.

Pick the **Parameter Favorites** button in the **Inventor Studio** panel. The **Parameter Favorites** dialog box is displayed, **Figure 20-27.** It contains all of the user-defined parameters in the assembly. In this case, there is only one: AssemblyDistance. This parameter controls the distance between the plates. In order to animate a parameter, it must be added to the Animation Favorites branch in the **Browser**. To do this, check the check box in the **Favorites** column in the **Parameter Favorites** dialog box. Then, pick the **OK** button to close the dialog box.

Now, expand the Animation Favorites branch in the **Browser**. The parameter you just added to the favorites appears in the tree. Right-click on the parameter name and select **Animate Parameters** in the pop-up menu or pick the **Animate Parameters** button in the **Inventor Studio** panel. The **Animate Parameters** dialog box is displayed.

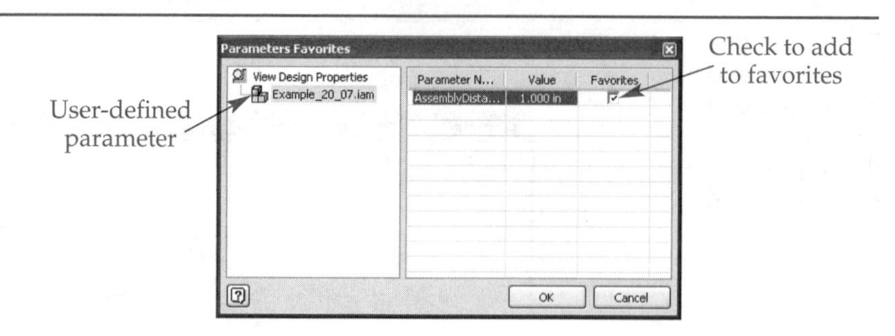

Figure 20-27.
To animate a parameter, it must first be added to the Animation Favorites branch in the **Browser** using the **Parameters Favorites** dialog box.

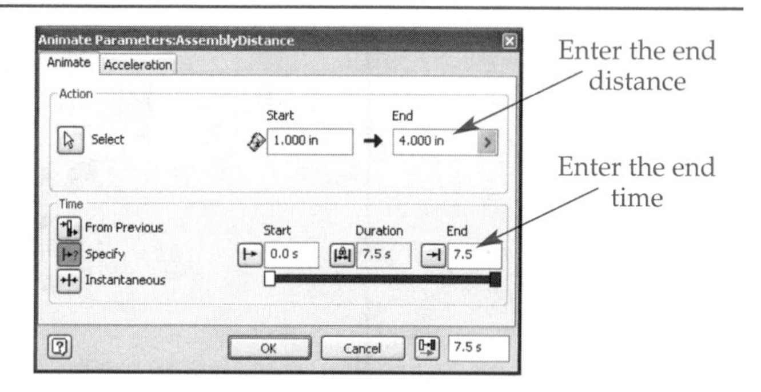

Figure 20-28.
Animating a parameter.

See **Figure 20-28.** If you right-clicked on the parameter name, the parameter is automatically selected. Otherwise, pick **Select** button in the **Action** area and select the parameter. Also, enter 1 in the **Start** text box and 4 in the **End** text box in the **Action** area. In the **Time** area, pick the **Specify** button, enter 0 in the **Start** text box, and enter 7.5 in the **End** text box. Finally, enter 7.5 in the **New Action Increment** text box and pick the **Complete action and start new** action button.

Next, make sure the start time is 7.5 seconds and the end time is 15.0 seconds. Also, make sure the start distance is 4. Then, enter 1 in the **End** text box in the **Action** area and pick the **OK** button to complete the action. Now, test the animation. The animation preview may be slow or "jump" because Inventor is performing a lot of calculations for the adaptivity of the spring. Save and close the file.

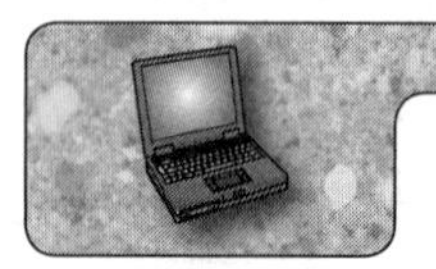

PROFESSIONAL TIP

Parameters can also be animated for individual parts.

Robot Hand

This next example really demonstrates how powerful parameters can be in creating an animation in Inventor Studio. Open Example_20_08.iam and examine the assembly and its components. This is a robot hand. Expand the tree in the **Browser** and look at the constraints. There is an angular constraint between each component in a finger (the "joint") and between the palm and arm. Those constraints are controlled by six assembly-level parameters—one parameter for each finger (the thumb being considered a finger) and the wrist joint. All of these individual constraints provide complete control over finger movement.

Switch to Inventor Studio. A camera has already been created, the animation length has been set, and a lighting style has been set up. Set the view to the camera and hide the camera. Also, double-click on the animation name in the **Browser** to enter animation mode. Now, the finger parameters need to be animated.

Pick the **Parameter Favorites** button in the **Inventor Studio** panel to display the **Parameter Favorites** dialog box. Add the six user-defined parameters to the Animation

Favorites branch in the **Browser** by checking the check box in the **Favorites** column for each. Then, pick the **OK** button to close the dialog box.

In the **Browser**, expand the Animation Favorites branch. Then, right-click on the IndexAngle parameter and select **Animate Parameters** from the pop-up menu. In the **Time** area of the **Animate Parameters** dialog box, pick the **Specify** button, enter 0 in the **Start** text box, and enter 5 in the **End** text box. In the **Action** area, enter 0 in the **Start** text box and 80 in the **End** text box. Then, enter 5 in the **New Action Increment** text box and pick the **Complete action and start new action** button. Make sure the start time is 5 and the end time is 10. Also, make sure the value in the **Start** text box in the **Action** area is 80. Then, enter 0 in the **End** text box in the **Action** area and pick the **OK** button to complete the action.

Test the animation. The index finger curls down to touch the palm. Apply the same settings to the MiddleAngle, RingAngle, LittleAngle, and ThumbAngle parameters. Also, animate the WristAngle parameter from 0 to 180 degrees over 10 seconds.

Test the animation again. Notice how all fingers move at the same time. If the motion is staggered, a more-human effect results. In the **Animation Timeline** dialog box, move the keys as follows. Remember, sometimes there are two keys, one on top of the other; move both keys.

- Move the first key for the middle finger to the 0.5 second mark, the second key to the 6.0 second mark, and leave the third key at the 10.0 second mark.
- Move the first key for the ring finger to the 1.0 second mark, the second key to the 7.0 second mark, and leave the third key at the 10.0 second mark.
- Move the first key for the little finger to the 1.5 second mark, the second key to the 8.0 second mark, and leave the third key at the 10.0 second mark.

Leave the keys for the index finger and thumb unchanged. Test the animation again. Notice how much better the motion appears with these small changes. See **Figure 20-29.** You may also want to start the wrist rotation later, such as around the eight second mark, to produce a much faster motion. Save and close the file.

Figure 20-29.
The animated robot hand.

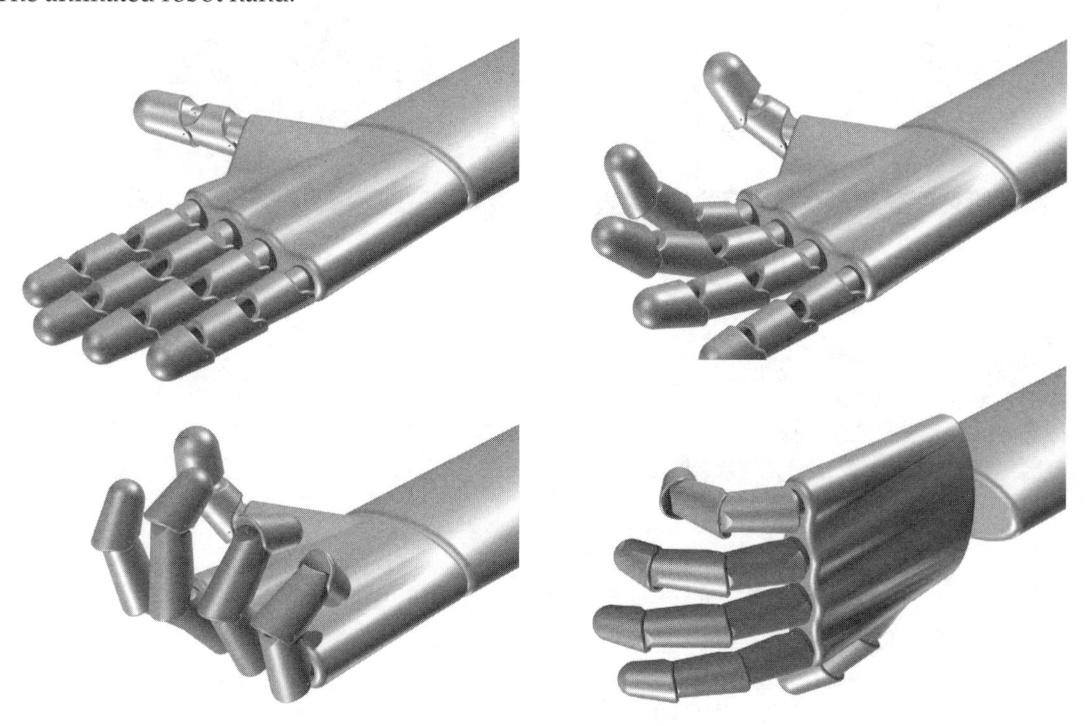

Chapter Test

Answer the following questions on a separate sheet of paper or complete the electronic chapter test on the Student CD.

1. What is a *rendering?*
2. Define *scene.*
3. What is a *surface style?*
4. Describe the process of rendering in Inventor.
5. Describe the process of animating in Inventor.
6. What are two ways in which a camera can be created in Inventor Studio?
7. List the three types of lights that can be added to a lighting style.
8. How are the width and height dimensions of a rendered still image set?
9. In which dialog box are settings made to animate a component?
10. What is a *key?*
11. How are keys moved?
12. Describe the simplest way to animate a camera.
13. Define *fade.*
14. When animating a parameter, where are the parameters located in the **Browser**?
15. How is the **Render Animation** dialog box displayed?

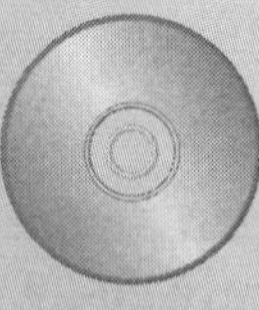

Chapter Exercises

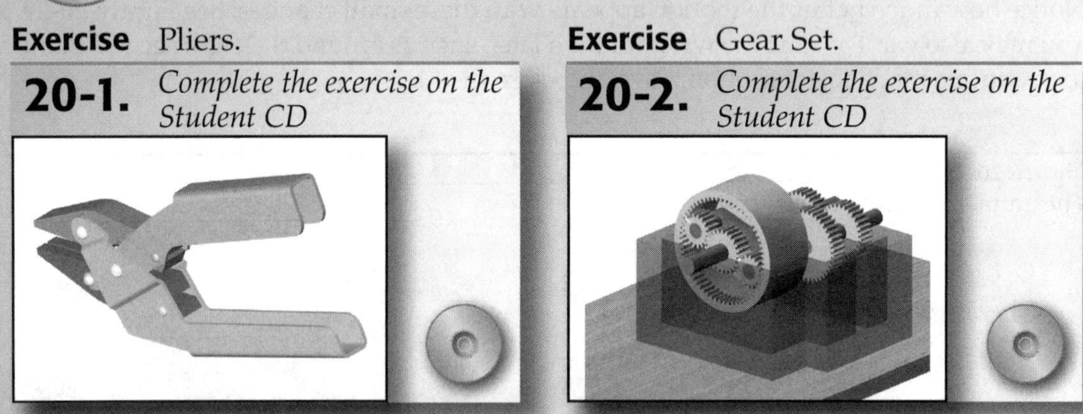

Exercise 20-1. Pliers.
Complete the exercise on the Student CD

Exercise 20-2. Gear Set.
Complete the exercise on the Student CD

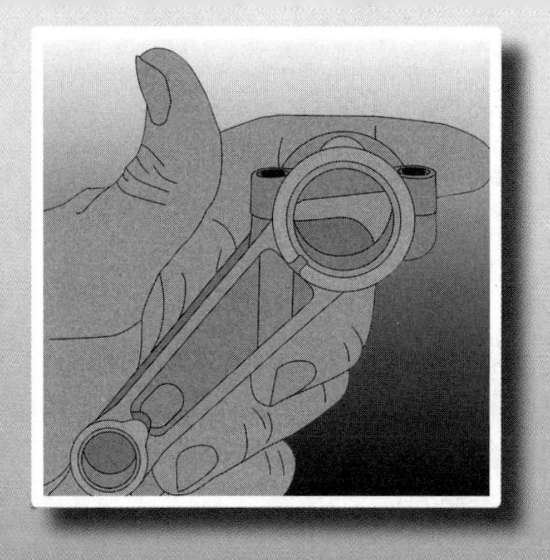

Index

2D drawings, 22
2D path, 236
2D Sketch Panel, 39, 68–69, 71–73, 76–79, 89, 94
3D sweeps, 241–244
  editing, 246

A

Adaptive parts, 284–288
  location, 284
  size, 286, 288
Adaptive spring, 447–448
Aligned section view, 194–195
Ambient light, 429
Ambiguous profiles, 77, 87–88
Analyze Faces button, 38
Angle assembly constraint, 278–279
Animate Components dialog box, 439–441
Animate Constraints dialog box, 443
Animate Fade dialog box, 445
Animate tool, 376–378
Animating,
  camera, 444–445
  components, 436–442
  constraints, 442–443
  parameters, 447–449
Animating components, 436–442
  creating a camera, 437
  editing animation segments, 441–442
  entering animation mode, 438
  setting the animation time, 438–439
  specifying the movement, 439–440
  testing the animation, 441
Animation, 376–378, 426
  editing segments, 441–442
  rendering, 446
  testing, 441
Animation dialog box, 376–378
Animation Options dialog box, 438
Animation process, 426–427
Animation Timeline dialog box, 438–444
Annotations, 355–366
  balloons, 359–363
  centerlines, 355
  dimensions, 355–358
  layers, 213–214
  leader text, 359
  preparing to annotate a drawing layout, 211
  repositioning, 203
Antialiasing, 433
Applications pull-down menu, 57
Arcs,
  and constraints, 77–79
  drawing from within the **Line** tool, 79–80
Aspect ratio, 171–172
Assembly, 21, 251
  adaptive parts, 284–288
  angle constraint, 278–279
  building, 254–265
  constraining centerlines, 257–258
  constraining edges of parts, 265–266
  constraining work planes and axes, 280–281
  constraints and sketches, 282–283
  creating parts in assembly view, 275–277
  insert constraint, 259
  parameters, 321–330
  part colors in, 386
  removing degrees of freedom, 256–257
  standard parts, 262–265
  working with, 273–291
Assembly drawings, 347–369
  annotations, 355–366
  break out views, 351–354
  creating drawing views using design view
    representations, 367–369
  overlay views, 369
  printing, 367
  section views, 349–350
  sliced and offset section views, 350–351
  views, 347–349

Assembly modeling, 21
Assembly Panel, 39–40
AutoCAD drawings, working with, 207–208
Auto Dimension dialog box, 71
Auto Explode dialog box, 373–374
Automated Centerlines dialog box, 355
Auxiliary views, 197–198

B

Background light, 429
Balloons, 359–363
  editing placement, 362–363
  placing, 360–361
  style, 362
Base view, 189–190
Bending, 392
Bend Part tool, 180
Bend radius, 396
Bitmap, 171–173
Border, 187–188
Boundary patch, 338–339
Break dialog box, 198–200
Break out view, 192, 351–354
Break Out View dialog box, 353–354
Broken views, 198, 200–201
Browser, 18, 41–45, 288
  changing dimensions, 43
  pop-up menu, 43–45
  renaming features, 43
  working with features, 45

C

Caching, 52
Camera, 426
  animating, 444–445
  creating, 428–429, 437, 442
  moving, 431
Camera flyout, 37
Centerlines, 214–218, 257–258, 355
Chamfers, 150, 407–408
  adding, 150–152
Circles, 76–77
Circular pattern, 153–155
Clearance holes, 135–136
Closed-loop loft, 248–249
Closed path, 238
Collision detector, 298–299
Color style, editing, 49–50
Common View tool, 34–35
Component Opacity flyout, 38
Composite iMate, 203, 304
Constraint bar, 71
Constraints, 66, 70–71
  and arcs, 77–79
  animating, 442–443
  assembly and sketches, 282–283
  driving, 279–280
  horizontal, 76–77
  iMate, 307–308, 310
  iPart, 306–307
  motion, 293
  review of, 83
  rotation, 294–296
  rotation-translation, 296, 299–300
  tangent, 76–77, 80
  work planes and axes, 280–281
Construction geometry, 93–96
Consumed sketch, 122
Contact Solver, 277–278, 367
Content Center, 262
Contour Flange dialog box, 412–413
Contour flanges, 412–415
  complex patterns as, 414–415
  creating, 412
  extending from an edge, 413–414
Convert pull-down menu, 57
Coordinate system indicator (CSI), 167
Coordinate system, moving on a sketch plane, 167–169
Corner Seam tool, 418–420
Counterbored holes, 136–137
Countersunk holes, 136–137
Create iMate dialog box, 311
Cross-sectional profile, 236
Cross section analysis, 343–344
Curvature analysis, 342
Custom iPart, 303
Cut tool, 400–401
Cutouts, 87

D

Decal tool, 171–175
Degrees of freedom, 256–257
Delete View dialog box, 198–199
Derived parts, 268–271
Design Assistant, 24
Design view representations, 288, 290–291
  creating drawing views, 367–369
Design View Representations dialog box, 290–291
Design views, 288, 290–291
Detail views, 198
Diagnose Sketch dialog box, 82
Diametric dimensions, 97
Dimension Properties dialog box, 91
Dimensioning,
  adding hole or thread notes, 221–222
  adding text, 226
  auxiliary view, 220
  dimension styles, 219–220
  feature control frames, 223–226
  in Inventor, 218–232
  parametric dimensions, 218–219
  revision table and revision tag, 228–229
  sketched symbols, 227–228
  surface texture symbols, 222
  weld symbols, 230–232
Dimensions, 71–72
  annotation, 355–358
  changing, 43
  using d0 model parameters in equations, 89–92
Dimension styles, 219–220
Dimension text, 226
  add text to, 356, 358
  completely replace, 358
Display flyout, 36

Draft, 176
Draft analysis, 341–342
Draft angle, 176
Draft views, 201
Drafting standards, 211–218
  centerlines, 214–218
  layers, 213
Drawing, 392
Drawing Review Panel, 207
Drawings,
  changing view orientation, 205–206
  creating views, 189–201
  editing a view's lines, 203
  editing views, 201–205
  Inventor's options for, 206–207
  projecting views, 191–192
Drawing views, creating using design view representations, 367–369
Drawing View dialog box, 189–191, 202–203
Drawing Views Panel, 39
Driven dimensions, 101
Driving constraints, 279–280

E

Edge Properties dialog box, 203–204
Edit Coordinate System tool, 167
Edit Dimension dialog box, 72, 90, 92, 218, 224, 358
Edit Task & Sequences dialog box, 384–385
Emboss, 169–171
End Fill dialog box, 231
Engineer's notebook, 23, 59–60
Engrave, 169
Equations, 89–93
Exploded views, 371
  multiple, 378–381
Extrude dialog box, 73–74, 80–82, 88, 110–113, 120–121, 331–332
Extrude tool,
  Intersect options, 121–122
  operation and extents options, 110–113

F

Face, 395
  analyzing, 340–344
Face dialog box, 395, 397
Face Draft dialog box, 177–178
Face drafts, 176–180
  fixed edge, 177–178
  fixed plane, 178
  other, 178–179
Face Properties dialog box, 107–108
Face tool, 395–398
  adding a second face, 395–397
  adding a third face, 397–398
Factory, 303, 318
Fade, 445
Family of parts, 303
Feature, 18–19
  adding, 131–156, 159–184
  adding nonsketch features to the part, 131
  renaming, 43
  working with, 45
Feature-based modeling, 18–19
Feature control frames, 223–226
  adding additional, 225–226
  defining datums, 224–225
Feature patterns, 153–154
File Open Options dialog box, 372–373
File pull-down menu, 46–47
File Save Options dialog box, 29
Fillet dialog box, 141–149
  Setbacks tab, 149
Fillets, 141–149
  all edges, 143–144
  applying, 141–143
  Fillet tool options, 145–146
  full round, 149
  setbacks, 148–149
  variable-radius, 147–149
Fill light, 429
Fix constraints, 72–73
Flanges, 402–405
  applying to a portion of an edge, 405–406
Flat pattern, 399–400
  projecting into a sketch, 401
Flush assembly constraint, 256
Fold tool, 415–418, 422
Format pull-down menu, 47–52
  Active Standard... option, 47
  Style and Standard Editor... option, 48–52
Format Text dialog box, 169–170
Free-rotate mode, 33
Full round fillets, 149
Full section view, 192
Fully constrained sketch, 19

G

General Dimension tool, 89–90
Grips, 75
Grounded, 255

H

Half-section view, 192
Help pull-down menu, 58–59
Hem dialog box, 408
Hems, 408–409
Hole centers, 133
Hole notes, adding, 221–222
Holes, creating without a sketch plane, 137–138
Hole tool, 132–138

I

iMates, 304
  constraining, 307–308, 310
  creating, 311–313
  glyphs, 308–310
Inline work features, 244–246
Insert assembly constraint, 259
Insert pull-down menu, 47
Instance, 251
Intelligent objects, 65
Interval, 376
Inventor Precise Input toolbar, 88–89, 99–103
Inventor project wizard, 252

Inventor Standard toolbar, 27–38, 93–94
Inventor Studio, 425–449
  animating a camera, 444–445
  animating components, 436–442
  animating constraints, 442–443
  animating fade, 445
  animating parameters, 447–449
  animation process, 426–427
  rendering components, 427–436
  rendering process, 426
iPart Author dialog box, 313–318
iParts, 303
  assembling, 304–311
  authoring, 313–317
  constraining, 306–307
  creating from scratch, 311–318
  custom, 303
  modeling the part, 311
  placed, 307
  placing, 304, 306
  placing the remaining parts, 310–311
  standard, 303
  testing in the factory, 318
Isometric views, 196–197
  redefining, 35

K

Key, 441, 444
Key column, 315
Keyframe, 441
Key light, 429

L

Layers, 191
  drafting standards, 213
Layout drawing, creating, 187–188
Leader text, 359
Lighting style, 426
  creating, 429–431
  editing, 50–51
Lighting Styles dialog box, 430–431
Line tool, drawing arcs, 79–80
Loft dialog box,
  Conditions tab, 246–247
  Curves tab, 335
Lofting, 246
Lofts, 246–249
  basic loft, 246–247
  closed-loop, 248–249
  editing, 249
Look At button, 36

M

Mate assembly constraint, 257
Material style, editing, 51
Menu Bar, 46–59
Mirror Feature tool, 156
Mirror Pattern dialog box, 156
Mirror tool, 96–97
Model, changing, 205
Modeling motion, 22
Motion constraint, 293

O

Offset section view, 351
Open dialog box, 25–26, 62
Open-loop loft, 248
Origin, 69
Overlay views, 369

P

Pan button, 32
Panel Bar, 38–41
  panels, 39–40
  resizing, 41
Parameters, 321–330
  animating, 447–449
  controlling a part with a spreadsheet, 322–324
  flanged pipe run example, 328–329
  using a spreadsheet to control an assembly, 324–329
Parameters dialog box, 92, 322–324
Parameters Favorites dialog box, 448–449
Parametric dimensions, 218–219
Parametric modeling, 19–21
Part colors in assemblies, 386
Part drawings, creating, 187–208
Part Features panel, 39–40
Parts, creating, 65–67
Parts list, 364–366
Parts List dialog box, 364–366
Part tree, 18
Path, 236
  closed, 238
Pattern the entire part, 154–155
Place Component dialog box, 304–305, 307–308
Place Constraint dialog box, 293–294, 306
Place Custom iPart dialog box, 307
Placed features, 19, 41
Place Standard iPart dialog box, 308
Pop-up menu, 43–45
Precise input, 88–89, 99–103
Presentation, 22
Presentation files, 371–388
  complex and rotational tweaks, 382–385
  creating multiple exploded views, 378–381
  lighting, 386–387
  styles, 385–388
Presentation Panel, 39–40
Printing the drawing, 367
Projects, 60–63, 251
  creating, 63
  creating with folders, 252–254
  selecting, 61–62
  setting additional paths in, 273–275
Projects dialog box, 252–254
Properties dialog box,
  Physical tab, 46
  Project tab, 254–255
Punching, 392, 420
Punch Tool, 420–422

Q

Quilt, 334, 337

R

Rail, 235, 246
Rectangle tool, 88–89
Rectangular pattern, 153–155
Reference dimensions, 218
Relative input, 102
Relief, 396–397
Remnant, 396
Renaming features, 43
Render Animation dialog box, 446
Render Image dialog box, 432–433
Rendering, 426
  an animation, 446
  creating a camera, 428–429
  creating a lighting style, 429–431
  creating a surface style, 434–436
  components, 427–436
  process, 426
  moving lights and cameras, 431
  rendering the scene, 432–433
  scene containing threads, 436
Resolution, 432–433
Return button, 30–31
Revision table, 228–229
Revision tag, 229
Revolve tool, 97–99, 415
Rib, 163
Rib tool, 163–166
Rotate button, 32–36
  Common View tool, 34–35
  redefining the isometric view, 35
  rotating around a large part, 35–36
Rotation constraint, 294–296
Rotation-translation constraint, 296, 299–300
Rounds, 141–149, 407–408
  all edges, 143–144
  applying, 141–143

S

Save As dialog box, 29
Scene styles, 426
Secondary sketch planes and features, creating, 107–109
Section View dialog box, 192–194
Section views, 349–350
  based on sketch geometry, 196
  sliced and offset, 350–351
  with depth, 195
Select Assembly dialog box, 372–373
Select flyout, 30
Setbacks, 148–149
Shadow draft, 179–180
Shadow flyout, 37–38
Shared environment, 252
Sheet, 187
Sheet Metal Features panel, 39–40
Sheet metal parts, 391–422
  contour flanges, 412–415
  creating a flat pattern, 399–400
  Cut tool, 400–401
  Face tool, 395–398
  flanges, 402–406
  Fold tool, 415–418
  hems, 408–409
  holes and cuts across bends, 409–411
  overriding styles, 398–399
  projecting a flat pattern into a sketch, 401
  Punch Tool, 420–422
  rounds and chamfers, 407–408
Sheet Metal Style tool, 393–396, 399
Sheet size, 188
Shell tool, 159–163
Silhouette curves, 114–121
Sketch coordinate system (SCS), 167
Sketch Doctor wizard, 81–82
Sketched features, 19, 41
Sketched symbols, 227–228
Sketching, 19, 65–80
  arcs and more constraints, 77–79
  circles, tangent and horizontal constraints, and trimming, 76–77
  constraints, 70–71
  dimensions, 71–72
  drawing an arc from within the **Line** tool, 79–80
  editing the feature and the sketch, 74–76
  extruding the part, 73–74
  fix constraints, 72–73
  things that can go wrong, 80–82
Sketch patterns, 155
Sliced section view, 350–351
Solids, creating from open sketches, 167
Split tool, 180–184
Spreadsheet,
  controlling a part with, 322–324
  editing, 326–328
  using to control an assembly, 324–329
Stamping, 420
Standard iPart, 303
Standard parts, 262–265
Storyboard, 382
Style and Standard Editor dialog box, 48–52
  balloon style, 362
  caching styles, 52
  changing text styles, 226–227
  colors, 388
  color style, 49–50
  dimension styles, 219–220
  importing and exporting styles, 52
  layers, 191–192
  lighting, 386–387
  lighting style, 50–51
  material style, 51
  parts list standards, 366
  reloading styles, 51
  standards, 212–213
Styles, 47, 385–388
  caching, 52
  importing and exporting, 52
  overriding, 398–399
  purging, 52
  reloading, 51
  saving to style library, 52
  updating, 52
Surface analysis, 343

Surface Analysis dialog box, 343
Surfaces, 331–344
  analyzing, 340–344
  as construction geometry, 337
  display, 332
  extruded, 333
  in 2D drawings, 339
  lofted, 333
  replace face, 335–336
  revolved, 333
  stitched, 337
  swept, 333–334
Surface style, 426
  creating, 434–436
Surface Texture dialog box, 222–223
Surface texture symbols, 222
Sweep dialog box, 236–237
Sweep path, creating, 236, 238–239
Sweeps, 235–249
  creating, 236–238
  profile not perpendicular to the path, 237–238
Symbols dialog box, 227–228

T

Tangent constraint, 266
Taper thread holes, 136
Text, dimension, 226, 356, 358
Text tool, 169–171
Thicken/Offset tool, 334–335
Threaded holes, 134–135
Thread notes, adding, 221–222
Thread tool, 139–140
Title block, 188
Tolerance, 358
Tools pull-down menu, 53–57
  application options, 53–57
  document settings, 53
Top view, 189
Trails, 373
  removing, 375
Transitional constraint, 300–301
Triad, 439
Trimming, 77
Tweaks, 373
  adding, 374–376
  complex and rotational, 382–385
  editing, 376–378
Two Point Rectangle button, 89

U

Unconsumed sketch, 122
Undo and **Redo** buttons, 30
Unitless, 92
Update button, 31–32
User interface, 25–63
  Inventor Standard toolbar, 27–38
  Menu Bar, 46–59
  opening an existing part file, 25
  overview, 26–27
  Panel Bar, 38–41

V

Variable-radius fillet, 147–148
View pull-down menu, 47
Views, 347–349, 384–385
  break out, 351–354
  changing orientation, 205–206
  creating, 189–201
  editing, 201–205
  editing lines, 203
Visibility, 288, 290–291, 385

W

Web, 163
Web pull-down menu, 58
Weld annotations, applying, 231–232
Weld Caterpillars dialog box, 232
Welding Symbol dialog box, 231
Weld symbols, 230–232
Window pull-down menu, 57
Work axis, 127
  constraining, 280–281
Work envelope, 21
Work features, 19
Work planes,
  angled from a face or existing work plane, 124
  between edges or points, 127
  constraining, 280–281
  creating and using, 122–127
  midway between two parallel faces, 125–126
  offset from an existing face or work plane, 123–124
  offset through a point, 124
  parallel to a face or plane and tangent to a curved face, 126
  perpendicular to a line at its endpoint, 126
  perpendicular to an axis and through a point, 127
  tangent to a curved face through an edge, 127
  through two parallel work axes, 127

Z

Zebra analysis, 340
Zebra Analysis dialog box, 340–341
Zoom All button, 32
Zoom button, 32
Zoom Selected button, 32
Zoom Window button, 32